Graduate Texts in Mathematics 240

Graduate Texts in Mathematics

(continued after index)

Henri Cohen

Number Theory

Volume II:
Analytic and Modern
Tools

 Springer

Henri Cohen
Université Bordeaux I
Institut de Mathématiques de Bordeaux
351, cours de la Libération
33405, Talence cedex
France
Henri.Cohen@math.u-bordeaux1.fr

Mathematics Subject Classification (2000): 11-xx 11-01 11Dxx 11Rxx 11Sxx

ISBN-13: 978-1-4419-2388-2   eISBN-13: 978-0-387-49894-2

Printed on acid-free paper.

9 8 7 6 5 4 3 2 1

springer.com

Preface

This book deals with several aspects of what is now called "explicit number theory," not including the essential algorithmic aspects, which are for the most part covered by two other books of the author [Coh0] and [Coh1]. The central (although not unique) theme is the solution of Diophantine equations, i.e., equations or systems of polynomial equations that must be solved in integers, rational numbers, or more generally in algebraic numbers. This theme is in particular the central motivation for the modern theory of arithmetic algebraic geometry. We will consider it through three of its most basic aspects.

The first is the *local* aspect: the invention of p-adic numbers and their generalizations by K. Hensel was a major breakthrough, enabling in particular the simultaneous treatment of congruences modulo prime powers. But more importantly, one can do *analysis* in p-adic fields, and this goes much further than the simple definition of p-adic numbers. The local study of equations is usually not very difficult. We start by looking at solutions in *finite fields*, where important theorems such as the Weil bounds and Deligne's theorem on the Weil conjectures come into play. We then *lift* these solutions to local solutions using *Hensel lifting*.

The second aspect is the *global* aspect: the use of number fields, and in particular of class groups and unit groups. Although local considerations can give a considerable amount of information on Diophantine problems, the "local-to-global" principles are unfortunately rather rare, and we will see many examples of failure. Concerning the global aspect, we will first require as a prerequisite of the reader that he or she be familiar with the standard basic theory of number fields, up to and including the finiteness of the class group and Dirichlet's structure theorem for the unit group. This can be found in many textbooks such as [Sam] and [Marc]. Second, and this is less standard, we will always assume that we have at our disposal a computer algebra system (CAS) that is able to compute rings of integers, class and unit groups, generators of principal ideals, and related objects. Such CAS are now very common, for instance Kash, magma, and Pari/GP, to cite the most useful in algebraic number theory.

The third aspect is the theory of zeta and L-functions. This can be considered a *unifying theme*[3] for the whole subject, and it embodies in a beautiful way the local and global aspects of Diophantine problems. Indeed, these functions are defined through the local aspects of the problems, but their analytic behavior is intimately linked to the global aspects. A first example is given by the Dedekind zeta function of a number field, which is defined only through the splitting behavior of the primes, but whose leading term at $s = 0$ contains at the same time explicit information on the unit rank, the class number, the regulator, and the number of roots of unity of the number field. A second very important example, which is one of the most beautiful and important conjectures in the whole of number theory (and perhaps of the whole of mathematics), the Birch and Swinnerton-Dyer conjecture, says that the behavior at $s = 1$ of the L-function of an elliptic curve defined over $\mathbb{Q}$ contains at the same time explicit information on the rank of the group of rational points on the curve, on the regulator, and on the order of the torsion group of the group of rational points, in complete analogy with the case of the Dedekind zeta function. In addition to the purely *analytical* problems, the theory of L-functions contains beautiful results (and conjectures) on *special values*, of which Euler's formula $\sum_{n \geqslant 1} 1/n^2 = \pi^2/6$ is a special case.

This book can be considered as having four main parts. The first part gives the tools necessary for Diophantine problems: equations over finite fields, number fields, and finally local fields such as $\mathfrak{p}$-adic fields (Chapters 1, 2, 3, 4, and part of Chapter 5). The emphasis will be mainly on the theory of $\mathfrak{p}$-adic fields (Chapter 4), since the reader probably has less familiarity with these. Note that we will consider function fields only in Chapter 7, as a tool for proving Hasse's theorem on elliptic curves. An important tool that we will introduce at the end of Chapter 3 is the theory of the Stickelberger ideal over cyclotomic fields, together with the important applications to the Eisenstein reciprocity law, and the Davenport–Hasse relations. Through Eisenstein reciprocity this theory will enable us to prove Wieferich's criterion for the first case of Fermat's last theorem (FLT), and it will also be an essential tool in the proof of Catalan's conjecture given in Chapter 16.

The second part is a study of certain basic Diophantine equations or systems of equations (Chapters 5, 6, 7, and 8). It should be stressed that even though a number of general techniques are available, each Diophantine equation poses a new problem, and it is difficult to know in advance whether it will be easy to solve. Even without mentioning *families* of Diophantine equations such as FLT, the congruent number problem, or Catalan's equation, all of which will be stated below, proving for instance that a specific equation such as $x^3 + y^5 = z^7$ with x, y coprime integers has no solution with $xyz \neq 0$ seems presently out of reach, although it has been proved (based on a deep theorem of Faltings) that there are only finitely many solutions; see [Dar-Gra]

---

[3] Expression due to Don Zagier.

and Chapter 14. Note also that it has been shown by Yu. Matiyasevich (after a considerable amount of work by other authors) in answer to Hilbert's tenth problem that there cannot exist a general algorithm for solving Diophantine equations.

The third part (Chapters 9, 10, and 11) deals with the detailed study of analytic objects linked to algebraic number theory: Bernoulli polynomials and numbers, the gamma function, and zeta and L-functions of Dirichlet characters, which are the simplest types of L-functions. In Chapter 11 we also study p-adic analogues of the gamma, zeta, and L-functions, which have come to play an important role in number theory, and in particular the Gross–Koblitz formula for Morita's p-adic gamma function. In particular, we will see that this formula leads to remarkably simple proofs of Stickelberger's congruence and the Hasse–Davenport product relation. More general L-functions such as Hecke L-functions for Grössencharacters, Artin L-functions for Galois representations, or L-functions attached to modular forms, elliptic curves, or higher-dimensional objects are mentioned in several places, but a systematic exposition of their properties would be beyond the scope of this book.

Much more sophisticated techniques have been brought to bear on the subject of Diophantine equations, and it is impossible to be exhaustive. Because the author is not an expert in most of these techniques, they are not studied in the first three parts of the book. However, considering their importance, I have asked a number of much more knowledgeable people to write a few chapters on these techniques, and I have written two myself, and this forms the fourth and last part of the book (Chapters 12 to 16). These chapters have a different flavor from the rest of the book: they are in general not self-contained, are of a higher mathematical sophistication than the rest, and usually have no exercises. Chapter 12, written by Yann Bugeaud, Guillaume Hanrot, and Maurice Mignotte, deals with the applications of Baker's explicit results on linear forms in logarithms of algebraic numbers, which permit the solution of a large class of Diophantine equations such as Thue equations and norm form equations, and includes some recent spectacular successes. Paradoxically, the similar problems on elliptic curves are considerably less technical, and are studied in detail in Section 8.7. Chapter 13, written by Sylvain Duquesne, deals with the search for rational points on curves of genus greater than or equal to 2, restricting for simplicity to the case of hyperelliptic curves of genus 2 (the case of genus 0—in other words, of quadratic forms—is treated in Chapters 5 and 6, and the case of genus 1, essentially of elliptic curves, is treated in Chapters 7 and 8). Chapter 14, written by the author, deals with the so-called super-Fermat equation $x^p + y^q = z^r$, on which several methods have been used, including ordinary algebraic number theory, classical invariant theory, rational points on higher genus curves, and Ribet–Wiles type methods. The only proofs that are included are those coming from algebraic number theory. Chapter 15, written by Samir Siksek, deals with the use of Galois representations, and in particular of Ribet's level-lowering theorem and Wiles's and Taylor–Wiles's theorem proving the modularity conjecture. The main application is to equations of "abc" type, in other words, equations of the form $a + b + c = 0$ with a, b, and c highly composite, the "easiest" application of this method being the proof of FLT. The author of this chapter has tried to hide all the sophisticated mathematics and to present the method as a black box that can be used without completely understanding the underlying theory. Finally, Chapter 16, also written by the author, gives the complete proof of Catalan's conjecture by P. Mihăilescu. It is entirely based on notes of Yu. Bilu, R. Schoof, and especially of J. Boéchat and M. Mischler, and the only reason that it is not self-contained is that it will be necessary to assume the validity of an important theorem of F. Thaine on the annihilator of the plus part of the class group of cyclotomic fields.

Warnings

Since mathematical conventions and notation are not the same from one mathematical culture to the next, I have decided to use systematically unambiguous terminology, and when the notations clash, the French notation. Here are the most important:

— We will systematically say that a is strictly greater than b, or greater than or equal to b (or b is strictly less than a, or less than or equal to a), although the English terminology a is greater than b means in fact one of the two (I don't remember which one, and that is one of the main reasons I refuse to use it) and the French terminology means the other. Similarly, positive and negative are ambiguous (does it include the number 0)? Even though the expression "x is nonnegative" is slightly ambiguous, it is useful, and I *will* allow myself to use it, with the meaning $x \geqslant 0$.

— Although we will almost never deal with noncommutative fields (which is a contradiction in terms since in principle the word field implies commutativity), we will usually not use the word field alone. Either we will write explicitly commutative (or noncommutative) field, or we will deal with specific classes of fields, such as finite fields, $\mathfrak{p}$-adic fields, local fields, number fields, etc., for which commutativity is clear. Note that the "proper" way in English-language texts to talk about noncommutative fields is to call them either skew fields or division algebras. In any case this will not be an issue since the only appearances of skew fields will be in Chapter 2, where we will prove that finite division algebras are commutative, and in Chapter 7 about endomorphism rings of elliptic curves over finite fields.

— The GCD (respectively the LCM) of two integers can be denoted by (a, b) (respectively by $[a, b]$), but to avoid ambiguities, I will systematically use the explicit notation $\gcd(a, b)$ (respectively $\operatorname{lcm}(a, b)$), and similarly when more than two integers are involved.

- An open interval with endpoints a and b is denoted by (a, b) in the English literature, and by $]a, b[$ in the French literature. I will use the French notation, and similarly for half-open intervals $(a, b]$ and $[a, b)$, which I will denote by $]a, b]$ and $[a, b[$. Although it is impossible to change such a well-entrenched notation, I urge my English-speaking readers to realize the dreadful ambiguity of the notation (a, b), which can mean either the ordered pair (a, b), the GCD of a and b, the inner product of a and b, or the open interval.
- The trigonometric functions $\sec(x)$ and $\csc(x)$ do not exist in France, so I will not use them. The functions $\tan(x)$, $\cot(x)$, $\cosh(x)$, $\sinh(x)$, and $\tanh(x)$ are denoted respectively by $\operatorname{tg}(x)$, $\operatorname{cotg}(x)$, $\operatorname{ch}(x)$, $\operatorname{sh}(x)$, and $\operatorname{th}(x)$ in France, but for once to bow to the majority I will use the English names.
- $\Re(s)$ and $\Im(s)$ denote the real and imaginary parts of the complex number s, the typography coming from the standard TEX macros.

Notation

In addition to the standard notation of number theory we will use the following notation.

- We will often use the practical self-explanatory notation $\mathbb{Z}_{>0}$, $\mathbb{Z}_{\geqslant 0}$, $\mathbb{Z}_{<0}$, $\mathbb{Z}_{\leqslant 0}$, and generalizations thereof, which avoid using excessive verbiage. On the other hand, I prefer not to use the notation $\mathbb{N}$ (for $\mathbb{Z}_{\geqslant 0}$, or is it $\mathbb{Z}_{>0}$?).

- If a and b are nonzero integers, we write $\gcd(a, b^\infty)$ for the limit of the ultimately constant sequence $\gcd(a, b^n)$ as $n \to \infty$. We have of course $\gcd(a, b^\infty) = \prod_{p \mid \gcd(a,b)} p^{v_p(a)}$, and $a/\gcd(a, b^\infty)$ is the largest divisor of a coprime to b.
- If n is a nonzero integer and $d \mid n$, we write $d\|n$ if $\gcd(d, n/d) = 1$. Note that this is *not* the same thing as the condition $d^2 \nmid n$, except if d is prime.
- If $x \in \mathbb{R}$, we denote by $\lfloor x \rfloor$ the largest integer less than or equal to x (the *floor* of x), by $\lceil x \rceil$ the smallest integer greater than or equal to x (the *ceiling* of x, which is equal to $\lfloor x \rfloor + 1$ if and only if $x \notin \mathbb{Z}$), and by $\lfloor x \rceil$ the nearest integer to x (or one of the two if $x \in 1/2 + \mathbb{Z}$), so that $\lfloor x \rceil = \lfloor x + 1/2 \rfloor$. We also set $\{x\} = x - \lfloor x \rfloor$, the *fractional part* of x. Note that for instance $\lfloor -1.4 \rfloor = -2$, and not -1 as almost all computer languages would lead us to believe.
- For any α belonging to a field K of characteristic zero and any $k \in \mathbb{Z}_{\geqslant 0}$ we set
$$\binom{\alpha}{k} = \frac{\alpha(\alpha - 1) \cdots (\alpha - k + 1)}{k!}.$$
In particular, if $\alpha \in \mathbb{Z}_{\geqslant 0}$ we have $\binom{\alpha}{k} = 0$ if $k > \alpha$, and in this case we will set $\binom{\alpha}{k} = 0$ also when $k < 0$. On the other hand, $\binom{\alpha}{k}$ is *undetermined* for $k < 0$ if $\alpha \notin \mathbb{Z}_{\geqslant 0}$.

- Capital italic letters such as K and L will usually denote number fields.
- Capital calligraphic letters such as $\mathcal{K}$ and $\mathcal{L}$ will denote general $\mathfrak{p}$-adic fields (for specific ones, we write for instance $K_{\mathfrak{p}}$).
- Letters such as $\mathbb{E}$ and $\mathbb{F}$ will always denote finite fields.
- The letter $\mathbb{Z}$ indexed by a capital italic or calligraphic letter such as $\mathbb{Z}_K$, $\mathbb{Z}_L$, $\mathbb{Z}_{\mathcal{K}}$, etc., will always denote the ring of integers of the corresponding field.
- Capital italic letters such as A, B, C, G, H, S, T, U, V, W, or lowercase italic letters such as f, g, h, will usually denote polynomials or formal power series with coefficients in some base ring or field. The coefficient of degree m of these polynomials or power series will be denoted by the corresponding letter indexed by m, such as A_m, B_m, etc. Thus we will always write (for instance) $A(X) = A_d X^d + A_{d-1} X^{d-1} + \cdots + A_0$, so that the ith elementary symmetric function of the roots is equal to $(-1)^i A_{d-i}/A_d$.

Acknowledgments

A large part of the material on local fields has been taken with little change from the remarkable book by Cassels [Cas1], and also from unpublished notes of Jaulent written in 1994. For p-adic analysis, I have also liberally borrowed from work of Robert, in particular his superb GTM volume [Rob1]. For part of the material on elliptic curves I have borrowed from another excellent book by Cassels [Cas2], as well as the treatises of Cremona and Silverman [Cre2], [Sil1], [Sil2], and the introductory book by Silverman–Tate [Sil-Tat]. I have also borrowed from the classical books by Borevich–Shafarevich [Bor-Sha], Serre [Ser1], Ireland–Rosen [Ire-Ros], and Washington [Was]. I would like to thank my former students K. Belabas, C. Delaunay, S. Duquesne, and D. Simon, who have helped me to write specific sections, and my colleagues J.-F. Jaulent and J. Martinet for answering many questions in algebraic number theory. I would also like to thank M. Bennett, J. Cremona, A. Kraus, and F. Rodriguez-Villegas for valuable comments on parts of this book. I would especially like to thank D. Bernardi for his thorough rereading of the first ten chapters of the manuscript, which enabled me to remove a large number of errors, mathematical or otherwise. Finally, I would like to thank my copyeditor, who was very helpful and who did an absolutely remarkable job.

It is unavoidable that there still remain errors, typographical or otherwise, and the author would like to hear about them. Please send e-mail to

Henri.Cohen@math.u-bordeaux1.fr

Lists of known errors for the author's books including the present one can be obtained on the author's home page at the URL

http://www.math.u-bordeaux1.fr/~cohen/

Table of Contents

Part III

Analytic Tools

9. Bernoulli Polynomials and the Gamma Function

We now begin our study of analytic methods in number theory. This is of course a vast subject, but we will *not* deal with what is usually called "analytic number theory," but with the methods that are related to the study of L-functions, which we will study in the next chapter. This essentially involves Bernoulli numbers and polynomials, the Euler–MacLaurin summation formula, and the gamma function and related functions.

9.1 Bernoulli Numbers and Polynomials

9.1.1 Generating Functions for Bernoulli Polynomials

We start by recalling some properties of Bernoulli numbers and polynomials.

Definition 9.1.1. *We define the Bernoulli polynomials $B_k(x)$ and their exponential generating function $E(t, x)$ by*

$$E(t, x) = \frac{te^{tx}}{e^t - 1} = \sum_{k \geqslant 0} \frac{B_k(x)}{k!} t^k ,$$

and the Bernoulli numbers B_k by $B_k = B_k(0)$.

The first few polynomials are $B_0(x) = 1$, $B_1(x) = x - 1/2$, $B_2(x) = x^2 - x + 1/6$, and $B_3(x) = x^3 - 3x^2/2 + x/2$. Note that most of the results that we give in this section for Bernoulli *polynomials* also apply to Bernoulli numbers by specializing to 0 the variable x.

The reader will notice as we go along that more natural numbers would be B_k/k instead of B_k. However, it is impossible to change a definition that is centuries old.

Proposition 9.1.2. *We have the following properties:*

(1) $B_k'(x) = kB_{k-1}(x)$.

(2) $B_k(x)$ *is a monic polynomial of degree k.*

(3) *For $k \neq 1$ we have $B_k(1) = B_k(0) = B_k$, while for $k = 1$ we have $B_1(1) = 1/2 = B_1(0) + 1$. In other words, if we set $\delta_{k,1} = 1$ if $k = 1$ and $\delta_{k,1} = 0$ otherwise, we have $B_k(1) = B_k + \delta_{k,1}$.*

(4) $B_k = 0$ *if k is odd and $k \geqslant 3$.*
(5) *We have*

$$B_k(x) = \sum_{j=0}^{k} \binom{k}{j} B_j x^{k-j} \; .$$

Proof. All these results are immediate consequences of the definition: (1) is equivalent to $\frac{\partial E(t,x)}{\partial x} = tE(t,x)$, (2) follows by induction, (3) is equivalent to $E(t,1) - E(t,0) = t$, (4) to the fact that $E(t,0) + t/2 = (t/2)\coth(t/2)$ is an even function, and (5) by formal multiplication of the power series for e^{tx} by $E(t,0)$.   □

It is immediate to check that (1) and (3) together with $B_0(x) = 1$ in fact *characterize* Bernoulli polynomials (Exercise 1).

In addition to the initial values $B_0 = 1$ and $B_1 = -1/2$, the first few nonzero values are $B_2 = 1/6$, $B_4 = -1/30$, $B_6 = 1/42$, $B_8 = -1/30$, $B_{10} = 5/66$, $B_{12} = -691/2730$, $B_{14} = 7/6$, $B_{16} = -3617/510$. For instance, every time that you meet the (prime) number 691, you must immediately think of the Bernoulli number B_{12}.

Further immediate properties of Bernoulli polynomials are the following.

Proposition 9.1.3. *We have*

$$B_k(x+1) = B_k(x) + kx^{k-1} \; ,$$
$$B_k(-x) = (-1)^k (B_k(x) + kx^{k-1}) \; ,$$
$$B_k(1-x) = (-1)^k B_k(x) \; ,$$

$$\sum_{j=0}^{k} \binom{k}{j} y^{k-j} B_j(x) = B_k(x+y) \; , \quad \text{and in particular}$$

$$\sum_{j=0}^{k-1} \binom{k}{j} B_j(x) = kx^{k-1} \; , \quad \text{hence}$$

$$\sum_{j=0}^{k-1} \binom{k}{j} B_j = 0 \text{ for } k \neq 1 \; ,$$

$$\sum_{0 \leqslant j < N} B_k\left(x + \frac{j}{N}\right) = \frac{B_k(Nx)}{N^{k-1}} \quad \text{for } N \in \mathbb{Z}_{\geqslant 1} \; .$$

Proof. It is immediate that these formulas are equivalent respectively to the trivial identities $E(t, x+1) = E(t,x) + te^{tx}$, $E(-t,-x) = e^t E(t,x) = E(t,x) + te^{tx}$, $E(-t,1-x) = E(t,x)$, $E(t,x+y) = e^{ty}E(t,x)$, $(e^t-1)E(t,x) = te^{tx}$, and $\sum_{0 \leqslant j < N} E(Nt, x + j/N) = NE(t, Nx)$.   □

Note that the formula for $B_k(-x)$ generalizes the fact that $B_k = 0$ for $k \geqslant 3$ odd. Like all formulas involving j/N for $0 \leqslant j < N$, the last formula is called the *distribution formula* for Bernoulli polynomials.

Bernoulli numbers and polynomials are by definition Taylor coefficients of certain power series. Thus they occur in the Taylor expansion of a number of classical functions, as follows.

Proposition 9.1.4. *We have the following Taylor series expansions with radii of convergence R indicated in parentheses:*

$$\coth(t) = \frac{1}{t} + \sum_{k \geqslant 1} 2^{2k} \frac{B_{2k}}{(2k)!} t^{2k-1} \qquad (R = \pi) \,,$$

$$\cot(t) = \frac{1}{t} - \sum_{k \geqslant 1} (-1)^{k-1} 2^{2k} \frac{B_{2k}}{(2k)!} t^{2k-1} \qquad (R = \pi) \,,$$

$$\tanh(t) = \sum_{k \geqslant 1} 2^{2k}(2^{2k} - 1) \frac{B_{2k}}{(2k)!} t^{2k-1} \qquad (R = \pi/2) \,,$$

$$\tan(t) = \sum_{k \geqslant 1} (-1)^{k-1} 2^{2k}(2^{2k} - 1) \frac{B_{2k}}{(2k)!} t^{2k-1} \qquad (R = \pi/2) \,,$$

$$\frac{1}{\sinh(t)} = \frac{1}{t} - \sum_{k \geqslant 1} 2(2^{2k-1} - 1) \frac{B_{2k}}{(2k)!} t^{2k-1} \qquad (R = \pi) \,,$$

$$\frac{1}{\sin(t)} = \frac{1}{t} + \sum_{k \geqslant 1} (-1)^{k-1} 2(2^{2k-1} - 1) \frac{B_{2k}}{(2k)!} t^{2k-1} \qquad (R = \pi) \,,$$

$$\frac{2}{e^t + 1} = 1 - \sum_{k \geqslant 1} 2(2^{2k} - 1) \frac{B_{2k}}{(2k)!} t^{2k-1} \qquad (R = \pi) \,.$$

Proof. By definition

$$\coth(t) = \frac{\cosh(t)}{\sinh(t)} = \frac{e^t + e^{-t}}{e^t - e^{-t}} = \frac{e^{2t} + 1}{e^{2t} - 1} = 1 + \frac{1}{t}\frac{2t}{e^{2t} - 1} \,,$$

and since $\cot(t) = i \coth(it)$ the first two formulas follow. Next, we note the trigonometric identity $\tan(t) = \cot(t) - 2\cot(2t)$, which immediately leads to the expansion for $\tan(t)$, and the one for $\tanh(t)$ follows from $\tanh(t) = \tan(it)/i$. Next, we note that $1/(e^t - e^{-t}) = 1/(e^t - 1) - 1/(e^{2t} - 1)$, giving the formula for $1/\sinh(t)$, hence for $1/\sin(t)$, and we also note that $1/(e^t + 1) = 1/(e^t - 1) - 2/(e^{2t} - 1)$, giving the last formula. The statements about the radii of convergence can be proved either directly from the asymptotic estimate for Bernoulli numbers that we will give below (Corollary 9.1.22), or from the fact that it is equal to the distance from the origin of the nearest singularity. □

Corollary 9.1.5. *We have*

$$B_k\left(\frac{1}{2}\right) = -\left(1 - \frac{1}{2^{k-1}}\right)B_k,$$

and in particular the polynomial $B_k(x)$ is divisible by $x(x-1/2)(x-1)$ when k is odd and $k \geqslant 3$.

Proof. For the first formula we note that $te^{t/2}/(e^t - 1) = (t/2)/\sinh(t/2)$, and the second statement follows from the vanishing of B_k for $k \geqslant 3$ odd and the fact that $B_k(1) = B_k(0)$ for $k \neq 1$.   □

Definition 9.1.6. *We define the* tangent numbers T_k *for $k \geqslant 0$ by*

$$T_k = 2^{k+1}(2^{k+1} - 1)\frac{B_{k+1}}{k+1}.$$

Thus $\tanh(t) = \sum_{k \geqslant 1} T_{2k-1}t^{2k-1}/(2k-1)!$ and similarly for $\tan(t)$. We have $T_0 = -1$, $T_{2k} = 0$ for $k \geqslant 1$, and the first few values of T_k for k odd are $T_1 = 1$, $T_3 = -2$, $T_5 = 16$, $T_7 = -272$, $T_9 = 7936$.

Corollary 9.1.7. *The tangent numbers satisfy the recurrence*

$$\sum_{j=1}^{k}\binom{2k-1}{2j-1}T_{2j-1} = 1 \text{ for } k > 0,$$

and in particular $T_{2k-1} \in \mathbb{Z}$ for all $k \geqslant 1$.

Proof. This immediately follows from the identity $\cosh(t)\tanh(t) = \sinh(t)$, and the details are left to the reader.   □

The fact that $T_{2k-1} \in \mathbb{Z}$ also follows from the Clausen–von Staudt theorem that we will prove below (Exercise 59).

Definition 9.1.8. *We define the* Euler numbers E_k *for $k \geqslant 0$ by setting $E_{2k+1} = 0$ for $k \geqslant 0$ and*

$$E_{2k} = -4^{2k+1}\frac{B_{2k+1}(1/4)}{2k+1}.$$

The first few values are $E_0 = 1$, $E_2 = -1$, $E_4 = 5$, $E_6 = -61$, $E_8 = 1385$, so once again if you meet the prime 61 in a computation, you may suspect that it comes from E_6.

Proposition 9.1.9. *We have*

$$B_{2k}(1/4) = B_{2k}(3/4) = \frac{B_{2k}(1/2)}{2^{2k}} = -\frac{1}{2^{2k}}\left(1 - \frac{1}{2^{2k-1}}\right)B_{2k},$$

and the Taylor series expansions

$$\frac{1}{\cosh(t)} = \sum_{k \geq 0} \frac{E_{2k}}{(2k)!} t^{2k} \qquad (R = \pi/2) \,,$$

$$\frac{1}{\cos(t)} = \sum_{k \geq 0} (-1)^k \frac{E_{2k}}{(2k)!} t^{2k} \qquad (R = \pi/2) \,.$$

Proof. Multiplying the identity $1/(e^t + 1) = 1/(e^t - 1) - 2/(e^{2t} - 1)$ given above by $e^{t/2}$ and replacing t by $2t$, we obtain

$$\frac{1}{\cosh(t)} = \frac{2e^t}{e^{2t} + 1} = -\sum_{k \geq 1} 2^k \frac{2^k B_k(1/4) - B_k(1/2)}{k!} t^{k-1} \,.$$

Since $\cosh(t)$ is an even function, we first deduce that $2^{2k} B_{2k}(1/4) = B_{2k}(1/2)$, and since $B_{2k}(1 - x) = B_{2k}(x)$, we obtain the first formula. Furthermore, since $B_{2k+1}(1/2) = 0$ for $k \geq 0$ by the above corollary, we have

$$\frac{1}{\cosh(t)} = -\sum_{k \geq 0} 4^{2k+1} \frac{B_{2k+1}(1/4)}{(2k+1)!} t^{2k} \,,$$

giving the formula for $1/\cosh(t)$, the last formula following by changing t into it. $\qquad \square$

Corollary 9.1.10. *The Euler numbers satisfy the recurrence*

$$\sum_{j=0}^{k} \binom{2k}{2j} E_{2j} = 0 \text{ for } k > 0 \,,$$

and in particular $E_{2k} \in \mathbb{Z}$ for all k.

Proof. This immediately follows from the identity $\cosh(t)(1/\cosh(t)) = 1$. It also follows from the second formula of Proposition 9.1.14 below applied to $x = y = 1/4$. We thus have $E_{2k} = -\sum_{0 \leq j < k} \binom{2k}{2j} E_{2j}$, from which we deduce by induction that E_{2k} is an integer for all k. $\qquad \square$

Remark. Although Bernoulli numbers satisfy the recurrence $\sum_{j=0}^{k-1} \binom{k}{j} B_j = 0$, which is very similar to the one for E_k if we replace k by $2k$ and B_j by 0 when $j > 1$ is odd, the main difference is that this recurrence leads to

$$B_{k-1} = -\frac{1}{k} \sum_{j=0}^{k-2} \binom{k}{j} B_j$$

for $k \geq 2$, and the denominator $1/k$ implies that the B_k are not necessarily integers (we will study some of their arithmetic properties in Section 9.5, and in particular we will see that the only integral B_{2k} is $B_0 = 1$).

Interestingly enough, although the natural generating function for Bernoulli polynomials is the *exponential generating function* $E(t,x) = \sum_{k \geqslant 0} B_k(x)t^k/k!$, it is also possible to consider the ordinary generating function

$$S(t,x) = \sum_{k \geqslant 0} \frac{B_k(x)}{t^{k+1}} ,$$

and I thank D. Zagier for pointing this out to me. We could of course consider the generating function $\sum_{k \geqslant 0} B_k(x)t^k = S(1/t,x)/t$, but the corresponding formulas would be slightly more complicated.

It is easy to check that the series $S(t,x)$ does not converge for any value of t, but as a formal power series it makes sense, and we will also see that even though the series is divergent we can assign to it a specific value. Note, however, that in Chapter 11 we will see that it converges for all *p-adic* values of t such that $|t| > 1$, and that $S(t,x) = \psi_p'(t - x + 1)$ (which follows immediately from Proposition 11.5.2 (2)), to be compared with Corollary 9.1.13, which is formally identical.

Proposition 9.1.11. *We have*

$$S(-t,-x) = -S(t,x) - \frac{1}{(t-x)^2} ,$$

$$S(t,x+1) = S(t,x) + \frac{1}{(t-x)^2} ,$$

$$S(t-y,x) = S(t,x+y) ,$$

and in particular

$$S(t-1,x) = -S(-t,-x) = S(t,x) + \frac{1}{(t-x)^2} \quad and$$

$$S(t,x) = S(t-x,0) .$$

Proof. Using the formula for $B_n(-x)$ mentioned above we have

$$S(-t,-x) = \sum_{k \geqslant 0} (-1)^{k+1} \frac{B_k(-x)}{t^{k+1}} = -\sum_{k \geqslant 0} \frac{B_k(x) + kx^{k-1}}{t^{k+1}}$$

$$= -S(t,x) - \frac{1}{t^2} \frac{1}{(1-x/t)^2} = -S(t,x) - \frac{1}{(t-x)^2} ,$$

proving the first formula, and the second follows similarly from the formula for $B_k(x+1)$ (or from the first and the formula for $B_k(1-x)$).

For the third, we use the formula for $B_k(x+y)$, which gives

$$S(t, x + y) = \sum_{k \geqslant 0} t^{-(k+1)} \sum_{j=0}^{k} \binom{k}{j} y^{k-j} B_j(x)$$

$$= \sum_{j \geqslant 0} B_j(x) t^{-(j+1)} \sum_{k \geqslant j} \binom{k}{j} (y/t)^{k-j}$$

$$= \sum_{j \geqslant 0} B_j(x) t^{-(j+1)} (1 - y/t)^{-(j+1)} ,$$

proving the third formula. The final two follow from this and the first two.
□

Proposition 9.1.12. *As a formal power series in t, $S(t, x)$ is the Laplace transform of $E(t, x)$; in other words, we have formally*

$$S(t, x) = \int_0^\infty e^{-tu} E(u, x) \, du .$$

Furthermore, for $t > x - 1$ the above integral converges absolutely.

Proof. The first statement is clear by expanding $E(u, x)$ as a power series in u since

$$\int_0^\infty e^{-tu} u^k \, du = \frac{k!}{t^{k+1}} ,$$

and the second follows since the integrand is continuous everywhere and is asymptotic to $ue^{u(x-1-t)}$ as $u \to \infty$.
□

Corollary 9.1.13. *For $t > x - 1$ we have*

$$S(t, x) = \psi'(t - x + 1) = \psi'(t - x) - \frac{1}{(t - x)^2} ,$$

where $\psi = \Gamma'/\Gamma$ is the logarithmic derivative of the gamma function (see Definition 9.6.13).

Proof. From Corollary 9.6.43 below we have

$$\psi'(s + 1) = \int_0^\infty \frac{ve^{-sv}}{e^v - 1} \, dv ,$$

so the result follows from the proposition.
□

See also Theorems 9.6.48 and 9.6.49 for continued fraction expansions of $S(t, x)$.

9.1.2 Further Recurrences for Bernoulli Polynomials

There are a great many useful recurrences for Bernoulli numbers and polynomials. We begin with the following.

Proposition 9.1.14. *For $k \geqslant 0$ we have*

$$\sum_{j=0}^{k} \binom{k}{j} y^{k-j} \frac{B_{j+1}(x)}{j+1} = \frac{B_{k+1}(x+y) - y^{k+1}}{k+1} ,$$

$$\sum_{j=0}^{\lfloor k/2 \rfloor} \binom{k}{2j} \frac{1}{y^{2j}} \frac{B_{2j+1}(x)}{2j+1} = \frac{B_{k+1}(x+y) + (-1)^k B_{k+1}(x-y)}{(2k+2)y^k} ,$$

$$\sum_{j=1}^{\lceil k/2 \rceil} \binom{k}{2j-1} \frac{1}{y^{2j}} \frac{B_{2j}(x)}{2j} = \frac{B_{k+1}(x+y) + (-1)^{k-1} B_{k+1}(x-y) - 2y^{k+1}}{(2k+2)y^{k+1}} .$$

Proof. We could give a proof of the first formula directly from the generating function, as we did for Proposition 9.1.3. It is however instructive to give an alternative proof. After all, if we integrate with respect to x the formula for $B_k(x+y)$ given in Proposition 9.1.3 and use $B'_{j+1}(x) = (j+1)B_j(x)$, we obtain the result *up to addition of a function of y*, which is not easy to determine. This approach almost never works. What does almost always work is to use trivial transformations of binomial coefficients. Here we note that for $j \geqslant 1$ we have $\binom{k}{j} = (k/j)\binom{k-1}{j-1}$, so that

$$y^k + \sum_{j=1}^{k} \frac{k}{j} \binom{k-1}{j-1} B_j(x) = B_k(x+y) ,$$

from which the first formula follows by dividing by k and changing j into $j+1$ and k into $k+1$. The other two formulas follow by computing the sum and difference of the first formula applied to y and to $-y$. $\square$

Corollary 9.1.15. *For $k \geqslant 0$ We have*

$$\sum_{j=0}^{k-1} \binom{k}{j} \frac{B_{j+1}(x)}{j+1} = x^k - \frac{1}{k+1} ,$$

$$\sum_{j=0}^{k-1} \binom{2k}{2j} \frac{B_{2j+1}(x)}{2j+1} = \frac{x^{2k} - (x-1)^{2k}}{2} ,$$

$$\sum_{j=1}^{k} \binom{2k}{2j-1} \frac{B_{2j}(x)}{2j} = \frac{x^{2k} + (x-1)^{2k}}{2} - \frac{1}{2k+1} ,$$

$$\sum_{j=1}^{k} \binom{2k}{2j-1} 2^{2j} \frac{B_{2j}(x)}{2j} = 2^{2k} \left(x - \frac{1}{2} \right)^{2k} - \frac{1}{2k+1} ,$$

$$\sum_{j=0}^{k} \binom{2k+1}{2j} \frac{B_{2j+1}(x)}{2j+1} = \frac{x^{2k+1} + (x-1)^{2k+1}}{2} ,$$

$$\sum_{j=1}^{k} \binom{2k+1}{2j-1} \frac{B_{2j}(x)}{2j} = \frac{x^{2k+1} - (x-1)^{2k+1}}{2} - \frac{1}{2k+2} ,$$

$$\sum_{j=0}^{k} \binom{2k+1}{2j} 2^{2j} \frac{B_{2j+1}(x)}{2j+1} = 2^{2k} \left(x - \frac{1}{2} \right)^{2k+1} .$$

Proof. These formulas are obtained by suitable specializations to $y = 1$ or $y = 1/2$ of the formulas of the proposition. □

Remarks. (1) If we want formulas involving $B_j(x)$ itself instead of $B_j(x)/j$, we simply differentiate with respect to x the formulas of the proposition and of the corollary. We can of course differentiate several times. Inversely, if we want formulas involving $B_j(x)/(j(j+1))$ for instance, we must in principle integrate the given formulas, but as explained above this will not give the constant term, so we simply use as above the relation $\binom{k}{j} = (k/j)\binom{k-1}{j-1}$ for $j \geqslant 1$; see Exercise 23.

(2) Since $B_k(x+1)$ and $B_k(1-x)$ have simple expressions in terms of $B_k(x)$, if we want to specialize again the above formulas (or their derivatives), we may as well restrict to $0 \leqslant x \leqslant 1/2$. Using the formulas $B_k(0) = B_k$, $B_k(1/2) = -(1-1/2^{k-1})B_k$, $B_{2k}(1/4) = -(1/2^{2k})(1-1/2^{2k-1})B_{2k}$, and the analogous formulas for $B_{2k}(1/3)$ and $B_{2k}(1/6)$ given by Exercise 10, we obtain in this way a very large number of recurrence relations for Bernoulli numbers. We can obtain even more such relations by replacing directly x and y in the formulas of Proposition 9.1.14, for instance $x = y = 1/4$ in the third formula. We also obtain the standard relation for Euler numbers given in Corollary 9.1.10 by choosing $x = y = 1/4$ in the second formula. It is to be noted, however, that all these formulas have approximately k terms; in other words, they express B_{2k} in terms of all the B_{2j} for $1 \leqslant j < k$. We are going to see that we can reduce this by a factor of 2.

The second type of recurrence that we are going to study is not well known, although it is essentially due to Seidel in 1877, and Lucas soon afterward. It has the advantage of having half as many terms in the sum, and smaller binomial coefficients. I thank my colleague C. Batut for having pointed it out to me.

Proposition 9.1.16. *For any k and m in $\mathbb{Z}_{\geqslant 0}$ we have*

$$\sum_{j=1}^{\max(k,m)} \left(\binom{k}{j} + (-1)^{j+1} \binom{m}{j} \right) \frac{B_{k+m+1-j}(x)}{k+m+1-j}$$

$$= x^k (x-1)^m + \frac{(-1)^{m+1}}{(k+m+1)\binom{k+m}{k}},$$

$$\sum_{j=1}^{\max(k,m)} \left(\binom{k}{j} + (-1)^{j+1} \binom{m}{j} \right) B_{k+m-j}(x)$$

$$= x^{k-1}(x-1)^{m-1}((k+m)x - k),$$

$$\sum_{j=0}^{\lfloor (k-1)/2 \rfloor} \binom{k}{2j+1} \frac{B_{2k-2j}(x)}{2k-2j} = \frac{x^k(x-1)^k}{2} + \frac{(-1)^{k+1}}{(4k+2)\binom{2k}{k}},$$

$$\sum_{j=0}^{\lfloor k/2 \rfloor} \left(\binom{k}{2j+1} + \binom{k+1}{2j+1} \right) \frac{B_{2k+1-2j}(x)}{2k+1-2j} = x^k(x-1)^k(x-1/2)$$

(in all the above we recall that when $k \in \mathbb{Z}_{\geqslant 0}$ we have $\binom{k}{j} = 0$ if $j < 0$ or if $j > k$).

Proof. Consider x as a fixed parameter and set

$$F_x(t) = \frac{E(t,x)}{t} = \frac{e^{tx}}{e^t - 1} = \frac{1}{t} + \sum_{k \geqslant 0} \frac{B_{k+1}(x)}{k+1} \frac{t^k}{k!},$$

let $D = d/dt$ be the differentiation operator with respect to t, and let I be the identity operator. We begin with the following lemma.

Lemma 9.1.17. *With the above notation we have*

$$(e^t D^m (D+I)^k - D^k (D-I)^m) F_x(t) = x^k (x-1)^m e^{xt}.$$

Proof. For simplicity write F_x instead of $F_x(t)$. Leibniz's rule can be written in operator notation

$$D^N (e^{at} F_x) = \left(\sum_{j=0}^{N} \binom{N}{j} a^{N-j} e^{at} D^j \right) F_x = e^{at} (D + aI)^N F_x.$$

If we apply D^k to the defining identity $e^t F_x - F_x = e^{xt}$ we thus obtain $e^t (D+I)^k F_x - D^k F_x = x^k e^{xt}$, so multiplying by e^{-t} we have

$$(D+I)^k F_x - e^{-t} D^k F_x = x^k e^{t(x-1)},$$

and finally applying D^m we obtain

$$D^m (D+I)^k F_x - e^{-t}(D-I)^m D^k F_x = x^k (x-1)^m e^{t(x-1)},$$

proving the lemma after multiplication by e^t. □

Proof of the proposition. Denote by $G[0]$ (*not* $G(0)$) the constant term in a Laurent series $G(t)$. Taking constant terms in the lemma we obtain

$$\sum_{j=0}^{k}\binom{k}{j}(e^t D^{k+m-j}F_x(t))[0]+\sum_{j=0}^{m}(-1)^{j+1}\binom{m}{j}(D^{k+m-j}F_x(t))[0] = x^k(x-1)^m .$$

By definition we have $(D^N F_x)[0] = B_{N+1}(x)/(N+1)$, and

$$D^N F_x(t) = \frac{(-1)^N N!}{t^{N+1}} + \sum_{j\geqslant 0}\frac{B_{j+N+1}(x)}{j+N+1}\frac{t^j}{j!} ,$$

$$(e^t D^N F_x(t))[0] = \frac{(-1)^N N!}{(N+1)!} + \frac{B_{N+1}(x)}{N+1} = \frac{(-1)^N}{N+1} + \frac{B_{N+1}(x)}{N+1} ,$$

so replacing in the formula we obtain

$$\sum_{j=0}^{k}\binom{k}{j}\frac{(-1)^{k+m-j}}{k+m-j+1} + \sum_{j=1}^{\max(k,m)}\left(\binom{k}{j}+(-1)^{j+1}\binom{m}{j}\right)\frac{B_{k+m+1-j}}{k+m+1-j}$$
$$= x^k(x-1)^m ,$$

where the second sum starts at $j = 1$, since for $j = 0$ the binomial coefficients cancel. Furthermore, we have

$$\sum_{j=0}^{k}(-1)^j\binom{k}{j}\frac{1}{k+m+1-j} = \int_0^1 t^m(t-1)^k\,dt$$

$$= (-1)^k\int_0^1 t^m(1-t)^k\,dt = \frac{(-1)^k}{(k+m+1)\binom{k+m}{k}} ,$$

since it is easily shown by induction on k that

$$\int_0^1 t^m(1-t)^k\,dt = \frac{k!\,m!}{(k+m+1)!} = \frac{1}{(k+m+1)\binom{k+m}{k}}$$

(see Proposition 9.6.39 below for a more general formula). Replacing gives the first formula of the proposition. The second immediately follows by differentiating with respect to x, the third follows by choosing $m = k$ in the first formula, and the fourth by choosing $m = k+1$ in the first formula and subtracting the third. □

Corollary 9.1.18. *For any k and m in $\mathbb{Z}_{\geqslant 0}$ we have*

$$\sum_{j=0}^{\max(k,m)}\left(\binom{k}{j}+(-1)^{k+m}\binom{m}{j}\right)\frac{B_{k+m+1-j}}{k+m+1-j} = \frac{(-1)^{m+1}}{(k+m+1)\binom{k+m}{k}} ,$$

$$\sum_{j=0}^{\max(k,m)} \left(\binom{k}{j} + (-1)^{k+m+1} \binom{m}{j} \right) B_{k+m-j} = 0 \,,$$

for $k \geqslant 1$

$$\sum_{j=0}^{\lfloor (k-1)/2 \rfloor} \binom{k}{2j+1} \frac{B_{2k-2j}}{2k-2j} = \frac{(-1)^{k+1}}{(4k+2)\binom{2k}{k}} \,,$$

for $k \geqslant 2$

$$\sum_{j=0}^{\lfloor k/2 \rfloor} \left(\binom{k}{2j+1} + \binom{k+1}{2j+1} \right) B_{2k-2j} = 0$$

and

$$\sum_{j=0}^{\lfloor k/2 \rfloor} \left(\binom{k}{2j+1} + \binom{k+1}{2j+1} \right) 4^{2j} E_{2k-2j} = (-3)^k \,.$$

Proof. The first four formulas follow by taking $x = 0$ in the proposition and using the formulas for the odd Bernoulli numbers. The replacement of $(-1)^j$ by $\pm(-1)^{k+m}$ and the fact that we begin at $j = 0$ removes the special cases. The details are left to the reader. The last formula is obtained by taking $x = 1/4$ in the last formula of the proposition.   □

A restatement of the fourth formula is the following:

Corollary 9.1.19. *For $k \geqslant 2$ we have*

$$B_{2k} = -\frac{1}{(k+1)(2k+1)} \sum_{j=1}^{\lfloor k/2 \rfloor} (2k-2j+1) \binom{k+1}{2j+1} B_{2k-2j} \,.$$

We could of course restate in the same way the last formula to obtain a shorter recurrence for Euler numbers, but it is not certain that this would be any better than the standard one since the recurrence would involve nonintegral rational numbers.

Thus, as mentioned above, we obtain a recurrence giving B_{2k} as a linear combination of the preceding B_{2k-2j}, but only those with $2k - 2j \geqslant 2\lceil k/2 \rceil$, hence half as many as the formulas obtained using the more standard recurrences. Furthermore, the coefficients of the linear combinations are smaller binomial coefficients since (forgetting the simple factor $(2k - 2j + 1)$) they have the form $\binom{k+1}{j}$ instead of $\binom{2k}{j}$.

9.1.3 Computing a Single Bernoulli Number

If we want to compute a *table* of Bernoulli numbers up to a desired limit, the above recurrence or others are suitable. But if we want to compute a *single* value of a Bernoulli number B_k for k even, computing all the preceding B_j

up to B_k using recurrences is a waste of time and space since there exist more efficient direct methods. We assume of course that k is even. The first method is based on a direct formula for B_k given in Exercise 26. The second method is based on two results that we shall prove below (Corollary 9.1.21 and Theorem 9.5.14). One is the well-known formula

$$B_k = (-1)^{k/2-1} \frac{2 \cdot k!}{(2\pi)^k} \sum_{m \geqslant 1} \frac{1}{m^k} \, ,$$

which gives a very precise asymptotic estimate on the size of B_k. The other is the Clausen–von Staudt congruence, which gives the exact denominator D_k of the rational number B_k:

$$D_k = \prod_{(p-1)|k} p \, ,$$

where the product is over prime numbers p such that $(p - 1) \mid k$. It is thus sufficient to compute an approximation A_k to $D_k B_k$ such that $|A_k - D_k B_k| < 1/2$, and the numerator of B_k will then be equal to the nearest integer to A_k. This indeed gives a very efficient method to compute an individual value of B_k.

Note that the implementation of this method should be done with care. We first compute the denominator D_k and $k!$ in a naïve way. We must then estimate the number of decimal digits d with which to perform the computation, and the number N of terms to take in the zeta series. A cursory analysis shows that one can take

$$d = 3 + \lceil d_1/ \log(10) \rceil \quad N = 1 + \lceil \exp((d_1 - \log(k - 1))/(k - 1)) \rceil \, ,$$

where

$$d_1 = \log(D_k) + (k + 1/2) \log(k) - k(\log(2\pi) + 1) + \log(2\pi)/2 + \log(2) + 0.1 \, .$$

Thanks to Stirling's formula the reader will recognize that d_1 is close to $\log(D_k|B_k|)$, and the 3+ and 1+ are safety precautions. Note that the above computations should be done to the lowest possible accuracy, since at this point we only want integers d and N.

The computation of π can be done using many different methods, but since anyway you will have to use a CAS for the multiprecision operations, this is always built in. Of course π^k is computed using a binary powering method.

When k is large all this takes only a small fraction of the time, almost all the time being spent in the computation of $\zeta(k) = \sum_{m \geqslant 1} m^{-k}$ to the desired number of decimal digits d. Note that since k is large, $\zeta(k)$ is very close to 1. Once again there are several methods to do this computation, but in the author's opinion the best method is as follows. First, instead of computing the *series* $\zeta(k)$, we compute the *Euler product* $1/\zeta(k) = \prod_p (1 - 1/p^k)$, the product being over all prime numbers up to the precomputed limit N. Second, the multiplication of the current product P by $1 - 1/p^k$ is not done naïvely as $P(1 - 1/p^k)$ but as $P - t(P, d - k\log(p))/p^k$, where $t(P, d')$ is equal to P truncated to the accuracy d'. Indeed, contrary to most computations in numerical analysis, here we need *absolute* and not relative accuracy. Although this is a technical remark it can in itself gain a factor of 3 or 4.

Note that when suitably implemented the above method is so efficient that it can even be faster than the method using recurrences for computing a *table*.

To give an idea of the speed, on a Pentium 4 at 3 Ghz the computation of B_{10000} requires 33 seconds using the formula of Exercise 26, but only 0.3 seconds using the above method. The computation of all Bernoulli numbers up to B_{5000} requires 205 seconds using the standard recurrences given above and 26 seconds using the above method that computes each number *individually*, which is indeed considerably faster.

9.1.4 Bernoulli Polynomials and Fourier Series

In this section we give a direct link between Bernoulli polynomials and certain Fourier series. This will later be useful for computing special values of Dirichlet L-functions (see Section 10.3).

It is important to compute the Fourier series corresponding to the functions $B_k(x)$ for $k \geqslant 1$ (for $k = 0$ it is trivial), more precisely to the functions obtained by extending by periodicity of period 1 the kth Bernoulli polynomial on the interval $[0, 1[$. We will denote by $\{x\}$ the *fractional part* of x, in other words the unique real number in $[0, 1[$ such that $x - \{x\} \in \mathbb{Z}$, i.e., $\{x\} = x - \lfloor x \rfloor$. The function $B_k(\{x\})$ is evidently periodic of period 1. The result is as follows.

Theorem 9.1.20. (1) *For $n \geqslant 2$ even we have*

$$\sum_{k \geqslant 1} \frac{\cos(2\pi kx)}{k^n} = \frac{(-1)^{n/2+1}}{2} \frac{(2\pi)^n B_n(\{x\})}{n!} .$$

(2) *For $n \geqslant 1$ odd we have*

$$\sum_{k \geqslant 1} \frac{\sin(2\pi kx)}{k^n} = \frac{(-1)^{(n+1)/2}}{2} \frac{(2\pi)^n B_n(\{x\})}{n!} ,$$

except for $n = 1$ and $x \in \mathbb{Z}$, in which case the left-hand side is evidently equal to 0.

(3) *For $x \notin \mathbb{Z}$ we have*

$$\sum_{k \geqslant 1} \frac{\cos(2\pi kx)}{k} = -\log(2|\sin(\pi x)|) .$$

Proof. (1) and (2). Since $B_n(1) = B_n(0)$ for $n \neq 1$, the function $B_n(\{x\})$ is piecewise C^∞ and continuous for $n \geqslant 2$, with simple discontinuities at the integers if $n = 1$. If $n \geqslant 2$ we thus have

$$B_n(\{x\}) = \sum_{k \in \mathbb{Z}} c_{n,k} e^{2i\pi kx} \;,$$

with

$$c_{n,k} = \int_0^1 B_n(t) e^{-2i\pi kt}\, dt \;.$$

For $n = 1$, the same formula is valid for $x \notin \mathbb{Z}$, and for $x \in \mathbb{Z}$ we must replace $B_1(\{x\})$ by $(B_1(1^-) + B_1(0^+))/2 = 0$.

Using the definitions and the formulas $B_n'(x) = nB_{n-1}(x)$ and $B_n(1) = B_n(0)$ for $n \neq 1$, by integration by parts we obtain for $k \neq 0$

$$c_{n,k} = \frac{n}{2i\pi k} c_{n-1,k} \quad \text{and} \quad c_{1,k} = -\frac{1}{2i\pi k} \;,$$

hence by induction

$$c_{n,k} = -\frac{n!}{(2i\pi k)^n} \;.$$

On the other hand, we clearly have

$$c_{n,0} = \frac{B_{n+1}(1) - B_{n+1}(0)}{n+1} = 0$$

as soon as $n \geqslant 1$. Thus, with the above interpretation for $x \in \mathbb{Z}$ when $n = 1$, we obtain that for $n \geqslant 1$ we have

$$B_n(\{x\}) = -\frac{n!}{(2i\pi)^n} \sum_{k \neq 0} \frac{e^{2i\pi kx}}{k^n} \;.$$

Separating the cases n even and n odd, and grouping the terms k and $-k$ proves (1) and (2).

For (3) we proceed differently. We have

$$\sum_{k \geqslant 1} \frac{\cos(2\pi kx)}{k} = \Re\left(\sum_{k \geqslant 1} \frac{e^{2i\pi kx}}{k} \right) = -\Re\left(\log(1 - e^{2i\pi x})\right)$$

$$= -\log\left(|1 - e^{2i\pi x}|\right) = -\log(2|\sin(\pi x)|) \;,$$

proving the theorem.   □

Corollary 9.1.21. *For $n \geqslant 1$ we have*

$$\sum_{k \geqslant 1} \frac{1}{k^{2n}} = \frac{(-1)^{n-1}}{2} \frac{(2\pi)^{2n} B_{2n}}{(2n)!} \;,$$

and for $n \geqslant 0$ we have

$$\sum_{k \geqslant 0} \frac{(-1)^k}{(2k+1)^{2n+1}} = \frac{(-1)^{n-1}}{2} \frac{(2\pi)^{2n+1} B_{2n+1}(1/4)}{(2n+1)!} = \frac{(-1)^n}{2} \frac{(\pi/2)^{2n+1} E_{2n}}{(2n)!}.$$

In particular, the sign of B_{2n} is equal to $(-1)^{n-1}$ for $n \geqslant 1$ and the sign of E_{2n} is equal to $(-1)^n$ for $n \geqslant 0$, so both have alternating signs.

Proof. This is a direct consequence of the theorem by choosing $x = 0$ for n even and $x = 1/4$ for n odd. Note that $\sum_{k \geqslant 0} (-1)^k/(2k+1)^{2n+1} > 0$ since it is an alternating series with decreasing terms. □

Note that these are special cases of Theorem 10.3.1, which we will prove in the next chapter. Conversely, we can give an alternative proof of this theorem using Theorem 9.1.20; see Exercise 35 of Chapter 10.

Corollary 9.1.22. *As n tends to infinity, we have*

$$B_{2n} \sim (-1)^{n-1} \frac{2(2n)!}{(2\pi)^{2n}} \quad and$$

$$E_{2n} \sim (-1)^n \frac{2(2n)!}{(\pi/2)^{2n+1}}.$$

Proof. Clear since $\sum_{k \geqslant 1} 1/k^{2n}$ and $\sum_{k \geqslant 0} (-1)^k/(2k+1)^{2n+1}$ tend to 1 as $n \to \infty$. □

This corollary shows that, as already mentioned, most asymptotic expansions involving Bernoulli numbers or Euler numbers will *diverge*, since $(2n)!$ grows much faster than any power of n. Only rare expansions which have an expression such as $(2n)!$ in the denominator may converge.

Examples.

$$\sum_{k \geqslant 1} \frac{1}{k^2} = \frac{\pi^2}{6}, \quad \sum_{k \geqslant 1} \frac{1}{k^4} = \frac{\pi^4}{90}, \quad \sum_{k \geqslant 1} \frac{1}{k^6} = \frac{\pi^6}{945}, \quad \sum_{k \geqslant 1} \frac{1}{k^8} = \frac{\pi^8}{9450},$$

$$\sum_{k \geqslant 0} \frac{(-1)^k}{2k+1} = \frac{\pi}{4}, \quad \sum_{k \geqslant 0} \frac{(-1)^k}{(2k+1)^3} = \frac{\pi^3}{32}, \quad \sum_{k \geqslant 0} \frac{(-1)^k}{(2k+1)^5} = \frac{5\pi^5}{1536},$$

$$\sum_{k \geqslant 0} \frac{(-1)^k}{(2k+1)^7} = \frac{61\pi^7}{184320}, \quad \sum_{k \geqslant 0} \frac{(-1)^k}{(2k+1)^9} = \frac{277\pi^9}{8257536}.$$

Note also the following corollary, which is very useful for giving upper bounds on the remainder terms in the Euler–MacLaurin summation formula.

Corollary 9.1.23. *If n is even we have*

$$\sup_{x \in \mathbb{R}} |B_n(\{x\})| = |B_n|$$

and if n is odd we have

$$\sup_{x \in \mathbb{R}} |B_n(\{x\})| \leqslant \frac{7|B_{n+1}|}{n+1} \, .$$

Proof. The first statement immediately follows from Theorem 9.1.20 and the fact that $|\cos(2\pi kx)| \leqslant 1$, with equality for all k if $x = 0$. This proof is not valid for n odd. For $n = 1$ we have $B_1(x) = x - 1/2$, hence $\sup_{x \in \mathbb{R}} |B_1(\{x\})| = 1/2 < 7|B_2/2|$, since $B_2 = 1/6$. For $n \geqslant 3$ odd, we have

$$\left| \frac{B_n(\{x\})}{B_{n+1}} \right| \leqslant \frac{2(n!)}{(2\pi)^n} \zeta(n) \frac{(2\pi)^{n+1}}{2((n+1)!)\zeta(n+1)} = \frac{2\pi}{n+1} \frac{\zeta(n)}{\zeta(n+1)} \, .$$

It is easily checked that for $s \geqslant 3$ the function $\zeta(s)/\zeta(s+1)$ is decreasing, so it attains its maximum value for $s = 3$, and the second result follows since $2\pi\zeta(3)/\zeta(4) < 7$. Note that one can prove the same result with 2π instead of 7, and that 2π is the optimal constant, but we do not need this for applications since we only want to give a reasonable upper bound for the error terms. □

9.2 Analytic Applications of Bernoulli Polynomials

Even though for us the main use of Bernoulli numbers is of number-theoretic nature, as we shall see for special values of L-functions (we have already seen some examples above) and, as we shall see in Chapter 11, in congruence properties leading to the definition of p-adic zeta and L-functions, it is important to note that they are also essential for purely analytic reasons, mainly because of the Euler–MacLaurin summation formula.

In addition to the above section on generating functions and recurrences, we will thus devote four sections to Bernoulli polynomials. The present section and the next deal with the analytic properties, i.e., essentially those linked to the Euler–MacLaurin formula, Section 9.4 deals with χ-Bernoulli polynomials, and Section 9.5 deals with the arithmetic properties of Bernoulli numbers.

9.2.1 Asymptotic Expansions

We begin by recalling the definition of an asymptotic expansion. Even though we can define this in a more general setting, we will assume that we deal with asymptotic expansions at infinity.

Definition 9.2.1. *Let u_N be a sequence of complex numbers. We will say that a sequence (a_n) is the sequence of coefficients of an* asymptotic expansion *(at infinity) of u_N if for every $k \geqslant 0$, as $N \to \infty$ we have*

$$u_N = a_0 + \frac{a_1}{N} + \frac{a_2}{N^2} + \cdots + \frac{a_k}{N^k} + o(1/N^k) ,$$

where we recall that $f(N) = o(1/N^k)$ means that $N^k f(N)$ tends to 0 as $N \to \infty$.

It is easy to see by induction on k that an asymptotic expansion, if it exists, is unique. However, I emphasize the fact that in practice it is quite rare that the corresponding power series $\sum_{j \geqslant 0} a_j/N^j$ converges; in other words, the power series $\sum_{j \geqslant 0} a_j x^j$ usually has a radius of convergence equal to 0. Nevertheless, by abuse of notation we will write

$$u_N = a_0 + \frac{a_1}{N} + \frac{a_2}{N^2} + \cdots ,$$

when it is understood that it is an asymptotic expansion in the above sense, and not a convergent power series.

Even though the series converges nowhere in general, we can usually use an asymptotic expansion to compute u_N numerically to quite high accuracy, by bounding the error term $o(1/N^k)$. We will see below as applications of the Euler–MacLaurin summation formula many examples of asymptotic expansions, of bounds on the error terms, and of numerical computations. For the moment consider the following example.

Example. Let u_N be defined by

$$u_N = e^N \int_N^\infty \frac{e^{-t}}{t} dt$$

(this is equal to $e^N E_1(N)$, see Section 8.5.3). Successive integration by parts shows by induction that

$$u_N = \frac{0!}{N} - \frac{1!}{N^2} + \frac{2!}{N^3} - \cdots + (-1)^{k-1}\frac{(k-1)!}{N^k} + (-1)^k k! e^N \int_N^\infty \frac{e^{-t}}{t^{k+1}} dt .$$

It is easy to show that this defines an asymptotic expansion in the above sense, so we will write

$$e^N \int_N^\infty \frac{e^{-t}}{t} dt = \frac{0!}{N} - \frac{1!}{N^2} + \frac{2!}{N^3} - \cdots + (-1)^{k-1}\frac{(k-1)!}{N^k} + \cdots ,$$

knowing that this expansion converges for *no* value of N. From the explicit expression of the remainder term it is however clear that u_N is always between two consecutive terms (this is very frequently the case in asymptotic expansions), and in particular the error is less than the absolute value of the first neglected term. If for instance we choose $N = 40$, taking $k = 40$ we see that the error is less than $40!/40^{41} < 2 \cdot 10^{-18}$, so that we can compute very accurately the value of u_{40} (we obtain $u_{40} = 0.024404115079628577\ldots$).

In practice we generalize the notion of asymptotic expansion in two ways: first by allowing a *finite* number of auxiliary functions of N such as positive powers of N, powers of logarithms or exponentials, etc., either additively or multiplicatively; second by allowing the expansion to be in powers of some other function of N than $1/N$, most frequently $1/N^{1/2}$ or more generally $1/N^{\alpha}$ for some $\alpha > 0$.

9.2.2 The Euler–MacLaurin Summation Formula

The Euler–MacLaurin summation formula is a simple but powerful tool that enables us to solve (for instance) the following problems:

- Find the asymptotic expansion of the Nth partial sum of a divergent series.
- Find the asymptotic expansion of the Nth remainder of a convergent series, and consequently considerably accelerate the convergence of the series.
- Find the asymptotic expansion of the difference between a definite integral and corresponding Riemann sums, which allows us to compute much more accurately and much faster the numerical value of the integral.
- Determine whether a given series converges by comparison with the corresponding integral.

We will see several examples of all of this. The general Euler–MacLaurin formula is not complicated, and is easy to prove, but this does not prevent it from being very useful. Taylor's formula is of a similar kind, and in fact Bourbaki calls formulas analogous to Euler–MacLaurin generalized Taylor expansions.

Although we could directly state and prove the formula, we prefer to begin with some preliminary remarks. We have seen above that $B_n(x+1) - B_n(x) = nx^{n-1}$. This should be compared with the identity $(x^n)' = nx^{n-1}$. Here the operation is derivation, and the antiderivative of nx^{n-1} is x^n. In our case, the operation is close to the derivation, it is the *difference operator* $f(x+1) - f(x)$, and the "antidifference" of nx^{n-1} is $B_n(x)$. This is why Bernoulli polynomials (and numbers) are so important in everything having to do with sums, as we will see in the Euler–MacLaurin summation formula.

The aim of this summation formula is to give an asymptotic expansion for a general sum of the type $\sum_{0 \leqslant m \leqslant N-1} f(m)$, where f is a regular function (for instance real analytic) on $\mathbb{R}$. Before giving a formal and rigorous proof, we will use a heuristic argument that is useful in other contexts. Denote by D the derivation operator d/dt. If f is an entire function, we have by Taylor's expansion

$$f(m) = \sum_{k \geqslant 0} m^k \frac{f^{(k)}(0)}{k!} = \left(\sum_{k \geqslant 0} \frac{m^k D^k}{k!} \right)(f)(0) = (e^{mD} f)(0) \,,$$

in a reasonable operator sense. Thus

$$\sum_{0\leqslant m\leqslant N-1} f(m) = \left(\sum_{0\leqslant m\leqslant N-1} e^{mD} f\right)(0) = \left(\frac{e^{ND}-1}{e^D-1}f\right)(0)$$

$$= \left(\frac{1}{e^D-1}f\right)(N) - \left(\frac{1}{e^D-1}f\right)(0) ,$$

since all power series operators in D commute and once again by Taylor we have $e^{ND}f(0) = f(N)$. By definition $1/(e^D-1) = 1/D + \sum_{j\geqslant 1}(B_j/j!)D^{j-1}$. The operator $1/D$ is of course the antiderivative operator, i.e., the integral, hence the above formal reasoning leads to the formula

$$\sum_{0\leqslant m\leqslant N-1} f(m) = \int_0^N f(t)\,dt + \sum_{j\geqslant 1}\frac{B_j}{j!}(f^{(j-1)}(N) - f^{(j-1)}(0)) .$$

This heuristic reasoning is essentially correct, but we have not taken into account the convergence conditions since in general the series that we have obtained is not convergent. In fact, the goal of the Euler–MacLaurin summation formula is to give an asymptotic expansion of the left-hand side, not an exact formula. The precise theorem is as follows, which we give in a more general form.

Theorem 9.2.2 (Euler–MacLaurin). *Let a and b be two real numbers such that $a \leqslant b$, and assume that $f \in C^k([a,b])$ for some $k \geqslant 1$. Then*

$$\sum_{\substack{a<m\leqslant b \\ m\in\mathbb{Z}}} f(m) = \int_a^b f(t)\,dt + \sum_{j=1}^k \frac{(-1)^j}{j!}\left(B_j(\{b\})f^{(j-1)}(b) - B_j(\{a\})f^{(j-1)}(a)\right)$$

$$+ \frac{(-1)^{k-1}}{k!}\int_a^b f^{(k)}(t)B_k(\{t\})\,dt .$$

Proof. We give a clean proof, using (very little) the language of distributions, and explain very briefly afterward how to avoid it.

By the basic properties of Bernoulli polynomials, we know that $B_k'(\{t\}) = kB_{k-1}(\{t\})$ for $k \geqslant 2$, except for $k = 2$ on a set of measure zero (the integers). Furthermore, $B_1'(\{t\}) = B_0(\{t\}) - \delta_{\mathbb{Z}}(t)$, where $\delta_{\mathbb{Z}}(t)$ is the Dirac distribution concentrated on $\mathbb{Z}$. Thus, if we set

$$R_k = \frac{(-1)^{k-1}}{k!}\int_a^b f^{(k)}(t)B_k(\{t\})\,dt ,$$

integration by parts gives for $k \geqslant 2$

$$R_k = \frac{(-1)^{k-1}}{k!}\left(B_k(\{b\})f^{(k-1)}(b) - B_k(\{a\})f^{(k-1)}(a)\right) + R_{k-1} .$$

For $k = 1$ we first assume that $a \notin \mathbb{Z}$ and $b \notin \mathbb{Z}$. Integration by parts gives

$$R_1 = B_1(\{b\})f(b) - B_1(\{a\})f(a) - \int_a^b f(t)\,dt + \sum_{a<m\leqslant b} f(m)\,.$$

Furthermore, we note that $R_1 = R_1(a,b)$ is a *continuous* function of a and b, and it is easily checked by letting a or b tend to integers that the right-hand side of this formula is also continuous, so it is valid for all a and b, integral or not. Using the recurrence on R_k we thus obtain

$$R_k = \sum_{j=1}^{k} \frac{(-1)^{j-1}}{j!} \left(B_j(\{b\})f^{(j-1)}(b) - B_j(\{a\})f^{(j-1)}(a) \right)$$

$$- \int_a^b f(t)\,dt + \sum_{a<m\leqslant b} f(m)\,,$$

proving the theorem.

To avoid the (very elementary) use of the Dirac distribution, we proceed as follows. Setting $a_0 = \lfloor a \rfloor$ and $b_0 = \lfloor b \rfloor$, we split the integral into the sum of an integral from a to $a_0 + 1$, of integrals from $a_0 + i$ to $a_0 + i + 1$ for $1 \leqslant i \leqslant b_0 - a_0 - 1$, and of an integral from b_0 to b. We then perform the same integrations by parts as above on each individual integral, and putting everything together we of course obtain the same result. $\square$

The following corollary gives three alternative forms of the Euler–MacLaurin formula, which for simplicity we give only for $b = a + N$ with $N \in \mathbb{Z}_{\geqslant 0}$, so that $\{b\} = \{a\}$.

Corollary 9.2.3. *Let $a \in \mathbb{R}$, $N \in \mathbb{Z}_{\geqslant 0}$, and $k \in \mathbb{Z}_{\geqslant 1}$.*

(1) *If $f \in C^k([a, N+a])$ we have*

$$\sum_{m=0}^{N-1} f(m+a) = \int_a^{N+a} f(t)\,dt + \sum_{j=1}^{k} \frac{B_j}{j!} \left(f^{(j-1)}(N+a) - f^{(j-1)}(a) \right) + R_k(f,N)\,,$$

$$\text{with}\quad R_k(f,N) = \frac{(-1)^{k-1}}{k!} \int_a^{N+a} f^{(k)}(t) B_k(\{t-a\})\,dt\,.$$

(2) *If $f \in C^{2k}([a, N+a])$ we have*

$$\sum_{m=0}^{N} f(m+a) = \int_a^{N+a} f(t)\,dt + \frac{f(N+a) + f(a)}{2}$$

$$+ \sum_{j=1}^{k} \frac{B_{2j}}{(2j)!} \left(f^{(2j-1)}(N+a) - f^{(2j-1)}(a) \right) + R_{2k}(f,N)\,,$$

$$\text{with}\quad R_{2k}(f,N) = -\frac{1}{(2k)!} \int_a^{N+a} f^{(2k)}(t) B_{2k}(\{t-a\})\,dt\,.$$

(3) *If $f \in C^{2k}([1, N])$ we have*

$$\sum_{m=1}^{N} f(m) = \int_{1}^{N} f(t)\, dt + \frac{f(N) + f(1)}{2}$$

$$+ \sum_{j=1}^{k} \frac{B_{2j}}{(2j)!} \left(f^{(2j-1)}(N) - f^{(2j-1)}(1) \right) + R_{2k}(f, N) ,$$

$$\text{with} \quad R_{2k}(f, N) = -\frac{1}{(2k)!} \int_{1}^{N} f^{(2k)}(t) B_{2k}(\{t\})\, dt .$$

Note that the main term of the last formula is equivalent to the one that we have obtained by our heuristic reasoning. Note also that we use the notation $R_k(f, N)$ with slightly different meanings.

Proof. For (1) we apply the theorem to $a = 0$ and $b = N$, replace the function $f(t)$ by $f(t + a)$, and subtract $f(N + a) - f(a)$. Formula (2) follows by changing k into $2k$ and using the values of the odd Bernoulli numbers. Formula (3) follows from (2) by choosing $a = 1$ and changing N into $N - 1$. ☐

Corollary 9.2.4. *Let $f \in C^{2k}([a, \infty[)$ for some $a \in \mathbb{R}$. Assume that both the series $\sum_{m \geq a} f(m)$ and the integral $\int_{a}^{\infty} f(t)\, dt$ converge, and that the derivatives $f^{(2j-1)}(N)$ tend to 0 as $N \to \infty$ for $1 \leq j \leq k$. Then*

$$\sum_{m=N+1}^{\infty} f(m) = \int_{N}^{\infty} f(t)\, dt - \frac{f(N)}{2} - \sum_{j=1}^{k} \frac{B_{2j}}{(2j)!} f^{(2j-1)}(N) + R_{2k}(f, N) ,$$

with

$$R_{2k}(f, N) = -\frac{1}{(2k)!} \int_{N}^{\infty} f^{(2k)}(t) B_{2k}(\{t\})\, dt .$$

Proof. Immediate and left to the reader (Exercise 52). ☐

Remark. If it is inconvenient to compute the successive derivatives of the function f, we could hope to replace them for instance by the iterated forward differences obtained by iterating $(\delta f)(t) = f(t+1) - f(t)$ (or by the centered differences $f(t+1/2) - f(t-1/2)$ if preferred). In this case, we would need a formula involving this operator instead of the derivative operator $D = d/dt$. This is where the heuristic reasoning made at the beginning comes in handy: by Taylor we have $\delta = e^{D} - 1$, hence $D = \log(1 + \delta)$. Thus the operator $1/(e^{D} - 1)$ that is involved in the Euler–MacLaurin formula can be rewritten formally as

$$\frac{1}{e^{D} - 1} = \frac{1}{D} + \left(\frac{1}{e^{D} - 1} - \frac{1}{D} \right) = \frac{1}{D} + \left(\frac{1}{\delta} - \frac{1}{\log(1 + \delta)} \right) .$$

Thus if we define the δ-Bernoulli numbers b_k by

$$\frac{t}{\log(1+t)} = \sum_{k \geqslant 0} \frac{b_k}{k!} t^k \;,$$

we have

$$\frac{1}{e^D - 1} = \frac{1}{D} - \sum_{k \geqslant 1} \frac{b_k}{k!} \delta^{k-1} \;.$$

From the formal Euler–MacLaurin formula we can thus deduce an asymptotic expansion for $\sum_{0 \leqslant m \leqslant N-1} f(m+a)$ involving the antiderivative operator $1/D$ and the iterates of the forward difference operator instead of the iterates of the derivative operator. An analogous argument also holds for similar operators (everywhere-convergent power series in D with zero constant term).

9.2.3 The Remainder Term and the Constant Term

To use the Euler–MacLaurin formula (usually in the form of Corollary 9.2.3) we must give an estimate for the remainder term. We give it for the second formula of the corollary, the third being obtained by replacing $[a, N+a]$ with $[1, N]$.

Proposition 9.2.5. *Assume that $f \in C^{2k+2}([a, N + a])$ and denote by $T_{2k+2}(f, N)$ the first "neglected term" in Corollary 9.2.3 (2), in other words $T_{2k+2}(f, N) = (B_{2k+2}/(2k + 2)!)(f^{(2k+1)}(N + a) - f^{(2k+1)}(a))$. Assume that $f^{(2k+2)}(t)$ has constant sign on $[a, a + N]$. Then*

(1) *The remainder term $R_{2k}(f, N)$ has the same sign as $T_{2k+2}(f, N)$ and satisfies $|R_{2k}(f, N)| \leqslant 2(1 - 2^{-2k-2})|T_{2k+2}(f, N)|$.*
(2) *If, in addition, $f \in C^{2k+4}([a, a + N])$ and $f^{(2k+4)}(t)$ are also of constant sign on $[a, a + N]$ then $|R_{2k}(f, N)| \leqslant |T_{2k+2}(f, N)|$; in other words, the remainder term is in absolute value smaller than the first neglected term.*

The term $T_{2k+2}(f, N)$ is of course not to be confused with the tangent numbers T_{2k-1}.

Proof. (1). For notational simplicity set $K = 2k + 2$, and let $\varepsilon = \pm 1$ be such that $\varepsilon f^{(K)}(t) \geqslant 0$ for $t \in [a, N + a]$. Since $|B_K(t)| \leqslant |B_K|$ for $t \in [0, 1]$ (Corollary 9.1.23) we have

$$|R_K(f, N)| \leqslant \frac{|B_K|}{K!} \int_a^{N+a} |f^{(K)}(t)| \, dt \leqslant \frac{|B_K|}{K!} \varepsilon \int_a^{N+a} f^{(K)}(t) \, dt$$

$$\leqslant \frac{|B_K|}{K!} \varepsilon (f^{(K-1)}(N + a) - f^{(K-1)}(a)) \leqslant |T_K(f, N)| \;.$$

On the other hand, applying Corollary 9.2.3 to k and to $k + 1$ it is clear that $R_{2k}(f, N) = T_K(f, N) + R_K(f, N)$. Since we have just proved that $|R_K(f, N)| \leqslant |T_K(f, N)|$ it follows that $R_{2k}(f, N)$ has the same sign as

page 26 9. Bernoulli Polynomials and the Gamma Function

$T_K(f, N)$ and also that $|R_{2k}(f, N) \leqslant 2T_K(f, N)$. To obtain the slightly stronger inequality given in the proposition we write

$$T_K(f, N) = \frac{B_K}{K!} \int_a^{N+a} (f^{(K)}(t) \, dt \ ,$$

so that

$$R_{2k}(f, N) = T_K(f, N) + R_K(f, N) = \frac{1}{K!} \int_a^{N+a} f^{(K)}(t)(B_K - B_K(\{t-a\})) \, dt \ .$$

By Exercise 15 we have $|B_K - B_K(\{t - a\})| \leqslant 2(1 - 2^{-K})|B_K|$, proving (1).

(2). Applying (1) to $2k + 2$ instead of $2k$, for $t \in [a, N + a]$ we have

$$\mathrm{sign}(R_{2k+2}(f, N)) = \mathrm{sign}(T_{2k+4}(f, N)) = \mathrm{sign}(B_{2k+4}) \, \mathrm{sign}(f^{2k+4}(t))$$
$$= -\mathrm{sign}(B_{2k+2}) \, \mathrm{sign}(f^{(2k+2)}(t)) = -\mathrm{sign}(T_{2k+2}(f, N)) \ ,$$

where $\mathrm{sign}(0)$ agrees with any value of ± 1. Since $R_{2k}(f, N) = T_{2k+2}(f, N) + R_{2k+2}(f, N)$ it follows that $|R_{2k}(f, N)| \leqslant |T_{2k+2}(f, N)|$ as claimed. □

The following is another useful form of the Euler–MacLaurin formula, where we introduce the notion of "constant term," used by Ramanujan without any justification.

Corollary 9.2.6. *Let $k \geqslant 1$, and let $f \in C^k([a, \infty[)$.*

(1) *Assume that the sign of $f^{(k)}(t)$ is constant on $[a, \infty[$ and that $f^{(k-1)}(t)$ tends to 0 as $t \to \infty$. There exists a constant $z_k(f, a)$ such that*

$$\sum_{m=0}^{N-1} f(m+a) = z_k(f, a) + \int_a^{N+a} f(t) \, dt + \sum_{j=1}^{k-1} \frac{B_j}{j!} f^{(j-1)}(N+a) + R_k(f, N) \ ,$$

where

$$R_k(f, N) = \frac{(-1)^k}{k!} \int_{N+a}^{\infty} f^{(k)}(t)(B_k(\{t - a\}) - B_k) \, dt$$

tends to 0 as $N \to \infty$.

(2) *Let $k_0 \geqslant 1$ be an integer. If the sign of $f^{(k)}(t)$ is constant and $f^{(k-1)}(t)$ tends to 0 as $t \to \infty$ for all $k \geqslant k_0$, then for $k \geqslant k_0$ the constant $z_k(f, a)$ is independent of k. It will be simply denoted by $z(f, a)$ and called the constant term of the formula, and we have the following identity, valid for any fixed $k \geqslant k_0$:*

$$z(f, a) = -\sum_{j=1}^{k} \frac{B_j}{j!} f^{(j-1)}(a) + \frac{(-1)^{k-1}}{k!} \int_a^{\infty} f^{(k)}(t) B_k(\{t - a\}) \, dt \ .$$

Proof. By the Euler–MacLaurin formula, we have for all $k \geqslant 1$

$$\sum_{m=0}^{N-1} f(m+a) = z_k(f,a,N) + \int_a^{N+a} f(t)\,dt + \sum_{j=1}^{k} \frac{B_j}{j!} f^{(j-1)}(N+a)\,,$$

where

$$z_k(f,a,N) = -\sum_{j=1}^{k} \frac{B_j}{j!} f^{(j-1)}(a) + \frac{(-1)^{k-1}}{k!} \int_a^{N+a} f^{(k)}(t) B_k(\{t-a\})\,dt\,.$$

Since the sign of $f^{(k)}(t)$ is constant it follows that the above integral is bounded in absolute value by $\sup_{t \in [0,1]} |B_k(t)| \|f^{(k-1)}(N+a) - f^{(k-1)}(a)|$. Since by assumption $f^{(k-1)}(t)$ tends to 0 as $t \to \infty$, it is in particular bounded, and it follows that the integral $\int_a^\infty f^{(k)}(t) B_k(\{t-a\})\,dt$ is absolutely convergent. Thus $z_k(f,a,N) = z_k(f,a) + I_k(f,N)$ with

$$z_k(f,a) = -\sum_{j=1}^{k} \frac{B_j}{j!} f^{(j-1)}(a) + \frac{(-1)^{k-1}}{k!} \int_a^\infty f^{(k)}(t) B_k(\{t-a\})\,dt$$

and

$$I_k(f,N) = \frac{(-1)^k}{k!} \int_{N+a}^\infty f^{(k)}(t) B_k(\{t-a\})\,dt\,.$$

Since the integral of $f^{(k)}(t) B_k(\{t-a\})$ converges (absolutely) at infinity, $I_k(f,N)$ tends to 0 as $N \to \infty$. Finally, by assumption $f^{(k-1)}(N+a)$ also tends to 0 and we have

$$\frac{B_k}{k!} f^{(k-1)}(N+a) = -\frac{B_k}{k!} \int_{N+a}^\infty f^{(k)}(t)\,dt\,,$$

proving (1).

If in addition we assume that $f^{(k)}(t)$ has constant sign and that $f^{(k-1)}(t)$ tends to 0 as $t \to \infty$ for all $k \geqslant k_0$, then subtracting (1) for k from (1) for $k+1$ we obtain

$$0 = z_{k+1}(f,a) - z_k(f,a) + \frac{B_k}{k!} f^{(k-1)}(N+a) + o(1)\,,$$

so letting $N \to \infty$ and using the fact that $f^{(k-1)}(t)$ tends to 0, we deduce that $z_{k+1}(f,a) = z_k(f,a)$, hence that $z_k(f,a)$ is indeed independent of k, proving (2). $\square$

9.2.4 Euler–MacLaurin and the Laplace Transform

See Section 9.7.4 for more details on the Laplace transform. Recall the following definition:

Definition 9.2.7. *Let g be a piecewise continuous function on $[0, \infty[$ such that for all $a > 0$, the function $g(t)e^{-at}$ tends to 0 as $t \to \infty$. We define the Laplace transform $\mathcal{L}(g)$ of g by the formula*

$$\mathcal{L}(g)(x) = \int_0^\infty e^{-tx} g(t)\, dt \ .$$

Thanks to the assumptions on g it is clear that $\mathcal{L}(g)$ is well defined and defines a holomorphic function on $\Re(x) > 0$, and that

$$\mathcal{L}(g)^{(k)}(x) = (-1)^k \int_0^\infty e^{-tx} t^k g(t)\, dt = (-1)^k \mathcal{L}(t^k g)(x) \ .$$

In addition, note that by Fubini's formula we have

$$\int_a^b \mathcal{L}(g)(x)\, dx = \int_0^\infty \frac{e^{-at} - e^{-bt}}{t} g(t)\, dt \ .$$

Applying this with $g(t) = 1$, hence $\mathcal{L}(x) = 1/x$, and $a = 1$ gives the well-known and important formula

$$\log(x) = \int_0^\infty \frac{e^{-t} - e^{-xt}}{t}\, dt \ .$$

The relation between the Laplace transform and the Euler–MacLaurin formula is clear: if $f(x) = \mathcal{L}(g)(x)$, then for instance

$$\sum_{1 \leqslant m \leqslant N} f(m) = \int_0^\infty \frac{1 - e^{-Nt}}{e^t - 1} g(t)\, dt \ .$$

From this we obtain both a formula for the sum of the infinite series if it converges (or more generally for the constant term $z(f, 1)$ defined above) by letting N tend to infinity, and a formula for the remainder term in Euler–MacLaurin by expanding $1/(e^t - 1)$ in terms of Bernoulli numbers and using the formula given above for $f^{(k)}(N)$.

Since there are many forms of the Euler–MacLaurin formula, there are as many expressions for the remainder term and the constant term. We give the following:

Proposition 9.2.8. *Keep the above assumptions on g and set $f(x) = \mathcal{L}(g)(x)$. For $k \geqslant 1$ we have*

(1)

$$\frac{(-1)^{k-1}}{k!} \int_a^{N+a} f^{(k)}(t) B_k(\{t - a\})\, dt$$

$$= \int_0^\infty g(t) \left(\frac{1}{e^t - 1} - \sum_{j=0}^k \frac{B_j}{j!} t^{j-1} \right) \left(e^{-at} - e^{-(N+a)t} \right) dt \ .$$

(2) *If f satisfies the assumptions of Corollary 9.2.6 (1) then*

$$R_k(f, N) = \frac{(-1)^k}{k!} \int_{N+a}^{\infty} f^{(k)}(t)(B_k(\{t - a\}) - B_k) \, dt$$

$$= -\int_0^{\infty} g(t) \left(\frac{1}{e^t - 1} - \sum_{j=0}^{k-1} \frac{B_j}{j!} t^{j-1} \right) e^{-(N+a)t} \, dt \ .$$

Proof. Immediate from the above remarks and left to the reader (Exercise 31). Note that the result is false for $k = 0$. $\qquad\square$

Corollary 9.2.9. *Keep the above assumptions on g and assume that f and all its derivatives have constant sign and tend to 0 as $t \to \infty$. With the notation of Corollary 9.2.6 we have*

$$z(f, a) = \frac{f(a)}{2} + \int_a^{\infty} f'(t) \left(\{t - a\} - \frac{1}{2} \right) dt$$

$$= \int_0^{\infty} g(t) \left(\frac{1}{1 - e^{-t}} - \frac{1}{t} \right) e^{-at} \, dt \ .$$

Proof. Simply take $k = 1$ in the proposition. $\qquad\square$

Examples. As examples of the proposition and its corollary, we give the following formulas. The functions $\Gamma(x)$, $\psi(x)$, and $\zeta(s, x)$ will be defined and studied in more detail below.

Proposition 9.2.10. (1) *For $\Re(s) > 0$ and $x > 0$ we have*

$$\zeta(s, x + 1) = \frac{x^{1-s}}{s - 1} - s \int_x^{\infty} \frac{\{t - x\}}{t^{s+1}} \, dt = \frac{1}{\Gamma(s)} \int_0^{\infty} \frac{t^{s-1} e^{-xt}}{e^t - 1} \, dt \ .$$

(2) *In particular, for $\Re(s) > 0$ we have*

$$\zeta(s) = \frac{1}{s - 1} + 1 - s \int_1^{\infty} \frac{\{t\}}{t^{s+1}} \, dt = \frac{1}{\Gamma(s)} \int_0^{\infty} \frac{t^{s-1}}{e^t - 1} \, dt \ .$$

(3) *For $x > 0$ we have*

$$\psi(x + 1) = \log(x) + \int_x^{\infty} \frac{\{t - x\}}{t^2} \, dt$$

$$= \log(x) + \int_0^{\infty} \left(\frac{1}{t} - \frac{1}{e^t - 1} \right) e^{-tx} \, dt = \int_0^{\infty} \left(\frac{e^{-t}}{t} - \frac{e^{-tx}}{e^t - 1} \right) dt \ .$$

(4) *In particular,*

$$\gamma = 1 - \int_1^{\infty} \frac{\{t\}}{t^2} \, dt = \int_0^{\infty} \left(\frac{1}{e^t - 1} - \frac{e^{-t}}{t} \right) dt \ .$$

(5) *For $x > 0$ we have*

$$\log(\Gamma(x+1)) = \left(x + \frac{1}{2}\right)\log(x) - x + \frac{1}{2}\log(2\pi) - I(x) \, ,$$

where

$$I(x) = \int_x^\infty \frac{\{t-x\} - 1/2}{t}\, dt = \int_0^\infty \left(\frac{1}{t} - \frac{1}{e^t - 1} - \frac{1}{2}\right)\frac{e^{-xt}}{t}\, dt \, ,$$

so that

$$\log(\Gamma(x+1)) = \int_0^\infty \left(x\frac{e^{-t}}{t} - \frac{1 - e^{-xt}}{t(e^t - 1)}\right)\, dt \, .$$

(6) *In particular,*

$$\frac{1}{2}\log(2\pi) = 1 + \int_1^\infty \frac{\{t\} - 1/2}{t}\, dt = \int_0^\infty \left(\frac{1}{t} - \frac{e^{-t}}{2} - \frac{1}{e^t - 1}\right)\frac{dt}{t} \, .$$

Proof. All the results except (5) and (6) are direct consequences of the definitions and of the proposition and its corollary. For (5) and (6), the formulas involving fractional parts come from Euler–MacLaurin, the formula for $\log(2\pi)/2$ coming from Stirling's formula (see below). It is to be noted that the integrals are only conditionally convergent. If you are uncomfortable with this, do an integration by parts to obtain formulas involving $B_2(\{t\})$, which will be absolutely convergent.

For the Laplace-type formulas we integrate the formula for $\psi(x+1)$, use Stirling's formula, do some rearrangements, and use the Laplace formula for $\log(x)$ seen above. The details will be seen below when we study the gamma function (Proposition 9.6.29 and Corollary 9.6.31). □

If we assume only that f is a holomorphic function of x for $\Re(x) > 0$, but not necessarily given as a Laplace transform, we have the following.

Proposition 9.2.11 (Abel–Plana). *Assume that f is a holomorphic function on $\Re(z) > 0$, that $f(z) = o(\exp(2\pi|\Im(z)|))$ as $|\Im(z)| \to \infty$ uniformly in vertical strips of bounded width, and that f and all its derivatives have constant sign and tend to 0 as $x \to \infty$ in $\mathbb{R}$.*

(1) *If $a > 0$ we have*

$$z(f, a) = \frac{f(a)}{2} + i\int_0^\infty \frac{f(a+it) - f(a-it)}{e^{2\pi t} - 1}\, dt \, .$$

(2) *If $a > 1/2$ we have*

$$z(f, a) = \int_0^{1/2} f(a - 1/2 + t)\, dt$$
$$- i\int_0^\infty \frac{f(a - 1/2 + it) - f(a - 1/2 - it)}{e^{2\pi t} + 1}\, dt \, .$$

Proof. See Exercise 33.   □

Remarks. (1) If $f(x) = \mathcal{L}(g)(x)$ this proposition implies Corollary 9.2.9; see Exercise 29.

(2) The formula is usually given in the first form. However, the second one is better suited for numerical computation since there is no problem in computing the integrand close to $t = 0$, while with the first form we must use some sort of Taylor expansion to obtain the result accurately.

9.2.5 Basic Applications of the Euler–MacLaurin Formula

As already mentioned, the Euler–MacLaurin formula has many applications. We begin with the easiest.

Proposition 9.2.12. *For every $k \geqslant 1$ we have*

$$\sum_{m=1}^{N} m^k = \frac{1}{k+1}\left(N^{k+1} + \frac{k+1}{2}N^k + \sum_{j=2}^{k}\binom{k+1}{j}B_j N^{k+1-j}\right)$$

$$= N^k + \frac{B_{k+1}(N) - B_{k+1}(0)}{k+1} = \frac{B_{k+1}(N+1) - B_{k+1}(0)}{k+1},$$

and more generally

$$\sum_{0 \leqslant m < N}(m+x)^k = \frac{B_{k+1}(N+x) - B_{k+1}(x)}{k+1}.$$

Proof. Immediate application of Euler–MacLaurin with $f(t) = t^k$. The proposition is also easily proved directly using Proposition 9.1.3.   □

Examples.

$$\sum_{m=1}^{N} m = \frac{N(N+1)}{2}, \quad \sum_{m=1}^{N} m^2 = \frac{N(N+1)(2N+1)}{6},$$

$$\sum_{m=1}^{N} m^3 = \frac{N^2(N+1)^2}{4}, \quad \sum_{m=1}^{N} m^4 = \frac{N(N+1)(2N+1)(3N^2+3N-1)}{30}.$$

Proposition 9.2.13. *Let $\alpha \in \mathbb{C}$ be different from -1.*

(1) *For every $k > \Re(\alpha) + 1$ such that $k \geqslant 1$ we have*

$$\sum_{m=1}^{N} m^\alpha = \zeta(-\alpha) + \frac{N^{\alpha+1}}{\alpha+1} + \frac{N^\alpha}{2} + \sum_{j=2}^{k}\binom{\alpha}{j-1}\frac{B_j}{j}N^{\alpha-j+1} + R_k(\alpha, N),$$

where

$$R_k(\alpha, N) = (-1)^k \binom{\alpha}{k} \int_N^\infty t^{\alpha-k} B_k(\{t\}) \, dt \ .$$

When k is even we have

$$|R_k(\alpha, N)| \leqslant \left| \binom{\alpha}{k+1} \frac{B_{k+2}}{k+2} N^{\alpha-k-1} \right| \ ;$$

in other words, $|R_k(\alpha, N)|$ *is smaller than the modulus of the first omitted term, and in particular* $R_k(\alpha, N)$ *tends to 0 as* $N \to \infty$.

(2) *With the same assumptions, we have the formula*

$$\zeta(-\alpha) = -\frac{1}{\alpha+1} - \sum_{j=1}^k \binom{\alpha}{j-1} \frac{B_j}{j} + (-1)^{k-1} \binom{\alpha}{k} \int_1^\infty t^{\alpha-k} B_k(\{t\}) \, dt \ .$$

Proof. The first statement follows directly from the Euler–MacLaurin formula and Proposition 9.2.5, apart from the determination of the constant. Fix some integer $k_0 > \Re(\alpha) + 1$ such that $k_0 \geqslant 1$, and let $f_\alpha(t) = t^\alpha$. For all $k \geqslant k_0$ the sign of $f_\alpha^{(k)}(t)$ is constant and $f_\alpha^{(k-1)}(t)$ tends to 0 as $t \to \infty$, so that we can apply Corollary 9.2.6. The first formula applied to $f_\alpha(t)$, $a = 1$, and N replaced by $N - 1$ gives

$$\sum_{m=0}^{N-2} (m+1)^\alpha = z(f_\alpha, 1) + \frac{N^{\alpha+1} - 1}{\alpha+1} - \frac{N^\alpha}{2} + \sum_{j=2}^k \binom{\alpha}{j-1} \frac{B_j}{j} N^{\alpha-j+1} + o(1) \ .$$

Adding N^α to both sides shows that the constant is equal to $z(f_\alpha, 1) - 1/(\alpha + 1)$.

Now using again Corollary 9.2.6 (2), we obtain that for any fixed $k \geqslant k_0$ we have

$$z(f_\alpha, 1) = -\sum_{j=1}^k \binom{\alpha}{j-1} \frac{B_j}{j} + (-1)^{k-1} \binom{\alpha}{k} \int_1^\infty t^{\alpha-k} B_k(\{t\}) \, dt \ .$$

From this formula it is immediately obvious that for all $\alpha \in \mathbb{C}$ the function $z(f_\alpha, 1)$ is a complex differentiable function of α, hence a holomorphic function. On the other hand, for $\alpha < -1$ in the formula that we have proved we may choose $k_0 = 1$, hence

$$\sum_{m=1}^N m^\alpha = z(f_\alpha, 1) - \frac{1}{\alpha+1} + \frac{N^{\alpha+1}}{\alpha+1} + \frac{N^\alpha}{2} + o(1) = z(f_\alpha, 1) - \frac{1}{\alpha+1} + o(1) \ ,$$

and since the left-hand side converges to $\zeta(-\alpha)$, we deduce that $z(f_\alpha, 1) = \zeta(-\alpha) + 1/(\alpha + 1)$. Since this is true on an open subset of $\mathbb{C}$ and both sides are meromorphic functions, it is true for all $\alpha \in \mathbb{C}$ such that $\alpha \neq -1$, proving (1). Statement (2) follows by taking $N = 1$. □

On the other hand, for $\alpha = -1$ we have the following.

Proposition 9.2.14. (1) *For $k \geqslant 1$ we have*

$$\sum_{m=1}^{N} \frac{1}{m} = \log N + \gamma + \frac{1}{2N} - \sum_{j=2}^{k} \frac{B_j}{jN^j} + R_k(-1, N) ,$$

where γ is Euler's constant and

$$R_k(-1, N) = \int_{N}^{\infty} t^{-1-k} B_k(\{t\}) \, dt .$$

When k is even we have

$$|R_k(-1, N)| \leqslant \frac{|B_{k+2}|}{(k+2)N^{k+2}} ;$$

in other words, $|R_k(-1, N)|$ is smaller than the modulus of the first omitted term.

(2) *For $k \geqslant 1$ we have*

$$\gamma = \frac{1}{2} + \sum_{j=2}^{k} \frac{B_j}{j} - \int_{1}^{\infty} t^{-1-k} B_k(\{t\}) \, dt .$$

(3) *We have $\lim_{s \to 1}(\zeta(s) - 1/(s-1)) = \gamma$.*

Proof. (1) is again a direct application of Euler–MacLaurin and the definition of γ, and (2) follows by choosing $N = 1$. If we choose $k = 1$ in (2) of the preceding proposition with $\alpha = -s$ we obtain

$$\zeta(s) = \frac{1}{s-1} + \frac{1}{2} - s \int_{1}^{\infty} t^{-s-1} B_1(\{t\}) \, dt ,$$

so by absolute convergence

$$\lim_{s \to 1} \left(\zeta(s) - \frac{1}{s-1} \right) = \frac{1}{2} - \int_{1}^{\infty} t^{-2} B_1(\{t\}) \, dt = \gamma$$

by (2). $\qquad\square$

Examples.

$$\sum_{m=1}^{N} \frac{1}{m} = \log N + \gamma + O(1/N) ,$$

$$\sum_{m=1}^{N} \frac{1}{\sqrt{m}} = 2\sqrt{N} + \zeta(1/2) + O(1/\sqrt{N}) ,$$

$$\sum_{m=1}^{N} \sqrt{m} = \frac{2}{3}N\sqrt{N} + \frac{1}{2}\sqrt{N} + \zeta(-1/2) + O(1/\sqrt{N}) .$$

An important practical question is how to compute Euler's constant (or $\zeta(-\alpha)$, or other constants of the type $z(f, a)$ occurring in the Euler–MacLaurin formula). It is out of the question to use the above definition since the convergence is much too slow. The whole point is that one can use the Euler–MacLaurin formula with suitably chosen parameters N and n.

As a toy example, assume for instance that we do not want to use Bernoulli numbers B_j for $j > 12$. We will thus use $k = 12$ in the formula. Since all the conditions of Proposition 9.2.5 are satisfied, we deduce that the modulus of the remainder r_{12} is bounded by

$$\frac{|B_{14}|}{14!} \frac{13!}{N^{14}} = \frac{1}{12N^{14}} .$$

Thus if we only take $N = 10$, we obtain 15 decimal digits of the correct result, using only the partial sum of the first ten terms plus a few corrective terms coming from the Euler–MacLaurin formula (we obtain $\gamma = 0.577215664901533\ldots$).

The same method can be used for many other sums or limits of the same kind. For instance, we can easily compute to 15 decimal digits any reasonable value of $\zeta(s)$; see Exercise 42.

If we choose $f(N) = \log N$, the summation formula immediately gives *Stirling's formula* in the following weak form:

$$\log(N!) = (N + 1/2) \log N - N + C + O(1/N)$$

for a certain constant C. As above, it is easy to compute C numerically. The asymptotic expansion given by Euler–MacLaurin is

$$\log(N!) = \left(N + \frac{1}{2}\right) \log N - N + C + \frac{B_2}{1 \cdot 2N} + \frac{B_4}{3 \cdot 4N^3} + \cdots .$$

However, the constant C can also be computed exactly. Classically this is done using Wallis's formulas (see Proposition 9.6.22). However, a more sophisticated, but more natural, way to compute it is to take the derivative with respect to α of the formulas of Proposition 9.2.13. Let us explain how this is done, since it can be used in other situations (see Exercise 44). Assume that $\Re(\alpha) < 1$, so that we can choose $k = 1$. We have

$$\sum_{m=1}^{N} m^\alpha = \zeta(-\alpha) + \frac{N^{\alpha+1}}{\alpha + 1} + \frac{N^\alpha}{2} - \alpha \int_N^\infty t^{\alpha-1}(\{t\} - 1/2)\, dt .$$

Differentiating with respect to α and setting $\alpha = 0$, we obtain

$$\sum_{m=1}^{N} \log m = -\zeta'(0) + N \log N - N + \frac{1}{2} \log N - \int_N^\infty \frac{\{t\} - 1/2}{t}\, dt .$$

Since the last integral tends to 0 as $N \to \infty$, we deduce that our constant C is equal to $-\zeta'(0) = \log(2\pi)/2$, as we will see in Section 10.2.4.

Remark. As already mentioned, if you are uncomfortable with the conditionally convergent integrals that occur in the above reasoning, simply choose $k = 2$ instead of $k = 1$. Everything will be absolutely convergent but the computations will of course be slightly longer.

The reader is strongly advised to solve Exercise 44 for analogous results.

The Euler–MacLaurin summation formula also permits the determination of the convergence behavior of certain series. We give the following example. Let x be a nonzero real number and let $\alpha \in \mathbb{R}$. We want to know the behavior (convergent or divergent) of the series

$$S = \sum_{m=1}^{\infty} \frac{\sin(x \log m)}{m^{\alpha}} .$$

Since $|\sin(x \log m)| \leqslant 1$, if $\alpha > 1$ the series trivially converges absolutely, and if $\alpha \leqslant 0$ the general term does not tend to 0, so the series diverges. We can thus assume that $0 < \alpha \leqslant 1$.

We use Euler–MacLaurin with $k = 2$, obtaining

$$\sum_{m=1}^{N} \frac{\sin(x \log m)}{m^{\alpha}} = \int_{1}^{N} \frac{\sin(x \log t)}{t^{\alpha}} \, dt + \frac{\sin(x \log N)}{2 N^{\alpha}} + r_1 ,$$

with

$$r_1 = \int_{1}^{N} B_1(\{t\}) \frac{x \cos(x \log t) - \alpha \sin(x \log t)}{t^{\alpha+1}} \, dt .$$

Since $B_1(\{t\})$ is bounded and $\alpha + 1 > 1$, it is clear that the integral defining r_1 is absolutely convergent as $N \to \infty$, and in particular has a limit. The term $\sin(t \log N)/(2N^{\alpha})$ tends to 0 as $N \to \infty$. It follows that our series has the same convergence properties as the integral. In the integral we make the change of variables $t = e^u$, and we obtain

$$\int_{1}^{N} \frac{\sin(x \log t)}{t^{\alpha}} \, dt = \int_{0}^{\log N} \sin(xu) e^{(1-\alpha)u} \, du .$$

It is now an easy exercise (for instance by explicit computation) to show that for $0 < \alpha \leqslant 1$ the integral does not converge as $N \to \infty$, and so neither does our series (there are, of course, other ways to prove this, for instance by grouping terms such that $\exp(k\pi/|x|) \leqslant m < \exp((k+1)\pi/|x|)$).

9.3 Applications to Numerical Integration

It is clear that in the opposite direction to the above examples, the Euler–MacLaurin formula gives approximations of integrals by sums, which often allows the numerical computation of these integrals. It is to be stressed from the start that our goal is to give *high-precision* approximations to integrals, not only 15 decimal digits, say.

9.3.1 Standard Euler–MacLaurin Numerical Integration

We begin with a direct application of Euler–MacLaurin. In the next subsections we will give little-known but very powerful methods for high-precision numerical integration.

Let $f \in C^r([a,b])$, where $[a,b]$ is a finite interval. When N is large the integral from a to b of f can be reasonably well approximated by the Riemann sum

$$\frac{b-a}{N} \sum_{m=0}^{N-1} f(a + m(b-a)/N) .$$

The Euler–MacLaurin formula allows us to state this much more precisely, and as usual gives us both an asymptotic expansion of the difference and an efficient method to compute the integral numerically.

Proposition 9.3.1. *Let $[a,b]$ be a finite closed interval, and assume that $f \in C^k([a,b])$ for some $k \geqslant 1$. Then for any integer $N \geqslant 1$, if we set $h = (b-a)/N$ we have*

$$\int_a^b f(t)\, dt = h \sum_{m=0}^{N-1} f(a + mh) + h \frac{f(b) - f(a)}{2}$$

$$- \sum_{j=2}^{k} \frac{B_j}{j!} h^j \left(f^{(j-1)}(b) - f^{(j-1)}(a) \right)$$

$$+ \frac{(-1)^k}{k!} h^k \int_a^b f^{(k)}(t) B_k(\{(t - a)/h\})\, dt .$$

Proof. For $t \in [0, N]$, set $g(t) = f(a + ht)$ and apply the formula to the function g on the interval $[0, N]$. Since $g^{(j)}(t) = h^j f^{(j)}(a + ht)$, we obtain

$$\sum_{m=0}^{N-1} f(a + hm) = \int_0^N f(a + ht)\, dt - \frac{f(b) - f(a)}{2}$$

$$+ \sum_{j=2}^{k} \frac{B_j}{j!} h^{j-1} \left(f^{(j-1)}(b) - f^{(j-1)}(a) \right)$$

$$+ \frac{(-1)^{k-1}}{k!} h^k \int_0^N f^{(k)}(a + ht) B_k(\{t\})\, dt ;$$

hence by making the change of variables $a + ht = t'$ in both integrals we obtain

$$\sum_{m=0}^{N-1} f(a+hm) = \frac{1}{h}\int_a^b f(t)\,dt - \frac{f(b)-f(a)}{2}$$

$$+ \sum_{j=2}^{k} \frac{B_j}{j!}h^{j-1}\left(f^{(j-1)}(b) - f^{(j-1)}(a)\right)$$

$$+ \frac{(-1)^{k-1}}{k!}h^{k-1}\int_a^b f^{(k)}(t)B_k(\{(t-a)/h\})\,dt\;;$$

hence, transferring to the left the integral that we want, we obtain the desired formula.   □

Remarks. (1) To avoid the term $h(f(b) - f(a))/2$, it is neater to replace the asymmetrical Riemann sum $\sum_{0\leqslant m\leqslant N-1}$ by the symmetrical sum $\sum'_{0\leqslant m\leqslant N}$, where the $'$ indicates that the extremal terms $m = 0$ and $m = N$ must be counted with coefficient $1/2$.

(2) In the sum from $j = 2$ to k we can of course restrict only to j even, and we may also choose k even (or k odd) as desired.

Example. Let us use the above formula to compute $\log 2$. We choose $f(t) = 1/(1+t)$, whose integral from 0 to 1 is equal to $\log 2$. We have $f^{(j)}(t) = (-1)^j j!/(1+t)^{j+1}$ and $f(m/N) = N/(N+m)$. We deduce that for all k and N we have

$$\log 2 = \sum_{m=1}^{N} \frac{1}{N+m} + \frac{1}{4N} - \sum_{j=2}^{k} \frac{B_j(1-2^{-j})}{jN^j} + \frac{1}{N^k}\int_0^1 \frac{B_k(\{Nt\})}{(1+t)^{k+1}}\,dt\,.$$

To bound the remainder, we choose as usual k even, and we can then bound $|B_k(\{Nt\})|$ by $|B_k|$. We bound $1/(1+t)^{k+1}$ by 1, and we deduce that the remainder is bounded by $|B_k|/N^k$. Choosing $k = 12$, we see that with $N = 10$ the remainder is bounded by $3\cdot10^{-13}$. Thus, as usual with Euler–MacLaurin, by dividing the interval of integration into only 10 subintervals, and adding a few corrective terms, we obtain 13 decimal digits of the result.

9.3.2 The Basic Tanh-Sinh Numerical Integration Method

We now consider a little-known but much more powerful method, due to Takahashi and Mori, see [Tak-Mor] and [Mori], which the author learned from [Bor-Bai-Gir]. Apart from the evident fact that this method is quite recent, the main reason that this method is not widespread is that the usual practitioners of numerical integration are engineers and numerical analysts, who in general do not need more than 15 correct decimal places. In contrast, in number theory we often want to *identify* certain integrals using linear dependence techniques (see Section 2.3.5). For this we often need hundreds if not thousands of decimal places, and the standard methods are totally unsuitable for that purpose.

Assume first that $F \in C^{\infty}(\mathbb{R})$ is such that for all $k \geqslant 0$ the derivatives $F^{(k)}(x)$ tend to 0 as $|x| \to \infty$ at least as fast as $1/|x|^{\alpha}$ for some $\alpha > 1$, and let h be a small positive parameter. Applying Corollary 9.2.3 with N replaced by $2N$ and $f(t)$ replaced by $F(h(t - N))$, we obtain the following formula valid for all $k \geqslant 1$:

$$h \sum_{m=-N}^{N} F(mh) = \int_{-N}^{N} F(t)\,dt + h\frac{F(N) + F(-N)}{2}$$

$$+ \sum_{j=1}^{k} \frac{B_{2j}h^{2j}}{(2j)!} \left(F^{(2j-1)}(N) - F^{(2j-1)}(-N) \right) + h^{2k+1}R_{2k}(F, N) \,,$$

with

$$R_{2k}(F, N) = -\frac{1}{(2k)!} \int_{-N}^{N} F^{(2k)}(t) B_{2k}(\{t\})\,dt\,.$$

From the assumptions on F we can let N tend to infinity and we obtain the simple estimate

$$\left| \int_{-\infty}^{\infty} F(t)\,dt - h \sum_{m=-\infty}^{\infty} F(mh) \right| \leqslant C_{2k}h^{2k+1}\,,$$

with

$$C_{2k} = \frac{|B_{2k}|}{(2k)!} \int_{-\infty}^{\infty} |F^{(2k)}(t)|\,dt\,.$$

In other words, for such functions F, as h tends to 0 the difference between the sum and the integral tends to 0 faster than any power of h. In actual practice the convergence is usually (although not always) at least as fast as $e^{-C/|h|}$ for some $C > 0$; see Exercise 47.

Now let $f \in C^{\infty}(]-1, 1[)$ be integrable on $[-1, 1]$ (it may have singularities at the endpoints). The fundamental trick is as follows. We introduce the magic function $\phi(t) = \tanh(\sinh(t))$. This function has the following evident properties: it is a one-to-one odd map from $\mathbb{R}$ to $]-1, 1[$, and as $t \to \pm\infty$ it tends to ± 1 *doubly exponentially fast*; more precisely, $\operatorname{sign}(t) - \tanh(\sinh(t))$ behaves approximately like $2/\exp(\exp(|t|))$. Thus the function $F(t) = f(\phi(t))\phi'(t)$ will certainly satisfy our assumptions above, and in fact its derivatives will tend to zero extremely rapidly (and in particular $F(t)$ will be in the so-called *Schwartz class*). Changing variables and applying the above remark based on Euler–MacLaurin we obtain

$$\int_{-1}^{1} f(x)\,dx = \int_{-\infty}^{\infty} f(\phi(t))\phi'(t)\,dt = h \sum_{m=-\infty}^{\infty} f(\phi(mh))\phi'(mh) + R(h)\,,$$

where the remainder term $R(h)$ tends to 0 very fast.

If f is a meromorphic function in $\mathbb{C}$, and not only a C^{∞} function on $\mathbb{R}$, it can be shown that $|R(h)| < e^{-C/|h|}$ for some $C > 0$, and that with N function evaluations we can reach an accuracy on the order of $\exp(-CN/\log(N))$ for some (other) $C > 0$.

Remark. It can also be shown that double-exponential convergence at infinity is optimal: choosing for instance functions giving triple-exponential convergence would give worse results. In particular, as we will see below, it is necessary to *adapt* the magic function $\phi(t)$ to the class of functions to be integrated.

Assume for instance that we want to compute the integral with an accuracy of approximately 500 decimal digits (this would be completely impossible with classical methods). We note that, due to the doubly exponential behavior, we have $\phi(t) < 10^{-500}$ for $t \geqslant 7.05$, so we sum only for $|m| < 7.05/h$. Although it is not easy to estimate $R(h)$ accurately we try successive values $h = 1/2^r$ for $r = 2, 3$, etc., until the value of the sum stabilizes. For instance, for 500 decimals, $h = 1/2^8$ is almost always sufficient.

Since this method is so useful we give an explicit algorithm essentially copied from [Bor-Bai-Gir] (the case in which we want to integrate on more general intervals than $[-1, 1]$ is studied in the next section). The algorithm needs to be given a small integral parameter r such that $h = 1/2^r$, which is found empirically by trying two or three values (for instance, as mentioned above we choose $r = 8$ for 500 decimal digits).

Algorithm 9.3.2 (Tanh-Sinh Numerical Integration) Given an integrable C^∞ function f on $]-1, 1[$, an accuracy ε, and a small integral parameter $r \geqslant 2$ as above, this algorithm computes an approximation to $\int_{-1}^{1} f(x)\, dx$ of order ε.

1. [Initialize] Set $h \leftarrow 1/2^r$, $e_1 \leftarrow e^h$, $e_2 \leftarrow 1$; $i \leftarrow 0$.
2. [Fill Arrays $x[]$ and $w[]$] Set $c \leftarrow e_2 + 1/e_2$, $s \leftarrow e_2 - c$, $e_3 \leftarrow 2/\left(e^{2s} + 1\right)$, $x[i] = 1 - e_3$, $w[i] \leftarrow ce_3(1 + x[i])$, $e_2 \leftarrow e_1e_2$. If $e_3 > \varepsilon$, set $i \leftarrow i + 1$ and go to Step 2. Otherwise set $w[0] \leftarrow w[0]/2$, $n \leftarrow i$, $S \leftarrow 0$, and $p \leftarrow 2^r$ (n will be the largest index i for which we have computed $x[i]$ and $w[i]$; it will never exceed $20 \cdot 2^r$).
3. [Outer Loop] Set $p \leftarrow p/2$ and $i \leftarrow 0$.
4. [Inner Loop] If $(2p) \nmid i$ or if $p = 2^{r-1}$ then set $S \leftarrow S + w[i](f(-x[i]) + f(x[i]))$. Set $i \leftarrow i + p$, and if $i \leqslant n$ go to Step 4.
5. [Terminate?] If $p \geqslant 2$ go to Step 3; otherwise, output $pS/2^r$ and terminate the algorithm.

Steps 1 and 2 should of course be done once and for all, independently of the function f.

9.3.3 General Doubly Exponential Numerical Integration

The above method computes $\int_{-1}^{1} f(x)\, dx$ in a quite general setting, but one in which f must be a C^∞ function and have at most reasonable type singularities at the endpoints ± 1. We now consider more general cases. We start by splitting the integral into a sum of integrals where the possible singularities are at the endpoints, so we assume that we are in this case. For integration on a finite interval $[a, b]$ we of course use the formula

$$\int_a^b f(x)\,dx = \frac{b-a}{2} \int_{-1}^1 f\left(\frac{b+a}{2} + \frac{b-a}{2}x\right)\,dx\,.$$

For integration on a semi-infinite interval $[a, \infty[$ (or symmetrically $]-\infty, a]$) we could use a change of variable $x \mapsto 1/x$, but this would give wild singularities at 0. Thus in these cases it is better to use other integration methods of "doubly exponential" type, similar to the tanh-sinh method. For instance, for an integral of the form $I = \int_0^\infty f(x)\,dx$, where f tends to 0 not too rapidly as $x \to \infty$ (for instance like $1/x^n$ for some $n > 1$). we may use the change of variable $x = \exp(K\sinh(t))$ for some constant $K > 0$, and write that I is very well approximated by

$$h \sum_{m \in \mathbb{Z}} f(\exp(K\sinh(mh)))K\cosh(mh)\exp(K\sinh(mh))\,.$$

On the other hand, if f tends to 0 exponentially fast as $x \to \infty$, say as $\exp(-x)$ for some $a > 0$, then we may use the change of variables $x = \exp(t - \exp(-t))$ and write that I is very well approximated by

$$h \sum_{m \in \mathbb{Z}} f(\exp(mh - \exp(-mh)))(1 + \exp(-t))\exp(t - \exp(-t))\,.$$

It is essential to adjust the change of variable to the rate at which f tends to 0 at infinity (if it does! but see below for oscillatory functions). For instance, if f tends to 0 as $\exp(-g(x))$ for some strictly increasing function g going to ∞ with x we should use the change of variables $x = g^{-1}(\exp(t - \exp(-t)))$. See Exercise 48 for $\int_{-\infty}^\infty f(x)\,dx$.

None of the above solutions is satisfactory for dealing with an oscillatory integral. Let us consider a typical example. Assume that we want to compute $\int_0^\infty \sin(x)/x\,dx$. The change of variable gives

$$\int_0^\infty \frac{\sin(x)}{x}\,dx = \int_0^1 \frac{\sin(x)}{x}\,dx + \int_0^1 \frac{\sin(1/x)}{x}\,dx\,,$$

and the singularity of the function $\sin(1/x)/x$ at $x = 0$ is too wild to be accessible to any integration method. We must therefore proceed differently. Once again there are two completely different methods: one consists in trying to save what we can from the tanh-sinh method; the other is to choose another function $\phi(t)$. We consider first what we can salvage of the tanh-sinh method.

Assume that we want to compute an integral of the form

$$\int_a^\infty s(x)f(x)\,dx\,,$$

where s and f are C^∞ functions satisfying the following additional properties (note that much weaker assumptions are possible):

(1) The function $f(x)$ is nonnegative and decreases monotonically to 0 as $x \to \infty$.

(2) There exists a "half-period" $P > 0$ such that $s(x + P) = -s(x)$ for all x (hence $s(x + 2P) = s(x)$), and such that $s(x) \geqslant 0$ for $x \in [0, P]$.

We can then write

$$\int_a^\infty s(x)f(x)\,dx = \int_a^0 s(x)f(x)\,dx + \sum_{k \geqslant 0} \int_{kP}^{(k+1)P} s(x)f(x)\,dx$$

$$= \int_a^0 s(x)f(x)\,dx + \sum_{k \geqslant 0} (-1)^k I_k ,$$

where

$$I_k = \int_0^P s(x)f(x + kP)\,dx .$$

Each individual integral can be computed using the tanh-sinh (or any other) method, and note that the values of $s(x)$ for $x \in [0, P]$ (more precisely of $s((P/2)(1 \pm x[i]))$ for $0 \leqslant i \leqslant n$ in the notation of the algorithm) should be computed and tabulated once and for all since they are used in all the integrals.

The infinite alternating sum will in general converge very slowly. However, there is a nice trick due to F. Rodriguez-Villegas, D. Zagier, and the author for accelerating alternating series in a very simple manner as follows (see [Coh-Vil-Zag]).

Algorithm 9.3.3 (Alternating Sums) Given an "alternating" series $S = \sum_{k \geqslant 0} (-1)^k I_k$ and an accuracy ε, this algorithm computes an approximation to S of order ε.

1. [Initialize] Set $n \leftarrow \lceil 0.57| \log(\varepsilon)| \rceil$, $d \leftarrow (1+\sqrt{2})^{2n}$, $d \leftarrow (d+1/d)/2$, $b \leftarrow -1$, $c \leftarrow -d$, $s \leftarrow 0$, $k \leftarrow 0$.

2. [Loop] Set $c \leftarrow b - c$, $s \leftarrow s + cI_k$, $b \leftarrow (k + n)(k - n)b/((k + 1/2)(k + 1))$, $k \leftarrow k + 1$. If $k \leqslant n - 1$ go to Step 2; otherwise, output s/d and terminate the algorithm.

The (easy) proof of the validity of this algorithm for a wide class of sums (not only alternating, and not necessarily convergent) is given in [Coh-Vil-Zag].

Applied to our specific problem, it gives a reasonably efficient method (although much slower than standard tanh-sinh integration) for integrating reasonable oscillatory functions on an infinite interval.

The second method consists in changing the function $\phi(t)$. Contrary to the preceding methods we must choose ϕ depending on the summation step

h. Assume that the integral has the form $\int_0^\infty \sin(x)f(x)\,dx$, where $f(x)$ is as above, and in particular nonoscillating. The above-mentioned authors show that the choice

$$x = \phi(t) = \frac{(\pi/h)t}{1 - \exp(-\sinh(t))}$$

leads to an excellent doubly exponential method. It should be stressed that this is specific to an integral from 0 to infinity, and with the specific oscillating function $s(x) = \sin(x)$. This leads to the following:

(1) For an integral whose lower bound is not 0, write

$$\int_a^\infty \sin(x)f(x)\,dx = \int_0^\infty \sin(x)f(x)\,dx - \int_0^a \sin(x)f(x)\,dx\,,$$

the second integral being computed by ordinary tanh-sinh integration.

(2) If $s(x) = \sin(kx)$, write

$$\int_0^\infty \sin(kx)f(x)\,dx = \frac{1}{k}\int_0^\infty \sin(x)f\left(\frac{x}{k}\right)\,dx\,.$$

(3) If $s(x) = \cos(kx)$, write

$$\int_0^\infty \cos(kx)f(x)\,dx = \frac{1}{k}\int_{\pi/2}^\infty \sin(x)f\left(\frac{x - \pi/2}{k}\right)\,dx\,,$$

and then use (1).

(4) If $s(x)$ is a general periodic function, compute its Fourier coefficients and apply the above.

Although more complicated to use practice, this is much faster than the use of alternating sums as above.

Examples. To compute our initial example $\int_0^\infty \sin(x)/x\,dx$ we apply directly the above method with $f(x) = 1/x$, $s(x) = \sin(x)$, and $P = \pi$.
To compute $\int_0^\infty \sin^2(x)/x^2\,dx$ we first write $\sin^2(x) = (1 - \cos(2x))/2$, so that

$$\int_0^\infty \frac{\sin^2(x)}{x^2}\,dx = \frac{1}{2}\int_0^{\pi/4} \frac{1 - \cos(2x)}{x^2}\,dx + \frac{1}{2}\int_{\pi/4}^\infty \frac{1}{x^2} - \frac{1}{2}\int_{\pi/4}^\infty \frac{\cos(2x)}{x^2}$$

$$= \frac{1}{2}\int_0^{\pi/4} \frac{1 - \cos(2x)}{x^2}\,dx + \frac{2}{\pi} + \frac{1}{2}\int_0^\infty \frac{\sin(2x)}{(x + \pi/4)^2}\,,$$

and this last integral is computed by the above method with $f(x) = 1/(x + \pi/4)^2$, $s(x) = \sin(2x)$, and $P = \pi/2$.
To the required accuracy we obtain the well-known values (see Proposition 9.6.38)

$$\int_0^\infty \frac{\sin(x)}{x}\,dx = \int_0^\infty \frac{\sin^2(x)}{x^2} = \frac{\pi}{2}\,.$$

9.4 χ-Bernoulli Numbers, Polynomials, and Functions

We now want to generalize the Euler–MacLaurin summation formula to more general sums of the form $\sum_j \chi(j) f(j)$, where χ is a periodic arithmetic function, for instance $\chi(j) = (-1)^j$, or χ a Dirichlet character (see the next chapter). For this we must generalize Bernoulli numbers and polynomials. Thus we let χ be a function from $\mathbb{Z}$ to $\mathbb{C}$ that is assumed to be m-periodic with $m \in \mathbb{Z}_{\geqslant 1}$, in other words such that $\chi(r+m) = \chi(r)$ for all $r \in \mathbb{Z}$, where m is not necessarily the minimal period. We do *not* necessarily assume that χ is a Dirichlet character.

9.4.1 χ-Bernoulli Numbers and Polynomials

There are essentially *four* possible definitions of the χ-Bernoulli numbers and polynomials, which differ only slightly from one another. Each definition has advantages and disadvantages. The most common one, found for instance in [Was], is mainly given for its application to p-adic L-functions, but is not well suited to the χ-Euler–MacLaurin formula or, for that matter, to special values of L-functions when χ is not necessarily a character. We will use a slight variant of that definition, which has the advantage of being more elegant in many formulas, and of exactly generalizing ordinary Bernoulli numbers and polynomials, but which gives slightly less uniform formulas for p-adic L-functions.

Definition 9.4.1. *We define the χ-Bernoulli polynomials $B_k(\chi, x)$ by*

$$E(\chi, t, x) = t e^{tx} \frac{\sum_{0 \leqslant r < m} \chi(r) e^{rt}}{e^{mt} - 1} = \sum_{k \geqslant 0} \frac{B_k(\chi, x)}{k!} t^k \ ,$$

and the χ-Bernoulli numbers $B_k(\chi)$ by $B_k(\chi) = B_k(\chi, 0)$.

The definition of the χ-Bernoulli numbers (hence implicitly of the χ-Bernoulli polynomials) used in [Was] and many other places consists in replacing the sum from 0 to $m - 1$ by the sum from 1 to m. This has the effect of replacing $B_k(\chi, x)$ by $B_k(\chi, x) + \chi(0) k x^{k-1}$, so that in particular the χ-Bernoulli numbers themselves differ from the present ones only when $k = 1$ and $\chi(0) \neq 0$, which in the context of Dirichlet characters means that χ is the trivial character.

Since there is an important alternative definition of Bernoulli numbers and polynomials, it is essential to introduce the following definition, which will enable us to use both definitions concurrently.

Definition 9.4.2. *For any function χ defined on $\mathbb{Z}$ we define the function χ^- by $\chi^-(n) = \chi(-n)$.*

Thus, if χ is a Dirichlet character we have $\chi^- = \chi(-1)\chi$, so that, contrary to $\overline{\chi}$, χ^- is *not* a Dirichlet character when $\chi(-1) = -1$, but only a periodic arithmetic function.

Lemma 9.4.3. *We have*

$$te^{tx}\frac{\sum_{1\leqslant r\leqslant m}\chi(r)e^{-rt}}{1 - e^{-mt}} = \sum_{k\geqslant 0}\frac{B_k(\chi^-, x)}{k!}t^k \ .$$

Proof. Immediate and left to the reader.   □

See also Proposition 9.4.9 below.

The alternative definition would be to use $B_k(\chi^-, x)$ as χ-Bernoulli polynomials. We will see that this is indeed the natural definition to use in many. applications.

Proposition 9.4.4. *We have* $B'_k(\chi, x) = kB_{k-1}(\chi, x)$.

Proof. Clear since $(d/dx)E(\chi, t, x) = tE(\chi, t, x)$.   □

Proposition 9.4.5. *We have the following formulas:*

$$B_k(\chi, x) = \sum_{j=0}^{k}\binom{k}{j}B_j(\chi)x^{k-j} = m^{k-1}\sum_{0\leqslant r<m}\chi(r)B_k\left(\frac{x+r}{m}\right)$$

$$= \sum_{j=0}^{k}\binom{k}{j}B_jm^{j-1}\sum_{0\leqslant r<m}\chi(r)(x+r)^{k-j} \ , \ and$$

$$B_k(\chi) = m^{k-1}\sum_{0\leqslant r<m}\chi(r)B_k\left(\frac{r}{m}\right) = \sum_{j=0}^{k}\binom{k}{j}B_jm^{j-1}S_{k-j}(\chi) \ ,$$

where $S_n(\chi) = \sum_{0\leqslant r<m}\chi(r)r^n$.

Proof. The first formula follows from the identity $E(\chi, t, x) = e^{tx}E(\chi, t, 0)$. The second follows from

$$E(\chi, t, x) = \frac{1}{m}\sum_{0\leqslant r<m}\chi(r)\frac{mte^{((x+r)/m)mt}}{e^{mt} - 1}$$

$$= \frac{1}{m}\sum_{0\leqslant r<m}\chi(r)\sum_{k\geqslant 0}B_k\left(\frac{x+r}{m}\right)\frac{m^k}{k!}t^k$$

by definition of Bernoulli polynomials. The third formula follows from $B_k(z) = \sum_{j=0}^{k}\binom{k}{j}B_jz^{k-j}$, and the last two are obtained by specializing to $x = 0$.   □

For example,

$$B_0(\chi) = \frac{1}{m} \sum_{0 \leqslant r < m} \chi(r) \,, \quad B_1(\chi) = \frac{1}{m} \sum_{0 \leqslant r < m} \chi(r) \left(r - \frac{m}{2} \right) \,, \quad \text{and}$$

$$B_2(\chi) = \frac{1}{m} \sum_{0 \leqslant r < m} \chi(r) \left(r^2 - mr + \frac{m^2}{6} \right) .$$

For future reference, note the following results.

Corollary 9.4.6. *If* $x \in \mathbb{Z}_{\geqslant 0}$ *we have*

$$m^{k-1} \sum_{0 \leqslant r < m} \chi(x + r) B_k \left(\frac{x + r}{m} \right) = B_k(\chi) + k \sum_{0 \leqslant r < x} \chi(r) r^{k-1} .$$

Proof. Indeed, for any function f and $x \in \mathbb{Z} \geqslant 0$ we have

$$\sum_{0 \leqslant r < m} f(x + r) = \sum_{x \leqslant r < m + x} f(r) = \sum_{0 \leqslant r < m} f(r) + \sum_{0 \leqslant r < x} (f(r + m) - f(r)) \,,$$

so the formula follows from $B_k(z + 1) - B_k(z) = kz^{k-1}$ and the proposition. □

Lemma 9.4.7. *If* $m \mid M$ *then*

$$B_k(\chi, x) = M^{k-1} \sum_{0 \leqslant r < M} \chi(r) B_k \left(\frac{x + r}{M} \right) .$$

Proof. Write $n = M/m$, and for $0 \leqslant r < M$ let $r = qm + s$ with $0 \leqslant s < m$ and $0 \leqslant q < n$. By the distribution formula for Bernoulli polynomials (Proposition 9.1.3) we have

$$M^{k-1} \sum_{0 \leqslant r < M} \chi(r) B_k \left(\frac{x + r}{M} \right) = M^{k-1} \sum_{0 \leqslant s < m} \chi(s) \sum_{0 \leqslant q < n} B_k \left(\frac{x + s}{M} + \frac{q}{n} \right)$$

$$= \frac{M^{k-1}}{n^{k-1}} \sum_{0 \leqslant s < m} \chi(s) B_k \left(\frac{n(x + s)}{M} \right)$$

$$= m^{k-1} \sum_{0 \leqslant s < m} \chi(s) B_k \left(\frac{x + s}{m} \right) = B_k(\chi, x)$$

as claimed. □

Proposition 9.4.8. *We have*

$$B_k(\chi, x + m) = B_k(\chi, x) + k \sum_{0 \leqslant r < m} \chi(r)(x + r)^{k-1} .$$

Proof. Follows from the formula

$$E(\chi, t, x + m) - E(\chi, t, x) = t \sum_{0 \leqslant r < m} \chi(r) e^{(x+r)t} .$$

$\square$

Proposition 9.4.9. (1) *We have*

$$B_k(\chi, -x) = (-1)^k (B_k(\chi^-, x) + \chi(0) k x^{k-1}) ,$$

or equivalently,

$$B_k(\chi^-, x) = (-1)^k B_k(\chi, -x) - \chi(0) k x^{k-1} .$$

In particular, $B_k(\chi^-) = (-1)^k B_k(\chi) - \chi(0) \delta_{k,1}$, *where we recall that* $\delta_{k,1} = 1$ *if* $k = 1$, *and* $\delta_{k,1} = 0$ *otherwise.*
(2) *In particular, if* χ *is an even function then* $B_k(\chi) = 0$ *for* $k \geqslant 3$ *odd and* $B_1(\chi) = -\chi(0)/2$, *while if* χ *is an odd function then* $B_k(\chi) = 0$ *for all* $k \geqslant 0$ *even.*

Proof. An easy computation shows that

$$E(\chi^-, t, x) - E(\chi, -t, -x) = -\chi(0) t e^{xt} ,$$

which is clearly equivalent to the first formula, and the other statements follow by specializing to $x = 0$. $\square$

The above proposition will be used in particular when χ is a Dirichlet character.

9.4.2 χ-Bernoulli Functions

Definition 9.4.10. *We define the* χ-*Bernoulli functions and we denote by* $B_k(\chi, \{x\}_\chi)$ *the functions defined for* $x \in \mathbb{R}$ *by*

$$B_k(\chi, \{x\}_\chi) = m^{k-1} \sum_{r \bmod m} \chi(r) B_k \left(\left\{ \frac{x+r}{m} \right\} \right) .$$

Note that since $\chi(r)$ and $\{(x+r)/m\}$ are periodic functions in r of period dividing m it is not necessary to specify the precise range of summation for r, so we simply write $r \bmod m$. It is clear that $B_k(\chi, \{x\}_\chi)$ generalizes the function $B_k(\{x\})$, and that $B_k(\chi, \{0\}_\chi) = B_k(\chi)$.

Proposition 9.4.11. *The* χ-*Bernoulli functions satisfy the following properties:*

(1) *We have* $B_0(\chi, \{x\}_\chi) = B_0(\chi) = S_0(\chi)/m$.

(2) *We have*

$$B_1'(\chi, \{x\}_\chi) = B_0(\chi, \{x\}_\chi) - \sum_{r \in \mathbb{Z}} \chi^-(r)\delta_r(x) \,,$$

where δ_r is the Dirac distribution concentrated at the point r.
(3) *We have $B_k'(\chi, \{x\}_\chi) = kB_{k-1}(\chi, \{x\}_\chi)$ for all $k \geqslant 1$ and all $x \notin \mathbb{Z}$.*
(4) *The function $B_k(\chi, \{x\}_\chi)$ is continuous for $k \geqslant 2$.*
(5) *We have $\int_0^m B_k(\chi, \{x\}_\chi) \, dx = 0$ for $k \geqslant 1$.*
(6) *If $n \in \mathbb{Z}$ we have*

$$\lim_{\substack{x \to n \\ x > n}} B_1(\chi, \{x\}_\chi) = B_1(\chi, \{n\}_\chi) \quad and$$

$$\lim_{\substack{x \to n \\ x < n}} B_1(\chi, \{x\}_\chi) = B_1(\chi, \{n\}_\chi) + \chi^-(n) \,.$$

(7) *On any interval $]r, r+1[$ with $r \in \mathbb{Z}$ the function $B_k(\chi, \{x\}_\chi)$ is a polynomial of degree less than or equal to k.*
(8) *For $k \geqslant 2$ we have $B_k(\chi, \{x\}_\chi) \in C^{k-2}(\mathbb{R})$.*
(9) *$B_k(\chi, \{x+m\}_\chi) = B_k(\chi, \{x\}_\chi)$ for all $x \in \mathbb{R}$ and $k \geqslant 0$.*

Conversely, the sequence of χ-Bernoulli functions is the only sequence satisfying properties (1) to (5) above.

Proof. All these properties are essentially clear from the definition and the basic properties of ordinary Bernoulli polynomials. For instance, let us prove (2) and (8). An easy exercise in distributions (Exercise 50) shows that

$$\left\{\frac{x+r}{m}\right\}' = \frac{1}{m} - \sum_{q \in \mathbb{Z}} \delta_{qm-r}(x) \,,$$

and (2) immediately follows. Property (8) follows from the fact that $B_k(\{x\}) \in C^{k-2}(\mathbb{R})$. The easy proofs of the other properties are left to the reader (Exercise 50).

Let us now prove the converse. Let $C_k(x)$ be another sequence of functions satisfying the first five properties above and set $D_k(x) = C_k(x) - B_k(\chi, \{x\}_\chi)$. We prove by induction that $D_k = 0$. This is clear for $k = 0$. For $k = 1$ we have $D_1'(x) = 0$ in the sense of distributions, so $D_1(x)$ is a constant, and this constant is 0 since $\int_0^m D_1(x) \, dx = 0$. Assume now $k \geqslant 2$ and that $D_{k-1} = 0$. By (3) the function $D_k(x)$ is constant on any interval $]r, r+1[$ with $r \in \mathbb{Z}$. But since $k \geqslant 2$, by (4) we know that $D_k(x)$ is continuous on $\mathbb{R}$. It follows that $D_k(x)$ is constant on $\mathbb{R}$, and as in the case $k = 1$ this constant is 0 since $\int_0^m D_k(x) \, dx = 0$, proving the proposition. $\qquad\square$

In the context of this proposition, which is the key to the χ-Euler–MacLaurin formula, it is clear that $B_k(\chi^-)$ would be a better definition of the χ-Bernoulli numbers. In fact, we have the following:

Proposition 9.4.12. *We have*

$$B_k(\chi, \{-x\}_\chi) = (-1)^k B_k(\chi^-, \{x\}_\chi) - \chi(x)\delta_{x,\mathbb{Z}}\delta_{k,1} ,$$

where $\delta_{x,\mathbb{Z}} = 1$ if $x \in \mathbb{Z}$ and $\delta_{x,\mathbb{Z}} = 0$ otherwise.

Proof. Left to the reader (Exercise 51). Note that this generalizes Proposition 9.4.9. $\square$

Proposition 9.4.13. *For $x \in \mathbb{R}_{\geq 0}$ we have*

$$B_k(\chi, \{x\}_\chi) = B_k(\chi, x) - k \sum_{1 \leq r \leq x} \chi^-(r)(x - r)^{k-1} .$$

Proof. Denote by $R_k(x)$ the right-hand side of this formula. We could show that $R_k(x)$ satisfies the first five conditions of Proposition 9.4.11, but it is easier to reason directly. Set $C_k(x) = \sum_{1 \leq r \leq x} \chi^-(r)(x - r)^{k-1}$. If $x \geq 0$ we have

$$C_k(x + m) = \sum_{1 \leq r \leq x+m} \chi^-(r)(x + m - r)^{k-1}$$

$$= \sum_{1 \leq r \leq m} \chi^-(r)(x + m - r)^{k-1} + \sum_{m+1 \leq r \leq x+m} \chi^-(r)(x + m - r)^{k-1}$$

$$= \sum_{1 \leq r \leq m} \chi^-(r)(x + m - r)^{k-1} + C_k(x) ,$$

so that

$$C_k(x + m) - C_k(x) = \sum_{0 \leq r < m} \chi(r)(x + r)^{k-1} = \frac{B_k(\chi, x + m) - B_k(\chi, x)}{k}$$

by Proposition 9.4.8. It follows that $R_k(x) = B_k(\chi, x) - kC_k(x)$ is periodic of period dividing m, as is the left-hand side of the equality to be proved, so we may assume that $0 \leq x < m$. In that case

$$B_k(\chi, \{x\}_\chi) = m^{k-1} \sum_{0 \leq r < m} \chi(r)B_k\left(\left\{\frac{x + r}{m}\right\}\right)$$

$$= m^{k-1}\left(\sum_{0 \leq r < m-x} \chi(r)B_k\left(\frac{x + r}{m}\right) + \sum_{m-x \leq r < m} \chi(r)B_k\left(\frac{x + r}{m} - 1\right)\right) ,$$

so by Proposition 9.1.3,

$$B_k(\chi, \{x\}_\chi) = m^{k-1} \sum_{0 \leq r < m} \chi(r)B_k\left(\frac{x + r}{m}\right) - k \sum_{m-x \leq r < m} \chi(r)(x + r - m)^{k-1}$$

$$= B_k(\chi, x) - k \sum_{1 \leq r \leq x} \chi^-(r)(x - r)^{k-1}$$

after changing r into $m - r$, proving the proposition. $\qquad\qquad\qquad$ $\square$

Note that even for $\chi = 1$, in other words for ordinary Bernoulli polynomials, the above proposition gives the not completely trivial formula

$$B_k(\{x\}) = B_k(x) - k \sum_{1 \leqslant r \leqslant x} (x - r)^{k-1} ,$$

which is of course an immediate consequence of the formula $B_k(x) = k(x - 1)^{k-1} + B_k(x - 1)$.

Since the χ-Bernoulli functions are the natural generalizations of the functions $B_n(\{x\})$, it is natural to compute their Fourier expansions as periodic functions of period m. Recall from Definition 2.1.38 that we have defined Gauss sums by the formula

$$\tau(\chi, a) = \sum_{r \bmod m} \chi(r) e^{2i\pi ar/m} .$$

Proposition 9.4.14. *For $n \geqslant 2$ we have the Fourier expansion*

$$B_n(\chi, \{x\}_\chi) = -\frac{n! m^{n-1}}{(2i\pi)^n} \sum_{k \in \mathbb{Z},\ k \neq 0} \frac{\tau(\chi, k)}{k^n} e^{2i\pi kx/m} .$$

For $n = 1$ this formula must be modified as follows: the right-hand side must be understood as a symmetrical summation, in other words as the limit as $N \to \infty$ of the sum for $|k| \leqslant N$, and the left-hand side must be changed to $B_1(\chi, \{x\}_\chi) + \chi(x)/2$ if $x \in \mathbb{Z}$.

Proof. We have seen above that the function $B_n(\chi, \{x\}_\chi)$ is piecewise C^∞ and continuous for $n \geqslant 2$, with simple discontinuities at the integers if $n = 1$, and periodic of period (dividing) m. Thus for $n \geqslant 2$ we have $B_n(\chi, \{x\}_\chi) = \sum_{k \in \mathbb{Z}} c_{n,k} \exp(2i\pi kx/m)$ with

$$c_{n,k} = \frac{1}{m} \int_0^m B_n(\chi, \{t\}_\chi) e^{-2i\pi kt/m} \, dt .$$

For $n = 1$ the same formula is valid for $x \notin \mathbb{Z}$, and for $x \in \mathbb{Z}$ we must replace $B_1(\chi, \{x\}_\chi)$ by $(B_1(\chi, x^+) + B_1(\chi, x^-))/2 = B_1(\chi, \{x\}_\chi) + \chi(x)/2$.

For $k = 0$ and $n > 0$, we have $c_{n,0} = 0$ by property (5) of Proposition 9.4.11. For $k \neq 0$ and $n \geqslant 2$, by integration by parts we compute that $c_{n,k} = (nm/(2i\pi k))c_{n-1,k}$, so that $c_{n,k} = (n! m^{n-1}/(2i\pi k)^{n-1})c_{1,k}$. For $k \neq 0$ and $n = 1$, by integration by parts we find that

$$c_{1,k} = -\frac{1}{2i\pi k} \int_{0+}^{m^+} e^{-2i\pi kt/m} \sum_{r \in \mathbb{Z}} \chi(r) \delta_{-r}(t) \, dt$$

$$= -\frac{1}{2i\pi k} \sum_{r \bmod m} \chi(r) e^{2i\pi kr/m} = -\frac{\tau(\chi, k)}{2i\pi k} ,$$

proving the proposition. As before, it is immediate to transform this proof into one not using distributions explicitly. □

9.4.3 The χ-Euler–MacLaurin Summation Formula

The χ-Euler–MacLaurin formula, which is the exact generalization of Theorem 9.2.2, is the following.

Proposition 9.4.15. *As above, let χ be a periodic arithmetic function of period (dividing) m, let a and b be two real numbers such that $a \leqslant b$, and assume that $f \in C^k([a,b])$ for some $k \geqslant 1$. Then*

$$\sum_{\substack{a<r\leqslant b \\ r\in\mathbb{Z}}} \chi(r)f(r) = B_0(\chi^-)\int_a^b f(t)\,dt$$

$$+ \sum_{j=1}^k \frac{(-1)^j}{j!}\left(B_j(\chi^-,\{b\}_\chi)f^{(j-1)}(b) - B_j(\chi^-,\{a\}_\chi)f^{(j-1)}(a)\right)$$

$$+ \frac{(-1)^{k-1}}{k!}\int_a^b f^{(k)}(t)B_k(\chi^-,\{t\}_\chi)\,dt\,.$$

Proof. The Bernoulli functions have been defined exactly in order for this proposition to be valid, and as can clearly be seen, in the present context it would have been much better to choose $B_k(\chi^-, x)$ as definition of χ-Bernoulli polynomials. By their basic properties, if we set

$$R_k(\chi, f) = \frac{(-1)^{k-1}}{k!}\int_a^b f^{(k)}(t)B_k(\chi^-,\{t\}_\chi)\,dt\,,$$

then for $k \geqslant 2$ integration by parts gives

$$R_k(\chi, f) = \frac{(-1)^{k-1}}{k!}(B_k(\chi^-,\{b\}_\chi)f^{(k-1)}(b) - B_k(\chi^-,\{a\}_\chi)f^{(k-1)}(a))$$
$$+ R_{k-1}(\chi^-, f)\,.$$

For $k = 1$, as in the proof of Theorem 9.2.2 we first assume that a and b are not in $\mathbb{Z}$, in which case we obtain

$$R_1(\chi^-, f) = f(b)B_1(\chi^-,\{b\}_\chi) - f(a)B_1(\chi^-,\{a\}_\chi)$$

$$- B_0(\chi^-)\int_a^b f(t)\,dt + \sum_{\substack{a<r\leqslant b \\ r\in\mathbb{Z}}} \chi(r)f(r)\,.$$

We then note that $R_1(\chi^-, f) = R_1(\chi^-, f, a, b)$ is a continuous function of a and b, and that the right-hand side of the above equality is also continuous, since by Proposition 9.4.11 (6), for $n \in \mathbb{Z}$ we have $\lim_{x\to n,\ x>n} B_1(\chi^-,\{x\}_\chi) = B_1(\chi^-,\{n\}_\chi)$ and $\lim_{x\to n,\ x<n} B_1(\chi^-,\{x\}_\chi) = B_1(\chi^-,\{n\}_\chi)+\chi(n)$, and the result follows. □

Corollary 9.4.16. *If $N \in \mathbb{Z}_{\geq 0}$ is such that $m \mid N$ and if $f \in C^k([0,N])$ we have for $k \geq 1$*

$$\sum_{0 \leq r < N} \chi(r)f(r) = B_0(\chi) \int_0^N f(t)\, dt$$

$$+ \sum_{j=1}^k \frac{B_j(\chi)}{j!} \left(f^{(j-1)}(N) - f^{(j-1)}(0) \right) + R_k(\chi, f),$$

with

$$R_k(\chi, f) = \frac{(-1)^{k-1}}{k!} \int_0^N f^{(k)}(t) B_k(\chi^-, \{t\}_\chi)\, dt.$$

Proof. This follows from the above proposition and the formulas

$$B_j(\chi^-, \{0\}_\chi) = B_j(\chi^-) = (-1)^j B_j(\chi) - \chi(0)\delta_{j,1}$$

coming from Proposition 9.4.9.

Corollary 9.4.17. *If $m \mid N$ and $k \geq 0$ we have*

$$\sum_{0 \leq r < N} \chi(r)(x+r)^k = \frac{B_{k+1}(\chi, N+x) - B_{k+1}(\chi, x)}{k+1}.$$

Proof. Clear. □

This corollary exactly generalizes Proposition 9.2.12.

Corollary 9.4.18. *Assume in addition that $f \in C^k([1, \infty[)$, that both the series $\sum_{r \geq 1} \chi(r)f(r)$ and the integral $\int_1^\infty f(t)\, dt$ converge, and that the $f^{(j)}(N)$ tend to 0 as $N \to \infty$ for $0 \leq j \leq k-1$. Then for $k \geq 1$ we have*

$$\sum_{r \geq N} \chi(r)f(r) = B_0(\chi) \int_N^\infty f(t)\, dt - \sum_{j=1}^k \frac{B_j(\chi)}{j!} f^{(j-1)}(N)$$

$$+ \frac{(-1)^{k-1}}{k!} \int_N^\infty f^{(k)}(t) B_k(\chi^-, \{t\}_\chi)\, dt.$$

Proof. Immediate and left to the reader (Exercise 52). □

Examples. As a first example we choose $\chi(r) = (-1)^{r-1}$, so that $m = 2$ and $S_0(\chi) = 0$. By the explicit formulas for $B_k(\chi)$ and for $B_k(1/2)$ we find that for all k we have $B_k(\chi) = (-1)^k 2^{k-1}(B_k(1/2) - B_k(1)) = -(2^k - 1)B_k$. For instance, choosing $f(t) = 1/t$ we obtain the following asymptotic expansion for N even:

$$\sum_{r \geqslant N+1} \frac{(-1)^{r-1}}{r} = \log(2) - \sum_{1 \leqslant r \leqslant N} \frac{(-1)^{r-1}}{r}$$

$$= \frac{1}{2N} - \sum_{j \geqslant 1} (2^{2j}-1)\frac{B_{2j}}{2jN^{2j}} = \frac{1}{2N} - \sum_{j \geqslant 1} \frac{T_{2j-1}}{(2N)^{2j}},$$

where the $T_{2j-1} \in \mathbb{Z}$ are the tangent numbers.

As second example we choose $\chi(r) = \left(\frac{-4}{r}\right)$, in other words $\chi(r) = 0$ for r even and $\chi(r) = (-1)^{(r-1)/2}$ for r odd, so that $m = 4$ and $S_0(\chi) = 0$. By the explicit formulas for $B_k(\chi)$ and for the Euler numbers, we find that $B_k(\chi) = 0$ for k even and $B_k(\chi) = kE_{k-1}/2$ for k odd. Thus, for instance, choosing again $f(t) = 1/t$ and multiplying by 2 we obtain for $4 \mid N$ the asymptotic expansion

$$2\sum_{r \geqslant N/2+1} \frac{(-1)^{r-1}}{2r-1} = \frac{\pi}{2} - 2\sum_{1 \leqslant r \leqslant N/2} \frac{(-1)^{r-1}}{2r-1} = \sum_{j \geqslant 0} \frac{E_{2j}}{N^{2j+1}}.$$

The two formulas above are given in [Bor-Bai] as the explanation of an amusing numerical phenomenon. If we compute the sum S of the first 5000000 terms of the above series for $\pi/2$, in other words if we choose $N = 10^7$, and if we put below it the value of $\pi/2$, we find that

$S = 1.5707962267948966192323216916397513920985846996936529104874709\mathbf{11}\ldots,$

$\frac{\pi}{2} = 1.5707963267948966192313216916397514420985846996875529104874722\mathbf{96}\ldots.$

Thus, even though S and $\pi/2$ differ by approximately 10^{-7}, the digits of $\pi/2$ can still be recognized much further. In fact, we see that at 10^{-7} we must subtract $1 \cdot 10^{-7}$, at 10^{-21} we must subtract $-1 \cdot 10^{-21}$, at 10^{-35} we must subtract $5 \cdot 10^{-35}$, at 10^{-49} we must subtract $-61 \cdot 10^{-49}$, etc., and similarly for $\log(2)$. We thus recognize the Euler numbers and the asymptotic expansion given above.

9.5 Arithmetic Properties of Bernoulli Numbers

9.5.1 χ-Power Sums

The following notation will be essential.

Definition 9.5.1. (1) *We denote by $\mathcal{Z}_p$ the ring of p-adic integers of $\mathbb{C}_p$, in other words elements α such that $|\alpha| \leqslant 1$.*
(2) *If α, β, and γ are in $\mathcal{Z}_p$ with $\gamma \neq 0$, we write $\alpha \equiv \beta \pmod{\gamma}$ (or $\alpha \equiv \beta \pmod{\gamma\mathcal{Z}_p}$) if $(\alpha - \beta)/\gamma \in \mathcal{Z}_p$.*

If $k \geqslant 0$ and χ is a primitive character modulo f, we set

$$S_k(\chi) = \sum_{0 \leqslant r < f} \chi(r) r^k \,.$$

Lemma 9.5.2. (1) *For any $m \in \mathbb{Z}_{\geqslant 0}$ we have*

$$\sum_{0 \leqslant r < mf} \chi(r) r^k \equiv m S_k(\chi) \pmod{f} \,.$$

(2) *If, in addition, $\gcd(m, f) = 1$ and χ is nontrivial, then*

$$\sum_{0 \leqslant r < mf} \chi(r) r^k \equiv m S_k(\chi) \pmod{mf} \,.$$

Proof. Immediate by writing $r = qf + s$ or $r = qm + s$ respectively, and left to the reader (Exercise 55). □

The goal of this technical section is to give rather precise estimates on the divisibility properties of $S_k(\chi)$. We may of course assume that χ is a nontrivial character. In particular, $S_0(\chi) = 0$, so we may assume that $k \geqslant 1$. We will divide our study into three cases: the case that f is not a prime power, the case that f is an odd prime power, which is more difficult, and the case that f is a power of 2.

Theorem 9.5.3. *Let χ be a nontrivial primitive Dirichlet character of conductor f, and assume that f is not a prime power.*

(1) *We have $S_k(\chi) \equiv 0 \pmod{2f}$.*
(2) *If $\gcd(f, k^\infty) \neq f$ then $S_k(\chi) \equiv 0 \pmod{2f \gcd(k, f^\infty)}$.*

Proof. (1). Since f is not a prime power there exist coprime integers f_1 and f_2 such that $f = f_1 f_2$, with $f_1 > 1$ and $f_2 > 1$. By Proposition 2.1.34 there exist primitive characters χ_1 of conductor f_1 and χ_2 of conductor f_2 such that $\chi = \chi_1 \chi_2$. For $0 \leqslant r < f$ we can write in a unique way $r = f_1 r_2 + r_1$ with $0 \leqslant r_1 < f_1$ and $0 \leqslant r_2 < f_2$, so that

$$S_k(\chi) = \sum_{\substack{0 \leqslant r_1 < f_1 \\ 0 \leqslant r_2 < f_2}} \chi_1 \chi_2 (f_1 r_2 + r_1)(f_1 r_2 + r_1)^k = \sum_{0 \leqslant j \leqslant k} \binom{k}{j} f_1^j T_{k,j}(\chi) \,,$$

where

$$T_{k,j}(\chi) = \sum_{0 \leqslant r_1 < f_1} \chi_1(r_1) r_1^{k-j} \sum_{0 \leqslant r_2 < f_2} \chi_2(f_1 r_2 + r_1) r_2^j \,.$$

Since f_1 and f_2 are coprime we have

$$\sum_{0 \leqslant r_2 < f_2} \chi_2(f_1 r_2 + r_1) = \chi_2(f_1) \sum_{r_2 \bmod f_2} \chi_2(r_2 + r_1 f_1^{-1})$$

$$= \sum_{r_3 \bmod f_2} \chi_2(r_3) = 0$$

since χ_2 is a primitive, hence nontrivial, character modulo $f_2 > 1$, so that $T_{k,0}(\chi) = 0$. This shows that $S_k(\chi) \equiv 0 \pmod{f_1}$, and by symmetry $S_k(\chi) \equiv 0 \pmod{f_2}$, so $S_k(\chi) \equiv 0 \pmod{f}$ since f_1 and f_2 are coprime. To prove the stronger congruence modulo $2f$, without loss of generality we may assume that f_2 is odd.

Assume first that f_1 is even, hence that $4 \mid f_1$. Since $T_{k,0}(\chi) = 0$ we have $S_k(\chi) \equiv n f_1 T_{k,1}(\chi) \pmod{4f_1}$. If we write

$$T_{k,1}(\chi) = \sum_{\substack{0 \leqslant r_1 < f_1 \\ 0 \leqslant r_2 < f_2}} f(r_1, r_2) \quad \text{with} \quad f(r_1, r_2) = \chi_1(r_1)\chi_2(f_1 r_2 + r_1)r_1^{k-1} r_2,$$

we see that $f(f_1 - r_1, f_2 - r_2 - 1) \equiv f(r_1, r_2) \pmod 2$. Since the involution $(r_1, r_2) \mapsto (f_1 - r_1, f_2 - r_2 - 1)$ has a single fixed point $(r_1, r_2) = (f_1/2, (f_2 - 1)/2)$ and $f(r_1, r_2) = 0$ (since $\chi_1(f_1/2) = 0$), it follows that $T_{k,1}(\chi) \equiv 0 \pmod 2$, so we deduce that $S_k(\chi) \equiv 0 \pmod{2f_1}$ in this case, hence that $S_k(\chi) \equiv 0 \pmod{2f}$ since f_2 is coprime to $2f_1$, as claimed. Note for future reference that the same proof shows that $T_{k,j}(\chi) \equiv 0 \pmod 2$ when f_1 is even and f_2 is odd.

If f_1 is odd then f is odd, and since $k > 0$ we have

$$S_k(\chi) \equiv \sum_{0 \leqslant r < f, \ r \text{ odd}} \chi(r) \equiv 0 \pmod 2$$

by Corollary 2.1.37, since by assumption χ is primitive and f is not a prime power, proving (1) in general since we already know that $S_k(\chi) \equiv 0 \pmod f$.

(2). If f and k are coprime there is nothing more to prove, so we assume that $\gcd(f, k) > 1$. Here we choose specifically $f_1 = \gcd(f, k^\infty)$. By definition $f_1 \mid f$, $f_2 = f/f_1$ is coprime to f_1, and $\gcd(f, k) \mid f_1$, so in particular $f_1 > 1$, and finally by assumption we have $f_2 > 1$. From the proof of (1) we know that there exist algebraic integers $T_{k,j}$ such that $S_k(\chi) = \sum_{1 \leqslant j \leqslant k} \binom{k}{j} f_1^j T_{k,j}$. I claim that $\gcd(k, f^\infty) \mid \binom{k}{j} f_1^{j-1}$ for all $j \in [1, k]$. Indeed, by definition we have $\gcd(k, f^\infty) = \prod_{p \mid f} p^{v_p(k)}$; hence let p be a prime dividing f and k, so that $p \mid f_1$. By Lemma 4.2.8 we have

$$v_p\left(\binom{k}{j} f_1^{j-1}\right) \geqslant j - 1 + \max(v_p(k) - v_p(j), 0) .$$

If $v_p(k) \geqslant v_p(j)$ this is greater than or equal to $v_p(k) + j - 1 - v_p(j)$, and $j - 1 - v_p(j) \geqslant 0$ for all $j \geqslant 1$. If $v_p(k) < v_p(j)$ this is greater than or equal to $j - 1 = j - 1 - v_p(j) + v_p(j) > v_p(k)$ for the same reason, proving my claim.

We thus have $S_k(\chi) \equiv 0 \pmod{f_1 \gcd(k, f^\infty)}$, and since $S_k(\chi) \equiv 0 \pmod{f_2}$ and f_2 is coprime to kf_1, we deduce that $S_k(\chi) \equiv 0 \pmod{f \gcd(k, f^\infty)}$.

To prove the stronger congruence, we reason as follows. If k is odd, (1) and what we have just proved imply that $S_k(\chi) \equiv 0 \pmod{f \operatorname{lcm}(2, \gcd(k, f^\infty))}$, hence modulo $2f \gcd(k, f^\infty)$. We may therefore assume that k is even. Using the same notation as above, since f_2 is coprime to kf_1 it is odd, and we have $S_k(\chi) \equiv 0 \pmod{f_2}$. Since f_1 is divisible by 4, we have seen in the proof of (1) that $T_{k,j} \equiv 0 \pmod 2$ for all j, so that the formula for $S_k(\chi)$ given above implies that $S_k(\chi) \equiv 0 \pmod{2f_1 \gcd(k, f^\infty)}$, proving the theorem. $\qquad\square$

We now consider the more delicate case that $f = p^v$ is an odd prime power. We begin with the case $v = 1$.

Lemma 9.5.4. *Let χ be any character modulo p for some odd prime p, let $o(\chi) \mid (p-1)$ be the order of χ, and let $K = \mathbb{Q}(\chi) = \mathbb{Q}(\zeta_{o(\chi)})$ be the corresponding cyclotomic field.*

(1) *If $o(\chi) \neq (p-1)/\gcd(p-1, k)$ then $S_k(\chi) \equiv 0 \pmod p$.*
(2) *If $o(\chi) = (p-1)/\gcd(p-1, k)$ there exists a (necessarily unique) prime ideal $\mathfrak{p}$ of K above p such that $S_k(\chi) \equiv 0 \pmod{\mathfrak{q}}$ for all prime ideals $\mathfrak{q}$ above p with $\mathfrak{q} \neq \mathfrak{p}$, while $S_k(\chi) \equiv -1 \pmod{\mathfrak{p}}$.*

Proof. Let $L = \mathbb{Q}(\zeta_{p-1}) \supset K$. Since $p \equiv 1 \pmod{p-1}$, by Proposition 3.5.18 the prime p splits completely in L. Let $\mathfrak{P}$ be some prime ideal of L above p. By definition of the Teichmüller character (Definition 3.6.2) we have $\omega_{\mathfrak{P}}(x) \equiv x \pmod{\mathfrak{P}}$, so that

$$pS_k(\chi) = \sum_{0 \leqslant r < p} \chi(r) r^k \equiv \sum_{r \bmod p} (\chi \omega_{\mathfrak{P}}^k)(r) \pmod{\mathfrak{P}} .$$

Thus, by orthogonality of characters, if $\chi \neq \omega_{\mathfrak{P}}^{-k}$ we have $pS_k(\chi) \equiv 0 \pmod{\mathfrak{P}}$, and otherwise $pS_k(\chi) \equiv p - 1 \equiv -1 \pmod{\mathfrak{P}}$. Since $\omega_{\mathfrak{P}}$ has order $p-1$, $\omega_{\mathfrak{P}}^{-k}$ has exact order $(p-1)/\gcd(p-1, k)$, hence if this is not equal to $o(\chi)$ we deduce that $pS_k(\chi) \equiv 0 \pmod p$, proving (1). On the other hand, since all characters of order $p-1$ are of the form $\omega_{\mathfrak{P}}$ for some (unique) $\mathfrak{P}$, all characters of order $(p-1)/\gcd(p-1, k)$ are of the form $\omega_{\mathfrak{P}}^{-k}$ for some $\mathfrak{P}$. It follows that if $o(\chi) = (p-1)/\gcd(p-1, k)$ we have $\chi = \omega_{\mathfrak{P}}^{-k}$ for some $\mathfrak{P}$. By Lemma 3.6.3 (3), $\mathfrak{P}$ may not be unique, but the ideal $\mathfrak{p}$ of $K = \mathbb{Q}(\zeta_{o(\chi)})$ below $\mathfrak{P}$ is unique, and since p is totally split the result follows. $\qquad\square$

We can deduce from this the general case $v \geqslant 1$ and p odd, as follows.

Theorem 9.5.5. *Let p be an odd prime, let $v \in \mathbb{Z}_{\geqslant 1}$, let χ be a primitive Dirichlet character of conductor $f = p^v$, let $o(\chi)$ be the order of χ, and let $K = \mathbb{Q}(\chi) = \mathbb{Q}(\zeta_{o(\chi)})$ be the cyclotomic field generated by the values of χ.*

(1) *If either $o(\chi) \neq p^{v-1}(p-1)/\gcd(p-1, k)$ or $p \mid k$ and $v \geqslant 2$, we have $S_k(\chi) \equiv 0 \pmod f$.*

(2) If $o(\chi) = p^{v-1}(p-1)/\gcd(p-1,k)$, and $p \nmid k$ or $v = 1$, then $S_k(\chi) \equiv 0$ (mod f/p).

(3) More precisely, there exists a unique prime ideal $\mathfrak{p}$ of K above p such that $v_\mathfrak{q}(S_k(\chi)/f) = 0$ for all prime ideals $\mathfrak{q}$ of K above p different from $\mathfrak{p}$, and such that

$$v_\mathfrak{p}\left(\frac{S_k(\chi)}{f} + \frac{1}{p}\right) \geqslant 0 \quad if \; v = 1,$$

$$v_\mathfrak{p}\left(\frac{S_k(\chi)}{f} - \frac{k}{1 - \chi(1+p)}\right) \geqslant 0 \quad if \; v \geqslant 2.$$

(4) If χ is an even character then $S_k(\chi) \equiv 0$ (mod 2), so that $S_k(\chi) \equiv 0$ (mod $2f$) or (mod $2f/p$) depending on whether we are in case (1) or (2) above.

Proof. (1), (2), and (3). The case $v = 1$ is nothing else than Lemma 9.5.4, so assume $v \geqslant 2$. Let r be coprime to p. Writing $r \equiv r_1 r_2$ (mod p^v) with $r_1 = r^{p^{v-1}}$, it is clear that r_1 is unique modulo p and that $r_2 \equiv 1$ (mod p) (see Lemma 2.1.26). Dually, we can write $\chi = \chi_1 \chi_2$ with $\chi_1 = \chi^{p^{v-1}}$, and it is immediate to see that χ_1 is a (possibly trivial) character modulo p, and χ_2 is a primitive character of exact order p^{v-1}. Since $r_2 \equiv 1$ (mod p) we have $\chi_1(r_2) = 1$, and since r_1 is a p^{v-1}th power we have $\chi_2(r_1) = 1$. It follows that

$$\chi(r) = \chi_1(r_1)\chi_2(r_1)\chi_1(r_2)\chi(r_2) = \chi_1(r_1)\chi_2(r_2) .$$

Thus

$$S_k(\chi) = \sum_{0 \leqslant r < p^v} \chi(r) r^k$$

$$\equiv \sum_{r_1 \bmod p} \chi_1(r_1) r_1^k \sum_{\substack{r_2 \bmod p^v \\ r_2 \equiv 1 \; (\bmod \; p)}} \chi_2(r_2) r_2^k \equiv S_1 S_2 \; (\bmod \; p^v)$$

with evident notation. Since S_1 is the sum studied in Lemma 9.5.4, we now study S_2. By Lemma 2.1.26, for any $a \equiv 1$ (mod p) there exists a unique x modulo p^{v-1} such that $a \equiv (1+p)^x$ (mod p^v), so that

$$S_2 \equiv \sum_{x \; (\bmod \; p^{v-1})} \chi_2(1+p)^x (1+p)^{kx} \equiv \frac{(1+p)^{kp^{v-1}} - 1}{\chi_2(1+p)(1+p)^k - 1} \; (\bmod \; p^v) .$$

Since $v \geqslant 2$ we have $(1+p)^{kp^{v-1}} \equiv 1 + kp^v$ (mod p^{v+1}). Furthermore, $(1+p)^k \equiv 1$ (mod p), $\chi_2(1+p)$ is a primitive p^{v-1}th root of unity, so $v_p(\chi_2(1+p) - 1) = 1/\phi(p^{v-1}) < 1$ by Proposition 3.5.5. Since $\chi_1(1+p) = 1$, it follows finally that

$$S_2 \equiv \frac{kp^v}{\chi(1+p) - 1} \; (\bmod \; p^v) .$$

Since we want $S_1 S_2$ modulo p^v and we have $v_p(\chi(1+p) - 1) < 1$, it is enough to know S_1 modulo p. Since $\chi_1 = \chi^{p^{v-1}}$, we deduce from Lemma 9.5.4 that if $o(\chi) \neq p^{v-1}(p-1)/\gcd(p-1, k)$ then $S_1 \equiv 0 \pmod{p}$, so that $S_1 S_2 \equiv 0 \pmod{p^v}$, proving (1). Otherwise, there exists a unique prime ideal $\mathfrak{p}$ above p in $K_1 = \mathbb{Q}(\zeta_{(p-1)/\gcd(p-1,k)})$ such that $S_1 \equiv 0 \pmod{\mathfrak{q}}$ for any prime ideal $\mathfrak{q}$ of K_1 above p and $S_1 \equiv -1 \pmod{\mathfrak{p}}$. However, in the extension K/K_1 all the prime ideals above p are totally ramified; in other words, every prime ideal $\mathfrak{q}$ of K_1 above p splits as $\mathfrak{q}\mathbb{Z}_K = \mathfrak{Q}^{\phi(p^{v-1})}$, and in particular there is a single prime ideal of K above each prime ideal $\mathfrak{q}$ of K_1 that is above p. Furthermore, $\chi(1+p) - 1 \in \mathbb{Q}(\zeta_{p^{v-1}})$ has norm p by Proposition 3.5.5, so $v_{\mathfrak{Q}}(\chi(1+p)-1) = 1$ for every prime ideal $\mathfrak{Q}$ of K above p. If we denote by $\mathfrak{P}$ the one above $\mathfrak{p}$ we thus have $v_{\mathfrak{Q}}(S/p^v) \geqslant 0$ for $\mathfrak{Q} \neq \mathfrak{P}$, and $v_{\mathfrak{P}}(S/p^v - k/(1 - \chi(1+p))) \geqslant 0$, proving (3).

For (4), we simply note that

$$S_k(\chi) \equiv \sum_{0 \leqslant r < f} \chi(r) r^k \equiv \sum_{0 \leqslant r < f,\ r \text{ odd}} \chi(r) \equiv 0 \pmod{2}$$

by Corollary 2.1.37 (1), since we have assumed that χ is an even character (the result of (4) is trivially false if χ is an odd character). $\square$

We finally consider the last remaining case, that $f = 2^v$.

Theorem 9.5.6. *If χ is a primitive Dirichlet character of conductor $f = 2^v$ for some $v \geqslant 1$ we have*

$$S_k(\chi) \equiv \begin{cases} 0 & (\mathrm{mod}\ 2f) & \text{if } k \text{ is even,} \\ \dfrac{2f}{1 - \chi(5)} & (\mathrm{mod}\ 2f) & \text{if } k \text{ is odd and } v \geqslant 3, \\ 6 & (\mathrm{mod}\ 2f) & \text{if } k \text{ is odd and } v = 2. \end{cases}$$

Proof. Since there are no primitive characters modulo 2, we may assume that $v \geqslant 2$. The only primitive character modulo 4 is $\chi(r) = (-1)^{(r-1)/2}$ for r odd, so that $S_k(\chi) = 1 - 3^k$, and this is congruent to 0 modulo 8 if k is even, and to 6 modulo 8 if k is odd, proving the case $v = 2$. Thus assume that $v \geqslant 3$, so that if $0 \leqslant r < f$ satisfies $r \equiv 1 \pmod{4}$ there exists a unique $z \in [0, 2^{v-2}[$ such that $r \equiv 5^z \pmod{f}$. Thus

$$
S_k(\chi) = \sum_{\substack{0 \leqslant r < f \\ r \equiv 1\ (\mathrm{mod}\ 4)}} \chi(r) r^k + \sum_{\substack{0 \leqslant r < f \\ r \equiv 3\ (\mathrm{mod}\ 4)}} \chi(r) r^k
$$

$$
= \sum_{\substack{0 \leqslant r < f \\ r \equiv 1\ (\mathrm{mod}\ 4)}} \chi(r) r^k + \sum_{\substack{0 \leqslant r < f \\ r \equiv 1\ (\mathrm{mod}\ 4)}} \chi(f - r)(f - r)^k
$$

$$\equiv (1 + (-1)^k \chi(-1)) \sum_{\substack{0 \leqslant r < f \\ r \equiv 1 \ (\mathrm{mod}\ 4)}} \chi(r) r^k$$

$$+ fk(-1)^{k-1}\chi(-1) \sum_{\substack{0 \leqslant r < f \\ r \equiv 1 \ (\mathrm{mod}\ 4)}} \chi(r) r^{k-1} \ (\mathrm{mod}\ 4f)$$

by the binomial theorem, since $4 \mid f$. The second sum is easily treated: we have

$$\sum_{\substack{0 \leqslant r < f \\ r \equiv 1 \ (\mathrm{mod}\ 4)}} \chi(r) r^{k-1} \equiv \sum_{\substack{0 \leqslant r < f \\ r \equiv 1 \ (\mathrm{mod}\ 4)}} \chi(r) \equiv \sum_{0 \leqslant z < 2^{v-2}} \chi(5)^z \ (\mathrm{mod}\ 4) \ .$$

This is a geometric series, and $\chi(5) \neq 1$; otherwise $\chi(5^z) = 1$ for all z, so that χ would be the trivial character if $\chi(-1) = 1$, and equal to $\left(\frac{-4}{\cdot}\right)$ if $\chi(-1) = -1$, both of which are excluded since by assumption χ is a primitive character modulo $f \geqslant 8$. It follows that the last sum is equal to

$$\frac{\chi(5)^{2^{v-2}} - 1}{\chi(5) - 1} = \frac{\chi(5^{2^{v-2}}) - 1}{\chi(5) - 1} = 0$$

since $5^{2^{v-2}} \equiv 1 \ (\mathrm{mod}\ f)$. We have thus shown that

$$S_k(\chi) \equiv (1 + (-1)^k \chi(-1)) \sum_{\substack{0 \leqslant r < f \\ r \equiv 1 \ (\mathrm{mod}\ 4)}} \chi(r) r^k \ (\mathrm{mod}\ 4f) \ ,$$

so that $S_k(\chi) \equiv 0 \ (\mathrm{mod}\ 4f)$ if $\chi(-1) = (-1)^{k-1}$. Thus assume that $\chi(-1) = (-1)^k$. As above, we have

$$\sum_{\substack{0 \leqslant r < f \\ r \equiv 1 \ (\mathrm{mod}\ 4)}} \chi(r) r^k \equiv \sum_{0 \leqslant z < 2^{v-2}} \chi(5)^z 5^{kz} \equiv \frac{5^{k2^{v-2}} - 1}{5^k \chi(5) - 1} \ (\mathrm{mod}\ f) \ ,$$

where here the denominator trivially cannot vanish. Since $1 + (-1)^k \chi(-1) = 2$ we can thus write

$$S_k(\chi) \equiv (5^{k2^{v-2}} - 1) \frac{2}{5^k \chi(5) - 1} \ (\mathrm{mod}\ 2f) \ .$$

Since $v_2(\chi(5) - 1) \leqslant 1$, as in the case $p \geqslant 3$ we deduce that $2/(5^k \chi(5) - 1) \equiv 2/(\chi(5) - 1) \ (\mathrm{mod}\ 2\mathbb{Z}_2)$, so its 2-adic valuation is nonnegative. Finally, since $v_2(5^{2^{v-2}} - 1) = v$, we have $v_2(5^{k2^{v-2}} - 1) = v + v_2(k)$, so that the first factor is divisible by $2f$ if k is even, while if k is odd we deduce that $S_k(\chi)/f \equiv 2/(\chi(5) - 1) \equiv 2/(1 - \chi(5)) \ (\mathrm{mod}\ 2\mathbb{Z}_2)$, proving the theorem. $\square$

Corollary 9.5.7. *Let χ be a nontrivial primitive Dirichlet character of conductor f and of order $o(\chi)$.*

(1) *If either f is not a prime power, or if $f = p^v$ is an odd prime power and either $o(\chi) \neq p^{v-1}(p-1)/\gcd(p-1,k)$ or $p \mid k$ or $v = 1$, or if $f = 2^v$ with either $v \geqslant 3$ or k even, then $S_k(\chi) \equiv 0 \pmod{f}$.*

(2) *If either f is not a prime power, or if $f = p^v$ is an odd prime power and either $o(\chi) \neq p^{v-1}(p-1)/\gcd(p-1,k)$ or $p \mid k$ or $v \geqslant 2$, and χ is an even character, or if $f = 2^v$ with k even, then $S_k(\chi) \equiv 0 \pmod{2f}$.*

Proof. Clear. $\qquad\qquad\qquad\qquad\qquad\qquad\qquad\qquad\qquad\qquad$ □

Corollary 9.5.8. *Let D be the discriminant of a quadratic field, let $k \in \mathbb{Z}_{\geqslant 1}$, and set $S_k(D) = \sum_{0 \leqslant r < |D|} \left(\frac{D}{r}\right) r^k$.*

(1) *If $|D|$ is a prime p such that $k \equiv (p-1)/2 \pmod{p-1}$ then $S_k(D) \equiv -1 \pmod{D}$.*

(2) (a) *If $D = -4$, then for k odd we have $S_k(D) \equiv -2 \pmod{2D}$ and for k even we have $S_k(D) \equiv 0 \pmod{2D}$.*

   (b) *If $D = -8$, then for k odd we have $S_k(D) \equiv 8 \pmod{2D}$ and for k even we have $S_k(D) \equiv 0 \pmod{8D}$.*

   (c) *If $D = 8$, then for k odd we have $S_k(D) \equiv 0 \pmod{8D}$ and for k even we have $S_k(D) \equiv 0 \pmod{2D}$.*

(3) *In all other cases $S_k(D) \equiv 0 \pmod{D}$. More precisely, if we are not in case (1) or (2) then:*

   (a) *If $D \neq -p$ for an odd prime p then $S_k(D) \equiv 0 \pmod{2D}$.*

   (b) *If $D = -p$ then $S_k(D) \equiv D \pmod{2D}$*

   (c) *If $D \neq -p$ and $D \nmid 4k$ then $S_k(D) \equiv 0 \pmod{2D \gcd(k, D^\infty)}$.*

The fact that negative prime discriminants are singled out by this corollary corresponds to an important algebraic fact: for instance by Dirichlet's class number formula, for $D < -4$ the class number $h(D)$ of the imaginary quadratic field of discriminant D, which is of course an integer, is equal to $S_1(D)/D$. Statements (3) (a) and (b) mean that $h(D)$ is odd if and only if $D = -p$ for an odd prime p (and in addition for $D = -8$ and $D = -4$).

Corollary 9.5.9. *Let D be the discriminant of a quadratic field, and assume that either $D \equiv 0 \pmod{4}$ or $D > 0$.*

(1) *We have*

$$\sum_{0 \leqslant r < |D|} \left(\frac{D}{r}\right) r^2 \equiv 0 \pmod{4D},$$

except if $D = -4$, 5, or 8, in which case the left-hand side is equal respectively to -8, 4, or 16.

(2) *We have*

$$\sum_{0 \leqslant r < |D|} \left(\frac{D}{r}\right) r^4 \equiv 0 \pmod{8D},$$

except if $D = -4$ or 8, in which case the left-hand side is equal respectively to $-80 = -2^4 \cdot 5$ or $1696 = 2^5 \cdot 53$.

Proofs. Immediate consequences of the theorem and of the fact that when χ is an even character and $k \geqslant 2$ is even, then $S_k(\chi) \equiv 0 \pmod 8$, except when $k = 2$, in which case the congruence is only modulo 4 (Exercise 56). □

Corollary 9.5.10. *Let χ be a primitive character of conductor f such that $4 \mid f$, and let $k \in \mathbb{Z}_{\geqslant 0}$.*

(1) *If f is not a power of 2 we have*

$$\sum_{0 \leqslant r < f/2} \chi(r) r^k \equiv \begin{cases} 0 \pmod f & \text{if } \chi(-1) = (-1)^k , \\ 0 \pmod 4 & \text{if } \chi(-1) = (-1)^{k-1} . \end{cases}$$

(2) *If $f = 2^v$ with $v \geqslant 3$ we have*

$$\sum_{0 \leqslant r < f/2} \chi(r) r^k \equiv \begin{cases} \dfrac{f}{1 - \chi(5)} \pmod f & \text{if } \chi(-1) = (-1)^k , \\ \dfrac{4}{1 - \chi(5)} \pmod 4 & \text{if } \chi(-1) = (-1)^{k-1} . \end{cases}$$

If $f = 4$ we have of course $\sum_{0 \leqslant r < f/2} \chi(r) r^k = 1$.

Proof. Assume first that $\chi(-1) = (-1)^k$. We have

$$S_k(\chi) = \sum_{0 \leqslant r < f/2} (\chi(r) r^k + \chi(f - r)(f - r)^k)$$

$$\equiv 2 \sum_{0 \leqslant r < f/2} \chi(r) r^k - \chi(-1) k f \sum_{0 \leqslant r < f/2} \chi(r) r^{k-1} \pmod{f^2} .$$

Now it is clear that

$$\chi(-1) \sum_{0 \leqslant r < f/2} \chi(r) r^{k-1} \equiv \sum_{0 \leqslant r < f/2} \chi(r) \pmod 2 .$$

This last sum vanishes when χ is an even (nontrivial) character, and by Corollary 2.1.30 when χ is odd we have $\chi(f/2 - r) = \chi(r)$, so that $\sum_{0 \leqslant r < f/2} \chi(r) = 2 \sum_{0 \leqslant r < f/4} \chi(r) \equiv 0 \pmod 2$. It follows from Theorem 9.5.3 that if f is not a power of 2 then $2 \sum_{0 \leqslant r < f/2} \chi(r) r^k \equiv 0 \pmod{2f}$, and from Theorem 9.5.6 that if $f = 2^v$ with $v \geqslant 3$ then $2 \sum_{0 \leqslant r < f/2} \chi(r) r^k \equiv 2f/(1 - \chi(5)) \pmod{2f}$, proving the theorem when $\chi(-1) = (-1)^k$.

Assume now that $\chi(-1) = (-1)^{k-1}$. Since we only want congruences modulo 4 and since $r^2 \equiv 1 \pmod 8$ when $2 \nmid r$, we have

$$\sum_{0 \leqslant r < f/2} \chi(r) r^k \equiv \sum_{0 \leqslant r < f/2} \chi_1(r) \pmod 4 ,$$

where $\chi_1 = \chi$ if k is even and $\chi_1 = \left(\frac{-4}{\cdot}\right)\chi$ if k is odd, so that in both cases the character χ_1 is odd and its conductor is still equal to f (note that $f \neq 4$). By Corollary 2.1.30 once again, we have

$$\sum_{0 \leqslant r < f} \chi_1(r)r = \sum_{0 \leqslant r < f/2} (\chi_1(r)r + \chi_1(r+f/2)(r+f/2)) = -(f/2) \sum_{0 \leqslant r < f/2} \chi_1(r).$$

As before it follows from Theorems 9.5.3 and 9.5.6 that the left hand side is divisible by $2f$ when f is not a power of 2 and congruent to $2f/(1 - \chi(5))$ modulo $2f$ if $f = 2^v$ with $v \geqslant 3$, so that $\sum_{0 \leqslant r < f/2} \chi_1(r)$ is congruent to 0 or to $4/(\chi(5) - 1)$ modulo 4, proving the theorem when $\chi(-1) = (-1)^{k-1}$ (note that $4/(\chi(5) - 1) \equiv 4/(1 - \chi(5)) \pmod 4$). $\qquad\square$

9.5.2 The Generalized Clausen–von Staudt Congruence

Lemma 9.5.11. *Let χ be a primitive Dirichlet character of conductor f and let p be a prime number.*

(1) *If $f > 1$ and either f is not a power of p, or $p = 2$ and $f = 2^v$ with $v \geqslant 3$, then $v_p(B_k(\chi)) \geqslant 0$ for all $k \geqslant 0$.*
(2) *In all other cases, in other words if either $f = 1$, or $f = p^v$ with p odd, or $f = 4$, then $v_p(B_k(\chi)) \geqslant -1$ for all $k \geqslant 0$.*

Proof. Set $N = \mathrm{lcm}(f, p)$. By Corollary 9.4.17 we have

$$\sum_{0 \leqslant r < N} \chi(r)r^k = \frac{1}{k+1} \sum_{1 \leqslant j \leqslant k+1} \binom{k+1}{j} B_{k+1-j}(\chi)N^j$$

$$= \sum_{0 \leqslant j \leqslant k} \binom{k}{j} B_{k-j}(\chi)\frac{N^{j+1}}{j+1},$$

so we obtain the induction formula

$$B_k(\chi) = \frac{1}{N} \sum_{0 \leqslant r < N} \chi(r)r^k - \sum_{1 \leqslant j \leqslant k} \binom{k}{j} B_{k-j}(\chi)\frac{N^j}{j+1}.$$

Set $z = 0$ in case (1) and $z = -1$ in case (2). We prove by induction on k that $v_p(B_k(\chi)) \geqslant z$ for all $k \geqslant 0$. Let $k \geqslant 0$, and assume that we have shown that $v_p(B_j) \geqslant z$ for $j < k$, so that $v_p(B_{k-j}(\chi)) \geqslant z$ for $1 \leqslant j \leqslant k$. Since trivially $v_p(j + 1) \leqslant j$ for $j \geqslant 1$ and $p \mid N$, we have $v_p(N^j/(j+1)) \geqslant 0$, so the valuation of the second sum in the induction formula is also greater than or equal to z. For the first term we consider three cases.

Case 1: $p \nmid f$ **and** $f > 1$. Then $N = pf$, and writing $r = qp + s$ we have

$$\sum_{0 \leqslant r < N} \chi(r)r^k = \sum_{0 \leqslant s < p} \sum_{0 \leqslant q < f} \chi(qp + s)(qp + s)^k$$

$$\equiv \sum_{0 \leqslant s < p} s^k \sum_{0 \leqslant q < f} \chi(qp + s) \pmod p.$$

Since p is coprime to f the map $q \mapsto qp + s$ is a bijection from $(\mathbb{Z}/f\mathbb{Z})^*$ onto itself, and since χ is a nontrivial character, $\sum_{0 \leqslant q < f} \chi(qp + s) = 0$, so that $\sum_{0 \leqslant r < N} \chi(r)r^k \equiv 0 \pmod{p}$. Thus the p-adic valuation of the first term in our induction is nonnegative, so is greater than or equal to z, proving the result in this case.

Case 2: $f = 1$. This case is immediate: we have $N = p$, so the valuation of the first term of our induction formula is greater than or equal to $z = -1$.

Case 3: $p \mid f$. In this case we have $N = f$, so the first term of our induction formula is equal to $S_k(\chi)/f$, with the usual notation $S_k(\chi) = \sum_{0 \leqslant r < f} \chi(r)r^k$. Note that since $p \mid f$, if f is a prime power it must be a power of p. Thus if f is not a power of p or if $p = 2$ and $f = 2^v$ with $v \geqslant 3$, it follows from Theorems 9.5.3 and 9.5.6 that $v_p(S_k(\chi)/f) \geqslant 0 > z$. On the other hand, if f is an odd prime power or if $f = 4$, it follows from Theorems 9.5.5 and 9.5.6 that $v_p(S_k(\chi)/f) \geqslant -1 = z$, proving the lemma. $\square$

Lemma 9.5.12. *Let χ be a primitive Dirichlet character of conductor f and let p be a prime number. Assume that we are in case (2) of the preceding lemma, in other words that either $f = 1$, or $f = p^v$ with p odd or with $p = 2$ and $v = 2$. Then for all k such that $\chi(-1) = (-1)^k$ we have*

$$pB_k(\chi) \equiv \begin{cases} \dfrac{S_k(\chi)}{p^{v-1}} \pmod{p} & \text{if } f > 1, \\ \displaystyle\sum_{0 \leqslant r < p} r^k \pmod{p} & \text{if } f = 1. \end{cases}$$

Note that if $\chi(-1) = (-1)^{k-1}$ we have $B_k(\chi) = 0$ except if $k = 1$ and $f = 1$, in which case $B_k(\chi) = B_1 = -1/2$; see Proposition 9.4.9.

Proof. Keep the notation of the preceding proof. Since $p \mid N$ it is immediate to check that $v_p(N^{j-1}/(j+1)) \geqslant 0$ for $j \geqslant 2$, and also for $j = 1$ if $p \neq 2$. Since from the preceding lemma we know that $v_p(B_j(\chi)) \geqslant -1$, it follows that for $j \geqslant 1$ we have

$$v_p(B_{k-j}(\chi)N^j/(j+1)) = v_p(NB_{k-j}(\chi)N^{j-1}/(j+1)) \geqslant 0 \,,$$

except perhaps for $j = 1$ when $p = 2$. Thus by our induction formula we have

$$pB_k(\chi) \equiv \frac{p}{N} \sum_{0 \leqslant r < N} \chi(r)r^k - \frac{kpN}{2}B_{k-1}(\chi) \pmod{p} \,,$$

where the last term can be omitted if $p \neq 2$, and must be omitted if $k = 0$. Note, in addition, that by Proposition 9.4.9, since we have assumed that $\chi(-1) = (-1)^k$ we have $B_{k-1}(\chi) = 0$ except if χ is even and $k = 2$, but then $v_2(kpN/2) \geqslant 2$, so the last term can also be omitted in that case. We deduce that

$$pB_k(\chi) \equiv \frac{p}{N} \sum_{0 \leqslant r < N} \chi(r)r^k \pmod{p} \,.$$

When $f > 1$ we have $N = f = p^v$, and when $f = 1$ we have $N = p$, so the lemma follows.   □

From this lemma and the results of the preceding subsection, it is now immediate to deduce the generalized Clausen–von Staudt congruence.

Theorem 9.5.13. *Let χ be a primitive Dirichlet character of conductor f, denote as usual by $o(\chi)$ the order of χ, and let $K = \mathbb{Q}(\chi) = \mathbb{Q}(\zeta_{o(\chi)})$. For any $k \geqslant 0$ such that $\chi(-1) = (-1)^k$, the number $B_k(\chi)$ is an algebraic integer, with the following exceptions:*

(1) *When $f = 1$: if p is a prime such that $(p-1) \nmid k$ then $v_p(B_k) \geqslant 0$, and if $(p-1) \mid k$ we have $v_p(B_k + 1/p) \geqslant 0$.*
(2) *When $f = 4$: we have $B_k(\chi) + 1/2 \in \mathbb{Z}$.*
(3) *When $f = p^v$ with p an odd prime and $v \geqslant 1$, and $o(\chi) = p^{v-1}(p-1)/\gcd(p-1,k)$, and either $p \nmid k$ or $v = 1$: in this case, there exists a unique prime ideal $\mathfrak{p}$ of K above p such that $v_\mathfrak{q}(B_k(\chi)) \geqslant 0$ for any prime ideal $\mathfrak{q} \neq \mathfrak{p}$ of K (above p or not), and such that for $\mathfrak{q} = \mathfrak{p}$ we have*

$$v_\mathfrak{p}\left(B_k(\chi) + \frac{1}{p}\right) \geqslant 0 \quad \text{if } v = 1,$$

$$v_\mathfrak{p}\left(B_k(\chi) - \frac{k}{1 - \chi(1+p)}\right) \geqslant 0 \quad \text{if } v \geqslant 2.$$

Proof. All these results follow immediately from Lemma 9.5.12: for $f = 1$ we apply Lemma 9.5.4, for $f = 4$ we use the evaluation $S_k(\chi) = 1 - 3^k$, and for the other values of f we use Theorems 9.5.3, 9.5.5, and 9.5.6.   □

See Corollary 11.4.2 for the corresponding and stronger result for $B_k(\chi)/k$.

The special case $f = 1$ of the above theorem is the usual Clausen–von Staudt congruence, which we restate because of its importance:

Theorem 9.5.14 (Clausen–von Staudt). *For any even $k \in \mathbb{Z}_{>0}$ we have*

$$B_k \equiv - \sum_{(p-1)|k} \frac{1}{p} \pmod{1},$$

where it is understood that p is a positive prime number.

The Clausen–von Staudt theorem means that $v_p(B_k) = 0$ if $(p-1) \nmid k$ and $B_k \equiv -1/p \pmod{1}$ if $(p-1) \mid k$. We will see in Proposition 11.4.4 and especially in Corollary 11.4.7 that this can be strengthened. For instance, (when $(p-1) \mid k$) we have $B_k \equiv (1 - 1/p) \pmod{p}$ for $p = 2$, 5, and 13.

Write $B_k = N_k/D_k$ uniquely with N_k and D_k coprime and $D_k > 0$.

Corollary 9.5.15. *For even $k > 0$ we have*

$$D_k = \prod_{(p-1)|k} p \, .$$

Proof. Indeed, from the theorem it is clear that D_k divides $\prod_{(p-1)|k} p$, but conversely for any prime p such that $(p-1) \mid k$ the theorem implies that $v_p(B_k) = -1$, so that the product of such p divides D_k. $\qquad\square$

Corollary 9.5.16. (1) *For even $k > 0$ we have $6 \mid D_k$.*
(2) *If $k = 2q$, where q is a prime such that $2q+1$ is not prime, then $D_k = 6$.*
(3) *The number of even $k \leqslant X$ such that $D_k > 6$ is greater than or equal to $X/4$, and in particular has a strictly positive lower density.*

Proof. The fact that $6 \mid D_k$ is clear from the above corollary. If $k = 2q$ with q prime, the only divisors of k are 1, 2, q, and $2q$, so the only possible primes p are 2, 3, $q+1$, and $2q+1$. But $q+1$ is even (unless $q = 2$, which is excluded since $2q+1$ must not be prime) so is not prime, and $2q+1$ is not prime by assumption, so $D_k = 6$ as claimed. Finally, note that if for instance $4 \mid k$ then $30 \mid D_k$, so that at least half of even k's have $D_k > 6$. $\qquad\square$

This corollary applies for instance when q is a prime such that $q \equiv 1 \pmod 6$ or $q \equiv 7 \pmod{10}$.

It is, however, possible to prove a stronger result as follows (see [Erd-Wag]):

Theorem 9.5.17. *For every given D divisible by 6 the density of even positive integers k such that $D_k = D$ exists and is strictly positive.*

For instance, for $D_k = 6$ the proportion seems experimentally to be around 0.14 of all even numbers.

9.5.3 The Voronoi Congruence

The exposition of the this subsection and the next two is taken with little change from Ireland–Rosen [Ire-Ros]. As already mentioned, almost all of the results will be given in a stronger form in Chapter 11, using the expansion around $s = 1$ of p-adic L-functions (Theorem 11.3.21), although the tools used are not really any deeper.

Recall that we have written canonically $B_k = N_k/D_k$. By abuse of notation, since in the rest of this chapter we will no longer consider χ-power sums or χ-Bernoulli numbers, for $k \geqslant 1$ we will set

$$S_k(n) = \sum_{0 \leqslant r < n} r^k \, .$$

We begin with the following result.

Proposition 9.5.18. *For all even $k \geqslant 2$ and all $n \geqslant 1$ we have*

$$D_k S_k(n) \equiv N_k n \pmod{n^2} \,.$$

Proof. By the Euler–MacLaurin formula (which we used more generally in the proof of Lemma 9.5.11), we can write

$$S_k(n) = \sum_{j=0}^{k} A_{k,j} n^2 \quad \text{with} \quad A_{k,j} = \binom{k}{j} B_{k-j} \frac{n^{j-1}}{j+1} \,.$$

I first claim that for $j \geqslant 1$, if $p \mid n$ and $p \geqslant 5$, then $A_{k,j}$ is p-integral. Indeed, for $j = 1$ the result is trivial for $k > 2$ since $k - 1$ is odd, and for $k = 2$ since $A_{2,1} = -1/2$, and we exclude $p = 2$ for the moment. Similarly for $j = 2$, $n^{j-1}/(j+1) = n/3$ is divisible by p since $p \neq 3$, and we have shown that $p B_{k-2}$ is p-integral. For $j \geqslant 3$, note that since $p \geqslant 5$ it is easy to show that $p^{j-2}/(j+1)$ is p-integral, so that $v_p(n^{j-1}/(j+1)) \geqslant 1$, and since we have shown that $v_p(B_{k-j}) \geqslant -1$ my claim follows.

I now claim that for $j \geqslant 1$ then $v_2(A_{k,j}) \geqslant -1$ and $v_3(A_{k,j}) \geqslant -1$ when $3 \mid n$. Consider first the case $p = 2$. If $j = 1$ we have as usual $B_{k-1} = 0$ if $k > 2$, and $A_{2,1} = -1/2$ as already mentioned. For $j > 1$ we have $B_{k-j} = 0$ except if j is even or $j = k - 1$. But if j is even then $v_2(j+1) = 0$, so that $v_2(A_{k,j}) \geqslant v_2(B_{k-j}) \geqslant -1$, while for $j = k - 1$ we have $A_{k,k-1} = -n^{k-2}/2$, and hence $v_2(A_{k,k-1}) \geqslant -1$.

Consider now the case $p = 3$ and $p \mid n$. Since $A_{k,2} = \binom{k}{2} B_{k-2} n/3$ and $A_{k,3} = \binom{k}{3} B_{k-3} n^2/4$, it is clear that $v_3(A_{k,j}) => \geqslant -1$ for $j = 2$ and $j = 3$. For $j \geqslant 4$ we easily check that $3^{j-2}/(j+1)$ is 3-integral, so that $v_3(A_{k,j}) \geqslant 0$ in that case, proving my claim.

Summarizing my claims, we have proved that for all $j \geqslant 1$ the number $6A_{k,j}$ is p-integral for all $p \mid n$. If we write $6A_{k,j} = a_{k,j}/b_{k,j}$ with $a_{k,j}$ and $b_{k,j}$ coprime, this means that $\gcd(b_{k,j}, n) = 1$. Thus

$$S_k(n) = n B_k + \sum_{j=1}^{k} A_{k,j} n^2 = \sum_{j=1}^{k} \frac{a_{k,j}}{6 b_{k,j}} n^2 \,.$$

Let B be the LCM of the $b_{k,j}$, which is still coprime to n. Multiplying by $B D_k$ we obtain

$$B D_k S_k(n) = B n N_k + (D_k/6) \sum_{j=1}^{k} a_{k,j} (B/b_{k,j}) n^2 \equiv B n N_k \pmod{n^2}$$

since $6 \mid D_k$ by the Clausen–von Staudt congruence. Since B is coprime to n we can divide this congruence by B, proving the proposition. $\qquad \square$

Lemma 9.5.12 tells us that (for k even) $p B_k \equiv S_k(p) \pmod{p}$, and it is immediate to see that the above proposition together with the Clausen–von

Staudt theorem tells us that if $(p-1) \nmid k$ we even have $pB_k \equiv S_k(p) \pmod{p^2}$. In fact, not only is the restriction $(p-1) \nmid k$ unnecessary, but even more is true.

Corollary 9.5.19. *If $p \geqslant 5$ and k is even then*

$$pB_k \equiv S_k(p) \pmod{p^3},$$

except if $(p-1) \mid (k-2)$ and $p \nmid k(k-1)$, in which case the congruence is only modulo p^2.

Proof. Using again the expression for $S_k(p)$ used at the beginning of the proof of Proposition 9.5.18, the Clausen–von Staudt theorem, and $B_{k-1} = 0$ when $k \geqslant 4$ is even, we deduce that

$$S_k(p) = pB_k + \frac{k(k-1)p^3}{6} B_{k-2},$$

and the conclusion again follows from the Clausen–von Staudt theorem for $p \geqslant 5$. It is immediate to see that the congruence is true modulo p^2 for $k = 2$. $\square$

We can now state and prove the Voronoi congruences.

Proposition 9.5.20 (Voronoi). *For any even $k \geqslant 2$ and for all coprime integers a and n in $\mathbb{Z}_{>0}$ we have*

$$(a^k - 1)N_k \equiv ka^{k-1}D_k \sum_{m=1}^{n-1} m^{k-1} \left\lfloor \frac{ma}{n} \right\rfloor \pmod{n}.$$

Proof. For $1 \leqslant m \leqslant n-1$, write $ma = q_m n + r_m$ with $0 \leqslant r_m < n$, so $q_m = \lfloor ma/n \rfloor$. By the binomial theorem we have

$$(ma)^k \equiv r_m^k + knq_m r_m^{k-1} \equiv r_m^k + kn(ma)^{k-1} \left\lfloor \frac{ma}{n} \right\rfloor \pmod{n^2}.$$

However, since a and n are coprime, $r_1, \ldots, r_{n-1}$ is a permutation of $1, \ldots, n-1$. Thus, summing the above congruence for $1 \leqslant m \leqslant n-1$ gives

$$a^k S_k(n) \equiv S_k(n) + kna^{k-1} \sum_{m=1}^{n-1} m^{k-1} \left\lfloor \frac{ma}{n} \right\rfloor \pmod{n^2}.$$

Multiplying by D_k, using Proposition 9.5.18, and dividing by n gives the desired result. $\square$

Corollary 9.5.21. *Let p be a prime such that $p \equiv 3 \pmod 4$ and $p > 3$. Then*

$$2 \left(2 - \left(\frac{2}{p} \right) \right) B_{(p+1)/2} \equiv - \sum_{m=1}^{(p-1)/2} \left(\frac{m}{p} \right) \pmod{p}.$$

Proof. We set $k = (p+1)/2$, $a = 2$, and $n = p$ in the above proposition. Since for any a we have $a^{(p-1)/2} \equiv \left(\frac{a}{p}\right)$ (mod p), we obtain

$$\left(2\left(\frac{2}{p}\right) - 1\right) N_{(p+1)/2} \equiv \frac{p+1}{2}\left(\frac{2}{p}\right) D_{(p+1)/2} \sum_{m=1}^{p-1} \left(\frac{m}{p}\right) \left\lfloor \frac{2m}{p} \right\rfloor \pmod{p} .$$

Now $\lfloor 2m/p \rfloor$ is equal to 0 for $m < (p-1)/2$, and to 1 for $(p+1)/2 \leqslant m \leqslant p-1$. Since $v_p(D_{(p+1)/2}) = 0$ by Corollary 9.5.15 (since otherwise $(p-1) \mid (p+1)/2$, hence $p - 1 \leqslant (p + 1)/2$, which is possible only for $p = 3$, which we have excluded), it follows that

$$2\left(2\left(\frac{2}{p}\right) - 1\right) B_{(p+1)/2} \equiv \left(\frac{2}{p}\right) \sum_{m=(p+1)/2}^{p-1} \left(\frac{m}{p}\right) \pmod{p} .$$

The result follows from the fact that $\sum_{1 \leqslant m \leqslant p-1} \left(\frac{m}{p}\right) = 0$.   □

Corollary 9.5.22. *Let p be a prime such that $p \equiv 3$ (mod 4) and $p > 3$. If we denote by $h(-p)$ the class number of the imaginary quadratic field $\mathbb{Q}(\sqrt{-p})$, then*

$$h(-p) \equiv -2B_{(p+1)/2} \pmod{p} .$$

Proof. The classical Dirichlet class number formula gives for any fundamental discriminant $D < -4$ the identity

$$\left(2 - \left(\frac{D}{2}\right)\right) h(D) = \sum_{m=1}^{\lfloor |D|/2 \rfloor} \left(\frac{D}{m}\right) .$$

When $D = -p$ with $p \equiv 3$ (mod 4), thanks to the quadratic reciprocity law this can be rewritten

$$\left(2 - \left(\frac{2}{p}\right)\right) h(-p) = \sum_{m=1}^{(p-1)/2} \left(\frac{m}{p}\right) .$$

The corollary thus immediately follows from the preceding one.   □

Note that even though this is not a very practical method of computation of $h(-p)$, it does determine the value of $h(-p)$ exactly from that of $B_{(p+1)/2}$ modulo p since it is well known and easy to show that $h(-p) < p^{1/2} \log p/\pi < p$ for all $p > 3$.

9.5.4 The Kummer Congruences

We begin with the following result.

Proposition 9.5.23 (J. Adams). *If $(p-1) \nmid k$ then B_k/k is p-integral.*

Proof. By Theorem 9.5.14, we already know that B_k is p-integral. Write $k = p^e k_0$ with $p \nmid k_0$. Choosing $n = p^e$ in Proposition 9.5.20, we see that $(a^k - 1)N_k \equiv 0 \pmod{p^e}$. Take for a a primitive root modulo p. Since $(p-1) \nmid k$, we have $p \nmid a^k - 1$. Thus $N_k \equiv 0 \pmod{p^e}$, so $B_k/k = (N_k/k)/D_k$ is p-integral. $\square$

The main result of this section is the following theorem, essentially due to Kummer.

Theorem 9.5.24 (Kummer). *For any $k \geqslant 2$ even, set*

$$z(k) = (p^{k-1} - 1)B_k/k = (1 - p^{k-1})\zeta(1 - k).$$

Then if $(p-1) \nmid k$ and $k' \equiv k \pmod{\phi(p^e)}$ we have $z(k') \equiv z(k) \pmod{p^e}$.

Proof. If we set $s = v_p(k)$, the above proposition shows that $p^s \mid N_k$. In Proposition 9.5.20 we choose $n = p^{e+s}$. Since p^s divides both k and N_k, we can divide the congruence by p^s, and since k/p^s and D_k are coprime to p we can divide by both and we obtain the congruence

$$\frac{(a^k - 1)B_k}{k} \equiv a^{k-1} \sum_{m=1}^{p^{e+s}-1} m^{k-1} \left\lfloor \frac{ma}{p^{e+s}} \right\rfloor \pmod{p^e}.$$

We have

$$\sum_{1 \leqslant m \leqslant p^{e+s}-1} m^{k-1} \left\lfloor \frac{ma}{p^{e+s}} \right\rfloor = \sum_{\substack{1 \leqslant m \leqslant p^{e+s}-1 \\ p \nmid m}} m^{k-1} \left\lfloor \frac{ma}{p^{e+s}} \right\rfloor$$

$$+ p^{k-1} \sum_{1 \leqslant m \leqslant p^{e+s-1}-1} m^{k-1} \left\lfloor \frac{ma}{p^{e+s-1}} \right\rfloor.$$

Using the congruence that we have obtained above with e replaced by $e - 1$, since $k \geqslant 2$ we deduce that

$$\frac{p^{k-1}(a^k - 1)B_k}{k} \equiv p^{k-1}a^{k-1} \sum_{1 \leqslant m \leqslant p^{e+s}-1} m^{k-1} \left\lfloor \frac{ma}{p^{e+s-1}} \right\rfloor \pmod{p^e}.$$

Putting the two congruences and the above identity together, we obtain

$$\frac{(1 - p^{k-1})(a^k - 1)B_k}{k} \equiv a^{k-1} \sum_{\substack{1 \leqslant m \leqslant p^{e+s}-1 \\ p \nmid m}} m^{k-1} \left\lfloor \frac{ma}{p^{e+s}} \right\rfloor \pmod{p^e}.$$

Now note that when $p \nmid m$ as in the last sum, $k' \equiv k \pmod{\phi(p^e)}$ implies that $m^{k'-k} \equiv 1 \pmod{p^e}$ (Euler's theorem for the group $(\mathbb{Z}/p^e\mathbb{Z})^*$ of cardinality

$\phi(p^e)$), so that $m^{k'-1} \equiv m^{k-1} \pmod{p^e}$. It follows that the right-hand side of the above congruence is unchanged modulo p^e if we replace k by $k' \equiv k$ $\pmod{\phi(p^e)}$ (recall that a is coprime to $n = p^{e+s}$, hence not divisible by p). We deduce that

$$\frac{(1 - p^{k'-1})(a^{k'} - 1)B_{k'}}{k'} \equiv \frac{(1 - p^{k-1})(a^k - 1)B_k}{k} \pmod{p^e}$$

when $k' \equiv k \pmod{\phi(p^e)}$. We now choose for a a primitive root modulo p. Since $(p - 1) \nmid k$ (hence $(p - 1) \nmid k'$), it follows that $a^k - 1$ and $a^{k'} - 1$ are coprime to p, and as above, $a^{k'} - 1 \equiv a^k - 1 \pmod{p^e}$. We can thus divide the above congruence by $a^k - 1$, thus giving the congruence of the theorem. □

Corollary 9.5.25. *If k and k' are even with $\min(k, k') \geqslant e + 1$, and p is a prime such that $(p - 1) \nmid k$ and $k' \equiv k \pmod{\phi(p^e)}$, then $B_{k'}/k' \equiv B_k/k$ $\pmod{p^e}$.*

Proof. Clear from the above theorem since B_k/k is p-integral by Proposition 9.5.23. □

In Section 11.4.2 we will see that the Kummer congruences are closely related to the p-adic zeta function and L-functions. In fact, we will give a statement (Proposition 11.4.4) that includes the case $(p - 1) \mid k$. These congruences can also be used in connection with Fermat's last theorem, because of the following result, which we will prove in Chapter 11 (Theorem 11.4.10; see also [Was]):

Theorem 9.5.26. *An odd prime p is irregular if and only if it divides the numerator of some B_k for $k \leqslant p - 3$.*

Proposition 9.5.27. *The set of irregular primes is infinite.*

Proof. Let $\{p_1, \ldots, p_s\}$ be a nonempty set of irregular primes (this is possible since 37 is irregular). In the way of Euclid's proof of the infinitude of primes, we will construct an irregular prime that is not in this set, proving the proposition. Set $n = \prod_{1 \leqslant i \leqslant s}(p_i - 1)$. Since $p_i \geqslant 37$ for all i, we have $n \geqslant 36$, so by trivial estimates we have $|B_n/n| > 1$. It follows that there exists a prime p such that $v_p(B_n/n) > 0$. I claim that p is irregular and distinct from the p_i. Indeed, by the Clausen–von Staudt congruence we know that $(p - 1) \nmid n$, so that $p \neq p_i$ and $p \neq 2$. Furthermore, if r is the remainder of the Euclidean division of n by $p - 1$ we have $2 \leqslant r \leqslant p - 3$ and r even, and by Corollary 9.5.25 we have $B_n/n \equiv B_r/r \pmod{p}$. Since $v_p(B_n/n) > 0$ it follows that $v_p(B_r) = v_p(B_r/r) > 0$, so that p is irregular. □

As already mentioned in the section on FLT, a famous conjecture is that the set of *regular* primes is also infinite, with density among primes equal to $\exp(-1/2)$.

9.5.5 The Almkvist–Meurman Theorem

We will give two proofs of this theorem, the first one in this section, the second as Exercise 63. Furthermore, in Section 11.4.3, we will prove a stronger result. We begin with a lemma of independent interest.

Lemma 9.5.28 (Hermite). *Let p be a prime number and let $n \geqslant 1$ be an integer. We have the congruence*

$$\sum_{1 \leqslant j \leqslant (n-1)/(p-1)} \binom{n}{(p-1)j} \equiv 0 \pmod{p} .$$

Proof. By Lemma 2.5.1 we know that for $m \geqslant 1$ we have $\sum_{a \in \mathbb{F}_p} a^m = -\delta_{m,p-1}$, where $\delta_{m,p-1} = 1$ if $(p-1) \mid m$ and $\delta_{m,p-1} = 0$ otherwise, while $\sum_{a \in \mathbb{F}_p} a^0 = 0$, so in $\mathbb{F}_p$ we have

$$-\delta_{n,p-1} = \sum_{a \in \mathbb{F}_p} (a+1)^n = \sum_{a \in \mathbb{F}_p} \sum_{0 \leqslant m \leqslant n} \binom{n}{m} a^m = -\sum_{1 \leqslant m \leqslant n} \delta_{m,p-1} \binom{n}{m} .$$

It follows that $\sum_{1 \leqslant j \leqslant (n-1)/(p-1)} \binom{n}{(p-1)j} \equiv 0 \pmod{p}$, proving the lemma.

$\square$

See Exercise 62 and Proposition 11.4.11 for generalizations.

Theorem 9.5.29 (Almkvist–Meurman). *For any $n \geqslant 0$, $k \in \mathbb{Z}_{\geqslant 1}$, and $h \in \mathbb{Z}$ we have $k^n(B_n(h/k) - B_n) \in \mathbb{Z}$.*

Proof. If we set $\widetilde{B}_n(x) = B_n(x) - B_n$ then by Proposition 9.1.3 we have

$$\widetilde{B}_n(x+y) = \sum_{m=0}^{n} \binom{n}{m} \widetilde{B}_m(x) y^{n-m} + \widetilde{B}_n(y) ,$$

so by induction on h it is enough to prove the theorem for $h = 1$. In addition, it is trivially true by inspection for $n \leqslant 1$, so we may assume that $n \geqslant 2$. Set $b_n(k) = k^n \widetilde{B}_n(1/k)$. By the basic formula for Bernoulli polynomials we have

$$b_n(k) = 1 - \frac{nk}{2} + \sum_{2 \leqslant m \leqslant n-1,\ m \text{ even}} \binom{n}{m} B_m k^m .$$

Thus, by the Clausen–von Staudt Theorem 9.5.14, with the understanding that p is prime, we have

$$b_n(k) \equiv -\frac{nk}{2} - \sum_{2 \leqslant m \leqslant n-1,\ m \text{ even}} \binom{n}{m} k^m \sum_{(p-1)\mid m} \frac{1}{p}$$

$$\equiv \frac{nk}{2} - \sum_{p \leqslant n} \frac{1}{p} \sum_{\substack{2 \leqslant m \leqslant n-1 \\ \mathrm{lcm}(2,p-1)\mid m}} \binom{n}{m} k^m \pmod{1} .$$

If $p \mid k$ then $k^m \equiv 0 \pmod{p}$, and if $p \nmid k$ and $(p-1) \mid m$ we have $k^m \equiv 1 \pmod{p}$ by Fermat's little theorem. Thus

$$b_n(k) \equiv \frac{nk}{2} - \sum_{p \leqslant n,\ p \nmid k} \frac{1}{p} \sum_{\substack{2 \leqslant m \leqslant (n-1) \\ \mathrm{lcm}(2,p-1) \mid m}} \binom{n}{m} \pmod{1} .$$

Denote by $S(n,p)$ the inner sum. Assume first that $p \neq 2$, so that $p-1$ is even. Setting $m = (p-1)j$ we thus have

$$S(n,p) = \sum_{1 \leqslant j \leqslant (n-1)/(p-1)} \binom{n}{(p-1)j} \equiv 0 \pmod{p}$$

by Lemma 9.5.28. It follows that $b_n(k) \equiv 0 \pmod{1}$ (in other words, $b_n(k) \in \mathbb{Z}$) when k is even, and otherwise

$$b_n(k) \equiv \frac{n}{2} - \frac{1}{2} S(n,2) \pmod{1} .$$

But

$$S(n,2) = \sum_{1 \leqslant j \leqslant (n-1)/2} \binom{n}{2j} = \begin{cases} 2^{n-1} - 2 & \text{if } n \text{ is even,} \\ 2^{n-1} - 1 & \text{if } n \text{ is odd;} \end{cases}$$

hence $S(n,2) \equiv n \pmod{2}$ since $n \geqslant 2$, proving that $b_n(k) \equiv 0 \pmod{1}$ in all cases. $\qquad\square$

9.6 The Real and Complex Gamma Functions

Although the complex gamma function is quite classical, and we have already mentioned it in Chapter 8, since it occurs in all the functional equations of functions linked to number theory (and in particular in so-called "motivic" L-functions), it is essential to have a thorough understanding of this function. We give here a slightly nonstandard approach that emphasizes the formulas that we need and is well suited to generalizations. The impatient reader can jump directly to Section 9.6.2. The main idea of our approach is that most basic formulas involving the gamma function are in fact specializations of formulas for the Hurwitz zeta function, which we now study.

9.6.1 The Hurwitz Zeta Function

The Hurwitz zeta function is an important tool, not only for the definition of the gamma function, but also in the study of L-functions.

Definition 9.6.1. *We define the Hurwitz zeta function for $x \in \mathbb{R}_{>0}$ and $s \in \mathbb{C}$ with $\Re(s) > 1$ by*

$$\zeta(s,x) = \sum_{n \geqslant 0} \frac{1}{(n+x)^s} \; .$$

We could extend this definition to $x \in \mathbb{C} \setminus \mathbb{Z}_{\leqslant 0}$ by deciding that $1/(n+x)^s = \exp(-s\log(n+x))$, where we choose the principal determination of the logarithm. However, this would create too many determination problems, so we restrict to $x \in \mathbb{R}_{>0}$. The only exception to this is when $s \in \mathbb{Z}_{>1}$, in which case we can define $\zeta(s,x)$ unambiguously (see Exercise 66 for an application of this).

We will see below that it is very easy to give the analytic continuation of $\zeta(s,x)$ to the whole s-plane. Before that, we prove a number of simple formulas for which we assume implicitly that $\Re(s) > 1$, but which will automatically be valid for all s by analytic continuation.

Proposition 9.6.2. *We have the functional equation*

$$\zeta(s, x+1) = \zeta(s,x) - x^{-s} \; ,$$

the (partial) differential equation

$$\frac{\partial \zeta(s,x)}{\partial x} = -s\zeta(s+1, x) \; ,$$

the asymptotic formula $\zeta(s,x) \sim x^{-s}$ as $\Re(s) \to \infty$, and the special cases

$$\zeta(s, 1) = \zeta(s) \quad and \quad \zeta(s, 1/2) = (2^s - 1)\zeta(s) \; ,$$

where $\zeta(s)$ is the Riemann zeta function.

Proof. The first formula corresponds to changing n into $n+1$, and the next two formulas are immediate by normal convergence of the series and its derivative. By definition we have $\zeta(s, 1) = \zeta(s)$. Finally,

$$\zeta(s, 1/2) = 2^s \sum_{m \geqslant 0} \frac{1}{(2m+1)^s} = 2^s \left(\zeta(s) - \sum_{m \geqslant 1} \frac{1}{(2m)^s} \right) = 2^s \zeta(s) \left(1 - \frac{1}{2^s} \right) \; ,$$

proving the last formula.   □

Corollary 9.6.3. *We have the following series expansion valid for $|y| < x$:*

$$\zeta(s, x+y) = \sum_{k \geqslant 0} (-1)^k \binom{s+k-1}{k} y^k \zeta(s+k, x) \; .$$

In particular, for $0 < x < 1$ we have

$$\zeta(s,x) = x^{-s} + \sum_{k \geqslant 0}(-1)^k \binom{s+k-1}{k}x^k\zeta(s+k) \ \ and$$

$$\zeta(s,x) = \sum_{k \geqslant 0}(-1)^k \binom{s+k-1}{k}(x-1/2)^k(2^{s+k}-2)\zeta(s+k) \ .$$

Proof. Denote by RHS the right-hand side of the first formula, so that by definition of $\zeta(s,x)$ and of $\binom{a}{k}$ we have

$$\text{RHS} = \sum_{k \geqslant 0}\binom{-s}{k}y^k \sum_{n \geqslant 0}\frac{1}{(n+x)^{s+k}} \ .$$

Since $|y| < x$ the double sum is absolutely convergent, so we may interchange the order of summation and obtain

$$\text{RHS} = \sum_{n \geqslant 0}(n+x)^{-s}\sum_{k \geqslant 0}\binom{-s}{k}(y/(n+x))^k$$

$$= \sum_{n \geqslant 0}(n+x)^{-s}(1+y/(n+x))^{-s} = \sum_{n \geqslant 0}(n+x+y)^{-s} = \zeta(s,x+y) \ ,$$

proving the first formula. The second follows by choosing $x = 1$ and exchanging x and y, and the last by choosing $x = 1/2$, exchanging x and y, and using the formula for $\zeta(s,1/2)$. $\qquad\square$

Lemma 9.6.4. *For $|y| < x$ we have the formulas*

$$\zeta(s,x) = \frac{\zeta(s-1,x)-\zeta(s-1,x+y)}{y(s-1)} + \sum_{k \geqslant 1}(-1)^{k-1}\binom{s+k-1}{k}y^k\frac{\zeta(s+k,x)}{k+1} \ ,$$

$$\zeta(s,x) = \frac{\zeta(s-1,x-y)-\zeta(s-1,x+y)}{2y(s-1)} - \sum_{k \geqslant 1}\binom{s+2k-1}{2k}y^{2k}\frac{\zeta(s+2k,x)}{2k+1} \ ,$$

$$\zeta(s,x) = \frac{\zeta(s-2,x-y)-2\zeta(s-2,x)+\zeta(s-2,x+y)}{y^2(s-2)(s-1)}$$

$$-2\sum_{k \geqslant 1}\binom{s+2k-1}{2k}y^{2k}\frac{\zeta(s+2k,x)}{(2k+1)(2k+2)} \ .$$

Proof. The first formula is a simple rearrangement of terms of the first formula of the corollary. The second and third formulas follow by changing y into $-y$ in the first formula and computing the sum and the difference. $\qquad\square$

Corollary 9.6.5. *We have the following formulas, valid for $x > 1$, except for the fourth, which is also valid for $x > 1/2$:*

$$\zeta(s,x) = \frac{x^{1-s}}{s-1} + \sum_{k \geqslant 1}(-1)^{k-1}\binom{s+k-1}{k}\frac{\zeta(s+k,x)}{k+1},$$

$$\zeta(s,x) = \frac{(x-1)^{1-s}}{s-1} - \sum_{k \geqslant 1}\binom{s+k-1}{k}\frac{\zeta(s+k,x)}{k+1},$$

$$\zeta(s,x) = \frac{(x-1)^{1-s}+x^{1-s}}{2(s-1)} - \sum_{k \geqslant 1}\binom{s+2k-1}{2k}\frac{\zeta(s+2k,x)}{2k+1},$$

$$\zeta(s,x) = \frac{(x-1/2)^{1-s}}{s-1} - \sum_{k \geqslant 1}\frac{1}{2^{2k}}\binom{s+2k-1}{2k}\frac{\zeta(s+2k,x)}{2k+1},$$

$$\zeta(s,x) = \frac{(x-1)^{2-s}-x^{2-s}}{(s-1)(s-2)} - 2\sum_{k \geqslant 1}\binom{s+2k-1}{2k}\frac{\zeta(s+2k,x)}{(2k+1)(2k+2)}.$$

Proof. This follows by taking $y = 1$ and $y = -1$ in the first formula of the lemma, $y = 1$ and $y = 1/2$ in the second formula, and $y = 1$ in the third formula. □

Remarks. (1) If we want formulas valid for $x \in \mathbb{R}_{>0}$, we can apply the above formulas to $x + 1$ and use $\zeta(s,x) = \zeta(s,x+1) + x^{-s}$.

(2) We can of course obtain similar formulas for the Riemann zeta function by choosing for instance $x = 2$ (not $x = 1$) and using the formula $\zeta(s,2) = \zeta(s) - 1$. Only in the fourth formula can we directly set $x = 1$.

Proposition 9.6.6. *The parameter $x \in \mathbb{R}_{>0}$ being fixed, the function $\zeta(s,x)$ (hence in particular the function $\zeta(s)$) can be analytically continued to the whole complex plane to a meromorphic function with a single pole, at $s = 1$, which is simple with residue 1.*

Proof. To prove this proposition, we can use any of the formulas of the above corollary. First note that since $\zeta(s, x+1) = \zeta(s,x) - x^{-s}$, it is enough to prove analytic continuation when $x > 1$. In that case, since by Proposition 9.6.2 we know that $\zeta(s+k,x) \sim x^{-s}x^{-k}$ as $k \to \infty$, it follows that the first formula above (for instance) expresses $\zeta(s,x)$ as a geometrically convergent series involving only $\zeta(s+k,x)$ for $k \geqslant 1$. We can thus extend analytically $\zeta(s,x)$ by strips of width 1: first to $\Re(s) > 0$, then to $\Re(s) > -1$, and so on. The only polar part is obtained with the term $x^{1-s}/(s-1)$, hence at $s = 1$, which gives a pole, which is simple, with residue equal to 1. Note that the pole at $s = 1-k$ of $\zeta(s+k,x+1)$ is canceled by $\binom{s+k-1}{k} = s(s-1)\cdots(s-k+1)/k!$, which vanishes for $s = 1 - k$ when $k \geqslant 1$. □

Note that Proposition 10.2.2, which we will prove in the next chapter, also gives an easy proof of analytic continuation of $\zeta(s,x)$ and its values at negative integers, which we shall give in Corollary 9.6.10 below; see Exercise 17 of Chapter 10.

In Proposition 9.2.13 we have seen how to express $\zeta(-\alpha)$ for *any* $\alpha \in \mathbb{C} \setminus \{-1\}$ using the Euler–MacLaurin formula. Now that we know the analytic continuation of $\zeta(s, x)$ to the whole plane, exactly the same reasoning gives the following more general statements.

Proposition 9.6.7. *Let $\alpha \in \mathbb{C}$ be different from -1, and let $x \in \mathbb{R}_{>0}$.*

(1) *For every $k > \Re(\alpha) + 1$ such that $k \geqslant 1$ we have*

$$\sum_{m=0}^{N} (m + x)^{\alpha} = \zeta(-\alpha, x) + \frac{(N + x)^{\alpha+1}}{\alpha + 1} + \frac{(N + x)^{\alpha}}{2}$$

$$+ \sum_{j=2}^{k} \binom{\alpha}{j-1} \frac{B_j}{j} (N + x)^{\alpha-j+1} + R_k(\alpha, x, N) ,$$

where

$$R_k(\alpha, x, N) = (-1)^k \binom{\alpha}{k} \int_N^{\infty} (t + x)^{\alpha-k} B_k(\{t\}) \, dt .$$

When k is even we have

$$|R_k(\alpha, x, N)| \leqslant \left| \binom{\alpha}{k+1} \frac{B_{k+2}}{k+2} (N + x)^{\alpha-k-1} \right| ;$$

in other words, $|R_k(\alpha, x, N)|$ is smaller than the modulus of the first omitted term, and in particular $R_k(\alpha, x, N)$ tends to 0 as $N \to \infty$.

(2) *With the same assumptions, we have the formula*

$$\zeta(-\alpha, x) = -\frac{x^{\alpha+1}}{\alpha + 1} - \sum_{j=1}^{k} \binom{\alpha}{j-1} \frac{B_j}{j} x^{\alpha+1-j}$$

$$+ (-1)^{k-1} \binom{\alpha}{k} \int_0^{\infty} (t + x)^{\alpha-k} B_k(\{t\}) \, dt .$$

Similarly, for $\alpha = -1$ we have the following:

Proposition 9.6.8. *For $x \notin \mathbb{Z}_{\leqslant 0}$ define*

$$\psi(x) = -\lim_{N \to \infty} \left(\sum_{m=0}^{N} \frac{1}{m + x} - \log(N + x) \right) .$$

(1) *For $k \geqslant 1$ we have*

$$\sum_{m=0}^{N} \frac{1}{m + x} = -\psi(x) + \log(N + x) - \sum_{j=1}^{k} \frac{B_j}{j(N + x)^j} + R_k(-1, x, N) ,$$

where

$$R_k(-1, x, N) = \int_N^{\infty} \frac{B_k(\{t\})}{(t + x)^{k+1}} \, dt$$

and $|R_k(-1, x, N)| \leqslant |B_{k+2}/((k + 2)(N + x)^{k+2})|$ when k is even.

(2) *For $k \geqslant 1$ we have*

$$\psi(x) = \log(x) - \frac{1}{2x} - \sum_{j=2}^{k} \frac{B_j}{jx^j} + \int_0^\infty \frac{B_k(\{t\})}{(t+x)^{k+1}} \, dt .$$

(3) *We have $\lim_{s \to 1}(\zeta(s, x) - 1/(s-1)) = -\psi(x)$, in other words*

$$\zeta(s, x) = \frac{1}{s-1} - \psi(x) + O(s-1) .$$

We will study below the properties of the function $\psi(x)$, and in particular we will see that $\psi(x) = \Gamma'(x)/\Gamma(x)$ is the logarithmic derivative of the gamma function; see Definition 9.6.13. Indeed, since we will *define* the gamma function by $\log(\Gamma(x)) = \frac{\partial \zeta}{\partial s}(0, x) - \frac{\partial \zeta}{\partial s}(0, 1)$ and since $\zeta(s, x)$ is meromorphic in s, around $s = 1$ we have

$$\zeta(s-1, x) = \zeta(0, x) + (s-1)\log(\Gamma(x)) + \cdots$$
$$= 1/2 - x + (s-1)\log(\Gamma(x)) + \cdots$$

(using $\zeta(0, x) = 1/2 - x$, which is immediate from Proposition 9.6.7), so that by Proposition 9.6.2,

$$-(s-1)\zeta(s, x) = \frac{\partial \zeta}{\partial x}(s-1, x) = -1 + (s-1)\psi(x) + \cdots ,$$

as claimed in the proposition.

Corollary 9.6.9. *As $x \to \infty$ we have:*

(1) *For $\Re(s) \geqslant 1$ and $s \neq 1$,*

$$\zeta(s, x) = \frac{x^{1-s}}{s-1} + O(x^{-s}) .$$

(2) *For $\Re(s) < 1$,*

$$\zeta(s, x) = -\frac{x^{1-s}}{1-s} + \frac{x^{-s}}{2} - \sum_{j=1}^{p} \binom{-s}{2j} \frac{B_{2j}}{2j} x^{-s+1-2j} + O(x^{-1}) ,$$

where $p = \lfloor (3 - \Re(s))/2 \rfloor$.

Proof. Clear by Proposition 9.6.7. □

Corollary 9.6.10. *If $k \in \mathbb{Z}_{\geqslant 1}$ we have*

$$\zeta(1-k, x) = -\frac{B_k(x)}{k} ,$$

and in particular $\zeta(1-k) = -B_k/k - \delta_{k,1}$.

Proof. Setting $\alpha = k - 1$ in Proposition 9.6.7, we find that for $n \geqslant k$,

$$-k\zeta(1 - k, x) = x^k + \sum_{j=1}^{n} \binom{k}{j} B_j x^{k-j} = B_k(x) .$$

$\square$

The statement for $\zeta(1 - k)$ will be proved again in the next chapter using the functional equation of the zeta function. Historically it was the first indication of the existence of this functional equation, discovered by L. Euler.

Proposition 9.6.11. *As $x \to 0$ we have*

$$\zeta(s, x) = \begin{cases} x^{-s} + \zeta(s) + o(1) & \text{if } \Re(s) \geqslant 0 , \\ 1/2 + o(1) & \text{if } s = 0 , \\ \zeta(s) + o(1) & \text{if } \Re(s) < 0, \ s \neq -2k \text{ with } k \in \mathbb{Z}_{\geqslant 1} , \\ -B_{2k}x + O(x^3) & \text{if } s = -2k \text{ with } k \in \mathbb{Z}_{\geqslant 2} , \\ -B_2 x + x^2/3 + O(x^3) & \text{if } s = -2 . \end{cases}$$

Proof. For $s \neq -2k$ with $k \in \mathbb{Z}_{\geqslant 1}$ this immediately follows from

$$\zeta(s, x) = x^{-s} + \zeta(s, x + 1) = x^{-s} + \zeta(s) + o(1) .$$

For $s = -2k$, by the above corollary we have $\zeta(-2k, x) = -B_{2k+1}(x)/(2k+1)$, so the result follows from the explicit formula for $B_n(x)$. $\square$

Proposition 9.6.12. *We have the duplication formula*

$$\zeta(s, x) + \zeta\left(s, x + \frac{1}{2}\right) = 2^s \zeta(s, 2x)$$

and more generally for $N \in \mathbb{Z}_{\geqslant 1}$ the distribution formula

$$\sum_{0 \leqslant j < N} \zeta\left(s, x + \frac{j}{N}\right) = N^s \zeta(s, Nx) .$$

Proof. Follows from an easy rearrangement of terms and left to the reader (Exercise 64). $\square$

9.6.2 Definition of the Gamma Function

Since we have seen above that $\zeta(s, x)$ can be extended to the whole complex plane with a simple pole at $s = 1$, the following definition makes sense.

Definition 9.6.13. (1) *We define the* real *gamma function for $x \in \mathbb{R}_{>0}$ by the formula*

$$\log(\Gamma(x)) = \zeta'(0, x) - \zeta'(0, 1) = \zeta'(0, x) - \zeta'(0) \,,$$

where here and elsewhere the derivative is taken with respect to the first variable.
(2) *We define the real ψ function for $x \in \mathbb{R}_{>0}$ as the logarithmic derivative of $\Gamma(x)$; in other words, $\psi(x) = \Gamma'(x)/\Gamma(x)$.*

We will see later that in fact $\zeta'(0) = -\log(2\pi)/2$, but for the moment we do not need this result. We will also see how to generalize this definition to all $x \in \mathbb{C} \setminus \mathbb{Z}_{\leqslant 0}$.

As already mentioned, since the gamma function is very often used in conjunction with L-series, it is customary to use the variable s and not the variable x, hence to write $\Gamma(s)$. The reader should be aware that although this *will* be the variable used in zeta and L-functions, it is *not* the variable s of the Hurwitz zeta function used to define the gamma function. For the moment, since we handle simultaneously $\zeta(s, x)$ and the gamma function, we keep the variable x, but we will switch to the variable s later, after the introduction of the complex gamma function.

We will study later in great detail the properties of the function $\Gamma(x)$. For the moment we note the following basic results.

Proposition 9.6.14. *For all $x \in \mathbb{R}_{>0}$ we have $\Gamma(x+1) = x\Gamma(x)$ and when $n \in \mathbb{Z}_{\geqslant 1}$ we have $\Gamma(n) = (n-1)!$.*

Proof. Since $\zeta(s, x+1) = \zeta(s, x) - x^{-s}$ we obtain the first formula by derivation with respect to x. The second follows by induction since $\log(\Gamma(1)) = \zeta'(0, 1) - \zeta'(0, 1) = 0$. $\qquad\square$

Proposition 9.6.15. (1) *Let $u \in \mathbb{R}_{>0}$. For $|x| < u$ we have*

$$\log(\Gamma(x+u)) = \log(\Gamma(u)) + \psi(u)x + \sum_{k \geqslant 2}(-1)^k \frac{\zeta(k, u)}{k} x^k \,.$$

(2) *In particular, for $|x| < 1$ we have*

$$\log(\Gamma(x+1)) = \sum_{k \geqslant 1}(-1)^k \frac{\zeta(k)}{k} x^k \,,$$

where by convention we set $\zeta(1) = \gamma$, Euler's constant.

Proof. This follows by differentiating with respect to s the first and second formulas of Corollary 9.6.3, and using the fact that around $s = 1$ we have $\zeta(s, u) = 1/(s-1) - \psi(u) + O(s-1)$, and in particular $\zeta(s) = 1/(s-1) + \gamma + O(s-1)$. $\qquad\square$

Proposition 9.6.16. *For $x > 0$ we have for any $k \geqslant 1$,*

$$\zeta'(0,x) = \left(x - \frac{1}{2}\right)\log(x) - x + \sum_{j=1}^{k-1} \frac{B_{j+1}}{j(j+1)x^j} - \frac{1}{k}\int_0^\infty \frac{B_k(\{t\})}{(t+x)^k}\,dt\,,$$

and in particular

$$\zeta'(0,x) = \left(x - \frac{1}{2}\right)\log(x) - x - \int_0^\infty \frac{\{t\}-1/2}{t+x}\,dt$$

and

$$\log(\Gamma(x)) = \left(x - \frac{1}{2}\right)\log(x) - x + 1 + (x-1)\int_1^\infty \frac{\{t\}-1/2}{t(t+x-1)}\,dt\,.$$

Proof. This follows by derivation after a short computation from the formula for $\zeta(-\alpha, x)$ given in Proposition 9.6.7. □

Remark. As already noted in Section 9.2.5, the integral $\int_0^\infty B_1(\{t\})/(t+x)\,dt$ is convergent, albeit only conditionally. If you are uncomfortable with this, simply choose $k = 2$ instead of $k = 1$.

Proposition 9.6.17. *For all $x \in \mathbb{C} \setminus \mathbb{Z}_{\leqslant 0}$ set*

$$u_N(x) = \frac{N^{x-1}N!}{x(x+1)\cdots(x+N-1)}\,.$$

Then for all $x \in \mathbb{R}_{>0}$ we have

$$\Gamma(x) = \lim_{N\to\infty} u_N(x)\,.$$

Proof. By differentiating the first formula of Proposition 9.6.7 a short computation gives

$$\sum_{m=0}^{N-1} \log(m+x) = -\zeta'(0,x) + \left(N + x - \frac{1}{2}\right)\log(N+x)$$

$$- (N+x) - \int_N^\infty \frac{\{t\}-1/2}{t+x}\,dt\,,$$

hence in particular

$$\log(N!) = \sum_{m=0}^{N-1} \log(m+1)$$

$$= -\zeta'(0) + \left(N + \frac{1}{2}\right)\log(N+1) - (N+1) - \int_N^\infty \frac{\{t\}-1/2}{t+1}\,dt\,.$$

Since the integrals converge, expanding $\log(N+x)$ and $\log(N+1)$ we obtain

$$\log(\Gamma(x)) = \log(N!) - \sum_{m=0}^{N-1} \log(m+x) + (x-1)\log(N) + o(1) \,,$$

proving the proposition.   □

Recall from Definition 9.6.13 that we have denoted by $\psi(x)$ the logarithmic derivative of $\Gamma(x)$. By differentiating the formulas of the above propositions, it is easy to see that this definition of $\psi(x)$ coincides with the notation used in Proposition 9.6.8, and also proves (2) and (3) of that proposition. It is also immediate to give analogous formulas for the derivatives of $\psi(x)$; see Exercise 77.

Thanks to the above proposition, we can now define the complex gamma function in a more traditional manner:

Definition 9.6.18. *For $s \in \mathbb{C} \setminus \mathbb{Z}_{\leqslant 0}$ we define $\Gamma(s) = \lim_{N \to \infty} u_N(s)$.*

Proposition 9.6.19. *The above limit exists and defines a meromorphic function on $\mathbb{C}$ that generalizes the real gamma function defined above for $s \in \mathbb{R}_{>0}$. It has no zeros in $\mathbb{C}$, and it has simple poles on $\mathbb{Z}_{\leqslant 0}$, the residue at $s = -k$ being equal to $(-1)^k/k!$.*

Proof. Since

$$u_{N+1}(s)/u_N(s) = (1+1/N)^s/(1+s/N) = 1 + O(1/N^2) \,,$$

it is clear that the limit exists and that it converges uniformly on any compact subset of $\mathbb{C} \setminus \mathbb{Z}_{\leqslant 0}$. It follows that $\Gamma(s)$ is indeed a meromorphic function on $\mathbb{C}$ with simple poles on $\mathbb{Z}_{\leqslant 0}$, and by Proposition 9.6.17 that it does generalize the real gamma function. By the functional equation, as s tends to $-k$ with $k \in \mathbb{Z}_{\geqslant 0}$ we have

$$\Gamma(s) = \frac{\Gamma(s+k+1)}{\prod_{0 \leqslant i \leqslant k}(s+i)} \sim \frac{(-1)^k}{k!} \frac{1}{s+k} \,,$$

giving the residues. Finally, since

$$\Gamma(s) = u_1(s) \prod_{N \geqslant 1} \frac{u_{N+1}(s)}{u_N(s)} = \frac{1}{s} \prod_{N \geqslant 1} \frac{(1+1/N)^s}{1+s/N} \,,$$

the absolute convergence of this infinite product implies that it does not vanish anywhere.   □

It follows from this proposition that if we set $\psi(s) = \Gamma'(s)/\Gamma(s)$ then ψ is also a meromorphic function with the same poles, residues equal to -1, and that it generalizes the real ψ function defined above.

Corollary 9.6.20. *There exists a unique holomorphic function* $\operatorname{Log}\Gamma(s)$ *defined on the simply connected set* $\mathbb{C}\setminus\mathbb{R}_{\leqslant 0}$ *such that* $\exp(\operatorname{Log}\Gamma(s)) = \Gamma(s)$ *and* $\operatorname{Log}\Gamma(1) = 0$. *We have* $\operatorname{Log}\Gamma'(s) = \psi(s)$ *and the formulas*

$$\operatorname{Log}\Gamma(s) = \lim_{N\to\infty}\left((s-1)\log(N) + \log(N!) - \sum_{0\leqslant k\leqslant N-1}\log(s+k)\right)$$

$$= -\log(s) + \sum_{N\geqslant 1}\left(s\log(1+1/N) - \log(1+s/N)\right).$$

Proof. Let $\Omega = \mathbb{C}\setminus\mathbb{R}_{\leqslant 0}$. Since $\Gamma(s)$ has no zeros or poles on Ω and since Ω is simply connected, it follows that there exists a holomorphic function $\log(\Gamma(s))$ defined on Ω, which is unique if we specify its value at a single point, for instance at 1, where it can be any integral multiple of $2i\pi$. Specifically, if $\mathcal{C}_s$ is any contour from 1 to s and lying in Ω, we set $\operatorname{Log}\Gamma(s) = \int_{\mathcal{C}_s}\psi(z)\,dz$, and this does not depend on the contour since ψ is holomorphic in Ω. Furthermore, the two given formulas are clearly equivalent, and since by uniform convergence the sum

$$-\log(s) + \sum_{N\geqslant 1}\left(s\log(1+1/N) - \log(1+s/N)\right)$$

defines a holomorphic function on Ω, equal to 0 at $s = 1$ and whose exponential is equal to $\Gamma(s)$ by the proposition, it follows by uniqueness that it is equal to $\operatorname{Log}\Gamma(s)$. $\qquad\square$

Remark. If we denote by log the principal determination of the logarithm, we have evidently $\operatorname{Log}\Gamma(s) = \log(\Gamma(s)) + 2i\pi m(s)$ for some $m(s) \in \mathbb{Z}$, but $m(s)$ is not equal to 0 in general (and it can be estimated approximately when $\Re(s)$ or $\Im(s)$ is large; see Exercise 35).

9.6.3 Preliminary Results for the Study of $\Gamma(s)$

In the sequel we are going to study in some detail the gamma function. For this we need some classical undergraduate material that we recall here with proof.

Proposition 9.6.21.
$$\int_{-\infty}^{\infty} e^{-t^2}\,dt = \sqrt{\pi}\,,$$

and more generally for $a > 0$ *we have*

$$\int_{-\infty}^{\infty} e^{-at^2}\,dt = \sqrt{\pi/a}\,.$$

Proof. There are several classical proofs of this result, and we give two. The first is using polar coordinates. Set $I_N = \int_{-N}^{N} e^{-t^2}\,dt$. Then $I_N^2 =$

$\int_{S(N)} e^{-(t^2+u^2)} \, dt \, du$, where $S(N)$ is the square $[-N, N]^2$. If $D(R)$ is the disk centered at the origin of radius R, we clearly have $D(N) \subset S(N) \subset D(N\sqrt{2})$. On the other hand, by passing to polar coordinates we have

$$\int_{D(N)} e^{-(t^2+u^2)} \, dt \, du = \int_0^{2\pi} d\theta \int_0^N e^{-\rho^2} \rho \, d\rho = \pi(-e^{-\rho^2})\Big|_0^N = \pi(1 - e^{-N^2}) \, .$$

Since the function $e^{-(t^2+u^2)}$ is nonnegative, it follows that

$$\pi(1 - e^{-N^2}) \leqslant I_N^2 \leqslant \pi(1 - e^{-2N^2}) \, ,$$

proving the first result by letting $N \to \infty$. The second follows by making the evident change of variable $u = a^{1/2}t$. A second proof is given in Exercise 72.   □

Note that this result is still true when a is complex with $\Re(a) \geqslant 0$ and $a \neq 0$; see Lemma 10.2.9 in the next chapter.

Proposition 9.6.22 (Stirling's formula). *As $n \to \infty$ we have*

$$n! \sim n^n e^{-n} \sqrt{2\pi n} \, ,$$

or equivalently,

$$\log(n!) = \left(n + \frac{1}{2}\right) \log(n) - n + \frac{1}{2} \log(2\pi) + o(1) \, .$$

Proof. Once again there are several classical proofs. Certainly the most classical is as follows: if we set $u_n = \log(n!/(n^n e^{-n} \sqrt{n}))$ then

$$u_{n+1} - u_n = 1 - \left(n + \frac{1}{2}\right) \log\left(1 + \frac{1}{n}\right) \sim -\frac{1}{12n^2} \, ;$$

hence this is the general term of an absolutely convergent series, so as $n \to \infty$, u_n tends to some limit $\log(A)$, say (we could also apply the Euler–MacLaurin summation formula). To obtain A we can use Wallis's formulas. We let $C_n = \int_0^{\pi/2} \cos^n(t) \, dt$. By integrating by parts, it is immediate that for $n \geqslant 2$ we have $C_n = (n-1)(C_{n-2} - C_n)$, hence $C_n = ((n-1)/n)C_{n-2}$. Since $C_0 = \pi/2$ and $C_1 = 1$, we deduce that

$$C_{2k} = \frac{(2k)!}{2^{2k}(k!)^2} \frac{\pi}{2} \quad \text{and} \quad C_{2k+1} = \frac{2^{2k}(k!)^2}{(2k+1)!} \, .$$

On the other hand, the sequence C_n is clearly decreasing, so that in particular $C_{2k+1} \leqslant C_{2k} \leqslant C_{2k-1}$. If we replace C_n by its asymptotic value $n^n e^{-n} n^{1/2} A$ (where A is the unknown nonzero constant above) a short computation shows that $A^2 = 2\pi$, proving Stirling's formula. Of course the $o(1)$

in the expression for $\log(n!)$ can be given a complete asymptotic expansion by the Euler–MacLaurin summation formula; see Section 9.2.5.

A more sophisticated, but easier to generalize, way of finding the value of the constant A has been explained in Section 9.2.5: by derivation of the formulas of Proposition 9.2.13, which come from the Euler–MacLaurin summation formula, we find that

$$\log(n!) = (n + 1/2)\log(n) - n - \zeta'(0) + O(1/n) \ .$$

The value of $\zeta'(0)$ is immediate to compute from the functional equation for the zeta function, which itself is a simple application of the Poisson summation formula; see Section 10.2.4. Once again we urge the reader to study Exercise 44 for generalizations of this idea.   □

Corollary 9.6.23. *For any $m \geqslant 1$ we have*

$$\mathrm{Log}\Gamma(s) = \left(s - \frac{1}{2}\right)\log(s) - s + \frac{\log(2\pi)}{2} + \sum_{k=1}^{m} \frac{B_{2k}}{2k(2k-1)s^{2k-1}}$$

$$- \frac{1}{2m+1}\int_0^\infty \frac{B_{2m+1}(\{t\})}{(t+s)^{2m+1}}\,dt \ .$$

Proof. Clear from Stirling's formula and Proposition 9.6.16.   □

Proposition 9.6.24. *We have the following expansions:*

(1)

$$\pi\cot(\pi x) = \frac{1}{x} + 2x\sum_{n\geqslant 1}\frac{1}{x^2 - n^2} \ .$$

(2)

$$\left(\frac{\pi}{\sin(\pi x)}\right)^2 = \sum_{n\in\mathbb{Z}}\frac{1}{(x-n)^2} \ .$$

(3)

$$\sin(\pi x) = \pi x \prod_{n\geqslant 1}\left(1 - \frac{x^2}{n^2}\right) \ .$$

Proof. Let $a \notin \mathbb{Z}$ be a parameter, and define $f(x)$ to be the 2π-periodic function such that $f(x) = \cos(ax)$ for $-\pi \leqslant x \leqslant \pi$. This function is clearly continuous and piecewise differentiable. It is thus everywhere equal to the sum of its Fourier series. A short computation gives

$$f(x) = \frac{\sin(\pi a)}{\pi}\left(\frac{1}{a} + 2a\sum_{n\geqslant 1}\frac{(-1)^n\cos(nx)}{a^2 - n^2}\right),$$

and taking $x = \pi$ gives

$$\pi \cot(\pi a) = \frac{1}{a} + 2a \sum_{n \geqslant 1} \frac{1}{a^2 - n^2} \,,$$

proving (1). Note incidentally that this formula immediately implies the formulas for $\zeta(2k)$; see Exercise 74. (2) follows by differentiation, after writing $2x/(x^2 - n^2) = 1/(x - n) + 1/(x + n)$. For (3), consider the function $g(a) = \sin(\pi a)/(\pi a \prod_{n \geqslant 1}(1 - a^2/n^2))$. Clearly the product converges absolutely, so the function is defined for all $a \notin \mathbb{Z}$. In addition, as a tends to 0 it is clear that $g(a)$ tends to 1, and since by writing $(1 - a^2/n^2) = (1 - a/n)(1 + a/n)$ it is clear that $g(a)$ is a periodic function of period 1, it follows that $g(a)$ tends to 1 as a tends to any integer. Moreover, it is also clear that $g(a)$ is differentiable (in fact infinitely). If we compute the logarithmic derivative of $g(a)$ we find using (1) that

$$\frac{g'(a)}{g(a)} = \pi \cot(\pi a) - \left(\frac{1}{a} + 2a \sum_{n \geqslant 1} \frac{1}{a^2 - n^2} \right) = 0 \,.$$

It follows that $g(a)$ is a constant, and since $g(0) = 1$, that $g(a) = 1$ for all a, proving (3). $\qquad \square$

Proposition 9.6.25. *For all s such that $\Re(s) > 0$ we have*

$$\log(s) = \int_0^\infty \frac{e^{-t} - e^{-st}}{t} \, dt \,,$$

where the left-hand side is the principal determination of the logarithm. More generally, if $\Re(s_1) > 0$ and $\Re(s_2) > 0$ we have

$$\log(s_1/s_2) = \int_0^\infty \frac{e^{-s_2 t} - e^{-s_1 t}}{t} \, dt \,.$$

Proof. Let $I(s)$ be the first integral above. It is clearly absolutely convergent for $\Re(s) > 0$, and its (for the moment formal) derivative with respect to s is $\int_0^\infty e^{-st} \, dt$, which is normally convergent in the domain $\Re(s) \geqslant \varepsilon > 0$ for any fixed ε. It follows that the derivation under the integral sign is justified; hence $I'(s) = 1/s$, so that $I(s) = \log(s)$ with the principal determination of the logarithm, since clearly $I(1) = 0$. The second formula follows from $\log(s_1/s_2) = \log(s_1) - \log(s_2)$ when $\Re(s_i) > 0$, with the principal determinations. $\qquad \square$

9.6.4 Properties of the Gamma Function

With this out the way, we can now begin our detailed study of the gamma function. Recall that we have set $u_n(s) = n^{s-1} n!/(s(s + 1) \cdots (s + n - 1))$

and that we have defined $\Gamma(s) = \lim_{n\to\infty} u_n(s)$ in Definition 9.6.18. Since $\mathbb{R}_{>0}$ is not a discrete subset of $\mathbb{C} \setminus \mathbb{Z}_{\leqslant 0}$, an important remark is that all the identities that we prove on the real gamma function (usually as byproducts of corresponding results for the Hurwitz zeta function) will be automatically valid for the complex gamma function by analytic continuation. In particular, by Proposition 9.6.14 we have $\Gamma(s+1) = s\Gamma(s)$ for all $s \in \mathbb{C} \setminus \mathbb{Z}_{\leqslant 0}$.

Proposition 9.6.26 (Hadamard product). *We have*

$$\Gamma(s+1) = e^{-\gamma s} \prod_{n\geqslant 1} \frac{e^{s/n}}{1 + s/n} \, ,$$

where $\gamma = 0.57721\ldots$ is Euler's constant.

Note that this is the *Hadamard product* expansion of the entire function $1/\Gamma(s+1)$; see Theorem 10.7.6.

Proof. If we divide the numerator and the denominator of $u_n(s)$ by $n! = 1 \cdot 2 \cdots n$ we obtain

$$u_n(s+1) = \frac{n^s}{\prod_{1\leqslant k\leqslant n}(1 + s/k)} = n^s e^{-sH_n} \prod_{1\leqslant k\leqslant n} \frac{e^{s/k}}{1 + s/k} \, ,$$

where $H_n = \sum_{1\leqslant k\leqslant n} 1/k$ is the harmonic sum. Since

$$n^s e^{-sH_n} = e^{-s(H_n - \log(n))} \, ,$$

by definition of γ this tends to $e^{-\gamma s}$, proving the proposition. Note that this implies that the infinite product is convergent, which is clear directly by noting that the logarithm of its general term is $O(1/n^2)$.   □

Proposition 9.6.27 (Complex Stirling formula). *For any $s \in \mathbb{C}$ set $\rho(s) = \max(\Re(s), |\Im(s)|)$. Then as $\rho(s) \to \infty$ we have*

$$\mathrm{Log}\Gamma(s) = \left(s - \frac{1}{2}\right)\log(s) - s + \frac{1}{2}\log(2\pi) + O(1/\rho(s)) \, .$$

Proof. First note that the region $R_N = \{s \in \mathbb{C}/\ \rho(s) \geqslant N\}$ is a subset of $\mathbb{C} \setminus \mathbb{R}_{\leqslant 0}$, hence we can choose the principal determination of the logarithm, which will be analytic in that region. Furthermore, if s belongs to R_N then so does $s + k$ for $k \in \mathbb{Z}_{\geqslant 0}$. Now an immediate exercise in complex integration shows that if $z \in R_N$ we have $\int_0^z \log(u)\,du = z\log(z) - z$, where the principal determination is also chosen on the right-hand side. Note that this equality is true as such, and not only modulo $2i\pi$. Thus, using an exact version of the Euler–MacLaurin summation formula and this remark we obtain

$$\sum_{k=0}^{n-1} \log(s+k) = \left(s+n-\frac{1}{2}\right)\log(n+s) - (n+s) - \left(\left(s-\frac{1}{2}\right)\log(s)-s\right)$$
$$+ \frac{1}{12}\left(\frac{1}{s+n}-\frac{1}{s}\right) + \frac{1}{2}\int_0^n \frac{B_2(\{t\})}{(t+s)^2}\,dt\,,$$

where once again I emphasize that this is true not only modulo $2i\pi$, but as written. Thus, by Corollary 9.6.20 and the ordinary Stirling formula we have

$$\mathrm{Log}\,\Gamma(s) = (s-1)\log(n) + \log(n!) + o(1) - \sum_{k=0}^{n-1}\log(s+k)$$

$$= \left(s+n-\frac{1}{2}\right)\log(n) - n + \frac{1}{2}\log(2\pi) + o(1)$$

$$- \left(s+n-\frac{1}{2}\right)\log(n+s) + (n+s) + \left(\left(s-\frac{1}{2}\right)\log(s)-s\right)$$

$$- \frac{1}{12}\left(\frac{1}{s+n}-\frac{1}{s}\right) - \frac{1}{2}\int_0^n \frac{B_2(\{t\})}{(t+s)^2}\,dt\,.$$

Since $s - (s+n-1/2)\log(1+s/n) = O(1/n)$, letting $n \to \infty$ it follows that we have the following integral representation for $\mathrm{Log}\,\Gamma(s)$:

$$\mathrm{Log}\,\Gamma(s) = \left(s-\frac{1}{2}\right)\log(s) - s + \frac{1}{2}\log(2\pi) + \frac{1}{12s} - \frac{1}{2}\int_0^\infty \frac{B_2(\{t\})}{(t+s)^2}\,dt\,.$$

Since $B_2(\{t\})$ is bounded in absolute value by $B_2 = 1/6$, it follows that

$$\left|\int_0^\infty \frac{B_2(\{t\})}{(t+s)^2}\,dt\right| \leqslant \int_0^\infty \frac{1}{|t+s|^2}\,dt\,,$$

and an easy computation shows that

$$\int_0^\infty \frac{1}{|t+s|^2}\,dt = \frac{1}{|\Im(s)|}\,\mathrm{atan}\left(\frac{|\Im(s)|}{\Re(s)}\right)$$

(or $1/\Re(s)$ if $\Im(s) = 0$). Since $\mathrm{atan}(x) \leqslant x$ for $x \geqslant 0$, it follows that this is less than or equal to $(\pi/2)/|\Im(s)|$ and to $1/\Re(s) \leqslant (\pi/2)/\Re(s)$ when $\Re(s) > 0$, hence to $(\pi/2)/\rho(s)$, proving the proposition.   □

Remark. Once the slightly delicate estimate above is made, we can of course apply Euler–MacLaurin to any order and deduce that the asymptotic expansion for $\log(n!)$ obtained in Section 9.2.5 is valid more generally for $\mathrm{Log}\,\Gamma(s+1)$. Note also that the expansion of $\log(\Gamma(s+1))$ would involve an additional multiple of $2i\pi$.

Corollary 9.6.28. (1) *As $x \to \infty$ in $\mathbb{R}$ we have $\Gamma(x) \sim x^{x-1/2}e^{-x}(2\pi)^{1/2}$.*

(2) *Let $\sigma \in \mathbb{R}$ be fixed. Then as $|t| \to \infty$ we have*

$$|\Gamma(\sigma + it)| \sim |t|^{\sigma - 1/2} e^{-\pi|t|/2} (2\pi)^{1/2} .$$

Proof. The first statement is clear. By the above proposition we have

$$\begin{aligned}
\log(|\Gamma(\sigma + it)|) &= \Re(\log(\Gamma(\sigma + it))) \\
&= \Re((\sigma + it - 1/2)\log(\sigma + it) - \sigma - it) + \log(2\pi)/2 + o(1) \\
&= \Re((\sigma + it - 1/2)(\log(\sigma^2 + t^2)/2 + i \operatorname{atan} 2(\sigma, t))) \\
&\quad - \sigma + \log(2\pi)/2 + o(1) \\
&= (\sigma - 1/2)(\log(|t|) + o(1)) - t \operatorname{atan} 2(\sigma, t) \\
&\quad - \sigma + \log(2\pi)/2 + o(1) ,
\end{aligned}$$

where $\operatorname{atan} 2(x, y) = \operatorname{Arg}(x + iy)$ is the unique angle θ in $]-\pi, \pi]$ such that $\cos(\theta) = x/\sqrt{x^2 + y^2}$ and $\sin(\theta) = y/\sqrt{x^2 + y^2}$. Changing t into $-t$ does not change the above expression, so we may assume $t \geqslant 0$. Clearly $\operatorname{atan} 2(x, y) = \operatorname{atan}(y/x) + k\pi$, where $k = 0$ if $x > 0$, while $k = \operatorname{sign}(y)$ if $x < 0$. In both cases $\sigma > 0$ and $\sigma < 0$, we see that

$$\begin{aligned}
\log(|\Gamma(\sigma + it)|) &= (\sigma - 1/2)\log(|t|) - t(\pi/2 - \operatorname{atan}(\sigma/t)) \\
&\quad - \sigma + \log(2\pi)/2 + o(1) \\
&= (\sigma - 1/2)\log(|t|) - t\pi/2 + \log(2\pi)/2 + o(1) ,
\end{aligned}$$

proving the result, and the same proof gives the result when $\sigma = 0$. $\qquad \square$

Remark. This result, which shows that the gamma function tends to zero exponentially fast as $|t| \to \infty$ in bounded vertical strips of the complex plane (which may seem paradoxical compared to its behavior on $\mathbb{R}_{>0}$), is essential in all proofs dealing with L-functions with functional equations involving products of gamma functions, since it easily allows us to shift the contours of integration.

Proposition 9.6.29. *We have the following integral representation, valid for $\Re(s) > -1$:*

$$\operatorname{Log}\Gamma(s + 1) = \int_0^\infty \left(s\frac{e^{-t}}{t} - \frac{1 - e^{-st}}{t(e^t - 1)} \right) dt .$$

Proof. By Corollary 9.6.20, as $n \to \infty$ we have $\operatorname{Log}\Gamma(s + 1) = s\log(n) - \sum_{1 \leqslant k \leqslant n} \log((s + k)/k) + o(1)$, so by Proposition 9.6.25 we have

$$\begin{aligned}
\operatorname{Log}\Gamma(s + 1) &= o(1) + \int_0^\infty \left(s\frac{e^{-t} - e^{-nt}}{t} - \sum_{1 \leqslant k \leqslant n} \frac{e^{-kt} - e^{-(s+k)t}}{t} \right) dt \\
&= o(1) + \int_0^\infty \left(s\frac{e^{-t}}{t} - s\frac{e^{-nt}}{t} - \frac{1 - e^{-st}}{t} \frac{1 - e^{-nt}}{e^t - 1} \right) dt
\end{aligned}$$

and since

$$\int_0^\infty \frac{e^{-nt}}{t} \left(-s + \frac{1 - e^{-st}}{e^t - 1} \right) dt$$

converges normally it is clear that it tends to 0 as $n \to \infty$, proving the proposition. □

Corollary 9.6.30. *We have the following integral representations:*

$$\gamma = \int_0^\infty \left(\frac{1}{e^t - 1} - \frac{e^{-t}}{t} \right) dt$$

and for $k \geqslant 2$

$$\zeta(k) = \frac{1}{(k-1)!} \int_0^\infty \frac{t^{k-1}}{e^t - 1} \, dt \ .$$

Proof. This follows by expanding in powers of s the integrand of the proposition and comparing with Proposition 9.6.15. □

Note that these formulas can easily be proved directly by writing $1/(e^t - 1) = \sum_{k \geqslant 1} e^{-kt}$ (see also Corollary 10.2.3). In particular, we have more generally for $\Re(s) > 1$,

$$\zeta(s) = \frac{1}{\Gamma(s)} \int_0^\infty \frac{t^{s-1}}{e^t - 1} \, dt \ .$$

Corollary 9.6.31. *We have the integral representation*

$$\frac{1}{2} \log(2\pi) = \int_0^\infty \left(\frac{1}{t} - \frac{e^{-t}}{2} - \frac{1}{e^t - 1} \right) \frac{dt}{t} \ .$$

Proof. Integrating the first formula of Proposition 9.6.25 we obtain

$$s \log(s) - s = \int_0^\infty \left(s \frac{e^{-t}}{t} - \frac{1 - e^{-st}}{t^2} \right) dt \ .$$

Subtracting this and $\log(s)/2$ given by Proposition 9.6.25 from the integral representation of $\mathrm{Log}\Gamma(s+1)$, the result follows from Stirling's formula by letting $s \to \infty$ in the positive integers for instance. □

Corollary 9.6.32. *For $\Re(s) > 0$ we have*

$$\mathrm{Log}\Gamma(s) = \left(s - \frac{1}{2} \right) \log(s) - s + \frac{1}{2} \log(2\pi) + \int_0^\infty \frac{e^{-st}}{t} \left(\frac{1}{e^t - 1} - \frac{1}{t} + \frac{1}{2} \right) dt \ .$$

Proof. Immediate from Propositions 9.6.29 and the integral representations of $s \log(s) - s$, $\log(s)$, and $\log(2\pi)/2$ given above. □

By expanding into a power series the function $1/(e^t - 1) - 1/t + 1/2$ we once again recover Stirling's asymptotic expansion for $\mathrm{Log}\Gamma(s+1)$.

Proposition 9.6.33. *We have the duplication formula*

$$\Gamma(s)\Gamma(s+1/2) = 2^{1-2s}\pi^{1/2}\Gamma(2s)$$

and more generally for $N \in \mathbb{Z}_{\geqslant 1}$ the distribution formula

$$\prod_{0 \leqslant j < N} \Gamma\left(s + \frac{j}{N}\right) = N^{1/2-Ns}(2\pi)^{(N-1)/2}\Gamma(Ns) \, .$$

In particular, we have

$$\prod_{1 \leqslant j \leqslant N} \Gamma\left(\frac{j}{N}\right) = \frac{(2\pi)^{(N-1)/2}}{N^{1/2}} \, .$$

Proof. Differentiating with respect to s the formula of Proposition 9.6.12, setting $s = 0$, and using Definition 9.6.13 gives

$$\sum_{0 \leqslant j < N} \log(\Gamma(x + j/N)) = \log(\Gamma(Nx)) - (N-1)\zeta'(0) + \zeta(0, Nx)\log(N) \, ,$$

and since $\zeta'(0) = -\log(2\pi)/2$ and $\zeta(0, Nx) = -B_1(Nx) = -(Nx - 1/2)$ by Corollary 9.6.10, the result follows for $x \in \mathbb{R}_{>0}$, hence for all $x \in \mathbb{C}$ by analytic continuation.

Another proof is to use the alternative definition of the gamma function given by Proposition 9.6.17; see Exercise 85.

The last statement is immediate by setting $s = 1/m$ in the distribution formula. $\qquad \square$

Proposition 9.6.34. *We have the reflection formula*

$$\Gamma(s)\Gamma(1-s) = \frac{\pi}{\sin(\pi s)} \, .$$

Proof. By Propositions 9.6.26 and 9.6.24 we have

$$\Gamma(1+s)\Gamma(1-s) = \prod_{n \geqslant 1}(1 - s^2/n^2) = \pi s/\sin(\pi s) \, ,$$

so the result follows since $\Gamma(1+s) = s\Gamma(s)$. $\qquad \square$

Remarks. At this point, several remarks are in order.

(1) The function $\sin(\pi s)$ is a natural function because it is $\mathbb{Z}$-periodic. The above reflection formula shows that $\Gamma(s)$ is in a certain sense "one half" of the sine function. Another way of saying this is that formulas obtained by summing (or taking products) over $\mathbb{Z}$, such as the Poisson summation formula, will be simpler than formulas obtained by summing over $\mathbb{Z}_{>0}$.

A case in point is the zeta function: the natural definition would be $\zeta(s) = \sum_{n \in \mathbb{Z}} 1/n^s$. This is not defined for $n = 0$, so we have to exclude $n = 0$ from the sum. But this is not the important point: n^s is multivalued when $n < 0$, so we have to agree on some determination, except if $s \in \mathbb{Z}$. And then $\zeta(2k + 1) = 0$, which removes all difficulties for zeta values at odd integers!

(2) We have proved all the properties of the gamma function as consequences of corresponding properties of the Hurwitz zeta function (for $s \in \mathbb{Z}_{>0}$, and then deduced their validity by analytic continuation), with the exception of the reflection formula. It can also be obtained in this way, but the proof is much less natural; see Exercise 67.

(3) It is not difficult to prove that the functional equation and the distribution formulas extend naturally to the function $\mathrm{Log}\Gamma(s)$ (for instance we have $\mathrm{Log}\Gamma(s + 1) = \mathrm{Log}\Gamma(s) + \log(s)$ with the principal determination, and this would not be true with $\mathrm{Log}\Gamma(s)$ replaced by $\log(\Gamma(s))$). On the other hand, the extension of the reflection formula is slightly more subtle; see Exercise 68.

The reader may be surprised that I have not yet mentioned the most standard definition of the gamma function. Indeed, we have the following classical result, which is usually taken as the definition of the gamma function:

Proposition 9.6.35. *For $\Re(s) > 0$ we have*

$$\Gamma(s) = \int_0^\infty t^s e^{-t} \frac{dt}{t} \ .$$

Proof. Recall that for any function $f \in C^\infty([0,1])$ we have Taylor's formula (which is trivially proved by integration by parts)

$$f(1) = \sum_{0 \leqslant k \leqslant n} \frac{f^{(k)}(0)}{k!} + \frac{1}{n!} \int_0^1 (1 - t)^n f^{(n+1)}(t)\, dt \ .$$

Applying this to $f(t) = t^{n+s}$, we deduce since $\Re(s) > 0$ that

$$1 = \frac{s(s + 1) \cdots (s + n)}{n!} \int_0^1 (1 - t)^n t^{s-1}\, dt \ ,$$

so that

$$v_n(s) = n^s \int_0^1 (1 - t)^n t^{s-1}\, dt \ ,$$

where

$$v_n(s) = \frac{n^s n!}{s(s + 1) \cdots (s + n)} = \frac{n}{s + n} u_n(s)$$

clearly tends to $\Gamma(s)$ as $n \to \infty$. Changing t into t/n in the integral gives

$$v_n(s) = \int_0^n (1 - t/n)^n t^{s-1} \, dt \ .$$

Since $(1 - t/n)^n$ tends to e^{-t} as $n \to \infty$, this starts to look like the desired result. However, we must justify the limiting process. An easy exercise in real analysis (Exercise 93) shows that for $0 \leqslant t \leqslant n$ we have

$$\left(1 - \frac{t}{n}\right)^n = e^{-t} - \frac{\phi(t)}{n} \ ,$$

where $0 \leqslant \phi(t) \leqslant t^2 e^{-t}$. It follows that

$$v_n(s) = \int_0^n e^{-t} t^{s-1} \, dt - \frac{1}{n} \int_0^n \phi(t) t^{s-1} \, dt \ .$$

Since $|\phi(t)| \leqslant t^2 e^{-t}$, it follows that as $n \to \infty$ the rightmost integral tends to some finite limit, and since we divide by n the quotient tends to 0. Thus

$$\Gamma(s) = \lim_{n \to \infty} v_n(s) = \int_0^\infty e^{-t} t^{s-1} \, dt \ ,$$

proving the proposition.   □

Corollary 9.6.36. *Let a and s be two complex numbers such that $\Re(a) > 0$ and $\Re(s) > 0$, or $\Re(a) = 0$, $a \neq 0$, and $0 < \Re(s) < 1$. Then*

$$\int_0^\infty t^s e^{-at} \frac{dt}{t} = a^{-s} \Gamma(s) \ ,$$

where as usual $a^{-s} = e^{-s \log(a)}$ with the principal determination of the logarithm, i.e., $-\pi < \Im(\log(a)) < \pi$. In particular, if $x \in \mathbb{R}^*$ and $0 < \Re(s) < 1$ we have

$$\int_0^\infty t^s \sin(xt) \frac{dt}{t} = \frac{\Gamma(s)}{|x|^s} \sin\left(\frac{s\pi}{2}\right) \operatorname{sign}(x) \ and$$

$$\int_0^\infty t^s \cos(xt) \frac{dt}{t} = \frac{\Gamma(s)}{|x|^s} \cos\left(\frac{s\pi}{2}\right) \ .$$

Proof. The integral converges (absolutely) at $t = 0$ if and only if $\Re(s) > 0$. When $\Re(a) > 0$ it is clear that the integral converges absolutely at infinity. When $\Re(a) = 0$ and $a \neq 0$, then $f(t) = e^{-at}$ is such that $\int_A^B f(t) \, dt$ is bounded independently of A and B, so by integration by parts we see that the integral converges at infinity if and only if $\Re(s) < 1$.

Setting $u = at$, we obtain

$$\int_0^\infty t^s e^{-at} \frac{dt}{t} = a^{-s} \int_{L_a} u^s e^{-u} \frac{du}{u} \ ,$$

where L_a is the line going from 0 to $a\infty$. If we consider the natural contour going from $\varepsilon > 0$ to $R > \varepsilon$, then on the circle of radius R to $e^{i\theta}R$ (where θ is the argument of a), back on the line L_a to $a\varepsilon$ and on the circle of radius ε back to ε, the above convergence proof shows that the integral over the two circles will tend to 0 as ε tends to 0 and R tends to infinity. Since the integrand has no poles inside the contour, it follows that the integral over L_a is equal to the integral from 0 to ∞, proving the first formula. The other two are special cases.   □

Corollary 9.6.37. *For $0 < \Re(s) < 1$ we have*

$$\int_0^\infty t^{s-1} \cos(t)\, dt = \cos\left(\frac{\pi s}{2}\right)\Gamma(s) \quad and \quad \int_0^\infty t^{s-1}\sin(t)\, dt = \sin\left(\frac{\pi s}{2}\right)\Gamma(s) .$$

Proof. Clear.   □

The following proposition shows that the formula involving $\sin(xt)$ is still valid as a limiting case for $s = 0$, and that there exists a similar formula for $\cos(xt)$.

Proposition 9.6.38. *We have*

$$\int_0^\infty \frac{\sin(xt)}{t}\, dt = \frac{\pi}{2}\operatorname{sign}(x) ,$$

and for $a > 0$,

$$\int_0^a \frac{\cos(xt) - 1}{t}\, dt + \int_a^\infty \frac{\cos(xt)}{t}\, dt = -(\gamma + \log(|ax|)) .$$

Proof. By integration by parts we have for $s < 1$,

$$\int_1^\infty t^s e^{ixt} \frac{dt}{t} = -\frac{e^{ix}}{ix} + \frac{1-s}{ix}\int_1^\infty \frac{e^{ixt}}{t^{2-s}}\, dt .$$

This last integral is absolutely convergent, so that

$$\lim_{s \to 0} \int_1^\infty t^s e^{ixt} \frac{dt}{t} = \int_1^\infty e^{ixt} \frac{dt}{t} .$$

Furthermore, since $(e^{ixt} - 1)/t$ is a continuous function on the compact set $[0, 1]$ we also have

$$\lim_{s \to 0} \int_0^1 t^s (e^{ixt} - 1) \frac{dt}{t} = \int_0^1 (e^{ixt} - 1)\frac{dt}{t} .$$

To prove the proposition we may of course assume $x > 0$. By Corollary 9.6.36 we have for $0 < s < 1$,

$$\int_0^\infty t^s e^{ixt} \frac{dt}{t} = \frac{\Gamma(s+1)}{sx^s} e^{is\pi/2} \,,$$

hence

$$\int_0^1 t^s (e^{ixt} - 1) \frac{dt}{t} + \int_1^\infty t^s e^{ixt} \frac{dt}{t} = \frac{\Gamma(s+1)}{sx^s} e^{is\pi/2} - \frac{1}{s} \,.$$

By what we have proved above we can take the limit as $s \to 0^+$, so that

$$\int_0^1 \frac{e^{ixt} - 1}{t} \, dt + \int_1^\infty \frac{e^{ixt}}{t} \, dt = \lim_{s\to 0^+} \left(\frac{\Gamma(s+1)}{sx^s} e^{is\pi/2} - \frac{1}{s} \right) \,.$$

Using the expansion $\Gamma(s+1) = 1 - \gamma s + O(s^2)$ we immediately obtain the first formula of the proposition and the second for $a = 1$, and the general case follows by changing x into ax and t into t/a. □

A direct classical proof of the first formula is given in Exercise 95.

Proposition 9.6.39. *For $\Re(a) > 0$ and $\Re(b) > 0$ define the* beta function *$B(a, b)$ by*

$$B(a, b) = \int_0^1 t^{a-1}(1 - t)^{b-1} \, dt \,.$$

Then

$$B(a, b) = \frac{\Gamma(a)\Gamma(b)}{\Gamma(a + b)} \,.$$

Proof. By Proposition 9.6.35 we have

$$\Gamma(a)\Gamma(b) = \int_0^\infty \int_0^\infty t^{a-1} u^{b-1} e^{-(t+u)} \, dt \, du \,,$$

so that setting $v = t + u$ we obtain

$$\Gamma(a)\Gamma(b) = \int_0^\infty e^{-v} \left(\int_0^v t^{a-1}(v - t)^{b-1} \, dt \right) dv \,;$$

hence making the change of variable $t = vz$ in the inner integral gives

$$\Gamma(a)\Gamma(b) = \int_0^\infty v^{a+b-1} e^{-v} \left(\int_0^1 z^{a-1}(1 - z)^{b-1} \, dz \right) dv = B(a, b)\Gamma(a + b) \,.$$

□

Corollary 9.6.40. (1) *For $0 < \Re(a) < 2\Re(b)$ we have*

$$\int_0^\infty \frac{t^{a-1}}{(1 + t^2)^b} \, dt = \frac{1}{2} B(a/2, b - a/2) = \frac{\Gamma(a/2)\Gamma(b - a/2)}{2\Gamma(b)} \,.$$

(2) *For $\Re(a) > 0$ and $\Re(b) > 0$ we have*

$$\int_0^1 t^{a-1}(1-t^2)^{b-1} = \frac{1}{2} B(a/2, b) = \frac{\Gamma(a/2)\Gamma(b)}{2\Gamma(b+a/2)} .$$

(3) *For $\Re(s) > 0$ we have*

$$\int_0^1 (1-t^2)^{s-1} dt = 2^{2s-2} \frac{\Gamma(s)^2}{\Gamma(2s)} \quad and \quad \int_0^1 \frac{t^{s-1}}{(1+t^2)^s} dt = \frac{\Gamma(s/2)^2}{4\Gamma(s)} .$$

(4) *We have*

$$\int_0^\pi \sin(t)^s \, dt = \pi^{1/2} \frac{\Gamma((s+1)/2)}{\Gamma(s/2+1)} ,$$

$$\int_0^\pi \sin(t)^{-s} \, dt = \pi^{1/2} \tan(\pi s/2) \frac{\Gamma(s/2)}{\Gamma((s+1)/2)} ,$$

$$\int_0^\infty \sinh(t)^{-s} \, dt = \frac{\pi^{1/2}}{2\cos(\pi s/2)} \frac{\Gamma(s/2)}{\Gamma((s+1)/2)} ,$$

$$\int_0^\infty \cosh(t)^{-s} \, dt = \frac{\pi^{1/2}}{2} \frac{\Gamma(s/2)}{\Gamma((s+1)/2)} ,$$

where the first formula is valid for $\Re(s) > -1$, the second for $\Re(s) < 1$, the third for $0 < \Re(s) < 1$, and the fourth for $\Re(s) > 0$.

See Exercise 105 (e) for an interesting application of (4).

Proof. If we set $u = t^2/(1+t^2)$ then $1+t^2 = 1/(1-u)$, $t = (u/(1-u))^{1/2}$, and $dt = du/(2(1-u)^{3/2}u^{1/2})$, so

$$\int_0^\infty \frac{t^{a-1}}{(1+t^2)^b} dt = \frac{1}{2} \int_0^1 u^{a/2-1}(1-u)^{-a/2-1+b} du = \frac{B(a/2, b-a/2)}{2} ,$$

proving (1). (2) follows immediately from the proposition by making the change of variable $u = t^2$. The first formula of (3) is a consequence of (2) applied to $a = 1$ and of the duplication formula for the gamma function. For the second, the change of variable $u = 1/t$ gives

$$\int_1^\infty t^{s-1}/(t^2+1)^s \, dt = \int_0^1 u^{s-1}/(u^2+1)^s \, du ,$$

so the result follows from (1). For (4) we set $u = \tan(t/2)$, so that $\sin(t) = 2u/(1+u^2)$, $dt = 2du/(1+u^2)$; hence by (1) we have

$$\int_0^\pi \sin(t)^s \, dt = 2^{s+1} \int_0^\infty \frac{u^s}{(1+u^2)^{s+1}} du = 2^s \frac{\Gamma((s+1)/2)^2}{\Gamma(s+1)} ,$$

and we obtain the first formula using the duplication formula, and the second using the reflection formula. Note that we could obtain this result directly using the composite change of variable $u = \sin^2(t/2)$. For the third formula of (4) we set $u = \sinh(t/2)$, so that $\sinh(t) = 2u(1+u^2)^{1/2}$ and $dt = 2\,du/(1+u^2)^{1/2}$; hence we obtain by (1)

$$\int_0^\infty \sinh(t)^{-s}\,dt = 2^{1-s}\int_0^\infty \frac{u^{-s}}{(1+u^2)^{(s+1)/2}} = 2^{-s}\frac{\Gamma((1-s)/2)\Gamma(s)}{\Gamma((s+1)/2)}\,,$$

giving the third formula after applying the duplication and reflection formulas. For the fourth formula we set $u = \tanh^2(t)$, so that $\cosh(t) = 1/(1-u)$, $dt = du/(2u^{1/2}(1-u))$, and hence

$$\int_0^\infty \cosh(t)^{-s}\,dt = \frac{1}{2}\int_0^1 u^{-1/2}(1-u)^{s/2-1}\,du$$

$$= \frac{1}{2}B(s/2,1/2) = \frac{\pi^{1/2}}{2}\frac{\Gamma(s/2)}{\Gamma((s+1)/2)}\,.$$

$\square$

9.6.5 Specific Properties of the Function $\psi(s)$

Recall from Definition 9.6.13 that $\psi(s) = \Gamma'(s)/\Gamma(s)$. All the formulas that we have seen up to now can of course be logarithmically differentiated several times if necessary to give formulas for ψ and its derivatives:

Proposition 9.6.41. *Let $k \in \mathbb{Z}_{\geqslant 0}$.*

(1) *We have*

$$\psi(s) = \lim_{N\to\infty}\left(\log(N) - \sum_{n=0}^N \frac{1}{n+s}\right)$$

$$= -\gamma + (s-1)\sum_{n=0}^\infty \frac{1}{(n+1)(n+s)} = -\gamma + \sum_{n=0}^\infty\left(\frac{1}{n+1} - \frac{1}{n+s}\right)\,.$$

(2) *For $k \geqslant 1$,*

$$\psi^{(k)}(s) = (-1)^{k+1}k!\zeta(k+1,s) = (-1)^{k+1}k!\sum_{n\geqslant 0}\frac{1}{(n+s)^{k+1}}\,.$$

(3) *We have*

$$\psi^{(k)}(s+1) = \psi^{(k)}(s) + \frac{(-1)^k k!}{s^{k+1}}\,,$$

$$\psi^{(k)}(n) = (-1)^{k-1}k!\left(\zeta(k+1) - \sum_{j=1}^{n-1}\frac{1}{j^{k+1}}\right)\,,$$

where we set by convention $\zeta(1) = \gamma$.

(4) We have
$$\sum_{0 \leqslant j < m} \psi(s + j/m) = -m \log(m) + m\psi(ms) \,,$$

and for $k \geqslant 1$,

$$\sum_{0 \leqslant j < m} \psi^{(k)}(s + j/m) = m^{k+1} \psi^{(k)}(ms) \,.$$

(5) We have
$$\psi(s) - \psi(1 - s) = -\pi \cotan(\pi s) \,.$$

Proposition 9.6.42. As $s \to \infty$ we have $\psi(s) = \log(s) + O(1/s)$ and $\psi^{(k)}(s) = (-1)^{k-1}(k-1)!/s^{k-1} + O(1/s^k)$ for $k \geqslant 1$.

Proposition 9.6.43. We have for $\Re(s) > -1$,

$$\psi(s+1) = \int_0^\infty \left(\frac{e^{-t}}{t} - \frac{e^{-st}}{e^t - 1} \right) dt$$

$$= -\gamma + \int_0^\infty \frac{1 - e^{-st}}{e^t - 1} \, dt = -\gamma + \int_0^1 \frac{1 - (1-t)^s}{t} \, dt \,,$$

and for $k \geqslant 1$,

$$\psi^{(k)}(s+1) = (-1)^{k-1} \int_0^\infty \frac{t^k e^{-st}}{e^t - 1} \, dt \,.$$

Proofs. Immediate and left to the reader (Exercise 77). □

Remarks. (1) In Proposition 9.2.10 we have already given without proof the integral representations of Proposition 9.6.43, together with integral representations involving fractional parts.

(2) Thanks to the formulas of Proposition 9.6.41 we can express exactly in terms of complex values of the function ψ and its derivatives the sum of any infinite series of rational function values, and similarly for infinite products. More precisely, we have the following:

Proposition 9.6.44. Let f be a rational function, and let its decomposition into partial fractions be

$$f(x) = \sum_{\alpha \ pole} \sum_{1 \leqslant k \leqslant -v(\alpha)} \frac{a_{\alpha,k}}{(x - \alpha)^k} \,,$$

where α runs through the poles of f, $-v(\alpha) \geqslant 1$ denotes the order of the pole α, and $a_{\alpha,k} \in \mathbb{C}$. Assume that $x^2 f(x)$ is bounded when $x \to \infty$, in other words that $\sum_\alpha a_{\alpha,1} = 0$. Then

$$\sum_{n \geqslant 0} f(n) = \sum_{\alpha \ pole} \sum_{1 \leqslant k \leqslant v(\alpha)} (-1)^k \frac{a_{\alpha,k}}{(k-1)!} \psi^{(k-1)}(-\alpha) \,.$$

Proposition 9.6.45. *Let f be a rational function, and write*

$$f(x) = C \prod_{\alpha \text{ zero or pole}} (x - \alpha)^{v(\alpha)} \,,$$

where α runs through the zeros and poles of f, $v(\alpha) \in \mathbb{Z}$ is the order of α (positive for a zero, negative for a pole), and $C \in \mathbb{C}^*$. Assume that $x^2(f(x) - 1)$ is bounded as $x \to \infty$, in other words that $C = 1$, $\sum_\alpha v(\alpha) = 0$, and $\sum_\alpha \alpha v(\alpha) = 0$. Then

$$\prod_{n \geqslant 0} f(n) = \prod_{\alpha \text{ zero or pole}} \Gamma(-\alpha)^{-v(\alpha)} \,.$$

Proof. Again immediate and left to the reader (Exercise 78). $\qquad\square$

In addition to the above, the function ψ has specific properties that do not immediately follow by derivation from corresponding properties of $\mathrm{Log}\Gamma(x)$. Perhaps the most interesting is the fact that it can be evaluated exactly at rational arguments in terms of elementary functions:

Proposition 9.6.46. *Assume that $0 < r < m$ are integers, and set as usual $\zeta_m = \exp(2i\pi/m)$. We have*

$$\psi\left(\frac{r}{m}\right) = -\gamma - \log(m) + \sum_{1 \leqslant k \leqslant m-1} \zeta_m^{-rk} \log(1 - \zeta_m^k)$$

$$= -\gamma - \log(m) - \frac{\pi}{2} \cotan\left(\frac{\pi r}{m}\right) + \sum_{k=1}^{m-1} \cos\left(\frac{2\pi k r}{m}\right) \log\left(2\sin\left(\frac{\pi k}{m}\right)\right) \,.$$

Proof. By Proposition 9.6.41 we have

$$\psi(r/m) = -\gamma + \sum_{n \geqslant 0}\left(\frac{1}{n+1} - \frac{m}{mn+r}\right) \,.$$

By Abel's theorem on the continuity of power series on their circle of convergence, since $1/(n+1) - m/(mn+r) = O(1/n^2)$ we have $\psi(r/m) = -\gamma + \lim_{t \to 1^-} f(t)$, where

$$f(t) = \sum_{n \geqslant 0}\left(\frac{t^{mn}}{n+1} - \frac{m t^{mn}}{mn+r}\right) \,.$$

As $t \to 1^-$ we have

$$\sum_{n \geqslant 0} t^{mn}/(n+1) = -t^{-m}\log(1 - t^m) = -\log(1-t) - \log(m) + o(1) \,.$$

On the other hand, we have

$$-\sum_{0\leqslant k<m} \zeta_m^{-rk}\log(1-t\zeta_m^k) = \sum_{n\geqslant 1}\frac{t^n}{n}\sum_{0\leqslant k<m}\zeta_m^{k(n-r)} .$$

The inner sum vanishes unless $m \mid (n-r)$, in other words $n = qm + r$ for some $q \geqslant 0$, in which case it is equal to m. Note that we use here the fact that $0 < r < m$. Thus

$$-\sum_{0\leqslant k<m}\zeta_m^{-rk}\log(1-t\zeta_m^k) = m\sum_{q\geqslant 0}\frac{t^{qm+r}}{qm+r} .$$

It follows that as $t \to 1^-$ we have

$$f(t) = -\log(1-t) - \log(m) + t^{-r}\sum_{0\leqslant k<m}\zeta_m^{-rk}\log(1-t\zeta_m^k) + o(1)$$

$$= -\log(m) + \sum_{1\leqslant k\leqslant m-1}\zeta_m^{-rk}\log(1-\zeta_m^k) + o(1) ,$$

where once again we use Abel's theorem mentioned above, proving the first formula. For the second, we use the following trick: replacing r by $m-r$, adding, and dividing by 2, we have

$$\frac{\psi(r/m) + \psi(1-(r/m))}{2} = -\gamma - \log(m) + \sum_{1\leqslant k\leqslant m-1}\cos(2\pi rk/m)\log(1-\zeta_m^k) .$$

On the other hand, by the reflection formula for the ψ function (Proposition 9.6.41 (5)) we have $\psi(1-(r/m)) = \psi(r/m) + \pi\cotan(\pi r/m)$. Replacing in the above formula and taking the real part gives the desired result, since $\psi(r/m) \in \mathbb{R}$ and $\Re(\log(1-\zeta_m^k)) = \log(2\sin(\pi k/m))$.   □

Using the functional equation $\psi(x+1) = \psi(x)+1/x$, the above proposition gives the value of $\psi(\alpha)$ for any $\alpha \in \mathbb{Q}$.

In a similar manner we can prove the following result:

Proposition 9.6.47. *We have*

$$\sum_{1\leqslant r\leqslant m}\psi\left(\frac{r}{m}\right)e^{2i\pi ar/m} = \begin{cases} m\log\left(\left|2\sin\left(\dfrac{\pi a}{m}\right)\right|\right) +im\pi\left(\left\{\dfrac{a}{m}\right\}-\dfrac{1}{2}\right) & \textit{if } m \nmid a , \\ -m(\log(m)+\gamma) & \textit{if } m \mid a. \end{cases}$$

Proof. Left to the reader (Exercise 103).   □

We also have the following results, which we will not prove:

Theorem 9.6.48. *We have the continued fraction expansion*

$$\psi'(s) = \cfrac{2}{1(2s-1) + \cfrac{1^4}{3(2s-1) + \cfrac{2^4}{5(2s-1) + \cdots}}}$$

which converges when $s \neq 1/2$, and

$$\psi''(s) = -\cfrac{2}{1(2s^2 - 2s + P(1)) - \cfrac{1^6}{3(2s^2 - 2s + P(2)) - \cfrac{2^6}{5(2s^2 - 2s + P(3)) + \cdots}}}$$

where $P(n) = n^2 - n + 1$, which converges for all s.

These results come implicitly from the work of Apéry on the irrationality of $\zeta(2)$ and $\zeta(3)$, and were made explicit by the author, C. Batut, and M. Olivier (see [Bat-Oli] and [Coh4]). It has also been extended to $\zeta(4)$ by G. Rhin and the author (see [Coh-Rhi] and [Coh4]), so that there also exists a similar but more complicated continued fraction for ψ''', which however is only an asymptotic expansion. In a different form, some of these continued fractions can also be found in the work of Stieltjes.

Finally, note also the following continued fractions due to Bender (see [Bor-Bai-Gir], page 324):

Theorem 9.6.49. *We have the continued fraction expansions*

$$2s^2\psi'(s) = 1 + 2s + \cfrac{2}{6s + \cfrac{2^2(2^2-1)}{10s + \cfrac{3^2(3^2-1)}{14s + \cfrac{4^2(4^2-1)}{18s + \cdots}}}}$$

$$= 1 + 2s + \cfrac{1}{2s(1+1/2) + \cfrac{1}{2s(1/2+1/3) + \cfrac{1}{2s(1/3+1/4) + \cdots}}}$$

$$-s^3\psi''(s) = 1 + s + \cfrac{1}{2s + \cfrac{1^3 \cdot 2}{3s + \cfrac{1 \cdot 2^3}{4s + \cfrac{2^3 \cdot 3}{5s + \cfrac{2 \cdot 3^3}{6s + \cdots}}}}}$$

$$= 1 + s + \cfrac{1}{2s/1 + \cfrac{1}{s(1 + 1/2) + \cfrac{1}{2s/2 + \cfrac{1}{s(1/2 + 1/3) + \cfrac{1}{2s/3 + \cdots}}}}}$$

and these continued fractions converge for all $s \neq 0$.

9.6.6 Fourier Expansions of $\zeta(s,x)$ and $\log(\Gamma(x))$

We begin with the following.

Proposition 9.6.50. *Let $n \in \mathbb{Z}$, $x \in \mathbb{R}_{>0}$, $s \in \mathbb{C}$, and set*

$$C_n(x) = \int_x^{x+1} e^{2i\pi nt} \zeta(s,t)\, dt \ .$$

Then $C_0(x) = x^{1-s}/(s-1)$, and when $n \neq 0$ we have

$$C_n(x) = \begin{cases} -\int_0^x e^{2i\pi nt} t^{-s}\, dt + \dfrac{\Gamma(1-s)}{(2\pi|n|)^{1-s}} e^{i(1-s)\pi/2\,\mathrm{sign}(n)} & \text{for } \Re(s) < 1, \\ \int_x^\infty e^{2i\pi nt} t^{-s}\, dt & \text{for } \Re(s) > 0. \end{cases}$$

Proof. Assume first that $0 < \Re(s) < 1$ and set $F_n(x) = \int_0^x e^{2i\pi nt} \zeta(s,t)\, dt$, which converges since $\Re(s) < 1$. We have $F_n'(x) = e^{2i\pi nx} \zeta(s,x)$ and $C_n(x) = F_n(x+1) - F_n(x)$, so that

$$C_n'(x) = e^{2i\pi n(x+1)} \zeta(s, x+1) - e^{2i\pi nx} \zeta(s,x) = -e^{2i\pi nx} x^{-s} \ ,$$

and hence

$$C_n(x) = -\int_0^x e^{2i\pi nt} t^{-s}\, dt + C_n$$

for some constant $C_n = C_n(0)$ to be determined. Setting $t = x + u$ we have

$$C_n(x) = e^{2i\pi nx} \int_0^1 e^{2i\pi nu} \zeta(s, x+u)\, du \ .$$

By Taylor's formula to order 2 and the formula $(d/dx)\zeta(s,x) = -s\zeta(s+1,x)$ there exists $\theta = \theta(u,x) \in]0,1[$ such that

$$\zeta(s, x+u)) = \zeta(s,x) - us\zeta(s+1,x) + \frac{u^2}{2}s(s+1)\zeta(s+2, x+\theta) .$$

By Corollary 9.6.9, since $\Re(s) > 0$ we have $\zeta(s+2, x+\theta) = O(x^{-1-s})$ uniformly in u, $\zeta(s,x) = -x^{1-s}/(1-s)+x^{-s}/2+O(x^{-1-s})$, and $\zeta(s+1,x) = x^{-s}/s + O(x^{-1-s})$. Thus

$$C_n(x) = e^{2i\pi n x}\left(\left(-\frac{x^{1-s}}{1-s} + \frac{x^{-s}}{2}\right)\int_0^1 e^{2i\pi n u}\, du - x^{-s}\int_0^1 ue^{2i\pi n u}\, du\right)$$
$$+ O(x^{-1-s}) .$$

For $n = 0$ this gives $C_n(x) = -x^{1-s}/(1-s) + O(x^{-1-s})$, and since $C_0(x) = -\int_0^x t^{-s}\, dt + C_n = -x^{1-s}/(1-s) + C_0$ we deduce that $C_0 = 0$. Assume now that $n \neq 0$. Since $\int_0^1 ue^{2i\pi n u}\, du = 1/(2i\pi n)$, the above estimate gives $C_n(x) = -x^{-s}/(2i\pi n) + O(x^{-1-s}) = o(1)$ since $\Re(s) > 0$. On the other hand, $I_n = \int_0^\infty e^{2i\pi n t}t^{-s}\, dt$ converges, so $C_n(x) = -I_n + C_n + o(1)$. Comparing the two expressions and using Corollary 9.6.36 we deduce that

$$C_n = I_n = \int_0^\infty e^{2i\pi n t}t^{-s}\, dt = \frac{\Gamma(1-s)}{(2\pi|n|)^{1-s}}e^{i(1-s)\pi/2\,\mathrm{sign}(n)},$$

proving the proposition for $0 < \Re(s) < 1$. For general s such that $\Re(s) < 1$ we note that all the integrals converge absolutely, so both sides of the formula define analytic functions of s, so the formula is still valid by analytic continuation. For $\Re(s) > 1$ and for $n \neq 0$ we again have $C_n'(x) = -e^{2i\pi n x}x^{-s}$, and since $\zeta(s,x) = O(x^{1-s})$ tends to zero as $x \to \infty$ we deduce that $C_n(x) = \int_x^\infty e^{2i\pi n t}t^{-s}\, dt$. Once again we conclude by analytic continuation that the formula is valid for $\Re(s) > 0$. $\qquad\square$

Corollary 9.6.51. *For $x \in \mathbb{R} \setminus \mathbb{Z}$ and $\Re(s) < 1$ we have*

$$\zeta(s, \{x\}) = 2(2\pi)^{s-1}\Gamma(1-s)\sum_{n \geqslant 1}\frac{\sin(2\pi n x + s\pi/2)}{n^{1-s}} .$$

Proof. The variable s being fixed, the periodic function $\zeta(s, \{x\})$ is a piecewise C^∞ function with simple discontinuities at the integers, so the corollary follows by a simple computation from the proposition and the fundamental theorem on Fourier series, which implies that outside of the discontinuities we have $\zeta(s, \{x\}) = \sum_{n \in \mathbb{Z}} C_n(0)e^{-2i\pi n x}$. $\qquad\square$

This result has many important consequences, in particular the functional equation for the Riemann zeta function and for Dirichlet L-functions, which we will study in the next chapter. For now we give the following.

Corollary 9.6.52. *For $x \in \mathbb{R} \setminus \mathbb{Z}$ and $\Re(s) > 0$ we have*

$$\sum_{n \geqslant 1} \frac{\cos(2\pi n x)}{n^s} = \frac{(2\pi)^s}{4\Gamma(s)\cos(s\pi/2)}\left(\zeta(1-s,\{x\}) + \zeta(1-s,\{1-x\})\right) ,$$

$$\sum_{n \geqslant 1} \frac{\sin(2\pi n x)}{n^s} = \frac{(2\pi)^s}{4\Gamma(s)\sin(s\pi/2)}\left(\zeta(1-s,\{x\}) - \zeta(1-s,\{1-x\})\right) .$$

Proof. Immediate and left to the reader. $\qquad\square$

Corollary 9.6.53. *For all $x \in \mathbb{R} \setminus \mathbb{Z}$ the Fourier expansion of $\log(\Gamma(\{x\}))$ is given by*

$$\log(\Gamma(\{x\})) = \frac{1}{2}\log(2\pi) + \frac{1}{2}\sum_{n \geqslant 1}\frac{\cos(2\pi n x)}{n} + \frac{1}{\pi}\sum_{n \geqslant 1}(\log(2\pi n) + \gamma)\frac{\sin(2\pi n x)}{n}$$

$$= \frac{1}{2}\log\left(\frac{\pi}{|\sin(\pi x)|}\right) - (\log(2\pi) + \gamma)\left(\{x\} - \frac{1}{2}\right)$$

$$+ \frac{1}{\pi}\sum_{n \geqslant 1}\frac{\log(n)}{n}\sin(2\pi n x) .$$

Proof. Using for instance Abel summation, we note that for fixed $x \notin \mathbb{Z}$, the series $\sum_{n \geqslant 1} \log(n)e^{2i\pi n x}n^{-s}$ is uniformly convergent in any compact subset of the right half-plane $\Re(s) > 0$. It follows that we can differentiate termwise the series for $\zeta(s, \{x\})$ for $\Re(s) < 1$, so that

$$\zeta'(s, \{x\}) = 2(2\pi)^{s-1}\Gamma(1-s)\left((\log(2\pi) - \psi(1-s))\sum_{n \geqslant 1}\frac{\sin(2\pi n x + s\pi/2)}{n^{1-s}}\right.$$

$$\left. + \sum_{n \geqslant 1}\frac{\log(n)\sin(2\pi n x + s\pi/2) + (\pi/2)\cos(2\pi n x + s\pi/2)}{n^{1-s}}\right) .$$

Setting $s = 0$ and using

$$\log(\Gamma(\{x\})) = \zeta'(0, \{x\}) - \zeta'(0, 1) = \zeta'(0, \{x\}) + \log(2\pi)/2$$

we obtain the corollary. An equivalent proof is to compute directly the Fourier coefficients of $\log(\Gamma(\{x\}))$ by differentiating with respect to s the formulas of Proposition 9.6.50. We obtain the following result, whose proof is left to the reader (Exercise 98).

Corollary 9.6.54. *For $n \in \mathbb{Z}$ and $x > 0$ set*

$$C_n(x) = \int_x^{x+1} e^{2i\pi n t}\log(\Gamma(t))\,dt .$$

Then

$$C_n(x) = \begin{cases} x\log(x) - x + \dfrac{1}{2}\log(2\pi) & \text{for } n = 0\,, \\ \int_0^x e^{2i\pi nt}\log(t)\,dt + \dfrac{1}{4|n|} + \dfrac{i}{2\pi n}(\gamma + \log(2\pi|n|)) & \text{for } n \neq 0\,. \end{cases}$$

The case $n = 0$ of the above formula is called Raabe's formula.

Example. Setting $x = 1/4$ in the second formula of Corollary 9.6.53 we obtain

$$\sum_{k\geqslant 0}(-1)^k \frac{\log(2k+1)}{2k+1} = \pi\left(\log\left(\Gamma\left(\frac{1}{4}\right)\right) - \left(\frac{3}{4}\log(\pi) + \frac{\log(2)}{2} + \frac{\gamma}{4}\right)\right)\,.$$

We will see in the next chapter (Proposition 10.3.5) that this is a special case of a more general result giving $L'(\chi, 1)$ for an odd primitive character χ.

9.7 Integral Transforms

Before studying integral transforms, we recall three theorems of undergraduate real analysis, which although very classical, are not always sufficiently well known (see for instance [Rud]):

Theorem 9.7.1 (Monotone convergence theorem). *Let $X \subset \mathbb{R}$, and let f_n be a sequence of measurable functions on X such that $f(x) = \lim_{n\to\infty} f_n(x)$ exists for every $x \in X$. If for all $x \in X$ we have $0 \leqslant f_0(x) \leqslant f_1(x) \leqslant \cdots$ then $f(x)$ is measurable, and*

$$\lim_{n\to\infty}\int_X f_n(X)\,dx = \int_X f(x)\,dx\,.$$

Theorem 9.7.2 (Dominated convergence theorem). *Let $X \subset \mathbb{R}$, and let f_n be a sequence of measurable functions on X such that $f(x) = \lim_{n\to\infty} f_n(x)$ exists for every $x \in X$. If there exists a function $g \in L^1(X)$ such that $|f_n(x)| \leqslant g(x)$ for all $x \in X$, then $f \in L^1(X)$, and*

$$\lim_{n\to\infty}\int_X |f_n(x) - f(x)|\,dx = 0 \quad and \quad \lim_{n\to\infty}\int_X f_n(x)\,dx = \int_X f(x)\,dx\,.$$

Theorem 9.7.3 (Riemann–Lebesgue lemma). *Let $f \in L^1(\mathbb{R})$ be a periodic function of period 1, and let*

$$c_n = \int_0^1 f(t)e^{-2i\pi nt}\,dt$$

be the nth Fourier coefficient of f. Then as $n \to \pm\infty$ we have $c_n \to 0$.

9.7.1 Generalities on Integral Transforms

Let C be a contour in the complex plane, and in particular a real interval. An *integral transform* is a map that sends a function f belonging to some reasonable class to another function F defined by

$$F(x) = \int_C K(x,t)f(t)\,dt\ ,$$

where $K(x,t)$, called the *kernel function*, is also a reasonable function of two variables (which, however, may have mild singularities, for instance for $x = t$).

There are many reasons why integral transforms are important. For instance, they can transform some properties of the function f into some quite different property of the function F. Furthermore, linear operations on f can be transformed into the same operations on K, thus giving explicit formulas. Finally, useful integral transforms can be *inverted*; in other words, we can recover the function f from the function F through another integral transform, evidently called the inverse transform.

The three simplest and most important transforms used in number theory and elsewhere are the *Fourier transform*, the *Laplace transform*, and the *Mellin transform*. We will study each one, but we will see that they are closely related.

Intimately linked to integral transforms are *convolutions*. If F and G are the integral transforms of f and g respectively, the convolution of f and g is the function whose integral transform is FG. It can usually be expressed as an integral involving f and g, but not involving explicitly the kernel K.

9.7.2 The Fourier Transform

This is probably the most important, and the oldest of all integral transforms, and should be part of every undergraduate curriculum. We give here the main results that we need.

Definition 9.7.4. *The Fourier transform of a function f is defined by*

$$\mathcal{F}(f)(x) = \int_{-\infty}^{\infty} e^{-2i\pi xt} f(t)\,dt\ .$$

Although the "correct" context in which to study the Fourier transform is the space $L^2(\mathbb{R})$, we will usually assume that our functions are nicer than simply L^2. Since our goal is concreteness and not abstraction we focus on the formulas and not on the minimal assumptions. The following theorem summarizes all that we need to know.

Theorem 9.7.5. (1) (*Inversion formula.*) *Assume that both f and $\mathcal{F}(f)$ are in $L^1(\mathbb{R})$. Then*

$$f(x) = \int_{-\infty}^{\infty} e^{2i\pi xt} \mathcal{F}(f)(t) \, dt$$

for all x, where f is continuous. In other words, we have the formula
$\mathcal{F}(\mathcal{F}(f))(x) = f(-x)$.

(2) *(Convolution formula.)* If we set

$$(f * g)(x) = \int_{-\infty}^{\infty} f(t)g(x - t) \, dt \,,$$

then $\mathcal{F}(f * g) = \mathcal{F}(f)\mathcal{F}(g)$ and $\mathcal{F}(fg) = \mathcal{F}(f) * \mathcal{F}(g)$.

(3) If $f \in C^1(\mathbb{R})$ and $xf(x)$ tends to 0 as $|x| \to \infty$ we have $\mathcal{F}(f')(x) = 2i\pi x\mathcal{F}(f)(x)$ and $\mathcal{F}(f)'(x) = -2i\pi\mathcal{F}(tf(t))(x)$ (hence $\mathcal{F}(tf(t))(x) = -(1/(2i\pi))\mathcal{F}(f)'(x)$).

(4) Assume that $f \in C^k(\mathbb{R})$, that $f^{(j)}(x)$ tends to 0 as $|x| \to \infty$ for $0 \leqslant j \leqslant k - 1$, and that $f^{(j)}(x) \in L^1(\mathbb{R})$ for $0 \leqslant j \leqslant k$. Then $x^k\mathcal{F}(f)(x)$ tends to 0 as $x \to \infty$; in other words, $\mathcal{F}(f)(x) = o(x^{-k})$.

(5) Conversely, if $f(x) = o(x^{-k})$ then $\mathcal{F}(f) \in C^k(\mathbb{R})$.

(6) The functions $f(x) = e^{-\pi x^2}$ and $1/\cosh(\pi x)$ are invariant under Fourier transform.

Proof. Set for $T > 0$,

$$f_T(x) = \int_{-T}^{T} \left(1 - \frac{|t|}{T}\right) e^{2i\pi xt} \mathcal{F}(f)(t) \, dt \,.$$

Replacing $\mathcal{F}(f)(t)$ by its expression we find after an easy computation that

$$f_T(x) = \int_{-T}^{T} \left(1 - \frac{|t|}{T}\right) e^{2i\pi xt} \int_{-\infty}^{\infty} e^{-2i\pi tu} f(u) \, du$$

$$= \int_{-\infty}^{\infty} f(u) \left(\int_{-T}^{T} \left(1 - \frac{|t|}{T}\right) e^{2i\pi t(x-u)} \, dt\right) du$$

$$= \int_{-\infty}^{\infty} f(u) \frac{\sin^2(\pi T(u - x))}{\pi^2 T(u - x)^2} \, du$$

$$= \frac{1}{\pi} \int_{-\infty}^{\infty} \frac{\sin^2(t)}{t^2} f\left(x + \frac{t}{\pi T}\right) dt \,.$$

Using the well-known formula $\int_{-\infty}^{\infty} \sin^2(t)/t^2 \, dt = \pi$, which can be proved in a number of ways (see Exercise 97), we thus have

$$f_T(x) - f(x) = \frac{1}{\pi} \int_{-\infty}^{\infty} \frac{\sin^2(t)}{t^2} \left(f\left(x + \frac{t}{\pi T}\right) - f(x)\right) dt \,.$$

Let $\varepsilon > 0$ be given, and set $X = T^{1/2}$. Since f is continuous at x there exists $\eta > 0$ such that $|h| \leqslant \eta$ implies $|f(x + h) - f(x)| \leqslant \varepsilon$. Since $f \in L^1(\mathbb{R})$, for $X/(\pi T) \leqslant \eta$, in other words for $T \geqslant (\pi\eta)^{-2}$, we have

$$\pi|f_T(x) - f(x)| \leqslant \int_{|t|>X} \frac{\sin^2(t)}{t^2}|f(t)|\,dt + 2\varepsilon \int_0^X \frac{\sin^2(t)}{t^2}\,dt$$

$$\leqslant \int_{|t|>T^{1/2}} |f(t)|\,dt + \varepsilon\pi\ .$$

Since $f \in L^1(\mathbb{R})$, as $T \to \infty$ this upper bound tends to $\varepsilon\pi$, and since ε is arbitrary we thus have $\lim_{T\to\infty} f_T(x) = f(x)$. On the other hand, since $\mathcal{F}(f) \in L^1(\mathbb{R})$, by the dominated convergence theorem we have

$$\lim_{T\to\infty} f_T(x) = \int_{-\infty}^\infty e^{2i\pi xt}\mathcal{F}(f)(t)\,dt\ ,$$

proving (1).

(2). We have

$$\mathcal{F}(f * g)(x) = \int_{-\infty}^\infty e^{-2i\pi xt}\left(\int_{-\infty}^\infty f(u)g(t-u)\,du\right) dt$$

$$= \int_{-\infty}^\infty e^{-2i\pi xu} f(u)\left(\int_{-\infty}^\infty e^{-2i\pi x(t-u)}g(t-u)\,dt\right) du$$

$$= \mathcal{F}(f)(x)\mathcal{F}(g)(x)\ ,$$

proving the first formula of (2), and the second follows from the inversion formula (1).

(3). Follows immediately by differentiation under the integral sign and by integration by parts.

(4) and (5). By (3) we have $\mathcal{F}(f)(x) = \mathcal{F}(f^{(k)})(x)/(2i\pi x)^k$. We are thus reduced to proving (4) for $k = 0$, in other words that if $f \in L^1(\mathbb{R})$ is continuous, then $\mathcal{F}(f)(x)$ tends to 0 as $x \to \infty$. But this is exactly the statement of the Riemann–Lebesgue lemma. The converse follows immediately from this and the inversion formula, but can be proved directly if desired.

(6). The Fourier-invariance of $e^{-\pi x^2}$ is very classical and fundamental for the functional equation of theta and L-functions that we will study in the next chapter, and it has been proved in Proposition 9.6.21. The Fourier-invariance of $1/\cosh(\pi x)$ is less well known, although it is essentially equivalent to the functional equation of the *square* of the usual theta function. See Exercise 101 for the proof.   □

Remarks. (1) The fact that, up to a multiplicative constant, the Fourier transform converts derivatives into multiplication by x, hence more generally higher derivatives into multiplication by powers of x, is very useful in the fields of differential equations and partial differential equations, since it can transform them into polynomial equations. The same is true for the Laplace transform (see below), which is closely related.

(2) Because of (2), in other words of the formula $\mathcal{F}(\mathcal{F}(f))(x) = f(-x)$, it is reasonable to consider the Fourier transform as a *dualizing* operator (it is indeed self-dual on even functions). Thus, to emphasize this aspect we can also write $\widehat{f}(x)$ instead of $\mathcal{F}(f)(x)$, and this is the notation that we have used in Proposition 2.2.16. A deeper reason for this dualizing property is given by the Weil representation; see Exercise 108.

9.7.3 The Mellin Transform

Definition 9.7.6. *The Mellin transform of f is defined by the formula*

$$\mathcal{M}(f)(s) = \int_0^\infty t^{s-1} f(t)\, dt\,.$$

As a basic and crucial example for number theory, the Mellin transform of $f(x) = e^{-x}$ is the gamma function. There of course exist many other important examples useful for number theory, which we will see mainly in the next chapter. The Mellin transform is in fact a version of the Fourier transform, as the following proposition shows. However, its applications are slightly different.

Proposition 9.7.7. *Assume that f is continuous on $]0, \infty[$, that $f(t) = O(t^{-\alpha})$ for some $\alpha \in \mathbb{R}$ as $t \to 0$, and that $f(t)$ tends to 0 faster than any power of t as $t \to \infty$.*

(1) *The Mellin transform of f converges absolutely for $\Re(s) > \alpha$ and defines a holomorphic function in that right half-plane.*

(2) *If we let $s = \sigma + iT$ with $\sigma > \alpha$ and set*

$$g_\sigma(t) = e^{-2\pi\sigma t} f(e^{-2\pi t})$$

then

$$\mathcal{M}(f)(\sigma + iT) = 2\pi \mathcal{F}(g_\sigma)(T)\,.$$

(3) *We have the Mellin inversion formula, valid for all $\sigma > \alpha$: for all $x > 0$,*

$$f(x) = \frac{1}{2i\pi} \int_{\sigma-i\infty}^{\sigma+i\infty} x^{-s} \mathcal{M}(f)(s)\, ds\,.$$

Proof. (1). By our assumptions on f the integral converges absolutely in a neighborhood of infinity, and in any compact interval not containing 0. Since $|t^{s-1} f(t)| = O(t^{\Re(s)-\alpha-1})$, the integral converges absolutely also at 0 when $\Re(s) > \alpha$. Furthermore, since $|t^{s-1} \log(t) f(t)| = O(t^{\Re(s)-\alpha-1-\varepsilon})$ for all $\varepsilon > 0$, the integral of the derivative with respect to s also converges absolutely and normally on compact intervals for $\Re(s) > \alpha$, so by the theorem on differentiation of improper integrals we deduce that $\mathcal{M}(f)(s)$ is complex-differentiable for $\Re(s) > \alpha$, hence is holomorphic.

(2). Making the change of variable $t = e^{-2\pi u}$ in the defining integral we have

$$\mathcal{M}(f)(s) = 2\pi \int_{-\infty}^{\infty} e^{-2\pi s u} f(e^{-2\pi u}) \, du$$

$$= 2\pi \int_{-\infty}^{\infty} e^{-2i\pi T u} g_\sigma(u) \, du = 2\pi \mathcal{F}(g_\sigma)(T) \, ,$$

proving (1). For (2) the Fourier inversion formula tells us that for all $x \in \mathbb{R}$,

$$g_\sigma(x) = \int_{-\infty}^{\infty} e^{2i\pi x T} \mathcal{F}(g_\sigma)(T) \, dT = \frac{1}{2\pi} \int_{-\infty}^{\infty} e^{2i\pi x T} \mathcal{M}(f)(\sigma + iT) \, dT$$

$$= \frac{1}{2i\pi} \int_{\sigma - i\infty}^{\sigma + i\infty} e^{2\pi x (s - \sigma)} \mathcal{M}(f)(s) \, ds \, .$$

Thus

$$f(e^{-2\pi x}) = e^{2\pi \sigma x} g_\sigma(x) = \frac{1}{2i\pi} \int_{\sigma - i\infty}^{\sigma + i\infty} e^{2\pi x s} \mathcal{M}(f)(s) \, ds \, ,$$

and the Mellin inversion formula follows by setting $X = e^{-2\pi x} > 0$.   □

The Mellin transform evidently also has a convolution formula, which is immediately deduced from that for the Fourier transform, but we will not need it.

9.7.4 The Laplace Transform

Definition 9.7.8. *The Laplace transform of f is defined by the formula*

$$\mathcal{L}(f)(x) = \int_0^\infty e^{-tx} f(t) \, dt \, .$$

Proposition 9.7.9. *Assume that f is piecewise continuous on $]0, \infty[$, that $f(t) = O(t^{-\alpha})$ for some $\alpha < 1$ as $t \to 0$, and that $e^{-at} f(t)$ tends to 0 for all $a > 0$ as $t \to \infty$.*

(1) *The Laplace transform of f converges absolutely for $\Re(x) > 0$ and defines a holomorphic function in that right half-plane.*

(2) *We have $\mathcal{L}(f)'(x) = -\mathcal{L}(tf)(x)$ and if $f \in C^1[0, \infty[$ we have*

$$\mathcal{L}(f')(x) = x\mathcal{L}(f)(x) - f(0) \, .$$

(3) *If we let $g(t) = f(-\log(t))$ for $t \in [0, 1]$ and $g(t) = 0$ for $t > 1$ then*

$$\mathcal{L}(f)(x) = \mathcal{M}(g)(x) \, .$$

(4) *We have the Laplace inversion formula: for all $\sigma > 0$ and all $x > 0$,*

$$f(x) = \frac{1}{2i\pi} \int_{\sigma - i\infty}^{\sigma + i\infty} e^{xs} \mathcal{L}(f)(s)\, ds \ .$$

In addition, for $x < 0$ we have

$$\frac{1}{2i\pi} \int_{\sigma - i\infty}^{\sigma + i\infty} e^{xs} \mathcal{L}(f)(s)\, ds = 0 \ .$$

(5) *If we set $f * g(x) = \int_0^x f(t)g(x - t)\, dt$ we have the convolution formula*

$$\mathcal{L}(f * g)(x) = \mathcal{L}(f)(x)\mathcal{L}(g)(x) \ .$$

Proof. Thanks to the assumptions made on f, (1) and the first formula of (2) are clear, and the second is immediate by integration by parts. For (3) we make the change of variable $u = e^{-t}$, which gives for $x > 0$,

$$\mathcal{L}(f)(x) = \int_0^1 u^{x-1} f(-\log(u))\, du = \int_0^\infty u^{x-1} g(u)\, du = \mathcal{M}(g)(x) \ ,$$

proving (3). Thus by the Mellin inversion formula we have for all $\sigma > 0$ and $x > 0$,

$$g(x) = \frac{1}{2i\pi} \int_{\sigma - i\infty}^{\sigma + i\infty} x^{-s} \mathcal{L}(f)(s)\, ds \ .$$

When $x < 1$ we replace x by e^{-t} for $t > 0$, giving the inversion formula, and when $x > 1$ we obtain the other result since $g(x) = 0$. Note that this latter result can be proved directly by showing that it is legitimate to shift the path of integration infinitely to the right. Finally, (4) can be proved from the convolution formula for the Fourier transform, but better directly as we did for the Fourier transform.   □

Note that we have already met the Laplace transform in the context of the Euler–MacLaurin summation formula (Section 9.2.4). However, one of its main uses is in the theory of ordinary differential equations, because of property (2), which essentially says that the operator $\mathcal{L}$ transforms differentiation with respect to x into multiplication by x. This is of course nothing else than the corresponding property of the Fourier transform, which is also used in the context of differential equations.

9.8 Bessel Functions

9.8.1 Definitions

We refer to [Wats] and [Abr-Ste] for more details on Bessel functions. Although we will not need all the standard Bessel functions, it is convenient and not longer to define them all.

Proposition 9.8.1. *For $\nu \in \mathbb{C}$, let E_ν be the differential equation*

$$y'' + \frac{y'}{x} + \left(1 - \frac{\nu^2}{x^2}\right) y = 0 .$$

(1) *When $\nu \notin \mathbb{Z}$, a basis of the space of solutions of E_ν is given by the two linearly independent solutions $J_{\pm\nu}$ such that*

$$J_{\pm\nu}(x) = \frac{(x/2)^{\pm\nu}}{\Gamma(\pm\nu + 1)} S_{\pm\nu}(x) ,$$

where $S_{\pm\nu}(x)$ is a power series such that $S_{\pm\nu}(0) = 1$. Explicitly, we have the series expansion with infinite radius of convergence

$$J_{\pm\nu}(x) = (x/2)^{\pm\nu} \sum_{k \geqslant 0} \frac{(-1)^k (x/2)^{2k}}{k! \Gamma(\pm\nu + 1 + k)} .$$

(2) *For $\nu \notin \mathbb{Z}$ set*

$$Y_\nu(x) = \frac{J_\nu(x) \cos(\pi\nu) - J_{-\nu}(x)}{\sin(\pi\nu)} ,$$

and for $n \in \mathbb{Z}$ set

$$Y_n(x) = \lim_{\nu \to n, \, \nu \notin \mathbb{Z}} Y_\nu(x) .$$

For any ν a basis of the space of solutions of E_ν is given by the functions J_ν and Y_ν.

(3) *For $n = 0$ the function $Y_0(x) - (2/\pi)(\log(x/2) + \gamma) J_0(x)$ has a power series expansion around $x = 0$ with no constant term, where as usual γ is Euler's constant. More precisely, we have*

$$Y_0(x) = -\frac{2}{\pi} \sum_{k \geqslant 0} \frac{(-1)^k (x/2)^{2k}}{k!^2} (H_k - \gamma - \log(x/2)) ,$$

where $H_k = \sum_{1 \leqslant j \leqslant k} 1/j$ is the harmonic sum.

   Proof. This is a classical undergraduate exercise, so we only give a sketch (see Exercise 110). If $\nu \notin \mathbb{Z}$, we can set $y = x^{\pm\nu} \sum_{k \geqslant 0} a_k x^k$ with $a_0 \neq 0$. The differential equation gives $a_1 = 0$ and a simple recurrence for a_{k+2} in terms of a_k, which shows the existence of the power series $S_{\pm\nu}(x)$, the fact that its radius of convergence is infinite, and the explicit formula for $J_\nu(x)$, proving (1). For (2) the above procedure works for n, but not for $-n$, which gives the zero solution. On the other hand, the given expression $(J_\nu(x) \cos(\pi\nu) - J_{-\nu}(x))/\sin(\pi\nu)$ is evidently a solution of E_ν for $\nu \notin \mathbb{Z}$, and it is easily checked on the explicit expansions both that it has a limit when $\nu \to n$ and that this limit is indeed a solution of E_n, clearly independent of J_n since it is also easily seen that it has a logarithmic singularity at $x = 0$

(the fact that the limit exists will also follow from the integral representation that we give below). The same explicit expansions also give (3).   □

Similarly we have the following:

Proposition 9.8.2. *For $\nu \in \mathbb{C}$, let F_ν be the differential equation*

$$y'' + \frac{y'}{x} - \left(1 + \frac{\nu^2}{x^2}\right)y = 0 .$$

(1) *When $\nu \notin \mathbb{Z}$, a basis of the space of solutions of F_ν is given by the two linearly independent solutions $I_{\pm\nu}$ such that*

$$I_{\pm\nu}(x) = \frac{(x/2)^{\pm\nu}}{\Gamma(\pm\nu + 1)} T_{\pm\nu}(x) ,$$

where $T_{\pm\nu}(x)$ is a power series such that $T_{\pm\nu}(0) = 1$. Explicitly, we have the series expansion with infinite radius of convergence

$$I_{\pm\nu}(x) = (x/2)^{\pm\nu} \sum_{k\geqslant 0} \frac{(x/2)^{2k}}{k!\,\Gamma(\pm\nu + 1 + k)} .$$

(2) *For $\nu \notin \mathbb{Z}$ set*

$$K_\nu(x) = \frac{\pi}{2} \frac{I_{-\nu}(x) - I_\nu(x)}{\sin(\nu\pi)} ,$$

and for $n \in \mathbb{Z}$ set

$$K_n(x) = \lim_{\nu \to n,\ \nu\notin\mathbb{Z}} K_\nu(x) .$$

For any ν a basis of the space of solutions of F_ν is given by the functions I_ν and K_ν.

(3) *For $n = 0$ the function $K_0(x) + (\log(x/2) + \gamma)I_0(x)$ has a power series expansion around $x = 0$ with no constant term. More precisely, we have*

$$K_0(x) = \sum_{k\geqslant 0} \frac{(x/2)^{2k}}{k!^2} \left(H_k - \gamma - \log(x/2)\right) .$$

Proof. Exactly the same proof as the preceding proposition. Note that $T_{\pm\nu}(x) = S_{\pm\nu}(ix)$.   □

Definition 9.8.3. *The functions $J_\nu(x)$ and $I_\nu(x)$ are called the Bessel functions of the first and second kind respectively, and the functions $Y_\nu(x)$ and $K_\nu(x)$ the modified Bessel functions of the first and second kind.*

Remarks. (1) As will become clear from the asymptotic expansions given below, the reader should think of the functions $J(x)$ and $Y(x)$ as the functions $\cos(x)$ and $\sin(x)$ respectively, and of the functions $I(x)$ and

$K(x)$ as the functions e^x and e^{-x}. In particular, the functions J, Y, and K often occur in expansions, but almost never the function I since it is exponentially large.

(2) The normalization of the functions J and I is natural. That of the functions K is canonical up to multiplication by a constant, since it is the only solution to the differential equation that tends exponentially fast to zero at infinity. On the other hand, the normalization of the function Y (which is sometimes denoted by N) is less natural, but we have chosen the one occurring in the literature.

x

Proposition 9.8.4. *We have*

$$J_{\nu-1}(x) + J_{\nu+1}(x) = \frac{2\nu}{x} J_\nu(x) , \quad J_{\nu-1}(x) - J_{\nu+1}(x) = 2J_\nu'(x) ,$$

$$Y_{\nu-1}(x) + Y_{\nu+1}(x) = \frac{2\nu}{x} Y_\nu(x) , \quad Y_{\nu-1}(x) - Y_{\nu+1}(x) = 2Y_\nu'(x) ,$$

$$I_{\nu-1}(x) - I_{\nu+1}(x) = \frac{2\nu}{x} I_\nu(x) , \quad I_{\nu-1}(x) + I_{\nu+1}(x) = 2I_\nu'(x) ,$$

$$K_{\nu+1}(x) - K_{\nu-1}(x) = \frac{2\nu}{x} K_\nu(x) , \quad K_{\nu-1}(x) + K_{\nu+1}(x) = -2K_\nu'(x) .$$

Proof. Immediate from the series expansions and the definitions of Y and K, using $\Gamma(\nu + k + 1) = (\nu + k)\Gamma(\nu + k)$, and left to the reader (Exercise 116). □

Proposition 9.8.5. *When $\nu \in (1/2) + \mathbb{Z}$ the four Bessel functions are elementary functions. More precisely:*

(1) *We have*

$$J_{1/2}(x) = Y_{-1/2}(x) = \sqrt{\frac{2}{\pi x}} \sin(x), \quad J_{-1/2}(x) = -Y_{1/2}(x) = \sqrt{\frac{2}{\pi x}} \cos(x) ,$$

$$I_{1/2}(x) = \sqrt{\frac{2}{\pi x}} \sinh(x), \quad I_{-1/2}(x) = \sqrt{\frac{2}{\pi x}} \cosh(x) ,$$

$$K_{1/2}(x) = K_{-1/2}(x) = \sqrt{\frac{\pi}{2x}} e^{-x} .$$

(2) *More generally, there exist polynomials $P_n(X)$ and $Q_n(X)$ satisfying $\deg(P_n) = \deg(Q_n) = n$, $P_n(-X) = (-1)^n P_n(X)$, $Q_n(-X) = (-1)^n Q_n(X)$, and such that for $k \in \mathbb{Z}_{\geq 0}$ we have*

$$J_{k+1/2}(x) = \sqrt{\frac{2}{\pi x}} \left(P_k(1/x)\sin(x) - Q_{k-1}(1/x)\cos(x) \right),$$

$$J_{-k-1/2}(x) = (-1)^k \sqrt{\frac{2}{\pi x}} \left(P_k(1/x)\cos(x) + Q_{k-1}(1/x)\sin(x) \right),$$

$$Y_{k+1/2}(x) = (-1)^{k-1} J_{-k-1/2}(x), \quad Y_{-k-1/2} = (-1)^k J_{k+1/2}(x),$$

$$I_{k+1/2}(x) = \sqrt{\frac{2}{\pi x}} \left(i^k P_k(i/x)\sinh(x) + i^{k-1} Q_{k-1}(i/x)\cosh(x) \right),$$

$$I_{-k-1/2}(x) = \sqrt{\frac{2}{\pi x}} \left(i^k P_k(i/x)\cosh(x) + i^{k-1} Q_{k-1}(i/x)\sinh(x) \right),$$

$$K_{k+1/2}(x) = K_{-k-1/2}(x) = \sqrt{\frac{\pi}{2x}} \left(i^k P_k(1/(ix)) + i^{k-1} Q_{k-1}(1/(ix)) \right) e^{-x}.$$

Proof. The formulas for $J_{\pm 1/2}(x)$ and $I_{\pm 1/2}(x)$ follow immediately from the power series expansion, using the formula

$$\Gamma(k + 3/2) = (2k + 1)! \sqrt{\pi}/(k! 2^{2k+1}),$$

which is an immediate consequence of the duplication formula of the gamma function. The formulas for Y and K then follow from the definition. Finally, the assertions of (2) follow from (1) and the recurrences of Proposition 9.8.4. The details are left to the reader (Exercise 117).   □

9.8.2 Integral Representations and Applications

Apart from the power series expansions around $x = 0$, which are readily found, the only results that we need are given in the following propositions.

Proposition 9.8.6. *We have the integral representations*

$$J_\nu(x) = \frac{1}{\pi} \int_0^\pi \cos(x\sin(t) - \nu t)\, dt - \frac{\sin(\pi\nu)}{\pi} \int_0^\infty e^{-x\sinh(t) - \nu t}\, dt,$$

$$Y_\nu(x) = \frac{1}{\pi} \int_0^\pi \sin(x\sin(t) - \nu t)\, dt - \frac{1}{\pi} \int_0^\infty e^{-x\sinh(t)} \left(e^{\nu t} + \cos(\pi\nu)e^{-\nu t} \right) dt,$$

$$I_\nu(x) = \frac{1}{\pi} \int_0^\pi e^{x\cos(t)} \cos(\nu t)\, dt - \frac{\sin(\nu\pi)}{\pi} \int_0^\infty e^{-x\cosh(t) - \nu t}\, dt,$$

$$K_\nu(x) = \int_0^\infty e^{-x\cosh(t)} \cosh(\nu t)\, dt.$$

Proof. We first prove the formula for $J_\nu(x)$. By Proposition 9.8.1 we have

$$J_\nu(x) = \sum_{k \geqslant 0} \frac{(-1)^k (x/2)^{\nu+2k}}{k! \Gamma(\nu + 1 + k)}.$$

On the other hand, by Exercise 99, for all $z \in \mathbb{C}$ and all $\varepsilon > 0$ we have

$$\frac{1}{\Gamma(z)} = \frac{1}{2i\pi} \int_C t^{-z} e^t \, dt \,,$$

where C is any contour coming from $-\infty$, turning in the positive direction around 0, and going back to $-\infty$. Since the radius of convergence of the series is infinite, we deduce that

$$J_\nu(x) = \frac{(x/2)^\nu}{2i\pi} \int_C \sum_{k \geq 0} \frac{(-1)^k (x/2)^{2k} t^{-\nu-k-1}}{k!} e^t \, dt$$

$$= \frac{(x/2)^\nu}{2i\pi} \int_C t^{-\nu-1} e^{t - x^2/(4t)} \, dt \,,$$

so setting $t = (x/2)u$ we obtain

$$J_\nu(x) = \frac{1}{2i\pi} \int_{C'} u^{-\nu-1} e^{(x/2)(u-1/u)} \, du$$

for some other contour C' of the same type.

We now make the change of variable $u = e^w$. We choose as contour C_1 the rectangular contour with vertices $\infty - i\pi$, $-i\pi$, $i\pi$, $\infty + i\pi$. It is clear that as w goes along this contour, $u = e^w$ goes from $-\infty$ to -1, around the trigonometric circle back to -1, and then returns to $-\infty$, hence is (the limit of) a suitable contour C. Thus

$$J_\nu(x) = \frac{1}{2i\pi} \int_{C_1} e^{-\nu w} e^{x \sinh(w)} \, dw \,,$$

which gives the desired integral representations after splitting the contour C_1 into its three sides and making the evident necessary changes of variable.

It is now immediate to deduce the integral for $Y_\nu(x)$ from the definition: we have

$$\sin(\nu\pi) Y_\nu(x) = \cos(\nu\pi) J_\nu(x) - J_{-\nu}(x) = \frac{I_1}{\pi} - \frac{\sin(\nu\pi)}{\pi} I_2 \,,$$

where

$$I_1 = \cos(\nu\pi) \int_0^\pi \cos(x \sin(t) - \nu t) \, dt - \int_0^\pi \cos(x \sin(t) + \nu t) \, dt \quad \text{and}$$

$$I_2 = \int_0^\infty e^{-x \sinh(t)} (\cos(\nu\pi) e^{-\nu t} + e^{\nu t}) \, dt \,.$$

Now since

$$\cos(\nu\pi) \cos(x \sin(t) - \nu t) = \cos(x \sin(t) + \nu(\pi - t)) + \sin(\nu\pi) \sin(x \sin(t) - \nu t)$$

and

$$\int_0^\pi \cos(x \sin(t) + \nu(\pi - t)) \, dt = \int_0^\pi \cos(x \sin(t) + \nu t) \, dt \,,$$

we have

$$I_1 = \sin(\nu\pi) \int_0^\pi \sin(x\sin(t) - \nu t)\, dt .$$

Combining this with the formula for I_2 we obtain the integral representation of Y_ν for $\nu \notin \mathbb{Z}$, hence for all ν by continuity.

For $I_\nu(x)$ the proof is identical to that of $J_\nu(x)$ since the series expansion is obtained by removing the factor $(-1)^k$, so that

$$I_\nu(x) = \frac{1}{2i\pi} \int_C u^{-\nu-1} e^{(x/2)(u+1/u)}\, du .$$

Finally, the formula for $K_\nu(x)$ immediately follows from the definition $K_\nu(x) = (\pi/(2\sin(\nu\pi)))(I_{-\nu}(x) - I_\nu(x))$, even more simply than for $Y_\nu(x)$.

$\square$

Note that the above integral representations give another proof of the existence of Y_n and K_n when $n \in \mathbb{Z}$.

Proposition 9.8.7. *As $x \to \infty$ in $\mathbb{R}$ we have*

$$J_\nu(x) \sim (\pi x/2)^{-1/2} \cos(x - \pi/4 - \nu\pi/2),$$
$$Y_\nu(x) \sim (\pi x/2)^{-1/2} \sin(x - \pi/4 - \nu\pi/2),$$
$$I_\nu(x) \sim (2\pi x)^{-1/2} e^x, \quad K_\nu(x) \sim (2x/\pi)^{-1/2} e^{-x} .$$

Proof. We prove this in the reverse order of the formulas. For K_ν we make the bijective change of variable $u = 2x^{1/2}\sinh(t/2)$. An easy calculation gives

$$K_\nu(x) = \frac{e^{-x}}{x^{1/2}} \int_0^\infty e^{-u^2/2} \cosh(2\nu \sinh^{-1}(u/(2x^{1/2}))) \frac{du}{(1 + u^2/(4x))^{1/2}} .$$

By normal convergence it is clear that as $x \to \infty$ the integral tends to $\int_0^\infty e^{-u^2/2}\, du = (\pi/2)^{1/2}$ by Proposition 9.6.21, proving the result. For I_ν we first note that $\int_0^\infty \exp(-x\cosh(t) - \nu t)\, dt$ tends to 0 exponentially fast, so we need only consider the first integral. We split it into an integral from 0 to $\pi/2$ and an integral from $\pi/2$ to π. Since $\cos(t) \leqslant 0$ in this second interval, the second integral is bounded. Thus, setting $u = 2x^{1/2}\sin(t/2)$ in the first integral we obtain

$$I_\nu(x) = O(1) + \frac{e^x}{\pi x^{1/2}} \int_0^{(2x)^{1/2}} e^{-u^2/2} \cos(2\nu \sin^{-1}(u/(2x^{1/2}))) \frac{du}{(1 - u^2/4x)^{1/2}} .$$

Once again we have normal convergence, proving the result for I_ν.

As for I_ν, the integrals from 0 to ∞ occurring in the integral representations of J_ν and Y_ν tend to 0 exponentially fast, so they can be ignored. Thus for any $A > 0$,

$$J_\nu(x) + iY_\nu(x) = O(x^{-A}) + \frac{1}{\pi} \int_0^\pi e^{i(x\sin(t) - \nu t)} \, dt \ .$$

Simple changes of variables give

$$J_\nu(x) + iY_\nu(x) = O(x^{-A}) + \frac{2e^{-i\pi\nu/2}}{\pi}(I_1(x) + I_2(x)) \ ,$$

where

$$I_1(x) = \int_0^{\pi/3} e^{ix\cos(t)} \cosh(\nu t) \, dt \quad \text{and} \quad I_2(x) = \int_{\pi/3}^{\pi/2} e^{ix\cos(t)} \cosh(\nu t) \, dt \ .$$

We first consider $I_2(x)$. We make the change of variables $\cos(t) = u$ and obtain

$$I_2(x) = \int_0^{1/2} e^{ixu} \phi(u) \, du \ ,$$

where

$$\phi(u) = \frac{\cosh(\nu \cos^{-1}(u))}{(1 - u^2)^{1/2}} \ .$$

By integration by parts we have

$$I_2(x) = \frac{e^{ixu}}{ix}\phi(u)\Big|_0^{1/2} - \frac{1}{ix}\int_0^{1/2} e^{ixu}\phi'(u) \, du = O(x^{-1})$$

since $u \leqslant 1/2$ stays away from the singularities at 1 of $(1-u^2)^{-1/2}$ and $\phi'(u)$. For $I_1(x)$ we set $u = (2x)^{1/2}\sin(t/2)$ and we obtain

$$I_1(x) = \frac{2^{1/2}e^{ix}}{x^{1/2}} \int_0^{(x/2)^{1/2}} e^{-iu^2} \cosh(2\nu \sin^{-1}(u/(2x)^{1/2})) \frac{du}{(1 - u^2/(2x))^{1/2}} \ .$$

Once again we have normal convergence, so we obtain that as $x \to \infty$,

$$I_1(x) \sim 2^{1/2}e^{ix}x^{1/2} \int_0^\infty e^{-it^2} \, dt \ .$$

Now it is well known, and we will prove in the next chapter, that

$$\int_0^\infty e^{-it^2} \, dt = \frac{\pi^{1/2}}{2}e^{-i\pi/4} \ .$$

Thus

$$J_\nu(x) + iY_\nu(x) \sim \frac{2^{1/2}e^{i(x - \pi/4 - \pi\nu/2)}}{\pi^{1/2}x^{1/2}} \ ,$$

proving the asymptotic formulas for J_ν and Y_ν. $\qquad\qquad\qquad \square$

Remark. For the reader having some knowledge of numerical analysis, the proof that we have given for K_ν and I_ν is essentially the method of *steepest descent*, while the proof for J_ν and Y_ν is the method of *stationary phase*.

Proposition 9.8.8. *For $0 < \Re(s) < 1$ we have the following Mellin transforms:*

$$\int_0^\infty t^{s-1} J_0(t)\, dt = \frac{2^{s-1}}{\pi} \sin\left(\frac{\pi s}{2}\right) \Gamma(s/2)^2,$$

$$\int_0^\infty t^{s-1} Y_0(t)\, dt = -\frac{2^{s-1}}{\pi} \cos\left(\frac{\pi s}{2}\right) \Gamma(s/2)^2,$$

$$\int_0^\infty t^{s-1} K_0(t)\, dt = 2^{s-2}\Gamma(s/2)^2 .$$

Proof. All of these formulas immediately follow from the corresponding integral representations, so we simply prove the first. Exchanging the orders of integration, which is legal since $0 < \Re(s) < 1$, and making the change of variable $y = x \sin(t)$, we have

$$\int_0^\infty t^{s-1} J_0(t)\, dt = \frac{1}{\pi} \int_0^\pi \int_0^\infty x^{s-1} \cos(x \sin(t))\, dx\, dt$$

$$= \frac{1}{\pi} \int_0^\pi \sin(t)^{-s}\, dt \int_0^\infty y^{s-1} \cos(y)\, dy .$$

Thus, by Corollaries 9.6.40 and 9.6.37 we have

$$\int_0^\infty t^{s-1} J_0(t)\, dt = \pi^{-1/2} \frac{\Gamma((1-s)/2)}{\Gamma(1-s/2)} \cos(\pi s/2)\Gamma(s) ,$$

and the formula for J_0 follows by using the reflection and duplication formula for the gamma function.

The Mellin transforms of Y_0 and K_0 are obtained in a similar way, using all the formulas of Corollaries 9.6.40 (4) and 9.6.37, and the details are left to the reader; see Exercise 113. □

Finally, we prove an additional result on the function $K_\nu(x)$ that we will need in the next chapter. We change on purpose the index from ν to s (in fact to $s - 1/2$) since it will become a variable in the next chapter.

Theorem 9.8.9. *For $x > 0$ and $\Re(s) > 1/2$ we have*

$$\int_0^\infty \frac{\cos(xt)}{(t^2+1)^s}\, dt = \frac{\pi^{1/2}(x/2)^{s-1/2}}{\Gamma(s)} K_{s-1/2}(x) .$$

Proof. Set $I_s(x) = \int_0^\infty \cos(xt)/(t^2+1)^s \, dt$. Using the integral definition of $\Gamma(s)$ we have by Fubini's theorem

$$\Gamma(s)I_s(x) = \int_0^\infty \cos(xt) \left(\int_0^\infty u^{s-1} e^{-u(t^2+1)} \, du \right) dt$$

$$= \int_0^\infty u^{s-1} e^{-u} \left(\int_0^\infty \cos(xt) e^{-ut^2} \, dt \right) du \ .$$

Since $\cos(xt) = (e^{ixt} + e^{-ixt})/2$ we have by Lemma 10.2.9,

$$\int_0^\infty \cos(xt) e^{-ut^2} \, dt = \frac{1}{2} \left(\int_{-\infty}^\infty e^{-ut^2 + ixt} \, dt \right) = \frac{1}{2} \left(\frac{\pi}{u} \right)^{1/2} e^{-x^2/(4u)} \ .$$

Thus, making the change of variable $u = (x/2)e^v$ we obtain

$$\Gamma(s)I_s(x) = \frac{\pi^{1/2}}{2} \int_0^\infty u^{s-3/2} e^{-(u+x^2/(4u))} \, du$$

$$= \frac{\pi^{1/2}}{2} \left(\frac{x}{2} \right)^{s-1/2} \int_{-\infty}^\infty e^{-x \cosh(v)} e^{(s-1/2)v} \, dv \ ,$$

and since $e^{(s-1/2)v} = \cosh((s-1/2)v) + \sinh((s-1/2)v)$ and $\sinh((s-1/2)v)$ is an odd function of v, the result follows from Proposition 9.8.6.   □

9.9 Exercises for Chapter 9

1. Prove that the Bernoulli polynomials $B_n(x)$ are characterized by $B_n'(x) = nB_{n-1}(x)$, $B_0(x) = 1$, and $B_n(1) = B_n(0)$ for $n \neq 1$.

2.

(a) Let $P_n = (p_{i,j})_{0 \leqslant i,j \leqslant n-1}$ be the $n \times n$ matrix such that $p_{i,j} = \binom{i}{j}$, in other words the lower triangular Pascal triangle. Compute P_n^{-1}.

(b) Let $Q_n = (q_{i,j})_{0 \leqslant i,j \leqslant n-1}$ be such that $q_{i,j} = \binom{i+1}{j+1}$, in other words Pascal's triangle without the left column of 1's. Compute Q_n^{-1}.

(c) Let $R_n = (r_{i,j})_{0 \leqslant i,j \leqslant n-1}$ be such that $r_{i,j} = \binom{i+1}{j}$ for $j \leqslant i$ and $r_{i,j} = 0$ otherwise, in other words Pascal's triangle without the diagonal of 1's. Compute R_n^{-1} in terms of Bernoulli numbers.

3. Prove that

$$\sum_{0 \leqslant k \leqslant n} \binom{6n+3+a}{6k+a} B_{6k+a} = \begin{cases} 2n+1 & \text{when } a = 0 \ , \\ 2n+5/3 & \text{when } a = 2 \ , \\ -n-7/6 & \text{when } a = 4 \ . \end{cases}$$

4.

(a) By expressing $e^{t(x+y)}/(e^t - 1)^2$ in terms of the derivative of $e^{t(x+y)}/(e^t - 1)$, prove that

$$\sum_{0 \leqslant k \leqslant n} \binom{n}{k} B_k(x) B_{n-k}(y) = n(x + y - 1) B_{n-1}(x + y) - (n - 1) B_n(x + y) .$$

(b) Deduce that for $n \geqslant 3$ the Bernoulli numbers satisfy the quadratic recurrence

$$B_n = -\frac{1}{n+1} \sum_{2 \leqslant k \leqslant n-2} \binom{n}{k} B_k B_{n-k} .$$

(c) Let $m \in \mathbb{Z}_{\geqslant 1}$. Compute $\sum_{0 \leqslant j < m} j B_n(j/m)$ for n even, and for $n = 1$, 3, and 5.

(d) Generalizing (a), show that if for $n \geqslant 1$ we write

$$\frac{t^n}{(e^t - 1)^n} = \sum_{k \geqslant 0} c_{n,k} \frac{t^k}{k!} \quad \text{and} \quad \frac{t}{(e^t + 1)^n} = \sum_{k \geqslant 0} d_{n,k} \frac{t^k}{k!} ,$$

then $c_{n,k}$ and $d_{n,k}$ can each be expressed as an explicit linear combination of the Bernoulli numbers B_{k-j} for $0 \leqslant j < n$ involving Stirling numbers of the first kind (see the proof of Proposition 4.2.28 for the definition).

5. Prove the following reciprocity formula:

$$m! \sum_{j=0}^{m} \frac{B_{m-j}}{(m-j)!} \frac{B_{n+j+1}}{(j+1)!} + n! \sum_{j=0}^{n} \frac{B_{n-j}}{(n-j)!} \frac{B_{m+j+1}}{(j+1)!} = -B_{m+n} .$$

   See Exercise 3 of Chapter 11 for another proof.

6.

(a) Show that two of the identities of Corollary 9.1.18 can be restated as follows: if m and n are in $\mathbb{Z}_{\geqslant 0}$ not both zero, then

$$(-1)^m \sum_{j=0}^{m} \binom{m+1}{j} (n+j+1) B_{n+j} + (-1)^n \sum_{j=0}^{n} \binom{n+1}{j} (m+j+1) B_{m+j} = 0 ,$$

and for $n \in \mathbb{Z}_{\geqslant 1}$,

$$\sum_{j=0}^{n} \binom{n+1}{j} (n+j+1) B_{n+j} = 0 ,$$

formulas rediscovered by Momiyama and Kaneko respectively.

(b) Generalize the above formulas to Bernoulli polynomials.

7. Using the same method as that for evaluating Ramanujan sums (see Proposition 10.1.6), prove that

$$\sum_{\substack{0 \leqslant j < m \\ \gcd(j,m)=1}} B_k\left(\frac{j}{m}\right) = \frac{B_k}{m^{k-1}} \prod_{p \mid m} (1 - p^{k-1}) .$$

8. Let $I_{m,n} = \int_0^1 B_m(x)B_n(x)\,dx$. By integration by parts, show that when $m \geq 1$ and $n \geq 1$,

$$I_{m,n} = (-1)^{m-1}\frac{B_{n+m}}{\binom{m+n}{m}},$$

and that $I_{m,0} = 0$ when $m \geq 1$ and $I_{0,0} = 1$.

9.

(a) Show that any $P(X) \in \mathbb{R}[X]$ can be written in the form $P(X) = \sum_{k \geq 0} a_k B_k(X)$ for some unique $a_k \in \mathbb{R}$, and compute the a_k in terms of $\int_0^1 P(t)\,dt$ and the coefficients of the polynomial $P(X+1) - P(X)$.

(b) Apply this to the polynomial $P(X) = \sum_{0 \leq k \leq n} B_k(X)B_{n-k}(X)$, and deduce the identity

$$\sum_{0 \leq k \leq n} B_k(X)B_{n-k}(X) = \frac{2}{n+2}\sum_{k=0}^{n-2}\binom{n+2}{k}B_{n-k}B_k(X) + (n+1)B_n(X).$$

(c) Setting as usual $H_n = \sum_{1 \leq j \leq n} 1/j$, find a similar identity for the polynomial $P(X) = \sum_{0 \leq k \leq n} B_k(X)B_{n-k}(X)/(k(n-k))$.

10.

(a) Show that

$$B_{2k}(1/3) = B_{2k}(2/3) = -\frac{B_{2k}}{2}\left(1 - \frac{1}{3^{2k-1}}\right) \quad \text{and}$$

$$B_{2k}(1/6) = B_{2k}(5/6) = \frac{B_{2k}}{2}\left(1 - \frac{1}{2^{2k-1}}\right)\left(1 - \frac{1}{3^{2k-1}}\right).$$

(b) Compute in terms of $B_{2k+1}(1/3)$ and $B_{2k+1}(1/6)$ the Taylor series expansions of $1/(2\cosh(t)+1)$, $1/(2\cosh(t)-1)$, and $\cosh(t/2)/(2\cosh(t)+1)$ (hence also the corresponding ones where $\cosh(t)$ is replaced by $\cos(t)$).

11. Let $p \geq 5$ be a prime number, and assume that $k \in \mathbb{Z}$ is such that $1 \leq k < p/2$.

(a) Show that for all $m \in \mathbb{Z}_{\geq 1}$ we have

$$\sum_{1 \leq j < p/m} j^{2k-1} \equiv \frac{B_{2k}(\{p/m\}) - B_{2k}}{2k} \pmod{p}.$$

(b) Assume that $m = 4$ or 6. Show that

$$\sum_{1 \leq j < p/m} j^{2k-1} \equiv \frac{B_{2k}(1/m) - B_{2k}}{2k} \pmod{p}.$$

(c) Using the preceding exercise, deduce that for $1 \leq k < p/2$ we have

$$-(2^{p-2k} - 1)(3^{p-2k} - 2^{p-2k} - 1)\frac{B_{2k}}{4k} \equiv \sum_{p/6 < j < p/4} j^{2k-1} \pmod{p}.$$

Using this we can check whether p is a regular prime approximately six times faster than the naïve method (we must of course be careful when the factor in front of $B_{2k}/(4k)$ is divisible by p). There are, however, much faster FFT-based methods to make tables of irregular primes.

12. Compute explicitly the Taylor expansion of $\log\left((e^t-1)/t\right)$ around $t = 0$.

13. Find the Taylor expansion of the following functions in terms of Bernoulli and Euler numbers: $f(t) = e^t/(e^{2t}+1)$, $f(t) = 1/(\cos(t)+\sin(t))$, $f(t) = 1/\cos(t)^k$, and $f(t) = 1/\sin(t)^k$ for $2 \leqslant k \leqslant 4$.

14. (I thank V. Arnold for this exercise.) Define the *Bernoulli–Euler* triangle $A_{n,k}$ as follows. We set $A_{0,0} = 1$, and for $n \geqslant 1$ and $0 \leqslant k \leqslant n$ we set

$$A_{n,k} = \begin{cases} \sum_{j=k}^{n-1} A_{n-1,j} & \text{when } n \text{ is odd,} \\ \sum_{j=0}^{k-1} A_{n-1,j} & \text{when } n \text{ is even.} \end{cases}$$

(a) Compute the evident triangle for $n \leqslant 6$.
(b) Prove that $A(n,n) = |E_n|$ and that $A(n,0) = |T_n|$.

15. By induction show that for $x \in [0,1]$ we have the following:
(a) $B_{2k}(x) - B_{2k} = 0$ if and only if $x = 0$ or $x = 1$.
(b) For $k \geqslant 1$, $B_{2k+1}(x) - B_{2k+1} = 0$ if and only if $x = 0$, $x = 1/2$, or $x = 1$.
(c) $|B_{2k}(x)| \leqslant |B_{2k}|$ and $|B_{2k}(x) - B_{2k}| \leqslant 2(1 - 2^{-2k})|B_{2k}|$.

16.
(a) Prove that

$$\sum_{0 \leqslant j \leqslant k} (-1)^j \binom{2k}{2j} = \begin{cases} 0 & \text{if } k \text{ is odd,} \\ (-1)^{k/2} & \text{if } k \text{ is even.} \end{cases}$$

(b) Deduce from Corollary 9.1.10 that the Euler numbers E_{2k} are not only integers, but odd integers, and more precisely that $E_{2k} \equiv (-1)^k \pmod 4$.

17. Let p be a prime number such that $(p-1) \mid (2k)$. Show that

$$E_{2k} \equiv 1 - \left(\frac{-4}{p}\right) \pmod p .$$

18. Define the Euler polynomials $E_k(x)$ by the generating series

$$\frac{2e^{tx}}{e^t+1} = \sum_{k \geqslant 0} \frac{E_k(x)}{k!} t^k ,$$

so that the Euler numbers are given by $E_k = 2^k E_k(1/2)$. Express $E_k(x)$ as a linear combination of two Bernoulli polynomials and show that essentially all of the formulas given for Bernoulli polynomials have analogues for Euler polynomials.

19. Generalizing Corollary 9.1.13, prove that with a suitable integral definition of the left-hand sides, for $t > x - 1$ (and also formally) we have

$$\sum_{k \geqslant 0} \frac{B_{k+1}(x)}{(k+1)t^{k+1}} = -\psi(t - x + 1) + \log(|t|) \quad \text{and}$$

$$\sum_{k \geqslant 0} \frac{B_{k+2}(x)}{(k+2)(k+1)t^{k+1}} = \log(\Gamma(t - x + 1)) - \left(t - x + \frac{1}{2}\right)\log(|t|) + t - \frac{\log(2\pi)}{2} .$$

20. (D. Zagier.) The present exercise is due to D. Zagier. All the power series or Laurent series are formal. Let

$$T(u, x) = \sum_{k \geqslant 0} \frac{B_{k+1}(x)}{(k+1)u^{k+1}} ,$$

so that $\frac{\partial T}{\partial u}(u, x) = -(S(u, x) - 1/u)$, with the notation of Proposition 9.1.11 (see also Exercise 19). For $n \in \mathbb{Z}_{\geqslant 1}$ define

$$f(n, x) = \sum_{0 \leqslant r \leqslant n} \binom{n+r}{2r} \frac{B_r(x)}{n+r} \quad \text{and} \quad g(n, x) = \sum_{\substack{0 \leqslant r \leqslant n \\ r \not\equiv n \pmod 2}} \binom{n+r}{2r} \frac{B_r(x)}{n+r} ,$$

and let $F(t, x) = \sum_{n \geqslant 1} f(n, x)t^n$ and $G(t, x) = \sum_{n \geqslant 1} g(n, x)t^n$.

(a) Show that

$$T(u - 1, x) = T(u, x) + \log\left(1 - \frac{1}{u}\right) + \frac{1}{u - x} \quad \text{and} \quad T(-u, 1 - x) = T(u, x) .$$

(b) Show that

$$F(t, x) = \frac{1}{2} T\left(\frac{(1 - t)^2}{t}, x\right) - \log(1 - t) .$$

(c) Deduce that

$$2F(t, x) = T\left(t + \frac{1}{t}, x\right) - \log(1 + t^2) + \frac{t}{1 - tx + t^2} + \frac{t}{1 - (x + 1)t + t^2} .$$

(d) Conclude that

$$4G(t, x) = 2(F(t, x) - F(-t, 1 - x)) = \sum_{-2 \leqslant j \leqslant 1} \frac{t}{1 - (x + j)t + t^2} ,$$

in other words that $4g(n, x)$ is the coefficient of t^n in

$$\sum_{-2 \leqslant j \leqslant 1} t/(1 - (x + j)t + t^2) .$$

(e) Deduce for instance that if $0 < x < 1$ is fixed then $|g(n, x)|$ is bounded. We now specialize to $x = 0$ and write $f(n) = f(n, 0)$.

(f) Prove that for *odd* n we have

$$f(n) = \frac{1}{4}\left(\frac{-4}{n}\right) + \frac{1}{2}\left(\frac{-3}{n}\right) ,$$

and in particular that f is periodic of period 12 on odd integers.

(g) From now on we assume that n is even, and we set $\widetilde{B}_n = 2nf(n) - B_n$. Prove the following analogue of the Clausen–von Staudt congruence, for $n \geqslant 2$ even:

$$\widetilde{B}_n \equiv \sum_{\substack{(p+1)|n \\ p \text{ prime}}} \frac{1}{p} \pmod 1 .$$

(h) Still assuming n even, prove that as $n \to \infty$ we have

$$f(n) \sim (-1)^{n/2} \pi Y_n(4\pi) \sim (-1)^{n/2-1} \frac{(n-1)!}{(2\pi)^n} .$$

(i) More precisely, compute asymptotic estimates as $n \to \infty$ for $\widetilde{B}_n$, $f(n) - (-1)^{n/2} \pi Y_n(4\pi)$, and $f(n) - (-1)^{n/2-1}(n-1)!/(2\pi)^n$.

21. Prove the convergence of the continued fractions for $\psi'(x)$ and $\psi''(x)$ in the domains given by Theorem 9.6.48.

22. Compute explicitly

$$\sum_{k=0}^{n-1} \binom{2n-1}{2k-1} \frac{B_{2k+1}(x)}{2k+1} \quad \text{and} \quad B_{2n}(x) + \sum_{k=1}^{n-1} \binom{2n-1}{2k-2} \frac{B_{2k}(x)}{2k} .$$

23. Compute explicitly

$$\sum_{k=0}^{n-1} \binom{n}{k} y^{n-k} \frac{B_{k+2}(x)}{(k+1)(k+2)}$$

and generalize.

24. This exercise is indirectly related to the recurrences of Proposition 9.1.16.

(a) Show that for $k \in \mathbb{Z}_{\geqslant 0}$ we have

$$\sum_{k/2 \leqslant n \leqslant k} \binom{n}{2n-k} = F_{k+1} ,$$

where F_k is the Fibonacci sequence. In fact, show that this is true for all $k \in \mathbb{Z}$ with a suitable interpretation of both sides.

(b) Compute explicitly in its domain of absolute convergence the sum of the power series

$$\sum_{n \geqslant 0} \frac{(-1)^n t^{2n}}{(2n+1)\binom{2n}{n}} .$$

25. With the notation of Lemma 9.1.17, prove the identities

$$e^t (D^2 - I)^n F - D^n (D - 2I)^n F = x^n (x-2)^n e^{xt}$$

and

$$e^t D^n (D + 2I)^n F - (D^2 - I)^n F = (x^2 - 1)^n e^{xt}$$

and deduce from them identities analogous to those of Proposition 9.1.16.

26.

(a) By computing the mth derivative of $(1 - e^{-t})^n$ at $t = 0$, compute

$$\sum_{k=1}^{n} (-1)^k \binom{n}{k} k^m$$

for $0 \leqslant m \leqslant n$, and compute this quantity also for $m = -1$.

(b) By expanding the inner sum in terms of $B_m(x)$, prove that

$$B_n(x) = \sum_{k=1}^{n+1} \frac{(-1)^{k-1}}{k} \binom{n+1}{k} \sum_{j=0}^{k-1} (j+x)^n .$$

(c) Similarly, show that

$$B_n(x) = \sum_{k=1}^{n} \frac{(-1)^{k-1}}{k} \binom{n}{k} \sum_{j=0}^{k-1} (j+x)^n + \frac{(-1)^{n+1} n!}{n+1} .$$

In particular, if $x = 0$ these formulas give reasonable methods for computing a *single* value of B_n, but the method given in Section 9.1.3 is much more efficient for large n.

27. In the text and in the exercises we have given many methods and recurrences for Bernoulli numbers. Let N be a large integer, and assume that we want to compute exactly all the Bernoulli numbers B_{2k} for $2k \leqslant N$. Implement many different methods, and compare their efficiency. In particular, implement the method explained in Section 9.1.3 where each Bernoulli number is computed separately, and implement the recurrences given in Proposition 9.1.3, Corollary 9.1.19, and Exercise 3.

28. Show that for all $t \in \mathbb{R}$,

$$-2e^{2i\pi t} = B_0(\{t\}) + 2i\pi B_1(\{t\}) + \sum_{k \geqslant 0} \left(\frac{(2i\pi)^{k+2} B_{k+2}(\{t\})}{(k+2)!} - \frac{(2i\pi)^k B_k(\{t\})}{k!} \right) ,$$

and estimate the speed of convergence of the series.

29. Using the power series for $\sin(at)$, Proposition 9.2.10 (2), and Corollary 9.1.21, show that

$$\int_0^\infty \frac{\sin(at)}{e^{2\pi t} - 1} dt = \frac{1}{2} \left(\frac{1}{e^a - 1} - \frac{1}{a} + \frac{1}{2} \right) .$$

30. In connection with Proposition 9.2.8, let $m(k)$ be the maximum on $[0, \infty[$ of the absolute value of the function $(1/(e^t - 1) - \sum_{0 \leqslant j \leqslant k} (B_j/j!) t^{j-1}) e^{-t}$, which exists since the function is continuous and tends to 0 at infinity. Compute $m(k)$ for $k \leqslant 18$, but show that $m(k)$ tends to infinity as $k \to \infty$. Give an asymptotic estimate for the growth of $m(k)$.

31. Prove Proposition 9.2.8.

32. Assume that the hypotheses of Corollary 9.2.6 (1) and (2) are satisfied.

(a) Show that $z(f, a + 1) = z(f, a) - f(a) + \int_a^{a+1} f(t) dt$.

(b) Deduce that the quantity

$$z(f) = z(f, a) + \sum_{1 \leqslant m \leqslant a-1} f(m) - \int_1^m f(t) dt$$

is independent of $a \in \mathbb{Z}_{>0}$.

(c) Give an expression for $z(f)$, both in terms of Bernoulli polynomials and in terms of the inverse Laplace transform of f.

(d) Compute $z(f)$ for the usual functions f seen in the text.

33. Assume that f satisfies the assumptions of Proposition 9.2.11. Let C be the rectangle with vertices $\pm iT$ and $N \pm iT$, with a small indentation around $z = a$ and $z = N$ (including the points a and N), where T is large.

(a) By considering $\int_C \cotan(\pi t) f(t)\, dt$ and letting the indentations tend to zero and T tend to infinity, show that when $a \in \mathbb{Z}_{>0}$,

$$\sum_{m=a}^{N} f(m) = \int_a^N f(t)\, dt + \frac{f(N) + f(a)}{2}$$
$$+ i \int_0^\infty \frac{f(a+iy) - f(N+iy) - f(a-iy) + f(N-iy)}{e^{2\pi y} - 1}\, dy \, .$$

(b) Deduce the first formula for $z(f, a)$ given in Proposition 9.2.11 for $a > 0$, not necessarily integral.

(c) In a similar manner, prove the second formula.

34.

(a) Using the formula of the preceding exercise, prove that for $\Re(s) > 0$ we have the following integral representation due to Binet:

$$\mathrm{Log}\,\Gamma(s) = \left(s - \frac{1}{2}\right) \log(s) - s + \frac{1}{2}\log(2\pi) + 2 \int_0^\infty \frac{\mathrm{atan}(t/s)}{e^{2\pi t} - 1}\, dt \, .$$

(b) Using the second Abel–Plana formula, prove that we also have

$$\mathrm{Log}\,\Gamma\left(s + \frac{1}{2}\right) = s\log(s) - s + \frac{1}{2}\log(2\pi) - 2\int_0^\infty \frac{\mathrm{atan}(t/s)}{e^{2\pi t} + 1}\, dt \, .$$

Note that this formula can also be obtained directly from (a) by using the duplication formula for the gamma function.

(c) By differentiation under the integral sign, compute in the range of convergence the integral

$$f(x, y) = \int_0^\infty e^{-xt} \sin(yt)\, \frac{dt}{t} \, .$$

(d) Deduce from this another proof of (a), using Corollary 9.6.32 and the formula of Exercise 29.

35.

(a) Set $f(t) = 1/(e^t - 1) - 1/t + 1/2$. Show that for $t \in \mathbb{R}_{>0}$ we have $0 < f(t) < 1/12$.

(b) Set $s = x + iy$ with x and y in $\mathbb{R}$. Deduce from Corollary 9.6.32 that for $x = \Re(s) > 0$ we have

$$\Im(\mathrm{Log}\,\Gamma(s)) = \left(x - \frac{1}{2}\right) \mathrm{atan}\left(\frac{y}{x}\right) + \frac{y}{2}\log(x^2 + y^2) - y + R(s) \, ,$$

with $|R(s)| < y/(12(x^2 + y^2))$.

(c) Writing $\Im(\mathrm{Log}\,\Gamma(s)) = \Im(\log(\Gamma(s))) + 2\pi m(s)$ for some $m(s) \in \mathbb{Z}$, where $\log$ is the principal determination, deduce an approximate formula for $m(s)$ when x or y is large.

The goal of the following three exercises is to prove the results given in Exercise 38.

36. Define the *Stirling numbers of the second kind* by the formula

$$X^n = \sum_{k \geqslant 0} S(n, k) X(X - 1) \cdots (X - k + 1) \, .$$

(a) Show that $S(n,k) = 0$ for $k > n$, $S(n,0) = 0$ for $n \geqslant 1$, $S(n,n) = 1$, and the recurrence formula $S(n+1,k) = kS(n,k) + S(n,k-1)$ for $k \geqslant 1$, so that in particular $S(n,k) \in \mathbb{Z}_{\geqslant 0}$ for all n and k. Prove also the explicit formula

$$S(n,k) = \frac{1}{k!} \sum_{0 \leqslant j \leqslant k} (-1)^{k-j} \binom{k}{j} j^n .$$

(b) Let $D = d/dT$ be the differentiation operator with respect to T. Define the *Eulerian polynomials* $P_n(X)$ by

$$D^n \left(\frac{1}{e^T - 1} \right) = (-1)^n \frac{P_n(e^T)}{(e^T - 1)^{n+1}} .$$

Show that $P_n(X) \in \mathbb{Z}[X]$, prove the recurrence formula

$$P_{n+1}(X) = (n+1)XP_n(X) - (X-1)XP_n'(X) ,$$

and show that we have the explicit formula

$$P_n(X+1) = \sum_{0 \leqslant k \leqslant n} S(n+1, k+1)k! X^{n-k} .$$

(c) Show that if we define the *Eulerian numbers* $A(n,k)$ (not to be confused with the Euler numbers of Definition 9.1.8) by $P_n(X) = \sum_{0 \leqslant k \leqslant n} A(n,k)X^k$ we have the explicit formula

$$A(n,k) = \sum_{j=0}^{k} (-1)^j (k-j)^n \binom{n+1}{j} .$$

Show also that the $A(n,k)$ for $1 \leqslant k \leqslant n$ ($A(n,0) = 0$ being excluded) form a symmetrical array whose first rows are as follows:

				1				
			1		1			
		1		4		1		
	1		11		11		1	
1		26		66		26		1
1	57		302		302		57	1

37. Let $n \geqslant 0$, let $N \in \mathbb{Z}_{\geqslant 1}$, and set $S_n(N) = \sum_{0 \leqslant r < N} r^n$ (with the convention that $0^0 = 1$ for $n = 0$). Let p be a prime number dividing N (otherwise there is nothing to prove). Using the Clausen–von Staudt theorem, show that:

(a) If $p \geqslant 3$ we have

$$v_p(S_n(N)) \geqslant \begin{cases} v_p(N) & \text{if } 2 \nmid n, \text{ or } n = 0, \text{ or } (p-1) \nmid n, \\ v_p(N) - 1 & \text{if } (p-1) \mid n, \end{cases}$$

and that in the latter case, we have more precisely $v_p(S_n(N) + N/p) \geqslant v_p(N)$.

(b) If $p = 2$ we have $v_p(S_p(N)) = v_p(N) - 1$.

In particular, show that for all $v \geqslant 1$ we have

$$v_p \left(\sum_{0 \leqslant r < p^v} r^n \right) \geqslant v - 1 \,,$$

and that for any prime number p we have $v_p(p^n S_n(N)) \geqslant v_p(N)$.

38. Let $n \in \mathbb{Z}_{\geqslant 0}$, let $N \in \mathbb{Z}_{>1}$, let ζ be an Nth root of unity different from 1, let $o(\zeta)$ be the order of ζ, so that $o(\zeta) \mid N$ and $o(\zeta) > 1$, and set $\pi = \zeta - 1$ and $S_n(N) = \sum_{0 \leqslant r < N} r^n \zeta^r$.

(a) By considering the formal power series $\sum_{0 \leqslant r < N} r^n e^{rT}$ and Exercise 36, show that

$$S_n(N) = N \sum_{j=1}^{n} (-1)^{n-j} N^{j-1} \binom{n}{j} \sum_{0 \leqslant i \leqslant n-j} S(n-j+1, i+1) \frac{i!}{\pi^{i+1}} \,.$$

(b) Deduce that if $o(\zeta)$ is not a prime power we have $S_n(N) \equiv 0 \pmod{N \gcd(n, N^\infty)}$, where we recall that $\gcd(n, N^\infty) = \prod_{p \mid N} p^{v_p(n)}$.

(c) Show that if $o(\zeta) = p^k$ for some prime p and $k \geqslant 2$, then

$$S_n(N) \equiv 0 \pmod{(N/(1-\zeta)^p) \gcd(n, N^\infty) \mathbb{Z}[\zeta])} \,,$$

and in particular modulo $(N/p) \gcd(n, N^\infty) \mathbb{Z}[\zeta]$.

(d) Show that if $o(\zeta) = p$ for some prime p then

$$S_n(N) \equiv 0 \pmod{(N/(1-\zeta)^{n \bmod (p-1)}) \gcd(n, (N/p^{v_p(N)})^\infty) \mathbb{Z}[\zeta])} \,,$$

where $n \bmod (p-1)$ is defined as the unique integer congruent to n modulo $p-1$ in the interval $[1, p-1]$, and in particular the congruence is true modulo $(N/p) \gcd(n, (N/p^{v_p(N)})^\infty) \mathbb{Z}[\zeta]$.

39. Deduce from Proposition 9.2.11 that

$$\gamma = \frac{1}{2} + 2 \int_0^\infty \frac{t}{(1+t^2)(e^{2\pi t} - 1)} \, dt \,,$$

and more generally find analogous expressions for the functions and constants occurring in Proposition 9.2.10.

40. Define Catalan's constant G by the formula

$$G = \sum_{k \geqslant 0} \frac{(-1)^k}{(2k+1)^2} = 1 - \frac{1}{3^2} + \frac{1}{5^2} - \cdots$$

(we have

$$G = L(\chi_{-4}, 2) = 0.9159655941772190150546035149323841\ldots \,,$$

using a notation introduced in the next chapter). Find integral formulas for G analogous to those given for γ in Proposition 9.2.10 (4) and in Exercise 39.

41. Let $m \in \mathbb{Z}_{\geqslant 1}$.

(a) Show that we have the convergent series

$$m^s \zeta(s) = \sum_{k \geqslant 1} \binom{s+k-2}{k-1} \frac{\zeta(s+k-1)}{k} \frac{B_k(m) - B_k}{m^{k-1}} .$$

For instance, if $n \in \mathbb{Z}_{\geqslant 2}$ we have

$$\sum_{k \geqslant 1} \binom{n+k-2}{k} \zeta(n+k-1) \frac{B_k(m) - B_k}{m^{k-1}} = (n-1)m^n \zeta(n) .$$

(b) Do these series converge for any other value of m?

(c) Show that

$$\sum_{k \geqslant 1} \zeta(k+1) \frac{B_k(m) - B_k}{m^{k-1}} = \frac{\pi^2}{6} m^2 ,$$

$$\sum_{k \geqslant 2} \zeta(k) \frac{B_k(m) - B_k}{k m^{k-1}} = m \log(m) ,$$

$$\sum_{k \geqslant 3} \zeta(k-1) \frac{B_k(m) - B_k}{k(k-1)m^{k-1}} = \frac{1}{2}((m-1)(\log(2\pi) - \gamma) - \log(m)) ,$$

$$\sum_{k \geqslant 4} \zeta(k-2) \frac{B_k(m) - B_k}{k(k-1)(k-2)m^{k-1}} = \frac{1-\gamma}{6}\left(m - \frac{3}{2} + \frac{1}{2m}\right) + \frac{\log(m)}{12m}$$

$$+ \frac{m-1}{4}(\log(2\pi) - 1) + \zeta'(-1)\left(m - \frac{1}{m}\right) .$$

(d) Explain why the $\zeta'(-1)$ that occurs in this last formula is "the same" as the $\zeta'(-1)$ that occurs in Exercise 44; see also Exercise 71.

42. Using Proposition 9.2.13, compute to 15 decimal digits $\zeta(-3)$, $\zeta(1/2)$, and $\zeta(-1/2)$.

43. Using the general Euler–MacLaurin formula prove that for $\alpha \neq -1$ and $0 < x < 1$ we have

$$\sum_{m=0}^{N-1} (m+x)^\alpha = x^\alpha + \frac{N^{\alpha+1} - 1}{\alpha + 1} + \sum_{1 \leqslant j \leqslant n} \binom{\alpha}{j-1} \frac{B_j(x)}{j}(N^{\alpha-k+1} - 1)$$

$$+ (-1)^{n-1} \binom{\alpha}{n} \int_1^N t^{\alpha-n} B_n(\{t - x\}) \, dt .$$

Deduce from this that for $0 < x < 1$,

$$\zeta(-\alpha, x) = x^\alpha - \frac{1}{\alpha + 1} - \sum_{1 \leqslant j \leqslant n} \binom{\alpha}{j-1} \frac{B_j(x)}{j}$$

$$+ (-1)^n \binom{\alpha}{n} \int_1^\infty t^{\alpha-n} B_n(\{t - x\}) \, dt .$$

44.

(a) Using the idea explained in the proof of Stirling's formula given in Section 9.2.5, prove the following asymptotic estimate as $N \to \infty$:

$$\sum_{1 \leqslant m < N} m \log(m) = \frac{B_2(N)}{2} \log(N) - \frac{N^2}{4} + \frac{1}{12} - \zeta'(-1) + o(1) \ .$$

(b) Denoting as usual by $H_n = \sum_{1 \leqslant j \leqslant n} 1/j$ the harmonic sum, show more generally that when $r \in \mathbb{Z}_{\geqslant 0}$ we have

$$\sum_{1 \leqslant m < N} m^{r-1} \log(m) = \frac{B_r(N)}{r} (\log(N) + H_{r-1})$$

$$- \frac{1}{r} \sum_{k=0}^{r-1} \binom{r}{k} H_{r-k} B_k N^{r-k} - \zeta'(1-r) + o(1) \ .$$

(c) Using $\psi(t) = \lim_{N \to \infty} (\log(N) - \sum_{0 \leqslant n < N} 1/(t+n))$, the explicit computation of $\int_0^1 t^k/(t+n)\, dt$, and the preceding question, show that for $k \in \mathbb{Z}_{\geqslant 1}$ we have

$$\int_0^1 t^k \psi(t)\, dt = \sum_{j=0}^{k-1} (-1)^j \binom{k}{j} \left(\zeta'(-j) + H_j \zeta(-j) \right)$$

(note that the sum stops at $j = k - 1$).

(d) Generalizing Raabe's formula (Theorem 9.6.54) deduce that for $k \in \mathbb{Z}_{\geqslant 0}$ we have

$$\int_0^1 t^k \log(\Gamma(t))\, dt = \frac{1}{k+1} \sum_{j=0}^{k} (-1)^{j+1} \binom{k+1}{j} \left(\zeta'(-j) + H_j \zeta(-j) \right) \ .$$

See Exercise 105 (d) for an interesting consequence of this formula.

45. The aim of this exercise is to give another proof of the last formula of the preceding exercise, and to give more general results.

(a) Using generating functions, prove the identity

$$\sum_{j=0}^{m} \frac{(-1)^{j+1}}{s-j-1} \binom{m}{j} = B(m+1, 1-s) = \frac{m!}{(1-s)(2-s)\cdots(m+1-s)}$$

(here B is of course the beta function, not the Bernoulli polynomial).

(b) By integrating by parts, show that for $s \neq 1$ we have

$$\int_0^1 t^k \zeta(s, x+t)\, dt = \sum_{m=1}^{k} (-1)^{m+1} \frac{k!}{(k+1-m)!} \frac{\zeta(s-m, x+1)}{(1-s)(2-s)\cdots(m-s)}$$

$$+ (-1)^{k+1} \frac{k!}{(1-s)(2-s)\cdots(k+1-s)} x^{k+1-s} \ .$$

(c) Deduce the following formula, which is a generalization both of Raabe's formula (for $k = 0$) and of the formula of the preceding exercise (for $x = 0$):

$$\int_0^1 t^k \log(\Gamma(x+t))\, dt = (-1)^k \frac{x^{k+1}}{k+1}(\log(x) - H_{k+1}) + \frac{1}{k+1}\frac{\log(2\pi)}{2}$$
$$+ \frac{1}{k+1} \sum_{j=1}^{k} (-1)^{j+1} \binom{k+1}{j} \left(\frac{\partial \zeta}{\partial s}(-j, x+1) + H_j \zeta(-j, x+1) \right).$$

(d) By integration by parts, compute $\int_0^1 t^k \psi^{(m)}(x+t)\, dt$ for $k \geqslant m$, and deduce the value of $\int_0^1 t^k \zeta(m+1, x+t)\, dt$ for $m \in \mathbb{Z}$ such that $1 \leqslant m \leqslant k$.

46. If ω_1 and ω_2 are two nonzero complex numbers we define for $p \geqslant 0$,

$$C_p(\omega_1, \omega_2) = \sum_{k=0}^{p} \binom{p}{k} B_k B_{p-k} \omega_1^{k-1} \omega_2^{p-k-1},$$

so that by abuse of notation

$$C_p(\omega_1, \omega_2) = \frac{\omega_2^{p-1}}{\omega_1} C_p \left(\frac{\omega_1}{\omega_2} \right),$$

with for instance

$$C_0(z) = 1, \quad C_1(z) = -\frac{z+1}{2}, \quad C_2(z) = \frac{z^2 + 3z + 1}{6}, \quad C_3(z) = -\frac{z^2 + z}{4}.$$

(a) Let f be a complex function that is holomorphic in a suitable region of the complex plane, and let $z \in \mathbb{C}$. Prove the following complex generalization of the Euler–MacLaurin summation formula:

$$\sum_{0 \leqslant m_1, m_2 < N} f(z + m_1\omega_1 + m_2\omega_2) = \sum_{p=0}^{n} \frac{C_p(\omega_1, \omega_2)}{p!} \left(f^{(p-2)}(z + N(\omega_1 + \omega_2)) \right.$$
$$\left. - f^{(p-2)}(z + N\omega_1) - f^{(p-2)}(z + N\omega_2) + f^{(p-2)}(z) \right) + R_n(N),$$

where $R_n(N)$ is a suitable "remainder term," and by convention $f^{(-1)}(z)$ is an antiderivative of $f(z)$ and similarly for $f^{(-2)}(z)$.

(b) As an application, assume that $\Re(\omega_1) > 0$, $\Re(\omega_2) > 0$, and $\Re(z) > 0$, and that in what follows we choose the principal determination of the logarithm. Prove that there exists a function $F_{\omega_1,\omega_2}(z)$ such that as $N \to \infty$,

$$\sum_{0 \leqslant m_1, m_2 < N} \log(z + m_1\omega_1 + m_2\omega_2) = N^2 \log(N) - \frac{3}{2}N^2 + \Omega_2 \left(\frac{N^2 - N}{2} \right)$$
$$+ \frac{\log(\omega_1\omega_2)}{2} N - \frac{\omega_1^2 + 3\omega_1\omega_2 + \omega_2^2}{12\omega_1\omega_2} \log(N)$$
$$+ z \left(\Omega_1 N + \frac{\omega_1 + \omega_2}{2\omega_1\omega_2} \log(N) \right) - \frac{z^2}{2\omega_1\omega_2} \log(N) + F_{\omega_1,\omega_2}(z) + o(1),$$

where

$$\Omega_k = \frac{(\omega_1 + \omega_2)^k \log(\omega_1 + \omega_2) - \omega_1^k \log(\omega_1) - \omega_2^k \log(\omega_2)}{\omega_1\omega_2}.$$

(c) Prove that as $N \to \infty$,

$$\sum_{0 \leqslant m_2 < N} \log(z + N\omega_1 + m_2\omega_2) = N\log(N) - N$$

$$+ N\frac{(\omega_1 + \omega_2)\log(\omega_1 + \omega_2) - \omega_1\log(\omega_1)}{\omega_2}$$

$$+ \left(\frac{z}{\omega_2} - \frac{1}{2}\right)\log\left(\frac{\omega_1 + \omega_2}{\omega_1}\right) + o(1) .$$

(Warning: you cannot use Euler–MacLaurin directly.)

(d) Prove that the function $F_{\omega_1,\omega_2}(z)$ defined above satisfies the homogeneity property $F_{a\omega_1,a\omega_2}(az) = F_{\omega_1,\omega_2}(z)$ and the relation

$$F(z + \omega_1) = F(z) + \log\left(\Gamma\left(\frac{z}{\omega_2}\right)\right) - \frac{1}{2}\log(2\pi) + \left(\frac{z}{\omega_2} - \frac{1}{2}\right)\log\left(\frac{\omega_1 + \omega_2}{\omega_1}\right)$$

(and of course the symmetrical one obtained by exchanging ω_1 and ω_2).

(e) Relate this function F to Barnes's Γ_2 function defined in Exercise 71.

47. Compute an upper bound for

$$\left| \int_{-\infty}^{\infty} g(t)\,dt - h \sum_{m=-\infty}^{\infty} g(mh) \right|$$

for some standard C^∞ functions $g(t)$ satisfying the hypotheses of Section 9.3.2, for instance for $g(t) = 1/(1 + t^2)$ or $g(t) = \exp(-t^2)$.

48.

(a) Show experimentally that to compute $\int_{-\infty}^{\infty} f(x)\,dx$ using the doubly exponential integration method, one can use $x = \sinh(\sinh(t))$ if the function does not tend to zero exponentially fast as $x \to \pm\infty$, and $x = \sinh(t)$ if it does.

(b) Give two solutions in the case that $f(x)$ tends to zero exponentially fast as $x \to -\infty$, but not as $x \to +\infty$. (Hint: consider $\sinh(t)\exp(\exp(t))$.)

(c) Find an analytic function f satisfying the above assumptions, and compare the two solutions on f.

49. Assume that we want to compute $I = \int_0^\infty f(x)\sin(x)\,dx$, where f is a priori a nonoscillatory function tending sufficiently rapidly to 0 at infinity. Implement the change of variable $x = (2\pi/h)t/(1 - \exp(-K\sinh(t)))$ suggested in the text, and compare the efficiency with that of other methods.

50.

(a) Let ϕ be a test function in the Schwartz space. Show that

$$\int_{qm-r}^{(q+1)m-r} \left\{\frac{x+r}{m}\right\}' \phi(x)\,dx = \frac{1}{m}\int_{qm-r}^{(q+1)m-r} \phi(x)\,dx - \phi((q+1)m - r) ,$$

and deduce that, as claimed in the text, we have

$$\left\{\frac{x+r}{m}\right\}' = \frac{1}{m} - \sum_{q \in \mathbb{Z}} \delta_{qm-r}(x) .$$

(Warning: you cannot simply replace $\{(x+r)/m\}'$ by $1/m$.)

(b) Complete the proof of Proposition 9.4.11.

51. Prove Proposition 9.4.12.

52. Prove Corollaries 9.2.4 and 9.4.18.

53. Let χ be a primitive character of conductor $f > 1$, and for all $k \geqslant 0$ set
$Q_k = \left(\sum_{0 \leqslant r < f/2} \chi(r) r^k \right) / f^k$ and $R_k = (2^k - \overline{\chi}(2)) Q_k$.

(a) By splitting in two different ways the formulas for $B_k(\chi)$ given in Proposition 9.4.5 and separating the cases f odd and f even, show that if $\chi(-1) = (-1)^k$ we have

$$\frac{B_k(\chi)}{f^{k-1}} = 2 \sum_{j=0}^{k} \binom{k}{j} B_j Q_{k-j} \quad \text{and}$$

$$(2^k - \overline{\chi}(2)) \frac{B_k(\chi)}{f^{k-1}} = 2^{k+1} \sum_{j=1}^{k} \binom{k}{j} \left(1 - \frac{1}{2^j} \right) B_j Q_{k-j} .$$

(b) Show that if $\chi(-1) = (-1)^{k-1}$ we have

$$R_{k-1} = \frac{2}{k} \sum_{j=1}^{\lfloor k/2 \rfloor} \binom{k}{2j} (2^{2j} - 1) B_{2j} R_{k-2j} .$$

(c) Deduce that when χ is an odd primitive character we have

$$(2 - \overline{\chi}(2)) B_1(\chi) = -Q_0, \qquad (8 - \overline{\chi}(2)) \frac{B_3(\chi)}{f^2} = -12 Q_2 + 3 \frac{1 - \overline{\chi}(2)}{2 - \overline{\chi}(2)} Q_0 ,$$

$$(2 - \overline{\chi}(2)) Q_1 = \frac{1}{2}(1 - \overline{\chi}(2)) Q_0, \ (8 - \overline{\chi}(2)) Q_3 = \frac{3}{2}(4 - \overline{\chi}(2)) Q_2 - \frac{1}{4}(1 - \overline{\chi}(2)) Q_0 ,$$

and when χ is an even nontrivial primitive character we have $Q_0 = 0$,

$$(4 - \overline{\chi}(2)) \frac{B_2(\chi)}{f} = -4 Q_1, \qquad (16 - \overline{\chi}(2)) \frac{B_4(\chi)}{f^3} = -32 Q_3 + 24 \frac{2 - \overline{\chi}(2)}{4 - \overline{\chi}(2)} Q_1 ,$$

$$(4 - \overline{\chi}(2)) Q_2 = (2 - \overline{\chi}(2)) Q_1, \quad (16 - \overline{\chi}(2)) Q_4 = 2(8 - \overline{\chi}(2)) Q_3 - (2 - \overline{\chi}(2)) Q_1 .$$

54. Using the χ-Euler–MacLaurin formula, compute $\sum_{0 \leqslant m < 2N} (-1)^{m-1} m^k$ and $\sum_{0 \leqslant m < 2N} (-1)^{m-1} (2m+1)^k$ in completely factored form for $0 \leqslant k \leqslant 5$.

55. Prove Lemma 9.5.2.

56.

(a) Prove Corollaries 9.5.8 and 9.5.9.

(b) Generalize Corollary 9.5.9 to exponents 6 and 8 by showing that, under the same conditions on D, we have

$$\sum_{0 \leqslant r < |D|} \left(\frac{D}{r} \right) r^6 \equiv 0 \pmod{4D} ,$$

except for $D = -4, 5, 8,$ and 13, and

$$\sum_{0 \leqslant r < |D|} \left(\frac{D}{r} \right) r^8 \equiv 0 \pmod{16D} ,$$

except for $D = -4, 8,$ and 17, and compute the value of the left-hand side for the excluded values of D.

57. (Balog–Darmon–Ono). Let $p \geqslant 5$ be prime, let $N > 0$ be such that $\left(\frac{-N}{p}\right) = 1$, assume that $D = (-1)^{(p+1)/2}pN$ is the discriminant of a quadratic field, and recall that we set $S_n(D) = \sum_{0 \leqslant r < |D|} \left(\frac{D}{r}\right)r^n$. Corollary 9.5.8 tells us that $S_n(D) \equiv 0 \pmod{D}$. Prove that in fact $S_{(p+1)/2}(D) \equiv 0 \pmod{pD}$ (for help and several other results of the same type, see [Bal-Dar-Ono]).

58. Prove that for $k \in 2\mathbb{Z}_{>0}$ the numerator of $|B_k/k|$ is equal to 1 if and only if $k = 2$, 4, 6, 8, 10, and 14. Note that these are exactly the (strictly positive) values of k for which there are no modular cusp forms of weight k over $\mathrm{SL}_2(\mathbb{Z})$. Indeed, if there *are* such cusp forms, then one can prove that there exists one that is congruent to an Eisenstein series modulo a prime factor of the numerator of $|B_k/k|$. For instance, $\tau(n) \equiv \sigma_{11}(n) \pmod{691}$, where $\tau(n)$ is Ramanujan's tau function (see Section 10.1.3).

59. Using the Clausen–von Staudt theorem give another proof that the tangent numbers T_{2k-1} are integral (see Definition 9.1.6).

60. By Corollary 9.1.10, the Euler numbers $E_{2k} = -4^{2k+1}B_{2k+1}(1/4)/(2k+1)$ are in $\mathbb{Z}$, which is a slightly stronger statement than what the Almkvist–Meurman Theorem 9.5.29 asserts. More generally, show that if $q = 2^m$ with $m \geqslant 1$, then for any $p \in \mathbb{Z}$ we have $q^{2k+1}B_{2k+1}(p/q)/(2k+1) \in \mathbb{Z}$ (see Theorem 11.4.12 for a more general statement). What happens for $m = 0$?

61. Using the Voronoi congruences (Proposition 9.5.20) prove the following congruence, due to Kummer. Let $e \geqslant 1$, k an even integer such that $k \geqslant e + 1$, and p a prime such that $(p-1) \nmid k$. Then

$$\sum_{j=0}^{e}(-1)^j\binom{e}{j}\frac{B_{k+j(p-1)}}{k+j(p-1)} \equiv 0 \pmod{p^e}.$$

62. Generalizing Hermite's Lemma 9.5.28, prove the following congruence due to Glaisher: for $1 \leqslant r \leqslant p-1$ and $n \geqslant 1$ we have

$$\sum_{\substack{1 \leqslant m \leqslant n \\ m \equiv r \pmod{p-1}}}\binom{n}{m} \equiv \binom{n \bmod (p-1)}{r} \pmod{p},$$

where $n \bmod p - 1$ is the unique integer congruent to n modulo $p-1$ in the interval $[1, p-1]$ (not $[0, p-2]$). (Hint: use a similar proof, but now with expressions of the form $\sum_{a \in \mathbb{F}_p} a^k(a+1)^n$ for a suitable k.)

63. The aim of this exercise is to give an alternative proof of Theorem 9.5.29. As in the proof given in the text, we may assume that $h = 1$ and we must show that $a_n = b_n(k) = k^n \tilde{B}_n(1/k)$ is an integer.

(a) Compute explicitly the exponential generating series $\sum_{n \geqslant 0} a_n t^n/n!$, and by multiplying by $e^{kt} - 1$ or by $\sum_{0 \leqslant j \leqslant k-1} e^{jt}$, prove that the a_n satisfy the following two recurrences:

$$(n+1)(1 - a_n) = \sum_{j=1}^{n-1}\binom{n+1}{j}a_j k^{n-j} \quad \text{and} \quad -ka_n = \sum_{j=1}^{n-1}\binom{n}{j}a_j s_{n-j},$$

where $s_m = \sum_{1 \leqslant j \leqslant k-1} j^m$ for $m \geqslant 1$.

(b) We prove that $a_n \in \mathbb{Z}$ by induction, the result being clear for $n \leqslant 1$. Assume $n \geqslant 2$ and that $a_j \in \mathbb{Z}$ for $1 \leqslant j \leqslant n - 1$. Write $k = \prod_{1 \leqslant i \leqslant g} p_i^{v_i}$, where the p_i are distinct primes and $v_i \geqslant 1$, and $n + 1 = q \prod_{1 \leqslant i \leqslant g} p_i^{w_i}$ with $\gcd(q, k) = 1$ and $w_i \geqslant 0$. Using the recurrences, prove that $p_i^{w_i} \mid (n + 1)a_n$ for all i. (Hint: for $w_i = 0$ this is trivial, and for $w_i \geqslant 1$ prove that $v_{p_i}(\binom{n+1}{j} k^{n-j}) \geqslant w_i$ for $1 \leqslant j \leqslant n - 1$ by separating the cases $j \leqslant n - w_i$ and $n - w_i + 1 \leqslant j \leqslant n - 1$ and using Lemma 4.2.8.)

(c) Keeping the above notation, deduce that $(n + 1)a_n$ is divisible by $(n + 1)/q$, and then that $a_n \in \mathbb{Z}$.

64. Prove the distribution relation for the Hurwitz zeta function (Proposition 9.6.12).

65. Prove that for $x \notin \mathbb{Z}$ we have

$$\sum_{k \geqslant 2} (\zeta(k, 1 - x) + (-1)^k \zeta(k, x))t^k = \frac{\pi t \sin(\pi t)}{\sin(\pi x) \sin(\pi(x + t))} .$$

66. Prove that for $\Re(s) > 0$ we have the convergent series expansion

$$\operatorname{Log}\Gamma(s) = \left(s - \frac{1}{2}\right)\log(s) - s + \frac{1}{2}\log(2\pi) + \frac{1}{2}\sum_{k \geqslant 2} \frac{k - 1}{k(k + 1)}\zeta(k, s + 1) ,$$

where $\zeta(k, s+1)$ can be defined for complex s by the usual series, since $k \in \mathbb{Z}_{\geqslant 2}$.

67. The attentive reader will have noticed that we have proved all the functional equations of the gamma function as corollaries of corresponding formulas for the Hurwitz zeta function, with the exception of the reflection formula (Proposition 9.6.34).

(a) As a consequence of the first formula of Corollary 9.6.52, prove that for $x \in \mathbb{R}_{\geqslant 0} \setminus \mathbb{Z}_{\geqslant 0}$ and $\Re(s) > 0$ we have

$$\zeta(1 - s, x) + \zeta(1 - s, 1 - x) = 4(2\pi)^{-s}\Gamma(s)\cos(s\pi/2)\sum_{n \geqslant 1} \frac{\cos(2\pi n x)}{n^s}$$

$$+ \left(e^{i\pi(s-1)} - 1\right)\sum_{0 \leqslant j \leqslant x-1} (j + \{x\})^{s-1} .$$

(b) Deduce from this the reflection formula for the gamma function.

(c) What can one deduce in the same manner from the *second* formula of Corollary 9.6.52?

68.

(a) Prove that the function $\operatorname{Log}\Gamma(s)$ satisfies the functional equation $\operatorname{Log}\Gamma(s+1) = \operatorname{Log}\Gamma(s) + \log(s)$ and the distribution formula

$$\sum_{0 \leqslant j < n} \operatorname{Log}\Gamma\left(s + \frac{j}{n}\right) = \left(\frac{1}{2} - ns\right)\log(n) + \frac{n - 1}{2}\log(2\pi) + \operatorname{Log}\Gamma(ns) .$$

(b) Write $s = x + iy$ with x and y in $\mathbb{R}$. Prove that the reflection formula for the function $\operatorname{Log}\Gamma$ is given for $s \notin \mathbb{R}$ by

$$\operatorname{Log}\Gamma(s) + \operatorname{Log}\Gamma(1 - s) = \log(\pi) - \log(\sin(\pi s)) + 2i\pi k(s) ,$$

where $k(s)$ is an integer given for $y \neq 0$ by

$$k(s) = \text{sign}(y) \left\lfloor \frac{x + 1/2}{2} \right\rfloor .$$

69. For this exercise you will first need to study the elementary properties of Dirichlet L-functions, in particular Corollary 10.3.2 and Proposition 10.3.5. Let $m \in \mathbb{Z}_{\geqslant 2}$ and let χ be a nontrivial (but not necessarily primitive) character modulo m.

(a) For $x \in \mathbb{R}_{\geqslant 0}$ and $\Re(s) > 1$ set $\zeta_\chi(s, x) = \sum_{n \geqslant 1} \chi(n)(x + n)^{-s}$. Compute $\zeta_\chi(s, x)$ in terms of the ordinary Hurwitz zeta function, and deduce that it can be extended to a holomorphic function of $s \in \mathbb{C}$. Note that in particular $\zeta_\chi(s, 0) = L(\chi, s)$.

(b) Show that $\zeta_\chi(0, x) = L(\chi, 0)$ and $\zeta'_\chi(0, 0) = L'(\chi, 0)$, where here and below ζ'_χ denotes derivation with respect to the first variable s. For simplicity of notation, set

$$C(\chi) = L'(\chi, 0) + \log(m)L(\chi, 0) = \sum_{1 \leqslant r < m} \chi(r) \log \left(\Gamma \left(\frac{r}{m} \right) \right) .$$

(c) In analogy with the ordinary gamma function, define

$$\Gamma_\chi(x) = \exp(\zeta'_\chi(0, x) - \zeta'_\chi(0, 0)) = \exp(\zeta'_\chi(0, x) - L'(\chi, 0)) ,$$

so that in particular $\Gamma_\chi(0) = 1$. Show that

$$\Gamma_\chi(x) = \exp \left(-C(\chi) + \sum_{0 \leqslant r < m} \chi(r) \log \left(\Gamma \left(\frac{r + x}{m} \right) \right) \right) .$$

(d) Deduce the analogues of all the standard formulas for the ordinary gamma function such as the functional equation, the reflection formula, the distribution formula, Raabe's formula, the Hadamard product expansion, the power series expansion of its logarithm, and Stirling's formula.

70. Prove that

$$\int_0^\infty \frac{\cosh(t/2) - 1}{t(e^t - 1)} \, dt = \frac{1}{2} \log \left(\frac{\pi}{2} \right) .$$

71. Let $r \in \mathbb{Z}_{\geqslant 1}$. Define Barnes's multiple gamma function $\Gamma_r(x)$ (which has no relation with the function defined in Chapter 8) by the following formula analogous to that used to define $\Gamma(x) = \Gamma_1(x)$:

$$\log(\Gamma_r(x)) = \zeta'(1 - r, x) - \zeta'(1 - r, 0) .$$

In particular, we have $\Gamma_r(x + 1) = x^{(x^{r-1})}\Gamma_r(x)$, so when $N \in \mathbb{Z}_{>0}$ we have $\Gamma_r(N) = \prod_{1 \leqslant m < N} m^{(m^{r-1})}$ (see Exercise 44). Prove as many results as you can that generalize those for $\Gamma(s)$, such as a distribution formula, a reflection formula, and so on. You may also want to study the properties of the modified function

$$\sum_{i=1}^r (-1)^{r-i} \binom{x - 1}{r - i} \log(\Gamma_i(x)) .$$

72. Set $K = \int_0^\infty e^{-t^2}\, dt$ and for $x \geqslant 0$ set

$$f(x) = 2K \int_0^x e^{-t^2}\, dt + \int_0^\infty \frac{e^{-x^2(1+t^2)}}{1+t^2}\, dt \ .$$

Prove that $f'(x) = 0$, then that $f(x) = \pi/2$ for all $x \geqslant 0$, and deduce that $K = \sqrt{\pi}/2$.

73.

(a) Show that under reasonable assumptions on a function f we have

$$\int_{-\infty}^\infty f(t - 1/t)\, dt = \int_{-\infty}^\infty f(t)\, dt$$

(you must give a sufficient condition on f for this to be valid).

(b) For example, deduce from Proposition 9.6.21 that

$$\int_{-\infty}^\infty e^{-t^2 - 1/t^2}\, dt = \frac{\sqrt{\pi}}{e^2} \ .$$

74. Using the proof of Proposition 9.6.24, give another proof of the formula

$$\zeta(2k) = (-1)^{k-1} \frac{2^{2k-1} \pi^{2k} B_{2k}}{(2k)!} \ .$$

75. For $k \in \mathbb{Z}_{\geqslant 2}$ let

$$P(k) = \prod_{n \geqslant 2} \frac{n^k + 1}{n^k - 1} \ ,$$

which is clearly a convergent product for $k \geqslant 2$.

(a) Using Proposition 9.6.24 show that $P(2) = \sinh(\pi)/\pi$.
(b) By decomposing $x^3 \pm 1$, compute explicitly by induction $\prod_{2 \leqslant n \leqslant N}(n^3 + 1)/(n^3 - 1)$ and deduce that $P(3) = 3/2$.
(c) Compute $P(k)$ for general $k \in \mathbb{Z}_{\geqslant 2}$ in terms of a finite product of values of the gamma function at complex arguments.
(d) Compute explicitly $P(k)$ for k even in terms of trigonometric and hyperbolic functions.

76. Prove that in a suitable domain of the complex plane we have

$$s = \int_0^\infty \left(\frac{1 - e^{-st}}{t^2} - s\frac{e^{-st}}{t} \right) dt \ .$$

77. Prove Propositions 9.6.41, 9.6.42, and 9.6.43.
78. Prove Propositions 9.6.44 and 9.6.45.
79.

(a) Show that

$$\psi'(x) - \frac{1}{x} - \frac{1}{x^2} = \int_0^\infty e^{-xt} \left(\frac{t + 1 - e^t}{e^t - 1} \right) dt \quad \text{and}$$

$$\psi'(x) - \frac{1}{x} - \frac{1}{2x^2} = \int_0^\infty e^{-xt} \left(\frac{t/2 + 1 + (t/2 - 1)e^t}{e^t - 1} \right) dt \ .$$

(b) Deduce that for all $x > 0$ we have

$$\frac{1}{x} + \frac{1}{2x^2} < \psi'(x) < \frac{1}{x} + \frac{1}{x^2} .$$

80. This exercise has nothing to do with the topics studied in this book, but serves as a motivation for the next one, and I thank B. Conrey for it. Its aim is to prove a weak form of the *large sieve inequality*, sufficient for most applications. The optimal form will be given below in Exercise 83.

Let f be a continuously differentiable periodic function of period 1 on $\mathbb{R}$.

(a) Using integration by parts, prove that for all $x \in [0, 1]$ we have

$$f(x) = \int_0^1 f(t)\, dt + \int_0^x t f'(t)\, dt + \int_x^1 (t - 1) f'(t)\, dt .$$

(b) Deduce that

$$f(1/2) \leqslant \int_0^1 |f(t)|\, dt + \frac{1}{2} \int_0^1 |f'(t)|\, dt .$$

(c) Deduce that more generally, for any $\alpha \in [0, 1]$ and any $\delta > 0$ we have

$$|f(\alpha)| \leqslant \frac{1}{\delta} \int_{\alpha - \delta/2}^{\alpha + \delta/2} |f(t)|\, dt + \frac{1}{2} \int_{\alpha - \delta/2}^{\alpha + \delta/2} |f'(t)|\, dt .$$

(d) For $x \in \mathbb{R}$ define $\|x\| = \min_{n \in \mathbb{Z}} |x - n|$; in other words, $\|x\|$ is the distance from x to the nearest integer, and let $x_1, \ldots, x_R$ be real numbers such that $\|x_r - x_s\| \geqslant \delta$ for all $r \neq s$ (such numbers are said to be δ-spaced), where $0 < \delta \leqslant 1/2$ is given. Deduce from the preceding inequality that

$$\sum_{r=1}^R |f(x_r)| \leqslant \frac{1}{\delta} \int_0^1 |f(t)|\, dt + \frac{1}{2} \int_0^1 |f'(t)|\, dt .$$

(e) Let $a_1, \ldots, a_N$ be arbitrary complex numbers and let

$$S(\alpha) = \sum_{n=1}^N a_n e^{2i\pi n\alpha} .$$

Applying the above inequality to the function $f(\alpha) = e^{-2i\pi N\alpha} S(\alpha)^2$, and using Parseval's equality and the Cauchy–Schwarz inequality, prove the following large sieve inequality:

$$\sum_{r=1}^R |S(x_r)|^2 \leqslant \left(\frac{1}{\delta} + \pi N \right) \sum_{n=1}^N |a_n|^2 .$$

We will see in Exercise 83 that π can be replaced by 1.

81. (I thank J. Rivat for the following exercises.) Define the *Beurling–Selberg function* $H(z)$ by the formula

$$H(z) = \left(\frac{\sin(\pi z)}{\pi} \right)^2 \left(\psi'(-z) - \psi'(z) + \frac{2}{z} \right) ,$$

set

$$H_N(z) = \left(\frac{\sin(\pi z)}{\pi}\right)^2 \left(\sum_{n=-N}^{N} \frac{\text{sign}(n)}{(z-n)^2} + \frac{2}{z}\right)$$

(with $\text{sign}(0) = 0$), and finally set also

$$K(z) = \left(\frac{\sin(\pi z)}{\pi z}\right)^2 \quad \text{and} \quad B(z) = H(z) + K(z) .$$

(a) Prove that although the function $H(z)$ is a priori defined only for $z \notin \mathbb{Z}$, it can be extended to $z \in \mathbb{Z}$ into a holomorphic function. Compute $H(z)$ for $z \in \mathbb{Z}$, and draw a picture of its graph for real values of z (between -4 and 4 for instance). You will of course need to be careful around integral values of z.

(b) Show that $\lim_{N\to\infty} H_N(z) = H(z)$ and $\lim_{N\to\infty} H'_N(z) = H'(z)$ uniformly on $\mathbb{R}$.

(c) Compute the Fourier transform of the function with compact support $f(x) = \max(1 - |x|, 0)$, and deduce from the Fourier inversion formula the Fourier transform of the function $K(x)$.

(d) Using the preceding exercise, show that for all $x \in \mathbb{R}$ we have

$$|H(x)| \leqslant 1, \quad |\text{sign}(x) - H(x)| \leqslant K(x), \quad \text{and} \quad \int_{-\infty}^{\infty} (B(t) - \text{sign}(t))\, dt = 1 .$$

82. (Continuation of the preceding exercise.) The aim of this exercise is to compute the Fourier transform of the function $H(x)$.

(a) Using the preceding exercise, show that

$$H_N(z) = \sum_{n=-N}^{N} \text{sign}(n) K(z - n) + 2z K(z) ,$$

and deduce that

$$H_N(z) = \int_{-1}^{1} \left((1 - |t|)\left(\cot\!an(\pi t) - \frac{\cos((2N+1)\pi t)}{\sin(\pi t)}\right) + \frac{\text{sign}(t)}{\pi}\right) \frac{e^{2i\pi zt}}{i}\, dt .$$

(b) We would now like to apply the Riemann–Lebesgue lemma, but this is not possible because of the singularity of the integrand at $t = 0$. We can make this singularity disappear by computing the derivative with respect to z. Thus, compute $H'(z)$ as a Fourier integral, and using the Fourier inversion formula deduce the Fourier transform of $H'(x)$. In particular, show that it vanishes outside $[-1, 1]$.

(c) Finally, by a careful integration by parts compute the Fourier transform of the function $H(x) - \text{sign}(x)$, and in particular show that it is equal to $i/(\pi x)$ when $|x| > 1$ and to 0 when $x = 0$.

(d) Let a, b, and δ be fixed real numbers such that $a \leqslant b$ and $\delta > 0$, and set

$$F(x) = \frac{1}{2}(B(\delta(x - a)) + B(\delta(b - x))) .$$

Prove that $F(x) \geqslant 0$ for all $x \in \mathbb{R}$, that $F(x) \geqslant 1$ for $x \in [a, b]$, that $\widehat{F}(x) = 0$ for $|x| \geqslant \delta$, and that $\widehat{F}(0) = b - a + 1/\delta$, where as usual $\widehat{F}$ denotes the Fourier transform of F.

83. (Continuation of the preceding exercises.) The aim of this exercise is to show how the Beurling–Selberg function gives a reasonably simple proof of the optimal form of the discrete *large sieve inequality*, improving on Exercise 80, and which was initially obtained by more complicated means. We keep the notation of that exercise, we set by convention $a_j = 0$ for $j \leqslant 0$ and $j > N$, and we let $F(x)$ be the function defined in the preceding exercise with $a = 1$ and $b = N$.

(a) Using the Poisson summation formula and the properties of the function $\widehat{F}$, show that

$$\sum_{n \in \mathbb{Z}} F(n)e^{-2i\pi n(x_r - x_s)} = \begin{cases} \widehat{F}(0) & \text{if } r = s, \\ 0 & \text{otherwise.} \end{cases}$$

(b) By expanding

$$\sum_{n \in \mathbb{Z}} \left| \frac{a_n}{\sqrt{F(n)}} \widehat{F}(0) - \sum_{r=1}^{R} \sqrt{F(n)} S(x_r) e^{-2i\pi n x_r} \right|^2 ,$$

deduce the large sieve inequality

$$\sum_{r=1}^{R} |S(x_r)|^2 \leqslant \left(\frac{1}{\delta} + N - 1 \right) \sum_{n=1}^{N} |a_n|^2 ,$$

where as above $S(\alpha) = \sum_{1 \leqslant n \leqslant N} a_n e^{2i\pi n\alpha}$.

Note that this improves on the inequality obtained in Exercise 80, and it is not difficult to show that it is optimal. Also see any good book on analytic number theory such as [Iwa-Kow] for numerous number-theoretic applications of large sieve inequalities.

84. By expanding $1/(e^t - 1)$ in powers of e^{-t}, show directly that

$$\Gamma(s)\left(\zeta(s) - \frac{1}{s-1} \right) = \int_0^\infty \left(\frac{1}{e^t - 1} - \frac{e^{-t}}{t} \right) t^{s-1}\, dt$$

(see also Corollary 10.2.3 (2)). Deduce from this another proof of the formula $\lim_{s \to 1}(\zeta(s) - 1/(s-1)) = \gamma$ seen in Proposition 9.2.14.

85. Give an alternative proof of Proposition 9.6.33 using Proposition 9.6.17 and Stirling's formula.

86.

(a) Using the change of variables $(x_1, y_1) = (x, -\log(xy))$, compute in terms of the gamma function

$$\int_0^1 \int_0^1 x^k y^\ell (-\log(xy))^s\, dx\, dy ,$$

for $k \in \mathbb{Z}_{\geqslant 0}$, $\ell \in \mathbb{Z}_{\geqslant 0}$, and $s \in \mathbb{C}$, and specify for which s it converges. You may assume $k \geqslant \ell$, and should separate the cases $k > \ell$ and $k = \ell$.

(b) Deduce the value of

$$\int_0^1 \int_0^1 \frac{x^k y^\ell (-\log(xy))^s}{1 - xy}\, dx\, dy .$$

(c) Deduce Sondow's formula

$$\gamma = \int_0^1 \int_0^1 \frac{1-x}{(1-xy)(-\log(xy))} \, dx \, dy \ .$$

87. Compute $\sum_{n\geqslant 1} \frac{n}{2^{4n}} \zeta(2n+1)$ in terms of Catalan's constant G defined in Exercise 40.

88. For $x \in \mathbb{C}$ with $\Re(x) > 0$, or for $x \in \mathbb{C}_p$ with $|x| > 1$, set

$$S(x) = \sum_{n\geqslant 0} \frac{1}{n+1} \begin{bmatrix} n \\ x \end{bmatrix} \quad \text{and} \quad T(x) = \sum_{n\geqslant 0} \frac{1}{n+x} \begin{bmatrix} n \\ x \end{bmatrix} ,$$

where $\begin{bmatrix} n \\ x \end{bmatrix} = n!/(x(x+1)\cdots(x+n))$ is the inverse binomial symbol introduced in Exercise 37 of Chapter 4.

(a) Show that these series converges absolutely and that

$$S(x) - S(x+1) = \frac{1}{x^2} \quad \text{and} \quad T(x) + T(x+1) = \frac{2}{x^2}$$

(use Exercise 37 (a) of Chapter 4).

(b) Deduce from Proposition 9.6.41 that for $x \in \mathbb{C}$ with $\Re(x) > 0$ we have $S(x) = \psi'(x)$ and

$$T(x) = 2 \sum_{n\geqslant 0} \frac{(-1)^n}{(x+n)^2} = \psi'(x/2) - 2\psi'(x) ,$$

in other words that

$$\psi'(x) = \sum_{n\geqslant 0} \frac{1}{n+1} \begin{bmatrix} n \\ x \end{bmatrix} \quad \text{and} \quad \psi'(x/2) - 2\psi'(x) = \sum_{n\geqslant 0} \frac{1}{n+x} \begin{bmatrix} n \\ x \end{bmatrix}$$

(the analogues of these results in the p-adic case are proved in Exercises 21 and 23 of Chapter 11).

(c) Similarly, show that

$$\sum_{n\geqslant 0} \frac{1}{n+2} \begin{bmatrix} n \\ x \end{bmatrix} = (1-x)\psi'(x) + 1 .$$

(d) Deduce for instance that in $\mathbb{R}$ we have the equalities

$$\sum_{n\geqslant 0} \frac{1}{n+1} \begin{bmatrix} n \\ 1/2 \end{bmatrix} = \frac{\pi^2}{2}, \quad \sum_{n\geqslant 0} \frac{1}{n+2} \begin{bmatrix} n \\ 1/2 \end{bmatrix} = \frac{\pi^2}{4}+1, \quad \sum_{n\geqslant 0} \frac{1}{n+1/2} \begin{bmatrix} n \\ 1/2 \end{bmatrix} = 8G ,$$

where G is Catalan's constant.

(e) (Harder.) More generally, show that for $k \in \mathbb{Z}_{\geqslant 1}$ we have

$$\sum_{n\geqslant 0} \frac{1}{n+k} \begin{bmatrix} n \\ x \end{bmatrix} = (-1)^{k-1} \binom{x-1}{k-1} (\psi'(x) + h_k(x)), \quad \text{where}$$

$$h_k(x) = \sum_{1\leqslant j\leqslant k-1} \left(\frac{1}{(x-k+j)^2} + \frac{(-1)^j(x-k+2j)}{j^2(x-k+j)\binom{x-k+j}{j}} \right) .$$

Note for experts: $h_k(x)$ is essentially the function that occurs in the Padé approximation table used by Apéry in his proof of the irrationality of $\zeta(2)$.

89. (Continuation of the preceding exercise.)

(a) By decomposing $\begin{bmatrix} n \\ x \end{bmatrix}$ into partial fractions and computing its asymptotic expansion as $x \to \infty$, prove that

$$B_m = \sum_{n=0}^{m} \frac{1}{n+1} \sum_{j=0}^{n} (-1)^j \binom{n}{j} j^m .$$

(b) More generally, prove directly that

$$B_m(z) = \sum_{n=0}^{m} \frac{1}{n+1} \sum_{j=0}^{n} (-1)^j \binom{n}{j} (j+z)^m .$$

Show also directly that this is equivalent to the identity of Exercise 26 (b).

(c) By decomposing $\begin{bmatrix} n \\ x \end{bmatrix}/(n+x)$ into partial fractions, prove that as $x \to \infty$ we have the asymptotic expansion

$$\psi'(x/2) - 2\psi'(x) = \sum_{m \geqslant 0} \frac{(-1)^{m+1}}{x^{m+1}} \sum_{n=0}^{m-1} \sum_{j=0}^{n} (-1)^j \frac{n^m - j^m}{n-j} \binom{n}{j} ,$$

and where $(n^m - j^m)/(n-j)$ is to be interpreted as mn^{m-1} if $j = n$.

(d) Deduce that for all $m \geqslant 0$ we have the following formula for Bernoulli numbers:

$$-2(2^m - 1)B_m = \sum_{n=0}^{m-1} \sum_{j=0}^{n} (-1)^j \frac{n^m - j^m}{n-j} \binom{n}{j} .$$

(e) More generally, prove directly that

$$-2(2^m B_m(z/2) - B_m(z)) = \sum_{n=0}^{m-1} \sum_{j=0}^{n} (-1)^j \frac{(n+z)^m - (j+z)^m}{n-j} \binom{n}{j} ,$$

where $((n+z)^m - (j+z)^m)/(n-j)$ is to be interpreted as $m(n+z)^{m-1}$ if $j = n$.

90. Let $(b_n)_{n \geqslant 0}$ be a sequence. In analogy with Exercise 38 of Chapter 4, define its 2-Stirling transform as the sequence $(a_n)_{n \geqslant 0}$ given by the formal identity

$$e^{T/2} \sum_{n \geqslant 0} (-1)^n \frac{n!^2 2^{2n+1}}{(2n+1)!} b_n (\sinh(T/2))^{2n+1} = \sum_{n \geqslant 0} a_n \frac{T^n}{n!} .$$

(a) Prove that we have the Taylor series expansion

$$\frac{\sinh^{-1}(x)}{\sqrt{1+x^2}} = \sum_{n \geqslant 0} (-1)^n n!^2 2^{2n} \frac{x^{2n+1}}{(2n+1)!} .$$

(b) Prove that under suitable conditions on x and the sequence b_n, either in $\mathbb{C}$ or in $\mathbb{C}_p$, we have the Laurent series expansion

$$\sum_{n \geqslant 0} b_n \begin{bmatrix} n \\ x \end{bmatrix} \begin{bmatrix} n \\ 1-x \end{bmatrix} = \sum_{n \geqslant 0} \frac{a_n}{x^{n+1}} ,$$

where a_n is the 2-Stirling transform of b_n.

(c) Deduce from this another proof of the formula for $T(x)$ given in Exercise 88.

(d) Show that for suitable values of $x \in \mathbb{C}$ we have

$$\sum_{n \geqslant 0} \frac{1}{n+1} \begin{bmatrix} n \\ x \end{bmatrix} \begin{bmatrix} n \\ 1-x \end{bmatrix} = \psi''(x)$$

(the p-adic analogue of this result is proved in Exercise 24 of Chapter 11).

(e) Set $f(n,x) = \begin{bmatrix} n \\ x \end{bmatrix} \begin{bmatrix} n \\ 1-x \end{bmatrix}$. Prove that

$$f(n,x) - f(n,x+1) = 2 \begin{bmatrix} n \\ 1-x \end{bmatrix} \begin{bmatrix} n+1 \\ x \end{bmatrix},$$

and deduce from this a more direct proof of the formula of the preceding question. (Hint: find a sequence $u_n = u_n(x)$ such that $(f(n,x) - f(n,x+1))/(n+1) = u_{n+1} - u_n$.)

(f) Generalize as much as you can all the results of Exercises 88 and 89. In particular, the experts should recover the Padé approximation table used by Apéry for $\zeta(3)$.

91.

(a) Using the duplication formula, prove that the Taylor expansion of $\mathrm{Log}\Gamma(s)$ around $s = 1/2$ is given by

$$\mathrm{Log}\Gamma(s) = \frac{\log(\pi)}{2} - (2\log(2)+\gamma)\left(s - \frac{1}{2}\right) + \sum_{k \geqslant 2}(-1)^k(2^k - 1)\frac{\zeta(k)}{k}\left(s - \frac{1}{2}\right)^k,$$

with radius of convergence $1/2$.

(b) Deduce from this the value of $\psi^{(k)}(1/2)$ and more generally of $\psi^{(k)}(n+1/2)$ for $n \in \mathbb{Z}$.

92. Recall that the harmonic sum H_m is defined as $H_m = \sum_{1 \leqslant r \leqslant m} 1/r$.

(a) Show that

$$\sum_{1 \leqslant r \leqslant m} \frac{\psi(1+r/m)}{r} = -(\gamma H_m + I_m) \quad \text{with} \quad I_m = \int_0^1 \frac{(1-x^m)\log(1-x^m)}{1-x}\,dx\,.$$

(b) Show that the asymptotic expansion as $m \to \infty$ of I_m is given by

$$I_m = -K + \sum_{k \geqslant 1} \frac{B_k}{m^k}\left(1 - \frac{1}{k}\sum_{2 \leqslant j \leqslant k+1} \zeta(j)\right),$$

where

$$K = \int_0^1 \frac{\psi(1+x) + \gamma}{x}\,dx = \int_0^1 \frac{(1-x)\log(1-x)}{x\log(x)}\,dx$$

$$= \sum_{n \geqslant 1} \frac{(-1)^{n-1}}{n}\zeta(n+1) = -\sum_{n \geqslant 2}\zeta'(n) = \sum_{k \geqslant 1}\frac{\log(k+1)}{k(k+1)} = \sum_{k \geqslant 1}\frac{\log(1+1/k)}{k}$$

$$= 1.2577468869443696300098998304958815285115408905088848689775\ldots\,.$$

I do not know if this constant can be given more "explicitly", for instance by a formula similar to that of I in Exercise 104.

93. Prove that for $n \in \mathbb{Z}_{\geqslant 1}$ and $0 \leqslant t \leqslant n$ we have the double inequality

$$e^{-t}(1 - t^2/n) \leqslant (1 - t/n)^n \leqslant e^{-t} .$$

(Hint: for the inequality on the left, show that the derivative of the auxiliary function $f(t) = n \log(1-t/n)+t-\log(1-t^2/n)$ is nonnegative for $0 \leqslant t < n^{1/2}$.)

94.

(a) Using complex exponentials (i.e., de Moivre's formulas) prove that

$$\sin^{2k+1}(x) = \frac{(-1)^k}{2^{2k}} \sum_{j=0}^{k}(-1)^j \binom{2k+1}{j} \sin(2k+1-2j)(x) .$$

(b) Using Proposition 9.6.38, deduce that

$$\int_0^\infty \frac{\sin^{2k+1}(x)}{x} \, dx = \frac{\pi}{2^{2k+1}} \binom{2k}{k} .$$

95. For $y \geqslant 0$, set

$$f(y) = \int_0^\infty e^{-yt} \frac{\sin(t)}{t} \, dt .$$

Show that for $y > 0$ it is legal to differentiate under the integral sign, compute $f'(y)$, then $f(y)$ for $y > 0$. Finally, show that $f(y)$ tends to $f(0)$ as y tends to 0 from above, and deduce the value of $\int_0^\infty \sin(xt)/t \, dt$.

96. Prove the formulas

$$\int_0^\infty \frac{\sin(x)}{x^s} \, dx = \cos(\pi s/2)\Gamma(1 - s) = \frac{\pi}{2 \sin(s\pi/2)\Gamma(s)} ,$$

$$\int_0^\infty \frac{1 - \cos(x)}{x^s} \, dx = - \sin(\pi s/2)\Gamma(1 - s) = -\frac{\pi}{2 \cos(s\pi/2)\Gamma(s)} ,$$

$$\int_0^\infty \frac{\sin^2(x)}{x^s} \, dx = -2^{s-2} \sin(\pi s/2)\Gamma(1 - s) = -2^{s-3}\frac{\pi}{\cos(s\pi/2)\Gamma(s)} ,$$

the first one for $0 < \Re(s) < 2$, and the next two for $1 < \Re(s) < 3$, so that in particular

$$\int_0^\infty \frac{\sin^2(x)}{x^2} \, dx = \frac{\pi}{2} .$$

97. Prove the formula $\int_{-\infty}^\infty \sin^2(t)/t^2 \, dt = \pi$:

(a) By solving Exercise 96.
(b) By using a similar method to that of Exercise 95.
(c) By integrating along a suitable contour in the complex plane.

98. Prove Corollary 9.6.54, both directly and by differentiating the formulas of Proposition 9.6.50.

99. The following two exercises are taken from [Bor-Bai] and [Bor-Bai-Gir]. Let C_ε be the contour in the complex plane going from $-\infty - i\varepsilon$ to $-\varepsilon - i\varepsilon$, then around the circle of radius $\varepsilon\sqrt{2}$ to $\varepsilon + i\varepsilon$ and finally to $-\infty + i\varepsilon$, where $\varepsilon > 0$. Set

$$I(z) = \int_{C_\varepsilon} t^{-z} e^{-t}\, dt\;,$$

where $t^{-z} = \exp(-z\log(t))$ and we choose the principal determination of the logarithm $-\pi < \Im(\log(t)) \leqslant \pi$.

(a) Show that for $\Re(z) < 1$ the integral $I(z)$ is independent of $\varepsilon > 0$, and by letting ε tend to 0, show that $I(z) = 2i\sin(\pi z)\Gamma(1 - z)$.

(b) Deduce that for *all* $z \in \mathbb{C}$ and for all $\varepsilon > 0$ we have

$$\frac{1}{\Gamma(z)} = \frac{1}{2i\pi}\int_{C_\varepsilon} t^{-z} e^{t}\, dt\;.$$

(c) Deduce that for all $\varepsilon > 0$,

$$\int_0^\infty \frac{dx}{\Gamma(x)} = \frac{1}{2i\pi}\int_{C_\varepsilon} \frac{e^t}{\log(t)}\, dt\;.$$

(d) By choosing $\varepsilon > 1$ (explain why this is necessary), prove finally the formula

$$\int_0^\infty \frac{dx}{\Gamma(x)} = e + \int_0^\infty \frac{e^{-t}}{\log^2(t) + \pi^2}\, dt\;.$$

(e) In a similar manner, show that

$$\int_0^\infty \frac{x^{-1}}{\Gamma(x)}\, dx = \int_0^\infty \frac{dx}{\Gamma(x+1)} = e - \int_0^\infty \frac{e^{-t}}{t(\log^2(t) + \pi^2)}\, dt$$

and compute $\int_0^\infty x^k/\Gamma(x)\, dx$ for small positive integral values of k.

100. An alternative way to prove the above results, not using complex integration, is as follows. Set

$$I(a,t) = \int_{-t}^\infty \frac{a^x}{\Gamma(x+1)}\, dx + \int_0^\infty \frac{e^{-ax}x^{t-1}}{\pi^2 + \log^2(x)}\left(\cos(\pi t) - \frac{\sin(\pi t)}{\pi}\log(x)\right)\, dx\;.$$

(a) Prove that this integral converges absolutely for $a > 0$ and $t \geqslant 0$, and that its derivative with respect to t vanishes, hence that it is a function $I(a)$ of a alone.

(b) Prove that $I'(a) = I(a)$.

(c) By letting $a \to 0^+$ and using the change of variable $x = \exp(-t)$, deduce the value of $I(a) = I(a,t)$.

(d) In particular, prove that for $k \geqslant 0$,

$$\int_0^\infty \frac{x(x-1)\cdots(x-k+1)a^x}{\Gamma(x+1)}\, dx = a^k e^a + (-1)^{k-1} a^k \int_0^\infty \frac{x^{k-1}e^{-ax}}{\pi^2 + \log^2(x)}\, dx\;,$$

so that in particular,

$$\int_0^\infty \frac{a^x x^{-1}}{\Gamma(x)}\, dx = e^a - \int_0^\infty \frac{e^{-ax}}{x(\pi^2 + \log^2(x))}\, dx \quad \text{and}$$

$$\int_0^\infty \frac{a^x}{\Gamma(x)}\, dx = ae^a + a\int_0^\infty \frac{e^{-ax}}{\pi^2 + \log^2(x)}\, dx\;.$$

101.

(a) Prove that for $x \notin \mathbb{Z}$,

$$\sum_{m=0}^{\infty} \frac{(-1)^m}{m+x} = \psi(x) - \psi(x/2) - \log(2) \quad \text{and}$$

$$\sum_{m \in \mathbb{Z}} \frac{(-1)^m}{m+x} = \frac{\pi}{\sin(\pi x)} .$$

(b) By splitting the integral at $t = 1$, show that for all s such that $\Re(s) > 0$ we have

$$\int_0^{\infty} \frac{t^{s-1}}{t^2+1} \, dt = \frac{\pi}{2\sin(\pi s/2)} .$$

(c) By making the change of variable $u = e^{\pi t}$, deduce from this that the function $1/\cosh(\pi x)$ is equal to its Fourier transform.

102.

(a) Prove the following formulas for $\Re(s) > 0$:

$$\int_0^1 \frac{t^{s-1}}{1+t} \, dt = \psi(s) - \psi(s/2) - \log(2) ,$$

$$\int_0^1 t^{s-1} \log(1+t) \, dt = \frac{1}{s}\left(\psi(s) - \psi(s/2) - \frac{1}{s}\right) .$$

(b) Set

$$F(a) = \sum_{j \geq 1} \frac{(-1)^{j-1}}{j} \psi\left(\frac{j}{a}\right) \quad \text{and} \quad G(a) = \sum_{i,j \geq 1} \frac{(-1)^{j-1}}{i(ai+j)} .$$

Deduce from (a) that

$$L(a) = \int_0^1 \frac{\log(1+t^a)}{1+t} \, dt = -a\frac{\pi^2}{12} + F(a) - F(2a) ,$$

and show that $F(a) = -\gamma \log(2) - a\pi^2/12 + G(a)$, so that $L(a) = G(a) - G(2a)$.

(c) Show that

$$L(a) = \sum_{i,j \geq 1} \frac{(-1)^{i+j}}{i(ai+j)} \quad \text{and} \quad G(a) = -\int_0^1 \frac{\log(1-t^a)}{1+t} \, dt .$$

See Exercise 60 of Chapter 10 for the sequel.

103. Prove Proposition 9.6.47.

104. Set

$$I = \int_0^1 \left(\psi^2(x) - \frac{1}{x^2} - \frac{2\gamma}{x}\right) dx .$$

Show that I is a convergent integral, and using Proposition 9.6.41 show that

$$I = 1 - \frac{\pi^2}{3} + 2\gamma_1, \quad \text{with} \quad \gamma_1 = \lim_{N \to \infty} \left(\sum_{1 \leq j \leq N} \frac{\log(j)}{j} - \frac{\log^2(N)}{2}\right)$$

(see Section 10.3.5 for γ_1). Using Exercise 49 of Chapter 10, deduce that

$$I = -2.4354998246638063226660030850418530167133724790822727806691\ldots.$$

105. For a and b in $\mathbb{Z}_{\geqslant 0}$ set

$$I(a,b) = \int_0^\pi t^a \log^b(2\sin(t))\, dt \; .$$

The aim of this exercise is to compute $I(a,b)$ explicitly for certain values of a and b in terms of usual quantities, including $\zeta(k)$ for $k \in \mathbb{Z}_{\geqslant 2}$. Note that evidently $I(a,0) = \pi^{a+1}/(a+1)$.

(a) Prove the identity

$$\prod_{1 \leqslant k \leqslant m-1} \sin\left(\frac{\pi k}{m}\right) = \frac{m}{2^{m-1}} \; ,$$

and using Riemann sums deduce that $I(0,1) = I(1,1) = 0$.

(b) Using Proposition 9.6.46 or Proposition 9.6.47 show that

$$\sum_{1 \leqslant r \leqslant m} \psi^2\left(\frac{r}{m}\right) = \frac{\pi^2}{12}(m-1)(m-2)$$

$$+ m\left((\gamma + \log(m))^2 + \sum_{1 \leqslant k \leqslant m-1} \log^2\left(2\sin\left(\frac{\pi k}{m}\right)\right)\right) \; .$$

(c) Using Riemann sums and (a), deduce that

$$I(0,2) = \frac{\pi^3}{12} \quad \text{and} \quad I(1,2) = \frac{\pi^4}{24}$$

(see the proof of Proposition 10.3.17 (3) for help).

(d) Using Exercise 44 (d), the reflection formula, and Exercise 19 of Chapter 10, show that

$$I(k,1) = \sum_{j=1}^{\lfloor k/2 \rfloor} \frac{(-1)^j}{2^{2j}} \frac{k!}{(k+1-2j)!} \pi^{k+1-2j} \zeta(2j+1) \; .$$

(e) By considering the exponential generating series $\sum_{k \geqslant 0} (I(0,k)/k!)x^k$ and using Corollary 9.6.40 (4), show that $I(0,k)$ satisfies the following recurrence for $k \geqslant 1$:

$$I(0,k) = \sum_{j=0}^{k-2} \frac{(k-1)!}{j!}(k-j)a_{k-j}I(0,j), \quad \text{with} \quad a_k = (-1)^k\left(1 - \frac{1}{2^{k-1}}\right)\frac{\zeta(k)}{k}$$

(recall that $I(0,0) = \pi$ and $I(0,1) = 0$). Note also that trivially $I(1,k) = (\pi/2)I(0,k)$. We thus have for instance

$$I(0,3) = -\frac{3}{2}\pi\zeta(3) \quad \text{and} \quad I(1,3) = -\frac{3}{4}\pi^2\zeta(3) \; .$$

(f) Using the LLL algorithm and the numerical integration methods of Section 9.3.2, the author has found the following experimental equalities, but has been too lazy to prove them. Do it for him:

$$I(2,2) = \frac{13}{4}\pi\zeta(4) = \frac{13}{360}\pi^5, \quad I(2,3) = -\left(\frac{3}{2}\pi\zeta(5) + \frac{5}{8}\pi^3\zeta(3)\right),$$

$$I(2,5) = -\left(\frac{45}{4}\pi\zeta(7) + \frac{35}{4}\pi^3\zeta(5) + \frac{71}{96}\pi^5\zeta(3)\right),$$

$$I(4,3) = \frac{45}{4}\pi\zeta(7) - \frac{21}{8}\pi^3\zeta(5) - \frac{5}{8}\pi^5\zeta(3).$$

(g) Deduce that

$$I(3,2) = 3\pi^2\zeta(4) = \frac{\pi^6}{30}, \quad I(3,3) = -\left(\frac{9}{4}\pi^2\zeta(5) + \frac{9}{16}\pi^4\zeta(3)\right),$$

$$I(3,5) = -\left(\frac{135}{8}\pi^2\zeta(7) + \frac{15}{2}\pi^4\zeta(5) + \frac{51}{64}\pi^6\zeta(3)\right),$$

$$I(5,3) = \frac{225}{8}\pi^2\zeta(7) - \frac{45}{16}\pi^4\zeta(5) - \frac{3}{4}\pi^6\zeta(3).$$

106. For $k \in \mathbb{Z}_{\geqslant 1}$ set

$$S_k = \int_0^\infty (\log(2\sinh(t)) - t)^k \, dt \quad \text{and} \quad C_k = \int_0^\infty (\log(2\cosh(t)) - t)^k \, dt.$$

(a) By a series of successive changes of variable, or using Proposition 9.6.43 in a manner similar to the previous exercise, show that $S_k = (-1)^k k! \zeta(k+1)/2$.
(b) Show that $C_1 = \zeta(2)/4 = \pi^2/24$ and $C_2 = \zeta(3)/8$ (note that for $k \geqslant 3$ the expression for C_k involves $\mathrm{Li}_{k+1}(1/2)$, which is believed not to have any "explicit" form).

107. Let $k \in \mathbb{Z}_{\geqslant 1}$. Generalizing Proposition 9.6.46, show that when $0 < r < m$ we have

$$\psi^{(k)}\left(\frac{r}{m}\right) = (-1)^{k-1}k!m^k\left(\zeta(k+1) + \sum_{1\leqslant j\leqslant m-1} \zeta_m^{-jr}\,\mathrm{Li}_{k+1}(\zeta_m^j)\right),$$

where Li_{k+1} is the polylogarithm function defined in Exercise 22 of Chapter 4.

108.

(a) For any nice function f defined on $\mathbb{R}$ and tending to zero sufficiently rapidly at $\pm\infty$, and for any $\theta \notin \pi\mathbb{Z}$, set

$$r_\theta(f)(x) = \frac{\lambda(\theta)}{|\sin(\theta)|^{1/2}} \int_{-\infty}^{+\infty} f(t)\exp\left(i\pi\left((x^2 + t^2)\cot\!an(\theta) - \frac{2xt}{\sin(\theta)}\right)\right) dt,$$

where

$$\lambda(\theta) = \exp\left(\frac{i\pi}{2}\left(B_1\left(\left\{\frac{\theta}{\pi}\right\}\right)\right)\right) = \exp\left(\frac{i\pi}{2}\left(\left\{\frac{\theta}{\pi}\right\} - \frac{1}{2}\right)\right).$$

Show that $r_{\theta'}(r_\theta(f)) = r_{\theta'+\theta}(f)$ (it may be useful to use Lemma 10.2.9 proved in the next chapter; see also the proof of Theorem 9.7.5 (1)).

(b) Show that $\lim_{\theta \to 0} r_\theta(f)(x)$ and $\lim_{\theta \to \pi} r_\theta(f)(x)$ exist and are equal to $f(x)$ and $f(-x)$ respectively. (Hint: one way of doing this is to first show convergence in the sense of distributions.)

(c) Noting that $\mathcal{F}(f)(x) = r_{\pi/2}(f)(x)$, deduce the Fourier inversion formula $\mathcal{F}(\mathcal{F}(f))(x) = f(-x)$.

(d) If $f(t) = e^{-\pi t^2}$, show that $r_\theta(f) = f$ for all θ.

Remark. The map $\exp(i\theta) \mapsto r_\theta(f)$ is therefore a *representation* of the group S^1 of complex numbers of modulus 1 to functions, called the Weil representation. The Fourier transform is thus only a special case. As for the Fourier transform, it is easily generalized to $\mathbb{R}^n$.

109. Compute (in a suitable range of the variable s) the Mellin transforms of the functions $\cos(a(x \pm 1/x)/2)$ and $\sin(a(x \pm 1/x)/2)$, where a is a fixed parameter.

110. Fill in the details of the proof of Propositions 9.8.1 and 9.8.2, and in particular find explicitly the expansions of all the Bessel functions around $x = 0$.

111.

(a) Show directly on the power series expansion that for fixed x, as $\nu \to +\infty$ we have
$$J_\nu(x) \sim \frac{(x/2)^\nu}{\Gamma(\nu+1)} \sim \frac{1}{\sqrt{2\pi\nu}}\left(\frac{ex}{2\nu}\right)^\nu ,$$
hence tends to 0 very fast.

(b) Again using directly the power series expansion, deduce that for all $t \in \mathbb{C}^*$ and $x \in \mathbb{C}$ we have the absolutely convergent Laurent generating series due to Schlömilch:
$$\sum_{n \in \mathbb{Z}} t^n J_n(x) = e^{(x/2)(t-1/t)} .$$

(c) By multiplying this series with the one in which t is changed into $1/t$, prove the following identities, valid for all $x \in \mathbb{C}$. For all $N \in \mathbb{Z}_{\neq 0}$:
$$\sum_{n \in \mathbb{Z}} J_n(x) J_{n+N}(x) = 0 ,$$
and
$$J_0^2(x) + 2\sum_{n \geq 1} J_n^2(x) = 1 .$$

This shows in particular that for all $x \in \mathbb{R}$ we have $|J_0(x)| \leq 1$ and $|J_n(x)| \leq 1/\sqrt{2}$ for $n \in \mathbb{Z}_{\neq 0}$.

112.

(a) Similarly to the first question of the preceding exercise, show that for fixed x, as $\nu \to +\infty$ we have
$$Y_\nu(x) \sim -\frac{(x/2)^{-\nu}\Gamma(\nu)}{\pi} \sim -\sqrt{\frac{2}{\pi\nu}}\left(\frac{ex}{2\nu}\right)^{-\nu} .$$

(b) Find the corresponding results for the two other Bessel functions $I_\nu(x)$ and $K_\nu(x)$.

113. Fill in the details of the proof of Proposition 9.8.8 in the case of Y_0 and K_0.

114. Compute $\sum_{k \geqslant 1} \Gamma'(k)/\Gamma^3(k)$ in terms of K-Bessel functions.

115. For $c \notin \mathbb{Z}_{\leqslant 0}$ define Gauss's hypergeometric series $F(a, b, c; x)$ by

$$F(a, b, c; x) = \sum_{n \geqslant 0} \frac{\binom{-a}{n}\binom{-b}{n}}{\binom{-c}{n}}(-x)^n$$

$$= \sum_{n \geqslant 0} \frac{a(a+1)\cdots(a+n-1)b(b+1)\cdots(b+n-1)}{c(c+1)\cdots(c+n-1)}\frac{x^n}{n!} .$$

Note that $F(b, a, c; x) = F(a, b, c; x)$, so in all of the formulas that we will obtain below we can exchange a and b.

(a) Compute the radius of convergence of this series in $\mathbb{C}$, and determine the set of triples $(a, b, c) \in \mathbb{C}^3$ for which the series converges absolutely at $x = 1$.

(b) Show that in a suitable range of the parameters we have the integral representation

$$F(a, b, c; x) = \frac{\Gamma(c)}{\Gamma(c-a)\Gamma(a)} \int_0^1 t^{a-1}(1-t)^{c-a-1}(1-xt)^{-b}\, dt .$$

(Hint: expand in powers of x and use Proposition 9.6.39.)

(c) Deduce that in a suitable range of the parameters, we have the following evaluation, also due to Gauss:

$$F(a, b, c; 1) = \frac{\Gamma(c-a-b)\Gamma(c)}{\Gamma(c-a)\Gamma(c-b)} .$$

(d) Using Corollary 9.6.40 (2), prove that in a suitable range of the parameters (in particular with $\Re(b) < 0$), we have

$$F(a, b, a-b+1; -1) = \frac{\Gamma(a-b+1)\Gamma(a/2)}{2\Gamma(a)\Gamma(a/2-b+1)} .$$

(e) Prove the following *contiguity relation*:

$$(c-b)F(a, b, c+1; x) + bF(a, b+1, c+1; x) = cF(a, b, c; x) .$$

See Exercise 19 of Chapter 11 for the p-adic analogue of this exercise.

116. Prove Proposition 9.8.4.

117.

(a) Fill in the details of the proof of Proposition 9.8.5.

(b) Compute explicitly the polynomials P_n and Q_n.

118. Prove that $K_{1/2}(x) = \sqrt{\pi/(2x)}e^{-x}$ by making the change of variable $u = \sinh(t/2)$ in the integral representation of Proposition 9.8.6.

119. Prove the following Mellin transform formulas and give their range of validity:

$$\int_0^\infty \frac{t^{s-1}}{1+t}\, dt = \frac{\pi}{\sin(\pi s)}\,,$$

$$\int_0^\infty \log(1+t) t^{s-1}\, dt = \frac{\pi}{s\sin(\pi s)}\,,$$

$$\int_0^\infty e^{i(x/2)(t-1/t)} t^{s-1}\, dt = 2K_s(x) e^{i\pi s/2}\,,$$

$$\int_0^\infty e^{i(x/2)(t+1/t)} t^{s-1}\, dt = \pi e^{i\pi s/2}(iJ_s(x) - Y_s(x))\,.$$

10. Dirichlet Series and L-Functions

This chapter deals with the analytic and arithmetic properties of Dirichlet series and in particular of L-functions, of which the Riemann zeta function is the prototypical example. In a sense it is analytic number theory, but it would be inappropriate to use this expression since it now means a part of number theory that extensively uses tools from real and complex analysis, while our purpose is slightly different. Perhaps more appropriate would be "elementary number theory," which deals with elementary number-theoretic functions, but which is also a misnomer since in no way should it be understood as "easy" number theory. In fact, the Riemann hypothesis, one of the most famous number-theoretical conjectures, can be considered as elementary number theory since it can be stated in "elementary" terms, for instance through the use of the Möbius function.

10.1 Arithmetic Functions and Dirichlet Series

An *arithmetic function* $a(n)$ is simply a complex-valued function defined on $\mathbb{Z}_{>0}$ (or sometimes only on a subset). By extension, we will also use the term to denote functions defined on integral ideals of number fields for instance. Almost all of the definitions and properties given in this section extend to this more general setting.

To an arithmetic function $a(n)$ is associated a *formal Dirichlet series* of a variable s:[1]

$$L(a, s) = \sum_{n=1}^{\infty} \frac{a(n)}{n^s} .$$

Many manipulations on arithmetic functions need only the formal aspect of these Dirichlet series, while others need convergence. In that case, s is a complex number, and $L(a, s)$ is the complex number equal to the sum of the series, when it converges. We start by studying only the formal aspects.

---

[1] Most authors write $L(s, a)$ instead of $L(a, s)$. I believe, however, that it is better to put the fixed parameters, here the arithmetic function a, at the beginning, and the variables after. After all, the usual notation for the L-function associated with an elliptic curve (or a more general algebro-geometric object) E and to a modular form f is $L(E, s)$ and $L(f, s)$. Thus, later I will write $L(\chi, s)$ instead of the more usual $L(s, \chi)$.

10.1.1 Operations on Arithmetic Functions

If $a(n)$ and $b(n)$ are two arithmetic functions and α and β are complex numbers, then $c(n) = \alpha a(n) + \beta b(n)$ is an arithmetic function, and clearly $L(c, s) = \alpha L(a, s) + \beta L(b, s)$. Thus arithmetic functions form naturally a $\mathbb{C}$-vector space (of infinite dimension).

Much more important is the multiplicative aspect. Although $a(n)b(n)$ is an arithmetic function in its own right, it is usually uninteresting. The interesting product is the so-called *arithmetic convolution* as follows.

Proposition 10.1.1. *Let $a(n)$, $b(n)$, and $c(n)$ be three arithmetic functions. The following two conditions are equivalent:*

(1)
$$L(c, s) = L(a, s)L(b, s) .$$

(2) *For all $n \geqslant 1$,*

$$c(n) = \sum_{d|n} a(d)b(n/d) = \sum_{d|n} a(n/d)b(d) .$$

If these conditions are satisfied, we say that the function c is the arithmetic convolution (or simply the convolution) of the functions a and b, and we sometimes write $c = a * b$.

Note that when we write $d \mid n$, we mean that d is a *positive* divisor of n.
Proof. We simply write

$$L(a, s)L(b, s) = \sum_{n=1}^{\infty} \frac{a(n)}{n^s} \sum_{m=1}^{\infty} \frac{b(m)}{m^s} = \sum_{n,m \geqslant 1} \frac{a(n)b(m)}{(nm)^s}$$

$$= \sum_{N \geqslant 1} \frac{1}{N^s} \sum_{\substack{nm=N \\ n,m \geqslant 1}} a(n)b(m) = \sum_{N \geqslant 1} \frac{1}{N^s} \sum_{n|N} a(n)b(N/n) .$$

The last equality of the proposition comes of course from the symmetry $d \mapsto n/d$ among divisors of n. $\qquad\square$

The following proposition is now clear:

Proposition 10.1.2. *The set $\mathcal{A}$ of (complex-valued) arithmetic functions together with the natural $\mathbb{C}$-vector space structure and arithmetic convolution as multiplication forms a commutative algebra with unit, the unit being the function $\delta(n)$ defined by $L(\delta, s) = 1$, in other words $\delta(1) = 1$ and $\delta(n) = 0$ if $n > 1$.*

We will denote by **1** the arithmetic function defined by $\mathbf{1}(n) = 1$ for all n (*not* to be confused with the function δ), so that by definition

$$L(\mathbf{1}, s) = \sum_{n=1}^{\infty} \frac{1}{n^s} = \zeta(s) .$$

This defines the (formal for the moment) Riemann zeta series.

The following proposition is easily proved by induction and left to the reader.

Proposition 10.1.3. *An arithmetic function $a(n)$ is invertible in $\mathcal{A}$ if and only if $a(1) \neq 0$.*

Definition 10.1.4. *We denote by $\mu(n)$ the inverse in $\mathcal{A}$ of the function $\mathbf{1}$, so that $\sum_{d|n} \mu(d) = \delta(n)$, or equivalently, $L(\mu, s) = 1/\zeta(s)$.*

Proposition 10.1.5 (Möbius inversion formula). *Let a and b be arithmetic functions.*

(1) *(First form.) Assume that $b(n) = \sum_{d|n} a(d)$. Then*

$$a(n) = \sum_{d|n} \mu(d) b(n/d) = \sum_{d|n} \mu(n/d) b(d) .$$

(2) *(Second form.) Assume that $b(n) = \sum_{k=1}^{\infty} a(kn)$, where all the series are absolutely convergent. Then*

$$a(n) = \sum_{k=1}^{\infty} \mu(k) b(kn) .$$

Proof. By definition, b is the arithmetic convolution of a with $\mathbf{1}$; hence $L(b, s) = L(a, s)\zeta(s)$, so that $L(a, s) = L(b, s)/\zeta(s) = L(\mu, s)L(b, s)$, and (1) follows.

For (2), we have

$$\sum_{k \geqslant 1} \mu(k) b(kn) = \sum_{k \geqslant 1} \mu(k) \sum_{d \geqslant 1} a(dkn) = \sum_{N \geqslant 1} a(Nn) \sum_{k|N} \mu(k) = a(n) ,$$

where the interchanges of summation are justified by absolute convergence, proving (2).   □

The Möbius inversion formula is useful in many contexts. One of its frequent uses is to replace the "rigid" function δ by the more tractable convolution of μ with $\mathbf{1}$. For example, a summation of the type $S(b) = \sum_{a, \, \gcd(a,b)=1} f(a, b)$ is very often advantageously replaced by

$$S(b) = \sum_{a} \sum_{d|\gcd(a,b)} \mu(d) f(a, b) = \sum_{d|b} \mu(d) \sum_{a_1} f(da_1, b) ,$$

and we see that in the inner sum the GCD condition has disappeared.

The following proposition gives an application of this:

Proposition 10.1.6 (Evaluation of Ramanujan sums). *We have*

$$R(m,a) = \sum_{\substack{x \bmod m \\ \gcd(x,m)=1}} \exp(2i\pi a x/m) = \sum_{d|\gcd(m,a)} \mu(m/d)d \ .$$

Proof. Using the above argument, setting $x = dy$ we have

$$R(m,a) = \sum_{d|m} \mu(d) \sum_{y \bmod m/d} \exp(2i\pi ay/(m/d)) \ .$$

The inner sum is now an honest geometric series that vanishes if $(m/d) \nmid a$ and is equal to m/d otherwise. Thus

$$R(m,a) = \sum_{d|m,\ (m/d)|a} \mu(d)(m/d) = \sum_{d|\gcd(m,a)} \mu(m/d)d$$

after changing d into m/d, giving the proposition.   □

Remark. The reader should understand in this example what we mean by "evaluation" or "explicit computation" of an expression. One could argue that the formula that we have obtained for $R(m,a)$ is barely simpler than the defining expression. But there is a huge difference, which is perhaps best seen in algorithmic terms: using the initial definition, we need to sum essentially m terms, which is extremely long if $m > 10^9$, say. On the other hand, using the result of the proposition, we need to sum only on the divisors of (m,a), of which there are very few, even if m and a are large. Even the (necessary) work of factoring m is small compared to the defining expression.

10.1.2 Multiplicative Functions

Most useful arithmetic functions a have a fundamentally number-theoretic property called *multiplicativity*:

Definition 10.1.7. *A (nonzero) arithmetic function a is said to be multiplicative if for all coprime integers n and m we have $a(nm) = a(n)a(m)$. It is said to be* completely *multiplicative if this is true for all n and m, not necessarily coprime.*

The crucial point about multiplicative functions is the following easy proposition.

Proposition 10.1.8. *A function a is multiplicative if and only if $L(a,s)$ has a formal Euler product, i.e., can be written formally as*

$$L(a,s) = \prod_p L_p(a,s), \quad where \ L_p(a,s) = 1 + \sum_{k=1}^{\infty} \frac{a(p^k)}{p^{ks}} \ .$$

Here and in the sequel, it is understood that a product such as $\prod_p$ is over all prime numbers.

In addition, the function a is completely multiplicative if and only if for all p we have

$$L_p(a, s) = \left(1 - \frac{a(p)}{p^s}\right)^{-1}.$$

(Note that $L_p(a, s)$ has no relation to the p-adic L-functions that we will study later.)

Proof. If we expand formally the product $L_p(a, s)$, we obtain a formal Dirichlet series of the form $\sum_{n \geqslant 1} b(n)/n^s$, where

$$b(n) = \prod_{p^k \| n} a(p^k).$$

The notation $p^k \| n$ (read: p^k exactly divides n) means that $p^k \mid n$ and $p^{k+1} \nmid n$. More generally, it will be useful to write $d \| n$ if $d \mid n$ and $\gcd(d, n/d) = 1$. This is the same as the previous definition when d is a prime power.

Resuming our proof, since a is multiplicative and the $p^k \| n$ are pairwise coprime, we obtain that $b(n) = a(n)$, so $\prod_p L_p(a, s) = L(a, s)$, as claimed. Furthermore, if a is completely multiplicative, then $\sum_{k \geqslant 0} a(p^k)/p^{ks} = (1 - a(p)/p^s)^{-1}$, proving the second statement. It is clear that the converse statements are also true. $\qquad\square$

Corollary 10.1.9. *If a and b are multiplicative functions, then so is the arithmetic convolution of a and b, and if a is invertible, then its inverse is also multiplicative.*

Proof. Clear from the above interpretation of multiplicativity in terms of formal Euler products. Of course this can also be proved directly. $\qquad\square$

Note the important fact that the arithmetic convolution of two completely multiplicative functions is almost never a completely multiplicative function. Indeed, since the Euler factors $L_p(a, s)$ of completely multiplicative functions are inverses of polynomials of degree at most 1 in p^{-s}, the Euler factor of the arithmetic convolution of two such functions will be the inverse of a polynomial of degree at most 2 in p^{-s}, but usually not of degree 1.

On the other hand, the ordinary product (which is rarely used) of two multiplicative (respectively completely multiplicative) functions is clearly multiplicative (respectively completely multiplicative).

10.1.3 Some Classical Arithmetical Functions

We give a list of the most important arithmetic functions, with their Dirichlet series when appropriate, their multiplicativity properties, and corresponding

Euler products. We also give some important relations between them. It should be emphasized that the list given here covers the large majority of the functions that are used in practice. Furthermore, the proofs of the given results can always be trivially obtained either from the Dirichlet series or from the Euler products when they exist, so the proofs are omitted and left as excellent exercises for the reader. Remember that for now all Dirichlet series and Euler products are *formal*, so there are no convergence problems. We will come to these problems later.

Completely Multiplicative Functions

- The function δ: $a(n) = \delta(n)$, $L(a, s) = 1$, $L_p(a, s) = 1$ for all p.
- The function **1**: $a(n) = 1$, $L(a, s) = \zeta(s)$, $L_p(a, s) = (1 - 1/p^s)^{-1}$ for all p.
- More generally, for any complex number t the function n^t: $a(n) = n^t$, $L(a, s) = \zeta(s - t)$, $L_p(a, s) = (1 - p^t/p^s)^{-1}$.
- Dirichlet characters χ modulo m: $a(n) = \chi(n)$, $L(a, s) = L(\chi, s)$ by definition, $L_p(a, s) = (1 - \chi(p)/p^s)^{-1}$.
  For any integer n, we will denote by $\omega(n)$ the number of *distinct* prime factors of n, and by $\Omega(n)$ the number of prime factors of n counted with multiplicity. In other words, if $n = \prod_{1 \leqslant i \leqslant g} p_i^{k_i}$ is the decomposition of n into powers of distinct primes, then we set $\omega(n) = g$ and $\Omega(n) = \sum_{1 \leqslant i \leqslant g} k_i$.
- For any complex number z the function $a(n) = z^{\Omega(n)}$ is completely multiplicative, and $L_p(a, s) = (1 - z/p^s)^{-1}$.

Elementary Multiplicative Functions

- The Möbius function μ: $L(\mu, s) = 1/\zeta(s)$, $L_p(\mu, s) = 1 - 1/p^s$. In particular:

Proposition 10.1.10. *The Möbius function is uniquely defined by $\mu(n) = 0$ if n is divisible by p^2 for some prime p (we say in this case that n is not squarefree), and otherwise $\mu(n) = (-1)^{\omega(n)}$.*

- If z is any complex number, then $a_z(n) = z^{\omega(n)}$ is a multiplicative function with

$$L_p(a_z, s) = 1 + \frac{z}{p^s - 1} = \frac{1 - (1 - z)/p^s}{1 - 1/p^s} .$$

The most important such function that occurs in practice is the function $2^{\omega(n)}$, which is equal to the number of divisors of n when n is squarefree (see the function $d(n)$ below), and we have $L(a_2, s) = \zeta(s)^2/\zeta(2s)$.
- The Euler totient function ϕ: $L(\phi, s) = \zeta(s - 1)/\zeta(s)$, $L_p(\phi, s) = (1 - 1/p^s)/(1 - p/p^s)$. In particular:

Proposition 10.1.11(1) *We have*

$$\phi(n) = n \prod_{p|n} \left(1 - \frac{1}{p} \right) ,$$

where the product is over primes dividing n.
(2) *We have the identity*

$$\sum_{d|n} \phi(d) = n .$$

(3) *We have $\phi(n) = |(\mathbb{Z}/n\mathbb{Z})^*|$, the number of invertible elements of the ring $\mathbb{Z}/n\mathbb{Z}$ of integers modulo n.*

Proof. By expanding the formal power series we have

$$\frac{1 - 1/p^s}{1 - p/p^s} = 1 + \frac{p-1}{p^s} + \frac{p^2 - p}{p^{2s}} + \cdots = 1 + \sum_{k \geqslant 1} \frac{p^k(1 - 1/p)}{p^{ks}} ;$$

hence for $k \geqslant 1$ we have $\phi(p^k) = p^k(1 - 1/p)$, so (1) follows by multiplicativity. (2) is trivial since it corresponds to the Dirichlet series identity $L(\phi, s)\zeta(s) = \zeta(s - 1)$. Finally, by the Chinese remainder theorem we know that $|(\mathbb{Z}/n\mathbb{Z})^*|$ is a multiplicative function (in our sense), and clearly $|(\mathbb{Z}/p^k\mathbb{Z})^*| = p^k - p^{k-1} = p^k(1 - 1/p)$, proving (3). □

Note that we have already proved and used the last two results of this proposition in Section 2.4.1.

Corollary 10.1.12. *We have*

$$\phi(n) = \sum_{d|n} \mu(d)(n/d) = \sum_{d|n} \mu(n/d)d .$$

Proof. Simply apply the Möbius inversion formula to (2). □

– Let t be a fixed complex number. The tth power divisor sum function $\sigma_t(n)$ is defined by $\sigma_t(n) = \sum_{d|n} d^t$. Then $L(\sigma_t, s) = \zeta(s - t)\zeta(s)$, $L_p(\sigma_t, s) = ((1 - p^t/p^s)(1 - 1/p^s))^{-1}$, so for $t \neq 0$ we have

$$\sigma_t(n) = \prod_{p^k \| n} \frac{p^{(k+1)t} - 1}{p^t - 1} ,$$

while for $t = 0$ we have

$$\sigma_0(n) = \prod_{p^k \| n} (k + 1) = \prod_{p|n} (v_p(n) + 1) .$$

Special cases: the number of divisors function $\sigma_0(n)$ is denoted by $d(n)$ (analytic number theorists often use the notation $\tau(n)$, but this is also used to denote another multiplicative function, the Ramanujan τ function), and we have $L(d, s) = \zeta(s)^2$. The sum of divisors function $\sigma_1(n)$ is simply denoted by $\sigma(n)$ and we have $L(\sigma, s) = \zeta(s - 1)\zeta(s)$.

Very precise results are known about the size of the functions $\phi(n)$ and $\sigma_t(n)$ which do not concern us here. The main thing to remember in practice about these functions is the following proposition:

Proposition 10.1.13. (1) *There exists a constant $c > 0$ such that for all n,*

$$c\,n/\log(\log(n)) \leqslant \phi(n) \leqslant n\,.$$

(2) *There exists a constant $c > 0$ such that for all n,*

$$n \leqslant \sigma(n) \leqslant c\,n\log(\log(n))\,.$$

(3) *For any $t > 1$ and for all n we have*

$$n^t \leqslant \sigma_t(n) \leqslant \zeta(t)\,n^t\,,$$

where $\zeta(t) > 1$ is Riemann's zeta function at t.
(4) *There exists a constant $c > 0$ such that for all n,*

$$0 \leqslant \omega(n) \leqslant \log(n)/\log(\log(n))\,.$$

(5) *There exists a constant $c > 0$ such that for all n,*

$$1 \leqslant d(n) \leqslant \exp(c\log(n)/\log(\log(n)))\,.$$

In particular, for any $\varepsilon > 0$ there exists $c_\varepsilon > 0$ such that for all n, $d(n) \leqslant c_\varepsilon n^\varepsilon$.

Remarks. (1) All these results are easy consequences of very weak forms of the prime number theorem that can be proved much more simply than the strong versions that we will prove in Section 10.7: we need to know only the existence of strictly positive constants C_1 and C_2 such that $C_1 x/\log(x) \leqslant \pi(x) \leqslant C_2 x/\log(x)$, where $\pi(x)$ is the number of primes less than or equal to x.
(2) The results of this proposition are all best possible, apart from the determination of the best constants. In particular, for any k there exist infinitely many n such that $d(n) > \log(n)^k$.
(3) The result for $\omega(n)$ is evidently false for $\Omega(n)$ since $\Omega(2^k) = k$.

Since $\mu(n) = 0$ or ± 1, there is nothing to say about the size of the Möbius function. It follows from the prime number theorem (and in fact is equivalent to it) that $\mu(n) = \pm 1$ with equal probability, in other words that if we let

$M(x) = \sum_{1 \leqslant n \leqslant x} \mu(n)$ then $M(x) = o(x)$, or equivalently, $M(x)/x \to 0$ as $x \to \infty$. More precise results are known, corresponding to more precise forms of the prime number theorem. What is completely conjectural, however, is the size of $M(x)$ as $x \to \infty$. In fact, it is conjectured that for all $\varepsilon > 0$ we have $M(x) = O(x^{1/2+\varepsilon})$ as $x \to \infty$, and it can be shown that this is equivalent to the Riemann hypothesis.

The von Mangoldt Function

Although not multiplicative, this function deserves a short study. We define the von Mangoldt function $\Lambda(n)$ by $L(\Lambda, s) = -\zeta'(s)/\zeta(s)$.

Proposition 10.1.14. (1) *We have* $\Lambda(n) = 0$ *if* n *is not a prime power, and* $\Lambda(n) = \log(p)$ *if* $n = p^k$ *is a power of a prime* p *with* $k \geqslant 1$.
(2) *We have*

$$\sum_{d|n} \Lambda(d) = \log(n) \quad and \quad \Lambda(n) = \sum_{d|n} \mu(n/d) \log(d) .$$

Proof. (1) is immediate since

$$-\frac{\zeta'(s)}{\zeta(s)} = \sum_{p \text{ prime}} \frac{\log(p)}{p^s(1 - 1/p^s)} = \sum_{p \text{ prime}} \sum_{k \geqslant 1} \frac{\log(p)}{p^{ks}} ,$$

and the two formulas of (2) are equivalent to the equality $\zeta(s)L(\Lambda, s) = -\zeta'(s)$. $\qquad \square$

Nonelementary Multiplicative Functions

There are of course many other interesting multiplicative functions in number theory. However, a particular class deserves to be mentioned, although it is outside the scope of this book: functions coming from the theory of modular forms. The sum of divisors functions $\sigma_t(n)$ are in fact of this type when t is an odd positive integer since they are the Fourier coefficients of holomorphic Eisenstein series. The most famous of the nonelementary functions is certainly Ramanujan's τ function defined by the formal expansion

$$q \prod_{m=1}^{\infty} (1 - q^m)^{24} = \sum_{n=1}^{\infty} \tau(n)q^n .$$

Of course, for someone not at all familiar with the theory of modular forms, this looks like a very artificial definition: for instance, why take the exponent 24? In any case, it was proved by Ramanujan and Mordell that $\tau(n)$ is indeed a multiplicative function (the proof is not difficult), and in addition that τ is equal to the arithmetic convolution of two (noncanonical) completely multiplicative functions α and β. This means that $L(\tau, s) = \prod_p L_p(\tau, s)$ and that $L_p(\tau, s)$ is the inverse of a second-degree polynomial in p^{-s}, and in fact $L_p(\tau, s) = (1 - \tau(p)/p^s + p^{11}/p^{2s})^{-1}$.

A deep conjecture of Ramanujan, which was proved by Deligne only in 1970 using all the machinery of modern algebraic geometry, is that the completely multiplicative functions α and β have modulus exactly equal to $n^{11/2}$, or equivalently, that they are conjugate (since their product is equal to n^{11}).

10.1.4 Numerical Dirichlet Series

As we have seen above, in many cases it is sufficient to consider formal Dirichlet series and formal Euler products. We now consider convergence problems, so that in this section s is a complex variable. The term Dirichlet series will thus denote numerical (or functional) Dirichlet series, and no longer formal ones.

For power series, the domain of convergence is a disk (possibly reduced to a point or infinite), where the behavior is a priori undetermined on the boundary. For Dirichlet series, we have a similar result:

Proposition 10.1.15. *Let* $f(s) = \sum_{n \geqslant 1} a(n)/n^s$ *be a Dirichlet series. There exists a* $\sigma \in [-\infty, +\infty]$ *(i.e., a real number or* $\pm\infty$*), called the ab-scissa of absolute convergence of the series, such that f converges absolutely in the half-plane* $\Re(s) > \sigma$*, and does not converge absolutely for* $\Re(s) < \sigma$*. Furthermore, for any* $\varepsilon > 0$*, the series* $f(s)$ *converges normally, hence uniformly, in the closed half-plane* $\Re(s) \geqslant \sigma + \varepsilon$*.*

Before proving this result, we note that $\sigma = +\infty$ means that the series never converges absolutely, while $\sigma = -\infty$ means that it converges absolutely for all $s \in \mathbb{C}$. Contrary to the case of power series, note that in number-theoretical practice, these situations do not occur, although of course they are possible (see below).

Proof. Assume that $f(s_0)$ converges absolutely for some $s_0 \in \mathbb{C}$. Then since

$$\frac{|a(n)|}{|n^s|} = \frac{|a(n)|}{|n^{s_0}|} |n^{s_0 - s}| = \frac{|a(n)|}{|n^{s_0}|} n^{\Re(s_0 - s)} ,$$

it follows that the series $f(s)$ is dominated in absolute value by the series $f(s_0)$ as soon as $\Re(s) \geqslant \Re(s_0)$. Thus, denote by σ the infimum in $[-\infty, +\infty]$ of the real parts of s such that $f(s)$ converges absolutely. It follows that if $\Re(s) > \sigma$ then $\Re(s) \geqslant \Re(s_0) > \sigma$ for some s_0 such that $f(s_0)$ converges absolutely, hence that $f(s)$ converges absolutely. The domination inequality that we have shown also proves normal hence uniform convergence in $\Re(s) \geqslant \Re(s_0)$. In addition, by definition if $\Re(s) < \sigma$ then $f(s)$ does not converge absolutely, proving the proposition. $\square$

It follows in particular from this proposition that the series $f(s)$ defines an analytic function (which by abuse of notation we will again denote by $f(s)$) on the half-plane $\Re(s) > \sigma$.

Another useful result is the following.

Proposition 10.1.16. *Let $f(s) = \sum_{n \geq 1} a_n/n^s$ be a Dirichlet series with nonnegative coefficients, i.e., such that $a_n \geq 0$ for all n, let σ_0 be the abscissa of convergence of f, and assume that $\sigma_0 \neq \pm\infty$. Then f cannot be analytically continued into a* holomorphic *function in the neighborhood of $s = \sigma_0$; in other words, $s = \sigma_0$ is a singularity of f (pole or otherwise).*

Proof. Assume the contrary. Then for $\varepsilon > 0$ sufficiently small, f is holomorphic in a circle centered at $\sigma_0 + 1$ of radius $1 + 2\varepsilon$, so inside this circle it is equal to the sum of its power series expansion

$$f(s) = \sum_{k \geq 0} \frac{(s - \sigma_0 - 1)^k}{k!} f^{(k)}(\sigma_0 + 1) \quad \text{with} \quad f^{(k)}(\sigma_0 + 1) = \sum_{n \geq 1} \frac{a_n (-\log(n))^k}{n^{\sigma_0 + 1}} .$$

Thus for instance we have

$$f(\sigma_0 - \varepsilon) = \sum_{k \geq 0} \frac{(1 + \varepsilon)^k}{k!} \sum_{n \geq 1} \frac{a_n (\log(n))^k}{n^{\sigma_0 + 1}} .$$

In this convergent double sum all the terms are nonnegative, so we can interchange the order of summation and obtain

$$f(\sigma_0 - \varepsilon) = \sum_{n \geq 1} \frac{a_n}{n^{\sigma_0 + 1}} \sum_{k \geq 0} \frac{(1 + \varepsilon)^k \log(n)^k}{k!} = \sum_{n \geq 1} \frac{a_n}{n^{\sigma_0 - \varepsilon}}$$

since

$$\sum_{k \geq 0} \frac{(1 + \varepsilon)^k \log(n)^k}{k!} = \exp((1 + \varepsilon) \log(n)) = n^{1 + \varepsilon} .$$

Thus the series $\sum_{n \geq 1} a_n/n^s$ converges (absolutely of course) for $s = \sigma_0 - \varepsilon < \sigma_0$, a contradiction since σ_0 is the abscissa of convergence. $\square$

Corollary 10.1.17. *Let $f(s) = \sum_{n \geq 1} a_n/n^s$ be a Dirichlet series with nonnegative coefficients and abscissa of convergence different from $+\infty$. If $f(s)$ can be holomorphically continued to $\Re(s) > \sigma$ then the series $f(s)$ converges for $\Re(s) > \sigma$.*

Proof. Indeed, if $\sigma_0 < \infty$ is the abscissa of convergence of $f(s)$ then either $\sigma_0 = -\infty$ and there is nothing to prove, or $\sigma_0 \neq \pm\infty$ and by the above proposition we know that $f(s)$ has a singularity at $s = \sigma_0$. In particular, $f(s)$ cannot be holomorphically continued at σ_0, so by assumption, $\sigma_0 \leq \sigma$, and hence $f(s)$ converges for $\Re(s) > \sigma$. $\square$

Examples. – Clearly $\sum_{n \geq 1} 2^n/n^s$ does not converge anywhere, so $\sigma = +\infty$ in this case. In the opposite direction, $\sum_{n \geq 1} 2^{-n}/n^s$ converges for any value of s, so that $\sigma = -\infty$ in this case. However, as already mentioned, in almost all number-theoretic applications of Dirichlet series we have $\sigma \in \mathbb{R}$, i.e., not equal to $\pm\infty$, so there really is a half-plane of convergence.

- The Riemann zeta function $\zeta(s) = \sum_{n\geqslant 1} 1/n^s$ has abscissa of absolute convergence $\sigma = 1$, for example because of the comparison theorem between series and integrals. On the line $\sigma = 1$, the series diverges (both absolutely and not; see Exercise 10). The corresponding Euler product $\zeta(s) = \prod_p (1 - 1/p^s)^{-1}$ also converges absolutely only for $\Re(s) > 1$.

- It follows from the deep result of Deligne mentioned above that the abscissa of absolute convergence of $L(\tau, s) = \sum_{n\geqslant 1} \tau(n)/n^s$ is equal to $\sigma = 13/2$ (one can easily prove that $\sigma \leqslant 7$, and with a little more difficulty that $\sigma \leqslant 7 - 1/4$).

The reader will have noticed that we have only mentioned absolute convergence. There are two reasons for this. The first one is that only absolutely convergent series are safe for computations; others should be avoided whenever possible. The second reason is more subtle: in the case of power series, the radius of absolute convergence and the radius of convergence are the same (although on the circle of convergence itself there may be differences). For Dirichlet series, the situation is different: the abscissa of convergence (which also exists; see Exercise 7) may be smaller than the abscissa of absolute convergence. To give an example, the abscissa of convergence of $\sum_{n\geqslant 1}(-1)^n/n^s$ is $\sigma = 0$, while its abscissa of absolute convergence is that of the Riemann zeta function, i.e., $\sigma = 1$; see Exercise 11. It can (easily) be proved, however, that the difference between the two abscissas is less than or equal to 1, and 1 is best possible as this example shows; see Exercise 8.

For a much deeper example, the abscissa of absolute convergence of the Dirichlet series $1/\zeta(s) = \sum_{n\geqslant 1} \mu(n)/n^s$ is equal to 1 (see Exercise 9), but nobody knows its abscissa of convergence σ. It is trivial to prove that $1/2 \leqslant \sigma \leqslant 1$, but even the proof that $\sigma < 1$ would be a major accomplishment worthy of the Fields medal plus a million US dollar Clay prize. The Riemann hypothesis is equivalent to the strongest possible statement that $\sigma = 1/2$. An equivalent formulation is that $M(x) = \sum_{1\leqslant n\leqslant x} \mu(n)$ satisfies $M(x) = O(x^{1/2+\varepsilon})$ for all $\varepsilon > 0$. The best currently known result in that direction (using sophisticated techniques of trigonometric sums in analytic number theory) is $M(x) = O(x\exp(-c\log(x)^{3/5}\log(\log(x))^{-1/5}))$ for some $c > 0$, and this result has not been improved upon for half a century; see the remarks after Theorem 10.7.8 below.

10.2 The Analytic Theory of L-Series

Let χ be a Dirichlet character modulo m. Recall that for $\Re(s) > 1$ we have the absolutely convergent series and Euler products (see the footnote at the beginning of Section 10.1)

$$L(\chi, s) = \sum_{n=1}^{\infty} \frac{\chi(n)}{n^s} = \prod_p \left(1 - \frac{\chi(p)}{p^s}\right)^{-1}.$$

Lemma 10.2.1. *Let χ be a character modulo m, let $f \mid m$ be the conductor of χ, and χ_f the corresponding primitive character modulo f. We have*

$$L(\chi, s) = L(\chi_f, s) \prod_{p \mid m} \left(1 - \frac{\chi_f(p)}{p^s} \right).$$

In particular, if $m = p^k$ is a power of a prime p and χ is a nontrivial character we have $L(\chi, s) = L(\chi_f, s)$.

Proof. Clear.   □

Since a finite Euler product is easy to study, it suffices to limit our study to *primitive* characters. Thus, in the rest of this section, we will usually assume that χ is a primitive character modulo m.

The basic results that we need about Dirichlet L-functions are their *analytic continuation*, *functional equation*, and their *special values*, either at negative integers, or at positive integers of suitable parity. It is important to note that Dirichlet L-functions are very simple objects compared to more complicated L-functions such as Dedekind zeta functions, Artin L-functions, or L-functions attached to elliptic curves or to modular forms. Thus we can use tools that are difficult if not impossible to generalize to these more general contexts. We will thus first give the simplest possible proofs. We will also give the more complicated proofs, so that the reader can have an idea of how to generalize to more complicated L-functions.

To obtain analytic continuation and special values at negative integers, we are going to see that a very simple approach based on integration by parts is sufficient, and in that case it is not necessary to assume that χ is a character. To obtain the functional equation and the special values at positive integers, we must assume that χ is a Dirichlet character, and the tool that we will use is the fundamental theorem on Fourier series. To treat more general L-series we would need a generalization of this tool, when it exists.

10.2.1 Simple Approaches to Analytic Continuation

We begin with the following general result. Recall that a function f *tends rapidly to 0 at infinity* if for any $k \geqslant 0$, as $x \to +\infty$ the function $x^k f(x)$ tends to 0.

Proposition 10.2.2. *Let f be a C^∞ function on $[0, \infty[$ tending rapidly to 0 at infinity, and for $\Re(s) > 0$ define*

$$L(f, s) = \frac{1}{\Gamma(s)} \int_0^\infty f(t) t^s \frac{dt}{t}.$$

(1) *For any $k \in \mathbb{Z}_{\geqslant 0}$ we have for $\Re(s) > -k$,*

$$\Gamma(s+k)L(f,s) = (-1)^k \int_0^\infty \frac{d^k f}{dt^k}(t)t^{s+k}\frac{dt}{t} .$$

(2) *The function $L(f,s)$ can be analytically continued to the whole of $\mathbb{C}$ into a holomorphic function.*

(3) *For any $n \in \mathbb{Z}_{\geqslant 0}$ we have*

$$L(f,-n) = (-1)^n \frac{d^n f}{dt^n}(0) = (-1)^n n! a_n ,$$

where $\sum_{n \geqslant 0} a_n t^n$ is the formal Taylor expansion of f around 0.

Proof. Assume first that all the derivatives of f also tend rapidly to 0 at infinity. Integrating by parts the definition of $L(f,s)$ and using the fact that f tends rapidly to 0 at infinity, we obtain

$$L(f,s) = -\frac{1}{s\Gamma(s)} \int_0^\infty f'(t)t^{s+1}\frac{dt}{t} = -L(f',s+1) ,$$

so (1) follows by induction. Since $\Gamma(s+k) \neq 0$ for all s and k, and the integral in (1) defines an analytic function for $\Re(s+k) > 0$, it follows that $L(f,s)$ can be analytically continued into a holomorphic function for $\Re(s) > -k$, hence in the whole of $\mathbb{C}$ since k is arbitrary, proving (2). Finally, (1) applied to $s = 1-k$ gives

$$L(f,1-k) = (-1)^k \int_0^\infty \frac{d^k f}{dt^k}(t)\,dt = (-1)^{k-1}\frac{d^{k-1}f}{dt^{k-1}}(0) ,$$

which is equivalent to the two formulas of (3).

We now assume only that f is a C^∞ function that tends rapidly to 0 at infinity. We must show that we can reduce to the case in which all its derivatives also do. In fact, we are going to show that we can reduce to the case in which f has *compact support*, which is stronger than what we need. Let ϕ be an auxiliary C^∞ function equal to 1 on $[0,1]$ and to 0 on $[2,\infty[$. We can evidently write $f = \phi f + (1-\phi)f$, so that $L(f,s) = L(\phi f,s) + L((1-\phi)f,s)$. Since $(1-\phi)f$ vanishes in a neighborhood of 0 and tends rapidly to 0 at infinity, it is clear that the integral defining $L((1-\phi)f,s)$ converges absolutely for all s and hence defines a holomorphic function on $\mathbb{C}$. Furthermore, since $1/\Gamma(-n) = 0$ for $n \in \mathbb{Z}_{\geqslant 0}$ we also have $L((1-\phi)f,-n) = 0$. It follows that we may replace f by ϕf, in other words by a C^∞ function with compact support, as claimed, thus finishing the proof. $\square$

Applying this to L-functions attached to arithmetic functions, we obtain the following.

Corollary 10.2.3. *Let χ be any arithmetic function of period dividing m, and recall that $B_0(\chi) = s_0(\chi)/m = (\sum_{0 \leqslant r < m} \chi(r))/m$.*

(1) *For any $k \in \mathbb{Z}_{\geqslant 0}$ and $\Re(s) > -k$ we have*

$$\Gamma(s+k)\left(L(\chi,s) - \frac{B_0(\chi)}{s-1}\right) = (-1)^k \int_0^\infty \frac{d^k F_\chi}{dt^k}(t) t^{s+k} \frac{dt}{t} ,$$

where

$$F_\chi(t) = \frac{\sum_{1 \leqslant r \leqslant m} \chi(r) e^{-rt}}{1 - e^{-mt}} - B_0(\chi) \frac{e^{-t}}{t} .$$

(2) *In particular, for any $k \in \mathbb{Z}_{\geqslant 0}$ and $\Re(s) > -k$ we have*

$$\Gamma(s+k)\left(\zeta(s) - \frac{1}{s-1}\right) = (-1)^k \int_0^\infty \frac{d^k F}{dt^k}(t) t^{s+k} \frac{dt}{t} ,$$

with

$$F(t) = \frac{1}{e^t - 1} - \frac{e^{-t}}{t} .$$

(3) *The function $L(\chi, s)$ can be analytically continued to the whole complex plane into a holomorphic function if $B_0(\chi) = 0$ (in particular if χ is a nontrivial character modulo m), and otherwise to a meromorphic function with a single pole, at $s = 1$, with residue $B_0(\chi)$.*

(4) *If $B_0(\chi) = 0$ the series $\sum_{n \geqslant 1} \chi(n)/n$ converges and its sum is equal to $L(\chi, 1)$.*

(5) *For $k \in \mathbb{Z}_{\geqslant 1}$ we have*

$$L(\chi, 1-k) = -\frac{B_k(\chi)}{k} - \chi(0)\delta_{k,1} .$$

(see Definition 9.4.1).

Proof. (1), (2), and (3). The integral definition of the gamma function immediately implies that

$$\int_0^\infty e^{-nt} t^s \frac{dt}{t} = \frac{\Gamma(s)}{n^s} .$$

It follows by absolute convergence that for $\Re(s) > 1$,

$$\Gamma(s)L(\chi,s) = \int_0^\infty G_\chi(t) t^{s-1}\, dt ,$$

where

$$G_\chi(t) = \sum_{n=1}^\infty \chi(n) e^{-nt} = \sum_{r=1}^m \chi(r) \sum_{q=0}^\infty e^{-(r+qm)t} = \frac{\sum_{1 \leqslant r \leqslant m} \chi(r) e^{-rt}}{1 - e^{-mt}} .$$

We cannot yet apply the proposition to $f(t) = G_\chi(t)$, since f is not defined at 0 except if $\sum_{1 \leqslant r \leqslant m} \chi(r) = m B_0(\chi) = 0$. However, for $\Re(s) > 1$ we have

$$\int_0^\infty t^{s-1} \frac{e^{-t}}{t} \, dt = \Gamma(s-1) = \frac{\Gamma(s)}{s-1} ,$$

hence

$$\Gamma(s) \left(L(\chi, s) - \frac{B_0(\chi)}{s-1} \right) = \int_0^\infty F_\chi(t) t^{s-1} \, dt ,$$

with $F_\chi(t) = G_\chi(t) - B_0(\chi) e^{-t}/t$. In the above formula, we could replace $B_0(\chi)$ by any constant, but the point of choosing specifically $B_0(\chi)$ is that now $F_\chi(t)$ is defined at $t = 0$, and is in $\mathbb{C}^\infty([0, +\infty[)$, so that the proposition is applicable, and (1), (2), and (3) immediately follow.

(4). Since $s_0(\chi) = 0$ the quantity $\sum_{n \leqslant x} \chi(n)$ is bounded, and it follows by Abel summation that the series $S = \sum_{n \geqslant 1} \chi(n)/n$ converges. It also follows from Abel's theorem that $\sum_{n \geqslant 1} \chi(n)/n^s$ tends to S as s tends to 1, $\Re(s) > 1$, uniformly in any sector $|\operatorname{Arg}(s-1)| \leqslant \theta < \pi/2$. $\qquad \square$

(5). Keep the above notation. By Lemma 9.4.3 and Proposition 9.4.9, we have

$$F_\chi(t) = \sum_{k \geqslant 0} \frac{B_k(\chi^-) - (-1)^k B_0(\chi)}{k!} t^{k-1}$$

$$= -\chi(0) + \sum_{k \geqslant 1} (-1)^k \frac{B_k(\chi) - B_0(\chi)}{k!} t^{k-1} .$$

Thus, if we write $F_\chi(t) = \sum_{n \geqslant 0} a_n t^n$ we have $(-1)^n n! a_n = -\chi(0) \delta_{n,0} - (B_{n+1}(\chi) - B_0(\chi))/(n+1)$, so we deduce from the proposition that for $k \geqslant 1$ we have

$$L(\chi, 1-k) + B_0(\chi)/k = -\chi(0)\delta k, 1 - B_k(\chi)/k + B_0(\chi)/k ,$$

proving (5). $\qquad \square$

Remarks. (1) The above approach does not give the *functional equation* of the L-functions. The essential reason is that we use only the *periodicity* of $\chi(n)$, and not its multiplicativity, which implies, through the use of Gauss sums, that its finite Fourier transform is a constant times its conjugate (see Proposition 2.1.39).

(2) A deeper reason for which the above approach *exists* is the fact that Dirichlet characters, or more generally periodic arithmetic functions, are intimately linked to *Abelian extensions* of $\mathbb{Q}$ via the Kronecker–Weber theorem asserting that any Abelian extension is a subextension of a cyclotomic field. Another way of stating this is that the Dedekind zeta function of an Abelian extension of $\mathbb{Q}$ splits as a product of L-functions of Dirichlet characters. Thus if we consider L or zeta functions attached to non-Abelian extensions (such as non-Galois cubic fields), no elementary

method of the above type is known even to prove analytic continuation. This is the reason for that Artin's conjecture, claiming essentially that all such L-functions are holomorphic, is so difficult to prove. Continuing with this language, Dirichlet characters are in one-to-one correspondence with characters of the group $GL_1(\mathbb{Z}/m\mathbb{Z}) = (\mathbb{Z}/m\mathbb{Z})^*$. To be able to work with more general extensions it is necessary to understand characters of more general groups such as $GL_n(\mathbb{Z}/m\mathbb{Z})$ and generalizations: this is the main thrust of the *Langlands program*.

Using the Fourier expansion of the χ-Bernoulli functions, we can also easily give the values of $L(\chi, s)$ at positive integers of suitable parity. In accordance with remark (1) above, here it is necessary to restrict to Dirichlet characters, and for simplicity we will even restrict to primitive characters.

Proposition 10.2.4. *Let χ be a primitive Dirichlet character modulo m, let $W(\chi)$ be the root number given by Definition 2.2.25, let $k \in \mathbb{Z}_{\geqslant 1}$ be such that $\chi(-1) = (-1)^k$, and let $e = 0$ or 1 be such that $k \equiv e \pmod 2$. We have*

$$L(\chi, k) = (-1)^{k-1+(k+e)/2} W(\chi) \frac{2^{k-1}\pi^k \overline{B_k(\chi)}}{m^{k-1/2}k!} .$$

Proof. Applying Proposition 9.4.14 to $\overline{\chi}$ and $x = 0$ (with k and n exchanged) we obtain

$$B_k(\overline{\chi}) = -\frac{m^{k-1}k!}{(2i\pi)^k} \sum_{n \in \mathbb{Z},\ n \neq 0} \frac{\tau(\overline{\chi}, n)}{n^k} .$$

This is a priori valid only for $k \geqslant 2$, but for $k = 1$ the corrective term is $\chi(0)/2 = 0$ since $\chi(-1) = -1$ in that case, so χ cannot be a trivial character.

Since χ is a primitive character, by Corollary 2.1.42 and Proposition 2.1.45 we have

$$\tau(\overline{\chi}, n) = \chi(n)\tau(\overline{\chi}) = \chi(-n)\overline{\tau(\chi)} = (-1)^k \chi(n) \frac{m}{\tau(\chi)} .$$

Furthermore, we clearly have $B_k(\overline{\chi}) = \overline{B_k(\chi)}$. Thus

$$\overline{B_k(\chi)} = (-1)^{k-1} 2 \frac{m^k k!}{(2i\pi)^k \tau(\chi)} \sum_{n \geqslant 1} \frac{\chi(n)}{n^k} ,$$

proving the proposition after $\tau(\chi)$ is replaced by $i^e m^{1/2} W(\chi)$. $\square$

We will see below other proofs of the above proposition. In particular, it is a special case of the functional equation of $L(\chi, s)$.

10.2.2 The Use of the Hurwitz Zeta Function $\zeta(s, x)$

Another simple approach to the study of L-functions of Dirichlet characters is based on the use of the Hurwitz zeta function $\zeta(s, x)$, which we have studied in detail in the preceding chapter. We are going to see that this approach, which again is quite simple and works because we have periodic functions, gives not only analytic continuation and values at negative integers, but also the functional equation in a painless way.

Proposition 10.2.5. *Let χ be any arithmetic function of period dividing m.*

(1) *We have*

$$L(\chi, s) = \frac{1}{m^s} \sum_{1 \leqslant r \leqslant m} \chi(r) \zeta(s, r/m) .$$

(2) *The function $L(\chi, s)$ has an analytic continuation to the whole complex plane, to a holomorphic function if $B_0(\chi) = 0$, and to a meromorphic function with a single pole at $s = 1$, simple with residue $B_0(\chi)$ otherwise.*

(3) *For any $k \in \mathbb{Z}_{\geqslant 1}$ we have*

$$L(\chi, 1 - k) = -\frac{B_k(\chi)}{k} - \chi(0)\delta_{k,1} = -\frac{m^{k-1}}{k} \sum_{1 \leqslant r \leqslant m} \chi(r) B_k\left(\frac{r}{m}\right) .$$

(4) *If $B_0(\chi) = 0$ we have*

$$L(\chi, 1) = -\frac{1}{m} \sum_{1 \leqslant r \leqslant m} \chi(r) \psi\left(\frac{r}{m}\right) ,$$

where as usual $\psi(x)$ is the logarithmic derivative of $\Gamma(x)$.

Proof. Since

$$L(\chi, s) = \sum_{n \geqslant 1} \frac{\chi(n)}{n^s} = \sum_{1 \leqslant r \leqslant m} \chi(r) \sum_{q \geqslant 0} \frac{1}{(qm + r)^s} = \frac{1}{m^s} \sum_{1 \leqslant r \leqslant m} \chi(r) \zeta(s, r/m) ,$$

the first formula is clear, and analytic continuation, residue, and special values at negative or zero integers follow from the corresponding properties of $\zeta(s, x)$ seen in Proposition 9.6.6 and Corollary 9.6.10, together with the formula for $B_k(\chi)$ given by Proposition 9.4.5. The formula for $L(\chi, 1)$ follows from Proposition 9.6.8 (3) since $B_0(\chi) = 0$. □

The functional equation for L-functions is in fact an immediate consequence of the Fourier expansion of $\zeta(s, \{x\})$ that we computed in the previous chapter (Corollary 9.6.51):

Theorem 10.2.6. *For $\Re(s) > 1$ set $Z(s, x) = \sum_{n \geqslant 1} e^{2i\pi n x}/n^s$.*

(1) *For $\Re(s) > 1$ we have the functional equation*

$$\zeta(1 - s, \{x\}) = (2\pi)^{-s}\Gamma(s)\left(e^{-is\pi/2}Z(s, x) + e^{is\pi/2}Z(s, 1 - x)\right).$$

(2) *For all s we have the functional equation*

$$\zeta(1 - s) = 2(2\pi)^{-s}\Gamma(s)\cos(s\pi/2)\zeta(s).$$

(3) *More generally let χ be any primitive character modulo m. For all s we have the functional equation*

$$L(\chi, 1 - s) = 2W(\chi)m^{s-1/2}(2\pi)^{-s}\Gamma(s)\cossin(s\pi/2)L(\overline{\chi}, s),$$

where $\cossin(x) = \cos(x)$ when $\chi(-1) = 1$ and $\cossin(x) = \sin(x)$ when $\chi(-1) = -1$, and $W(\chi)$ is as in Definition 2.2.25.

Proof. Statement (1) is a simple rephrasing of Corollary 9.6.51 seen in the previous chapter. For (2), assume first that $\Re(s) > 1$. Then by Proposition 9.6.11 the function $\zeta(1 - s, \{x\})$ is an everywhere *continuous* function of x, including at the integers, where it takes the value $\zeta(1 - s)$. Thus letting x tend to 0, we obtain

$$\zeta(1-s) = (2\pi)^{-s}\Gamma(s)\left(e^{-s\pi/2}\zeta(s) + e^{s\pi/2}\zeta(s)\right) = 2(2\pi)^{-s}\Gamma(s)\cos(s\pi/2)\zeta(s).$$

By analytic continuation this equation is valid for all s (as usual interpreting suitably the values at the poles). The same proof is valid for (3) except that we do not have to worry about integers: for $\Re(s) > 1$ we have

$$L(\chi, 1 - s) = \frac{1}{m^{1-s}}\sum_{r=1}^{m}\chi(r)\zeta(1 - s, r/m)$$

$$= m^{s-1}(2\pi)^{-s}\Gamma(s)\sum_{r=1}^{m}\chi(r)\left(e^{-is\pi/2}Z\left(s, \frac{r}{m}\right) + e^{is\pi/2}Z\left(s, -\frac{r}{m}\right)\right)$$

$$= m^{s-1}(2\pi)^{-s}\Gamma(s)\left(e^{-is\pi/2}\sum_{n\geqslant 1}\frac{\tau(\chi, n)}{n^s} + e^{is\pi/2}\sum_{n\geqslant 1}\frac{\tau(\chi, -n)}{n^s}\right)$$

$$= m^{s-1}(2\pi)^{-s}\Gamma(s)\tau(\chi)L(\overline{\chi}, s)\left(e^{-is\pi/2} + \chi(-1)e^{is\pi/2}\right)$$

by Corollary 2.1.42, which is applicable since χ is a primitive character, proving (3) after separating the cases $\chi(-1) = 1$ and $\chi(-1) = -1$ and extending to all s by analytic continuation. $\qquad\square$

10.2.3 The Functional Equation for the Theta Function

Perhaps the most classical way of proving the functional equation of Dirichlet L-series is to apply the Poisson summation formula (Proposition 2.2.16) to closely related series called *theta functions*. We will prove in fact more than is necessary.

Definition 10.2.7. *Let* $z \in \mathbb{C}^*$ *be any nonzero complex number.*

(1) *We define the principal determination of the argument of* z, *and denote by* $\operatorname{Arg}(z)$, *with a capital A, the unique* $\theta \in \,]-\pi, \pi]$ *such that* $z = |z| \exp(i\theta)$.

(2) *We define the principal determination of the square root of* z *by the formula* $z^{1/2} = |z|^{1/2} \exp(i \operatorname{Arg}(z)/2)$, *in other words as the unique complex number* w *such that* $w^2 = z$ *with* $\operatorname{Arg}(w) \in \,]-\pi/2, \pi/2]$.

(3) *If* $k \in \mathbb{Z}$, *we define* $z^{k/2}$ *by the formula* $z^{k/2} = (z^{1/2})^k$.

Note that in this definition we allow negative real numbers.

Warning. When k is even, we evidently have $(z^{1/2})^k = z^{k/2}$ in the usual sense. On the other hand, when k is odd, we do *not* in general have $z^{k/2} = (z^k)^{1/2}$.

The following lemma is very easy and left to the reader.

Lemma 10.2.8. *Let* x *and* y *be two nonzero complex numbers. Then*

(1) $\operatorname{Arg}(xy) = \operatorname{Arg}(x) + \operatorname{Arg}(y) - 2k\pi$, *where* $k = \lceil (\operatorname{Arg}(x) + \operatorname{Arg}(y) - \pi)/(2\pi) \rceil$.

(2) $(xy)^{1/2} = \varepsilon x^{1/2} y^{1/2}$, *where* $\varepsilon = \pm 1$, *and* $\varepsilon = 1$ *if and only if* $\operatorname{Arg}(x) + \operatorname{Arg}(y) \in \,]-\pi, \pi]$.

(3) $\operatorname{Arg}(1/x) = -\operatorname{Arg}(x)$, *except if* $x \in \mathbb{R}_{<0}$, *in which case* $\operatorname{Arg}(1/x) = \operatorname{Arg}(x) = \pi$.

(4) $x^{1/2}(1/x)^{1/2} = 1$ *except if* $x \in \mathbb{R}_{<0}$, *in which case* $x^{1/2}(1/x)^{1/2} = -1$.

We also need the following integral evaluation.

Lemma 10.2.9. *Assume that* $\Im(a) \geqslant 0$, *and that when* $\Im(a) = 0$ *we have* $a \neq 0$ *and* $\Im(b) = 0$. *Then*

$$\int_{-\infty}^{+\infty} \exp(i\pi(at^2 + bt + c))\,dt = \left(\frac{i}{a}\right)^{1/2} \exp\left(i\pi \frac{4ac - b^2}{4a}\right),$$

where as usual we use the principal determination of the square root.

Proof. Note that when $\Im(a) = 0$, we need the condition $a \neq 0$ and $\Im(b) = 0$ to ensure the convergence of the integral.

Since a/i is not a negative real number, by Lemma 10.2.8, we have $(a/i)^{1/2}(i/a)^{1/2} = 1$. If we set $u = (t + b/(2a))(a/i)^{1/2}$, the integral thus becomes

$$\left(\frac{i}{a}\right)^{1/2} \exp\left(i\pi \frac{4ac - b^2}{4a}\right) \int_C \exp(-\pi u^2)\,du\,,$$

where C is the line $(t + b/(2a))(a/i)^{1/2}$ as t goes from $-\infty$ to $+\infty$. Since $\operatorname{Arg}((a/i)^{1/2}) \in [-\pi/4, \pi/4]$, it is easy to show that we can modify the contour of integration C to the horizontal line $\mathbb{R}$ without changing the integral, and since $\int_{-\infty}^{+\infty} \exp(-\pi u^2)\,du = 1$, the result follows. $\square$

Proposition 10.2.10. *Let χ be a primitive character modulo m and let $e = 0$ if $\chi(-1) = 1$ and $e = 1$ if $\chi(-1) = -1$. For $\Im(\tau) > 0$ and any $z \in \mathbb{C}$, set*

$$\Theta(\chi, \tau, z) = \sum_{n \in \mathbb{Z}} \chi(n) \exp\left(\frac{i\pi n^2 \tau + 2i\pi nz}{m}\right).$$

We have the functional equation

$$\Theta\left(\chi, -\frac{1}{\tau}, \frac{z}{\tau}\right) = \frac{W(\chi)}{i^e} \left(\frac{\tau}{i}\right)^{1/2} \exp\left(\frac{i\pi z^2}{m\tau}\right) \Theta(\overline{\chi}, \tau, z),$$

where $W(\chi) = \tau(\chi)/(i^e m^{1/2})$ is as in Definition 2.2.25.

Proof. We have

$$\Theta(\chi, \tau, z) = \sum_{r \bmod m} \chi(r) \sum_{k \in \mathbb{Z}} \exp\left(\frac{i\pi(km+r)^2\tau + 2i\pi(km+r)z}{m}\right).$$

Since $\Im(\tau) > 0$, the function $f(x) = \exp(i\pi((xm+r)^2\tau + 2(xm+r)z)/m)$ tends rapidly to 0 at infinity, so we may apply the Poisson summation formula (Corollary 2.2.17) to the function $f(x)$. We thus apply Lemma 10.2.9 to $a = m\tau$, $b = 2(r\tau + y + z)$, and $c = (r^2\tau + 2rz)/m$, and we obtain

$$\int_{-\infty}^{+\infty} f(t) \exp(2i\pi yt)\, dt = \left(\frac{i}{m\tau}\right)^{1/2} \exp\left(-i\pi\left(\frac{2ry}{m} + \frac{(y+z)^2}{m\tau}\right)\right).$$

Thus

$$\Theta(\chi, \tau, z) = \left(\frac{i}{\tau}\right)^{1/2} m^{-1/2} \sum_{r \bmod m} \chi(r) \sum_{n \in \mathbb{Z}} \exp\left(-i\pi\left(\frac{2rn}{m} + \frac{(n+z)^2}{m\tau}\right)\right)$$

$$= \left(\frac{i}{\tau}\right)^{1/2} m^{-1/2} \sum_{n \in \mathbb{Z}} \exp\left(-i\pi\frac{n^2}{m\tau}\right) \sum_{r \bmod m} \chi(r) \exp\left(-2i\pi\frac{rn}{m}\right).$$

By Corollary 2.1.42, since χ is primitive we have

$$\sum_{r \bmod m} \chi(r) \exp\left(-2i\pi\frac{rn}{m}\right) = \overline{\chi}(-n)\tau(\chi) = \chi(-1)\overline{\chi}(n)\tau(\chi).$$

Thus

$$\Theta(\chi, \tau, z) = \left(\frac{i}{\tau}\right)^{1/2} \chi(-1)\tau(\chi)m^{-1/2} \sum_{n \in \mathbb{Z}} \overline{\chi}(n) \exp\left(-i\pi\frac{(n+z)^2}{m\tau}\right),$$

giving the result after changing τ into $-1/\tau$ and z into z/τ.   □

Definition 10.2.11. *We say that χ is an even (respectively odd) character if $\chi(-1) = 1$ (respectively $\chi(-1) = -1$), and as in the proposition we let $e = 0$ or 1 such that $\chi(-1) = (-1)^e$. Finally, we define*

$$\theta(\chi, \tau) = \sum_{n \in \mathbb{Z}} n^e \chi(n) \exp\left(\frac{i\pi n^2 \tau}{m}\right) .$$

Corollary 10.2.12. *The function $\theta(\chi, \tau)$ satisfies the functional equation*

$$\theta\left(\chi, -\frac{1}{\tau}\right) = W(\chi) \left(\frac{\tau}{i}\right)^{(2e+1)/2} \theta(\overline{\chi}, \tau) ,$$

where $(\tau/i)^{(2e+1)/2}$ is given by Definition 10.2.7.

Proof. If $e = 0$, i.e., if χ is even, the corollary immediately follows from the proposition by setting $z = 0$. Assume now that $e = 1$. We clearly have

$$\frac{\partial \Theta}{\partial z}(\chi, \tau, 0) = \frac{2i\pi}{m} \theta(\chi, \tau) .$$

On the other hand, the above proposition implies that

$$\frac{\partial \Theta}{\partial z}\left(\chi, -\frac{1}{\tau}, 0\right) = \frac{W(\chi)}{i} \tau \left(\frac{\tau}{i}\right)^{1/2} \frac{\partial \Theta}{\partial z}(\overline{\chi}, \tau, 0)$$

and the corollary follows.   □

This corollary shows that the function $\theta(\chi, \tau)$ is a *modular form* of weight $e + 1/2$ on a suitable congruence subgroup of $\mathrm{SL}_2(\mathbb{Z})$. In fact, we need the transformation formula only when $\tau = it$ with $t > 0$ real.

Corollary 10.2.13. *Let t be a real positive number. As $t \to +\infty$, then $\theta(\chi, it) = \chi(0) + O(\exp(-\pi t/m))$, and as $t \to 0^+$, then $\theta(\chi, it) = t^{-e-1/2}(\chi(0) + O(\exp(-\pi/(tm))))$.*

Proof. As $t \to +\infty$, this follows immediately from the definition. As $t \to 0^+$, we have by the functional equation

$$\theta(\chi, it) = W(\chi)t^{-e-1/2}\theta(\overline{\chi}, i/t) = W(\chi)t^{-e-1/2}(\chi(0) + O(\exp(-\pi/(tm)))) .$$

Now $\chi(0) \neq 0$ if and only if $\chi = 1$, so that $W(\chi) = 1$, giving the corollary.   □

10.2.4 The Functional Equation for Dirichlet L-Functions

Recall from Definition 8.5.9 that the incomplete gamma function is defined for $x > 0$ and $s \in \mathbb{C}$ by $\Gamma(s, x) = \int_x^\infty t^s e^{-t} \, dt/t$.

Theorem 10.2.14 (Functional Equation for Dirichlet L-Functions).
Let χ be a primitive character modulo m, let $e = 0$ (respectively $e = 1$) if χ is an even (respectively odd) character, and set

$$\gamma(s) = \pi^{-s/2}\Gamma(s/2) \quad and \quad \Lambda(\chi, s) = m^{(s+e)/2}\gamma(s+e)L(\chi, s) \ .$$

Then $\Lambda(\chi, s)$ and $L(\chi, s)$ can be analytically continued to meromorphic functions in the whole complex plane and satisfy the functional equation

$$\Lambda(\chi, 1 - s) = W(\chi)\Lambda(\overline{\chi}, s) \ .$$

In addition, these functions are holomorphic except when $\chi = 1$, in which case $\Lambda(\chi, s)$ has two poles at $s = 0$ and $s = 1$, which are simple with residues -1 and 1 respectively, and $L(\chi, s) = \zeta(s)$ has a single pole at $s = 1$, which is simple with residue 1.

Furthermore, we have the following rapidly convergent formula valid for all $A > 0$:

$$\Gamma\left(\frac{s+e}{2}\right)L(\chi, s) = \chi(0)\pi^{s/2}\left(\frac{A^{(s-1)/2}}{s-1} - \frac{A^{s/2}}{s}\right)$$

$$+ \sum_{n\geqslant 1}\frac{\chi(n)}{n^s}\Gamma\left(\frac{s+e}{2}, \frac{\pi n^2 A}{m}\right)$$

$$+ W(\chi)\left(\frac{\pi}{m}\right)^{s-1/2}\sum_{n\geqslant 1}\frac{\overline{\chi}(n)}{n^{1-s}}\Gamma\left(\frac{1-s+e}{2}, \frac{\pi n^2}{Am}\right) \ .$$

Proof. Set

$$I(\chi, s) = \int_0^\infty t^{(s+e)/2}(\theta(\chi, it) - \chi(0))\frac{dt}{t} \ .$$

By the above corollary, this integral converges (exponentially fast in fact) when t is large, while when t is close to 0, either we have $\chi \neq 1$, in which case we also have exponential convergence, or we have $\chi = 1$, and then by the corollary the integrand is asymptotic to $t^{(s-3)/2}$ as $t \to 0^+$. Thus, for $\Re(s) > 1$ the integral is absolutely convergent.

An easy computation gives

$$I(\chi, s) = 2\sum_{n\geqslant 1}n^e\chi(n)\int_0^\infty t^{(s+e)/2}e^{-\pi n^2 t/m}\frac{dt}{t}$$

$$= 2\sum_{n\geqslant 1}n^e\chi(n)\frac{m^{(s+e)/2}}{\pi^{(s+e)/2}n^{s+e}}\int_0^\infty u^{(s+e)/2}e^{-u}\frac{du}{u}$$

$$= 2m^{(s+e)/2}\gamma(s+e)L(\chi, s) = 2\Lambda(\chi, s) \ .$$

The exchange of summation is justified by absolute and normal convergence for $\Re(s) > 1$, and it also proves that $\Lambda(\chi, s)$ is holomorphic in that region.

Let $A > 0$ be arbitrary. We can write, still for $\Re(s) > 1$,

$$I(\chi, s) = \int_A^\infty t^{(s+e)/2}(\theta(\chi, it) - \chi(0))\frac{dt}{t} + \int_0^A t^{(s+e)/2}(\theta(\chi, it) - \chi(0))\frac{dt}{t}$$

$$= \int_A^\infty t^{(s+e)/2}(\theta(\chi, it) - \chi(0))\frac{dt}{t} + \int_0^A t^{(s+e)/2}\theta(\chi, it)\frac{dt}{t} - \chi(0)\frac{A^{s/2}}{s/2} \ ,$$

since $\chi(0) \neq 0$ implies $e = 0$. In the second integral we change t into $1/t$. Since by Proposition 10.2.10 we have for all real $t > 0$,

$$\theta(\chi, i/t) = W(\chi)t^{1/2+e}\theta(\overline{\chi}, it) \ ,$$

we obtain

$$\int_0^A t^{(s+e)/2}\theta(\chi, it)\frac{dt}{t} = W(\chi)\int_{1/A}^\infty t^{(1-s+e)/2}\theta(\overline{\chi}, it)\frac{dt}{t}$$

$$= W(\chi)\int_{1/A}^\infty t^{(1-s+e)/2}(\theta(\overline{\chi}, it) - \chi(0))\frac{dt}{t} + \chi(0)\frac{A^{(s-1)/2}}{(s-1)/2} \ ,$$

since $\chi(0) \neq 0$ implies $e = 0$ and $W(\chi) = 1$. We thus obtain the the preliminary formula

$$2\Lambda(\chi, s) = \chi(0)\left(\frac{A^{(s-1)/2}}{(s-1)/2} - \frac{A^{s/2}}{s/2}\right) + \int_A^\infty t^{(s+e)/2}(\theta(\chi, it) - \chi(0))\frac{dt}{t}$$

$$+ W(\chi)\int_{1/A}^\infty t^{(1-s+e)/2}(\theta(\overline{\chi}, it) - \chi(0))\frac{dt}{t} \ .$$

This has three consequences. First of all, since $\theta(\chi, it)-\chi(0)$ and $\theta(\overline{\chi}, it)-\chi(0)$ tend to zero exponentially as $t \to \infty$, the integrals converge normally in any compact subset of $\mathbb{C}$, so $\Lambda(\chi, s)$ has a meromorphic continuation to $\mathbb{C}$. Furthermore, its only possible poles can occur when $\chi(0) \neq 0$, i.e., $\chi = 1$, in which case they are at $s = 0$ and $s = 1$, are simple, with residues -1 and 1 respectively, as claimed. Furthermore, since $L(\chi, s) = \Lambda(\chi, s)/(m^{(s+e)/2}\gamma(s+e))$ and since $m^{(s+e)/2}\gamma(s + e)$ never vanishes, it follows that $L(\chi, s)$ is also holomorphic on the whole of $\mathbb{C}$, except perhaps at $s = 0$ and $s = 1$ when $\chi = 1$. But if $\chi = 1$, then $m^{(s+e)}\gamma(s + e) = \pi^{-s/2}\Gamma(s/2)$ has a simple pole with residue 2 at $s = 0$, which cancels the simple pole of $\Lambda(\chi, s)$ at $s = 0$, and proves in fact that $\zeta(0) = L(\chi, 0) = -1/2$. On the other hand, for $s = 1$, $\pi^{-s/2}\Gamma(s/2) = 1$, so that $\zeta(s) = L(\chi, s)$ has a simple pole with residue 1, as claimed.

Second, since by definition

$$\tau(\overline{\chi}) = \overline{\tau(\chi, -1)} = \chi(-1)\overline{\tau(\chi)}$$

and $|\tau(\chi)|^2 = m$ and $\chi(-1) = (-1)^e$, it follows that

$$W(\chi)W(\overline{\chi}) = (-1)^e \frac{\tau(\chi)\tau(\overline{\chi})}{m} = 1 \,.$$

Thus, if we change simultaneously s into $1-s$ and A into $1/A$, our preliminary formula immediately gives the functional equation $\Lambda(\chi, 1-s) = W(\chi)\Lambda(\overline{\chi}, s)$.

Third, by definition of $\Gamma(s, x)$, we have

$$\int_x^\infty t^s e^{-\pi n^2 t/m} \frac{dt}{t} = (m/\pi)^s n^{-2s} \Gamma(s, \pi n^2 x) \,.$$

Thus, replacing explicitly the theta functions in our preliminary formula and exchanging summation and integration (which is justified since the integrals converge exponentially fast), we obtain

$$m^{(s+e)/2} \gamma(s+e) L(\chi, s) = \chi(0) \left(\frac{A^{(s-1)/2}}{s-1} - \frac{A^{s/2}}{s} \right)$$

$$+ \left(\frac{m}{\pi}\right)^{(s+e)/2} \sum_{n \geqslant 1} \frac{\chi(n)}{n^s} \Gamma\left(\frac{s+e}{2}, \pi n^2 A\right)$$

$$+ W(\chi) \left(\frac{m}{\pi}\right)^{(1-s+e)/2} \sum_{n \geqslant 1} \frac{\overline{\chi}(n)}{n^{1-s}} \Gamma\left(\frac{1-s+e}{2}, \frac{\pi n^2}{A}\right) \,.$$

We obtain the final formula of the proposition by multiplying with $(\pi/m)^{(s+e)/2}$.
□

Remarks. (1) When $\chi = 1$, in other words $L(\chi, s) = \zeta(s)$, we will simply write $\Lambda(s)$ instead of $\Lambda(1, s)$.

(2) Although the additional formula of the theorem seems to be a "plus" compared to the functional equation, this is not so: it is not difficult to prove (see [Coh1], Appendix A) that the functional equation alone in turn implies the formula. We have already mentioned this phenomenon in Proposition 8.5.10 in the context of L-functions attached to elliptic curves.

(3) The point of the formula given above is not the formula itself, which is not very pretty, but its use for the algorithmic computation of $L(\chi, s)$: since for fixed s, $\Gamma(s, x)$ behaves roughly like e^{-x} as $x \to \infty$, we thus have a formula that converges exponentially fast to $L(\chi, s)$. In addition, it is not difficult to give rapidly convergent methods for the computation of $\Gamma(s, x)$; see for instance Sections 8.5.4 and 8.5.5.

(4) The formula converges fastest when $A = 1/A$, i.e., when $A = 1$. However, it is essential to use the formula with a variable value of A for at least three reasons. First of all, it gives an excellent check on the correctness of the implementation, since the result must be independent of A. Second, although $W(\chi)$ can be computed directly from the definition of $\tau(\chi)$, this takes $O(m)$ operations, hence is very costly when m is large. We can use

the formula with a given value of s and two slightly different values of A to compute $W(\chi)$, and this will be in only $O(m^{1/2}\log(m))$ operations. Third, the formula can be applied to *complex* values of A (using Definition 8.5.9 to define $\Gamma(s,x)$), and gives in this case the so-called *approximate functional equation*.

(5) Another even faster way to compute $W(\chi)$ that avoids the expensive computations of $\Gamma(s,x)$ is to use directly the functional equation of the theta function: by Proposition 10.2.10, for any $t > 0$ such that $\theta(\overline{\chi}, it) \neq 0$ we have

$$W(\chi) = \frac{\theta(\chi, i/t)}{t^{e+1/2}\theta(\overline{\chi}, it)} \ .$$

The optimal value of t in terms of convergence is $t = 1$, and it seems that we always have $\theta(\overline{\chi}, i) \neq 0$ for all χ, so this can be applied. If this is not the case, we can simply use a value of t close to 1 for which $\theta(\overline{\chi}, it) \neq 0$.

Corollary 10.2.15. (1) *If χ is a primitive character modulo m and $e = 0$ or 1 such that $\chi(-1) = (-1)^e$, the functional equation may be rewritten in the form*

$$L(\chi, s) = (-i)^e \tau(\chi) \left(\frac{2\pi}{m}\right)^s \frac{L(\overline{\chi}, 1-s)}{\cos(\pi/2(s-e))\Gamma(s)} \ .$$

(2) *If χ is any character, primitive or not, the function $L(\chi, s)$ does not vanish for $\Re(s) > 1$, and in the domain $\Re(s) < 0$ it vanishes if and only if $s = e - 2k$ for $k \in \mathbb{Z}_{\geqslant 1}$, where it vanishes to order 1.*

(3) *The function $L(\chi, s)$ vanishes at $s = 0$ if and only if χ is a nontrivial even character.*

Proof. (1) immediately follows from the functional equation and the formulas $\Gamma(s)\Gamma(1-s) = \pi/\sin(s\pi)$ and $\Gamma(s/2)\Gamma((s+1)/2) = \pi^{1/2}2^{1-s}\Gamma(s)$. For (2) we note that for $\Re(s) > 1$ we have the convergent Euler product $L(\chi, s) = \prod_p (1 - \chi(p)/p^s)^{-1}$, so that $L(\chi, s) \neq 0$. Furthermore, if χ is a nonprimitive character of conductor $f \mid m$, and if χ_f is the primitive character equivalent to χ then

$$L(\chi, s) = L(\chi_f, s) \prod_{p \mid m, p \nmid f} \left(1 - \frac{\chi_f(p)}{p^s}\right) \ .$$

Thus $L(\chi, s) = 0$ if and only if either $L(\chi_f, s) = 0$ or $p^s = \chi_f(p)$ for some $p \mid m$, $p \nmid f$, in other words $s = (\mathrm{Log}\,(\chi_f(p)) + 2ik\pi)/\log(p)$ for some $k \in \mathbb{Z}$, where Log denotes the principal determination of the complex logarithm. Since $|\chi_f(p)| = 1$, these latter values (infinite in number) are such that $\Re(s) = 0$, in other words are on the imaginary axis, so we do not need to deal with them for now. We can thus restrict to primitive characters. In that case for $\Re(s) < 0$ we have $L(\overline{\chi}, 1-s) \neq 0$ by what we have just said, so (1) implies that in that region $L(\chi, s) = 0$ if and only if $\cos(\pi/2(s-e))\Gamma(s)$ has a pole, which evidently occurs if and only if $s \in \mathbb{Z}_{\leqslant 0}$ is such that $s \equiv e \pmod 2$, and since $\Gamma(s)$ has only simple poles, these zeros have order 1. Finally, if χ is the trivial character then $L(\overline{\chi}, 1 - s)$ has a pole at $s = 0$ that cancels the pole of $\Gamma(s)$ (and it is immediate to compute $L(\chi_0, 0)$; see below), and otherwise $L(\overline{\chi}, 1 - s)$ does not have a pole at $s = 0$, while $\cos(\pi/2(s - e))\Gamma(s)$ has one if and only if $e = 0$. $\qquad\qquad$ □

Note that (1) is equivalent to the form of the functional equation that we obtained in Theorem 10.2.6 (3) using $\zeta(s, x)$.

Definition 10.2.16. *Denote by* Log *the principal determination of the complex logarithm. The zeros* $s = (\mathrm{Log}\,(\chi_f(p)) + 2ik\pi)/\log(p)$ *for* $k \in \mathbb{Z}$, $p \mid m$, *and* $p \nmid f$ *of the function* $L(\chi, s)$ *for a nonprimitive character* χ *are called* extraneous zeros. *The zeros* $s = e - 2k$ *for* $k \in \mathbb{Z}_{\geqslant 1}$, *and in addition* $s = 0$ *if* χ *is a nontrivial even character, are called* trivial zeros. *All other zeros are called* nontrivial.

Note that if $m = p^k$ with $k \geqslant 1$ is a power of a prime number p and if χ is a nontrivial character then all primes dividing m also divide the conductor $f = p^j$ for some j such that $1 \leqslant j \leqslant k$, so in that case $L(\chi, s) = L(\chi_f, s)$. In particular, $L(\chi, s)$ will have a clean functional equation and no extraneous zeros.

It follows from the above corollary that the problem of localizing the zeros of L-functions is reduced to the strip $0 \leqslant \Re(s) \leqslant 1$, and in fact as we will see in Section 10.5.7, to the strip $0 < \Re(s) < 1$. Furthermore, thanks to the functional equation and the elementary property $\overline{L(s, \chi)} = L(\overline{s}, \overline{\chi})$ we may even restrict the study to the smaller strip $1/2 \leqslant \Re(s) < 1$ and $\Im(s) \geqslant 0$.

10.2.5 Generalized Poisson Summation Formulas

We have given several proofs of the functional equation of Dirichlet L-functions. Although perhaps the longest, the most "natural" such proof is via the functional equation of the theta function, itself an immediate consequence of the Poisson summation formula (Proposition 2.2.16). In fact, it is quite easy to see that this summation formula (which of course is quite simple) can be deduced from the functional equation of Dirichlet L-functions. We will do this in quite a general context as follows.

Let $a(n)$ be an arithmetic function and $f(x)$ a nice function as occurs for instance in the Poisson summation formula (we will make this precise later). We would like to give an exact Poisson-style formula for $\sum_{n \geqslant 1} a(n) f(n)$, the Poisson formula itself corresponding to the case $a(n) = 1$ and $f(x)$ even. To simplify we will first assume that f is in the Schwartz space, in other words that $f \in C^{\infty}(\mathbb{R})$ and that $f(x)$ and all its derivatives tend to 0 faster than any power of $|x|$ as $|x| \to \infty$, and will mention below how to prove the result under much milder assumptions.

To obtain the desired formula we recall that the L-function associated with the arithmetic function $a(n)$ is defined as usual by $L(a,s) = \sum_{n \geqslant 1} a(n)n^{-s}$. We make the following assumptions, which are essential, contrary to the assumptions on $f(x)$. We assume that $L(a,s)$ converges for $\Re(s) > 1$, that it has an analytic continuation to the whole complex plane with a possible single pole at 1 of exact order $r \geqslant 0$, say, and that it has a functional equation of the type $\Lambda(a,s) = w\Lambda(a^*, 1-s)$, where $a^*(n)$ is some other arithmetic function, $|w| = 1$, $\Lambda(a,s) = \gamma(s)L(a,s)$, where $\gamma(s) = A^s \prod_{1 \leqslant i \leqslant g} \Gamma(a_i s + b_i)$, for strictly positive constants A and a_i, and similarly for a^*. Note that this is the form of the functional equation of L-series of Dirichlet characters, and up to a shift of the variable s it is the form of essentially all global functional equations occurring in number theory and algebraic geometry.

The result is as follows.

Theorem 10.2.17. *Keep all the above notation and assumptions. For any σ such that $0 < \sigma < 1$ and $x > 0$ set*

$$K(x) = \frac{1}{2i\pi} \int_{\Re(s) = \sigma} \frac{\gamma(s)}{\gamma(1-s)} x^{-s}\, ds \quad and \quad g(x) = \int_0^\infty f(t)K(xt)\, dt\,.$$

(1) *We have*

$$\int_0^\infty t^{s-1}K(t)\, dt = \frac{\gamma(s)}{\gamma(1-s)}\,, \quad \int_0^\infty g(t)K(xt)\, dt = f(x)\,, \quad and$$

$$\int_0^\infty K(xt)K(yt)\, dt = \delta(x-y)\,,$$

where $\delta(x)$ is the Dirac distribution.

(2) *If we set $a(0) = -L(a,0)$ we have the summation formula*

$$\sum_{n \geqslant 0} a(n)f(n) = \mathrm{Res}_{s=1}\left(L(a,s)\int_0^\infty t^{s-1}f(t)\, dt\right) + w\sum_{n \geqslant 1} a^*(n)g(n)\,.$$

Proof. (1). Recall from Section 9.7.3 that the Mellin transform of f is defined by $M(f)(s) = \int_0^\infty f(t)t^{s-1}\, dt$. Since f is in the Schwartz space, $M(f)(s)$ converges for $\Re(s) > 0$ and can be analytically continued to the whole complex plane with possible poles only in $\mathbb{Z}_{\leqslant 0}$, since by integration by parts for $\Re(s) > 0$ we have

$$M(f)(s) = f(t)\frac{t^s}{s}\Big|_0^\infty - \frac{1}{s}\int_0^\infty f'(t)t^s\, dt = -\frac{M(f')(s+1)}{s}\,.$$

The Mellin inversion formula (Proposition 9.7.7) tells us that for all $\sigma > 0$ and $x > 0$ we have

$$f(x) = \frac{1}{2i\pi} \int_{\Re(s)=\sigma} M(f)(s) x^{-s} \, ds \; ,$$

so the first formula of (1) is clear. For the second we have for $0 < \Re(s) < 1$,

$$\int_0^\infty x^{-s} g(x) \, dx = \int_0^\infty x^{-s} \int_0^\infty f(t) K(xt) \, dt = \int_0^\infty t^{s-1} f(t) \int_0^\infty x^{-s} K(x) \, dx$$
$$= M(f)(s) M(K)(1-s) = (\gamma(1-s)/\gamma(s)) M(f)(s)$$

by the first formula. Thus for $0 < \sigma < 1$,

$$\int_0^\infty g(t) K(xt) \, dt = \frac{1}{2i\pi} \int_{\Re(s)=\sigma} \frac{\gamma(s)}{\gamma(1-s)} x^{-s} \left(\int_0^\infty t^{-s} g(t) \, dt \right) ds$$
$$= \frac{1}{2i\pi} \int_{\Re(s)=\sigma} x^{-s} M(f)(s) \, ds = f(x)$$

by Mellin inversion, proving the second formula. For the third, let $\phi(x)$ be a function in the Schwartz space, and let $\psi(x) = \int_0^\infty \phi(t) K(xt) \, dt$, so that $\phi(x) = \int_0^\infty \psi(t) K(xt) \, dt$ by what we have just proved. We have

$$\int_0^\infty \left(\int_0^\infty K(xt) K(yt) \, dt \right) \phi(y) \, dy = \int_0^\infty K(xt) \left(\int_0^\infty \phi(y) K(yt) \, dy \right) dt$$
$$= \int_0^\infty K(xt) \psi(t) \, dt = \phi(x) \; ,$$

proving the last formula of (1).

(2). For simplicity of notation set $F = M(f)$. By Mellin inversion we have for all $\sigma > 1$,

$$\sum_{n \geqslant 1} a(n) f(n) = \frac{1}{2i\pi} \int_{\Re(s)=\sigma} F(s) \sum_{n \geqslant 1} a(n) n^{-s} \, ds = \frac{1}{2i\pi} \int_{\Re(s)=\sigma} F(s) L(a, s) \, ds \; ,$$

where here and elsewhere all the interchanges of summation are justified by the fact that f is in the Schwartz space. We now shift the line of integration from $\Re(s) = \sigma > 1$ to $\Re(s) = -1/2$, say. By assumption $L(a, s)$ has at most a pole at $s = 1$. From the formula $F(s) = M(f)(s) = -M(f')(s+1)/s$ we deduce that $F(s)$ may have a pole only at $s = 0$, which is simple with residue $-M(f')(1) = -\int_0^\infty f'(t) \, dt = f(0)$ (hence no pole at all if $f(0) = 0$). Thus, applying the functional equation for $L(a, s)$ we obtain

$$\sum_{n \geqslant 1} a(n) f(n) = \mathrm{Res}_{s=1}(L(a, s) F(s)) + f(0) L(a, 0) + I \; ,$$

where

$$I = \frac{1}{2i\pi} \int_{\Re(s)=-1/2} F(s)L(a,s)\,ds$$

$$= \frac{w}{2i\pi} \int_{\Re(s)=-1/2} F(s)\frac{\gamma(1-s)}{\gamma(s)}L(a^*,1-s)\,ds$$

$$= \frac{w}{2i\pi} \int_{\Re(s)=3/2} F(1-s)\frac{\gamma(s)}{\gamma(1-s)}L(a^*,s)\,ds .$$

Thus, if we set $G(s) = F(1-s)\gamma(s)/\gamma(1-s)$ then as above we have

$$\sum_{n\geqslant 1} a(n)f(n) = \mathrm{Res}_{s=1}(L(a,s)F(s)) + f(0)L(a,0) + w\sum_{n\geqslant 1} a^*(n)g(n) ,$$

where for $\sigma > 1$,

$$g(x) = \frac{1}{2i\pi} \int_{\Re(s)=\sigma} G(s)x^{-s}\,ds = \frac{1}{2i\pi} \int_{\Re(s)=\sigma} x^{-s}\frac{\gamma(s)}{\gamma(1-s)}F(1-s)\,ds .$$

We choose $\sigma = 3/2$ so that we can use the convergent formula

$$F(1-s) = -\frac{1}{1-s}M(f')(2-s) = -\frac{1}{1-s}\int_0^\infty f'(t)t^{1-s}\,dt .$$

If we set for $0 < \sigma < 1$,

$$K_1(x) = \frac{1}{2i\pi} \int_{\Re(s)=\sigma} \frac{\gamma(s)}{(1-s)\gamma(1-s)}x^{-s}\,ds$$

(no connection with Bessel functions), then

$$g(x) = -\frac{1}{2i\pi} \int_{\Re(s)=\sigma} x^{-s}\frac{\gamma(s)}{(1-s)\gamma(1-s)} \int_0^\infty f'(t)t^{1-s}\,dt\,ds$$

$$= -\int_0^\infty f'(t)\frac{K_1(xt)}{x}\,dt = -f(t)\frac{K(xt)}{x}\Big|_0^\infty + \int_0^\infty f(t)K_1'(xt)\,dt$$

$$= \int_0^\infty f(t)K(xt)\,dt ,$$

proving the theorem. $\qquad\square$

Remarks. (1) The second formula of (1) means that $K(x)$ is a *self-dual* integration kernel, generalizing the same fact for the cosine Fourier transform. This is essentially equivalent to the fact that its Mellin transform $\gamma(s)/\gamma(1-s)$ changes into its inverse when s is changed into $1-s$.

(2) Note that if around $s=1$ we have $L(a,s) = \sum_{k\geqslant -r}\lambda(k)(s-1)^k$, then since $t^{s-1} = \exp((s-1)\log(t))$ we have

$$\mathrm{Res}_{s=1}\left(L(a,s)\int_0^\infty f(t)t^{s-1}\,dt\right) = \sum_{k=0}^{r-1}\frac{\lambda(-k-1)}{k!}\int_0^\infty f(t)\log(t)^k\,dt .$$

(3) Although we have proved the theorem for a function f belonging to the Schwartz space, it is not difficult to show by approximation techniques that it is still valid if f is only piecewise C^∞ and piecewise monotonic (Exercise 32). In particular, we can multiply f by the characteristic function of a finite interval $[A, B]$ and obtain a summation formula for $\sum_{A \leqslant n \leqslant B} a(n)f(n)$, the integrals from 0 to ∞ being replaced by integrals from A to B (see for instance Proposition 2.2.16).

Examples. (1) As already mentioned at the beginning, the basic example is $a(n) = 1$, so that $L(a, s) = \zeta(s)$, $\gamma(s) = \pi^{-s/2}\Gamma(s/2)$. By Corollary 9.6.37 we have

$$\int_0^\infty \cos(2\pi t) t^{s-1}\, dt = (2\pi)^{-s} \cos(\pi s/2)\Gamma(s) .$$

On the other hand, the reflection and duplication formulas for the gamma function give

$$\frac{\pi^{-s/2}\Gamma(s/2)}{\pi^{-(1-s)/2}\Gamma((1-s)/2)} = 2(2\pi)^{-s} \cos(\pi s/2)\Gamma(s) .$$

It follows that $K(x) = 2\cos(2\pi x)$ and the theorem reads

$$\sum_{n \geqslant 1} f(n) + \frac{f(0)}{2} = \int_0^\infty f(t)\, dt + 2\sum_{n \geqslant 1} \int_0^\infty f(t) \cos(2\pi n t)\, dt ,$$

which is the Poisson summation formula for an even function f.

(2) We now choose $a(n) = r_2(n)$, where $r_2(n)$ is the number of decompositions of n as a sum of two squares. By Corollary 10.5.8 and Proposition 10.5.5, which we will prove below, we have $L(a, s) = 4\zeta_{\mathbb{Q}(i)}(s)$, $w = 1$, and $\gamma(s) = (2\pi)^{-s}\Gamma(s)$. From Proposition 9.8.8 we deduce that

$$\int_0^\infty t^{s-1} J_0(4\pi t^{1/2})\, dt = \frac{1}{2\pi} \int_0^\infty \left(\frac{u}{4\pi}\right)^{2s-1} J_0(u)\, du$$

$$= (2\pi)^{-2s} \frac{\Gamma(s)}{\Gamma(1-s)} = \frac{1}{2\pi} \frac{\gamma(s)}{\gamma(1-s)} .$$

It follows that $K(x) = 2\pi J_0(4\pi x^{1/2})$, and since (for instance by Corollary 10.2.3) we know that $L(\chi_{-4}, 1) = \pi/4$ and $L(\chi_{-4}, 0) = 1/2$, the theorem gives the summation formula

$$\sum_{n \geqslant 0} r_2(n) f(n) = \pi \int_0^\infty f(t)\, dt + 2\pi \sum_{n \geqslant 1} r_2(n) \int_0^\infty f(t) J_0(4\pi(nt)^{1/2})\, dt ,$$

where we set $r_2(0) = 1$.

(3) Finally, we choose $a(n) = d(n)$, where $d(n)$ is the number of (positive) divisors of n. We have seen that $L(a,s) = \zeta(s)^2$, so that $\gamma(s) = \pi^{-s}\Gamma(s/2)^2$. From the computation made for $\zeta(s)$ itself we have

$$\gamma(s)/\gamma(1-s) = 4(2\pi)^{-2s}\cos(s\pi/2)^2\Gamma(s)^2 .$$

On the other hand, Proposition 9.8.8 gives

$$\int_0^\infty t^{s-1}Y_0(4\pi t^{1/2})\,dt = -\frac{1}{\pi}(2\pi)^{-2s}\cos(s\pi)\Gamma(s)^2 ,$$

$$\int_0^\infty t^{s-1}K_0(4\pi t^{1/2})\,dt = \frac{1}{2}(2\pi)^{-2s}\Gamma(s)^2 .$$

Since $4\cos(s\pi/2)^2 = 2(\cos(s\pi)+1)$ we have

$$K(x) = 4K_0(4\pi x^{1/2}) - 2\pi Y_0(4\pi x^{1/2}) ,$$

and using the expansion $\zeta(s) = 1/(s-1) + \gamma + O(s-1)$ the theorem gives the summation formula

$$\sum_{n\geqslant 1} d(n)f(n) = \int_0^\infty f(t)(\log(t) + 2\gamma)\,dt + f(0)/4$$

$$+ \sum_{n\geqslant 1} d(n)\int_0^\infty f(t)(4K_0(4\pi(nt)^{1/2}) - 2\pi Y_0(4\pi(nt)^{1/2}))\,dt .$$

The above summation formulas are due to Voronoi and are used to give error estimates in the *circle problem* (estimate $\Delta(X) = \sum_{1\leqslant n\leqslant X} r_2(n) - \pi X$) and in the *divisor problem* (estimate $\Delta(X) = \sum_{1\leqslant n\leqslant X} d(n) - (X\log(X) + (2\gamma - 1)X)$). In both cases we have the "trivial" estimate $\Delta(X) = O(X^{1/2})$ (see Exercise 33), and from Voronoi's formulas it is not difficult to obtain $\Delta(X) = O(X^{1/3})$, which we will prove below for the circle problem (Theorem 10.2.18). It is also not too difficult to show that we cannot have $\Delta(X) = O(X^\alpha)$ for $\alpha \leqslant 1/4$. Several mathematicians have succeeded in obtaining values of α such that $\alpha < 1/3$, but at the price of considerable additional effort, and it is an ongoing race. The present record, due to Huxley, is $\alpha = 131/416 + \varepsilon$ for any $\varepsilon > 0$.

10.2.6 Voronoi's Error Term in the Circle Problem

We are now going to show that Voronoi's summation formula leads to a quite simple but powerful estimate for the error term in the circle problem. Recall that the *circle problem* consists in computing a precise estimate for the number of points with integral coordinates in the closed disk of radius $X^{1/2}$, in other words in estimating $\sum_{0\leqslant n\leqslant X} r_2(n)$. A heuristic shows that this should be close to the area of the disk, in other words to πX, and a rigorous and easy argument shows that more precisely it is $\pi X + O(X^{1/2})$, see Exercise 33. We will prove the following stronger result.

Theorem 10.2.18 (Voronoi). *As $X \to \infty$ we have*

$$\sum_{0 \leqslant n \leqslant X} r_2(n) = \pi X + O(X^{1/3}) .$$

Proof. By Voronoi's summation formula that we have seen above we have for all reasonable functions f (for instance piecewise continuous and tending to zero sufficiently rapidly at infinity)

$$\sum_{n \geqslant 0} r_2(n) f(n) = \pi \int_0^\infty f(t) \, dt + 2\pi \sum_{n \geqslant 1} r_2(n) \int_0^\infty f(t) J_0(4\pi(nt)^{1/2}) \, dt .$$

If we apply this formula to $f(t) = 1$ for $0 \leqslant t \leqslant X$ and $f(t) = 0$ for $t > X$ we obtain

$$\sum_{0 \leqslant n \leqslant X} r_2(n) = \pi X + 2\pi \sum_{n \geqslant 1} r_2(n) \int_0^X J_0(4\pi(nt)^{1/2}) \, dt .$$

Now from Proposition 9.8.4 (2) we have $J_1'(x) + J_1(x)/x = J_0(x)$, from which a short computation shows that

$$(t^{1/2} J_1(4\pi(at)^{1/2}))' = 2\pi a^{1/2} J_0(4\pi(at)^{1/2}) ,$$

giving the formula

$$\sum_{0 \leqslant n \leqslant X} r_2(n) = \pi X + X^{1/2} \sum_{n \geqslant 1} \frac{r_2(n)}{n^{1/2}} J_1(4\pi(nX)^{1/2}) .$$

At first sight this formula seems quite nice since by Proposition 9.8.7 we have

$$J_1(4\pi(nX)^{1/2}) \sim \frac{X^{-1/4}}{\pi 2^{1/2}} \frac{\cos(4\pi(nX)^{1/2} - 3\pi/4)}{n^{1/4}} ,$$

and we could even easily strengthen this asymptotic estimate into one with a negligible remainder term. However, estimating the series

$$\sum_{n \geqslant 1} \frac{r_2(n)}{n^{3/4}} \cos(4\pi(nX)^{1/2} - 3\pi/4)$$

is not a trivial task.

Thus we prefer to avoid taking a function f having a brutal cutoff at X, and the simplest method is to choose the function f defined in the the following way for $t \geqslant 0$. We set $Y = X^{1/3}$ (which will be seen in the course of the proof to give the best results), and define

$$f(t) = \begin{cases} 1 & \text{for } 0 \leqslant t \leqslant X , \\ 1 - (t - X)/Y & \text{for } X \leqslant t \leqslant X + Y , \\ 0 & \text{for } t \geqslant X + Y , \end{cases}$$

in other words $f(t) = \min(1 - (t - X)/Y, 1)$ for $0 \leqslant t \leqslant X + Y$ and $f(t) = 0$ elsewhere. For this function f, Voronoï's summation formula given above reads

$$\sum_{0 \leqslant n \leqslant X} r_2(n) + \sum_{X < n \leqslant X+Y} r_2(n)(1 - (n - X)/Y)$$

$$= \pi(X + Y/2) + 2\pi \sum_{n \geqslant 1} r_2(n) h_X(n) \,,$$

where

$$h_X(n) = \int_0^{X+Y} \min(1 - (t - X)/Y, 1) J_0(4\pi(nt)^{1/2}) \, dt \,.$$

We can thus write

$$\sum_{0 \leqslant n \leqslant X} r_2(n) = \pi X + \pi Y/2 + 2\pi S_1 + 2\pi S_2 - S_3$$

with

$$S_1 = \sum_{1 \leqslant n \leqslant X^{1/3}} r_2(n) h_X(n), \quad S_2 = \sum_{n > X^{1/3}} r_2(n) h_X(n), \quad \text{and}$$

$$S_3 = \sum_{X < n \leqslant X+Y} r_2(n)(1 - (n - X)/Y) \,.$$

We first estimate S_1 and S_2. From Proposition 9.8.4 (2) we easily deduce that

$$\frac{1}{2\pi n^{1/2}} \left(t^{(\nu+1)/2} J_{\nu+1}(4\pi(nt)^{1/2}) \right)' = t^{\nu/2} J_\nu(4\pi(nt)^{1/2}) \,.$$

Thus if we integrate by parts once, then another time, we obtain

$$h_X(n) = \frac{\min(1 - (t - X)/Y, 1)}{2\pi n^{1/2}} t^{1/2} J_1(4\pi(nt)^{1/2}) \bigg|_0^{X+Y}$$

$$+ \frac{1}{2\pi n^{1/2} Y} \int_X^{X+Y} t^{1/2} J_1(4\pi(nt)^{1/2}) \, dt$$

$$= \frac{1}{2\pi n^{1/2} Y} \int_X^{X+Y} t^{1/2} J_1(4\pi(nt)^{1/2}) \, dt = \frac{1}{4\pi^2 nY} t J_2(4\pi(nt)^{1/2}) \bigg|_X^{X+Y}$$

$$= \frac{1}{4\pi^2 nY} \left((X + Y) J_2(4\pi(n(X+Y))^{1/2}) - X J_2(4\pi(nX)^{1/2}) \right) \,.$$

We are going to use both of these last two formulas.

(S_1). We use the last formula involving J_1. Using the asymptotic estimate $J_1(x) = O(x^{-1/2})$ given by Proposition 9.8.7 we obtain

$$h_X(n) = O(X^{1/2} n^{-1/2} X^{-1/4} n^{-1/4}) = O(X^{1/4} n^{-3/4}) \,.$$

Let $R(N) = \sum_{1 \leqslant n \leqslant N} r_2(n)$ be the summatory function of $r_2(n)$, so that by a trivial estimate we know that $R(N) = O(N)$. By Abel summation we thus have

$$S_1 = O(X^{1/4}) \sum_{1 \leqslant n \leqslant X^{1/3}} \frac{R(n) - R(n-1)}{n^{3/4}}$$

$$= O(X^{1/4}) \left(\sum_{1 \leqslant n \leqslant X^{1/3}} R(n)(n^{-3/4} - (n+1)^{-3/4}) \right.$$

$$\left. + O(R(X^{1/3})X^{-1/4}) \right)$$

$$= O(X^{1/4}) \sum_{1 \leqslant n \leqslant X^{1/3}} R(n)n^{-7/4} + O(X^{1/3}) = O(X^{1/3}) \,.$$

(S_2). Here we use the formula involving J_2 and the asymptotic estimate $J_2(x) = O(x^{-1/2})$. Since $Y = X^{1/3}$ it follows that

$$h_X(n) = O(X^{-1/3}n^{-1}X^{3/4}n^{-1/4}) = O(X^{5/12}n^{-5/4}) \,,$$

so that we obtain similarly

$$\sum_{n > X^{1/3}} r_2(n) h_X(n) = O(X^{5/12}) \sum_{n > X^{1/3}} \frac{R(n) - R(n-1)}{n^{5/4}}$$

$$= O(X^{5/12}) \left(\sum_{n > X^{1/3}} R(n)(n^{-5/4} - (n+1)^{-5/4}) \right.$$

$$\left. + O(R(X^{1/3})X^{-5/12}) \right)$$

$$= O(X^{5/12}) \sum_{n > X^{1/3}} R(n)n^{-9/4} + O(X^{1/3}) = O(X^{1/3}) \,.$$

Estimating S_3 directly is not so easy, although it follows immediately from Exercise 18 of Chapter 5 or directly that $S_3 = O_\varepsilon(X^{1/3+\varepsilon})$, which is not quite sufficient. Thus, we use only the trivial fact that $S_3 \geqslant 0$, and deduce that $\sum_{n \leqslant X} r_2(n) \leqslant \pi X + O(X^{1/3})$.

To obtain an inequality in the other direction is completely analogous: we now use the function $f(t)$ defined by $f(t) = \min((X-t)/Y, 1)$ for $t \leqslant X$ and $f(t) = 0$ elsewhere. Here Voronoi's summation formula gives

$$\sum_{0 \leqslant n \leqslant X} r_2(n) = \pi(X - Y/2) + 2\pi S_1 + 2\pi S_2 + S_3 \,,$$

where S_1 and S_2 are defined as before (for the new function $f(t)$), and

$$S_3 = \sum_{X-Y < n \leqslant X} r_2(n)(1 - (X - n)/Y) .$$

The computations made above are valid verbatim, so that $S_1 = O(X^{1/3})$, $S_2 = O(X^{1/3})$, and evidently $S_3 \geqslant 0$, so that $\sum_{n \leqslant X} r_2(n) \geqslant \pi X + O(X^{1/3})$, finishing the proof.   □

As already mentioned above it is much harder but possible to obtain an error term in $O(X^\alpha)$ with $\alpha < 1/3$.

10.3 Special Values of Dirichlet L-Functions

10.3.1 Basic Results on Special Values

The aim of this section is to give still another proof of the following theorem.

Theorem 10.3.1. *Let χ be any periodic arithmetic function with period dividing m and let $k \geqslant 1$ be an integer.*

(1) *We have*

$$L(\chi, 1 - k) = -\frac{B_k(\chi)}{k} - \chi(0)\delta_{k,1} .$$

(2) *If, in addition, χ is a primitive character modulo m, then*

$$L(\chi, k) = (-1)^{e+k+1} W(\chi) \frac{2^{k-1}\pi^k \overline{B_k(\chi)}}{m^{k-1/2}k! \cos((\pi/2)(k - e))} ,$$

where we recall that we have set $e = 0$ or 1 so that $\chi(-1) = (-1)^e$. In other words, when $k \not\equiv e \pmod 2$, then $L(\chi, 1 - k) = B_k(\chi) = 0$, except when $k = 1$ and $m = 1$, and when $k \equiv e \pmod 2$ we have

$$L(\chi, k) = (-1)^{k-1+(k+e)/2} W(\chi) \frac{2^{k-1}\pi^k \overline{B_k(\chi)}}{m^{k-1/2}k!} .$$

Note that of course, this theorem does not tell us anything about $L(\chi, k)$ when $k \not\equiv e \pmod 2$, $k \geqslant 2$ (otherwise we would know the value of $\zeta(3)$), and that we have already proved the vanishing of the higher χ-Bernoulli numbers in that case.

Proof. Thanks to the functional equation (for instance from Corollary 10.2.15) it is clear that (2) follows from (1), so it is enough to prove (1). We have already given two proofs of (1), one using Corollary 10.2.3, the other using $\zeta(s, x)$ in Proposition 10.2.5. We will give a third proof, which is complex analytic, and which can be generalized to other contexts. This proof can be skipped.

Let ρ and ε be real numbers with $0 < \varepsilon < \rho$, and let $C_{\rho,\varepsilon}$ be the contour in the complex plane starting at $+\infty + i\varepsilon$ and proceeding to the circle of radius

ρ centered at the origin, then following this circle in the counterclockwise direction, and then from the circle to $+\infty - i\varepsilon$. We set $z^{s-1} = \exp((s-1)\log(z))$ (any determination of the logarithm, but continuous on the contour), and

$$I(s) = \int_{C_{\rho,\varepsilon}} t^{s-1} \exp(-t)\, dt \ .$$

Since $t^{s-1}\exp(-t)$ has no pole and is single-valued in the difference of two contours, it is clear that $I(s)$ is independent of ρ and ε, whence the notation. If $\Re(s) > 0$, then it is clear that the integral around the circle tends to 0 as $\rho \to 0$. As ρ and ε tend to 0, the integral from $+\infty + i\varepsilon$ to the circle tends to $-\int_0^\infty t^{s-1}\exp(-t)\,dt$, and since the argument of the logarithm has increased by 2π as we go around the circle, the integral from the circle to $+\infty - i\varepsilon$ tends to $\exp(2i\pi(s-1))\int_0^\infty t^{s-1}\exp(-t)$. Thus, for $\Re(s) > 0$ we have

$$I(s) = 2i\exp(i\pi s)\sin(\pi s)\Gamma(s) = 2i\pi\frac{\exp(i\pi s)}{\Gamma(1-s)}$$

by the reflection formula for the gamma function.

Let $z \in \mathbb{C}$ be such that $|z| < 1$, and consider

$$J(s,z) = \int_{C_{\rho,\varepsilon}} t^{s-1}\frac{\sum_{1\leqslant r\leqslant m}\chi(r)z^r\exp(-rt)}{1 - z^m\exp(-mt)}\,dt \ .$$

Note that this integral converges absolutely for $\Re(s) > 1$, and that as above, it is independent of ρ and ε, at least for ρ sufficiently small. If we choose $\rho < -\log(|z|)$, then $|z^m\exp(m\rho)| < 1$ so that $|z^m\exp(-mt)| < 1$ for all t on the contour, uniformly in t. We can thus expand $1/(1 - z^m\exp(-mt))$ as a power series in $z^m\exp(-mt)$, and the exchange of summation and integration will be justified. We thus obtain for $\Re(s) > 1$,

$$J(s,z) = \int_{C_{\rho,\varepsilon}} t^{s-1}\sum_{k\geqslant 0}\sum_{1\leqslant r\leqslant m}\chi(r)z^{r+km}\exp(-(r+km)t)\,dt$$

$$= \int_{C_{\rho,\varepsilon}} t^{s-1}\sum_{n=1}^{\infty}\chi(n)z^n\exp(-nt)\,dt$$

$$= \sum_{n=1}^{\infty}\frac{\chi(n)z^n}{n^s}\int_{C_{n\rho,n\varepsilon}} t^{s-1}\exp(-t)\,dt = 2i\pi\frac{\exp(i\pi s)}{\Gamma(1-s)}L(\chi,s,z) \ ,$$

where we have set

$$L(\chi,s,z) = \sum_{n=1}^{\infty}\frac{\chi(n)z^n}{n^s} \ .$$

Although this expression is valid only for $\Re(s) > 1$, since the integral $J(s,z)$ converges for all complex s, this in fact gives a meromorphic continuation of $L(\chi,s,z)$ to the whole complex plane. Furthermore, its only possible poles are those of $\Gamma(1-s)$, i.e., at $\mathbb{Z}_{>0}$. But by absolute convergence, it is clear that these s are not poles, except perhaps for $s = 1$ and $|z| = 1$, and we check that $s = 1$ is a pole if and only if $z = 1$ and χ is a trivial character.

I claim that for $\Re(s) > 1$, both sides of the equality that we have proved are continuous as z tends to 1, $z < 1$ real, and tend to the expected limits. Indeed, for $J(s, z)$ on the two straight lines of the contour there is no problem since $|\exp(-t)| \leqslant \exp(-\rho) < 1$. On the circle, since $\Re(s) > 1$, it is easy to see that the same result holds. Finally, the result holds also for $L(\chi, s, z)$ and $\Re(s) > 1$ by normal convergence. Thus, we have the identity

$$J(s) = \int_{C_{\rho,\varepsilon}} t^{s-1} \frac{\sum_{1 \leqslant r \leqslant m} \chi(r) \exp(-rt)}{1 - \exp(-mt)} \, dt = 2i\pi \frac{\exp(i\pi s)}{\Gamma(1-s)} L(\chi, s) \ .$$

A priori this identity is valid only for $\Re(s) > 1$, but since both sides have analytic continuation to $\mathbb{C}$, the identity is valid for all s.

Set

$$f(t) = \frac{\sum_{1 \leqslant r \leqslant m} \chi(r) \exp(-rt)}{1 - \exp(-mt)} \ ,$$

so that by Lemma 9.4.3,

$$f(t) = \sum_{k=-1}^{\infty} \frac{B_{k+1}(\chi^-)}{(k+1)!} t^k \ .$$

By Cauchy's formula we deduce that for $k \geqslant 1$ integral we have

$$\frac{B_k(\chi^-)}{k!} = \frac{1}{2i\pi} \int_{C_{\rho,\varepsilon}} \frac{f(t)}{t^k} \, dt = \frac{1}{2i\pi} J(1-k) = \frac{(-1)^{k-1}}{(k-1)!} L(\chi, 1-k) \ ,$$

proving (1) thanks to Proposition 9.4.9, and hence the theorem. $\square$

Corollary 10.3.2. *Let χ be a character modulo m. Then*

$$L(\chi, 0) = \begin{cases} 0 & \text{if } \chi(-1) = 1 \text{ and } m > 1 \ , \\ -\dfrac{1}{2} & \text{if } m = 1 \ , \\ -\dfrac{1}{m} \sum_{r=1}^{m-1} \chi(r) r & \text{if } \chi(-1) = -1 \ . \end{cases}$$

In particular, if $\chi(-1) = -1$ and χ is a primitive character we have

$$L(\chi, 1) = -W(\chi) \frac{\pi}{m^{3/2}} \sum_{r=1}^{m-1} \overline{\chi}(r) r \ .$$

Corollary 10.3.3. *Let χ be a nontrivial character modulo m.*

(1) *If χ is even we have $L(\chi, 1-k) = 0$ for $k \geqslant 1$ odd, and otherwise*

$$L(\chi, -1) = -\frac{1}{2m} \sum_{r=1}^{m-1} \chi(r) r^2 \,,$$

$$L(\chi, -3) = -\frac{1}{4m} \sum_{r=1}^{m-1} \chi(r) r^2 (r^2 - 2m^2) \,.$$

(2) *If χ is odd we have $L(\chi, 1-k) = 0$ for $k \geqslant 2$ even, and otherwise*

$$L(\chi, -2) = -\frac{1}{3m} \sum_{r=1}^{m-1} \chi(r) r (r^2 - m^2) \,,$$

$$L(\chi, -4) = -\frac{1}{15m} \sum_{r=1}^{m-1} \chi(r) r (r^2 - m^2)(3r^2 - 7m^2) \,.$$

Corollary 10.3.4. *We have $\zeta(0) = -1/2$, and for $k \geqslant 1$ we have $\zeta(-2k) = 0$, $\zeta(1 - 2k) = -B_{2k}/(2k)$, and*

$$\zeta(2k) = (-1)^{k-1} \frac{2^{2k-1} \pi^{2k} B_{2k}}{(2k)!} \,.$$

Proofs. Left to the reader (Exercise 12). □

Examples.

$$\zeta(2) = 1 + \frac{1}{2^2} + \frac{1}{3^2} + \cdots = \frac{\pi^2}{6} \,,$$

$$\zeta(4) = 1 + \frac{1}{2^4} + \frac{1}{3^4} + \cdots = \frac{\pi^4}{90} \,,$$

$$L\left(\left(\frac{-3}{\cdot}\right), 1\right) = 1 - \frac{1}{2} + \frac{1}{4} - \frac{1}{5} + \cdots = \frac{\pi}{3^{3/2}} \,,$$

$$L\left(\left(\frac{-4}{\cdot}\right), 1\right) = 1 - \frac{1}{3} + \frac{1}{5} - \cdots = \frac{\pi}{4} \,,$$

$$L\left(\left(\frac{-4}{\cdot}\right), 3\right) = 1 - \frac{1}{3^3} + \frac{1}{5^3} - \cdots = \frac{\pi^3}{32} \,.$$

See Exercise 41 for a general formula giving $L\left(\left(\frac{-4}{\cdot}\right), 2k+1\right)$.

It is often very useful to know the value of $L(\chi, 1)$ in all cases. This is given by the above theorem when χ is an odd character. However, when χ is an even character, we can still give an explicit expression.

Proposition 10.3.5. *Let χ be a character modulo m. Then*

(1)

$$L'(\chi,0) = \begin{cases} -\frac{1}{2}\sum_{r=1}^{m-1} \chi(r)\log\left(\sin\left(\frac{r\pi}{m}\right)\right) & \text{if } \chi \text{ is even and nontrivial}, \\ \sum_{r=1}^{m-1} \chi(r)\log\left(\Gamma\left(\frac{r}{m}\right)\right) - \log(m)L(\chi,0) & \text{if } \chi \text{ is odd}, \\ -\frac{1}{2}\Lambda(m) & \text{if } \chi \text{ is trivial and } m > 1, \\ -\frac{1}{2}\log(2\pi) & \text{if } m = 1, \end{cases}$$

where $\Lambda(m)$ is the von Mangoldt function (see Proposition 10.1.14).

(2) If χ is an even primitive character and $m > 1$, we have

$$L(\chi,1) = -\frac{W(\chi)}{m^{1/2}} \sum_{r=1}^{m-1} \overline{\chi(r)}\log\left(\sin\left(\frac{r\pi}{m}\right)\right).$$

(3) If χ is an odd primitive character we have

$$L'(\chi,1) = -\frac{\pi W(\chi)}{m^{1/2}}\left(\sum_{r=1}^{m-1} \overline{\chi}(r)\log\left(\Gamma\left(\frac{r}{m}\right)\right) - (\log(2\pi) + \gamma)L(\overline{\chi},0)\right).$$

The arithmetic function $\Lambda(m)$ should of course not be confused with the meromorphic function $\Lambda(\chi,s)$.

Proof. We have seen in Proposition 10.2.5 that we have

$$L(\chi,s) = m^{-s}\sum_{1\leqslant r\leqslant m} \chi(r)\zeta(s,r/m),$$

where $\zeta(s,x)$ is the Hurwitz zeta function. Since by Definition 9.6.13 we have $\zeta'(0,x) = \log(\Gamma(x)) + \zeta'(0)$ and by the functional equation we find that $\zeta'(0) = -(1/2)\log(2\pi)$, we thus have

$$L'(\chi,0) = \sum_{1\leqslant r\leqslant m} \chi(r)\frac{d}{ds}\zeta(s,r/m)\Big|_{s=0} - \log(m)L(\chi,0)$$

$$= -\log(m)L(\chi,0) + \sum_{1\leqslant r\leqslant m} \chi(r)\left(\log(\Gamma(r/m)) - \frac{1}{2}\log(2\pi)\right).$$

We consider the four cases of the proposition, using Corollary 10.3.2. If χ is a nontrivial even character, we have

$$L'(\chi, 0) = \sum_{1 \leqslant r \leqslant m-1} \chi(r) \log(\Gamma(r/m))$$

$$= \frac{1}{2} \sum_{1 \leqslant r \leqslant m-1} (\chi(r) \log(\Gamma(r/m)) + \chi(m-r) \log(\Gamma((m-r)/m)))$$

$$= \frac{1}{2} \sum_{1 \leqslant r \leqslant m-1} \chi(r) \log(\pi/\sin(r\pi/m))$$

$$= -\frac{1}{2} \sum_{1 \leqslant r \leqslant m-1} \chi(r) \log(\sin(r\pi/m))$$

using the reflection formula for the gamma function, giving the first case. If χ is an odd character, then

$$L'(\chi, 0) = \frac{\log(m)}{m} \sum_{1 \leqslant r \leqslant m-1} \chi(r) r + \sum_{1 \leqslant r \leqslant m-1} \chi(r) \log(\Gamma(r/m)) ,$$

giving the second case.

If χ is a trivial character, we can use the above formula, but it is easier to work directly. In that case,

$$L(\chi, s) = \prod_{p \mid m} \left(1 - \frac{1}{p^s}\right) \zeta(s) ,$$

so if $m > 1$,

$$L'(\chi, 0) = -\frac{1}{2} \sum_{p \mid m} \log(p) \prod_{q \mid m, \, q \neq p} 0 .$$

Thus, if m is not a prime power we have $L'(\chi, 0) = 0$, while if $m = p^k$ we have $L'(\chi, 0) = -\log(p)/2$, which gives the third case. The fourth case is the formula for $\zeta'(0)$.

Finally, consider $L(\chi, 1)$ when χ is a nontrivial character (otherwise there is a pole at $s = 1$). For χ even, the functional equation is easily seen to give

$$L(\chi, 1) = \frac{2W(\chi)}{m^{1/2}} L'(\overline{\chi}, 0) ,$$

and the formula follows. For χ odd, the functional equation gives

$$-\frac{1}{2} \frac{\Gamma'(1/2)}{\Gamma(1/2)} - \frac{L'(\chi, 0)}{L(\chi, 0)} = \log\left(\frac{m}{\pi}\right) + \frac{1}{2} \frac{\Gamma'(1)}{\Gamma(1)} + \frac{L'(\overline{\chi}, 1)}{L(\overline{\chi}, 1)} ,$$

and using the values $\Gamma'(1) = -\gamma$ and $\Gamma'(1/2) = -\Gamma(1/2)(2\log(2) + \gamma)$ coming from the duplication formula (see Exercise 91 of Chapter 9) the last formula follows. $\qquad\square$

Remarks. (1) The reader will have noticed that the formula for $L'(\chi, 0)$
when χ is odd is considerably more complicated than when χ is even. This
is due to the fact that when χ is even we have $L(\chi, 0) = 0$, so that $L'(\chi, 0)$,
if nonzero, is the first significant coefficient in the Taylor expansion of
$L(\chi, s)$ around $s = 0$. This is a general philosophy for special values in
number theory: $f'(k)$ will almost certainly have a nicer expression when
$f(k) = 0$ than when $f(k) \neq 0$, and similarly for higher derivatives. See
Corollary 8.5.14 and Section 10.6, and see Exercise 39 for an example.

(2) When χ is a nontrivial even character, and if we let $\zeta_m = e^{2i\pi/m}$ be this
specific primitive mth root of unity, then

$$\log(1 - \zeta_m^r) = \log(\sin(\pi r/m)) + i\pi r/m + \log(2) - i\pi/2 \,,$$

and since $\sum_{1 \leqslant r < m} \chi(r) = 0$ and $\sum_{1 \leqslant r < D} r\chi(r) = 0$, it follows that

$$\sum_{r=1}^{m-1} \chi(r) \log\left(\sin\left(\frac{r\pi}{m}\right)\right) = \sum_{r=1}^{m-1} \chi(r) \log(1 - \zeta_m^r) \,.$$

(3) We can give similar formulas for $L'(\chi, 1 - k)$ for any k; see for example
Exercise 40, and remark (1) is still valid if we replace the ordinary logarithm by the polylogarithm of order k, and the ordinary gamma function
by Barnes's gamma functions of higher order.

Corollary 10.3.6. *Let $D > 1$ be a fundamental discriminant, and set as
usual $\chi_D = \left(\frac{D}{\cdot}\right)$.*

(1) *We have*

$$L(\chi_D, 1) = \frac{2\log(\varepsilon_D)}{D^{1/2}} \,,$$

   where

$$\varepsilon_D = \prod_{r=1}^{\lfloor D/2 \rfloor} \sin(r\pi/D)^{-\chi_D(r)} \,,$$

   and ε_D is the fundamental unit of $K = \mathbb{Q}(\sqrt{D})$ such that $\varepsilon_D > 1$.
(2) *If $\zeta = e^{2i\pi u/D}$ is any primitive Dth root of unity we have*

$$\prod_{1 \leqslant r < D} (1 - \zeta^r)^{\chi_D(r)} = \varepsilon_D^{-2\chi_D(u)} \,.$$

Proof. The formula of (1) follows immediately from the above proposition
since we know that $W\left(\left(\frac{D}{\cdot}\right)\right) = 1$ and χ_D, and the fact that ε_D is the fundamental unit of K follows from Dirichlet's class number formula (Proposition
3.4.5). Note that this formula tells us that $L(\chi_D, 1) > 0$, so that $\varepsilon_D > 1$, but
this can be proved directly; see Corollary 10.5.6 below. The formula of (2)
follows from Remark (2) above when $u = 1$. For general u coprime to D we
have

$$\sum_{1\leqslant r<D} \left(\frac{D}{r}\right) \log(1 - \zeta_D^{ur}) = \left(\frac{D}{u}\right) \sum_{1\leqslant s<D} \left(\frac{D}{s}\right) \log(1 - \zeta_D^s) \,,$$

so the result follows in general.   □

For instance, we have

$$L(\chi_5, 1) = \frac{2\log((1 + \sqrt{5})/2)}{5^{1/2}} \,,$$

$$L(\chi_8, 1) = \frac{2\log(1 + \sqrt{2})}{8^{1/2}} \,.$$

The fact that $L(\chi_D, 1)$ (or $L'(\chi_D, 0)$) is equal to a simple factor times the logarithm of an algebraic unit is a special (proved!) case of an important conjecture due to H. Stark that essentially states that the same thing will happen for much more general L-functions.

10.3.2 Special Values of L-Functions and Modular Forms

This section assumes some knowledge of modular forms of integral and half-integral weight and can be skipped on first reading.

There are several close links between special values of L-functions and modular forms. We will give without proof two related types of examples, referring to the literature for more details.

The first type of examples comes from *Hilbert modular forms* attached to a totally real number field K. We refer for instance to [Fre] for the (easy) definition. As usual in the theory of modular forms it is not too difficult to construct explicitly *Eisenstein series*, a construction due in this case to Hecke. Also as usual, the generalized Fourier coefficients of these Eisenstein series are given by simple formulas generalizing the divisor function in the one-variable case. Also, we can restrict a Hilbert modular form to the diagonal, thus obtaining an ordinary modular form whose weight is equal to n times the weight k of the initial Hilbert modular form. The remarkable fact about the restrictions of the Hecke–Eisenstein series is that their *constant term* is essentially the value of the Dedekind zeta function of the number field K at the negative integer $1 - k$. Finally, a nice argument due to Siegel, which amounts to the finite dimensionality of spaces of modular forms, shows that the constant term of a modular form in a given space can be expressed as a universal linear combination of a small finite number of the nonconstant terms. Applying this to the restrictions of the Hecke–Eisenstein series, we thus obtain explicit formulas for $\zeta_K(1 - k)$.

In the special case thatn $K = \mathbb{Q}(\sqrt{D})$ is a real quadratic field, the formulas are especially easy to state, and have been given in many places, see [Coh2] for a comprehensive treatment. The simplest occur when the dimension of the corresponding space of modular forms is equal to 1, and this happens only for $k = 2$ and 4. Using $\zeta_{\mathbb{Q}(\sqrt{D})}(s) = \zeta(s)L((\frac{D}{\cdot}), s)$ and the functional equations, we obtain the following theorem, which is thus essentially due to Siegel.

Theorem 10.3.7. *Recall that we denote by $\sigma_k(n)$ the sum of the kth powers of the (positive) divisors of n. Let D be the discriminant of a real quadratic field.*

(1)

$$L\left(\left(\frac{D}{\cdot}\right), -1\right) = -\frac{1}{5} \sum_{\substack{s \in \mathbb{Z}, \ s^2 < D \\ s \equiv D \ (\mathrm{mod}\ 2)}} \sigma_1\left(\frac{D - s^2}{4}\right),$$

and by the functional equation

$$L\left(\left(\frac{D}{\cdot}\right), 2\right) = -\frac{2\pi^2}{D^{3/2}} L\left(\left(\frac{D}{\cdot}\right), -1\right).$$

(2)

$$L\left(\left(\frac{D}{\cdot}\right), -3\right) = \sum_{\substack{s \in \mathbb{Z}, \ s^2 < D \\ s \equiv D \ (\mathrm{mod}\ 2)}} \sigma_3\left(\frac{D - s^2}{4}\right),$$

and by the functional equation

$$L\left(\left(\frac{D}{\cdot}\right), 4\right) = \frac{4\pi^4}{3D^{7/2}} L\left(\left(\frac{D}{\cdot}\right), -3\right).$$

Remarks. (1) As already mentioned, such formulas exist for all even k, not only $k = 2$ and 4.

(2) This gives a fast $O(D^{1/2+\varepsilon})$ method for computing special values. Note that the explicit formulas given in the preceding section are $O(D)$, hence much slower. The main point was to show that such explicit formulas exist, but they are not really practical for actual computation, except for small conductors. The rapidly convergent explicit formula coming from the functional equation given in Theorem 10.2.14, which is valid for any s, special or not, also gives a $O(D^{1/2+\varepsilon})$ method, but is nonetheless slower and more complicated because of the need to compute the incomplete gamma function. In fact, it is only very recently that algorithms have been given that compute $L((\frac{D}{\cdot}), -1)$ (or equivalently $L((\frac{D}{\cdot}), 2)$) to reasonably high accuracy, using K-theory, see [Bel-Gan].

(3) The above formulas show that $L((\frac{D}{\cdot}), -3) \in \mathbb{Z}$, which is not completely trivial. We prove this directly, and in fact a little more.

Proposition 10.3.8. *Let $D \neq 1$ be a fundamental discriminant.*

(1) *If $D \neq 5$ and $D \neq 8$ then $L((\frac{D}{\cdot}), -1) \in 2\mathbb{Z}$, and $L((\frac{5}{\cdot}), -1) = -2/5$ and $L((\frac{8}{\cdot}), -1) = -1$.*

(2) *If* $D \neq 8$ *then* $L((\frac{D}{\cdot}), -3) \in 2\mathbb{Z}$, *and* $L((\frac{8}{\cdot}), -3) = 11$.

Proof. By Corollary 10.3.3, for $D < 0$ we have $L((\frac{D}{\cdot}), -1) = L((\frac{D}{\cdot}), -3) = 0$. For $D > 0$ we have

$$L\left(\left(\frac{D}{\cdot}\right), -1\right) = -\frac{1}{2D} \sum_{r=1}^{D-1} \left(\frac{D}{r}\right) r^2,$$

so by Corollary 9.5.9 we see that $L((\frac{D}{\cdot}), -1) \in 2\mathbb{Z}$, except if $D = 5$ or 8, in which case it is equal to $-2/5$ or -1 respectively.

Similarly, for $D > 0$ we have

$$L\left(\left(\frac{D}{\cdot}\right), -3\right) = -\frac{1}{4D} \sum_{r=1}^{D-1} \left(\frac{D}{r}\right) r^2(r^2 - 2D^2),$$

and again by Corollary 9.5.9 we see that $L((\frac{D}{\cdot}), -3) \in 2\mathbb{Z}$, except perhaps for $D = 5$ or 8, but a direct check shows that $L((\frac{5}{\cdot}), -3) = 2$, while $L((\frac{8}{\cdot}), -3) = 11$. $\square$

Note that if we combine this proposition with Theorem 10.3.7 we obtain the following result.

Corollary 10.3.9. *If* $D \neq 5$ *and* $D \neq 8$ *are positive fundamental discriminants, we have*

$$\sum_{\substack{s \in \mathbb{Z}, \ s^2 < D \\ s \equiv D \ (\mathrm{mod} \ 2)}} \sigma_1\left(\frac{D - s^2}{4}\right) \equiv 0 \ (\mathrm{mod} \ 10).$$

We have an analogous result for $D < 0$:

Proposition 10.3.10. *Let* $D \neq 1$ *be a fundamental discriminant.*

(1) *If* $D \neq -3, -4, -7,$ *and* -8 *then* $L((\frac{D}{\cdot}), -2) \in 2\mathbb{Z}$, *and* $L((\frac{-3}{\cdot}), -2) = -2/9$, $L((\frac{-4}{\cdot}), -2) = -1/2$, $L((\frac{-7}{\cdot}), -2) = -16/7$, *and* $L((\frac{-8}{\cdot}), -2) = -3$.

(2) *If* $D \neq -3, -4, -8,$ *and* -11 *then* $L((\frac{D}{\cdot}), -4) \in 2\mathbb{Z}$, *and* $L((\frac{-3}{\cdot}), -4) = 2/3$, $L((\frac{-4}{\cdot}), -4) = 5/2$, $L((\frac{-8}{\cdot}), -4) = 57$, *and* $L((\frac{-11}{\cdot}), -4) = 2550/11$.

Proof. Left as an exercise for the reader (Exercise 13). $\square$

We will generalize the above two propositions to all negative arguments in the next chapter (Corollary 11.4.3).

The second link between special values of L-functions and modular forms is provided by the theory of modular forms of *half-integral weight*. The theory of modular forms of half-integral weight was invented by G. Shimura, and developed by several people such as Waldspurger, Kohnen–Zagier, and the author. The link with special values of L-functions was found by Shimura and systematically explored by the author.

Remarkably enough, the process is very similar to the case of Hilbert modular forms. One defines in a natural way Eisenstein series of half-integral weight $k + 1/2$ for $k \in \mathbb{Z}_{\geq 2}$, and as usual one finds that it is possible to compute their Fourier coefficients explicitly. They happen to be simply expressible in terms of the special values of L-functions of quadratic characters at k (or equivalently at $1-k$). Thus, contrary to the case of Hecke–Eisenstein series where the special values occur as constant terms, here they occur all together (for a given value k) in a single Eisenstein series of half-integral weight. We refer to [Coh3] for details.

To state the theorem, we first need a definition.

Definition 10.3.11. *Let $k \geq 1$ be an integer. For any $n \geq 1$ such that $(-1)^k n \equiv 0$ or 1 modulo 4, write $(-1)^k n = Df^2$, where $f \in \mathbb{Z}$ and D is a fundamental discriminant (including 1). We define the functions $H_k(n)$ by the formula*

$$H_k(n) = L\left(\left(\frac{D}{\cdot}\right), 1-k\right) \sum_{d \mid f} \mu(d)\left(\frac{D}{d}\right) d^k \sigma_{2k-1}(f/d) ,$$

and we also set by convention $H_k(0) = \zeta(1 - 2k)$.

The theorem is then as follows.

Theorem 10.3.12. *For $k \geq 2$ the Fourier series*

$$\mathcal{H}_k(\tau) = \sum_{n \geq 0} H_k(n) q^n$$

is a modular form of weight $k + 1/2$ on the congruence subgroup $\Gamma_0(4)$, where as usual $q = \exp(2i\pi\tau)$.

Since the space of modular forms is finite-dimensional, it is then an easy matter to identify precisely a given form from its first few Fourier coefficients, given a specific basis.

It is easy to show that the function

$$\theta(\tau) = \sum_{n \in \mathbb{Z}} q^{n^2} = 1 + 2 \sum_{n \geq 1} q^{n^2}$$

(of weight $1/2$) and the function

$$\theta^4(\tau + 1/2) = \left(\sum_{n \in \mathbb{Z}} (-1)^n q^{n^2}\right)^4 = \left(1 + 2 \sum_{n \geq 1} (-1)^n q^{n^2}\right)^4$$

(of weight 2) generate the *algebra* of all modular forms of half-integral weight on $\Gamma_0(4)$. In other words, any modular form of integral or half-integral weight on $\Gamma_0(4)$ is an isobaric polynomial in these two functions. A little computation gives the following corollary, which is very useful for the computation of special values when many of them are needed.

Corollary 10.3.13. *We have*

$$\mathcal{H}_2(\tau) = \frac{5\theta(\tau)\theta^4(\tau+1/2) - \theta^5(\tau)}{480} ,$$

$$\mathcal{H}_3(\tau) = -\frac{7\theta^3(\tau)\theta^4(\tau+1/2) + \theta^7(\tau)}{2016} ,$$

$$\mathcal{H}_4(\tau) = \frac{\theta(\tau)\theta^8(\tau+1/2) + 14\theta^5(\tau)\theta^4(\tau+1/2) + \theta^9(\tau)}{3840} .$$

Remarks. (1) Since the θ function is lacunary, even applied naïvely these formulas give a very efficient method for computing large batches of special values of L-functions of quadratic characters. However, it is still $O(D^{1/2+\varepsilon})$ on average. On the other hand, if we use FFT-based techniques for multiplying power series, we can compute large numbers of coefficients even faster, and go down to $O(D^\varepsilon)$ on average.

(2) The above formulas are essentially equivalent to those that we have given in Theorem 5.4.16.

(3) Because Hilbert modular forms exist only for totally real number fields, the method using Hecke–Eisenstein series is applicable for computing special values of *real* quadratic characters only, while the present method is applicable both to real and to imaginary quadratic characters.

The formulas obtained by the above two methods are in fact closely related. For instance, if we set classically

$$E_2(\tau) = 1 - 24 \sum_{n \geqslant 1} \sigma_1(n) q^n ,$$

which is not quite a modular form, it is easy to check directly that

$$-\frac{\theta'(\tau)/(2i\pi)}{20} + \frac{E_2(4\tau)\theta(\tau)}{120}$$

is a true modular form of weight $5/2$, and the first coefficients show that it is equal to $\mathcal{H}_2(\tau)$. Similarly it is not difficult to check that $\mathcal{H}_4(\tau) = E_4(4\tau)\theta(\tau)/240$, where

$$E_4(\tau) = 1 + 240 \sum_{n \geqslant 1} \sigma_3(n) q^n .$$

This gives the following formulas, which generalize to arbitrary $N > 0$ (and not only discriminants of real quadratic fields) Siegel's formulas coming from Hecke–Eisenstein series:

Proposition 10.3.14. *By convention set $\sigma_k(0) = \zeta(-k)/2$ (so that $\sigma_1(0) = -1/24$ and $\sigma_3(0) = 1/240$). We have*

$$H_2(N) = -\frac{1}{5} \sum_{\substack{s\in\mathbb{Z},\ s^2\leqslant N \\ s\equiv N \ (\mathrm{mod}\ 2)}} \sigma_1\left(\frac{N-s^2}{4}\right) - \frac{N}{10}\delta(\sqrt{N}),$$

$$H_4(N) = \sum_{\substack{s\in\mathbb{Z},\ s^2\leqslant N \\ s\equiv N \ (\mathrm{mod}\ 2)}} \sigma_3\left(\frac{N-s^2}{4}\right),$$

where $\delta(\sqrt{N}) = 1$ if N is a square and 0 otherwise.

Remarks. (1) There also exist similar formulas for $H_3(N)$ and $H_5(N)$ involving modified σ_2 functions; see Exercise 52.

(2) Since the formulas coming from modular forms of half-integral weight include those coming from Hilbert modular forms, the reader may wonder why we have included the latter. The main reason is that they also give explicit formulas for computing the special values of Dedekind zeta functions at negative integers of all totally real number fields, not only quadratic ones, and this is in fact how Siegel's Theorem 10.5.3 on the rationality of such values is proved.

(3) The reader will have noticed that we do not mention the function $H_1(N)$, which is essentially a class number, and the corresponding Fourier series $\mathcal{H}_1(\tau)$. The theory is here complicated by the fact that the latter is not quite a modular form of weight $3/2$ (analogous to but more complicated than the situation for $E_2(\tau)$). However, the theory can be worked out completely, and it gives beautiful formulas on class numbers, due to Hurwitz, Eichler, Zagier, and the author. We refer for instance to [Coh2] for details.

10.3.3 The Pólya–Vinogradov Inequality

In the next subsection we will give some bounds for $L(\chi, 1)$. For this, it is useful, although not essential, to have some good estimates on $\sum_{1\leqslant n\leqslant X} \chi(n)$. Such an estimate is the following *Pólya–Vinogradov inequality*:

Proposition 10.3.15 (Pólya–Vinogradov). *Let χ be a nontrivial character modulo m of conductor $f > 1$. For all $X \geqslant 0$ we have the inequality*

$$\left| \sum_{1\leqslant a\leqslant X} \chi(a) \right| \leqslant d(m/f) f^{1/2} \log(f),$$

where $d(n)$ denotes the number of positive divisors of n.

Proof. Assume first that χ is a *primitive* character and set $S(X) = \sum_{1 \leqslant a \leqslant X} \overline{\chi}(a)$. It is clear that $S(X) = S(\lfloor X \rfloor)$, so we may assume that $X = N \in \mathbb{Z}_{\geqslant 0}$. By Corollary 2.1.42 and the fact that $\chi(x) = 0$ when $\gcd(x, m) > 1$ we have

$$
\begin{aligned}
\tau(\chi)S(N) &= \sum_{1 \leqslant a \leqslant N} \tau(\chi, a) = \sum_{1 \leqslant a \leqslant N} \sum_{x \bmod m} \chi(x)e^{2i\pi a x/m} \\
&= \sum_{x \bmod m} \chi(x) \sum_{1 \leqslant a \leqslant N} e^{2i\pi a x/m} \\
&= \sum_{x \bmod m,\ \gcd(x,m)=1} \chi(x) \frac{e^{2i\pi(N+1)x/m} - e^{2i\pi x/m}}{e^{2i\pi x/m} - 1} .
\end{aligned}
$$

Note that the denominator does not vanish since $\gcd(x, m) = 1$ and $m > 1$. We bound this crudely as follows:

$$
\begin{aligned}
|\tau(\chi)S(N)| &\leqslant \sum_{1 \leqslant x \leqslant m-1,\ x \neq m/2} \frac{1}{\sin(\pi x/m)} \\
&\leqslant 2 \sum_{1 \leqslant x \leqslant (m-1)/2} \frac{1}{\sin(\pi x/m)} \leqslant m \sum_{1 \leqslant x \leqslant (m-1)/2} \frac{1}{x} ,
\end{aligned}
$$

using the high-school inequality $\sin(t) \geqslant (2/\pi)t$ for $t \in [0, \pi/2]$. Now since $1/x$ is a convex function, we have the inequality

$$
\int_{x-1/2}^{x+1/2} \frac{dt}{t} > \frac{1}{x}
$$

(see Exercise 43). Thus

$$
\sum_{1 \leqslant x \leqslant (m-1)/2} \frac{1}{x} < \int_{1/2}^{m/2} \frac{dt}{t} = \log(m) .
$$

Since $|\tau(\chi)| = m^{1/2}$ by Proposition 2.1.45, the result follows for primitive characters.

Now let χ be any nontrivial character modulo m, let f be the conductor of χ, and let χ_f be the character modulo f equivalent to χ. Since $\gcd(a, f) = 1$ and $\gcd(a, m/f) = 1$ implies $\gcd(a, m) = 1$, using the definition of the Möbius function we have

$$
\begin{aligned}
\sum_{1 \leqslant a \leqslant X} \chi(a) &= \sum_{\substack{1 \leqslant a \leqslant X \\ \gcd(a,m/f)=1}} \chi_f(a) = \sum_{1 \leqslant a \leqslant X} \chi_f(a) \sum_{d | \gcd(a,m/f)} \mu(d) \\
&= \sum_{d | m/f} \mu(d) \chi_f(d) \sum_{1 \leqslant b \leqslant X/d} \chi_f(b) .
\end{aligned}
$$

Thus using the bound for primitive characters and the fact that $|\mu(d)| \leqslant 1$ and $|\chi_f(d)| \leqslant 1$ we deduce that $|\sum_{1 \leqslant a \leqslant X} \chi(a)| \leqslant d(m/f) f^{1/2} \log(f)$, proving the proposition in general. $\square$

Remark. It is easy to improve the bound to $Kd(m/f) f^{1/2} \log(f)$ for some $K < 1$; see Exercise 43. On the other hand, it is much more difficult to improve on the factor $\log(f)$. More precisely, assuming the extended Riemann hypothesis (ERH) for all Dirichlet L-functions, Montgomery and Vaughan showed in [Mon-Vau] that it can be replaced by $O(\log(\log(f)))$ for an explicit O-constant. Very recently, Granville and Soundararajan have shown in [Gra-Sou] that without the assumption of ERH it is nonetheless possible to improve on the factor $\log(f)$ for characters of *odd* order. More precisely, they show that we may replace it by $\log(f)^{1-\delta_g}$ for some $\delta_g > 0$, by $\log(f)^{2/3+\varepsilon}$ for $g = 3$, and by $\log(\log(f))^{1-\delta_g}$ under the ERH.

10.3.4 Bounds and Averages for $L(\chi, 1)$

Although we have given reasonably explicit formulas for $L(\chi, 1)$, these formulas do not lead to any reasonable estimate on the *size* of $L(\chi, 1)$. Finding lower bounds is quite difficult, and in fact we will prove in Section 10.5.5 the important but very weak result that $L(\chi, 1) \neq 0$. On the other hand, finding upper bounds is quite easy (although the best bounds, which we will not mention, rely on the extended Riemann hypothesis). Such a result is as follows.

Proposition 10.3.16. *Let χ be a nontrivial character modulo m of conductor $f > 1$.*

(1) *We have*

$$|L(\chi, 1)| \leqslant \frac{1}{2} \log(f) + \log(\log(f)) + \log(d(m/f)) + 2.8 \ .$$

(2) *Let $\beta \geqslant 1/2$. As $m \to \infty$ we have*

$$L(\chi, 1) = \sum_{n=1}^{m^\beta} \frac{\chi(n)}{n} + O(m^{1/2-\beta} \log(m)) \ .$$

Proof. This proof is fundamentally based on partial (or Abel) summation. For $X \geqslant 0$ set $S(X) = \sum_{1 \leqslant n \leqslant X} \chi(n)$. For any integers M and N such that $N \geqslant M \geqslant 1$ we have

$$\sum_{n=M+1}^{N} \frac{\chi(n)}{n} = \sum_{n=M+1}^{N} \frac{S(n) - S(n-1)}{n} = \sum_{n=M+1}^{N} \frac{S(n)}{n} - \sum_{n=M}^{N-1} \frac{S(n)}{n+1}$$

$$= \frac{S(N)}{N+1} - \frac{S(M)}{M+1} + \sum_{n=M+1}^{N} S(n) \left(\frac{1}{n} - \frac{1}{n+1} \right) \ .$$

Since χ is periodic of period m and since $S(m) = 0$ by orthogonality, S is also a periodic function on $\mathbb{Z}_{\geqslant 0}$, and in particular it is bounded by some constant B, say, which for the moment we do not specify. Letting $N \to \infty$ gives

$$\left| \sum_{n \geqslant M+1} \frac{\chi(n)}{n} \right| \leqslant \frac{|S(M)|}{M+1} + B \sum_{n \geqslant M+1} \left(\frac{1}{n} - \frac{1}{n+1} \right) \leqslant \frac{2B}{M} \ .$$

On the other hand, by Euler–MacLaurin or any other method it is clear that

$$\left| \sum_{1 \leqslant n \leqslant M} \frac{\chi(n)}{n} \right| \leqslant \sum_{1 \leqslant n \leqslant M} \frac{1}{n} \leqslant \log(M) + 1 \ .$$

We thus obtain for any $M \in \mathbb{Z}_{\geqslant 1}$,

$$|L(\chi, 1)| \leqslant \log(M) + 1 + \frac{2B}{M} \ .$$

By differentiating, we see that the optimal choice of M is $M = 2B$, but since this is not necessarily an integer we choose instead $M = 2B + \theta$ with $0 \leqslant \theta < 1$. An immediate computation shows that for this choice of M we obtain

$$|L(\chi, 1)| \leqslant \log(2B) + 2 + \frac{1}{4B^2} \ .$$

By Proposition 10.3.15, we can choose $B = d(m/f) f^{1/2} \log(f)$, and since $f \geqslant 3$ (why?), we have $1/(4B^2) < 1/12$, and replacing proves (1).

For (2), we use the bound obtained above for $\sum_{n \geqslant M+1} \chi(n)/n$ with $M = m^\beta$. By Corollary 10.2.3 (4) we thus have

$$\left| L(\chi, 1) - \sum_{n=1}^{m^\beta} \frac{\chi(n)}{n} \right| \leqslant \frac{2d(m/f) f^{1/2} \log(f)}{m^\beta} \leqslant \frac{2d(m/f)(f/m)^{1/2} \log(m)}{m^{\beta - 1/2}} \ .$$

The result follows since $d(n)/n^\alpha$ is bounded for any $\alpha > 0$, and in particular for $\alpha = 1/2$. $\qquad \square$

Remark. Evidently the constant 2.8, and even the term $\log(\log(f))$, are unimportant. On the other hand, the main term $\log(f)/2$ is difficult to improve. As mentioned above, the best results are obtained assuming the extended Riemann hypothesis, and the main term is then $O(\log(\log(f)))$.

As usual with analytic techniques, we can obtain much better results if we want only *average* (as opposed to individual) estimates for $L(\chi, 1)$.

Proposition 10.3.17. *By convention set $L(\chi_0, 1) = \gamma \phi(m)/m$ for the trivial character χ_0 modulo m, which is the constant term in the expansion of $L(\chi_0, s)$ around $s = 1$, and let $A(m) = \sum_{p \mid m} \log(p)/(p-1)$.*

(1) *The average of $L(\chi, 1)$ over all characters modulo m is equal to*

$$-\frac{1}{m}\left(\psi\left(\frac{1}{m}\right) + \log(m) + A(m)\right) = 1 - \frac{\log(m) + A(m) - \gamma}{m} + O\left(\frac{1}{m^2}\right).$$

(2) *As $m \to \infty$ we have $A(m) = O(\log(\log(m)))$.*

(3) *The average of $|L(\chi, 1)|^2$ over all characters modulo m is equal to*

$$\frac{1}{m^2}\sum_{\substack{1 \leqslant r \leqslant m \\ \gcd(r,m)=1}} \psi^2\left(\frac{r}{m}\right) - \frac{\phi(m)}{m^2}((\log(m) + A(m) + \gamma)^2 - \gamma^2)$$

$$= \frac{\pi^2}{6}\prod_{p|m}\left(1 - \frac{1}{p^2}\right) - \frac{\phi(m)}{m^2}((\log(m) + A(m))^2 + C) + O\left(\frac{1}{m^2}\right),$$

where

$$C = 1 - 2\gamma^2 - \int_0^1\left(\psi^2(x) - \frac{1}{x^2} - \frac{2\gamma}{x}\right)dx = 2\left(\frac{\pi^2}{6} - \gamma^2 - \gamma_1\right)$$

$$= 2.769143977048368974\ldots,$$

with

$$\gamma_1 = \lim_{N\to\infty}\left(\sum_{1\leqslant j\leqslant N}\frac{\log(j)}{j} - \frac{\log^2(N)}{2}\right)$$

(see Section 10.3.5).

Proof. (1). By Proposition 10.2.5 (4) and orthogonality of characters we have

$$\sum_{\chi\neq\chi_0} L(\chi, 1) = -\frac{1}{m}\sum_{1\leqslant r\leqslant m}\psi(r/m)\sum_{\chi\neq\chi_0}\chi(r)$$

$$= -\frac{1}{m}\left(\phi(m)\psi(1/m) - \sum_{\substack{1\leqslant r\leqslant m \\ \gcd(r,m)=1}}\psi(r/m)\right).$$

Using the Möbius function as explained after Proposition 10.1.5, we have

$$\sum_{\substack{1\leqslant r\leqslant m \\ \gcd(r,m)=1}}\psi(r/m) = \sum_{1\leqslant r\leqslant m}\psi(r/m)\sum_{d|(r,m)}\mu(d)$$

$$= \sum_{d|m}\mu(d)\sum_{1\leqslant k\leqslant m/d}\psi(k/(m/d)).$$

If we differentiate logarithmically the distribution formula for the gamma function (Proposition 9.6.33) we obtain $\sum_{0\leqslant j\leqslant m-1}\psi(s+j/m) = m\psi(ms) - m\log(m)$, hence $\sum_{1\leqslant r\leqslant m}\psi(r/m) = -m(\log(m) + \gamma)$. Thus

$$\sum_{\substack{1\leqslant r\leqslant m \\ \gcd(r,m)}} \psi(r/m) = -m\sum_{d\mid m}(\mu(d)/d)(\log(m) - \log(d) + \gamma) \ .$$

By multiplicativity (Corollary 10.1.12 and Exercise 45) we have the identities

$$\sum_{d\mid m}\mu(d)/d = \phi(m)/m, \quad \sum_{d\mid m}\mu(d)\log(d)/d = -(\phi(m)/m)\sum_{p\mid m}\log(p)/(p-1) \ ,$$

so that

$$\sum_{\substack{1\leqslant r\leqslant m \\ \gcd(r,m)=1}} \psi(r/m) = -\phi(m)(\log(m) + A(m) + \gamma) \ ,$$

proving the first formula of (1) after adding the contribution $\gamma\phi(m)/m$ of the trivial character to the average. The second formula follows from the functional equation $\psi(x+1) = \psi(x) + 1/x$ and Proposition 9.6.15, which tell us that $\psi(1/m) = -m - \gamma + O(1/m)$ as $m \to \infty$.

(2). Denote by p_j the jth prime number and let $m = \prod_{1\leqslant j\leqslant k} p_{i_j}^{v_j}$ be the prime-power decomposition of m with $v_j \geqslant 1$. We have $p_{i_j} \geqslant p_j$, and since the function $\log(p)/(p-1)$ is decreasing it follows that $A(m) \leqslant \sum_{1\leqslant j\leqslant k}\log(p_j)/(p_j - 1)$. Using the estimate $p_j = O(j\log(j))$, which is much weaker than the prime number theorem and is very easy to prove (see the remarks following Proposition 10.1.13), we deduce that $A(m) = O(\sum_{1\leqslant j\leqslant k} 1/j) = O(\log(k))$. But $k = \omega(m)$ and trivially $m \geqslant 2^k$; hence $k = O(\log(m))$ (in fact the above-mentioned proposition tells us that $k = O(\log(m)/\log(\log(m)))$ but we do not need this), so that we obtain $A(m) = O(\log(\log(m)))$, proving (2).

(3). For simplicity denote by $X_2(m)$ the average of $|L(\chi, 1)|^2$ over all characters modulo m, and set

$$S_2(m) = \sum_{\substack{\chi \bmod m \\ \chi\neq\chi_0}} |L(\chi, 1)|^2 \ ,$$

so that $X_2(m) = (S_2(m) + \gamma^2\phi(m)^2/m^2)/\phi(m)$. As in (1) we find that

$$S_2(m) = \frac{1}{m^2} \sum_{1\leqslant r,s\leqslant m} \psi(r/m)\psi(s/m) \sum_{\chi\neq\chi_0} \chi(r)\overline{\chi}(s) \ .$$

Since we restrict to r and s coprime to m, we have $\chi(r)\overline{\chi}(s) = \chi(rs^{-1})$; therefore once again by orthogonality of characters the inner sum is equal to -1 unless $r = s$, and otherwise is equal to $\phi(m) - 1$. Thus

$$S_2(m) = \frac{\phi(m)}{m^2} \sum_{\substack{1\leqslant r\leqslant m \\ \gcd(r,m)=1}} \psi^2(r/m) - \frac{1}{m^2}\left(\sum_{\substack{1\leqslant r\leqslant m \\ \gcd(r,m)=1}} \psi(r/m)\right)^2 \ ,$$

so that by the formula proved in (1) we obtain

$$S_2(m) = \phi(m)B_2(m) - \frac{\phi(m)^2}{m^2}(\log(m) + A(m) + \gamma)^2 ,$$

where $B_2(m) = (1/m^2)\sum_{1 \leqslant r \leqslant m, \ \gcd(r,m)=1} \psi^2(r/m)$, proving the first formula of (3). For the second formula, using Möbius inversion we have

$$B_2(m) = \frac{1}{m^2} \sum_{1 \leqslant r \leqslant m} \psi^2(r/m) \sum_{d | \gcd(r,m)} \mu(d)$$

$$= \frac{1}{m^2} \sum_{d|m} \mu(d) \sum_{1 \leqslant s \leqslant m/d} \psi^2(s/(m/d)) = \sum_{d|m} \frac{\mu(d)}{d^2} B\left(\frac{m}{d}\right) ,$$

where we set $B(m) = (1/m^2)\sum_{1 \leqslant r \leqslant m} \psi^2(r/m)$. It remains to estimate $B(m)$ (see Exercise 105 (b) of Chapter 9 for an "explicit" formula for $B(m)$). As in (1), when x is small we have $\psi(x) = -1/x - \gamma + (\pi^2/6)x + O(x^2)$, so that $\psi(x)^2 = 1/x^2 + 2\gamma/x + (\gamma^2 - \pi^2/3) + O(x)$. Thus if we set $f(x) = \psi(x)^2 - 1/x^2 - 2\gamma/x$ then $f(x) \in C^4([0,1])$, say (in fact $f(x) \in C^\infty(]-1,\infty[)$, but we do not need this). Thus the Riemann sum $(1/m)\sum_{1 \leqslant r \leqslant m} f(r/m)$ tends to a limit $I = \int_0^1 f(t)\,dt$ as $m \to \infty$. More precisely, by Euler–MacLaurin it is easy to see that $(1/m)\sum_{1 \leqslant r \leqslant m} f(r/m) = I + C_1/m + C_2/m^2 + O(1/m^3)$ as $m \to \infty$ for some constants C_1 and C_2. It follows that

$$\sum_{1 \leqslant r \leqslant m} \psi^2(r/m) = \sum_{1 \leqslant r \leqslant m} m^2/r^2 + 2\gamma \sum_{1 \leqslant r \leqslant m} m/r + I \cdot m + C_1 + C_2/m + O(1/m^2) ,$$

so using the standard Euler–MacLaurin expansions we obtain

$$B(m) = \frac{1}{m^2} \sum_{1 \leqslant r \leqslant m} \psi^2\left(\frac{r}{m}\right) = \frac{\pi^2}{6} + \frac{2\gamma \log(m)}{m} + \frac{C_3}{m} + \frac{C_4}{m^2} + \frac{C_5}{m^3} + O\left(\frac{1}{m^4}\right) ,$$

with $C_3 = I + 2\gamma^2 - 1$ and some other constants C_4 and C_5. We could of course push this expansion further if desired, but this is sufficient. Indeed, note that

$$\sum_{d|m} \left|\frac{\mu(d)}{d^2} \frac{d^4}{m^4}\right| \leqslant \frac{1}{m^4} \sum_{d|m} d^2 = \frac{\sigma_2(m)}{m^4} = O\left(\frac{1}{m^2}\right)$$

by Proposition 10.1.13, so the term $O(1/m^4)$ in the expansion of $B(m)$ contributes $O(1/m^2)$ to the average $X_2(m)$ (if we had stopped the expansion at $O(1/m^3)$ we would have obtained a superfluous factor of $\log(\log(m))$). Furthermore, by multiplicativity we have

$$\sum_{d|m} \frac{\mu(d)}{d^2} \frac{d^k}{m^k} = \frac{1}{m^k} \sum_{d|m} \mu(d)d^{k-2} = \frac{1}{m^k} \prod_{p|m}(1 - p^{k-2}) .$$

Since $|\prod_{p|m}(1-p)| \leqslant \prod_{p|m} p \leqslant m$, the term C_5/m^3 also contributes $O(1/m^2)$ to $X_2(m)$. For $k = 2$ the above expression vanishes (except for $m = 1$), so the term C_4/m^2 does not contribute. For $k = 1$, the term C_3/m contributes $C_3\phi(m)/m^2$, and for $k = 0$, the term $\pi^2/6$ contributes $(\pi^2/6)\prod_{p|m}(1-1/p^2)$. Finally, there remains to consider the term $2\gamma \log(m)/m$. Using Exercise 45 once again we obtain

$$\sum_{d|m} \frac{\mu(d)}{d^2} \frac{\log(m/d)}{m/d} = \frac{1}{m} \sum_{d|m} \frac{\mu(d)}{d}(\log(m) - \log(d))$$

$$= \frac{1}{m}\left(\frac{\log(m)}{m}\phi(m) - \sum_{d|m}\frac{\mu(d)}{d}\log(d)\right)$$

$$= \frac{\phi(m)}{m^2}\left(\log(m) + \sum_{p|m}\frac{\log(p)}{p-1}\right).$$

Putting everything together proves the second formula of (3) with $C = -C_3$. The second expression for the constant C in terms of γ_1 is proved in Exercise 104 of Chapter 9. $\square$

10.3.5 Expansions of $\zeta(s)$ Around $s = k \in \mathbb{Z}_{\leqslant 1}$

In this subsection we give for completeness some expansions of $\zeta(s)$ around $s = k \in \mathbb{Z}_{\leqslant 1}$. In practice, only the leading term is really useful, but around the special points $s = 0$ and 1 it is sometimes useful to have extra terms or even the whole expansion.

We begin with the following definitions.

Definition 10.3.18. *For $m \geqslant 0$ we define*

$$\gamma_m = \lim_{N \to \infty}\left(\sum_{k=1}^{N} \frac{\log(k)^m}{k} - \frac{\log(N)^{m+1}}{m+1}\right)$$

and constants δ_m for $m \geqslant 1$ by the recurrence formula

$$\delta_{m+1} = (m+1)\frac{\gamma_m}{m!} + \sum_{k=0}^{m-1}\frac{\gamma_k \delta_{m-k}}{k!}.$$

Note that by the Euler–MacLaurin summation formula the limit defining γ_m exists and that in fact we have

$$\sum_{k=1}^{N}\frac{\log(k)^m}{k} = \frac{\log(N)^{m+1}}{m+1} + \gamma_m + O\left(\frac{\log(N)^m}{N}\right),$$

or even more precisely an asymptotic expansion that as usual enables us to compute γ_m to any desired number of decimal digits.

For example we have $\delta_1 = \gamma_0 = \gamma$ and $\delta_2 = 2\gamma_1 + \gamma^2$.

Proposition 10.3.19. (1) *For s around 1 we have*

$$\zeta(s) = \frac{1}{s-1} + \sum_{m \geqslant 0} (-1)^m \frac{\gamma_m}{m!} (s-1)^m = \frac{1}{s-1} + \gamma + O(s-1) .$$

(2) *For s around 0 we have*

$$\log(-2\zeta(s)) = \log(2\pi) s + \sum_{m \geqslant 2} \frac{a_m}{m} s^m ,$$

where

$$a_m = \zeta(m) \left(1 - \frac{1 + (-1)^m}{2^m} \right) - \delta_m .$$

In particular,

$$\zeta(s) = -\frac{1}{2} - \frac{1}{2} \log(2\pi) s + O(s^2) .$$

(3) *For s around* $-2k$ *with* $k \in \mathbb{Z}_{\geqslant 1}$*, we have*

$$\zeta(s) = (-1)^k \frac{(2k)!}{2 \cdot (2\pi)^{2k}} \zeta(2k+1)(s+2k) + O(s+2k)^2 .$$

Proof. (1). Let us restrict to $s \in \mathbb{R}$, and for $k \geqslant 1$ set

$$u_k(s) = \frac{1}{k^s} - \int_{k-1}^{k} \frac{dt}{t^s} .$$

By Taylor's formula to order 2 it is clear that as $s \to \infty$ we have $u_k(s) \sim (s/2)k^{-s-1}$, so that the series $\sum_{k \geqslant 2} u_k(s)$ converges for $s > 0$. For $s > 1$ we have

$$\sum_{k \geqslant 2} u_k(s) = \sum_{k \geqslant 2} \frac{1}{k^s} - \int_{1}^{\infty} \frac{dt}{t^s} = \zeta(s) - 1 - \frac{1}{s-1} .$$

Since the series on the left converges normally for $s \geqslant \varepsilon > 0$, it follows by analytic continuation of $\zeta(s)$ that this equality is still valid for $s > 0$.

On the other hand, expanding around $s = 1$ it is easy to see that we have the following power series in $s - 1$ with infinite radius of convergence:

$$u_k(s) = \sum_{m \geqslant 0} \frac{(-1)^m}{m!} \left(\frac{\log(k)^m}{k} - \frac{\log(k)^{m+1} - \log(k-1)^{m+1}}{m+1} \right) (s-1)^m .$$

Thus by absolute convergence we can reorder the terms in the double sum $\sum_{k \geqslant 2} u_k(s)$ and obtain

$$\zeta(s) - 1 - \frac{1}{s-1} = \sum_{m \geqslant 0} \frac{(-1)^m}{m!} \gamma'_m (s-1)^m ,$$

where

$$\gamma'_m = \sum_{k \geqslant 2} \left(\frac{\log(k)^m}{k} - \frac{\log(k)^{m+1} - \log(k-1)^{m+1}}{m+1} \right)$$

$$= \lim_{N \to \infty} \left(\sum_{k=2}^{N} \frac{\log(k)^m}{k} - \frac{\log(N)^{m+1}}{m+1} \right) ,$$

and by definition we have $\gamma'_m = \gamma_m$ for $m \geqslant 1$, while $\gamma'_0 = \gamma_0 - 1$, proving (1).

(2). By Corollary 10.2.15 we can rewrite the functional equation in the form $-2\zeta(s) = (2\pi)^s(-s\zeta(1-s))/(\cos(\pi s/2)\Gamma(s+1))$. Taking formally the logarithm of both sides, we see that we can obtain the expansion of $\log(-2\zeta(s))$ around $s = 0$ as soon as we know the expansions of the logarithms of the factors occurring on the right-hand side. Evidently $\log((2\pi)^s) = s\log(2\pi)$. We note that the derivative of $\log(\cos(x))$ is $-\tan(x)$, whose expansion is given in Proposition 9.1.4 in terms of Bernoulli numbers, which can of course be translated in terms of $\zeta(2k)$ thanks to Corollary 9.1.21, so by integration we obtain that of $\log(\cos(x))$. Finally, the expansion of $\log(\Gamma(s+1))$ is given by Proposition 9.6.15. We thus need only to compute the expansion of $\log(-s\zeta(1-s))$ around 0. By (1) we know that $-s\zeta(1-s) = 1 - \sum_{m \geqslant 0}(\gamma_m/m!)s^{m+1}$. It follows that if we set $\log(-s\zeta(1-s)) = -\sum_{k \geqslant 1}(\delta_k/k)s^k$ we have $(-s\zeta(1-s))'/(-s\zeta(1-s)) = -\sum_{k \geqslant 1}\delta_k s^{k-1}$; hence

$$\sum_{m \geqslant 0}(m+1)\frac{\gamma_m}{m!}s^m = \left(1 - \sum_{m \geqslant 0}(\gamma_m/m!)s^{m+1}\right)\sum_{k \geqslant 1}\delta_k s^{k-1} ,$$

and identifying the coefficients of s^m on both sides gives

$$\delta_{m+1} = (m+1)\frac{\gamma_m}{m!} + \sum_{k=0}^{m-1} \frac{\gamma_k \delta_{m-k}}{k!}$$

as claimed. We leave the detailed computations to the reader (Exercise 47).

(3). This immediately follows from the functional equation. □

For the reader's convenience we give a small table of the constants γ_m; see Exercise 48 (the constants δ_m, a_m and the Taylor expansion of $\zeta(s)$ itself around $s = 0$ are immediate to compute from the γ_m). Note that this table gives the wrong impression that the γ_m are small. In fact, it can be shown that the γ_m are unbounded; for instance

$$\gamma_{50} = 126.8236026513227165967252536 48657\ldots .$$

m	γ_m
0	0.5772156649015328606065120900082
1	−0.0728158454836767248605863758755
2	−0.0096903631928723184845303860350
3	0.0020538344203033458661600465430
4	0.0023253700654673000574681701780
5	0.0007933238173010627017533348770
6	−0.0002387693454301996098724218420
7	−0.0005272895670577510460740975050
8	−0.0003521233538030395096020521650
9	−0.0000343947744180880481779146240
10	0.0002053328149090647946837222890

Table of γ_m for $0 \leqslant m \leqslant 10$

10.3.6 Numerical Computation of Euler Products and Sums

Functions such as Dirichlet L-functions (and in particular the Riemann zeta function) or more general L-functions can be expressed in two quite different ways, related by the underlying number theory: they are both Dirichlet series and Euler products. To compute *numerically* an L-function to reasonably high accuracy (say 28 decimal digits), we must use the Dirichlet series and not the Euler product since there are several available methods to accelerate the convergence of the series, such as for instance methods based on the Euler–MacLaurin summation formula, or on the functional equation. In this section we make the important remark that conversely, any reasonable Euler product or sum (i.e., a product or sum over prime numbers) can be computed to high accuracy using Dirichlet series. We first give a useful notation.

Definition 10.3.20. *Let $Z(s) = \prod_p Z_p(s)$ be an Euler product. We denote by $Z_{p>N}$ the Euler product*

$$Z_{p>N}(s) = \prod_{p>N} Z_p(s) = \frac{Z(s)}{\prod_{p \leqslant N} Z_p(s)} .$$

The following proposition gives a basic example of the computation of sums over primes.

Proposition 10.3.21. *Let $A(x) = \sum_{m \geqslant 1} a(m)x^m$ be a power series with radius of convergence strictly larger than $1/2$ and such that $A(0) = A'(0) = 0$ (hence with $a(1) = 0$), and set $S(A) = \sum_p A(1/p)$. Define*

$$c(n) = \sum_{d|n} \frac{\mu(d)}{d} a\left(\frac{n}{d}\right) .$$

Then for all $N \geqslant 1$ we have

$$S(A) = \sum_{p \leqslant N} A\left(\frac{1}{p}\right) + \sum_{n \geqslant 2} c(n) \log(\zeta_{p>N}(n)) \,.$$

Proof. Using the Euler product for the zeta function we have for $n \geqslant 2$,

$$\log(\zeta_{p>N}(n)) = \sum_{p>N} \sum_{k \geqslant 1} \frac{1}{kp^{kn}} = \sum_{k \geqslant 1} \frac{n}{kn} \sum_{p>N} \frac{1}{p^{kn}} \,.$$

By the second form of the Möbius inversion formula (Proposition 10.1.5 (2)) we thus have

$$\frac{1}{n} \sum_{p>N} \frac{1}{p^n} = \sum_{k \geqslant 1} \mu(k) \frac{\log(\zeta_{p>N}(kn))}{kn} \,,$$

so that

$$\begin{aligned}
S(A) &= \sum_{p \leqslant N} A\left(\frac{1}{p}\right) + \sum_{p>N} \sum_{m \geqslant 2} a(m) \frac{1}{p^m} \\
&= \sum_{p \leqslant N} A\left(\frac{1}{p}\right) + \sum_{m \geqslant 2} a(m) \sum_{k \geqslant 1} \mu(k) \frac{\log(\zeta_{p>N}(mk))}{k} \\
&= \sum_{p \leqslant N} A\left(\frac{1}{p}\right) + \sum_{n \geqslant 1} \log(\zeta_{p>N}(n)) \sum_{k \mid n} \frac{\mu(k)}{k} a(n/k) \,,
\end{aligned}$$

where the interchanges of summations are justified by absolute convergence, proving the proposition. □

Example. In milliseconds, we can in this way compute that

$$\sum_{p} \frac{1}{p^2} = 0.45224742004106549850654336483224793417323134323989\ldots \,,$$

although the sum is over primes, which do not display a regular behavior.

This basic example can be generalized in many different ways; see Exercise 53. A particularly important application is to the computation of Euler *products* as follows.

Corollary 10.3.22. *Let $B(x) = 1 + \sum_{m \geqslant 1} b(m)x^m$ be a power series with radius of convergence strictly greater than $1/2$, and such that $B(0) = 1$ and $B'(0) = 0$ (hence with $b(1) = 0$), and set $P(B) = \prod_p B(1/p)$. Write $\log(B(x)) = \sum_{m \geqslant 1} a(m)x^m$ and define*

$$c(n) = \sum_{d \mid n} \frac{\mu(d)}{d} a\left(\frac{n}{d}\right) \,.$$

Then for all $N \geqslant 1$ we have

$$P(B) = \prod_{p \leqslant N} B\left(\frac{1}{p}\right) \prod_{n \geqslant 2} \zeta_{p>N}(n)^{c_n} .$$

In addition, $c(n)$ satisfies the recurrence

$$c(n) = b(n) - \frac{1}{n} \sum_{1 \leqslant k \leqslant n-1} kc(k) \sum_{1 \leqslant q \leqslant n/k} b(n - qk) .$$

Proof. If we write $A(X) = \log(B(x))$ we have $P(B) = e^{S(A)}$, so the first formula follows from the proposition. For the second we note that by definition $nc(n) = \mu * na(n)$, where $*$ is the arithmetic convolution, so that $\sum_{d|n} dc(d) = na(n)$ (we are simply reversing Möbius inversion). The last formula now follows from

$$B'(x) = \sum_{m \geqslant 1} mb(m)x^{m-1} = B(x)\frac{B'(x)}{B(x)} = B(x) \sum_{m \geqslant 1} ma(m)x^{m-1}$$

after computing explicitly the product of the power series.   □

Remarks. (1) The coefficients $c(n)$ are simply the unique integers such that we have the formal expansions

$$B(x) = 1 + \sum_{m \geqslant 1} b(m)x^m = \prod_{n \geqslant 1} \left(\frac{1}{1 - x^n}\right)^{c(n)} .$$

(2) It is usually better to use the definition of $c(n)$. However, in some cases the function $a(n)$ is not easy to compute directly, and it is then necessary to use the recurrence for $c(n)$.

Example. Again in milliseconds we compute that

$$\prod_{p} \left(1 - \frac{1}{p(p-1)}\right) = 0.3739558136192022880547280543464164151116\ldots .$$

I refer to the author's unpublished and unfinished preprint at the URL
http://www.math.u-bordeaux1.fr/~cohen/hardylw.dvi
for many more examples and details on this subject.

10.4 Epstein Zeta Functions

These are other types of zeta functions that are also useful and quite beautiful. Before defining and studying them we introduce a nonholomorphic Eisenstein series that is very useful in many contexts.

10.4.1 The Nonholomorphic Eisenstein Series $G(\tau, s)$

Although properly speaking the study of this function belongs to the realm of modular forms, we study it completely independently. However, the most interesting properties of $G(\tau, s)$ are linked to its modularity properties.

Definition 10.4.1. *Let $\tau = x + iy$ be a complex number such that $y = \Im(\tau) > 0$. For $\Re(s) > 1$ we set*

$$G(\tau, s) = \frac{1}{2} \sum_{(m,n)\in\mathbb{Z}^2}' \frac{y^s}{|m\tau + n|^{2s}} \,,$$

where $\sum'$ means that the term $(m, n) = (0, 0)$ must be omitted.

Recall that $\mathrm{SL}_2(\mathbb{Z})$ is the group of 2×2 integral matrices with determinant 1.

Proposition 10.4.2. *The above series converges for $\Re(s) > 1$, and for any $\left(\begin{smallmatrix} a & b \\ c & d \end{smallmatrix}\right) \in \mathrm{SL}_2(\mathbb{Z})$ we have*

$$G\left(\frac{a\tau + b}{c\tau + d}, s\right) = G(\tau, s) \,;$$

in other words, $G(\tau, s)$ is a (nonholomorphic) modular form of weight 0 on $\mathrm{SL}_2(\mathbb{Z})$.

Proof. Left to the reader (Exercise 55). □

The main result that we need is the Fourier expansion of $G(\tau, s)$.

Theorem 10.4.3. *Let $\tau = x + iy$ with $y > 0$. For $\Re(s) > 1$ we have*

$$G(\tau, s) = \zeta(2s)y^s + \frac{\pi^{1/2}\Gamma(s - 1/2)}{\Gamma(s)}\zeta(2s - 1)y^{1-s}$$

$$+ 2\frac{\pi^s}{\Gamma(s)} \sum_{n\geqslant 1} n^{s-1}\sigma_{1-2s}(n)F_{s-1/2}(2\pi ny)\cos(2\pi nx) \,,$$

where $\sigma_z(n) = \sum_{d|n} d^z$ is the sum of the zth powers of the divisors of n and $F_{s-1/2}(z) = (2z/\pi)^{1/2}K_{s-1/2}(z)$, where $K_{s-1/2}(z)$ is the K-Bessel function.

Proof. First a friendly word to the reader: this result is quite technical; however, its proof is instructive and completely straightforward. Set

$$S(\tau, s) = \sum_{n\in\mathbb{Z}} \frac{y^s}{|\tau + n|^{2s}} \,.$$

By the Poisson summation formula (Corollary 2.2.17) applied to the function $f(t) = |iy + t|^{2s} = (t^2 + y^2)^s$ we have

$$S(\tau, s) = y^s \sum_{k \in \mathbb{Z}} c_k(y) e^{2i\pi kx} \quad \text{with} \quad c_k(y) = \int_{-\infty}^{\infty} \frac{e^{-2i\pi kt}}{(t^2 + y^2)^s} \, dt \; .$$

Since $(t^2 + y^2)^s$ is an even function and $y > 0$, making the change of variable $t = yu$ we have

$$c_k(y) = y^{1-2s} I_s(2\pi|k|y), \quad \text{with} \quad I_s(a) = \int_{-\infty}^{\infty} \frac{\cos(at)}{(t^2 + 1)^s} \, dt \; .$$

By Corollary 9.6.40 we have $I_s(0) = \pi^{1/2}\Gamma(s - 1/2)/\Gamma(s)$, while for $a > 0$ we have by Theorem 9.8.9,

$$I_s(a) = \frac{2\pi^{1/2}(a/2)^{s-1/2}}{\Gamma(s)} K_{s-1/2}(a) \; .$$

It follows that

$$S(\tau, s) = \frac{\pi^{1/2}\Gamma(s - 1/2)}{\Gamma(s)} y^{1-s} + \frac{4\pi^s}{\Gamma(s)} y^{1/2} \sum_{k \geqslant 1} k^{s-1/2} K_{s-1/2}(2\pi ky) \cos(2\pi kx)$$

$$= \frac{\pi^{1/2}\Gamma(s - 1/2)}{\Gamma(s)} y^{1-s} + 2\frac{\pi^s}{\Gamma(s)} \sum_{k \geqslant 1} k^{s-1} F_{s-1/2}(2\pi ky) \cos(2\pi kx) \; .$$

The proof of the theorem is now immediate: separating the terms with $m = 0$, $m < 0$, and $m > 0$ we have

$$G(\tau, s) = \zeta(2s)y^s + \sum_{m \geqslant 1} \frac{S(m\tau, s)}{m^s} = \zeta(2s)y^s + \frac{\pi^{1/2}\Gamma(s - 1/2)}{\Gamma(s)} \zeta(2s - 1) y^{1-s}$$

$$+ 2\frac{\pi^s}{\Gamma(s)} \sum_{k,m \geqslant 1} \frac{k^{s-1}}{m^s} F_{s-1/2}(2\pi kmy) \cos(2\pi kmx) \; .$$

For a given $N = km \geqslant 1$, we have

$$\sum_{km=N} \frac{k^{s-1}}{m^s} = \sum_{km=N} \frac{(km)^{s-1}}{m^{2s-1}} = N^{s-1}\sigma_{1-2s}(N) \; ,$$

and replacing gives the desired expansion of $G(\tau, s)$.   □

This technical theorem has a large number of corollaries.

Corollary 10.4.4. *The function $G(\tau, s)$ has a meromorphic continuation to the whole complex s-plane, with a single pole, at $s = 1$, which is simple with residue $\pi/2$ (in particular which is independent of τ). Moreover, if we set*

$$\mathcal{G}(\tau, s) = \pi^{-s}\Gamma(s)G(\tau, s)$$

we have the functional equation $\mathcal{G}(\tau, 1 - s) = \mathcal{G}(\tau, s)$.

Proof. Indeed, by Proposition 9.8.6 the function $K_{s-1/2}(z)$ is a holomorphic function of $s \in \mathbb{C}$ that tends to zero faster than any power of z as $z \to \infty$. It follows from the known analytic continuation of the other functions occurring in the expansion of $G(\tau, s)$ that this function can also be analytically continued to the whole complex plane into a meromorphic function. If as usual we set $\Lambda(s) = \pi^{-s/2}\Gamma(s/2)\zeta(s)$, by the functional equation of the zeta function we have $\Lambda(1 - s) = \Lambda(s)$ so the theorem gives

$$\mathcal{G}(\tau, s) = \pi^{-s}\Gamma(s)G(\tau, s) = \Lambda(2s)y^s + \Lambda(2 - 2s)y^{1-s}$$
$$+ 2\sum_{n \geq 1} n^{s-1}\sigma_{1-2s}(n)F_{s-1/2}(2\pi ny)\cos(2\pi nx) .$$

The possible poles are thus those of the Λ functions, hence at $s = 0$, $1/2$, and 1. The residue at $s = 1/2$ is equal to $y^{1/2}/2 - y^{1/2}/2 = 0$, so there is no pole. At $s = 0$ the function $G(\tau, s)$ clearly has no pole. At $s = 1$ the function $\Lambda(2 - 2s) = \Lambda(2s - 1)$ has a simple pole with residue $1/2$, so $G(\tau, s)$ has a simple pole with residue $\pi/2$, as claimed.

Finally, note that by definition $K_s(x)$ is an even function of s, so that $K_{s-1/2}(x)$ is invariant under the change s into $1 - s$, and similarly

$$n^{s-1}\sigma_{1-2s}(n) = \sum_{d_1 d_2 = n} (d_1 d_2)^{s-1}d_1^{1-2s} = \sum_{d_1 d_2 = n} d_1^{-s}d_2^{s-1}$$

is also invariant under the change s into $1-s$, proving the functional equation. $\square$

Corollary 10.4.5. *We have $G(\tau, 0) = -1/2$ and $G(\tau, -k) = 0$ for all $k \in \mathbb{Z}_{\geq 1}$.*

Proof. Immediate from the preceding corollary and left to the reader. $\square$

Note that, as for Dirichlet L-series, we may also ask for the value at positive integers s of $G(\tau, s)$. This can be done in the case that τ is imaginary quadratic, in other words when a, b, c are integers, and is a part of the theory of *complex multiplication*, which we will not study in this book for lack of space, although it is a very beautiful part of number theory.

10.4.2 The Kronecker Limit Formula

By Corollary 10.4.4 we know that when s is close to 1 we have $G(\tau, s) = (\pi/2)(1/(s-1) + C(\tau) + O(s-1))$ for a certain constant $C(\tau)$. The goal of *Kronecker's limit formula* is to give an explicit formula for $C(\tau)$. The result is as follows.

Theorem 10.4.6 (Kronecker's limit formula). *Define*

$$q = e^{2i\pi\tau} = e^{-2\pi y}e^{2i\pi x} \quad \text{and} \quad \eta(\tau) = e^{i\pi\tau/12}\prod_{n \geq 1}(1 - q^n) .$$

(1) *Around $s = 1$ we have*

$$G(\tau, s) = \frac{\pi}{2} \left(\frac{1}{s-1} + C(\tau) + O(s-1) \right) ,$$

where

$$C(\tau) = 2 \left(\gamma - \log(2) - \log \left(\Im(\tau)^{1/2} |\eta(\tau)|^2 \right) \right) .$$

(2) *We have $G(\tau, 0) = -1/2$ and*

$$G'(\tau, 0) = -\log(2\pi) - \log \left(\Im(\tau)^{1/2} |\eta(\tau)|^2 \right) ,$$

where of course G' denotes the derivative with respect to s.

Proof. We must give the Taylor expansions around $s = 1$ up to terms in $O(s-1)$ of all the terms occurring in the Fourier expansion of $G(\tau, s)$. For notational simplicity set $G(s) = \pi^{1/2} \Gamma(s - 1/2) \zeta(2s - 1) y^{1-s} / \Gamma(s)$. By Proposition 9.6.15 and Exercise 91 of Chapter 9 we have

$$\log(G(s)) = \frac{1}{2} \log(\pi) + \log(\Gamma(1/2)) + \frac{\Gamma'(1/2)}{\Gamma(1/2)}(s-1) + O(s-1)^2$$

$$+ \log \left(\frac{1}{2(s-1)} + \gamma + O(s-1) \right) - \log(y)(s-1) - \frac{\Gamma'(1)}{\Gamma(1)}(s-1)$$

$$= \log(\pi) - (2\log(2) + \gamma)(s-1) - \log(2) - \log(s-1)$$

$$+ 2\gamma(s-1) - \log(y)(s-1) + \gamma(s-1) + O(s-1)^2 ,$$

hence

$$\frac{\pi^{1/2} \Gamma(s-1/2) \zeta(2s-1) y^{1-s}}{\Gamma(s)} = \frac{\pi}{2} \left(\frac{1}{s-1} + 2\gamma - 2\log(2) - \log(y) \right)$$

$$+ O(s-1) .$$

It follows from Proposition 9.8.5 (1) that $F_{1/2}(z) = e^{-z}$, hence that

$$\frac{\pi}{2} C(\tau) = \frac{\pi^2}{6} y + \frac{\pi}{2} (2\gamma - 2\log(2) - \log(y)) + 2\pi S(\tau) ,$$

where

$$S(\tau) = \Re \left(\sum_{n \geq 1} \sigma_{-1}(n) q^n \right) \quad \text{with} \quad q = e^{2i\pi\tau} .$$

Now note that

$$\log \left(\prod_{n \geq 1} (1 - q^n) \right) = -\sum_{n \geq 1} \sum_{k \geq 1} \frac{q^{nk}}{k} = -\sum_{N \geq 1} q^N \sum_{k \mid N} k^{-1} = -\sum_{N \geq 1} \sigma_{-1}(N) q^N .$$

It follows that

$$S(\tau) = -\Re\left(\log\left(\prod_{n\geq 1}(1-q^n)\right)\right) = -\log\left(\left|\prod_{n\geq 1}(1-q^n)\right|\right)$$

$$= -\left(\log(|\eta(\tau)|) + \frac{\pi y}{12}\right),$$

proving (1) after simplifications, and (2) immediately follows from the functional equation. □

We have seen that $G(\gamma(\tau), s) = G(\tau, s)$ for all $\gamma = \left(\begin{smallmatrix} a & b \\ c & d \end{smallmatrix}\right) \in \mathrm{SL}_2(\mathbb{Z})$, where $\gamma(\tau) = (a\tau+b)/(c\tau+d)$. It follows from the above theorem that $\Im(\tau)^{1/2}|\eta(\tau)|^2$ is invariant under the change τ into $\gamma(\tau)$. Since $\Im(\gamma(\tau))^{1/2} = \Im(\tau)^{1/2}/|c\tau+d|$, this means that $|\eta(\gamma(\tau))| = |c\tau+d|^{1/2}|\eta(\tau)|$. In fact, the function $\eta(\tau)$ is the well-known Dedekind eta modular function of weight $1/2$, which satisfies $\eta(-1/\tau) = (\tau/i)^{1/2}\eta(\tau)$, and as a consequence $\eta(\gamma(\tau)) = v(\gamma)(c\tau+d)^{1/2}\eta(\tau)$ for an explicit 24th root of unity $v(\gamma)$.

Definition 10.4.7. *Let $Q(x,y) = ax^2 + bxy + cy^2$ be a positive definite quadratic form with real coefficients (in other words $a > 0$, $c > 0$, and $b^2 - 4ac < 0$). The Epstein zeta function attached to Q is the function defined for $\Re(s) > 1$ by*

$$\zeta_Q(s) = \frac{1}{2}\sideset{}{'}\sum_{(m,n)\in\mathbb{Z}^2}\frac{1}{Q(m,n)^s},$$

where here and elsewhere $\sum'$ means that we omit the term $(m,n) = (0,0)$.

If for any $n \geq 1$ we write $r_Q(n)$ for the number of representations of n by the form Q, in other words the number of pairs $(x,y) \in \mathbb{Z}^2$ such that $Q(x,y) = n$, then we clearly have

$$\zeta_Q(s) = \frac{1}{2}\sum_{n\geq 1}\frac{r_Q(n)}{n^s}.$$

Corollary 10.4.8. *Let Q be as above and set $\tau = (-b + i\sqrt{4ac - b^2})/(2a)$, so that $\Im(\tau) > 0$, and let $D = b^2 - 4ac$ be the discriminant of Q. Then*

(1) $\zeta_Q(s) = (|D|/4)^{-s/2}G(\tau, s)$.

(2) $\zeta_Q(s)$ *can be analytically continued to the whole complex plane with a simple pole, at $s = 1$, with residue $\pi/|D|^{1/2}$, and satisfies the functional equation $\Lambda_Q(1 - s) = \Lambda_Q(s)$, where*

$$\Lambda_Q(s) = \left(\frac{2\pi}{|D|^{1/2}}\right)^{-s}\Gamma(s)\zeta_Q(s).$$

(3) *Around $s = 1$ we have the expansion*

$$\zeta_Q(s) = \frac{\pi}{|D|^{1/2}}\left(\frac{1}{s-1} + C(Q) + O(s-1)\right),$$

where

$$C(Q) = 2\gamma - 2\log(2) - \frac{\log(|D|/4)}{2} - 2\log\left(\Im(\tau)^{1/2}|\eta(\tau)|^2\right).$$

(4) *We have* $\zeta_Q(0) = -1/2$ *and*

$$\zeta_Q'(0) = \frac{\log(|D|/4)}{4} - \log(2\pi) - \log\left(\Im(\tau)^{1/2}|\eta(\tau)|^2\right).$$

Proof. By definition we have

$$\zeta_Q(s) = \frac{a^{-s}}{2} \underset{(m,n)\in\mathbb{Z}^2}{\sum\nolimits'} \frac{1}{|m-n\tau|^{2s}} = (a\Im(\tau))^{-s}G(\tau,s) = (|D|/4)^{-s/2}G(\tau,s),$$

proving both (1) and (2) because of the corresponding properties of $G(\tau,s)$. Around $s = 1$ we have

$$(|D|/4)^{-s/2} = (|D|/4)^{-1/2}(1 - (s-1)\log(|D|/4)/2 + O(s-1)^2);$$

hence Kronecker's limit formula gives

$$\zeta_Q(s) = \frac{\pi}{\sqrt{|D|}}\left(\frac{1}{s-1} + C(Q) + O(s-1)\right),$$

with $C(Q) = C(\tau) - \log(|D|/4)/2$, proving (3), and (4) follows from the functional equation or directly from Theorem 10.4.6 (2). □

10.5 Dirichlet Series Linked to Number Fields

10.5.1 The Dedekind Zeta Function $\zeta_K(s)$

As we have seen above, the exact analytic translation of the existence and uniqueness of prime decomposition in $\mathbb{Z}$ is the fact that the Dirichlet series associated with a completely multiplicative function has an Euler product, and in particular the Riemann zeta function $\zeta(s) = \prod_p (1-p^{-s})^{-1}$. If K is a general number field and $\mathbb{Z}_K$ its ring of integers, the existence and uniqueness of prime decomposition are valid for *ideals*, as is the case for every Dedekind domain. Thus, it is natural to define the *Dedekind zeta function*

$$\zeta_K(s) = \sum_{\mathfrak{a}\subset\mathbb{Z}_K} \frac{1}{\mathcal{N}(\mathfrak{a})^s} = \prod_{\mathfrak{p}} \frac{1}{1-\mathcal{N}(\mathfrak{p})^{-s}},$$

where $\mathfrak{a}$ runs through all integral ideals of $\mathbb{Z}_K$ and $\mathfrak{p}$ through all prime ideals of $\mathbb{Z}_K$, and $\mathcal{N}$ denotes the absolute norm. The (formal) equality of the two definitions is the exact translation of existence and uniqueness of prime ideal decomposition. We of course have $\zeta_{\mathbb{Q}}(s) = \zeta(s)$, the ordinary Riemann zeta function.

Set $n = [K : \mathbb{Q}]$. Denoting as usual by p ordinary prime numbers, if s is a real number such that $s > 1$ we can write

$$\prod_{\mathfrak{p}}(1 - \mathcal{N}(\mathfrak{p})^{-s})^{-1} = \prod_{p}\prod_{\mathfrak{p}|p\mathbb{Z}_K}(1 - p^{-f(\mathfrak{p}/p)s})^{-1} \leqslant \prod_{p}(1 - p^{-s})^{-n} = \zeta(s)^n ,$$

since there are at most n prime ideals above p, so that $\zeta_K(s) \leqslant \zeta(s)^n$. In particular, this shows that $\zeta_K(s)$ converges for $s > 1$ real, hence converges absolutely (since it has nonnegative coefficients) for $\Re(s) > 1$. It follows that $\zeta_K(s)$ is a holomorphic function for $\Re(s) > 1$. This proof is only to show convergence, but we will soon see that $\zeta_K(s)$ has only a simple pole at $s = 1$, not a pole of order n.

The basic analytic properties of $\zeta_K(s)$ are summarized in the following theorem.

Theorem 10.5.1. *Let K be a number field of degree n and signature (r_1, r_2). Denote by $d(K)$, $h(K)$, $R(K)$, and $w(K)$ (standard notation) the discriminant, class number, regulator, and number of roots of unity in K. Then*

(1) *The function $\zeta_K(s)$ extends analytically to the whole complex plane to a meromorphic function having a single pole at $s = 1$, which is simple.*

(2) *It satisfies the functional equation $\Lambda_K(1 - s) = \Lambda_K(s)$, where*

$$\Lambda_K(s) = |d(K)|^{s/2}\gamma(s)^{r_1+r_2}\gamma(s+1)^{r_2}\zeta_K(s) ,$$

and $\gamma(s) = \pi^{-s/2}\Gamma(s/2)$ is as in Theorem 10.2.14.

(3) *If we set $r = r_1 + r_2 - 1$, which is the rank of the unit group of K, then $\zeta_K(s)$ has a zero at $s = 0$ of order r (no zero if $r = 0$ of course) and we have*

$$\lim_{s\to 0} s^{-r}\zeta_K(s) = -\frac{h(K)R(K)}{w(K)} .$$

(4) *Equivalently, by the functional equation, the residue of the pole at $s = 1$ is given by*

$$\lim_{s\to 1}(s - 1)\zeta_K(s) = 2^{r_1}(2\pi)^{r_2}\frac{h(K)R(K)}{w(K)|d(K)|^{1/2}} .$$

Proof. We will not prove this theorem, but we will make a number of remarks on the proof. There are two ways to prove the analytic continuation and functional equation of $\zeta_K(s)$ to the whole plane. One is Hecke's initial proof: he essentially copies the proof that we have given for the ordinary zeta function and Dirichlet L-series using the Poisson summation formula. For this, one must introduce theta functions in n variables, use a generalized Poisson summation formula, and so on. A large part of the difficulty, which does not occur for $K = \mathbb{Q}$, is the existence of an infinite unit group, which makes the domains of integration *noncompact*, so suitable regularization procedures have to be applied. The proof gives at the same time the residue at $s = 1$, hence the leading term at $s = 0$.

The other more recent proof is due to J. Tate. He gives an *adelic* proof, explaining each factor $(1 - \mathcal{N}(\mathfrak{p})^{-s})^{-1}$ as a $\mathfrak{p}$-adic integral, and the factors $\gamma(s)$ and $\gamma(s+1)$ as the factors corresponding to the places at infinity of K. This proof is more elegant, and more amenable to generalizations, but not really much shorter.

Finally, it is not difficult to prove the analytic continuation of $\zeta_K(s)$ to $\Re(s) > 1-1/n$ using quite simple means, and from that a volume computation gives the residue at $s = 1$. This is done for example in [Marc]. This is sufficient for many purposes, but does not give the functional equation.   □

Corollary 10.5.2. *For $k \in \mathbb{Z}_{\leqslant 0}$, the order of the (possible) zero at $s = k$ of $\zeta_K(s)$ is given by*

$$\begin{cases} r_1 + r_2 - 1 & \text{if } k = 0 \,, \\ r_1 + r_2 & \text{if } k < 0, \ k \equiv 0 \pmod 2 \,, \\ r_2 & \text{if } k < 0, \ k \equiv 1 \pmod 2 \,. \end{cases}$$

Proof. Follows immediately from the fact that $\gamma(s)$ has simple poles for all $s \in \mathbb{Z}_{\leqslant 0}$, and left to the reader.   □

In particular, we see that $\zeta_K(-2k) = 0$ for all number fields K when $k \geqslant 1$. Furthermore, if the field K is not totally real ($r_2 > 0$), we also have $\zeta_K(1-2k) = 0$ for all $k \geqslant 1$. Thus the only fields for which some of the values of $\zeta_K(-k)$ can be nonzero are *totally real fields*, whose complex embeddings are in fact all real. The field $K = \mathbb{Q}$ is of course the simplest example. We have seen that in that case the values $\zeta(1 - 2k)$ are in fact rational numbers, more precisely that $\zeta(1 - 2k) = -B_{2k}/(2k)$ (Corollary 10.3.4). For the Dedekind zeta function, there is a similar result, but the proof is more difficult and uses the Fourier expansion of the Hecke–Eisenstein series (see the discussion at the beginning of Section 10.3.2):

Theorem 10.5.3 (Siegel). *Let K be a totally real number field. For all $k \geqslant 1$ we have $\zeta_K(1 - 2k) \in \mathbb{Q}^*$.*

Corollary 10.5.4. *Let K be a totally real number field of degree n. For every $k \geqslant 1$ there exists $r_k \in \mathbb{Q}^*$ such that*

$$\zeta_K(2k) = r_k \frac{\pi^{2kn}}{|d(K)|^{1/2}} \,.$$

Proof. Clear by using the functional equation.   □

10.5.2 The Dedekind Zeta Function of Quadratic Fields

Another very important property of Dedekind zeta functions is that they can be factored into L-functions having a simpler functional equation. This is in fact linked to one of the most famous conjectures in number theory, Artin's conjecture on the analytic continuation of L-series. We will speak a little about this below, but note for now that everything is well understood for number fields that are Abelian extensions of $\mathbb{Q}$. We begin with the simplest nontrivial extensions of $\mathbb{Q}$: quadratic fields.

Proposition 10.5.5. *Let $K = \mathbb{Q}(\sqrt{D})$ be a quadratic field of discriminant D. We have*
$$\zeta_K(s) = \zeta(s)L(\chi_D, s) \,,$$
where as usual $\chi_D(n) = \left(\frac{D}{n}\right)$ is the Legendre–Kronecker character.

Proof. Indeed, if p is a prime number, then we know that p is inert, splits, or ramifies in $K/\mathbb{Q}$ according to whether $\left(\frac{D}{p}\right) = -1$, 1, or 0. Thus

$$\zeta_K(s) = \prod_p \prod_{\mathfrak{p}|p\mathbb{Z}_K} (1 - \mathcal{N}(\mathfrak{p})^{-s})^{-1}$$

$$= \prod_{\left(\frac{D}{p}\right)=-1} (1 - p^{-2s})^{-1} \prod_{\left(\frac{D}{p}\right)=1} (1 - p^{-s})^{-2} \prod_{\left(\frac{D}{p}\right)=0} (1 - p^{-s})^{-1}$$

$$= \zeta(s) \prod_p \left(1 - \left(\frac{D}{p}\right)p^{-s}\right)^{-1} = \zeta(s)L(\chi_D, s) \,,$$

proving the proposition. $\square$

Since we have proved the functional equation of L-functions of Dirichlet characters, this proposition implies Theorem 10.5.1 for quadratic fields.

Corollary 10.5.6. *Let D be a nonsquare integer congruent to 0 or 1 modulo 4, and let $\chi_D = \left(\frac{D}{\cdot}\right)$ be the corresponding Legendre–Kronecker symbol. Then $L(\chi_D, 1) > 0$.*

Proof. Write $D = D_0 f^2$, where D_0 is a fundamental discriminant. Since $L(\chi_D, 1) = \prod_{p|f}(1 - \chi_{D_0}(p)/p)L(\chi_{D_0}, 1)$ and $1 - \chi_{D_0}(p)/p > 0$, we may assume that D is a fundamental discriminant. In that case we note that by the above proposition we have $L(\chi_D, s) = \zeta_K(s)/\zeta(s)$, and that by definition $\zeta_K(s) > 1$ and $\zeta(s) > 1$ for $s \in \mathbb{R}_{>1}$. The result follows by letting s tend to 1. $\square$

In the special case of *imaginary* quadratic fields K, i.e., when $D < 0$, the function $\zeta_K(s)$ is closely related to the Epstein zeta functions that we have studied in Section 10.4. Indeed, for every ideal class $\mathcal{A} \in Cl(K)$ define the *partial zeta function* by the formula

$$\zeta_K(\mathcal{A}, s) = \sum_{\mathfrak{a} \subset \mathbb{Z}_K, \, \mathfrak{a} \in \mathcal{A}} \frac{1}{\mathcal{N}(\mathfrak{a})^s} \, .$$

We then have the following:

Proposition 10.5.7. *Let K be an imaginary quadratic field of discriminant D, and denote by $w(D)$ the number of roots of unity of K (so that $w(D) = 2$ if $D < -4$, $w(-4) = 4$, $w(-3) = 6$).*

(1) *We have the finite sum decomposition*

$$\zeta_K(s) = \sum_{\mathcal{A} \in Cl(K)} \zeta_K(\mathcal{A}, s) \, .$$

(2) *Let $\mathfrak{b}$ be an ideal of K such that $\mathfrak{b} \in \mathcal{A}^{-1}$ and $1 \in \mathfrak{b}$, and let $(1, \tau)$ be a $\mathbb{Z}$-basis of $\mathfrak{b}$. We have*

$$\zeta_K(\mathcal{A}, s) = \frac{|D|^{-s/2}}{w(D)/2} \zeta_{Q_{\mathcal{A}}}(s) \, ,$$

where

$$Q_{\mathcal{A}}(m, n) = \frac{1}{2\Im(\tau)} (m^2 - 2mn\Re(\tau) + n^2|\tau|^2)$$

and $\zeta_{Q_{\mathcal{A}}}(s)$ is the Epstein zeta function attached to the positive definite quadratic form $Q_{\mathcal{A}}$.

Proof. (1) is a trivial consequence of the definition of the class group $Cl(K)$. For (2), let $\mathfrak{b} \in \mathcal{A}^{-1}$ be an ideal belonging to the inverse class. Multiplying $\mathfrak{b}$ by a suitable principal ideal we may assume that $1 \in \mathfrak{b}$. Then $\mathfrak{a} \in \mathcal{A}$ if and only if $\mathfrak{a}\mathfrak{b} = \lambda \mathbb{Z}_K$ is a principal ideal, and in addition $\mathfrak{a} \subset \mathbb{Z}_K$ if and only if $\lambda \in \mathfrak{b}$, and by multiplicativity $\mathcal{N}(\mathfrak{a}) = \mathcal{N}(\mathfrak{b})^{-1}|\mathcal{N}(\lambda)| = \mathcal{N}(\mathfrak{b})^{-1}\mathcal{N}(\lambda)$ since $D < 0$. Finally, again since $D < 0$, there is a finite number $w(D)$ of possibilities for λ, so that

$$\zeta_K(\mathcal{A}, s) = \frac{\mathcal{N}(\mathfrak{b})^s}{w(D)} \sum_{\lambda \in \mathfrak{b}, \, \lambda \neq 0} \frac{1}{\mathcal{N}(\lambda)^s} \, .$$

Since $1 \in \mathfrak{b}$ we have $\mathbb{Z}_K \subset \mathfrak{b}$. I claim that $\mathcal{N}(\mathfrak{b}) = [\mathfrak{b} : \mathbb{Z}_K]^{-1}$: indeed, if m is any integer such that $m\mathfrak{b} \subset \mathbb{Z}_K$, then multiplication by m gives the equality $[\mathfrak{b} : \mathbb{Z}_K] = [m\mathfrak{b} : m\mathbb{Z}_K]$, and

$$m^2 = [\mathbb{Z}_K : m\mathbb{Z}_K] = [\mathbb{Z}_K : m\mathfrak{b}][m\mathfrak{b} : m\mathbb{Z}_K] = \mathcal{N}(m\mathfrak{b})[\mathfrak{b} : \mathbb{Z}_K]$$
$$= m^2 \mathcal{N}(\mathfrak{b})[\mathfrak{b} : \mathbb{Z}_K] \, ,$$

proving my claim. Since $1 \in \mathfrak{b}$ we have $\mathfrak{b} = \mathbb{Z} + \tau\mathbb{Z}$ for some $\tau \in K$, where by changing if necessary τ into $-\tau$ we may assume $\Im(\tau) > 0$. On the other hand, $\mathbb{Z}_K = \mathbb{Z} + \omega\mathbb{Z}$, where $\omega = (\delta + \sqrt{D})/2$ for any integer δ such that $\delta \equiv D$

(mod 2). Writing $\mathrm{Vol}(I)$ for the covolume in $\mathbb{C} \simeq \mathbb{R}^2$ of any fractional ideal I it follows that

$$\mathcal{N}(\mathfrak{b}) = [\mathfrak{b} : \mathbb{Z}_K]^{-1} = \frac{\mathrm{Vol}(\mathfrak{b})}{\mathrm{Vol}(\mathbb{Z}_K)} = \frac{\Im(\tau)}{\Im(\omega)} = \frac{2\Im(\tau)}{\sqrt{|D|}} \ .$$

Finally, we can write $\lambda = m - n\tau$ with $(m, n) \in \mathbb{Z}^2$, and $\mathcal{N}(\lambda) = m^2 - 2mn\Re(\tau) + n^2|\tau|^2$. We thus obtain

$$\zeta_K(\mathcal{A}, s) = \frac{|D|^{-s/2}}{w(D)} {\sum_{(m,n)\in\mathbb{Z}^2}}' \frac{1}{Q_{\mathcal{A}}(m,n)^s} = \frac{|D|^{-s/2}}{w(D)/2}\zeta_{Q_{\mathcal{A}}}(s) \ ,$$

where

$$Q_{\mathcal{A}}(m, n) = \frac{1}{2\Im(\tau)}(m^2 - 2mn\Re(\tau) + n^2|\tau|^2)$$

is a positive definite quadratic form, proving the proposition.   □

Corollary 10.5.8. *Let K be an imaginary quadratic field of discriminant D, let δ be any integer such that $D \equiv \delta \pmod 2$, and assume that the class number of K is equal to 1. Then*

$$\zeta_K(s) = \frac{1}{w(D)} {\sum_{(m,n)\in\mathbb{Z}^2}}' \frac{1}{(m^2 + mn\delta + ((\delta^2 - D)/4)n^2)^s} \ .$$

Proof. Clear.   □

Example. As in Theorem 5.4.15 denote by $r_2(n)$ the number of decompositions of n as a sum of two squares. Then

$$\zeta_{\mathbb{Q}(i)}(s) = \frac{1}{4} {\sum_{(m,n)\in\mathbb{Z}^2}}' \frac{1}{(m^2 + n^2)^s} = \frac{1}{4} \sum_{n\geq 1} \frac{r_2(n)}{n^s} \ ,$$

and the formula for $r_2(n)$ given in Theorem 5.4.15 is equivalent to the formula $\zeta_{\mathbb{Q}(i)}(s) = \zeta(s)L(\chi_{-4}, s)$.

Corollary 10.5.9. *Let K be an imaginary quadratic field of discriminant D. The statements of Corollary 10.4.8 (2), (3), and (4) are valid verbatim if we replace $\zeta_{\mathbb{Q}}(s)$ by $(\omega(D)/2)\zeta_K(\mathcal{A}, s)$; in other words:*

(1) *$\zeta_K(\mathcal{A}, s)$ can be analytically continued to the whole complex plane with a simple pole, at $s = 1$, with residue $2\pi/(\omega(D)|D|^{1/2})$, and satisfies the functional equation $\Lambda_K(\mathcal{A}, 1 - s) = \Lambda_K(\mathcal{A}, s)$, where*

$$\Lambda_K(\mathcal{A}, s) = \left(\frac{2\pi}{|D|^{1/2}}\right)^{-s} \Gamma(s)\zeta_K(\mathcal{A}, s) \ .$$

(2) *Around $s = 1$ we have the expansion*

$$\zeta(\mathcal{A}, s) = \frac{2\pi}{w(D)|D|^{1/2}}\left(\frac{1}{s-1} + C_K(\mathcal{A}) + O(s-1)\right),$$

where

$$C_K(\mathcal{A}) = 2\gamma - 2\log(2) - \frac{\log(|D|/4)}{2} - 2\log\left(\Im(\tau)^{1/2}|\eta(\tau)|^2\right).$$

(3) *We have $\zeta_K(\mathcal{A}, 0) = -1/w(D)$ and*

$$\zeta'_K(\mathcal{A}, 0) = \frac{1}{w(D)/2}\left(\frac{\log(|D|/4)}{4} - \log(2\pi) - \log\left(\Im(\tau)^{1/2}|\eta(\tau)|^2\right)\right).$$

Proof. This is an immediate consequence of the proposition and of Corollary 10.4.8: we note that the quadratic form $Q_{\mathcal{A}}(x, y)$ has determinant $(4\Re(\tau)^2 - 4|\tau|^2)/(4\Im(\tau)^2) = -1$, and we write a usual $|D|^{-s/2} = |D|^{-1/2}(1 - (s-1)\log(|D|)/2)$ around $s = 1$ and $|D|^{-s/2} = 1 - s\log(|D|)/2$ around $s = 0$. □

We can now *combine* this corollary with the decomposition $\zeta_K(s) = \zeta(s)L(\chi_D, s)$:

Proposition 10.5.10. *Let $D < 0$ be a fundamental discriminant, and denote by $h(D)$ the class number of $K = \mathbb{Q}(\sqrt{D})$. Denote by $\mathcal{Q}(D)$ the set of equivalence classes of quadratic numbers $\tau = (-b + \sqrt{D})/(2a)$ of discriminant D, modulo the natural action of $\mathrm{SL}_2(\mathbb{Z})$, which has cardinality $h(D)$. We have the following formulas:*

$$L(\chi_D, 1) = \frac{2\pi h(D)}{w(D)|D|^{1/2}}, \quad L(\chi_D, 0) = \frac{2h(D)}{w(D)},$$

$$\frac{L'(\chi_D, 1)}{L(\chi_D, 1)} = \gamma - \log(2) - \frac{\log(|D|)}{2} - \frac{2}{h(D)}\sum_{\tau \in \mathcal{Q}(D)}\log\left(\Im(\tau)^{1/2}|\eta(\tau)|^2\right),$$

$$\frac{L'(\chi_D, 0)}{L(\chi_D, 0)} = \log(4\pi) - \frac{\log(|D|)}{2} + \frac{2}{h(D)}\sum_{\tau \in \mathcal{Q}(D)}\log\left(\Im(\tau)^{1/2}|\eta(\tau)|^2\right).$$

Proof. First note that if $(1, \tau')$ is another $\mathbb{Z}$-basis of an ideal $\mathfrak{b}$ with $\Im(\tau') > 0$ we have $\tau' = \gamma(\tau) = (a\tau + b)/(c\tau + d)$ for some $\left(\begin{smallmatrix} a & b \\ c & d \end{smallmatrix}\right) \in \mathrm{SL}_2(\mathbb{Z})$ (and conversely), and that if we replace $\mathfrak{b}$ by another ideal $\alpha\mathfrak{b}$ in the same class the corresponding τ is unchanged. Thus the map $\mathfrak{b} \mapsto \tau$ induces a natural map from the ideal class group $Cl(K)$ to the set of $\tau = (-b + \sqrt{D})/(2a)$ of discriminant D, up to the action of $\mathrm{SL}_2(\mathbb{Z})$, in other words to the set $\mathcal{Q}(D)$. The above formulas are thus obtained by summing on ideal classes the corresponding formulas for $\zeta_K(\mathcal{A}, s)$ and using the factorization $\zeta_K(s) = \zeta(s)L(\chi_D, s)$. The details are left to the reader (Exercise 57). □

Note that thanks to the above results we have proved Theorem 10.5.1 in the special case of imaginary quadratic fields.

10.5.3 Applications of the Kronecker Limit Formula

We can combine the above proposition, which is an immediate consequence of Kronecker's limit formula, with the formula for the same quantities obtained in Proposition 10.3.5 in terms of the gamma function.

Proposition 10.5.11 (Lerch, Chowla–Selberg). *For any negative fundamental discriminant D we have the identity*

$$\prod_{\tau \in \mathcal{Q}(D)} \Im(\tau)|\eta(\tau)|^4 = (4\pi|D|^{1/2})^{-h(D)}\left(\prod_{r=1}^{|D|} \Gamma(r/|D|)^{\left(\frac{D}{r}\right)}\right)^{\frac{w(D)}{2}}.$$

Proof. Since $D < 0$ we have $L(\chi_D, 0) = 2h(D)/w(D)$, so Proposition 10.3.5 gives

$$\frac{L'(\chi_D, 0)}{L(\chi_D, 0)} = \frac{w(D)}{2h(D)} \sum_{r=1}^{|D|} \left(\frac{D}{r}\right) \log\left(\Gamma\left(\frac{r}{|D|}\right)\right) - \log(|D|) ;$$

hence comparing with Proposition 10.5.10 above we obtain the identity

$$\sum_{\tau \in \mathcal{Q}} \log\left(\Im(\tau)^{1/2}|\eta(\tau)|^2\right) = \frac{w(D)}{4} \sum_{r=1}^{|D|} \left(\frac{D}{r}\right) \log\left(\Gamma\left(\frac{r}{|D|}\right)\right)$$
$$- \frac{h(D)}{2}\log(4\pi|D|^{1/2}) ,$$

which gives the desired formula after doubling and exponentiation.   □

The above formula was obtained at the end of the nineteenth century by Lerch, and rediscovered by Chowla and Selberg in 1947.

It is also possible to generalize the Chowla–Selberg formula to nonfundamental discriminants, but with some difficulty; see [Nak-Tag]. We give the result without proof, but first need the following definition.

Definition 10.5.12. *Let $D < 0$ be congruent to 0 or 1 modulo 4. We denote by $\mathcal{Q}(D)$ the set of quadratic numbers $(-b + \sqrt{D})/(2a)$ with $b^2 - 4ac = D$ and $\gcd(a, b, c) = 1$, modulo the natural action of $\mathrm{SL}_2(\mathbb{Z})$. We write $h(D)$ for the cardinality of $\mathcal{Q}(D)$.*

Theorem 10.5.13. *Let $D < 0$ be congruent to 0 or 1 modulo 4, and write $D = D_0 f^2$, where D_0 is a fundamental discriminant, in other words the discriminant of the quadratic field $\mathbb{Q}(\sqrt{D})$. Then*

$$\prod_{\tau \in \mathcal{Q}(D)} \Im(\tau)|\eta(\tau)|^4 = \left(\frac{\prod_{p|f} p^{e(p)}}{4\pi|D|^{1/2}}\right)^{h(D)} \left(\prod_{r=1}^{|D_0|} \Gamma(r/|D_0|)^{\left(\frac{D_0}{r}\right)}\right)^{\frac{h(D)}{h(D_0)}\frac{w(D_0)}{2}},$$

where

$$e(p) = \frac{\left(1 - p^{-v_p(f)}\right)\left(1 - \left(\frac{D_0}{p}\right)\right)}{(1 - 1/p)\left(p - \left(\frac{D_0}{p}\right)/p\right)}.$$

Note that we have the following well-known formula, coming directly from Dirichlet's class number formula:

$$\frac{h(D)}{h(D_0)}\frac{w(D_0)}{2} = \frac{w(D)}{2}f\prod_{p|f}\left(1 - \left(\frac{D}{p}\right)/p\right).$$

Examples. When $h(D) = 1$ there is a single term on the left-hand side, and we can always choose $\tau = (-\delta + \sqrt{D})/2$, where $\delta \equiv D \pmod 2$. It follows from the definition of $\eta(\tau)$ that $\eta(\tau) = |\eta(\tau)|$ if $D \equiv 0 \pmod 4$ and $\eta(\tau) = e^{-i\pi/24}|\eta(\tau)|$ if $D \equiv 1 \pmod 4$. Using the reflection formula for the gamma function we obtain for instance the following formulas:

$$\eta\left(\frac{-1+\sqrt{-3}}{2}\right) = e^{-i\pi/24}2^{-1}3^{1/8}\pi^{-1}\Gamma(1/3)^{3/2},$$

$$\eta(\sqrt{-1}) = 2^{-1}\pi^{-3/4}\Gamma(1/4),$$

$$\eta\left(\frac{-1+\sqrt{-7}}{2}\right) = e^{-i\pi/24}2^{-1}7^{-1/8}\pi^{-1}\left(\Gamma(1/7)\Gamma(2/7)\Gamma(4/7)\right)^{1/2},$$

$$\eta(\sqrt{-2}) = 2^{-11/8}\pi^{-3/4}\left(\Gamma(1/8)\Gamma(3/8)\right)^{1/2}$$

$$\eta(\sqrt{-1}/2) = 2^{-7/8}\pi^{-3/4}\Gamma(1/4),$$

the last formula coming from the theorem for nonfundamental discriminants. As an example with $h(D) = 2$ we have for instance

$$\eta(\sqrt{-6})\eta(\sqrt{-6}/2) = 2^{-11/4}3^{-1/2}\pi^{-3/2}\left(\Gamma(1/24)\Gamma(5/24)\Gamma(7/24)\Gamma(11/24)\right)^{1/2}$$

(we will see below that these two eta values can be computed individually).

Corollary 10.5.14. *We have the formulas*

$$\prod_{n\geqslant 1}\left(1 + e^{-\pi n}\right) = 2^{-1/8}e^{\pi/24},$$

$$\prod_{n\geqslant 1}\tanh(\pi n/2) = (2\pi)^{-3/4}\Gamma(1/4),$$

$$\prod_{n\geqslant 1}\tanh(\pi n/\sqrt{2}) = 2^{-7/8}\pi^{-3/4}(\Gamma(1/8)\Gamma(3/8))^{1/2}.$$

Proof. These formulas are simply obtained by replacing the eta function by its infinite product expansion and using the special values. The details are left to the reader (Exercise 59).   □

When $h(D) > 1$ we can ask whether it is possible to compute all the $\eta(\tau)$ *individually*. The answer is yes, as follows.

Theorem 10.5.15. *For each* $\tau \in \mathcal{Q}(D)$ *there exists an algebraic number* $\alpha(\tau)$ *that can be given explicitly such that*

$$\Im(\tau)|\eta(\tau)|^4 = \frac{\alpha(\tau)}{4\pi|D|^{1/2}} \left(\prod_{r=1}^{|D|} \Gamma(r/|D|)^{\left(\frac{D}{r}\right)} \right)^{\frac{w(D)}{2h(D)}}.$$

Proof. This proof requires a basic knowledge of modular forms and complex multiplication. If τ_1 and τ_2 are in $\mathcal{Q}(D)$ there exists some integral matrix $\gamma = \left(\begin{smallmatrix} a & b \\ c & d \end{smallmatrix} \right)$ with nonzero determinant N such that $\tau_2 = (a\tau_1 + b)/(c\tau_1 + d)$. If as usual we set $\Delta(\tau) = \eta(\tau)^{24}$, it follows that $(\eta(\tau_2)/\eta(\tau_1))^{24} = \Delta(\gamma(\tau_1))/\Delta(\tau_1)$ is the value at the quadratic number τ_1 of the function $\Delta(\gamma(\tau))/\Delta(\tau)$. This is a modular function of weight 0 on a congruence subgroup of level N, so there is an algebraic relation with algebraic coefficients between this function and the modular invariant function $j(\tau)$. By the basic theorem of complex multiplication we know that $j(\tau_1)$ is algebraic, so we conclude that $\eta(\tau_2)/\eta(\tau_1)$ is algebraic. It follows that all the terms on the LHS of the Chowla–Selberg formula are equal to an algebraic number times one of them, proving the theorem. $\qquad\square$

One can in fact give an explicit formula for $\alpha(\tau)$; see [Poo-Wil]. We will give a special case below, and the reader can easily work out himself many other examples (Exercise 58).

The Kronecker limit formula for imaginary quadratic fields also has applications to *real* quadratic fields. We begin with the following lemma.

Lemma 10.5.16. *Let D_1 and D_2 be two coprime fundamental discriminants, and set $D = D_1 D_2$ and $K = \mathbb{Q}(\sqrt{D})$. Then:*

(1) *D is a fundamental discriminant.*
(2) *For any $\lambda \in \mathbb{Z}_K$ such that $\mathcal{N}(\lambda)$ is coprime to D_1 we have $\left(\frac{D_1}{\mathcal{N}(\lambda)} \right) = 1$.*
(3) *For any integral ideal $\mathfrak{a}$ of $\mathbb{Z}_K$ such that $\mathcal{N}(\mathfrak{a})$ is coprime to D we have $\left(\frac{D}{\mathcal{N}(\mathfrak{a})} \right) = 1$.*

Proof. (1). If D_1 and D_2 are squarefree, hence congruent to 1 modulo 4, then $D_1 D_2$ is squarefree and congruent to 1 modulo 4, so is fundamental. Otherwise, by symmetry we may assume that $D_1 = 4d_1$ with $d_1 \equiv 2$ or 3 modulo 4 and squarefree. It follows that D_2 is squarefree and is congruent to 1 modulo 4; hence $D_1 D_2 = 4d_1 D_2$ with $d_1 D_2$ squarefree and congruent to 2 or 3 modulo 4, so $D_1 D_2$ is fundamental.

(2). We can write $\lambda = (a + b\sqrt{D})/2$ with a and b integers such that $a \equiv bD$ (mod 2), so $\mathcal{N}(\lambda) = (a^2 - b^2 D)/4$. Assume first that $4 \nmid D_1$. Then

$$\left(\frac{D_1}{\mathcal{N}(\lambda)} \right) = \left(\frac{D_1}{a^2 - b^2 D} \right) = \left(\frac{D_1}{a^2} \right) = 1$$

as claimed, since $D_1 \mid D$ and the Kronecker symbol $\left(\frac{D_1}{n}\right)$ is periodic of period dividing D_1 when $D_1 \equiv 0$ or 1 modulo 4. Assume now that $D_1 = 4d_1$, hence that $D_2 \equiv 1 \pmod 4$. Then $D = 4d$ with $d = d_1 D_2$, and $\lambda = a_1 + b_1 \sqrt{d}$ for integers $a_1 = a/2$ and $b_1 = b$. Since we assume $\mathcal{N}(\lambda)$ coprime to D_1 we have

$$\left(\frac{D_1}{\mathcal{N}(\lambda)}\right) = \left(\frac{4d_1}{a_1^2 - b_1^2 d}\right) = \left(\frac{d_1}{a_1^2 - b_1^2 d}\right) = \left(\frac{d_1}{a_1^2}\right) = 1 \,,$$

again because $d_1 \mid d$ and by the periodicity property of the Kronecker symbol.

(3). Since $\mathfrak{a}$ is a product of prime ideals it is sufficient to prove this for prime ideals $\mathfrak{p}$ above p with $p \nmid D$. If p is inert then $\mathfrak{p} = p\mathbb{Z}_K$ and $\mathcal{N}(\mathfrak{p}) = p^2$, so that $\left(\frac{D}{\mathcal{N}(\mathfrak{p})}\right) = \left(\frac{D}{p^2}\right) = 1$. If p is split then $\left(\frac{D}{p}\right) = 1$, while $\mathcal{N}(\mathfrak{p}) = p$, so that $\left(\frac{D}{\mathcal{N}(\mathfrak{p})}\right) = \left(\frac{D}{p}\right) = 1$. $\square$

Corollary 10.5.17. *Let D_1 and D_2 be two coprime fundamental discriminants, and set $D = D_1 D_2$ and $K = \mathbb{Q}(\sqrt{D})$.*

(1) *For any ideal class $\mathcal{A} \in Cl(K)$ there exists an integral ideal $\mathfrak{a} \in \mathcal{A}$ such that $\gcd(\mathcal{N}(\mathfrak{a}), D_1) = 1$.*
(2) *The quantity*

$$\chi_{D_1}(\mathcal{N}(\mathfrak{a})) = \left(\frac{D_1}{\mathcal{N}(\mathfrak{a})}\right)$$

does not depend on the choice of the integral ideal $\mathfrak{a} \in \mathcal{A}$, as long as $\gcd(\mathcal{N}(\mathfrak{a}), D_1) = 1$, and by abuse of notation it will be written $\chi_{D_1}(\mathcal{A})$.
(3) *The map $\mathcal{A} \mapsto \chi_{D_1}(\mathcal{A})$ defines a nontrivial character of the finite abelian group $Cl(K)$.*
(4) *We have $\chi_{D_1}(\mathcal{A}) = \chi_{D_2}(\mathcal{A})$.*

Proof. (1). Let $\mathfrak{b} \in \mathcal{A}$ be any integral ideal. By the approximation theorem for Dedekind domains there exists $\alpha \in K$ such that $v_\mathfrak{p}(\alpha) = -v_\mathfrak{p}(\mathfrak{b})$ for every prime ideal $\mathfrak{p}$ above a prime number dividing $\mathfrak{p}$, and $v_\mathfrak{p}(\alpha) \geqslant 0$ for all other $\mathfrak{p}$. It is clear that $\mathfrak{a} = \alpha\mathfrak{b}$ is an integral ideal in $\mathcal{A}$ whose norm is coprime to D_1, proving (1). Note for future reference that we can in fact ask for $\mathcal{N}(\mathfrak{a})$ to be coprime to any fixed integer, not only to D_1.

(2). Let $\mathfrak{a}$ and $\mathfrak{b}$ be ideals in $\mathcal{A}$ such that $\mathcal{N}(\mathfrak{a})$ and $\mathcal{N}(\mathfrak{b})$ are coprime to D_1, so that $\mathfrak{a} = \lambda\mathfrak{b}$ for some $\lambda \in K$ such that $v_\mathfrak{p}(\lambda) = 0$ for all $\mathfrak{p}$ above a prime dividing D_1. Once again by the approximation theorem we can find α and β in $\mathbb{Z}_K$ such that $\lambda = \beta/\alpha$ with $\mathcal{N}(\alpha)$ and $\mathcal{N}(\beta)$ coprime to D_1, so that $\mathfrak{a}\alpha = \mathfrak{b}\beta$. It follows from the lemma that

$$\left(\frac{D_1}{\mathcal{N}(\mathfrak{a}\alpha)}\right) = \left(\frac{D_1}{\mathcal{N}(\mathfrak{a})}\right)\left(\frac{D_1}{\mathcal{N}(\alpha)}\right) = \left(\frac{D_1}{\mathcal{N}(\mathfrak{a})}\right)$$

and similarly $\left(\frac{D_1}{\mathcal{N}(\mathfrak{b}\beta)}\right) = \left(\frac{D_1}{\mathcal{N}(\mathfrak{b})}\right)$, proving (2).

(3). Since the Kronecker symbol is multiplicative it follows from (2) that χ_{D_1} is a character of $Cl(K)$. Let us show that it is nontrivial. For $i = 1$ and $i = 2$ let p_i be an inert prime in $\mathbb{Q}(\sqrt{D_i})$, in other words $\left(\frac{D_1}{p_1}\right) = \left(\frac{D_2}{p_2}\right) = -1$. Since D_1 and D_2 are coprime, by the Chinese remainder theorem there exists b such that $b \equiv p_1 \pmod{D_1}$ and $b \equiv p_2 \pmod{D_2}$. By Dirichlet's theorem on primes in arithmetic progression (which we will prove below; see Theorem 10.5.30) we can find a prime number p such that $p \equiv b \pmod{D}$. By periodicity of the Kronecker symbol we have

$$\left(\frac{D_1}{p}\right) = \left(\frac{D_1}{b}\right) = \left(\frac{D_1}{p_1}\right) = -1$$

and similarly for D_2, and

$$\left(\frac{D}{p}\right) = \left(\frac{D_1}{p}\right)\left(\frac{D_2}{p}\right) = 1 \ .$$

It follows from this last equation that p is split in $K = \mathbb{Q}(\sqrt{D})$, and if $\mathfrak{p}$ is an ideal above p then $\mathcal{N}(\mathfrak{p})$ is coprime to D, so that

$$\chi_{D_1}([\mathfrak{p}]) = \left(\frac{D_1}{\mathcal{N}(\mathfrak{p})}\right) = \left(\frac{D_1}{p}\right) = -1 \ ,$$

proving that the character χ_{D_1} is nontrivial.

(4). Choose as representative of $\mathcal{A}$ any integral ideal $\mathfrak{a}$ such that $\mathcal{N}(\mathfrak{a})$ is coprime to D, which is possible by (1), so that $\mathcal{N}(\mathfrak{a})$ is coprime to D_1 and D_2. Then by (2) we have

$$\chi_{D_1}(\mathcal{A})\chi_{D_2}(\mathcal{A}) = \left(\frac{D_1}{\mathcal{N}(\mathfrak{a})}\right)\left(\frac{D_2}{\mathcal{N}(\mathfrak{a})}\right) = \left(\frac{D}{\mathcal{N}(\mathfrak{a})}\right) = 1$$

by Lemma 10.5.16 (4).   □

Definition 10.5.18. *Let K be a number field and χ a character of the class group of K. For any ideal $\mathfrak{a}$ of K denote by $[\mathfrak{a}]$ its ideal class. We define the L-function $L_K(\chi, s)$ associated with χ by the formula*

$$L_K(\chi, s) = \sum_{\mathfrak{a} \subset \mathbb{Z}_K} \frac{\chi([\mathfrak{a}])}{\mathcal{N}(\mathfrak{a})^s} = \prod_{\mathfrak{p}} \frac{1}{1 - \chi([\mathfrak{p}])\,\mathcal{N}(\mathfrak{p})^{-s}} \ ,$$

where as usual $\mathfrak{a}$ runs through all integral ideals of $\mathbb{Z}_K$ and $\mathfrak{p}$ through all prime ideals of $\mathbb{Z}_K$.

It is clear that

$$L_K(\chi, s) = \sum_{\mathcal{A} \in Cl(K)} \chi(\mathcal{A})\zeta_K(\mathcal{A}, s) \ .$$

Proposition 10.5.19. *As above, let D_1 and D_2 be two coprime fundamental discriminants, $D = D_1 D_2$, $K = \mathbb{Q}(\sqrt{D})$, and let χ_{D_1} be the character of $Cl(K)$ defined in the above corollary. Then*

$$L_K(\chi_{D_1}, s) = L(\chi_{D_1}, s)L(\chi_{D_2}, s) \,,$$

where the L-functions on the right-hand side are the ordinary Dirichlet L-functions associated with the Dirichlet characters χ_{D_i}.

Proof. It is sufficient to show that the corresponding Euler factors are the same on both sides. Let p be a prime number. As usual we consider three cases. If p is inert in K there is a single prime ideal $\mathfrak{p} = p\mathbb{Z}_K$ above p that is a principal ideal, so that $\chi_{D_1}([\mathfrak{p}]) = 1$, and in fact

$$\chi_{D_1}([\mathfrak{p}]) = \left(\frac{D_1}{\mathcal{N}(\mathfrak{p})}\right) = \left(\frac{D_1}{p^2}\right) = 1 \,.$$

The Euler factor on the LHS is thus equal to $(1 - p^{-2s})^{-1}$. On the other hand, since p is inert we have $\left(\frac{D}{p}\right) = -1$, hence $\left(\frac{D_1}{p}\right) = -\left(\frac{D_2}{p}\right)$, so the Euler factor on the RHS is equal to $(1 - p^{-s})^{-1}(1 + p^{-s})^{-1} = (1 - p^{-2s})^{-1}$. If p is split in K we have two ideals $\mathfrak{p}$ and $\bar{\mathfrak{p}}$ above p, and

$$\chi_{D_1}([\mathfrak{p}]) = \chi_{D_1}([\bar{\mathfrak{p}}]) = \left(\frac{D_1}{\mathcal{N}(\mathfrak{p})}\right) = \left(\frac{D_1}{p}\right) \,,$$

so the Euler factor on the LHS is equal to $\left(1 - \left(\frac{D_1}{p}\right)p^{-s}\right)^{-2}$. On the other hand, since p is split we have $\left(\frac{D}{p}\right) = 1$ hence $\left(\frac{D_1}{p}\right) = \left(\frac{D_2}{p}\right)$, so the Euler factor on the RHS is equal to $\left(1 - \left(\frac{D_1}{p}\right)p^{-s}\right)^{-2}$. Finally, if p is ramified in K we have a single ideal $\mathfrak{p}$ above p, and $p \mid D$. Since $D = D_1 D_2$ with D_1 and D_2 coprime, p divides exactly one of the D_i. We consider both cases. If $p \mid D_2$ then $\mathcal{N}(\mathfrak{p}) = p$ is coprime to D_1; hence $\chi_{D_1}([\mathfrak{p}]) = \left(\frac{D_1}{p}\right)$, so the Euler factor on the LHS is equal to $\left(1 - \left(\frac{D_1}{p}\right)p^{-s}\right)^{-1}$, which is equal to the Euler factor on the RHS since $\left(\frac{D_2}{p}\right) = 0$. If $p \mid D_1$ then $p \nmid D_2$, and by Corollary 10.5.16 (4) we have $\chi_{D_1}([\mathfrak{p}]) = \chi_{D_2}([\mathfrak{p}]) = \left(\frac{D_2}{p}\right)$, and since $\left(\frac{D_1}{p}\right) = 0$ we conclude again that the Euler factors are equal. $\qquad\square$

We can now obtain the desired result on *real* quadratic fields.

Corollary 10.5.20. *Let D_1 and D_2 be two coprime fundamental discriminants, $D = D_1 D_2$, $K = \mathbb{Q}(\sqrt{D})$, and assume that $D_1 > 0$, $D_2 < 0$ hence $D < 0$. Then*

$$L(\chi_{D_1}, 1) = -\frac{w(D_2)}{h(D_2)D_1^{1/2}} \sum_{\mathcal{A} \in Cl(K)} \chi_{D_1}(\mathcal{A}) \log\left(\Im(\tau_{\mathcal{A}})^{1/2}|\eta(\tau_{\mathcal{A}})|^2\right) \,,$$

where $\tau_{\mathcal{A}}$ is the complex number corresponding to the ideal class $\mathcal{A}$ as above. Equivalently, if we denote by ε_{D_1} the fundamental unit greater than 1 of the real quadratic field $\mathbb{Q}(\sqrt{D_1})$ we have

$$\varepsilon_{D_1} = \left(\prod_{\mathcal{A} \in Cl(K)} \left(\Im(\tau_{\mathcal{A}})^{1/4} |\eta(\tau_{\mathcal{A}})| \right)^{\chi_{D_1}(\mathcal{A})} \right)^{-\omega(D_2)/(h(D_1)h(D_2))} .$$

Proof. By the proposition we have

$$L_K(\chi_{D_1}, s) = \sum_{\mathcal{A} \in Cl(K)} \chi_{D_1}(\mathcal{A}) \zeta_K(\mathcal{A}, s) = L(\chi_{D_1}, s) L(\chi_{D_2}, s) .$$

Note the trivial fact that $D_1 \geqslant 5$ and $|D_2| \geqslant 3$, so that $|D| \geqslant 15$ and $\omega(D) = 2$. By Kronecker's limit formula (here Corollary 10.5.9), around $s = 1$ we have

$$\zeta(\mathcal{A}, s) = \frac{\pi}{|D|^{1/2}} \left(\frac{1}{s-1} + C_K(\mathcal{A}) + O(s-1) \right) ,$$

where

$$C_K(\mathcal{A}) = 2\gamma - 2\log(2) - \frac{\log(|D|/4)}{2} - 2\log\left(\Im(\tau)^{1/2} |\eta(\tau)|^2 \right) .$$

Since χ_{D_1} is a nontrivial character on $Cl(K)$ we have $\sum_{\mathcal{A} \in Cl(K)} \chi_{D_1}(\mathcal{A}) = 0$, so $L_K(\chi_{D_1}, s)$ does not have a pole at $s = 1$ and we have

$$L_K(\chi_{D_1}, 1) = -\frac{2\pi}{|D|^{1/2}} \sum_{\mathcal{A} \in Cl(K)} \chi_{D_1}(\mathcal{A}) \log\left(\Im(\tau_{\mathcal{A}})^{1/2} |\eta(\tau_{\mathcal{A}})|^2 \right) .$$

On the other hand, by Proposition 10.5.10 (which is the simplest nontrivial case of Dirichlet's class number formula) we have

$$L(\chi_{D_2}, 1) = \frac{2\pi h(D_2)}{\omega(D_2)|D_2|^{1/2}} .$$

Thus we obtain the formula

$$L(\chi_{D_1}, 1) = -\frac{\omega(D_2)}{h(D_2)D_1^{1/2}} \sum_{\mathcal{A} \in Cl(K)} \chi_{D_1}(\mathcal{A}) \log\left(\Im(\tau_{\mathcal{A}})^{1/2} |\eta(\tau_{\mathcal{A}})|^2 \right) ,$$

proving the first formula of the corollary. The second immediately follows from Dirichlet's class number formula for real quadratic fields $L(\chi_{D_1}, 1) = 2h(D_1) \log(\varepsilon_{D_1})/D_1^{1/2}$.   □

Since ε_{D_1} is the fundamental solution of Pell's equation, this is called Kronecker's solution to Pell's equation, expressing an algebraic number as a combination of values of a transcendental function, which was part of Kronecker's Jugendtraum.

Example. Consider the case $D_1 = 8$, $D_2 = -3$, so that $D = -24$, $K = \mathbb{Q}(\sqrt{-6})$. There are two ideal classes in $\mathbb{Z}_K$, the corresponding $\tau_{\mathcal{A}}$ are $\sqrt{-6}$ and $\sqrt{-6}/2$, and the corresponding values of $\chi_{D_1}(\mathcal{A})$ are 1 (for the trivial class) and -1 (since otherwise χ_{D_1} would be a trivial character). Since $\varepsilon_8 = 1 + \sqrt{2}$ we obtain the formula

$$\frac{2\log(1+\sqrt{2})}{\sqrt{8}} = -\frac{6}{\sqrt{8}}\left(\log\left((\sqrt{6})^{1/2}|\eta(\sqrt{-6})|^2\right)\right.$$
$$\left. - \log\left((\sqrt{6}/2)^{1/2}\left|\eta(\sqrt{-6}/2)\right|^2\right)\right),$$

hence

$$\log(1+\sqrt{2}) = -3\left(\frac{\log(2)}{2} + \log\left(\frac{|\eta(\sqrt{-6})|^2}{|\eta(\sqrt{-6}/2)|^2}\right)\right).$$

Since by definition $\eta(\sqrt{-6})$ and $\eta(\sqrt{-6}/2)$ are positive real, this can also be written

$$\frac{\eta(\sqrt{-6})}{\eta(\sqrt{-6}/2)} = 2^{-1/4}(1+\sqrt{2})^{-1/6}.$$

Combining with the formula given above for $\eta(\sqrt{-6})\eta(\sqrt{-6}/2)$ coming from the Chowla–Selberg formula we obtain

$$\eta(\sqrt{-6}) = 2^{-3/2}3^{-1/4}(1+\sqrt{2})^{-1/12}\pi^{-3/4}\left(\Gamma(1/24)\Gamma(5/24)\Gamma(7/24)\Gamma(11/24)\right)^{1/4}$$
$$\eta(\sqrt{-6}/2) = 2^{-5/4}3^{-1/4}(1+\sqrt{2})^{1/12}\pi^{-3/4}\left(\Gamma(1/24)\Gamma(5/24)\Gamma(7/24)\Gamma(11/24)\right)^{1/4}.$$

These are special cases of Theorem 10.5.15. The reader is advised to work out for himself a few more examples (Exercise 58).

10.5.4 The Dedekind Zeta Function of Cyclotomic Fields

We now study the Dedekind zeta function of cyclotomic fields. We begin with the following.

Proposition 10.5.21. *Let m be an integer, and for all primes $p \nmid m$ denote by f_p the order of p modulo m (i.e., of the class of p in $(\mathbb{Z}/m\mathbb{Z})^*$), and set $g_p = \phi(m)/f_p$. Then*

$$\prod_{\chi \bmod m} L(\chi, s) = \prod_{p \nmid m}(1 - p^{-f_p s})^{-g_p},$$

where the product on the left is over all $\phi(m)$ Dirichlet characters modulo m.

Proof. Let G be the group of Dirichlet characters modulo m and H_p the group of f_pth roots of unity. If $\chi \in G$, then $\chi(p) \in H_p$. The map $\chi \mapsto \chi(p)$ is a group homomorphism from G to H_p. I claim that this homomorphism is surjective. Indeed, if $\zeta \in H_p$, define $\phi(\overline{p^k}) = \zeta^k$ for k modulo f_p, where $\overline{p^k}$ denotes the class of p^k in $(\mathbb{Z}/m\mathbb{Z})^*$. Since $\overline{p}$ has order f_p in this group, this is well defined and gives a character on the subgroup H of $(\mathbb{Z}/m\mathbb{Z})^*$ generated by $\overline{p}$. By Corollary 2.1.17, ϕ can be extended to a character of $(\mathbb{Z}/m\mathbb{Z})^*$, and the corresponding Dirichlet character χ will thus satisfy $\chi(p) = \zeta$, proving my claim.

It follows from this that the kernel and all the cosets of the map $\chi \mapsto \chi(p)$ have cardinality $g_p = \phi(m)/f_p$; in other words, for any $\zeta \in H_p$ there exist exactly g_p characters such that $\chi(p) = \zeta$. Thus

$$\prod_{\chi \bmod m} (1 - \chi(p)T) = \prod_{\zeta \in H_p} (1 - \zeta T)^{g_p} = (1 - T_p^f)^{g_p} .$$

Replacing T by p^{-s} and taking the product over p proves the result.   □

Theorem 10.5.22. *Let $\mathbb{Q}_m = \mathbb{Q}(\zeta_m)$ be the mth cyclotomic field. We have*

$$\zeta_{\mathbb{Q}_m}(s) = \prod_{\chi \bmod m} L(\chi_f, s) ,$$

where χ_f is the primitive character associated with χ. In particular, if m is a prime power we have $\zeta_{\mathbb{Q}_m}(s) = \prod_{\chi \bmod m} L(\chi, s)$. Furthermore, $|d(\mathbb{Q}_m)| = \prod_{\chi \bmod m} f(\chi)$, where $f(\chi)$ is the conductor of χ.

Proof. We know from Proposition 2.1.29 that if $m \equiv 2 \pmod 4$ the conductor of any character modulo m also divides $m/2$, and we also have $\mathbb{Q}(\zeta_m) = \mathbb{Q}(\zeta_{m/2})$, so we may assume that $m \not\equiv 2 \pmod 4$. Furthermore, the theorem is trivial for $m = 1$, so we may also assume that $m > 1$. In that case, $\mathbb{Q}_m$ is a totally complex field of degree $\phi(m)$; in other words, its signature is $(r_1, r_2) = (0, \phi(m)/2)$.

Recall that if $p \nmid m$, then the decomposition of $p\mathbb{Z}_{\mathbb{Q}_m}$ into prime ideals is $p\mathbb{Z}_{\mathbb{Q}_m} = \prod_{1 \leqslant i \leqslant g_p} \mathfrak{p}_i$, where $e(\mathfrak{p}_i/p) = f_p$, and f_p and g_p are as in the above proposition. Thus

$$\zeta_{\mathbb{Q}_m}(s) = \prod_{p|m} \prod_{\mathfrak{p}|p} (1 - \mathcal{N}(\mathfrak{p})^{-s})^{-1} \prod_{p \nmid m} (1 - p^{-f_p s})^{-g_p}$$

$$= a(s) \prod_{\chi \bmod m} L(\chi, s) = b(s) \prod_{\chi \bmod m} L(\chi_f, s) ,$$

where $a(s)$ and $b(s)$ are finite products and quotients of expressions of the form $1 - wp^{-fs}$, where $|w| = 1$ and $f \geqslant 1$ is an integer, and $p \mid m$. Now the point is that $\zeta_{\mathbb{Q}_m}(s)$ and all the $L(\chi_f, s)$ have functional equations when s goes to $1 - s$, so that $b(s)$ does also. More precisely, from above we know that

$$\Lambda_{\mathbb{Q}_m}(s) = |d(\mathbb{Q}_m)|^{s/2} \gamma(s)^{\phi(m)/2} \gamma(s+1)^{\phi(m)/2} \zeta_{\mathbb{Q}_m}(s)$$

is invariant by $s \mapsto 1 - s$, and that if we set $e(\chi) = 0$ if χ is even and $e(\chi) = 1$ if χ is odd, we know from Theorem 10.2.14 that if we let

$$\Lambda_1(\chi_f, s) = \Lambda(\chi_f, s) f^{-e(\chi)/2} = f^{s/2} \gamma(s + e(\chi)) L(\chi_f, s) \,,$$

then $\Lambda_1(\chi_f, 1 - s) = W(\chi_f) \Lambda(\overline{\chi_f}, s)$ for some complex number $W(\chi_f)$ of modulus 1. Furthermore, by orthogonality of characters, if $-1 \not\equiv 1 \pmod{m}$, i.e., if $m > 2$, which we have assumed, we have $\sum_{\chi \bmod m} \chi(-1) = 0$; in other words, there are exactly as many even as odd characters modulo m. Thus, if we set $g(s) = \Lambda_{\mathbb{Q}_m}(s) / \prod_{\chi \bmod m} \Lambda_1(\chi_f, s)$, we find that

$$g(s) = \frac{\zeta_{\mathbb{Q}_m}(s)}{\prod_{\chi \bmod m} L(\chi_f, s)} \left(\frac{|d(\mathbb{Q}_m)|}{\prod_{\chi \bmod m} f(\chi)} \right)^{s/2} = b(s) c(\mathbb{Q}_m)^{s/2} \,,$$

where $c(\mathbb{Q}_m) = |d(\mathbb{Q}_m)| / \prod_{\chi \bmod m} f(\chi)$. On the other hand, since $\chi \mapsto \overline{\chi}$ is an involution on the group of characters modulo m, the functional equations imply that $g(1 - s) = W g(s)$ for some $W \in \mathbb{C}$ with $|W| = 1$. Now the possible zeros and poles of $b(s)$, hence of $g(s)$, satisfy $p^{-fs} = 1/w = \exp(-it)$ for some real t, hence have the form $s = (t + 2k\pi) i / (f \log(p))$ for $k \in \mathbb{Z}$, in any case are purely imaginary (or 0). By the functional equation, these must also be zeros or poles of $g(1 - s)$, which is impossible since those satisfy $\Re(s) = 1$ instead. It follows that $b(s)$ cannot have zeros or poles, hence since it is equal to products or quotients of quantities of the form $1 - wp^{-fs}$, that $b(s) = 1$. This is nothing else than the first equality of the theorem. Thus $g(s) = c(\mathbb{Q}_m)^{s/2}$, but once more since $g(1 - s) = W g(s)$, this implies that $g(s) = c(\mathbb{Q}_m) = 1$, so that we obtain the second equality of the theorem. We also obtain that $W = 1$, but this is a trivial consequence of the fact that $\chi \mapsto \overline{\chi}$ is an involution of characters modulo m and the fact that $W(\chi) = 1$ for real characters as we have seen above because of Proposition 2.2.24. $\square$

Corollary 10.5.23. *If p is a prime then*

$$d(\mathbb{Q}_{p^k}) = \varepsilon p^{kp^k - (k+1)p^{k-1}} \,,$$

where $\varepsilon = -1$ if $p^k = 4$ or $p \equiv 3 \pmod{4}$ and $\varepsilon = 1$ otherwise.

Proof. The above result is trivially true for $p^k = 2$, so assume $p^k > 2$. By the above theorem and Proposition 2.1.29, whose notation we keep, we have

$$|d(K)| = \prod_{\chi \bmod p^k} f(\chi) = \prod_{f | p^k} f^{q(f)} = p^S \,,$$

where

$$S = \sum_{1 \leqslant j \leqslant k} j q(p^j) = -\frac{1}{p} + \left(1 - \frac{1}{p} \right)^2 \sum_{1 \leqslant j \leqslant k} j p^j = kp^k - (k+1)p^{k-1}$$

after a short computation, proving the corollary up to sign. Furthermore, we know that the sign of the discriminant of a number field is $(-1)^{r_2} = (-1)^{\phi(m)/2}$ for $m \geqslant 3$. Since $\phi(2^k)/2 = 2^{k-2}$ is odd if and only if $k = 2$ and for an odd prime p, $\phi(p^k)/2 = p^{k-1}(p-1)/2$ is odd if and only if $p \equiv 3$ (mod 4), the corollary follows.   □

An important theorem of Kronecker–Weber says that a number field K is an Abelian extension of $\mathbb{Q}$ if and only if it is a subfield of a cyclotomic field. Theorem 10.5.22 is in fact valid for such fields. More precisely, set the following definition:

Definition 10.5.24. *Let L be a subfield of $\mathbb{Q}(\zeta_m)$ and H the subgroup of $(\mathbb{Z}/m\mathbb{Z})^*$ corresponding to L by Galois theory and the canonical isomorphism with $\mathrm{Gal}(\mathbb{Q}(\zeta_m)/\mathbb{Q})$. The group of characters associated with L is the group of characters of $(\mathbb{Z}/m\mathbb{Z})^*$ (or equivalently, of Dirichlet characters modulo m) that are trivial on H, in other words the group of characters of $(\mathbb{Z}/m\mathbb{Z})^*/H$.*

We then have the following analogous theorem, which we give without proof, but is proved in the same way:

Theorem 10.5.25. *Let $L \subset \mathbb{Q}(\zeta_m)$ and let X be the group of characters associated with L. We have*

$$\zeta_L(s) = \prod_{\chi \in X} L(\chi_f, s)\,,$$

and $|d(L)| = \prod_{\chi \in X} f(\chi)$.

For instance, for a quadratic field this gives the easy decomposition $\zeta_{\mathbb{Q}(\sqrt{D})}(s) = \zeta(s)L(\chi, s)$, where $\chi(n) = \left(\frac{D}{n}\right)$.

Recall that we have defined h_{p^k} and $h_{p^k}^+$ to be the class numbers of $\mathbb{Q}(\zeta_{p^k})$ and $\mathbb{Q}(\zeta_{p^k})^+$ respectively, and that we have shown that $h_{p^k}^+ \mid h_{p^k}$ (see Section 3.5.4), so that $h_{p^k}^- = h_{p^k}/h_{p^k}^+ \in \mathbb{Z}$. Thanks to the above theorem and Dirichlet's class number formula, it is easy to give a reasonably efficient explicit formula for $h_{p^k}^-$.

Proposition 10.5.26. *Let $p \geqslant 3$ be a prime number, let $k \in \mathbb{Z}_{\geqslant 1}$, and set $N = \phi(p^k) = p^{k-1}(p-1)$. We have*

$$h_{p^k}^- = \frac{p^k}{2^{N/2-1}} \prod_{\chi \text{ odd}} L(\chi, 0) = (-1)^{N/2} \frac{p^k}{2^{N/2-1}} \prod_{\chi \text{ odd}} B_1(\chi)\,.$$

Proof. By Lemma 10.2.1, for all nontrivial characters χ we have $L(\chi_f, s) = L(\chi, s)$, so it follows from Theorem 10.5.22 that $\zeta_{\mathbb{Q}(\zeta_{p^k})}(s) = \zeta(s) \prod_{\chi \neq \chi_0} L(\chi, s)$. The characters associated with $\mathbb{Q}(\zeta_{p^k})^+$ correspond to those that are trivial on ι, in other words to the even characters, so Theorem 10.5.25 implies that $\zeta_{\mathbb{Q}(\zeta_{p^k})^+}(s) = \zeta(s)\prod_{\chi\neq\chi_0,\ \chi\ \text{even}} L(\chi, s)$. It follows that

$$\zeta_{\mathbb{Q}(\zeta_{p^k})}(s)/\zeta_{\mathbb{Q}(\zeta_{p^k})^+}(s) = \prod_{\chi\ \text{odd}} L(\chi, s)\ .$$

Now recall that Dirichlet's class number formula stated at $s = 0$ (which is much nicer than at $s = 1$) states that for any number field L we have

$$\zeta_L(s) \sim -\frac{h(L)R(L)}{w(L)} s^{r_1(L)+r_2(L)-1}\ ,$$

where $h(L)$, $R(L)$, and $w(L)$ denote the class number, regulator, and number of roots of unity of L, and $(r_1(L), r_2(L))$ is the signature of L. Setting as usual $K = \mathbb{Q}(\zeta_{p^k})$, by Theorem 3.5.20 we know that $U(K) = \langle\zeta_{p^k}\rangle U(K^+)$. The $N/2$ real embeddings of K^+ lift to $N/2$ pairs of complex embeddings of K. Since the elements of $U(K^+)$ are totally real and since the regulator matrix has order $N/2 - 1$, the regulator matrix of $U(K^+)$ considered in K will be equal to twice the regulator matrix of $U(K^+)$ considered in K^+, so that $R(K) = 2^{N/2-1}R(K^+)$. Finally, it is clear that $r_1(K) + r_2(K) = r_1(K^+) + r_2(K^+) = N/2$, $w(K) = 2p^k$, and $w(K^+) = 2$. Thus taking the limit as $s \to 0$ in the above quotient of Dedekind zeta functions we obtain

$$\frac{2^{N/2-1}h_{p^k}}{p^k h_{p^k}^+} = \prod_{\chi\ \text{odd}} L(\chi, 0)\ ,$$

proving the first formula, and the second follows from Corollary 10.2.3 and the fact that there are $N/2$ odd characters. $\square$

Corollary 10.5.27. *Let $p \geqslant 3$ be a prime number, let $g \in \mathbb{Z}$ be a primitive root modulo p^k, set $N = \phi(p^k) = p^{k-1}(p-1)$, and let $\zeta_N \in \mathbb{C}$ be a primitive Nth root of unity. Denote by P the polynomial*

$$P(X) = \sum_{0\leqslant j<N} (g^j \bmod p^k)X^j\ ,$$

where $(g^j \bmod p^k)$ denotes the unique integer congruent to g^j modulo p^k in the interval $[1, p^k[$. Then

$$h_{p^k}^- = -\frac{1}{(-2p^k)^{N/2-1}} \prod_{1\leqslant m\leqslant N/2} P(\zeta_N^{2m-1})$$

$$= -\frac{1}{(-2p^k)^{N/2-1}} \operatorname{Res}(P(X), X^{N/2} + 1)\ ,$$

where Res *denotes the resultant of the two polynomials.*

Proof. Indeed, note that it is easy to describe all characters modulo p^k: such a character is uniquely determined by its value on g; hence $\chi_1(g^j) = \zeta_N^j$ defines a character modulo p that generates the group of characters modulo p^k, so that any character is thus equal to some $\chi_n = \chi_1^n$ for a unique n such that $0 \leqslant n < N$. The character χ_n is odd if and only if

$$-1 = \chi_n(-1) = \chi_n(g^{N/2}) = \zeta_N^{nN/2} ,$$

hence if and only if n is odd. Finally, recall that since χ_n is nontrivial for $n \neq 0$ we have

$$B_1(\chi_n) = \frac{1}{p^k} \sum_{\substack{1 \leqslant r \leqslant p^k - 1 \\ p \nmid r}} r\chi_n(r) = \sum_{0 \leqslant j < N} (g^j \bmod p^k)\zeta_N^{nj} ,$$

proving the first formula, and the second is the definition of the resultant. □

Corollary 10.5.28. *With the same notation we have*

$$h_{p^k}^- = -\frac{1}{(-2p^k)^{N/2-1}} \prod_{d \mid N, \ d \nmid N/2} \mathrm{Res}(P(X), \Phi_d(X)) ,$$

where $\Phi_d(X)$ denotes the dth cyclotomic polynomial.

Proof. This immediately follows from the equality

$$X^{N/2} + 1 = \frac{X^N - 1}{X^{N/2} - 1} = \prod_{d \mid N, \ d \nmid N/2} \Phi_d(X)$$

together with the multiplicativity of the resultant. □

10.5.5 The Nonvanishing of $L(\chi, 1)$

One of the main easy results on $L(\chi, s)$, which will immediately imply a weak form of Dirichlet's theorem on primes in arithmetic progression, is that $L(\chi, 1) \neq 0$ for all nontrivial characters χ (otherwise $L(\chi, s)$ has a pole at $s = 1$). Considering the importance of this result we give three proofs, of which only two are really different.

Theorem 10.5.29. *For any nontrivial Dirichlet character χ we have $L(\chi, 1) \neq 0$.*

First Proof. By what we have seen in the preceding section, we have

$$\zeta_{\mathbb{Q}_m}(s) = \zeta(s) \prod_{\substack{\chi \bmod m \\ \chi \neq \chi_0}} L(\chi_f, s) ,$$

and since both $\zeta(s)$ and $\zeta_{\mathbb{Q}_m}(s)$ have simple poles at $s = 1$ and the $L(\chi_f, s)$ do not have poles for $\chi \neq \chi_0$, it follows that $L(\chi_f, 1) \neq 0$, so that $L(\chi, 1) \neq 0$ since both functions differ only by a finite nonvanishing Euler product.

The problem with this proof is that it assumes that we have proved that $\zeta_{\mathbb{Q}_m}(s)$ has a pole at $s = 1$, which is not difficult but still not completely trivial. It seems therefore appropriate to give another proof, which is in essence identical, but avoids assuming any results on $\zeta_{\mathbb{Q}_m}(s)$.

Second proof. Set $F_m(s) = \prod_{\chi \bmod m} L(\chi, s)$ (which is of course equal to $\zeta_{\mathbb{Q}_m}(s)$ up to a finite number of Euler factors, but the whole point is that we forget this). By Proposition 10.5.21 we have

$$F_m(s) = \prod_{p \nmid m}(1 - p^{-f_p s})^{-g_p} \,,$$

where $f_p \mid \phi(m)$ and $g_p = \phi(m)/f_p$. Since

$$(1 - p^{-f_p s})^{-g_p} = \sum_{k \geqslant 0}\binom{g_p + k - 1}{k}\frac{1}{p^{kf_p s}} \,,$$

it follows that $F_m(s)$ is a Dirichlet series with nonnegative coefficients. Furthermore, since the $L(\chi, s)$ can all be analytically continued to $\mathbb{C}$ (with a simple pole at $s = 1$ for $\chi = \chi_0$), so can the function $F_m(s)$.

Assume by contradiction that there exists a character $\chi \neq \chi_0$ such that $L(\chi, 1) = 0$. Since $L(\chi_0, s)$ has a *simple* pole at $s = 1$ it follows that $L(\chi, s)L(\chi_0, s)$, hence also $F_m(s)$, is holomorphic in the whole of $\mathbb{C}$. By Corollary 10.1.17 this implies in particular that the series $F_m(s)$ converges for all real $s > 0$, hence also the corresponding Euler product, which is a simple rearrangement of a series with positive terms (this would perhaps not be true if $s \notin \mathbb{R}$ or if $s \leqslant 0$). However, it is easy to see that this leads to a contradiction. Indeed, since $f_p \mid \phi(m)$ and $g_p \geqslant 1$, for $s > 1/\phi(m)$ we have

$$F_m(s) = \prod_{p \nmid m}(1 - p^{-f_p s})^{-g_p} \geqslant \prod_{p \nmid m}(1 - p^{-\phi(m)s})^{-1} = \zeta(\phi(m)s)\prod_{p \mid m}(1 - p^{-\phi(m)s}) \,,$$

and this is unbounded when s tends to $1/\phi(m)$ from above, contradicting the convergence of $F_m(s)$ for $s > 0$.

It is clear that this proof is a rephrasing of the preceding one that avoids any assumption about $\zeta_{\mathbb{Q}_m}$, and in fact that *proves* that $\zeta_{\mathbb{Q}_m}(s)$ has a simple pole at $s = 1$.

Third proof. This proof is a little different. We start again from the ubiquitous function $F_m(s)$ above (although it now seems natural to us, it was Dirichlet's important intuition to understand that it is simpler to treat L-functions modulo m all at once than individually). From its Euler product, or the fact that it is a Dirichlet series with nonnegative coefficients, the first one being equal to 1, it follows that for $s > 1$ real we have $F_m(s) \geqslant 1$.

Assume first that there exists a *nonreal* character χ such that $L(\chi, 1) = 0$. Then $\overline{\chi} \neq \chi$ and $\overline{\chi}$ is also a character modulo m. It follows that $L(\chi, s)L(\overline{\chi}, s)$ has at least a *double* zero at $s = 1$. Since $L(\chi_0, s)$ has only a simple pole, this implies that $F_m(s)$ would tend to 0 as s tends to 1 from above, contradicting $F_m(s) \geqslant 1$. This part of the proof shows that nonreal characters are *easy* to handle, and in fact it is immediate to deduce from this proof an explicit lower bound for $|L(\chi, 1)|$; see Exercise 30. All the difficulty comes from the real characters.

Thus it remains to show that $L(\chi, 1) \neq 0$ when χ is a real character modulo m. We could "cheat," and appeal to Dirichlet's results on such characters: we may of course assume that χ is primitive, since $L(\chi, 1)$ is equal to a finite nonvanishing Euler product times $L(\chi_f, 1)$, where χ_f is the primitive character equivalent to χ. By Theorem 2.2.15 such characters have the form $\left(\frac{D}{n}\right)$ for $D = \chi(-1)m$ a fundamental discriminant. Furthermore, Dirichlet's theorem gives explicitly the value of $L\left(\left(\frac{D}{n}\right), 1\right)$ in terms of a class number and a regulator, and implies immediately that it is nonzero.

But this cheat proof is not in the spirit of the proofs that we want to give, since Dirichlet's theorem, while not very difficult, is not trivial. A more proper proof is as follows. Let χ be a real character. Consider the function

$$g(s) = \frac{\zeta(s)L(\chi, s)}{\zeta(2s)}.$$

Although we will not need it, note that $\zeta(s)L(\chi, s)$ is equal to the Dedekind zeta function of the quadratic field $\mathbb{Q}(\sqrt{D})$.

It is immediately checked that

$$g(s) = \prod_p \left(1 + \sum_{k \geqslant 1} \frac{\chi(p)^k + \chi(p)^{k-1}}{p^{ks}}\right),$$

and since $\chi(n) = 0$ or ± 1 we have $\chi(p)^k + \chi(p)^{k-1} \geqslant 0$, so that $g(s)$ is a Dirichlet series with nonnegative coefficients and first coefficient equal to 1. As usual we apply Corollary 10.1.17. If $L(\chi, 1) = 0$ then $g(s)$ is holomorphic in the half-plane $\Re(s) > 1/2$, hence converges in that half-plane, so that $g(s) \geqslant 1$ for $s > 1/2$. On the other hand, when s tends to $1/2$ from above, $\zeta(s)L(\chi, s)$ stays bounded and $\zeta(2s)$ tends to ∞, so $g(s)$ tends to 0, a contradiction. $\square$

10.5.6 Application to Primes in Arithmetic Progression

Theorem 10.5.30 (Dirichlet). *Let a and m be coprime integers. There exist infinitely many primes p that are congruent to a modulo m. More precisely, the set of such primes has an analytic density $1/\phi(m)$, where the analytic density $d(P)$ of a set P of primes is defined, when it exists, by*

$$d(P) = \lim_{s \to 1+} \frac{\sum_{p \in P} p^{-s}}{\sum_p p^{-s}}.$$

Proof. First note that for $s > 1$ real,

$$\log(L(\chi, s)) = -\sum_p \log(1 - \chi(p)p^{-s}) = \sum_p \frac{\chi(p)}{p^s} - S(s) \, ,$$

where $S(s) = \sum_p (\log(1-\chi(p)p^{-s}) + \chi(p)p^{-s})$ converges absolutely for $\Re(s) > 1/2$. It follows that the analytic behavior at $s = 1$ of $\sum_p \chi(p)p^{-s}$ is the same as that of $\log(L(\chi, s))$. Thus, since we know by Theorem 10.5.29 that $L(\chi, s)$ can be analytically continued to $\mathbb{C}$ and that $L(\chi, 1) \neq 0$ (and of course $L(\chi, s) \neq 0$ for $\Re(s) > 1$ by the Euler product), it follows that

$$\sum_p \frac{\chi(p)}{p^s} = \begin{cases} O(1) & \text{if } \chi \neq \chi_0 \, , \\ -\log(s - 1) + O(1) & \text{if } \chi = \chi_0 \, . \end{cases}$$

Now by orthogonality of characters, we know that

$$\sum_{\chi \bmod m} \chi(a)^{-1}\chi(n) = \begin{cases} 0 & \text{if } n \not\equiv a \pmod{m} \, , \\ \phi(m) & \text{if } n \equiv a \pmod{m} \, . \end{cases}$$

Therefore

$$\sum_{p \equiv a \pmod{m}} \frac{1}{p^s} = \frac{1}{\phi(m)} \sum_{\chi \bmod m} \chi(a)^{-1} \sum_p \frac{\chi(p)}{p^s} = -\frac{\log(s - 1)}{\phi(m)} + O(1)$$

by what we have seen above, and the result follows by taking quotients. $\square$

Remark. It has been a long-standing conjecture that there exist arbitrarily long arithmetic progressions of prime numbers. In other words, for any N there should exist coprime integers a and b such that $ak + b$ is prime for *each* k such that $0 \leqslant k < N$ (note that this does not at all follow from Dirichlet's theorem). This was proved in 2004 by Green and Tao [Gre-Tao].

10.5.7 Conjectures on Dirichlet L-Functions

We have already seen, and will see again below, that there are many other types of L-functions than Dirichlet L-functions, and for those L-functions even basic questions such as analytic continuation are still conjectural. I would like to point out that even for Dirichlet L-functions some outstanding conjectures remain. Evidently the most famous one is the extended Riemann hypothesis (ERH), a generalization of the Riemann hypothesis for $\zeta(s)$: it states that the nontrivial zeros of $L(\chi, s)$ in the sense of Definition 10.2.16 are such that $\Re(s) = 1/2$. Since we know from Exercise 67 that $L(\chi, s) \neq 0$ for $\Re(s) = 1$, it follows that the nontrivial zeros are exactly those s such that $0 < \Re(s) < 1$.

The ERH can in fact be split into two parts, the first part stating that all *nonreal* zeros s such that $0 < \Re(s) < 1$ are such that $\Re(s) = 1/2$, and the second part stating that $L(\chi, s)$ does not vanish for $0 < s < 1$, except perhaps at $s = 1/2$. In fact a slightly stronger conjecture asserts that $L(\chi, s)$ does not vanish for $0 < s < 1$, including at $s = 1/2$. Note that this stronger conjecture is perhaps more rash since there do exist Artin L-series that vanish at $s = 1/2$.

On the first part, which exactly generalizes the RH for $\zeta(s)$, although a huge amount of work has been done on this conjecture, which is one of the most important of all mathematics, nothing much can be said (there are many reasons to believe that the conjecture is true, although some people believe the contrary). The second part, stating that $L(\chi, s) \neq 0$ for $0 < s < 1$, is of a different kind, although probably just as difficult. Such a real value of s (which probably does not exist) is called a *Siegel zero*.

First of all, for a given χ it is not difficult to check: indeed, it has been checked for *odd real* characters up to conductor $3 \cdot 10^8$ (see [Watk]), for *even real* characters up to conductor 10^6, and for all characters up to conductor 10^3 at least; see Exercise 31 for a simple approach.

Second, it has been shown by Conrey and Soundararajan in [Con-Sou] that a positive proportion of real characters χ are such that $L(\chi, s) \neq 0$ for $0 < s < 1$.

10.6 Science Fiction on L-Functions

I thank D. Zagier for considerable help in writing this section, but of course I am solely responsible for remaining errors or inaccuracies.

In my opinion, conjectures about special points and special values of L-functions are the most beautiful in all of mathematics. In this section, I would briefly like to describe the landscape, in very imprecise terms. Thus the reader is warned that nothing is defined, and even that what is defined is imprecise and/or misleading. The theory is however too beautiful to be overlooked, even in a graduate-level book such as this one.

First of all, we have to give some idea of what an L-function is. It is important to understand that there are (at least) two levels of L-functions, each with their own difficulties, although the second level is almost totally out of reach at present (but it is the most fascinating and important one).

10.6.1 Local L-Functions

To simplify, let us say that the first-level (or local) L-functions correspond to *finite fields*, or to a $\mathfrak{p}$-adic field whose residue field is the finite field in question. The typical example of such an L-function is the Hasse–Weil zeta function that we have described in Theorem 2.5.26. A more general example is the local zeta function of an arbitrary algebraic variety, which is defined in the same way as for curves. A special case is the *Euler factor* of the Dedekind zeta function of a number field, where p^{-s} is replaced by T. Indeed, let us keep the notation of Theorem 2.5.26, where ζ_C is defined in the same way for any algebraic variety.

Lemma 10.6.1. *Let $P(X) \in \mathbb{Z}[X]$ be an irreducible monic polynomial, and $K = \mathbb{Q}(\theta)$, where θ is a root of $P(X)$. If p is a prime number such that $p^2 \nmid \operatorname{disc}(P)$, then the Hasse–Weil zeta function of the 0-dimensional variety defined by $P(X) = 0$ over $\mathbb{F}_p$ is the Euler factor at p of $\zeta_K(s)$, where p^{-s} is replaced by T.*

Proof. If $p^2 \nmid \operatorname{disc}(P)$, then a fortiori p does not divide the index $[\mathbb{Z}_K : \mathbb{Z}[\theta]]$, so the factorization of $p\mathbb{Z}_K$ into prime ideals mimics that of the polynomial $P(X)$ modulo p, i.e., if $\overline{P}(X) = \prod_{1 \leq j \leq g} \overline{P_j}(X)^{e_j}$ with $\deg(P_j) = f_j$ and the P_j monic, then $p\mathbb{Z}_K = \prod_{1 \leq j \leq g} \mathfrak{p}_j^{e_j}$ with $f(\mathfrak{p}_j/p) = f_j$. It follows that the local factor at p of $\zeta_K(s)$ is equal to $\prod_{1 \leq j \leq g}(1 - p^{-f_j s})^{-1}$, and replacing p^{-s} by T gives $\prod_{1 \leq j \leq g}(1 - T^{f_j})^{-1}$.

On the other hand, we must compute the number $N(p^n)$ of solutions of $P(X) = 0$ in $\mathbb{F}_{p^n}$. Since $\overline{P}(X) = \prod_{1 \leq j \leq g} \overline{P_j}(X)^{e_j}$ and the $\overline{P_j}(X)$ are pairwise coprime, we have $N(p^n) = \sum_{1 \leq j \leq g} N_j(p^n)$, where $N_j(p^n)$ is the number of roots of $\overline{P_j}(X)$ in $\mathbb{F}_{p^n}$. Since $\overline{P_j}(X)$ is irreducible, the theory of finite fields tells us that $N_j(p^n) = 0$ if $f_j \nmid n$, and $N_j(p^n) = f_j$ if $f_j \mid n$. Therefore

$$N(p^n) = \sum_{\substack{1 \leq j \leq g \\ f_j \mid n}} f_j \; .$$

It follows that

$$\sum_{n \geq 1} \frac{N(p^n)}{n} T^n = \sum_{n \geq 1} \sum_{\substack{1 \leq j \leq g \\ f_j \mid n}} \frac{f_j T^n}{n} = \sum_{1 \leq j \leq g} \sum_{k \geq 1} \frac{T^{k f_j}}{k} = - \sum_{1 \leq j \leq g} \log(1 - T^{f_j}) \, ,$$

so that $\zeta_C(T) = \prod_{1 \leq j \leq g}(1 - T^{f_j})^{-1}$, proving the lemma. $\qquad\square$

Thus, to summarize, first-level zeta functions correspond to the *local* situation, where only one prime is involved (the characteristic of the finite field or of the residue field).

In this case, the definitive result is the extraordinary work of Deligne that proves the generalization of Theorem 2.5.26 to arbitrary nonsingular projective varieties. Thus, the first result, initially proved by Dwork, is that the local zeta function is always a *rational function* of T. The second result concerns the degree of its numerator and denominator, and the *local* functional equation that it satisfies when T is changed into $1/(q^d T)$ (note that if $T = q^{-s}$ then $1/(q^d T) = q^{-(d-s)}$, so the local functional equation relates s

and $d-s$). But by far the most difficult result to prove is the third, saying that the complex modulus of the reciprocal roots of the zeta function is exactly equal to $q^{k_i/2}$ for specific integers k_i. As we have seen with the Weil bounds, this is also the essential ingredient that we need to estimate the number of solutions of Diophantine equations over finite fields.

Deligne proved this by showing that, in accordance with the predictions made by Weil, these reciprocal roots have an interpretation as the eigenvalues of the "Frobenius map" $x \mapsto x^q$ acting on an appropriate cohomology group of the variety over $\mathbb{F}_q$.

The refinements of Deligne's result have mainly dealt with finding corresponding estimates for singular varieties (which are unfortunately unavoidable in actual practice), and also applications to other problems such as the proof of the Ramanujan conjecture, also by Deligne: he shows that it can be related to local L-functions of certain varieties.

In any case, even though many problems remain to be solved, the local theory for a given variety is well understood.

10.6.2 Global L-Functions

In rough terms, a global L-function is obtained by taking the product of local L-functions corresponding to all prime numbers p, with the variable T replaced by p^{-s} in the factor corresponding to p. The prototypical example is the Riemann zeta function $\zeta(s) = \prod_p (1 - p^{-s})^{-1}$ and more generally the Dedekind zeta function of a number field by the lemma that we have proved above.

Thus, we can also define the global L-function of a variety. However, many other L-functions of global type exist, not always "visibly" coming from a variety: for instance L-functions associated with modular forms as already mentioned above, Artin L-functions, etc. In the past decade, all this has been included in a vast theory of objects called "mixed motives" on which cohomology theories are defined, hence corresponding local and global L-functions.

In any case, just as there were three important conjectures concerning local L-functions (now all proved), there are now *four* conjectures concerning global L-functions, but except for some of these conjectures for specific classes of global L-functions, essentially *nothing* has been proved, as we shall see.

The first conjecture is the existence of *analytic continuation* to the whole complex plane of the L-series, with a possible finite number of poles at specific points. Indeed, all the L-series occurring in nature, say all motivic L-series, whatever this means, converge absolutely for $\Re(s)$ sufficiently large. The conjecture is that they can be meromorphically continued to the whole complex plane with a finite number of poles.

The second conjecture is that this continued L-function should satisfy a *functional equation* of a similar type to the ones that we have already seen:

after multiplication by a finite number of exponential functions and gamma factors of the form $\Gamma(as + b)$ with positive *rational* a (in fact almost always integral or half-integral) and complex b, one obtains a Λ-*function* that should satisfy an equation of the type $\Lambda(X, d - s) = W\Lambda(X^*, s)$, where X^* is "dual" to X in a certain sense, d is an integer, and W is a complex number of modulus 1 (see Theorem 10.2.14 for a typical example).

We have already proved these two simplest conjectures for the Riemann zeta function and for Dirichlet L-series, as easy consequences of the Poisson summation formula. But already for the Dedekind zeta function the proof is not at all easy, and in fact we have not given it in this book. As we have mentioned, there are two, related, proofs. The original one, due to Hecke, uses generalized theta functions and Poisson summation formulas, and is quite painful, although completely explicit. A more recent one due to Tate involves slightly simpler computations, and has the great advantage of explaining each of the local factors including the gamma factors individually as part of the global function, hence is considerably more elegant, although not that much shorter. In any case, the proof is not easy.

For Artin L-functions, the situation becomes already much more conjectural. A beautiful (and not too difficult) result of R. Brauer says that these L-functions can be analytically continued to the whole complex plane to *meromorphic* functions (with possibly an infinite number of poles) with the expected functional equation. On the other hand, apart from very special classes (including of course the Dirichlet L-functions, which are special cases), the fact that they can be *holomorphically* continued (with a finite number of known poles) is completely conjectural. A large body of theory called the *Langlands program* is intimately related to the Artin conjecture. In the past 30 years, only two (almost three) highly nontrivial cases of the Artin conjecture have been proved, using the complex machinery of the Langlands program: recall that a finite subgroup of $\mathrm{PSL}_2(\mathbb{C})$ is either cyclic, dihedral, or isomorphic to A_4, S_4, or A_5 (corresponding to the platonic solids); if the projective image of an odd irreducible 2-dimensional representation is cyclic or dihedral the Artin conjecture is easy to prove using the theory of induced characters. On the other hand, the A_4 and S_4 cases are considerably more difficult, and were solved only in the 1980s by Langlands and Tunnell. In addition, considerable progress was made in 2000 on the A_5 case by Buzzard et al., explaining the "almost three" above. This incredibly small number of solved cases shows the difficulty of the problem.

For *global* zeta functions of varieties of strictly positive dimension, the very simplest case is that of *elliptic curves* over $\mathbb{Q}$ (i.e., genus 1). The analytic continuation and functional equation of their L-series is one of the major achievements of the second half of the twentieth century: it is the remarkable work of Wiles, completed by Taylor–Wiles and proved in complete generality by Breuil, Conrad, Diamond, and Taylor in [BCDT].

Considering the difficulty of the above problem in the simplest case of elliptic curves, it goes without saying that for more general L-functions of varieties the problem is completely open.

On the other hand, some L-functions can naturally be extended analytically to the complex plane with a functional equation. This is for example the case of L-functions attached to *modular forms*. In that case the two above problems are easy. In fact, it is by showing that the L-function of an elliptic curve over $\mathbb{Q}$ is equal to the L-function of a modular form that Wiles et al. prove their result.

We come now to the third conjecture about global L-functions: the (global) Riemann hypothesis. It says the following. Define a Λ-function to be an L-function of one of the above types multiplied by its exponential and gamma factors so that it satisfies a functional equation when s goes to $1 - s$. The Riemann hypothesis states that the only zeros of a Λ-function are on the line $\Re(s) = 1/2$ (Λ-functions do not have any trivial zeros since they are canceled by the gamma factors; note also that the normalization of the functional equation to be $s \mapsto 1 - s$ is not natural. For example, the Λ-function of an elliptic curve satisfies a natural functional equation when $s \mapsto 2 - s$, and it is only by setting $\Lambda_1(s) = \Lambda(s + 1/2)$ that one recovers a functional equation when $s \mapsto 1 - s$.) Another way to say this is the following: if $\Lambda(\rho, d - s) = W\Lambda(\overline{\rho}, s)$, the interval $[0, d]$ is called the *critical strip*. Then the conjecture is that the zeros of Λ should be exactly in the middle of the critical strip, i.e., such that $\Re(s) = d/2$.

This is perhaps the most famous conjecture in mathematics, and you can earn 1 million US dollars by proving it. In my personal opinion, however, it is not as nice as the next one (which, by the way, can also earn you 1 million dollars, since these are two of the seven Clay prize problems). In any case, this global Riemann hypothesis is not known for even the simplest L-function, i.e., the Riemann zeta function. Note, however, that some other L-functions having an Euler product and a functional equation, the *Selberg zeta functions*, are known to satisfy the corresponding Riemann hypothesis, but unfortunately they do not shed any light on our Riemann hypothesis since there seems to be no relation between the two types of zeta function. One of the reasons is that Selberg zeta functions are complex functions of *order* 2, while usual L- and ζ-functions have order 1 (see definitions and examples preceding Lemma 10.7.5).

The fourth conjecture concerns *special values* of L-functions. Following Deligne we define a *special point* of an L- or Λ-function as follows. If X is some algebro-geometric object, for instance a representation, an algebraic variety, or more generally a motive, we assume that we have a natural global functional equation $\Lambda(X, d - s) = W\Lambda(X^*, s)$, where $\Lambda(X, s) = g(s)L(X, s)$, $|W| = 1$, X^* is another object "dual" to X, and $g(s)$ is equal to a product of exponential and gamma factors. A *special point* is then an integer $k \in \mathbb{Z}$ such that $g(k)$ and $g(d - k)$ (which are the gamma factors of $\Lambda(X, k)$ and

$\Lambda(X^*, d - k)$ respectively) do not have poles. In other words, they are the integers such that both $L(X, k)$ and $L(X^*, d - k)$ can be computed from the corresponding value of Λ by division by the gamma factor. For instance, for the Riemann zeta function the special points are the strictly positive even integers $2, 4, 6, \ldots$, and the negative odd integers $-1, -3, -5, \ldots$. The strictly negative *even* integers are *not* special points.

For any such X and each special point k Deligne associates a number $\omega_{X,k} \neq 0$, which is a *period*, in other words the integral of some algebraic differential form over some cycle, and he conjectures that $\Lambda(X, k)/\omega_{X,k}$ is an algebraic number. For instance, in the case of the Riemann zeta function we can choose $\omega_{X,k} = \pi^k$ if k is positive even, and $\omega_{X,k} = 1$ if k is negative and odd.

Deligne's conjecture has been proved in many cases, and has been verified numerically in many others, for instance for higher symmetric powers of L-functions associated with modular forms.

In some cases it is also possible to state precise conjectures giving explicitly the algebraic number occurring in Deligne's conjecture. This is for instance given by the Lichtenbaum conjectures; see more on this below.

An important generalization of the above is to compute the order of vanishing and the leading term of a "motivic" L-function at an integer $s \in \mathbb{Z}$, this time not necessarily special. A very general "conjecture" is as follows: if an L-function vanishes to order r, say, at some $s \in \mathbb{Z}$, there should exist a natural finitely generated abelian group of rank r closely related to the L-function, and the leading term in the expansion around s should also have some explanation in terms of this group and others. To understand this general philosophy, the best approach is to give three examples.

The first and most classical one is the behavior of the Dedekind zeta function $\zeta_F(s)$ of a number field F at $s = 0$. We have computed the order of the zeros at integers in Corollary 10.5.2, and we have seen that at $s = 0$ there is a zero of order $r_1 + r_2 - 1$, which is exactly the rank of the unit group of F, so that is the group in question. We have also given the leading term in Theorem 10.5.1.

A second example, which generalizes this, deals with the behavior of $\zeta_F(s)$ at negative integers: one can define so-called higher K-groups $K_n(F)$ for $n \geqslant 0$ such that in particular, $K_0(F) \simeq \mathbb{Z} \oplus Cl(F)$ and $K_1(F) \simeq U(F)$. A theorem of Borel tells us that for $k \geqslant 1$, the group $K_{2k}(F)$ is a finite abelian group, that $K_{2k-1}(F)$ is a finitely generated abelian group, and that the rank r of $K_{2k-1}(F)$ is exactly equal to the order of vanishing of $\zeta_F(s)$ at $s = 1 - k$, in other words $r_1 + r_2$ when $k \geqslant 3$ is odd, and r_2 when k is even. In addition, $K_{2k-1}(F) \otimes \mathbb{Q}$ is naturally a lattice in $\mathbb{R}^r$, and its covolume is equal to $\zeta_F(k)$. Finally, the Lichtenbaum conjectures state that if F is totally real, for $k \in \mathbb{Z}_{\geqslant 1}$ the rational number $|\zeta_F(1 - 2k)|$ is equal to $|K_{2k}(F)|/|K_{2k+1,\text{tors}}(F)|$, where we denote by $K_{2k+1,\text{tors}}(F)$ the torsion subgroup of $K_{2k+1}(F)$.

A third example is that of the global L-function attached to an elliptic curve over $\mathbb{Q}$. In that case we have $\Lambda(s) = (2\pi)^{-s}\Gamma(s)L(s)$, and the functional equation is $\Lambda(2 - s) = \pm\Lambda(s)$ for some sign $\pm$. At $s = 1$, which is the center of the critical strip, there may be a zero (there is always one if the sign of the functional equation is $-$), of order r, say. The beautiful Birch–Swinnerton-Dyer conjecture (see Conjecture 8.1.7), also a 1 million dollar problem, says that r should be equal to the rank of the group of rational points of the curve. As for the preceding examples, the BSD conjecture also gives a conjectural value for the leading term. After remarkable work by Coates–Wiles, Gross–Zagier, Kolyvagin, Rubin, and others, it can reasonably be said that we understand quite well what happens for $r = 0$ and $r = 1$ (see Theorem 8.1.8 and Corollary 8.1.9). On the other hand, for $r \geqslant 2$ not a single example has been proved, although the numerical evidence (which is very easy to compute; see Section 8.5) is absolutely overwhelming.

Work of Beilinson and Scholl made the above vague conjecture completely precise by defining an actual group and regulator map of which one can conjecture (and in some cases numerically check) that the rank is finite and equal to the order of vanishing of the L-function, and that the covolume corresponds to the leading term.

The conjectures mentioned above concerning special values, orders of vanishing, and leading terms, hence including the conjectures of Beilinson, Bloch–Kato, Birch–Swinnerton-Dyer, Stark, Zagier, and others, form in my opinion *the* most beautiful (and important) set of conjectures in the whole of mathematics.

For complete details on the material of this section, I refer to [Hul] and [Rap-Sch-Sch].

10.7 The Prime Number Theorem

The prime number theorem (PNT for short) states that the number $\pi(x)$ of prime numbers less than or equal to x is asymptotic to $x/\log x$. This was observed experimentally by Gauss and Legendre in the eighteenth and early nineteenth centuries, and a program to prove the result was put forward in a famous paper by Riemann on the zeta function in 1859. However, it was not before 1896 that the result was finally proved independently by Hadamard and de la Vallée Poussin, based on Riemann's remarkable insights, using similar methods of complex analysis. Since then many other proofs have been found, including a so-called "elementary" proof by P. Erdős and A. Selberg, i.e., one not using complex analysis, but it is much less natural than the complex-analytic ones. Furthermore, the PNT can be stated with an error term that can be reasonably estimated only with complex-analytic techniques. In this section we will present two proofs. The first one is essentially due to D. Newman [New], as rewritten by D. Zagier. It uses an original

"Tauberian theorem," and as with most proofs of this type, it cannot give an error term. The second proof is due to H. Iwaniec [Iwa-Kow], and gives a weak but nontrivial error term, and I thank E. Kowalski for showing it to me.

10.7.1 Estimates for $\zeta(s)$

All results on the PNT come from the knowledge of zero-free regions of $\zeta(s)$. In this subsection we give some elementary but useful estimates for $\zeta(s)$ in some regions of the complex plane, which are much more than we need for the versions of the PNT that we will prove. As always for $s \in \mathbb{C}$ we use the notation $s = \sigma + it$, where σ is the real part and t the imaginary part. Since all the meromorphic functions $f(s)$ that we use satisfy $f(\overline{s}) = \overline{f(s)}$ we may always assume $t \geqslant 0$, and to avoid trivial problems we will in fact implicitly always assume that t is sufficiently large.

Proposition 10.7.1. *For any fixed $k \geqslant 0$ we have $\zeta^{(k)}(s) = O(\log(t)^{k+1})$ uniformly in the region $1 - C/\log(t) \leqslant \sigma \leqslant 2$, and in particular for $s = 1 + it$ (recall that we also assume $t \geqslant t_0 > 0$).*

Proof. By the Euler–MacLaurin formula for $n = 1$ we have for $\sigma > 0$,

$$\zeta(s) = \sum_{m=1}^{N} \frac{1}{m^s} + \frac{N^{1-s}}{s-1} - \frac{N^{-s}}{2} - s \int_{N}^{\infty} \frac{B_1(\{t\})}{t^{s+1}}\, dt\ .$$

Differentiating k times (or applying Euler–MacLaurin directly to $(\log(m))^k/m^s$) we obtain

$$(-1)^k \zeta^{(k)}(s) = \sum_{m=1}^{N} \frac{(\log(m))^k}{m^s} - \frac{\log(N)^k}{2N^s} + k! N^{1-s} \sum_{j=0}^{k} \frac{\log(N)^j}{j!(s-1)^{k-j+1}}$$
$$+ \int_{N}^{\infty} \frac{\log(t)^{k-1} B_1(\{t\})}{t^{s+1}} (k - s\log(t))\, dt\ .$$

In the given region an easy estimate gives

$$(-1)^k \zeta^{(k)}(s) = \sum_{m=1}^{N} \frac{(\log(m))^k}{m^s} + O\left(\frac{N^{1-\sigma}(t\log(N))^k}{t^{k+1}} \right)$$
$$+ O\left(\frac{\log(N)^{k-1}(k/\sigma + \log(N))}{N^\sigma} \right)\ .$$

Finally, if $m \leqslant t$ we have $|m^{-s}| = m^{-\sigma} \leqslant m^{-(1-C/\log(t))} \leqslant K/m$ for some constant K. By choosing $N = \lfloor t \rfloor$, it is immediate that we obtain the desired estimate. $\qquad\square$

The basis of the initial proofs of the PNT is the first statement of the following lemma.

Lemma 10.7.2. *For all $\sigma > 1$ and $t \in \mathbb{R}$ we have*

$$\zeta(\sigma)^3 |\zeta(\sigma + it)|^4 |\zeta(\sigma + 2it)| \geqslant 1$$

and

$$-\Re \left(3 \frac{\zeta'(\sigma)}{\zeta(\sigma)} + 4 \frac{\zeta'(\sigma + it)}{\zeta(\sigma + it)} + \frac{\zeta'(\sigma + 2it)}{\zeta(\sigma + 2it)} \right) \geqslant 0 .$$

Proof. By expanding the logarithm of the Euler product defining $\zeta(s)$ it is clear that $\zeta(s) = \exp(\sum_p \sum_{k \geqslant 1} p^{-ks}/k)$, hence that

$$\log(|\zeta(s)|) = \sum_p \sum_{k \geqslant 1} \frac{\cos(kt \log(p))}{kp^{k\sigma}} .$$

The trick is to note that we have the positivity condition

$$3 + 4\cos(\theta) + \cos(2\theta) = 2(1 + \cos(\theta))^2 \geqslant 0 ,$$

so the logarithm of the first expression of the lemma is equal to

$$\sum_p \sum_{k \geqslant 1} \frac{3 + 4\cos(kt \log(p)) + \cos(2kt \log(p))}{kp^{k\sigma}} \geqslant 0 ,$$

proving the first inequality. The second follows in the same way since the logarithmic derivative of $\log |\zeta(s)|$ is equal to $\Re(\zeta'(s)/\zeta(s))$ and the derivative of p^{-ks} is $(-k \log(p))p^{-ks}$. $\qquad\square$

Corollary 10.7.3. *For $\sigma > 1$ we have*

$$\frac{1}{\zeta(\sigma + it)} = O \left(\frac{\log(t)^{1/4}}{(\sigma - 1)^{3/4}} \right) .$$

Proof. Since $\zeta(\sigma) = 1/(\sigma - 1) + O(1)$, by the lemma and Proposition 10.7.1 we have $|\zeta(\sigma + it)^{-4}| = O(\zeta(\sigma)^3 \log(t)) = O((\sigma - 1)^{-3} \log(t))$. $\qquad\square$

Corollary 10.7.4. *The function $\zeta(s)$ does not vanish in the closed half-plane $\Re(s) \geqslant 1$, and in particular on the line $\Re(s) = 1$.*

Proof. Since the Euler product is convergent for $\Re(s) > 1$ and none of its terms vanish, we know that $\zeta(s) \neq 0$ for $\Re(s) > 1$. Now assume by contradiction that $\zeta(1 + it_0) = 0$ for some $t_0 \in \mathbb{R}$. The function $\zeta(s + it_0)^4$ thus has a zero of order greater than or equal to 4 at $s = 1$. Since the function $\zeta(s)^3$ has a pole of order exactly equal to 3 and $\zeta(s + 2it)$ has no pole at $s = 1$, it follows that $\zeta(s)^3 \zeta(s + it_0)^4 \zeta(s + 2it_0)$ tends to 0 as s tends to 1, and this contradicts the first inequality of the lemma. $\qquad\square$

In fact, using the same method it is easy to show that $\zeta(s)$ has no zeros in a region of the form $\sigma > 1 - C/\log(t)^9$ and to give a uniform upper bound for $1/\zeta(s)$ in that region; see Exercise 64. However, for our purposes the above corollary is sufficient. In addition, it is easy to obtain much better estimates for $1/\zeta(s)$ (which we will not need for the versions of the PNT that we prove here), but for this we need to appeal to some additional complex analysis. Recall that an analytic function $f(s)$ is said to be of *order k* (where $k \in \mathbb{Z}_{\geqslant 0}$) if it is an entire function (i.e., holomorphic in the whole of $\mathbb{C}$, although it is trivial to allow a finite number of poles) such that for all $\varepsilon > 0$ we have $\log(|f(s)|) = O(|s|^{k+\varepsilon})$. The vast majority of functions in current use have order 0 or 1 (polynomials being of order 0), for instance the gamma function. Barnes's multiple gamma functions (see Exercise 71 of Chapter 9) and Selberg's zeta function mentioned above are examples of functions of higher order. For L-series with functional equation it is a general principle that they have the same order as the corresponding gamma factor. For instance:

Lemma 10.7.5. *The functions $(s-1)\zeta(s)$ and $s(1-s)\pi^{-s/2}\Gamma(s/2)\zeta(s)$ have order 1.*

Proof. Indeed, for $\sigma > 0$ by the integral representation we have for a suitable constant A, $|\Gamma(s/2)| \leqslant |\Gamma(\sigma/2)| = O(e^{A\sigma \log(\sigma)})$, and on the other hand, Euler–MacLaurin immediately gives for $\sigma \geqslant 1/2$, $|s-1| > A$, $\zeta(s) = O\left(|s| \int_1^\infty t^{-3/2}\, dt\right) + O(1) = O(|s|)$. It follows that $s(1-s)\pi^{-s/2}\Gamma(s/2)\zeta(s) = O(e^{A|s| \log |s|})$ for $\sigma \geqslant 1/2$, $|s-1| > A$, and since by the functional equation it is invariant under $s \mapsto 1-s$ and has no poles, it is an entire function of order 1. The result for $(s-1)\zeta(s)$ follows. $\square$

The result from complex function theory that we need is *Hadamard's factorization theorem*:

Theorem 10.7.6. *Let $f(s)$ be an entire function of order at most equal to $k \in \mathbb{Z}_{\geqslant 0}$. For all $s \in \mathbb{C}$ we have the absolutely convergent product*

$$f(s) = s^r e^{P_k(s)} \prod_{\rho} \left(1 - \frac{s}{\rho}\right) \exp\left(\sum_{1 \leqslant j \leqslant k} \frac{(s/\rho)^j}{j}\right),$$

where r is the order of f at $s = 0$ ($r = 0$ if $f(0) \neq 0$), $P_k(s)$ is a polynomial of degree less than or equal to k, and the product is over all zeros of $f(s)/s^r$ repeated with multiplicity.

Applying this to $\zeta(s)$ gives the following.

Corollary 10.7.7. *Set $b = \log(2\pi) - 1 - \gamma/2$. Then for all $s \in \mathbb{C}$ we have the convergent product*

$$\zeta(s) = \frac{e^{bs}}{s(s-1)\Gamma(s/2)} \prod_{\rho} \left(1 - \frac{s}{\rho}\right) e^{s/\rho},$$

the product being over all nontrivial zeros of $\zeta(s)$ (i.e., such that $0 \leqslant \Re(\rho) \leqslant 1$).

Proof. We apply Hadamard's theorem to the function

$$f(s) = s(1-s)\pi^{-s/2}\Gamma(s/2)\zeta(s) = 2(1-s)\pi^{-s/2}\Gamma(s/2+1)\zeta(s) .$$

Since the zeros of $\zeta(s)$ for $s = -2k$, $k \in \mathbb{Z}_{\geqslant 1}$, are killed by the poles of $\Gamma(s/2+1)$ and the pole of $\zeta(s)$ is killed by $1-s$, it follows that the zeros of $f(s)$ are the nontrivial zeros of $\zeta(s)$. Thus for suitable constants a_0 and a_1 we have

$$f(s) = a_0 e^{a_1 s} \prod_\rho \left(1 - \frac{s}{\rho}\right) e^{s/\rho} ,$$

so that

$$\zeta(s) = \frac{a_0 e^{bs}}{2(1-s)\Gamma(s/2+1)} \prod_\rho \left(1 - \frac{s}{\rho}\right) e^{s/\rho}$$

for $b = a_1 + \log(\pi)/2$. We deduce that $a_0 = 2\zeta(0) = -1$, and by logarithmic differentiation that

$$\frac{\zeta'(s)}{\zeta(s)} = b - \frac{1}{s-1} - \frac{\Gamma'(s/2+1)}{2\Gamma(s/2+1)} + \sum_\rho \left(\frac{1}{s-\rho} + \frac{1}{\rho}\right) ,$$

so that

$$\frac{\zeta'(0)}{\zeta(0)} = b + 1 - \frac{\Gamma'(1)}{\Gamma(1)} .$$

Using $\zeta'(0) = -\log(2\pi)/2$ and $\Gamma'(1) = -\gamma$ we obtain $b = \log(2\pi) - 1 - \gamma/2$. $\square$

We are now in a position to give a much better zero-free region than that given by Exercise 64.

Theorem 10.7.8. *There exists a constant $C > 0$ such that $\zeta(s) \neq 0$ for $t \geqslant t_0$ in the region*

$$\Re(s) > 1 - \frac{C}{\log(t)} .$$

Proof. Here we will use the second inequality of Lemma 10.7.2. Fix some $\sigma > 1$ (we will see at the end of the proof how to choose it appropriately). Since $\zeta(\sigma) = 1/(\sigma-1) + O(1)$ and $\zeta'(\sigma) = -1/(\sigma-1)^2 + O(1)$, we have $-\zeta'(\sigma)/\zeta(\sigma) < 1/(\sigma-1) + O(1)$. From the above corollary and trivial bounds on $\Gamma'(s)/\Gamma(s)$ we also have

$$-\frac{\zeta'(s)}{\zeta(s)} = O(\log(t)) - \sum_\rho \left(\frac{1}{s-\rho} + \frac{1}{\rho}\right) ,$$

so if we write $\rho = \beta + i\gamma$ with $0 \leqslant \beta \leqslant 1$ and $\gamma \in \mathbb{R}$ we have

$$-\Re\left(\frac{\zeta'(s)}{\zeta(s)}\right) = O(\log(t)) - \sum_{\rho}\left(\frac{\sigma-\beta}{(\sigma-\beta)^2+(t-\gamma)^2} + \frac{\beta}{\beta^2+\gamma^2}\right).$$

Since $\sigma > 1 \geqslant \beta \geqslant 0$, we deduce that for all $s \in \mathbb{C}$, $-\Re(\zeta'(s)/\zeta(s)) < O(\log(t))$ with $t = \Im(s)$. Now fix some nontrivial zero $\rho_0 = \beta_0 + i\gamma_0$. Then if $s = \sigma + i\gamma_0$ (same imaginary part but real part $\sigma > 1$) we evidently have the stronger inequality $-\Re(\zeta'(s)/\zeta(s)) < O(\log(\gamma_0)) - 1/(\sigma-\beta_0)$. Putting all this together in the second inequality of Lemma 10.7.2 applied to $t = \gamma_0$ we obtain

$$\frac{3}{\sigma-1} - \frac{4}{\sigma-\beta_0} + O(\log(\gamma_0)) \geqslant 0,$$

in other words $3/(\sigma-1) - 4/(\sigma-\beta_0) > -A\log(\gamma_0)$ for some constant A that we may choose strictly positive (since increasing A gives a worse estimate). Solving for $1-\beta_0$ gives

$$1-\beta_0 \geqslant \frac{1-(\sigma-1)A\log(\gamma_0)}{3/(\sigma-1)+A\log(\gamma_0)}.$$

Choosing for instance $\sigma - 1 = 1/(2A\log(\gamma_0))$ (this is why we must have $A > 0$), we obtain $1 - \beta_0 \geqslant 1/(14A\log(\gamma_0))$, proving the theorem. $\qquad\square$

Important Remarks. (1) Using a slight refinement of this proof, it is not difficult to show that in the given region we have $1/\zeta(s) = O(\log(t))$, and this zero-free region can be shown to lead to the PNT in the form

$$\pi(x) = \mathrm{Li}(x) + O(x\exp(-c\log(x)^{1/2}))$$

for some $c > 0$, where $\mathrm{Li}(x)$ is as defined before Corollary 10.7.20 below.
(2) With much more difficulty one can still improve the zero-free region hence the error term in the PNT. The best-known result is as follows. Set $g(t) = \log(t)^{2/3}\log(\log(t))^{1/3}$. There exists $C > 0$ such that $\zeta(s) = O(g(t))$ and $1/\zeta(s) = O(g(t))$ uniformly for $\sigma > 1 - C/g(t)$, and in particular $\zeta(s) \neq 0$ in that domain. This result is due to N. M. Korobov and I. M. Vinogradov, and is described for instance in [Ell]. It leads to the best known error term for the PNT:

$$\pi(x) = \mathrm{Li}(x) + O(x\exp(-c\log(x)^{3/5}\log(\log(x))^{-1/5}))$$

for some strictly positive constant c. This result has remained unchanged for almost half a century, and even the tiny $\log(\log(x))^{-1/5}$ factor has not been improved.

10.7.2 Newman's Proof

For $s \in \mathbb{C}$ and $x \in \mathbb{R}$ we set

$$\Phi(s) = \sum_p \frac{\log p}{p^s} \quad \text{and} \quad \theta(x) = \sum_{p \leqslant x} \log p .$$

The proof proceeds through a series of lemmas.

Lemma 10.7.9. *The function* $\Phi(s) - 1/(s-1)$ *is holomorphic in the closed half-plane* $\Re(s) \geqslant 1$.

Proof. It is clear that the series for $\Phi(s)$ converges absolutely for $\Re(s) > 1$ and normally for $\Re(s) \geqslant 1 + \varepsilon$ for any fixed $\varepsilon > 0$, hence defines an analytic function in $\Re(s) > 1$. For $\Re(s) > 1$ the absolutely convergent Euler product representation for $\zeta(s)$ implies that

$$-\frac{\zeta'(s)}{\zeta(s)} = \sum_p \frac{\log p}{p^s - 1} = \Phi(s) + \sum_p \frac{\log p}{p^s(p^s - 1)} .$$

The rightmost sum converges absolutely for $\Re(s) > 1/2$, proving that $\Phi(s)$ extends meromorphically to $\Re(s) > 1/2$ with poles only at the pole $s = 1$ of $\zeta(s)$ and at the zeros of $\zeta(s)$. At $s = 1$ we have a simple pole with residue 1. Furthermore, by Corollary 10.7.4 we know that $\zeta(s)$ does not vanish for $\Re(s) \geqslant 1$, so that $\Phi(s) - 1/(s-1)$ is holomorphic for $\Re(s) \geqslant 1$. □

Lemma 10.7.10. *We have* $\theta(x) = O(x)$.

Proof. For a positive integer n we have

$$2^{2n} = \sum_{0 \leqslant k \leqslant 2n} \binom{2n}{k} \geqslant \binom{2n}{n} \geqslant \prod_{n < p \leqslant 2n} p = e^{\theta(2n) - \theta(n)} .$$

Since $\theta(x)$ changes by $O(\log(x))$ when x changes by a bounded amount, we deduce that $\theta(x) - \theta(x/2) \leqslant Cx$ for any $C > \log 2$ and $x \geqslant x_0 = x_0(C)$. Summing this inequality for $x, x/2, \ldots, x/2^r$, where $x/2^r \geqslant x_0 > x/2^{r+1}$, we obtain $\theta(x) \leqslant 2Cx + O(1)$, proving the lemma. □

Lemma 10.7.11. *The integral*

$$\int_1^\infty \frac{\theta(x) - x}{x^2} \, dx$$

converges.

Proof. For $\Re(s) > 1$ we have by Stieltjes integration

$$\Phi(s) = \sum_p \frac{\log p}{p^s} = \int_1^\infty \frac{d\theta(x)}{x^s} = s \int_1^\infty \frac{\theta(x)}{x^{s+1}} \, dx = s \int_0^\infty e^{-st} \theta(e^t) \, dt .$$

A reader not familiar with Stieltjes integration can prove directly (but slightly more painfully) using Abel summation the equality $\Phi(s) = s \int_1^\infty \theta(x)/x^{s+1} \, dx$. The last equality above of course follows from the change of variable $x = e^t$.

Assume for the moment the following analytic theorem.

Theorem 10.7.12. *Let $f(t)$ be a bounded and locally integrable function for $t \geqslant 0$, and assume that the function $g(z) = \int_0^\infty f(t)e^{-zt}\,dt$ (defined for $\Re(z) > 0$) extends to a holomorphic function for $\Re(z) \geqslant 0$. Then $\int_0^\infty f(t)\,dt$ converges and is equal to $g(0)$.*

Consider the function $f(t) = \theta(e^t)e^{-t} - 1$. By Lemma 10.7.10, $f(t)$ is bounded, and it is clearly locally integrable. Furthermore, the corresponding function $g(z)$ is equal to $\Phi(z+1)/(z+1) - 1/z$ by the above formula. Lemma 10.7.9 tells us that $g(z)$ extends into a holomorphic function for $\Re(z) \geqslant 0$. Thus the hypotheses of the theorem are satisfied, so we deduce that $\int_0^\infty f(t)\,dt$ converges. Making the change of variable $x = e^t$ proves Lemma 10.7.11. $\square$

We will prove the above theorem later. We now have essentially all the ingredients to finish the proof.

Lemma 10.7.13. $\theta(x) \sim x$.

Proof. Assume that for some $\lambda > 1$ there exist arbitrary large x such that $\theta(x) \geqslant \lambda x$. Since $\theta(x)$ is nondecreasing, we have

$$\int_x^{\lambda x} \frac{\theta(t) - t}{t^2}\,dt \geqslant \int_x^{\lambda x} \frac{\lambda x - t}{t^2}\,dt = \int_1^\lambda \frac{\lambda - u}{u^2}\,du > 0$$

for such x, contradicting the convergence of the integral from 1 to ∞. Similar reasoning shows that the existence of $\lambda < 1$ such that there exist arbitrarily large x such that $\theta(x) \leqslant \lambda x$ leads to a contradiction. $\square$

Theorem 10.7.14 (Prime number theorem). *If $\pi(x)$ denotes the number of prime numbers less than or equal to x we have*

$$\pi(x) \sim \frac{x}{\log x} \ .$$

Proof. We have

$$\theta(x) = \sum_{p \leqslant x} \log p \leqslant \sum_{p \leqslant x} \log x = \pi(x) \log x \ .$$

On the other hand, for any $\varepsilon > 0$,

$$\theta(x) \geqslant \sum_{x^{1-\varepsilon} \leqslant p \leqslant x} \log p \geqslant \sum_{x^{1-\varepsilon} \leqslant p \leqslant x} (1-\varepsilon) \log x = (1-\varepsilon) \log x \left(\pi(x) + O(x^{1-\varepsilon})\right) \ .$$

It follows from the first inequality that

$$\liminf \pi(x) \frac{\log x}{x} \geqslant \liminf \frac{\theta(x)}{x} = \lim \frac{\theta(x)}{x} = 1$$

and from the second that

$$\limsup \pi(x)\frac{\log x}{x} \leqslant \frac{1}{1-\varepsilon} \, ,$$

and since $\varepsilon > 0$ is arbitrary, that

$$\limsup \pi(x)\frac{\log x}{x} \leqslant 1 \, ,$$

so that $\lim \pi(x)\log x/x$ exists and is equal to 1. □

It remains to prove the purely analytic Theorem 10.7.12, which is called a Tauberian theorem.

Proof of Theorem 10.7.12. For $T > 0$ set $g_T(z) = \int_0^T f(t)e^{-zt}\,dt$. This defines a holomorphic function for all z. We must show that $\lim_{T\to\infty} g_T(0) = g(0)$.

Let R be large and let C be the boundary of the region

$$\{z \in \mathbb{C}/\ |z| \leqslant R, \Re(z) \geqslant -\delta\} \, ,$$

where δ is chosen small enough (depending on R) so that $g(z)$ is holomorphic in and on C (δ exists since analyticity on $\Re(z) \geqslant 0$ implies analyticity on an open set containing $\Re(z) \geqslant 0$). Thus by the residue theorem

$$g(0) - g_T(0) = \frac{1}{2i\pi}\int_C (g(z) - g_T(z))e^{zT}(1 + z^2/R^2)\frac{dz}{z} \, .$$

Set $B = \sup_{t\geqslant 0}|f(t)|$, which exists since f is bounded. On the semicircle $C_+ = C \cap \{z/\ \Re(z) > 0\}$ we have

$$|g(z) - g_T(z)| = \left|\int_T^\infty f(t)e^{-zt}\,dt\right| \leqslant B\int_T^\infty |e^{-zt}|\,dt = \frac{Be^{-\Re(z)T}}{\Re(z)}$$

and

$$|e^{zT}(1 + z^2/R^2)/z| = e^{\Re(z)T}\frac{2\Re(z)}{R^2} \, ,$$

since $|1 + z^2/R^2| = 2\cos(\theta) = 2\Re(z)/R$ for $z = Re^{i\theta}$. Thus on C_+ the integrand is bounded (in absolute value) by $2B/R^2$, so the contribution to $g(0) - g_T(0)$ from the integral over C_+ is bounded by B/R. For the integral over $C_- = C \cap \{z/\ \Re(z) < 0\}$ we consider $g(z)$ and $g_T(z)$ separately. Since g_T is entire, the path of integration for the integral involving g_T can be replaced by the semicircle $C'_- = \{z \in \mathbb{C}/\ |z| = R,\ \Re(z) < 0\}$, so the contribution coming from the integral involving g_T over C'_- is bounded in absolute value by B/R exactly as before, since for $\Re(z) < 0$,

$$|g_T(z)| = \left|\int_0^T f(t)e^{-zt}\,dt\right| \leqslant B\int_{-\infty}^T |e^{-zt}|\,dt = \frac{Be^{-\Re(z)T}}{|\Re(z)|} \, .$$

Finally, the remaining integral involving $g(z)$ over C_- tends to 0 as $T \to \infty$ since the integrand is the product of the function $g(z)(1 + z^2/R^2)/z$, which is independent of T, by the function e^{zT}, which tends to zero rapidly and uniformly as $T \to \infty$ on compact subsets of the half-plane $\Re(z) < 0$. Hence $\limsup_{T\to\infty} |g(0) - g_T(0)| \leqslant 2B/R$. Since R is arbitrary, this finishes the proof of the theorem. $\square$

Remark. The above proof can be extended to Dirichlet L-functions: using exactly the same method as above and combining all L-functions corresponding to a given modulus, it is easy to prove that $L(\chi, s)$ does not vanish for all $s \in \mathbb{C}$ such that $\Re(s) = 1$. From this result, as above one can show the stronger statement that primes congruent to a modulo m have density $1/\phi(m)$ among all primes in the ordinary sense, i.e., when counting up to x; see Exercise 67.

10.7.3 Iwaniec's Proof

This proof has a different style from the above proof in that it does not use a Tauberian theorem, and as a consequence has the advantage of giving a nontrivial error term.

We begin by noting the following formula from elementary complex analysis:

Lemma 10.7.15. *For all $y > 0$ we have for any $\sigma > 1$,*

$$\max(\log(y), 0) = \frac{1}{2i\pi} \int_{\Re(s)=\sigma} \frac{y^s}{s^2}\, ds \ ,$$

the integral being on the vertical line $\Re(s) = \sigma$.

Proof. Indeed, the given integral is trivially less than $O(y^\sigma)$. Thus, if $y < 1$ it is immediate to check that we can shift the line of integration to the right without changing the value of the integral, and as σ tends to $+\infty$, y^σ tends to 0. On the other hand, if $y > 1$ we shift the line of integration to some $\sigma < 0$, catching a double pole at $s = 0$ with residue $\log(y)$. We now let σ tend to $-\infty$, and the residue formula tells us that the integral is equal to $\log(y)$. $\square$

We now introduce the following two functions, where $k \in \mathbb{Z}_{\geqslant 0}$ and $x \geqslant 0$:

$$G_k(s) = \sum_{m \geqslant 1} \frac{\mu(m)}{m^s} (\log(m))^k \ \text{and}$$

$$F_k(x) = \sum_{1 \leqslant m \leqslant x} \mu(m)(\log(m))^k \log(x/m)$$

$$= \sum_{m \geqslant 1} \mu(m)(\log(m))^k \max(\log(x/m), 0) \ .$$

The factor $\max(\log(x/m), 0)$ is a smoothing factor, and is a very common tool in analytic number theory (we have already used a similar ideal in the definition of the function $f(t)$ used to prove Voronoi's error term in the circle problem, see Section 10.2.6). We will remove it below.

Lemma 10.7.16. *Let $\sigma > 1$ be fixed. There exists s with $\Re(s) = \sigma$ such that*

$$|F_k(x)| \leqslant x^\sigma |G_k(s)||s|^{-1/2} .$$

Proof. From the above lemma we have

$$F_k(x) = \frac{1}{2i\pi} \int_{\Re(s)=\sigma} \frac{x^s G_k(s)}{s^2} \, ds ,$$

and since $|G_k(s)|$ is bounded by the convergent series $\sum_{m \geqslant 1} (\log(m))^k / m^\sigma$, we have

$$|F_k(x)| \leqslant \frac{1}{2\pi} x^\sigma \sup_{\Re(s)=\sigma} |G_k(s) s^{-1/2}| \int_{\Re(s)=\sigma} \frac{|ds|}{|s|^{3/2}} .$$

Since $|G_k(s)||s|^{-1/2}$ is a continuous and bounded function, the sup is attained. Furthermore,

$$\int_{\Re(s)=\sigma} \frac{|ds|}{|s|^{3/2}} = 2 \int_0^\infty \frac{dt}{(\sigma^2 + t^2)^{3/4}} \leqslant 2 \left(\sigma^{-3/2} \int_0^1 dt + \int_1^\infty \frac{dt}{t^{3/2}} \right) \leqslant 6 ,$$

proving the lemma. □

Lemma 10.7.17. *For $\sigma > 1$ and $\Re(s) = \sigma$ we have*

$$G_k(s) = O\left((\sigma - 1)^{-(3/4)(k+1)} \log(2|s|)^{(9k+1)/4} \right) ,$$

where the implied constant depends only on k.

Proof. By definition we have $G_k(s) = (-1)^k (1/\zeta(s))^{(k)}$. If s is close to 1, say $|s| \leqslant 2$ (still with $\Re(s) = \sigma > 1$), then $(1/\zeta(s))^{(k)}$ is bounded, so the result is trivial, so we may assume that $|s| > 2$. By explicitly expanding $(1/\zeta(s))^{(k)}$, we see that $\zeta(s)^{k+1} G_k(s)$ is a linear combination with coefficients depending only on k of monomials of the form $\prod_{j=0}^k (\zeta^{(j)}(s))^{a_j}$ with $\sum_j a_j = \sum_j j a_j = k$. Since we assume $|s| > 2$, by Proposition 10.7.1 we have $\zeta^{(j)}(s) = O(\log(t)^{j+1}) = O(\log(|s|)^{j+1})$, so that $\zeta(s)^{k+1} G_k(s) = O(\log(|s|)^m)$, with

$$m = \max_{(a_j)} \sum_j (j+1) a_j = \max_{(a_j)} \left(\sum_j j a_j + \sum_j a_j \right) = 2k .$$

Finally, by Corollary 10.7.3 we have $|1/\zeta(s)| = O((\sigma-1)^{-3/4} \log(|s|)^{1/4})$ and the lemma follows. □

Corollary 10.7.18. *For $x \geqslant 1$ we have*

$$F_k(x) = O\left(x\log(x)^{(3/4)(k+1)}\right) .$$

Proof. Combining Lemmas 10.7.16 and 10.7.17, and using the fact that any power of $\log(|s|)$ is negligible compared to $|s|^{1/2}$, we deduce that for all $\sigma > 1$ we have $F_k(x) = O(x^\sigma(\sigma - 1)^{-(3/4)(k+1)})$, so that choosing $\sigma = 1 + 1/\log(x)$ we obtain the desired conclusion. $\qquad\square$

Remark. If we estimated $F_k(x)$ crudely by bounding $\mu(m)$ by 1, we would obtain $F_k(x) = O(x\log(x)^k)$. The above bound is thus better as soon as $k > 3$.

We can now obtain the PNT in the following form.

Theorem 10.7.19. *For any $A > 0$ we have*

$$M(x) = \sum_{1 \leqslant m \leqslant x} \mu(m) = O\left(\frac{x}{\log(x)^A}\right) .$$

Proof. We introduce the function

$$H_k(x) = \sum_{1 \leqslant m \leqslant x} \mu(m)(\log(m))^k ,$$

which is the function $F_k(x)$ from which we have removed the smoothing factor $\log(x/m)$. It is easily related to $F_k(x)$ as follows:

$$
\begin{aligned}
F_k(x+y) - F_k(x) &= H_k(x)\log\left(\frac{x+y}{x}\right) \\
&\quad + \sum_{x < m \leqslant x+y} \mu(m)(\log(m))^k \log\left(\frac{x+y}{m}\right) \\
&= \left(H_k(x) + O(y\log(x)^k)\right)\log\left(\frac{x+y}{x}\right) ,
\end{aligned}
$$

as soon as $y = o(x)$, say. It follows from the above corollary that

$$H_k(x) = O(y\log(x)^k) + O\left((x^2/y)\log(x)^{(3/4)(k+1)}\right) .$$

The optimal choice of y makes the two terms of approximately equal size, and is thus $y = x\log(x)^{-A_k}$ with $A_k = (k-3)/8$, so that $H_k(x) = O(x\log(x)^{k-A_k})$.

Using partial (i.e., Abel) summation, we see that in the expression

$$M(x) = \sum_{1 \leqslant m \leqslant x} \mu(m) = \sum_{1 \leqslant m \leqslant x} \frac{H_k(m) - H_k(m-1)}{(\log(m))^k}$$

we may up to a multiplicative constant replace $H_k(x)$ by $x\log(x)^{k-A_k}$, hence $H_k(m) - H_k(m-1)$ by $\log(x)^{k-A_k}$, so that

$$|M(x)| = O\left(\sum_{1\leqslant m\leqslant x}(\log(m))^{-A_k}\right) = O(x\log(x)^{-A_k}).$$

Since $A_k = (k-3)/8$ tends to infinity with k and k is arbitrary, the theorem follows.   □

For the final result, we define a slight variation of the function $\theta(x)$ as follows:

$$\psi(x) = \sum_{1 < p^a \leqslant x}\log(p),$$

where the sum is over all nontrivial prime powers up to x (no relation to the logarithmic derivative of the gamma function). It is easy to see that $\psi(x) = \theta(x) + O(x^{1/2})$, so that estimating ψ and θ is essentially the same. Finally, we define

$$\mathrm{Li}(x) = \int_0^x \frac{dt}{\log(t)},$$

where the divergent integral is to be understood in the sense of the Cauchy principal value, in other words

$$\mathrm{Li}(x) = \lim_{\varepsilon\to 0+}\int_0^{1-\varepsilon} + \int_{1+\varepsilon}^x \frac{dt}{\log(t)}$$

(see Exercise 68). Note that this is completely unrelated to the polylogarithm functions $\mathrm{Li}_k(x)$ defined in Exercise 22 of Chapter 4. By successive integration by parts, we have

$$\mathrm{Li}(x) = \frac{x}{\log(x)}\left(\sum_{0\leqslant j\leqslant m}\frac{j!}{\log(x)^j} + O\left(\frac{1}{\log(x)^{m+1}}\right)\right).$$

Corollary 10.7.20. *For all $A > 0$ we have*

$$\psi(x) - x = O\left(\frac{x}{\log(x)^A}\right),$$

$$\theta(x) - x = O\left(\frac{x}{\log(x)^A}\right), \quad and$$

$$\pi(x) - \mathrm{Li}(x) = O\left(\frac{x}{\log(x)^A}\right).$$

Proof. (Sketch). Since this is very standard (in contrast to the above proof due to Iwaniec), we give only a sketch. Let $d(n)$ be the number of divisors of n and

$$\Delta(x) = \sum_{1 \leqslant n \leqslant x} (\log(n) - d(n) + 2\gamma) \ .$$

The standard application of the method of the hyperbola ($\sum_{1 \leqslant n \leqslant N} d(n)$ is equal to twice the number of integral points under the hyperbola $xy = N$ with $x \leqslant N^{1/2}$ minus the number of integral points in the square $[0, N^{1/2}]^2$) shows that $\Delta(x) = O(x^{1/2})$. In addition, Abel summation gives

$$\psi(x) - x + 2\gamma = \sum_{dk \leqslant x} \mu(d)(\log(k) - d(k) + 2\gamma)$$

$$= \sum_{k \leqslant x^{1/2}} (\log(k) - d(k) + 2\gamma)M(x/k) + \sum_{d \leqslant x^{1/2}} \mu(d)(\Delta(x/d) - \Delta(x^{1/2})) \ ,$$

and applying the estimate for $\Delta(x)$ as well as the estimate for $M(x)$ given by the theorem gives the estimate for $\psi(x) - x$. As mentioned, the estimate for $\theta(x)$ follows, and the estimate for $\pi(x)$ is obtained in a way similar to that used to obtain the PNT in the first proof.   □

10.8 Exercises for Chapter 10

1. Let $a \in \mathbb{Z}$ and $n \geqslant 1$.

(a) Assume that $n \geqslant 2$ and let p be a prime divisor of n. By writing $n = p^v n_1$ for $p \nmid n_1$, and similarly $d = p^w d_1$, prove that $\sum_{d|n} \mu(n/d)a^d \equiv 0 \pmod{p^v}$. Deduce that for all $n \geqslant 1$ we have $\sum_{d|n} \mu(n/d)a^d \equiv 0 \pmod{n}$ (note that this is a consequence of Corollary 2.4.14, but only when a is a prime power).

(b) Deduce that $\sum_{d|n} \phi(n/d)a^d \equiv 0 \pmod{n}$.

2. This exercise is a sequel to Exercise 32 of Chapter 2. Let K be a commutative field. For any $P \in K[X]$ different from 0, define the Möbius function as follows, analogously to Proposition 10.1.10: if P is not squarefree, set $\mu(P) = 0$; otherwise, set $\mu(P) = (-1)^{\omega(P)}$, where $\omega(P)$ is the number of irreducible monic divisors of P.

(a) Prove that μ is multiplicative, in other words that $\mu(PQ) = \mu(P)\mu(Q)$ when $\gcd(P, Q) = 1$.

(b) Let $N \in K[X]$, $N \neq 0$. Show that

$$\sum_{\substack{P|N \\ P \text{ monic}}} \mu(P) = \begin{cases} 1 & \text{if } \deg(N) = 0, \\ 0 & \text{if } \deg(N) > 0, \end{cases}$$

so that this is a perfect analogue of the Möbius function.

(c) (Stickelberger, Swan.) From now on, assume that $K = \mathbb{F}_q$, where q is *odd*. Denote by $\ell(P)$ the leading coefficient of P. Using Exercise 32 of Chapter 2, show that

$$\mu(P) = \begin{cases} (-1)^{\deg(P)} \left(\dfrac{\mathrm{disc}(P)}{q} \right) & \text{if } \left(\dfrac{\ell(P)}{q} \right) = 1, \\ (-1)^{\deg(P')+1} \left(\dfrac{\mathrm{disc}(P)}{q} \right) & \text{if } \left(\dfrac{\ell(P)}{q} \right) = -1. \end{cases}$$

For this, recall that $\mathrm{disc}(PQ) = \mathrm{disc}(P)\,\mathrm{disc}(Q)R(P,Q)^2$, where $R(P,Q)$ is the resultant of P and Q, and that $\mathrm{disc}(P) = (-1)^{\deg(P)(\deg(P)-1)/2} R(P,P')/\ell(P)$, so that $\mathrm{disc}(\ell P) = \ell^{\deg(P)+\deg(P')-1}\,\mathrm{disc}(P)$. Note also that we do not necessarily have $\deg(P') + 1 = \deg(P)$.

(d) (Conrad.) As an example, let $H \in \mathbb{F}_5[X]$. Show that either $H(0) = 0$, in which case $X^3 \mid (H(X)^5 + X^3)$, or

$$\mu(H(X)^5 + X^3) = \left(\frac{\ell(H)}{5} \right)^{\deg(H)}.$$

In particular, if H is monic or has even degree, the polynomial $H(X)^5 + X^3$ is never irreducible in $\mathbb{F}_5[X]$.

(e) Find a polynomial H of degree 3 such that $H(X)^5 + X^3$ is irreducible.

3. Show that $\sigma_{-t}(n) = n^{-t}\sigma_t(n)$ directly, and using formal Dirichlet series.

4.

(a) Find the formal Dirichlet series corresponding to $a(n) = d(n)^2$. What is the order of the pole at $s = 1$ of the corresponding numerical Dirichlet series?

(b) More generally, find the order of the pole of the numerical Dirichlet series corresponding to $a(n) = d(n)^k$ for $k \in \mathbb{Z}_{\geqslant 1}$.

(c) Same questions for $a(n) = d(n^k)$, and in addition prove the formula $d(n^z) = \sum_{m|n} z^{\omega(m)}$.

5.

(a) If $k \geqslant 1$ is a constant, find the formal Dirichlet series corresponding to $d(kn)$ in terms of $\zeta(s)$ and a finite Euler product depending on k.

(b) Same question for $R(k,n)d(n)$, where $R(k,n)$ is Ramanujan's sum defined in Proposition 10.1.6.

6. Similarly to the above exercise, using the definition and properties of the Ramanujan τ function given in the text, find the formal Dirichlet series corresponding to $\tau(n)^2$ in terms of the formal Dirichlet series corresponding to the convolution of the completely multiplicative functions α^2 and β^2.

7. Using similar reasoning to that of Proposition 10.1.15, show the existence of an abscissa of convergence for a Dirichlet series.

8. Assume that $\sum_{n \geqslant 1} a(n)/n^s$ converges (not necessarily absolutely). Show that $\sum_{n \geqslant 1} a(n)/n^{s'}$ converges absolutely when $\Re(s') > \Re(s) + 1$. (Hint: as in the power series case, use only the fact that $a(n)/n^s$ is bounded.) Deduce from this that the difference between the abscissas of absolute and ordinary convergence is less than or equal to 1.

9. Let f_1 and f_2 be two Dirichlet series with respective abscissas of absolute convergence σ_1 and σ_2. Show that when $\sigma_1 \neq \sigma_2$, the abscissa of absolute convergence

of $f_1 f_2$ is equal to $\min(\sigma_1, \sigma_2)$, while when $\sigma_1 = \sigma_2$, the abscissa of absolute convergence of $f_1 f_2$ is greater than or equal to this common value. In particular, if a Dirichlet series f is invertible, show that the abscissa of absolute convergence of its inverse is equal to that of f.

10. Show that the series defining $\zeta(s)$ diverges (absolutely and in the ordinary sense) for any s such that $\Re(s) = 1$.

11. Show that the series $\sum_{n \geqslant 1} (-1)^n / n^s$ converges for $\Re(s) > 0$, hence that its abscissa of convergence is $\sigma = 0$.

12. Prove Corollaries 10.3.2, 10.3.3, and 10.3.4.

13. Prove Proposition 10.3.10.

14. Generalize Propositions 10.3.8, 10.3.10, and Corollary 10.3.9, to the case that D is congruent to 0 or 1 modulo 4, not necessarily fundamental.

15.

(a) Prove that for all primes $p > 3$ we have $L\left(\left(\frac{\cdot}{p}\right), -(p+1)/2\right) \in \mathbb{Z}$, where $\left(\frac{n}{p}\right)$ is the Legendre symbol.

(b) Generalize to $L\left(\left(\frac{\cdot}{p}\right), -(p + 4k + 1)/2\right)$ for $k \in \mathbb{Z} \setminus \{-1\}$.

16. (J. Sondow.) Prove that for all $s \in \mathbb{C} \setminus \{1\}$ we have the convergent series

$$(1 - 2^{1-s})\zeta(s) = \sum_{n=0}^{\infty} \frac{1}{2^{n+1}} \sum_{k=0}^{n} (-1)^k \binom{n}{k} \frac{1}{(k+1)^s} ,$$

and estimate the speed of convergence of this series.

17. Find an integral representation for $\Gamma(s)\zeta(s, x)$ and deduce from it and Proposition 10.2.2 another proof of analytic continuation and special values at negative integers of $\zeta(s, x)$.

18. Using the Euler–MacLaurin summation formula, or directly, show that for $-1 < \Re(s) < 0$ we have

$$\zeta(s) = -s \int_0^{\infty} \frac{B_1(\{t\})}{t^{s+1}} \, dt .$$

Using Theorem 9.1.20 and Corollary 9.6.36, give another proof of the functional equation of $\zeta(s)$ (in fact this is a hidden way of using the Poisson summation formula).

19. Using the functional equation of $\zeta(s)$, show that for $k \in \mathbb{Z}_{\geqslant 1}$ we have

$$\zeta'(-2k) = (-1)^k \frac{(2k)!}{2} \frac{\zeta(2k+1)}{(2\pi)^{2k}} .$$

20. Let $\varepsilon = \pm 1$. Show that

$$\sum_{\substack{D \text{ fundamental} \\ \text{sign}(D) = \varepsilon}} \frac{1}{|D|^s} = \frac{1}{2\zeta(2s)} \left(\varepsilon L(\chi_{-4}, s) + \left(1 - \frac{1}{2^s} + \frac{2}{2^{2s}}\right) \zeta(s) \right) ,$$

where the sum is over fundamental discriminants (including 1) whose sign is equal to ε. For instance,

$$\sum_{D \text{ fundamental}} \frac{1}{D^2} = \frac{105}{8\pi^2} .$$

21. For $\tau \in \mathcal{H}$ set

$$R(\tau) = \sum_{n \in \mathbb{Z}} \frac{1}{\cos(\pi n \tau)} \ .$$

(a) Using Exercise 101 of Chapter 9, prove that $R(-1/\tau) = \tau R(\tau)$. Since clearly $R(\tau + 2) = R(\tau)$ this shows that R is a modular form of weight 1 on the same group as the function

$$\theta(\tau) = \theta(1, \tau) = \sum_{n \in \mathbb{Z}} e^{i\pi n^2 \tau} \ .$$

(b) Show that in fact $R(\tau) = \theta(1, \tau)^2$.

22. (D. Zagier.) Let $\chi_D = \left(\frac{D}{\cdot}\right)$ be the Kronecker symbol. Assume that D is odd and squarefree, and set $Z_D(n) = 0, 1, \left(\frac{D}{2}\right), -1, 0, 1, -\left(\frac{D}{2}\right), -1$ for $n \equiv 0, 1, 2, 3, 4, 5, 6, 7$ modulo 8. Prove that

$$L(\chi_D, s) = \frac{\zeta(2s)}{L(\chi_{-4}, s)} \sum_{n \geq 1} \frac{Z_D(n)}{n^s} \sum_{\substack{b \bmod n \\ b^2 \equiv -D \ (\mathrm{mod} \ n)}} 1$$

(note that the congruence is $b^2 \equiv -D \ (\mathrm{mod} \ n)$, not $b^2 \equiv D \ (\mathrm{mod} \ n)$).

23. Define $\chi(n) = (1 + \sqrt{5})/2, 1, 0, 0, 1$ when $n \equiv 0, 1, 2, 3, 4$ modulo 5. Note that χ is not a character. Prove that $L(\chi, s)$ satisfies the same functional equation as $L(\chi_5, s)$. In fact, show that χ and χ_5 form a basis for functions modulo 5 satisfying that functional equation.

24. Let χ be a nontrivial character modulo m. Prove that

$$L(\chi, 1) = \sum_{n \geq 1} \frac{\chi(n)}{n} = \prod_p \left(1 - \frac{\chi(p)}{p}\right)^{-1} \ ;$$

in other words, prove that the sum and product converge, and that they both converge to $L(\chi, 1)$.

25. Let $\chi(n) = \left(\frac{-12}{n}\right)$. Find all real numbers t such that $L(\chi, it) = 0$.

26. For a sufficiently nice function f define $D(f)$ by the formula

$$D(f)(x) = \frac{1}{2}\left(x f(x) - \frac{1}{2\pi} f'(x)\right) \ .$$

(a) If, as usual, $\mathcal{F}(f)$ denotes the Fourier transform of f, show that for all $n \geq 0$ we have $\mathcal{F}(D^n(f)) = i^n D^n(\mathcal{F}(f))$.

(b) Set $f_0(x) = e^{-\pi x^2}$, $f_n = D^n(f_0)$, and $\Lambda_n(s) = \mathcal{M}(f_n)(s)$, where $\mathcal{M}(f)$ denotes the Mellin transform of f. Show that $\Lambda_0(s) = (1/2)\pi^{-s/2}\Gamma(s/2)$ and that

$$\Lambda_{n+1} = \frac{1}{2}\left(\Lambda_n(s+1) + \frac{s-1}{2\pi}\Lambda_n(s-1)\right) \ .$$

(c) Deduce that there exist polynomials P_n and Q_n in $\mathbb{R}[X]$ such that

$$\Lambda_{2n}(s) = \frac{1}{2}\pi^{-s/2}\Gamma(s/2)P_n(s) \quad \text{and} \quad \Lambda_{2n+1}(s) = \pi^{-(s+1)/2}\Gamma((s+1)/2)Q_n(s) \ .$$

(d) Compute P_n and Q_n for $0 \leqslant n \leqslant 2$, and show that they satisfy the recurrences

$$P_{n+1}(s) = \frac{1}{8\pi}\left(sP_n(s+2) + (2s-1)P_n(s) + (s-1)P_n(s-2)\right) \quad \text{and}$$

$$Q_{n+1}(s) = \frac{1}{8\pi}\left((s+1)Q_n(s+2) + (2s-1)Q_n(s) + (s-2)Q_n(s-2)\right).$$

27. The goal of this exercise is to prove the Riemann hypothesis... for the functions Λ_n of the preceding exercise, which bear some resemblance to the functions $\Lambda(\chi, s)$ attached to a Dirichlet character.

 (a) Let $\alpha \in \mathbb{C}$ be such that $\Re(\alpha) > 0$. Prove that if $\Re(z) > 0$ we have $|z - \alpha| < |z + \overline{\alpha}|$, while if $\Re(z) < 0$ we have $|z - \alpha| > |z + \overline{\alpha}|$.
 (b) Let $P \in \mathbb{C}[X]$ be a nonconstant polynomial whose roots all have strictly positive real part. Deduce from (a) that all the roots of the polynomials $P(X) \pm \overline{P}(-X)$ are purely imaginary, where $\overline{P}$ denotes the polynomial obtained from P by complex conjugating all the coefficients.
 (c) Let $P \in \mathbb{R}[X]$ be a nonconstant polynomial that is either odd or even and whose roots are purely imaginary. Prove that the same is true for the polynomials $P(X - u) \pm P(X + u)$, where $u \in \mathbb{R}$.
 (d) Let $P \in \mathbb{R}[X]$ be a nonconstant polynomial that is either odd or even and whose roots are purely imaginary. Prove that the same is true for the polynomials $Q_u(X) = (X - u)P(X - 2) + 2XP(X) + (X + u)P(X + 2)$ for all $u > 0$. (Hint: apply (b) to the polynomial $(X - u)(P(X) + P(X - 2))$ and a suitable sign $\pm$.)
 (e) Deduce that for $n \geqslant 1$ all the zeros of the functions Λ_n of the preceding exercise are on the line $\Re(s) = 1/2$ (I thank D. Bump for asking this question).

28. Assume the Riemann hypothesis. Using similar methods to that of the preceding two exercises, but using also Hadamard's factorization theorem (Corollary 10.7.7), prove that the only nontrivial zeros of the functions

$$(s-1)\zeta(s+1) \pm 2\pi\zeta(s-1) \quad \text{and}$$

$$s(s+1)\zeta(s+1) \pm 2\pi(s-2)\zeta(s-1)$$

(as well as an infinite number of examples of the same type) are on the line $\Re(s) = 1/2$ (the trivial zeros being $s = -1 - 2n$ with $n \geqslant 1$ for the first two functions, and $n \geqslant 0$ for the last two). Surprisingly enough, one can show using the methods of [Tay] that these results are true unconditionally, in other words *without* assuming the Riemann hypothesis.

29. Let χ be a periodic function of period m, not necessarily a character, and assume that $\sum_{0 \leqslant r < m} \chi(r) = 0$. Prove that as t tends to 0 from above, then for all $N \geqslant 1$ we have

$$\sum_{n \geqslant 1} \chi(n)e^{-n^2 t} = \sum_{n=0}^{N-1}(-1)^n L(\chi, -2n)\frac{t^n}{n!} + O(t^N) \quad \text{and}$$

$$\sum_{n \geqslant 1} \chi(n)ne^{-n^2 t} = \sum_{n=0}^{N-1}(-1)^n L(\chi, -2n-1)\frac{t^n}{n!} + O(t^N).$$

30.

 (a) Show that the upper bound for $|L(\chi, 1)|$ given in Proposition 10.3.16 (1) is still valid for $|L(\chi, s)|$ if $s \in \mathbb{R}_{>1}$ (better bounds are possible, but we need one that is uniform in s).

(b) Using a completely similar method, find a uniform upper bound for $|L'(\chi, s)|$ for $s \in \mathbb{R}_{>1}$.
(c) Deduce from the third proof of the nonvanishing of $L(\chi, 1)$ given in the text that for any *nonreal* character χ modulo m and any $s \in \mathbb{R}_{>1}$ we have $|L(\chi, s)|^2 \zeta(s)| \geq C(m)$ for an explicit constant $C(m)$ not depending on s.
(d) Using (b), deduce an explicit lower bound for $|L(\chi, 1)|$ when χ is nonreal.

31. Let χ be a nontrivial character modulo m, as usual let $e = 0$ or 1 such that $\chi(-1) = (-1)^e$, and for all $k \geq 0$ set $S_k(\chi) = \sum_{1 \leq r \leq m} \chi(r)(m - 2r)^k$.
(a) Show that $S_k(\chi) = 0$ if $k \not\equiv e \pmod 2$ and if $k = 0$.
(b) Using Corollary 9.6.3, show that

$$L(\chi, s) = 2 \sum_{k \geq 1-e} \binom{s + 2k + e - 1}{2k + e} \frac{(2^{s+2k+e} - 2)\zeta(s + 2k + e)}{2^{2k+e}m^{s+2k+e}} S_{2k+e}(\chi),$$

and give the speed of convergence of this series.
(c) Assume now that χ is a nontrivial *real* character. Show that if $k \geq (m - 2)\log(2)/2$ and $k \equiv e \pmod 2$ then $S_k(\chi) > 0$.
(d) Deduce from this that if χ is a nontrivial real character and $S_k(\chi) \geq 0$ for all $k \equiv e \pmod 2$ such that $k < (m - 2)\log(2)/2$ then $L(s, \chi) > 0$ for $0 < s \leq 1$, and in particular $L(s, \chi)$ does not vanish in that interval.
(e) Using a small computer program, show that the only real character modulo m with $m \leq 100$ that does not satisfy the above condition occurs for $m = 68$, but show nonetheless that the corresponding L-function is strictly positive for $0 < s \leq 1$.
(f) Adapt the above method to characters that may be *nonreal*, by considering suitable expressions of the form $\Re(S_k(\chi)) + \lambda\Im(S_k(\chi))$, and show in this way that *all* L-functions of Dirichlet characters of conductor $m \leq 30$ do not vanish for $0 < s \leq 1$, except perhaps for a single character modulo 19 and its conjugate, and show by a specific argument that the result is also true for these two characters.

32. Using approximation techniques, prove that the results of Theorem 10.2.17 are still valid if we assume for instance that f has a finite number of simple discontinuities, that it is piecewise C^∞, piecewise monotonic, and that f and all its derivatives tend to 0 faster than any power of $|x|$ as $|x| \to \infty$.

33. With the notation of the examples following Theorem 10.2.17, prove the trivial bounds $\Delta(X) = O(X^{1/2})$.

34. Imitating the proof of Voronoi's Theorem 10.2.18, prove the estimate $\Delta(X) = O(X^{1/3})$ in the divisor problem explained in the examples given after Theorem 10.2.17.

35. Give an alternative proof of Theorem 10.3.1 using Theorem 9.1.20.

36. Fix $\tau \in \mathbb{C}$ such that $\Im(\tau) > 0$, and let $\Lambda = \mathbb{Z} + \mathbb{Z}\tau$ be the lattice generated by 1 and τ. For $z \notin \Lambda$ set

$$\wp(z) = \wp(z, \tau) = \frac{1}{z^2} + \sum_{w \in \Lambda}' \left(\frac{1}{(z - w)^2} - \frac{1}{w^2} \right),$$

where the sum is over all nonzero elements of Λ (this is of course the Weierstrass $\wp$-function for Λ), and for all $k \in \mathbb{Z}_{\geq 3}$ set

$$G_k(\tau) = {\sum_{w \in \Lambda}}' \frac{1}{w^k} = {\sum_{(m,n) \in \mathbb{Z}^2}}' \frac{1}{(m + n\tau)^k}$$

(these are the ordinary Eisenstein series of weight k).

(a) Prove that all the above series converge absolutely, that $G_k(\tau) = 0$ if $2 \nmid k$, and that around $z = 0$ we have the expansion

$$\wp(z) = \frac{1}{z^2} + \sum_{k \geq 1} (2k + 1) G_{2k+2}(\tau) z^{2k} .$$

(b) Show that for all $z \notin \Lambda$ and all $\alpha \in \Lambda$ we have $\wp(z + \alpha) = \wp(z)$.

(c) Set $g_2 = 60 G_4$ and $g_3 = 140 G_6$ and

$$f(z) = (\wp'(z))^2 - 4\wp(z)^3 - g_2 \wp(z) - g_3 ,$$

so that $f(z + \alpha) = f(z)$ for all $\alpha \in \Lambda$. Show that $f(z)$ is an entire bounded function in the whole complex plane that vanishes at $z = 0$, and deduce from Liouville's theorem that f is identically zero.

(d) Deduce that $\wp''(z) = 6\wp(z)^2 - 30 G_4$, and for $k \geq 4$ the recurrence

$$G_{2k}(\tau) = \frac{3}{(k-3)(4k^2 - 1)} \sum_{j=2}^{k-2} (2j - 1)(2k - 2j - 1) G_{2j}(\tau) G_{2k-2j}(\tau) .$$

37. (Sequel to the preceding exercise.) In this exercise we specialize to the cases $\tau = i$ and $\tau = \rho = (-1 + \sqrt{3}i)/2$, a primitive cube root of unity, which we treat together. We let w be the number of roots of unity in $\mathbb{Q}(\tau)$, so that $w = 4$ when $\tau = i$ and $w = 6$ when $\tau = \rho$.

(a) Show that $G_k(\tau) = 0$ if $w \nmid k$, and that $G_w(\tau) > 0$.

(b) Define $\Omega_w > 0$ by the formula $\Omega_w = (d_w G_w(\tau))^{1/w}$, with $d_4 = 15$ and $d_6 = 945$. Show that if we define the *Bernoulli–Hurwitz numbers* $H_{wk}^{(w)}$ by the formula

$$G_{wk}(\tau) = (-1)^{k-1} \frac{H_{wk}^{(w)}}{(wk)!} (2\Omega_w)^{wk} ,$$

the H_{wk} are rational numbers of alternating signs satisfying the recurrence

$$H_{wk}^{(w)} = -\frac{6}{(wk - 6)(w^2 k^2 - 1)} \sum_{j=1}^{k-1} (wj - 1)(wk - wj - 1) \binom{wk}{wj} H_{wj}^{(w)} H_{wk-wj}^{(w)}$$

for $k \geq 2$.

(c) Show that

$$H_4^{(4)} = \frac{1}{10}, \quad H_8^{(4)} = -\frac{3}{10}, \quad H_{12}^{(4)} = \frac{567}{130}, \quad H_{16}^{(4)} = -\frac{43659}{170}, \quad H_{20}^{(4)} = \frac{392931}{10},$$

$$H_6^{(6)} = \frac{1}{84}, \quad H_{12}^{(6)} = -\frac{25}{1092}, \quad H_{18}^{(6)} = \frac{1375}{1596}, \quad H_{24}^{(6)} = -\frac{257125}{1092},$$

and $\quad H_{30}^{(6)} = \frac{739234375}{2604} .$

(d) (Hard; for help, see [Kat2] and [Bar].) Find arithmetic properties of the $H_{wk}^{(w)}$ analogous to those of Bernoulli numbers (you may need to use results of Chapter 11 for this). More precisely:

- Prove that the denominator of $H_{wk}^{(w)}$ is equal to e_w times the product of primes $p \equiv 1 \pmod w$ such that $(p-1) \mid wk$, where $e_4 = 2$ and $e_6 = 12$. In particular, the denominator of $H_{4k}^{(4)}$ is always divisible by 10 and that of $H_{6k}^{(6)}$ is always divisible by 84.

- More precisely, prove the following analogue of the Clausen–von Staudt congruence: if $(p-1) \nmid wk$ then $H_{wk}^{(w)}$ is p-integral, while if $(p-1) \mid wk$ then $H_{wk}^{(w)} + a_p^{wk/(p-1)}$ is p-integral, where $p+1-a_p$ is the number of points on the elliptic curve $y^2 = x^3 - 4x$ for $w = 4$, or on the elliptic curve $y^2 = x^3 - 1$ for $w = 6$ (see Corollary 8.5.2 and Proposition 8.5.3 for the explicit formulas for the a_p).

- Prove the analogue of the Kummer congruences, which will also involve a_p.

(e) Show that the *numerator* also has interesting arithmetic properties, contrary to the numerator of Bernoulli numbers; in particular, give a precise statement implying that it is highly divisible by primes $p \equiv -1 \pmod w$. (Hint: its valuation at such primes is very close to $wkp/(p^2 - 1)$.)

(f) By computing $\wp(2/w)$, show that

$$\Omega_4 = 2 \int_0^1 \frac{1}{\sqrt{1 - t^4}}\, dt = \int_0^1 \frac{1}{\sqrt{t - t^3}}\, dt = \int_1^\infty \frac{1}{\sqrt{t^3 - t}}\, dt$$

$$\text{and} \quad \Omega_6 = 3 \int_0^1 \frac{1}{\sqrt{1 - t^3}}\, dt = \sqrt{3} \int_1^\infty \frac{1}{\sqrt{t^3 - 1}}\, dt = \int_{-\infty}^1 \frac{1}{\sqrt{1 - t^3}}\, dt$$

(see Section 7.3.2), and by using either the numerical integration methods given in Section 9.3.2 or, better, the formula given in Exercise 38 (b), show that

$$\Omega_4 = 2.62205755429211981046483958989111941368275495143 1623 \cdots$$
$$\text{and} \quad \Omega_6 = 4.20654631597636278352505723715088240638906661627 1958 \cdots$$

Remark. The fact that G_{2k} is equal to a rational number times the $2k$th power of a fixed "period" 2Ω is valid more generally when τ is a root of a quadratic polynomial with integral coefficients, in other words when τ is a CM point. This follows from the basic properties of *complex multiplication*, that we have already mentioned. However, even more generally we can also define Bernoulli–Hurwitz numbers as soon as g_2 and g_3 are rational, and they also satisfy Clausen–von Staudt and Kummer type congruences; see [Kat2] and [Bar].

38. Recall from Exercise 36 that for $k \in \mathbb{Z}_{\geqslant 3}$, $\Im(\tau) > 0$, and $\Lambda = \mathbb{Z} + \mathbb{Z}\tau$ we define

$$G_k(\tau) = \sum_{w \in \Lambda}' \frac{1}{w^k} = \sum_{(m,n) \in \mathbb{Z}^2}' \frac{1}{(m + n\tau)^k},$$

and as usual set $q = e^{2i\pi\tau}$, so that $|q| < 1$.

(a) By comparing the formula for $\pi \cot(\pi x)$ given in Proposition 9.6.24 and its Taylor expansion, prove that for $k \in \mathbb{Z}_{\geqslant 3}$ we have

$$\sum_{n \in \mathbb{Z}} \frac{1}{(n + \tau)^k} = (-1)^k \frac{(2i\pi)^k}{(k-1)!} \sum_{m \geqslant 1} m^{k-1} q^m.$$

(b) Deduce that for $k \in \mathbb{Z}_{\geqslant 2}$ we have

$$G_{2k}(\tau) = (-1)^{k-1}\frac{(2\pi)^{2k}}{(2k)!}B_{2k}\left(1 - \frac{4k}{B_{2k}}\sum_{n\geqslant 1}\sigma_{2k-1}(n)q^n\right)$$

$$= (-1)^{k-1}\frac{(2\pi)^{2k}}{(2k)!}B_{2k}\left(1 - \frac{4k}{B_{2k}}\sum_{n\geqslant 1}\frac{n^{2k-1}q^n}{1-q^n}\right).$$

(c) Deduce from this and Exercise 36 that for $k \in \mathbb{Z}_{\geqslant 1}$ we have the identity

$$\sum_{n=1}^{\infty}\frac{n^{4k+1}}{e^{2\pi n}-1} = \frac{B_{4k+2}}{8k+4}.$$

(d) (More difficult.) Using the function $G_2(\tau)$ defined as the right-hand side of the formula of (b), prove that

$$\sum_{n=1}^{\infty}\frac{n}{e^{2\pi n}-1} = \frac{B_2}{4} - \frac{1}{8\pi} = \frac{1}{24} - \frac{1}{8\pi},$$

so that the identity of (c) is still valid for $k = 0$ with the corrective term $1/(8\pi)$.

(e) Show that for $\Re(s) > 0$ we have

$$\int_0^{\infty}\frac{x^{s-1}}{e^{2\pi x}-1}\,dx = (2\pi)^{-s}\Gamma(s)\zeta(s).$$

(f) Deduce that for $k \in \mathbb{Z}_{\geqslant 0}$ we have

$$\int_0^{\infty}\frac{x^{4k+1}}{e^{2\pi x}-1}\,dx = \frac{B_{4k+2}}{8k+4}.$$

Note that the equality of the expressions in (c) and (f) is in the same vein as the well-known identity

$$\int_0^1 x^{-x}\,dx = \sum_{n\geqslant 1}n^{-n}.$$

39. Show that for $D = -3$ and $D = -4$, we have

$$L'(\chi_D, 0) = 2\log(\Gamma(1/|D|)) - \log(2\pi) + c_D\log(|D|)$$

with $c_{-3} = 1/6$ and $c_{-4} = -1/4$.

40. Let χ be a nontrivial character modulo m such that $\chi(-1) = -1$.

(a) Generalizing the technique used in the text for $L'(\chi, 0)$, compute $L'(\chi, -1)$ in terms of the function

$$S(z) = \int_0^z \log(\sin(t))\,dt.$$

See also Exercise 71 of Chapter 9.

(b) If in addition χ is primitive, find a formula for $L(\chi, 2)$.

(c) Compute explicitly $S(\pi/2)$ and $S(\pi)$.

41. Compute $L\left(\left(\frac{-4}{\cdot}\right), -2k\right)$ and

$$L\left(\left(\frac{-4}{\cdot}\right), 2k+1\right) = \sum_{n \geqslant 0} \frac{(-1)^n}{(2n+1)^{2k+1}}$$

in terms of the Euler numbers E_n (see Definition 9.1.8).

42.

(a) Using Corollary 9.1.21, show that

$$\prod_{p \equiv 1 \ (\mathrm{mod} \ 4)} \frac{p^3+1}{p^3-1} = \frac{105}{4} \frac{\zeta(3)}{\pi^3} \quad \text{and} \quad \prod_{p \equiv 3 \ (\mathrm{mod} \ 4)} \frac{p^3+1}{p^3-1} = 28 \frac{\zeta(3)}{\pi^3} \ .$$

(b) Similarly, if $G = L(\chi_{-4}, 2)$ denotes Catalan's constant introduced in Exercise 40 of Chapter 9, show that

$$\prod_{p \equiv 1 \ (\mathrm{mod} \ 4)} \frac{p^2+1}{p^2-1} = 12 \frac{G}{\pi^2} \quad \text{and} \quad \prod_{p \equiv 3 \ (\mathrm{mod} \ 4)} \frac{p^2+1}{p^2-1} = \frac{1}{8} \frac{\pi^2}{G} \ .$$

43.

(a) Let $f(t)$ be a C^2 convex function. Prove the inequality

$$f(x) \leqslant \int_{x-1/2}^{x+1/2} f(t) \, dt \ .$$

(b) Using this for the function $f(x) = 1/\sin(\pi x/m)$ with $x \in \ ]0, \pi/2[$, prove that the upper bound in the Pólya–Vinogradov inequality can be improved to

$$\frac{2}{\pi} d(m/f) f^{1/2} (\log(f) + \log(4e/(3\pi)))$$

(note that using Euler–MacLaurin instead of (a) would only very slightly improve the constant $\log(4e/(3\pi))$).

44. Let χ be a nontrivial character modulo m, not necessarily primitive, and let $f > 1$ be its conductor.

(a) Prove that

$$L(\chi, 1) = \sum_{n=1}^m \frac{\chi(n)}{n} + R \quad \text{with} \quad |R| \leqslant \frac{d(m/f) f^{1/2} \log(f)}{m} \ .$$

Note that this is a slightly more precise statement than Proposition 10.3.16 (2) for $\beta = 1$, and that there is no factor 2 in the bound for R.

(b) Deduce from this and Proposition 10.2.5 that

$$\left| \sum_{r=1}^m \chi(r) \psi(1 + r/m) \right| \leqslant d(m/f) f^{1/2} \log(f) \ .$$

Note that the right-hand side is exactly the one given by the Pólya–Vinogradov inequality.

45. Set $S(m) = \sum_{d|m} \mu(d) \log(d)/d$.

(a) Show that
$$\sum_{m\geqslant 1}\frac{S(m)}{m^s} = \frac{\zeta'(s+1)\zeta(s)}{\zeta(s+1)} .$$

(b) Show that
$$S(m) = -\frac{1}{m}\sum_{d|m}\phi(d)\Lambda(m/d) ,$$

where as usual $\Lambda(n)$ is the von Mangoldt function (see Proposition 10.1.14).

(c) Show that
$$S(m) = -\frac{\phi(m)}{m}\sum_{p|m}\frac{\log(p)}{p-1} = -\prod_{p|m}\left(1-\frac{1}{p}\right)\sum_{p|m}\frac{\log(p)}{p-1} ,$$

hence that $S(m) < 0$ for $m > 1$ and $S(m) = O(\log(\log(m)))$ by Proposition 10.3.17 (1).

(d) Using Mertens's theorem $\prod_{p\leqslant x}(1-1/p) \sim e^{-\gamma}/\log(x)$ prove that there exists a strictly positive constant C such that $|S(m)| > C\log(\log(m))$ for infinitely many m, and in particular that it is not bounded. (Hint: choose m equal to the product of all prime numbers between x and x^2 for large x.)

46. (I thank A. Granville for help on this exercise.)

(a) Prove that for all $\varepsilon > 0$ we have
$$\sum_{\substack{1\leqslant a,b,c,d\leqslant m \\ gcd(abcd,m)=1 \\ ab\equiv cd \ (mod\ m)}}\frac{1}{abcd} = \zeta(2)^4\zeta(4)\prod_{p|m}\left(1-\frac{1}{p^2}\right)^4\left(1-\frac{1}{p^4}\right)^{-1} + O\left(\frac{1}{m^{1/2-\varepsilon}}\right) .$$

(Hint: prove that the main contribution is due to the terms where $ab = cd$.)

(b) Deduce an asymptotic estimate for
$$\sum_{\substack{\chi \bmod m \\ \chi\neq\chi_0}}|L(\chi,1)|^4$$

analogous to those of Proposition 10.3.17.

47. Fill in the details of the proof of Proposition 10.3.19 (2) and (3).

48. Recall that the Stirling numbers of the first kind are defined by
$$X(X-1)\cdots(X-r+1) = \sum_{k=0}^{r}(-1)^{r-k}s(r,k)X^k$$

(see the proof of Proposition 4.2.28 for another occurrence).

(a) Show that the rth derivative of $\log(t)^m/t$ is given by the formula
$$\left(\frac{\log(t)^m}{t}\right)^{(r)} = \frac{m!}{t^{r+1}}\sum_{0\leqslant k\leqslant\min(r,m)}s(r+1,k+1)\frac{\log(t)^{m-k}}{(m-k)!} .$$

(b) Deduce an explicit Euler–MacLaurin-type formula for computing to a given accuracy the constants γ_m occurring in the expansion of $\zeta(s)$ around $s = 1$.

(c) Using this formula, compute the γ_m to 28 decimal digits for $0 \leqslant m \leqslant 10$ and for $m = 50$ (the values are given in the text).

49. Set $A_m = \sum_{k \geqslant 1} (-1)^{k-1} \log(k)^m / k$.

(a) Compute A_m as a linear combination of the γ_j for $0 \leqslant j \leqslant m - 1$.
(b) Deduce the following recurrence for the γ_m:

$$\gamma_m = \frac{\log(2)^{m+1}}{(m+1)(m+2)} - \frac{A_{m+1}}{(m+1)\log(2)} - \frac{1}{m+1} \sum_{j=0}^{m-1} \binom{m+1}{j} \log(2)^{m-j} \gamma_j .$$

(c) Using the built-in sumalt function of Pari/GP, which can compute efficiently sums of alternating series, compute values of γ_m and compare the efficiency of this computation with that of the preceding exercise.

50. By Proposition 10.3.19 and Cauchy's formula (or Fourier analysis), for any $\rho > 0$ we have

$$\gamma_m = \frac{(-1)^m m!}{2\pi \rho^m} \int_0^{2\pi} \zeta \left(1 + \rho e^{i\theta}\right) e^{-im\theta} \, d\theta ,$$

and this integral can be computed efficiently to high accuracy using Algorithm 9.3.2. Compare the efficiency of this method to that of the preceding two exercises.

51. Define $\gamma_m(x)$ by the formula

$$\gamma_m(x) = \lim_{N \to \infty} \left(\sum_{k=1}^{N} \frac{\log(k + x - 1)^m}{k + x - 1} - \frac{\log(N + x - 1)^{m+1}}{m+1} \right) .$$

(a) Show that this limit exists for $x \notin \mathbb{Z}_{\leqslant 0}$.
(b) Prove that $\gamma_0(x) = -\psi(x)$, where $\psi(x) = \Gamma'(x)/\Gamma(x)$ is the logarithmic derivative of the gamma function.
(c) Generalizing Proposition 10.3.19 (1), show that

$$\zeta(s, x) = \frac{1}{s-1} + \sum_{m \geqslant 0} (-1)^m \frac{\gamma_m(x)}{m!} (s - 1)^m .$$

(d) Deduce a formula for the mth derivative $L^{(k)}(\chi, 1)$ at $s = 1$ of the L-function of a Dirichlet character.
(e) Let K be a quadratic field of discriminant D, and denote by $\zeta_K^*(1)$ the residue at $s = 1$ of the Dedekind zeta function of K. Deduce a formula for $\lim_{s \to 1}(\zeta_K(s) - \zeta_K^*(1)/(s-1))$.

52. Set

$$\sigma_k^{(1)}(n) = \sum_{d|n} \left(\frac{-4}{d}\right) d^k \quad \text{and} \quad \sigma_k^{(2)}(n) = \sum_{d|n} \left(\frac{-4}{n/d}\right) d^k .$$

Give formulas for $H_3(N)$ and $H_5(N)$ analogous to those of Proposition 10.3.14, but involving the functions $\sigma_2^{(i)}$ and $\sigma_4^{(i)}$ respectively.

53.

(a) Using the methods of Section 10.3.6, compute to 28 decimal digits of accuracy the following sums and products over primes: $\sum_p 1/p^k$ for $k = 2$ and 3, $\sum_p 1/(p(p-1))$, $\sum_p 1/(p-1)^2$, $\prod_{p>2} p(p-2)/(p-1)^2$, and $\prod_p (1-1/(p(p-1)))$.

(b) Generalizing the above methods, compute to the same accuracy $\sum_p \log(p)/p^2$,
$\sum_p 1/(p\log(p))$, $\sum_p 1/(p^2 \log(p))$, and $\lim_{s\to 1+}(\sum_p 1/p^s - \log(\zeta(s)))$.

(c) Generalizing in a different direction, compute to the same accuracy

$$\prod_{p>2} \left(1 - \frac{\left(\frac{D}{p}\right)}{p-1} \right)$$

for $D = -3, -4, -7, 5$, and 8.

(d) Compute Landau's constant (see Proposition 5.4.10)

$$C = \frac{1}{\sqrt{2}} \prod_{p \equiv 3 \pmod 4} \left(1 - \frac{1}{p^2} \right)^{-1/2} .$$

For help see the URL
http://www.math.u-bordeaux1.fr/~cohen/hardylw.dvi

54. For $\Re(s) > 1$, set $P(s) = \sum_p p^{-s}$, where the sum is over all primes.

(a) Prove that $P(s)$ has a meromorphic continuation to $\Re(s) > 0$ with simple poles at the points $s = 1/n$ for $n \in \mathbb{Z}_{\geqslant 1}$.

(b) Using the method of Section 10.3.6, compute $P(2/3)$ to 28 decimal digits.

55. This exercise assumes knowledge of the theory of modular forms. Define

$$E(\tau, s) = \frac{1}{2} \sum_{(c,d)\in\mathbb{Z}^2, \ \gcd(c,d)=1} \frac{y^s}{|c\tau + d|^{2s}} .$$

Let $\mathcal{H} = \{\tau \in \mathbb{C}, \Im(\tau) > 0\}$ be the upper half-plane, and let $f(\tau) = \sum_{n\geqslant 1} a(n)q^n$ and $g(\tau) = \sum_{n\geqslant 1} b(n)q^n$ be two modular cusp forms of weight k on $\mathrm{SL}_2(\mathbb{Z})$, and set

$$L(f, g, s) = \int_{\mathcal{H}/\mathrm{SL}_2(\mathbb{Z})} E(\tau, s) f(\tau)\overline{g(\tau)} y^k \frac{dx\, dy}{y^2} .$$

(a) Show that the defining series for $E(\tau, s)$ converges for $\Re(s) > 1$ and that $E((a\tau + b)/(c\tau + d), s) = E(\tau, s)$ for any $\begin{pmatrix} a & b \\ c & d \end{pmatrix} \in \mathrm{SL}_2(\mathbb{Z})$.

(b) Show that $G(\tau, s) = \zeta(2s)E(\tau, s)$ (which gives another proof of (a)).

(c) Show that for $\Re(s) > 1$ the formula for $L(f, g, s)$ makes sense and that

$$L(f, g, s) = (4\pi)^{-(k+s-1)}\Gamma(k + s - 1) \sum_{n\geqslant 1} \frac{a(n)\overline{b(n)}}{n^{k+s-1}} .$$

(Hint: use the fact that an element $\begin{pmatrix} a & b \\ c & d \end{pmatrix} \in \mathrm{SL}_2(\mathbb{Z})$ can be written uniquely as $\begin{pmatrix} 1 & n \\ 0 & 1 \end{pmatrix}\begin{pmatrix} u & v \\ c & d \end{pmatrix}$, where u and v are fixed integers such that $ud - vc = 1$.)

(d) Deduce that $L(f, g, s)$ has an analytic continuation to the whole complex plane into a meromorphic function satisfying a functional equation, and give the poles of $\zeta(2s)L(f, g, s)$. Note that a much more difficult result due independently to Shimura and Zagier shows that $L(f, g, s)$ is holomorphic in $\mathbb{C}$.

56.

(a) Show that

$$\sum_{(m,n)\in\mathbb{Z}^2}' \frac{1}{(m^2+n^2)^3} = \frac{\pi^3\zeta(3)}{8} \quad \text{and} \quad \sum_{(m,n)\in\mathbb{Z}^2}' \frac{1}{(m^2+mn+n^2)^3} = \frac{8\pi^3\zeta(3)}{27\sqrt{3}}.$$

(b) Generalize to higher powers, both odd and even.

57. Perform the detailed computations giving the formulas of Proposition 10.5.10.

58.

(a) Compute $\eta((-b+\sqrt{D})/(2a))$ for all equivalence classes of quadratic numbers of negative discriminant D corresponding to those D such that $h(D) = 1$ or 2.

(b) Compute $\eta((-1+\sqrt{-23})/2)$ and $\eta((-1+\sqrt{-31})/2)$, the first two cases of class number 3.

59. Fill in the details of the proof of Corollary 10.5.14 and compute $\int_0^1 (1-t^2)^{1/4} \, dt$ and $\int_0^1 (1-t^2)^{1/8} \, dt$ in terms of $\prod_{m\geqslant 1} \tanh(\pi m/2)$ and $\prod_{m\geqslant 1} \tanh(\pi m/\sqrt{2})$ respectively.

60. (This is a research problem, and the author does not know the complete solution.) For $a > 0$ set

$$L(a) = \int_0^1 \frac{\log(1+t^a)}{1+t} \, dt \quad \text{and} \quad T(a) = \int_0^1 \frac{\operatorname{atan}(t^a)}{1+t^2} \, dt.$$

Using a modification of a partial Epstein zeta function associated with *real* quadratic fields, H. Muzzafar (unpublished) claims to have obtained a Chowla–Selberg-type formula that implies that when a is a unit in a real quadratic field certain values of $L(a)$ and $T(a)$ can be evaluated explicitly as rational linear combinations of $\log(2)^2$, $\log(2)\log(a)$, π^2, $\pi^2 a$, and similar quantities.

(a) Using numerical integration methods or expansions in terms of the derivatives of the gamma function (see for instance Exercise 102 of Chapter 9), compute $L(a)$ for $a = 1+\sqrt{2}$, $3+2\sqrt{2}$, $2+\sqrt{3}$, $2+\sqrt{5}$, and $4+\sqrt{17}$, compute $T(a)$ for $a = 3+2\sqrt{2}$ and $2+\sqrt{5}$, and show using a suitable linear dependence relation algorithm such as LLL that Muzzafar's claim is indeed correct, at least numerically. Show that $L((1+\sqrt{5})/2) + L((3+\sqrt{5})/2)$ is also of the same form.

(b) Read the paper of Herglotz [Her] on the Kronecker limit formula for *real* quadratic fields, and prove a result that is as general as possible and includes the relations that you have found experimentally, as well as for instance

$$L(4+\sqrt{15}) = -\frac{\pi^2}{12}(\sqrt{15}-2) + \log(2)\log(\sqrt{3}+\sqrt{5}) + \log((1+\sqrt{5})/2)\log(2+\sqrt{3})$$

(I am indebted to C. Meyer for this reference).

61. Let $p \geqslant 3$ be prime and let $M = (m_{i,j})_{1\leqslant i,j\leqslant (p-1)/2}$ be the $((p-1)/2)\times((p-1)/2)$ matrix such that $m_{i,j} = \lfloor (i+1)(j+1)/p \rfloor$. Prove that the determinant of M is equal to $\left(\frac{-8}{p}\right)h_p^-$ (see Corollary 10.5.27 and the remarks after the proof of Lemma 3.6.22).

62. Combine the proofs of Proposition 10.5.21 and of Theorem 10.5.22 to prove the following. Let m be an integer and p a prime number not necessarily prime to

m. Write $m = p^v m_0$ with $p \nmid m_0$ and $v = v_p(m)$, denote by $f_{p,0}$ the order of p modulo m_0, and set $g_{p,0} = \phi(m_0)/f_{p,0}$. Then we have

$$p\mathbb{Z}_{\mathbb{Q}_m} = \prod_{1 \leqslant j \leqslant g_{p,0}} \mathfrak{p}_j^{e_p} \quad \text{with} \quad f(\mathfrak{p}_j/p) = f_{p0} ,$$

where $e_p = \phi(p^v) = p^{v-1}(p-1)$.

63. Imitate the proof of Theorem 10.5.30 and use the quadratic reciprocity law to prove the following: if $a \in \mathbb{Z}$ is not a perfect square, the analytic density of primes p such that a is a square modulo p is equal to $1/2$.

64.

(a) Using Corollary 10.7.3 and the bound for $\zeta'(s)$ given by Proposition 10.7.1, show that

$$\frac{1}{\zeta(1+it)} = O(\log(t)^7) .$$

(b) Deduce that for a suitable strictly positive constant C we have $1/\zeta(s) = O(\log(t)^7)$ uniformly for $\sigma > 1 - C/\log(t)^9$, and in particular that $\zeta(s)$ has no zeros in this region (this is of course much weaker than Theorem 10.7.8, obtained using Hadamard's factorization theorem).

65. As in Corollary 10.7.7, let ρ be the nontrivial zeros of $\zeta(s)$.

(a) Compute $\prod_\rho (1 - s^2/\rho^2)$ in terms of $\zeta(s)$ and $\zeta(s+1)$. In particular, compute $\prod_\rho (1 - 1/\rho^2)$ and $\prod_\rho (1 - 4/\rho^2)$.

(b) Set

$$b_k = \zeta(k)\left(1 - \frac{1}{2^k}\right) - 1 - \delta_k ,$$

where δ_k is given by Definition 10.3.18. Prove that for $k \geqslant 2$ we have

$$\sum_\rho \frac{1}{\rho^k} = -b_k .$$

(c) Deduce from Proposition 10.3.19 that for $|s| < 1$ we have

$$\sum_{k \geqslant 2} b_k s^{k-1} = \frac{\zeta'(s)}{\zeta(s)} + \frac{1}{s-1} + \frac{\psi(s/2+1)}{2} + 1 + \frac{\gamma}{2} - \log(2\pi) .$$

(d) Show that we can make s tend to 1 from below in the above formula, and deduce that for $k = 1$ we have

$$\sum_\rho \frac{1}{\rho} = 1 + \frac{\gamma}{2} - \frac{\log(\pi)}{2} - \log(2) ,$$

where nontrivial zeros ρ and $1 - \rho$ are grouped together to make the left-hand side converge.

(e) Conclude that if we group zeros in the same way we have the very simple Hadamard product

$$s(s-1)\Lambda(s) = s(s-1)\pi^{-s/2}\Gamma(s/2)\zeta(s) = \prod_\rho (1 - s/\rho) ,$$

so that the term e^{bs} disappears in this form. Note that it has been shown by H. Stark in [Sta1] that this is in fact the case for all Dedekind zeta functions.

66. Let χ be a nontrivial primitive character modulo $m > 1$, and as usual set $e = 0$ or 1 such that $\chi(-1) = (-1)^e$. Generalizing Corollary 10.7.7, show that

$$L(\chi, s) = b_0(\chi) \frac{e^{b_1(\chi)s}}{\Gamma((s+e)/2)} \prod_\rho \left(1 - \frac{s}{\rho}\right) e^{s/\rho} ,$$

where

$$b_0(\chi) = W(\chi) m^{1/2} \pi^{-e/2} L(\overline{\chi}, 1) \quad \text{and}$$

$$b_1(\chi) = \log(2\pi) - e\log(2) - \frac{\log(m)}{2} + \frac{\gamma}{2} - \frac{L'(\overline{\chi}, 1)}{L(\overline{\chi}, 1)} .$$

67. As in the second proof of Theorem 10.5.29 set $F_m(s) = \prod_{\chi \bmod m} L(\chi, s)$.

(a) Similarly to Lemma 10.7.2, prove that

$$F_m(\sigma)^3 |F_m(\sigma + it)|^4 |F_m(\sigma + 2it)| \geqslant 1 .$$

(b) Deduce from this that $L(\chi, 1+it) \neq 0$ for all $t \in \mathbb{R}$ and every Dirichlet character χ (strictly speaking, set $L(\chi_0, 1) = \infty$).

(c) Using a similar method to that given in the text for the PNT, deduce the stronger following form of Dirichlet's theorem for primes in arithmetic progression:

$$\pi(x; m, a) \sim \frac{1}{\phi(m)} \frac{x}{\log(x)} ,$$

where $\gcd(a, m) = 1$ and $\pi(x; m, a)$ is the number of primes up to x congruent to a modulo m.

68.

(a) Using Corollary 9.6.30, prove that for $x > 1$ we have the two formulas

$$\mathrm{Li}(x) = \int_1^x \left(1 - \frac{1}{t}\right) \frac{dt}{\log(t)} + \log(\log(x)) + \gamma$$

$$= \int_0^x \left(\frac{1}{1-t} + \frac{1}{\log(t)}\right) dt + \log(x - 1) .$$

(b) Deduce from this the following two convergent series representations:

$$\mathrm{Li}(x) = \gamma + \log(\log(x)) + \sum_{n \geqslant 1} \frac{\log(x)^n}{n \cdot n!}$$

$$= \gamma + \log(\log(x)) + \sum_{n \geqslant 1} (n-1)! \frac{x - P_n(\log(x))}{(\log(x))^n} ,$$

where $P_n(u) = \sum_{0 \leqslant j \leqslant n} u^j / j!$ is the nth partial sum of the exponential series. In particular, estimate the speed of convergence of the second series. Comments?

11. p-adic Gamma and L-Functions

Independently of its intrinsic interest, one of the most fascinating aspects of the theory presented in this chapter is that, although completely "elementary" in the sense that it does not use any highbrow mathematical notions or results, it implies in quite a straightforward manner many results of classical (as opposed to p-adic) number theory, for instance strengthenings of almost all the arithmetic results on Bernoulli numbers seen in Section 9.5, of the Jacobstahl–Kazandzidis congruences (Corollary 11.6.22), of the Davenport–Hasse product relation (Theorem 3.7.3, which will be improved in Theorem 11.7.16), and a simpler proof of the Stickelberger congruence (Theorem 3.6.6). I would like to thank F. Rodriguez-Villegas for making me interested in the whole subject thanks to a GP script for computing Morita's p-adic gamma function, to F. Beukers and E. Friedman for very interesting discussions, and especially to P. Colmez for his help on proving the results of Section 11.5.3.

11.1 Generalities on p-adic Functions

11.1.1 Methods for Constructing p-adic Functions

There are many ways in which to define p-adic functions with nice properties (at least continuous, but usually analytic), and we have already seen a few in Chapter 4. These methods are of course interrelated. In this short section, we survey a little more systematically these methods, and in the rest of this chapter we will use them to define some p-adic analogues of the gamma, zeta, and L-functions seen in the previous chapters.

- Perhaps the most natural method is as follows. Let $(a_n)_{n \geqslant 0}$ be a sequence of integers. Since $\mathbb{Z}_{\geqslant 0}$ is dense in $\mathbb{Z}_p$ we can define a function f on $\mathbb{Z}_p$ thanks to the formula

$$f(x) = \lim_{n \to x,\ n \in \mathbb{Z}_{\geqslant 0}} a_n \, ,$$

where of course $n \to x$ in the p-adic topology. It is clear that $f(x)$ exists if and only if (a_n) is p-adically continuous, in other words if for all $k \in \mathbb{Z}_{\geqslant 0}$ there exists $j \in \mathbb{Z}_{\geqslant 0}$ such that

$$n \equiv m \pmod{p^j} \implies a_n \equiv a_m \pmod{p^k} \, .$$

By definition, $f(x)$ will then be a *continuous* function on $\mathbb{Z}_p$. This is of course the primary motivation for p-adic numbers: a sequence can be p-adically interpolated if and only if it satisfies *congruences* modulo arbitrarily high powers of p. In this chapter, we will see a prominent example of this in the definition of Morita's p-adic gamma function.

– An equivalent method of p-adic interpolation is the use of Mahler expansions seen in Section 4.2.4. Recall from Mahler's Theorem 4.2.26 that if we set (with different notation)

$$f(x) = \sum_{k \geqslant 0} c_k \binom{x}{k}, \quad \text{with} \quad c_k = \sum_{0 \leqslant m \leqslant k} (-1)^{k-m} \binom{k}{m} a_m \, ,$$

then $f(k) = a_k$, and f is p-adically continuous if and only if c_k tends to 0 as $k \to \infty$. By Corollary 4.2.27, this method is equivalent to the preceding one, but usually has some advantages.

We will see that Morita's p-adic gamma function has a very simple Mahler expansion that can serve as an alternative definition, and is, as far as I know, the most efficient method for *computing* it.

– A third method, familiar from complex analysis, is to define p-adic functions as sums of *power series*. The examples of the p-adic logarithm and exponential studied in Chapter 4 are certainly the most important. In the p-adic setting, however, power series are not as important as in the complex setting, mainly because of the impossibility of doing analytic continuation (because of the ultrametric topology), at least in a naïve manner. Also, as we have seen in Chapter 4 (see Proposition 4.2.28 and Exercise 17) it is not difficult to go back and forth between Mahler expansions and power series. We will see that all the p-adic functions that we will introduce in this chapter have rather simple power series expansions.

– A fourth method is the use of p-adic measures, in particular the Amice transform. This is explained in detail in Colmez's lectures; see [Colm] and a course on his web site. However, it needs some analytic preparation, so we will not use it, although it can quite easily prove the two main results that we will give in Section 11.5.3. Thus we will almost always use a more naïve method that is specific to the p-adic setting: the use of *Volkenborn integrals*, which we briefly study in the next section.

11.1.2 A Brief Study of Volkenborn Integrals

A detailed study of the Volkenborn integral is completely elementary but quite long, and will not be needed in this book, so we refer instead to the exposition of A. Robert in [Rob1]. We will simply give some definitions and results.

The simplest class of functions for which we can study classical Riemann integration is the class of continuous functions on a compact interval. In the p-adic case, we have to assume a much stronger property, that of strict differentiability.

Definition 11.1.1. (1) *We say that a function f is* strictly differentiable *at a point $a \in \mathbb{Z}_p$ if the function of two variables $\Phi f(x,y) = (f(x) - f(y))/(x - y)$ has a limit $\ell = f'(a)$ as $(x,y) \to (a,a)$, $x \neq y$.*
(2) *We say that f is* strictly differentiable *on some subset X of $\mathbb{Z}_p$, and write $f \in S^1(X)$, if f is strictly differentiable for all $a \in X$.*

It is easy to show that $f \in S^1(X)$ if and only if Φf can be extended to a continuous function on $X \times X$, if and only if there exists a continuous function ε defined on $X \times X$ such that $\varepsilon(x,x) = 0$ and satisfying $f(y) = f(x) + (y - x)f'(x) + (y - x)\varepsilon(x,y)$ for all $(x,y) \in X \times X$.

Theorem 11.1.2. *Let $f(x) = \sum_{k \geqslant 0} a_k \binom{x}{k}$ be the Mahler expansion of a continuous function f on $\mathbb{Z}_p$ (see Theorem 4.2.26).*

(1) *f is Lipschitz-continuous (in other words Φf is bounded) if and only if $k|a_k|$ is bounded. In that case,*

$$\|\Phi f\| = \sup_{x \neq y} |\Phi f(x,y)| = \sup_{k \geqslant 1} p^{\lfloor \log(k)/\log(p) \rfloor} |a_k| \,.$$

(2) *$f \in S^1(\mathbb{Z}_p)$ if and only if $k|a_k| \to 0$ as $k \to \infty$.*

Definition 11.1.3. *If f is Lipschitz-continuous we define the L^1-norm of f by the formula*

$$\|f\|_1 = \max(|f(0)|, \|\Phi f\|) \,,$$

which is indeed a norm.

We can now give the definition of the Volkenborn integral:

Definition 11.1.4. *Let g be a function from $\mathbb{Z}_p$ to $\mathbb{C}_p$. We define the Volkenborn integral of g on $\mathbb{Z}_p$, if it exists, by the formula*

$$\int_{\mathbb{Z}_p} g(t)\, dt = \lim_{r \to \infty} \frac{1}{p^r} \sum_{0 \leqslant n < p^r} g(n) \,.$$

If g is a function from $U_p = \mathbb{Z}_p^*$ to $\mathbb{C}_p$, we define similarly

$$\int_{\mathbb{Z}_p^*} g(t)\, dt = \lim_{r \to \infty} \frac{1}{p^r} \sum_{0 \leqslant n < p^r,\ p \nmid n} g(n) \,.$$

Note that if g is a function on $\mathbb{Z}_p^*$ and if we define g_0 to be the function on $\mathbb{Z}_p$ equal to g on $\mathbb{Z}_p^*$ and to 0 on $p\mathbb{Z}_p$ then evidently $\int_{\mathbb{Z}_p^*} g(t)\, dt = \int_{\mathbb{Z}_p} g_0(t)\, dt$. On the other hand, because of the p-adic topology it is clear that $g \in S^1(\mathbb{Z}_p^*)$ if and only if $g_0 \in S^1(\mathbb{Z}_p)$, so that we can always reduce an integral over $\mathbb{Z}_p^*$ to an integral over $\mathbb{Z}_p$ if desired. The following result, which we will not prove, ensures the existence of the Volkenborn integral of sufficiently regular functions; see [Rob1].

Proposition 11.1.5. *If $g \in S^1(\mathbb{Z}_p)$ then $\int_{\mathbb{Z}_p} g(t)\, dt$ exists, and similarly for* $\mathbb{Z}_p^*$.

We will thus be able to define p-adic functions by integrating functions of two variables, in other words by setting

$$ f(x) = \int_{\mathbb{Z}_p} g(x,t)\, dt \quad \text{or} \quad f(x) = \int_{\mathbb{Z}_p^*} g(x,t)\, dt \ . $$

We will see that all the functions that we will introduce in this chapter (the logarithm of Morita's p-adic gamma function, Diamond's p-adic log gamma function, and p-adic zeta and L-functions) have a simple definition in terms of Volkenborn integrals.

To avoid excessive technicalities, we will be a little sloppy, and often assume without any justification that we can differentiate under the integral sign. This is done in [Rob1] for integrals of the form $\int_{\mathbb{Z}_p} g(x+t)\, dt$, and otherwise it can be checked directly on the specific integral without appealing to general theorems.

Here are some basic properties of these integrals, which we will not need. We always assume that the functions f that occur are in $S^1(\mathbb{Z}_p)$.

Proposition 11.1.6. (1)

$$ \left| \int_{\mathbb{Z}_p} f(t)\, dt \right| \leqslant p\|f\|_1 \ . $$

(2) *If $\|f_n - f\|_1 \to 0$ (in other words if $f_n \to f$ in $S^1(\mathbb{Z}_p)$) then*

$$ \int_{\mathbb{Z}_p} f_n(t)\, dt \to \int_{\mathbb{Z}_p} f(t)\, dt \ . $$

(3)

$$ \int_{\mathbb{Z}_p} (f(t+1) - f(t))\, dt = f'(0) \ . $$

In particular, if $g(x) = \int_{\mathbb{Z}_p} f(x+t)\, dt$, then $g(x+1) - g(x) = f'(x)$.

(4) If $f(x) = \sum_{k \geqslant 0} a_k \binom{x}{k}$ then

$$\int_{\mathbb{Z}_p} f(t)\, dt = \sum_{k \geqslant 0} (-1)^k \frac{a_k}{k+1}\ .$$

(5) If f is an odd function ($f(-x) = -f(x)$), then

$$\int_{\mathbb{Z}_p} f(t)\, dt = -\frac{f'(0)}{2}\ .$$

Examples. (1) For $x \in \mathbb{C}_p$ such that $|x| < 1$, we have

$$\int_{\mathbb{Z}_p} (1+x)^t\, dt = \frac{\log_p(1+x)}{x}\ .$$

(2) For all $x \in \mathbb{Q}_p$ and $k \in \mathbb{Z}_{\geqslant 0}$ we have

$$\int_{\mathbb{Z}_p} (x+t)^k\, dt = B_k(x)\ .$$

We invite the reader to prove these formulas (Exercise 1). Since the second example above is essential, we give the proof of a more general result.

Lemma 11.1.7. *Let χ be a periodic function defined on $\mathbb{Z}$ of period a power of p, and let $k \in \mathbb{Z}_{\geqslant 0}$. For all $x \in \mathbb{C}_p$ we have*

$$\int_{\mathbb{Z}_p} \chi(t)(x+t)^k\, dt = B_k(\chi, x)\ .$$

In particular,

$$\int_{\mathbb{Z}_p} (x+t)^k\, dt = B_k(x) \quad \text{and} \quad \int_{\mathbb{Z}_p} \chi(t) t^k\, dt = B_k(\chi)\ .$$

Proof. By definition and Corollary 9.4.17 we have

$$\int_{\mathbb{Z}_p} \chi(t)(x+t)^k\, dt = \lim_{r \to \infty} \frac{1}{p^r} \sum_{0 \leqslant n < p^r} \chi(n)(n+x)^k$$

$$= \lim_{r \to \infty} \frac{B_{k+1}(\chi, p^r + x) - B_{k+1}(\chi, x)}{p^r(k+1)}$$

$$= \frac{B'_{k+1}(\chi, x)}{k+1} = B_k(\chi, x)$$

as soon as p^r is a multiple of the period of χ, by definition of the derivative and the fact that $B'_{k+1}(\chi, x) = (k+1)B_k(\chi, x)$. $\qquad\square$

11.2 The p-adic Hurwitz Zeta Functions

11.2.1 Teichmüller Extensions and Characters on $\mathbb{Z}_p$

Introduction.

Recall that in the complex case, our fundamental building block was the Hurwitz zeta function $\zeta(s, x)$, which enabled us first to motivate the definition of the gamma function and prove most of its properties as immediate consequences of the corresponding ones for $\zeta(s, x)$, and second to define the Dirichlet L-functions as a finite linear combination of $\zeta(s, x)$ for suitable rational values of x. We will proceed in exactly the same way in the p-adic case. We are going to see, however, that it is essential to distinguish between the cases $v_p(x) < 0$ and $v_p(x) \geqslant 0$.

Definition of q_p.

The prime number $p = 2$ is always annoying in number theory, and especially in $\mathfrak{p}$-adic theory: over a general $\mathfrak{p}$-adic field the annoying primes are those for which $e/(p-1) \geqslant 1$, in other words $e \geqslant p-1$. In the case of $\mathbb{Q}_p$, which is the main object of consideration in this chapter (although some variables will be in $\mathbb{C}_p$), the only annoying prime is $p = 2$ (the "oddest prime" as a famous saying goes). It is thus convenient to set the following notation, which we have met briefly in Proposition 4.4.47:

Definition 11.2.1. *We set $q_p = p$ when $p \geqslant 3$, and $q_2 = 4$. In addition, we define*
$$\mathrm{C}\mathbb{Z}_p = \{x \in \mathbb{Q}_p, \ v_p(x) \leqslant -v_p(q_p)\},$$
so that when $p \geqslant 3$ we have $\mathrm{C}\mathbb{Z}_p = \mathbb{Q}_p \setminus \mathbb{Z}_p$.

Extensions of the Teichmüller character.

Recall from Definition 4.3.10 that if $a \in \mathbb{Z}_p^*$ is a p-adic unit we let $\omega(a)$ be the Teichmüller representative of a. With the above notation it is the unique $\phi(q_p)$th root of unity such that $\langle a \rangle = a/\omega(a) \equiv 1 \pmod{q_p \mathbb{Z}_p}$. In particular, thanks to Corollary 4.2.18 we know that $\langle a \rangle^s = \exp_p(s \log_p(\langle a \rangle)) = \exp_p(s \log_p(a))$ is well defined by a power series that converges for $|s| < R_p = q_p/p^{1/(p-1)}$. Note the crucial fact that $R_p > 1$. It is essential to extend these functions to $\mathbb{Q}_p$, as follows.

Definition 11.2.2. (1) *We extend the notation $\langle \ \rangle$ to $\mathbb{Q}_p^*$ by setting $\langle a \rangle = \langle a/p^{v_p(a)} \rangle$.*

(2) *We extend the notation ω to $\mathbb{Z}_p$ as a Dirichlet character modulo p; in other words, we set $\omega(a) = 0$ for $a \in \mathbb{Z}_p \setminus \mathbb{Z}_p^* = p\mathbb{Z}_p$. More generally, for any $k \in \mathbb{Z}$ we let ω^k be the kth power of ω in the sense of characters, so that $\omega^k(a) = 0$ for $a \in p\mathbb{Z}_p$, even when $k \leqslant 0$.*

(3) *In particular, we set $\chi_0 = \omega^0$, which is the trivial character modulo p on $\mathbb{Z}_p$, equal to the characteristic function of $\mathbb{Z}_p^*$.*

(4) *If $a \in \mathbb{Q}_p^*$, we define $\omega_v(a)$ by the equivalent formulas*

$$\omega_v(a) = a/\langle a \rangle = p^{v_p(a)}\omega(a/p^{v_p(a)})$$

(the subscript v is simply to recall that the valuation is involved).

Remarks. (1) The use of the same notation $\langle a \rangle$ for $a \in \mathbb{Q}_p^*$ cannot lead to any confusion. On the other hand, it is essential to distinguish the two possible extensions of $\omega(a)$. It is reasonable to keep the same notation for the extension as a Dirichlet character, but the other extension must be given another notation, which I have chosen to be ω_v.
(2) In particular, if $a \in \mathbb{Z}_p$ we have $\omega(a) = \chi_0(a)\omega_v(a)$ with the notation χ_0 introduced above, and more generally for any $v \in \mathbb{Z}$ we have $\omega(a) = p^v \chi_0(a)\omega_v(a/p^v)$.
(3) It is clear that $\langle a \rangle$, $\omega(a)$, and $\omega_v(a)$ are still multiplicative functions, that by definition $\omega_v(a)\langle a \rangle = a$, and that $\log_p(\omega_v(a)) = 0$ since by definition of the Iwasawa p-adic logarithm we have set $\log_p(p) = 0$.
(4) By Proposition 4.4.44, the functions $\omega(a)$ and $\langle a \rangle$ can be canonically defined on the p-adic units of $\mathbb{C}_p$. However, we cannot naturally extend these symbols to $a \in \mathbb{C}_p^*$ since $p^{v_p(a)}$ is not uniquely defined (see the remarks preceding Proposition 4.4.44).

For future reference, we note the following formula:

Lemma 11.2.3. *We have*

$$\frac{\partial}{\partial x}\langle x \rangle^{1-s} = (1-s)\frac{\langle x \rangle^{1-s}}{x} = (1-s)\frac{\langle x \rangle^{-s}}{\omega_v(x)}.$$

Proof. Trivial and left to the reader.   □

Dirichlet characters on $\mathbb{Z}_p$.

Let χ be a Dirichlet character modulo p^v for some v. If $a \in \mathbb{Z}_p$ and a_n is a sequence of integers tending to a p-adically, we have $v_p(a_n - a_m) \geqslant v$ for n and m sufficiently large, so $\chi(a_n)$ is an ultimately constant sequence, and we of course set $\chi(a) = \chi(a_n)$ for $v_p(a - a_n) \geqslant v$. This is called a Dirichlet character on $\mathbb{Z}_p$, and it is clear that it has all the usual properties. In particular, it is multiplicative, and $\chi(a) = 0$ if and only if $a \in p\mathbb{Z}_p$. The characters ω^r that we have defined above are examples of such Dirichlet characters (with $v = v_p(q_p)$). Note that when the conductor of χ is not a power of p we cannot define an extension of χ to $\mathbb{Z}_p$.

11.2.2 The p-adic Hurwitz Zeta Function for $x \in C\mathbb{Z}_p$

Recall that $C\mathbb{Z}_p$ is the set of $x \in \mathbb{Q}_p$ such that $v_p(x) \leqslant -v_p(q_p)$.

Proposition 11.2.4. *Let $x \in C\mathbb{Z}_p$, and let $s \in \mathbb{C}_p$ be such that $|s| < R_p = q_p/p^{1/(p-1)}$ (in other words, $v_p(s) > 1/(p-1) - v_p(q_p)$).*

(1) *The Volkenborn integral $\int_{\mathbb{Z}_p} \langle x + t \rangle^{1-s} \, dt$ exists, and more precisely we have*

$$\int_{\mathbb{Z}_p} \langle x + t \rangle^{1-s} \, dt = \frac{1}{p^r} \sum_{0 \leqslant n < p^r} \langle n + x \rangle^{1-s} + O(p^{r-2}) ,$$

where we recall that $A = O(p^k)$ means that $v_p(A) \geqslant k$.

(2) *We have the convergent Laurent series expansion*

$$\int_{\mathbb{Z}_p} \langle x + t \rangle^{1-s} \, dt = \langle x \rangle^{1-s} \sum_{j \geqslant 0} \binom{1-s}{j} B_j x^{-j} .$$

Proof. Since $R_p > 1$, note first that $|s| < R_p$ is equivalent to $|1 - s| < R_p$, and furthermore by Proposition 4.4.47 we know that the series defining $\langle x \rangle^{1-s}$ converges for $|s| < R_p$, for all $x \in \mathbb{Q}_p$ with the extended definition of $\langle x \rangle$ that we have given. We can therefore write

$$\langle n + x \rangle^{1-s} = \langle x \rangle^{1-s}(1 + n/x)^{1-s} = \langle x \rangle^{1-s} \sum_{j \geqslant 0} \binom{1-s}{j} n^j x^{-j} ,$$

so that

$$\frac{1}{p^r} \sum_{0 \leqslant n < p^r} \langle n + x \rangle^{1-s} = \langle x \rangle^{1-s} \sum_{j \geqslant 0} \binom{1-s}{j} x^{-j} \frac{B_{j+1}(p^r) - B_{j+1}(0)}{(j+1)p^r}$$

by Euler–MacLaurin. In the proof of Lemma 11.1.7 we used the formula

$$\lim_{r \to \infty} \frac{B_{j+1}(p^r) - B_{j+1}(0)}{(j+1)p^r} = B_j .$$

Here we of course use this same formula, but since we have an infinite series of limits we must show that we can take the limit term by term. This can be done in great generality, but in this special case it is very easy: we have

$$\frac{B_{j+1}(p^r) - B_{j+1}(0)}{(j+1)p^r} = B_j + \sum_{2 \leqslant i \leqslant j+1} \binom{j}{i-1} B_{j+1-i} \frac{p^{r(i-1)}}{i} .$$

We have trivially $v_p(i) \leqslant i - 1$ and by Clausen–von Staudt $v_p(B_{j+1-i}) \geqslant -1$, so that for $i \geqslant 2$,

$$v_p(B_{j+1-i} p^{r(i-1)}/i) \geqslant (r-1)(i-1) - 1 \geqslant r - 2 .$$

It follows that there exist p-adic *integers* $A_j(r)$ such that

$$\frac{B_{j+1}(p^r) - B_{j+1}(0)}{(j+1)p^r} = B_j + p^{r-2} A_j(r) ,$$

so that

$$\sum_{j\geqslant 0}\binom{1-s}{j}x^{-j}\frac{B_{j+1}(p^r)-B_{j+1}(0)}{(j+1)p^r}$$

$$=\sum_{j\geqslant 0}\binom{1-s}{j}x^{-j}B_j+p^{r-2}\sum_{j\geqslant 0}\binom{1-s}{j}x^{-j}A_j(r)\,.$$

Since $|A_j(r)|\leqslant 1$ it follows that

$$\left|\sum_{j\geqslant 0}\binom{1-s}{j}x^{-j}A_j(r)\right|\leqslant\sum_{j\geqslant 0}\left|\binom{1-s}{j}x^{-j}\right|\,.$$

The series on the right-hand side is the (absolute value of the) power series expansion of $(1+1/x)^{1-s}$, which converges by Corollary 4.2.16 since $v_p(1/x)\geqslant v_p(q_p)>1/(p-1)$ and $v_p(1-s)>1/(p-1)-v_p(q_p)\geqslant 1/(p-1)-v_p(1/x)$, and in fact its absolute value is equal to 1. Thus the left-hand side is bounded independently of r, so we may indeed deduce that

$$\lim_{r\to\infty}\frac{1}{p^r}\sum_{0\leqslant n<p^r}\langle n+x\rangle^{1-s}=\langle x\rangle^{1-s}\sum_{j\geqslant 0}\binom{1-s}{j}x^{-j}B_j\,,$$

as well as the last statement.   □

By Proposition 9.6.6 and Lemma 11.1.7 we have for $x\in\mathbb{Q}$ and $k\in\mathbb{Z}_{\geqslant 1}$,

$$-\frac{1}{k}\int_{\mathbb{Z}_p}(x+t)^k\,dt=-\frac{B_k(x)}{k}=\zeta(1-k,x)\,.$$

Since $\mathbb{Z}_{\geqslant 1}$ is dense in $\mathbb{Z}_p$, this motivates the following definition, which makes sense thanks to the above proposition:

Definition 11.2.5. *For $s\in\mathbb{C}_p\setminus\{1\}$ such that $|s|<R_p$ and $x\in C\mathbb{Z}_p$ we define $\zeta_p(s,x)$ by the equivalent formulas*

$$\zeta_p(s,x)=\frac{1}{s-1}\int_{\mathbb{Z}_p}\langle x+t\rangle^{1-s}\,dt=\frac{\langle x\rangle^{1-s}}{s-1}\sum_{j\geqslant 0}\binom{1-s}{j}B_jx^{-j}\,.$$

Remarks. (1) In the case $p=2$ and $v_p(x)=-1$, which is not included in the above definition, we can still define $\zeta_p(s,x)$, but it is then necessary to restrict to $s\in\mathbb{Z}_p\setminus\{1\}$ and to slightly modify the formulas; see Exercise 5.
(2) We will see in the next section that there is an analogous definition for $v_p(x)\geqslant 0$, but this deserves a separate study.
(3) The first formula for $\zeta_p(s,x)$ is the most natural one, but for many purposes it will be simpler to use the second.

(4) As already mentioned we cannot extend $\langle\ \rangle$ to $\mathbb{C}_p$, so we must restrict to $x \in \mathbb{Q}_p$.

(5) Here and elsewhere, note that the p-adic analogue of a complex *infinite series* (here of $\sum_{n \geqslant 0}(n + x)^{-s}$) is, up to sign, the Volkenborn integral of the *integral* of the sum with respect to s, and not of the derivative as could be expected.

Proposition 11.2.6. *Assume that $x \in C\mathbb{Z}_p$.*

(1) *For any $k \in \mathbb{Z} \setminus \{0\}$ we have*

$$\zeta_p(1 + k, x) = \frac{\omega_v(x)^k}{k} \int_{\mathbb{Z}_p} \frac{dt}{(x + t)^k} \ .$$

(2) *For $k \in \mathbb{Z}_{\geqslant 1}$ we have*

$$\zeta_p(1 - k, x) = -\omega_v(x)^{-k}\frac{B_k(x)}{k} \ ,$$

and if in addition $x \in \mathbb{Q}$, then $\zeta_p(1 - k, x) = \omega_v(x)^{-k}\zeta(1 - k, x)$.

Proof. By definition of ω_v, if $x \in C\mathbb{Z}_p$ we have $\omega_v(1 + n/x) = 1$, so that $\omega_v(n+x) = \omega_v(x)$ for all $n \in \mathbb{Z}$. It follows that $\langle n+x \rangle^{-k} = (n+x)^{-k}\omega_v(x)^k$, hence

$$\zeta_p(1 + k, x) = \frac{\omega_v(x)^k}{k} \int_{\mathbb{Z}_p} (x + t)^{-k} \, dt \ ,$$

proving (1). For (2) we deduce from Lemma 11.1.7 that if $k \in \mathbb{Z}_{\geqslant 1}$ we have

$$\zeta_p(1 - k, x) = -\frac{\omega_v(x)^{-k}}{k} \int_{\mathbb{Z}_p} (x + t)^k \, dt = -\omega_v(x)^{-k}\frac{B_k(x)}{k} \ ,$$

and this is equal to $\omega_v(x)^{-k}\zeta(1 - k, x)$ when $x \in \mathbb{Q}$ by Corollary 9.6.10. $\square$

Corollary 11.2.7. *We have*

$$\frac{\partial \zeta_p}{\partial x}(s, x) = -\frac{s}{\omega_v(x)}\zeta_p(s + 1, x) \ .$$

Proof. Formally, this follows from the integral definition and Lemma 11.2.3, but we would need to justify the derivation under the integral sign. Instead, we use the series given by the proposition, since it is normally convergent for $x \in C\mathbb{Z}_p$. In that region we can therefore differentiate termwise. Since

$$\zeta_p(s, x) = \frac{1}{s - 1}\sum_{j \geqslant 0}\binom{1 - s}{j}B_j\langle x\rangle^{1-s-j}\omega_v(x)^{1-j}$$

and since $\omega_v(y) = \omega_v(x)$ if y is sufficiently close to x, it follows from Lemma 11.2.3 and the formula $(1 - s - j)\binom{1-s}{j} = (1 - s)\binom{-s}{j}$ that

$$\frac{\partial \zeta_p}{\partial x}(s,x) = \frac{1}{s-1}\sum_{j\geqslant 0}(1-s-j)\binom{1-s}{j}B_j\frac{\langle x\rangle^{-s-j}}{\omega_v(x)^j}$$

$$= -\frac{\langle x\rangle^{-s-1}}{\omega_v(x)}\sum_{j\geqslant 0}\binom{-s}{j}B_j x^{1-j} = -\frac{s}{\omega_v(x)}\zeta_p(s+1,x)\,,$$

proving the corollary. $\square$

Proposition 11.2.8. *For fixed $x \in C\mathbb{Z}_p$ the function $\zeta_p(s,x)$ is a p-adic meromorphic function on $|s| < R_p = q_p/p^{1/(p-1)}$, which in addition is analytic, except for a simple pole at $s = 1$ with residue 1.*

Proof. Since by definition $\langle x\rangle \equiv 1 \pmod{q_p\mathbb{Z}_p}$, we know that $\langle x\rangle^{1-s}$ is an analytic function on $|1-s| < R_p$, or equivalently, on $|s| < 1$ since $R_p > 1$, and in particular is defined and continuous on $\mathbb{Z}_p$, so that we need to consider only the infinite series.

By the Clausen–von Staudt Theorem 9.5.14 we have $v_p(B_j) \geqslant -1$, hence $v_p(x^{-j}B_j) \geqslant jv_p(q_p) - 1$. Applying Proposition 4.2.28 with $\alpha = v_p(q_p)$, $\alpha' = 0$, and $\alpha'' = -1$ we deduce that the series $\sum_{j\geqslant 0}\binom{s}{j}x^{-j}B_j$ has a radius of convergence greater than or equal to R_p, so that it defines an analytic function for $|s| < R_p$, with value 1 at $s = 0$, proving the proposition after changing s into $1-s$ and noting again that $|1-s| < R_p$ is equivalent to $|s| < R_p$. $\square$

Remarks. (1) We do not need any fancy definition of meromorphic p-adic functions: the analyticity and meromorphy statements simply mean that the function $f(s) = (s-1)\zeta_p(s,x)$ has a power series expansion that converges for $|s| < R_p$ and that $f(1) = 1$.

(2) As in the complex case (see the statements following Proposition 9.6.8), since we will *define* $\mathrm{Log}\Gamma_p(x)$ to be $\omega_v(x)\frac{\partial \zeta_p}{\partial s}(0,x)$, it follows from Corollary 11.2.7 that around $s = 1$ we have more precisely

$$\zeta_p(s,x) = \frac{1}{s-1} - \psi_p(x) + O(s-1)\,,$$

where $\psi_p(x) = (d/dx)(\mathrm{Log}\Gamma_p(x))$ (see Proposition 11.5.6).

Theorem 11.2.9. *Keep the above notation, and let $x \in C\mathbb{Z}_p$.*

(1) *For $k \in \mathbb{Z}_{\geqslant 1}$ we have*

$$\zeta_p(1-k,x) = -\omega_v(x)^{-k}\frac{B_k(x)}{k}\,,$$

which is also equal to $\omega_v(x)^{-k}\zeta(1-k,x)$ if $x \in \mathbb{Q}$.

(2) *If $x/u \in C\mathbb{Z}_p$ (and in particular if $u \in \mathbb{Z}_p$), we have the Laurent expansion*

$$\zeta_p(s,x+u) = \frac{\langle x\rangle^{1-s}}{s-1}\sum_{j\geqslant 0}\binom{1-s}{j}B_j(u)x^{-j}\,.$$

(3) *We have the functional equation*

$$\zeta_p(s, x+1) - \zeta_p(s, x) = -\frac{\langle x \rangle^{1-s}}{x} = -\omega_v(x)^{-1}\langle x \rangle^{-s} .$$

(4) *We have the reflection formula*

$$\zeta_p(s, 1-x) = \zeta_p(s, x) .$$

(5) *If $N \in \mathbb{Z}_{\geqslant 1}$ is such that $Nx \in C\mathbb{Z}_p$ we have the distribution formula*

$$\sum_{0 \leqslant j < N} \zeta_p\left(s, x + \frac{j}{N}\right) = \omega_v(N)\langle N \rangle^s \zeta_p(s, Nx) .$$

Proof. (1) Although we have already seen this in Proposition 11.2.6, which was in fact the motivation of our definition of $\zeta_p(s, x)$, we prove this from the series expansion. Indeed, by definition of the Bernoulli polynomials, for $k \geqslant 1$ we have

$$\zeta_p(1-k, x) = -\frac{\langle x \rangle^k}{k}\sum_{j \geqslant 0}\binom{k}{j}B_j x^{-j} = -\frac{\langle x \rangle^k}{k}x^{-k}B_k(x) = -\omega_v(x)^{-k}\frac{B_k(x)}{k} .$$

(2). Since $x/u \in C\mathbb{Z}_p$ we have $v_p(x+u) = v_p(x)$, hence

$$(x+u)/p^{v_p(x+u)} \equiv x/p^{v_p(x)} \pmod{q_p\mathbb{Z}_p} .$$

Thanks to our extended definition of ω_v and $\langle\ \rangle$ we have $\omega_v(x+u) = \omega_v(x)$ and $\langle x + u \rangle = (1 + u/x)\langle x \rangle$, hence

$$
\begin{aligned}
\zeta_p(s, x+u) &= \frac{\langle x+u \rangle^{1-s}}{s-1}\sum_{j \geqslant 0}\binom{1-s}{j}(x+u)^{-j}B_j \\
&= \frac{\langle x \rangle^{1-s}}{s-1}\sum_{j \geqslant 0}\binom{1-s}{j}x^{-j}(1+u/x)^{1-s-j}B_j \\
&= \frac{\langle x \rangle^{1-s}}{s-1}\sum_{j \geqslant 0}\binom{1-s}{j}x^{-j}B_j\sum_{k \geqslant 0}\binom{1-s-j}{k}u^k x^{-k} \\
&= \frac{\langle x \rangle^{1-s}}{s-1}\sum_{n \geqslant 0}\binom{1-s}{n}x^{-n}\sum_{j=0}^{n}\binom{n}{j}u^{n-j}B_j \\
&= \frac{\langle x \rangle^{1-s}}{s-1}\sum_{n \geqslant 0}\binom{1-s}{n}x^{-n}B_n(u) ,
\end{aligned}
$$

proving (2).

(3) and (4). Since $B_j(1) - B_j(0) = 0$ for $j \neq 1$, $B_1(1) - B_1(0) = 1$, and $B_j(1) = (-1)^j B_j$, (3) and (4) immediately follow from (2). Note that they clearly also follow from the definition of ζ_p as a Volkenborn integral.

(5). By Proposition 9.1.3 and (2) we have

$$\sum_{0\leqslant j<N} \zeta_p(s, x+j/N) = \frac{\langle x\rangle^{1-s}}{s-1} \sum_{n\geqslant 0} \binom{1-s}{n} x^{-n} \sum_{0\leqslant j<N} B_n(j/N)$$

$$= \frac{\langle x\rangle^{1-s}}{s-1} \sum_{n\geqslant 0} \binom{1-s}{n} x^{-n} \frac{B_n}{N^{n-1}} = \frac{N\langle x\rangle^{1-s}}{\langle Nx\rangle^{1-s}}\zeta_p(s, Nx)$$

$$= N\langle N\rangle^{s-1}\zeta_p(s, Nx) = \omega_v(N)\langle N\rangle^s \zeta_p(s, Nx)$$

since $\langle\ \rangle$ is multiplicative, proving (5). $\square$

Formulas (3) and (5) should be compared with the complex case (Propositions 9.6.2 and 9.6.12).

Statement (5) does not make sense when $v_p(Nx) \geqslant 0$, since $\zeta_p(s, Nx)$ is not defined, but since we will define $\zeta_p(s, x)$ (and more general functions) when $v_p(x) \geqslant 0$ below, there does exist a suitable generalization (Corollary 11.2.15).

We end this section with the following formula, which is a p-adic generalization of Raabe's formula (Proposition 9.6.50 and Corollary 9.6.54). I thank E. Friedman for having suggested that such a formula might exist; see [Coh-Fri].

Proposition 11.2.10. *For $|s| < R_p$ and $x \in C\mathbb{Z}_p$ we have*

$$\int_{\mathbb{Z}_p} \zeta_p(s, x+t)\, dt = s\zeta_p(s, x) + (x-1)\frac{\partial \zeta_p}{\partial x}(s, x)$$

$$= s\left(\zeta_p(s, x) - \langle x-1\rangle\zeta_p(s+1, x)\right) .$$

Proof. The series given by Theorem 11.2.9 (2) being normally convergent for $u \in \mathbb{Z}_p$ can thus be integrated term by term, so that using Exercise 3 (a) we obtain

$$\int_{\mathbb{Z}_p} \zeta_p(s, x+t)\, dt = \frac{\langle x\rangle^{1-s}}{s-1} \sum_{j\geqslant 0} \binom{1-s}{j} x^{-j} \int_{\mathbb{Z}_p} B_j(t)\, dt$$

$$= -\frac{\langle x\rangle^{1-s}}{s-1} \sum_{j\geqslant 0} \binom{1-s}{j}(jB_{j-1} + (j-1)B_j)x^{-j} .$$

Now by Corollary 11.2.7 we have

$$\sum_{j\geqslant 0} \binom{1-s}{j} jB_{j-1}x^{-j} = (1-s)\sum_{j\geqslant 0} \binom{-s}{j} B_j x^{-j-1}$$

$$= \frac{1-s}{x} s\langle x\rangle^s \zeta_p(s+1, x) = (s-1)\langle x\rangle^{s-1}\frac{\partial \zeta_p}{\partial x}(s, x) .$$

Furthermore, by a direct computation using Lemma 11.2.3 we have

$$x\frac{\partial \zeta_p}{\partial x}(s,x) = -\frac{\langle x\rangle^{1-s}}{s-1}\sum_{j\geq 0}\binom{1-s}{j}jB_jx^{-j} - (s-1)\zeta_p(s,x) \,,$$

hence

$$\frac{\langle x\rangle^{1-s}}{s-1}\sum_{j\geq 0}\binom{1-s}{j}(j-1)B_jx^{-j} = -x\frac{\partial \zeta_p}{\partial x}(s,x) - s\zeta_p(s,x) \,.$$

Replacing in the above formula for $\int_{\mathbb{Z}_p}\zeta_p(s,x+t)\,dt$ gives the first formula, and the second follows from Corollary 11.2.7. $\qquad\square$

Using Exercise 2 it is immediate to give a much simpler proof of this proposition, which generalizes; see Exercise 3.

11.2.3 The Function $\zeta_p(s,x)$ Around $s=1$

To finish the study of the function $\zeta_p(s,x)$ for $x \in C\mathbb{Z}_p$ we need more information on the coefficients of the power series expansion around $s=1$. We will see that this has important arithmetic applications. The main result is as follows.

Theorem 11.2.11. *Let* $x \in C\mathbb{Z}_p$.

(1) *We have* $1/(s-1)\sum_{j\geq 0}\binom{1-s}{j}B_jx^{-j} = 1/(s-1) + \sum_{j\geq 0}c_j(s-1)^j$ *with*

$$c_0 \equiv \frac{1}{2x} + \frac{1}{12x^2} \pmod{(q_p/x)\mathbb{Z}_p}\,, \quad c_1 \equiv \frac{1}{12x^2} \pmod{(q_p/x)\mathbb{Z}_p}\,, \text{ and}$$
$$c_j \equiv 0 \pmod{(q_p/x)\mathbb{Z}_p} \quad \text{for } j \geq 2.$$

(2) *We have* $\zeta_p(s,x) = 1/(s-1) + \sum_{j\geq 0}a_j(s-1)^j$ *with*

$$a_0 \equiv \frac{1}{2x} + \frac{1}{12x^2} - \log_p(\langle x\rangle) \pmod{(q_p/x)\mathbb{Z}_p}\,,$$
$$a_1 \equiv \frac{\log_p(\langle x\rangle)^2}{2} - \frac{\log_p(\langle x\rangle)}{2x} + \frac{1}{12x^2} \pmod{(q_p/x)\mathbb{Z}_p}\,, \text{ and}$$
$$a_j \equiv (-1)^{j+1}\frac{\log_p(\langle x\rangle)^{j+1}}{(j+1)!} \pmod{(q_p/x)\mathbb{Z}_p} \quad \text{for } j \geq 2.$$

(3) *If* $q_p \mid m$ *and* $p \nmid a$, *we have* $(m^{1-s}/m)\zeta_p(s,a/m) = 1/(m(s-1)) + \sum_{j\geq 0}b_j(s-1)^j$ *with*

$$b_0 \equiv \frac{1}{2a} + \frac{m}{12a^2} - \frac{\log_p(\langle a\rangle)}{m} \pmod{q_p\mathbb{Z}_p}\,,$$
$$b_1 \equiv \frac{\log_p(\langle a\rangle)^2}{2m} - \frac{\log_p(\langle a\rangle)}{2a} + \frac{m}{12a^2} \pmod{q_p\mathbb{Z}_p}\,, \text{ and}$$
$$b_j \equiv (-1)^{j+1}\frac{\log_p(\langle a\rangle)^{j+1}}{(j+1)!m} \pmod{q_p\mathbb{Z}_p} \quad \text{for } j \geq 2.$$

Remark. For $p \geqslant 5$ we have $1/(12x^2) \equiv 0 \pmod{(q_p/x)\mathbb{Z}_p}$, and for $p \geqslant 3$ we have $\log_p(\langle x \rangle)/(2x) \equiv 0 \pmod{(q_p/x)\mathbb{Z}_p}$.

Proof. For simplicity of notation, set $v = |v_p(x)| = -v_p(x) \geqslant v_p(q_p)$. By the Clausen–von Staudt theorem and Lemma 4.2.8, for $j \geqslant 1$ we have

$$v_p(B_j x^{-j}/j!) \geqslant -jv_p(x) - v_p(j!) - 1 \geqslant j(v - 1/(p-1)) - 1 + 1/(p-1) .$$

Since $v \geqslant v_p(q_p)$, this is an increasing function of j, so for $j \geqslant 6$ we have $v_p(B_j x^{-j}/j!) \geqslant 6(v - 1/(p-1)) - (p-2)/(p-1)$. We have $B_3 = B_5 = 0$, and $v_p(B_4 x^{-4}/4!) = 4v - v_p(720)$. Using $v \geqslant v_p(q_p)$, a case-by-case examination of $p = 2, 3, 5$ and $p \geqslant 7$ shows that $4v - v_p(720) < 6(v - 1/(p-1)) - (p-2)/(p-1)$. Since $R_p > 1$ we can transform the Mahler-type expansion of (1) into a power series in $(s - 1)$ with radius of convergence greater than or equal to R_p, so that

$$\sum_{j \geqslant 0} \binom{1-s}{j} B_j x^{-j} = 1 + \frac{s-1}{2x} + \frac{s(s-1)}{12x^2} + p^{4v - v_p(720)} F(s-1)$$

$$= 1 + (s-1)\left(\frac{1}{2x} + \frac{1}{12x^2}\right) + \frac{(s-1)^2}{12x^2} + p^{4v - v_p(720)} F(s-1) ,$$

for some power series $F \in \mathbb{Z}_p[[X]]$ with p-integral coefficients such that $F(0) = 0$. We check again on a case-by-case basis that $4v - v_p(720) \geqslant v + v_p(q_p) = v_p(q_p/x)$, proving (1).

For (2), since by definition $v_p(\langle x \rangle) \geqslant v_p(q_p)$, the series $\log_p(\langle x \rangle)$ converges, and by Proposition 4.2.14 we have $v_p(\log_p(\langle x \rangle)) = v_p(\langle x \rangle) \geqslant v_p(q_p)$, so we can write

$$\langle x \rangle^{1-s} = \exp_p((1-s)\log_p(\langle x \rangle)) = \sum_{j \geqslant 0} \frac{(1-s)^j}{j!} \log_p(\langle x \rangle)^j .$$

For simplicity of notation set $L = \log_p(\langle x \rangle)$. Since $\zeta_p(s,x) = (\langle x \rangle^{1-s}/(s-1)) \sum_{j \geqslant 0} \binom{1-s}{j} B_j x^{-j}$, it follows that

$$a_n = (-1)^{n+1} \frac{L^{n+1}}{(n+1)!} + \sum_{0 \leqslant j \leqslant n} \frac{(-1)^{n-j} L^{n-j}}{(n-j)!} c_j .$$

Note first that by Lemma 4.2.8 we have for $u \geqslant 1$,

$$v_p(L^u/u!) > uv_p(q_p) - u/(p-1) \geqslant u(v_p(q_p) - 1/(p-1)) .$$

In particular, we deduce that for $u \geqslant 2$ we have $v_p(L^u/u!) \geqslant v_p(q_p) + 1$, hence $v_p((L^u/u!)/(2x)) \geqslant v_p(q_p/x)$, and for $u \geqslant 1$ we have

$$v_p((L^u/u!)/(12x^2)) \geqslant v_p(q_p/x) + v_p(1/(12x)) \geqslant v_p(q_p/x) ,$$

since $v_p(1/x) \geqslant v_p(q_p) \geqslant v_p(12)$.

Since $c_j \equiv 0 \pmod{(q_p/x)\mathbb{Z}_p}$ for $j \geqslant 2$, we have

$$a_n \equiv (-1)^{n+1} \frac{L^{n+1}}{(n+1)!} + \sum_{0 \leqslant j \leqslant \min(n,1)} \frac{(-1)^{n-j} L^{n-j}}{(n-j)!} c_j \pmod{(q_p/x)\mathbb{Z}_p} .$$

By (1), this gives

$$a_0 \equiv -L + \frac{1}{2x} + \frac{1}{12x^2} \pmod{(q_p/x)\mathbb{Z}_p} ,$$

proving the first formula of (2), and for $n \geqslant 1$,

$$a_n \equiv (-1)^{n+1} \frac{L^{n+1}}{(n+1)!} + \frac{(-1)^n L^n}{n!} \left(\frac{1}{2x} + \frac{1}{12x^2} \right)$$
$$+ \frac{(-1)^{n-1} L^{n-1}}{(n-1)!} \frac{1}{12x^2} \pmod{(q_p/x)\mathbb{Z}_p} .$$

For $n \geqslant 2$, the inequalities proved above imply that $a_n \equiv (-1)^{n+1} L^{n+1}/(n+1)!$, and for $n = 1$ they give

$$a_1 \equiv \frac{L^2}{2} - \frac{L}{2x} + \frac{1}{12x^2} \pmod{(q_p/x)\mathbb{Z}_p} ,$$

proving (2).

(3). We have

$$\frac{\langle m \rangle^{1-s}}{m} \zeta_p \left(s, \frac{a}{m} \right) = \frac{\langle a \rangle^{1-s}}{m(s-1)} \sum_{j \geqslant 0} \binom{1-s}{j} B_j(a/m)^{-j} .$$

Since the only property of $L = \log_p(\langle x \rangle)$ that we used in the proof of (2) was that $v_p(L) \geqslant v_p(q_p)$, which is true for all x, it follows that we may replace $\langle x \rangle$ by $\langle a \rangle$ and x by a/m, thus obtaining a congruence modulo $q_p m \mathbb{Z}_p$, so that after division by m we obtain the given congruences modulo $q_p \mathbb{Z}_p$.   □

11.2.4 The p-adic Hurwitz Zeta Function for $x \in \mathbb{Z}_p$

The literature does not sufficiently emphasize that one can define $\zeta_p(s,x)$ also for $x \in \mathbb{Z}_p$, and that the corresponding function has important properties that nicely complement those for $x \in C\mathbb{Z}_p$ (see [Morit2]). We can in fact define more general functions depending on a Dirichlet character χ whose conductor is a power of p (see above). Note the important fact that if χ is defined modulo p^v for $v \geqslant 1$, and if χ' is the corresponding character modulo $p^{v'}$ for some $v' \geqslant v$, then in fact $\chi'(n) = \chi(n)$ for all n; in other words, the values of the characters χ and χ' are the same, so it is not necessary to specify the value of v, as long as $v \geqslant 1$.

Definition 11.2.12. *Let χ be a character modulo p^v with $v \geqslant 1$. If $x \in \mathbb{Z}_p$ and $s \in \mathbb{C}_p$ such that $|s| < R_p$ and $s \neq 1$ we define*

$$\zeta_p(\chi, s, x) = \frac{1}{s-1} \int_{\mathbb{Z}_p} \chi(x+t)\langle x+t \rangle^{1-s}\, dt \ ,$$

and by abuse of notation we will simply write $\zeta_p(s, x)$ instead of $\zeta_p(\chi_0, s, x)$, where χ_0 is the trivial character modulo p^v (where we recall that $\chi_0(a) = 0$ when $a \in p\mathbb{Z}_p$).

Even before showing that this definition makes sense, note the following:

Remarks. (1) As mentioned above it is clear that the definition of $\zeta_p(\chi, s, x)$ (and in particular of $\zeta_p(s, x)$) does not depend on the choice of $v \geqslant 1$ for which χ is defined modulo p^v.
(2) The only reason for which we restrict to characters modulo a power of p is that otherwise $\chi(x)$ does not make sense for $x \in \mathbb{Z}_p \setminus \mathbb{Z}$. However, if we restrict to $x \in \mathbb{Z}$, and in particular to $x = 0$, the above definition does make sense, and leads to p-adic L-functions, which we will study in great detail in Section 11.3.2 (the case $x \in \mathbb{Z}$ follows from the case $x = 0$ and Proposition 11.2.20 (1)).
(3) The function $\zeta_p(\chi, s, x)$ is the p-adic analogue of the function $\sum_{n \geqslant 0} \chi(n + x)(n + x)^{-s}$ (which has no specific name), as can be seen for instance in Proposition 11.2.20 (1).

To show that this definition makes sense, and in fact to relate it to the Hurwitz zeta function for $x \in \mathbb{C}\mathbb{Z}_p$, we first prove the following "change of variable" lemma in Volkenborn integrals.

Lemma 11.2.13. *Let χ be a character modulo p^v, let f be a function defined for $v_p(x) < -v$ such that for fixed x the function $f(x+t)$ is in $S^1(\mathbb{Z}_p)$, and set*

$$g(x) = \int_{\mathbb{Z}_p} f(x+t)\, dt \ .$$

Then for $x \in \mathbb{Z}_p$ we have

$$\frac{1}{p^v} \sum_{0 \leqslant j < p^v} \chi(x+j) g\left(\frac{x+j}{p^v}\right) = \int_{\mathbb{Z}_p} \chi(x+t) f\left(\frac{x+t}{p^v}\right) dt \ .$$

Proof. By definition, we have

$$\sum_{0 \leqslant j < p^v} \chi(x+j) g\left(\frac{x+j}{p^v}\right) = \lim_{r \to \infty} \frac{1}{p^r} \sum_{0 \leqslant j < p^v} \chi(x+j) \sum_{0 \leqslant a < p^r} f(a + (x+j)/p^v)$$

$$= \lim_{r \to \infty} \frac{1}{p^r} \sum_{0 \leqslant m < p^{v+r}} \chi(x+m) f((x+m)/p^v)$$

$$= p^v \int_{\mathbb{Z}_p} \chi(x+t) f((x+t)/p^v)\, dt \ ,$$

proving the lemma.   □

Corollary 11.2.14. *Definition 11.2.12 makes sense for $x \in \mathbb{Z}_p$ and $|s| < R_p$. More precisely, for any $M \in \mathbb{Z}_{\geqslant 1}$ such that $p^v \mid M$ we have*

$$\zeta_p(\chi, s, x) = \frac{\langle M \rangle^{1-s}}{M} \sum_{0 \leqslant j < M} \chi(x+j)\zeta_p\left(s, \frac{x+j}{M}\right) .$$

Proof. Applying the lemma to $f(x) = \langle x \rangle^{1-s}$ we obtain the given formula for $M = p^v$, which also shows the existence of $\zeta_p(\chi, s, x)$. For a general M, we write $M = Np^v$ and $j = p^v a + b$ with $0 \leqslant b < p^v$ and $0 \leqslant a < N$, so that

$$\sum_{0 \leqslant j < M} \chi(x+j)\zeta_p\left(s, \frac{x+j}{M}\right) = \sum_{0 \leqslant b < p^v} \chi(x+b) \sum_{0 \leqslant a < N} \zeta_p\left(s, \frac{x+b}{Np^v} + \frac{a}{N}\right)$$

$$= N\langle N \rangle^{s-1} \sum_{0 \leqslant b < p^v} \chi(x+b)\zeta_p\left(s, \frac{x+b}{p^v}\right) ,$$

using the distribution formula for ζ_p (Theorem 11.2.9 (5)), so the corollary follows in general since $\langle M \rangle^{1-s}/M = \langle N \rangle^{1-s}/(Np^v)$.   □

Thanks to this corollary, we can deduce most properties of $\zeta_p(\chi, s, x)$ for $x \in \mathbb{Z}_p$ from those of $\zeta_p(s, x)$ for $x \in C\mathbb{Z}_p$, and we will usually, although not always, choose $M = p^v$.

This corollary has many important consequences. For instance, we can now give a more general distribution formula that applies to the case $Nx \in \mathbb{Z}_p$ and $p^v \mid N$ (the case $p \nmid N$ will be considered in Proposition 11.2.20 below):

Corollary 11.2.15. *Let χ be a character modulo p^v. Then for any $x \in \mathbb{Q}_p$ and $N \in \mathbb{Z}_{\geqslant 1}$ such that $p^v \mid N$ and $Nx \in \mathbb{Z}_p$ we have*

$$\sum_{0 \leqslant j < N} \chi(Nx+j)\zeta_p\left(s, x + \frac{j}{N}\right) = \omega_v(N)\langle N \rangle^s \zeta_p(\chi, s, Nx) .$$

In particular,

$$\sum_{0 \leqslant j < N,\ p \nmid (Nx+j)} \zeta_p\left(s, x + \frac{j}{N}\right) = \omega_v(N)\langle N \rangle^s \zeta_p(s, Nx) ,$$

where we recall that we have defined $\zeta_p(s, x) = \zeta_p(\chi_0, s, x)$ when $x \in \mathbb{Z}_p$.

Proof. Follows from Corollary 11.2.14 applied to $M = N$ and x replaced by Nx.   □

We see that this is a perfect generalization of Theorem 11.2.9 to the case $Nx \in \mathbb{Z}_p$, and justifies the use of the same notation $\zeta_p(s, x)$ also when $x \in \mathbb{Z}_p$. We now study the functions $\zeta_p(\chi, s, x)$ for $x \in \mathbb{Z}_p$ in complete analogy with $x \in C\mathbb{Z}_p$.

Proposition 11.2.16. *If $x \in \mathbb{Z}_p$ We have*

$$\frac{\partial \zeta_p}{\partial x}(\chi, s, x) = -s\zeta_p(\chi\omega^{-1}, s+1, x) .$$

Note that in the above proposition $\chi\omega^{-1}$ is taken in the sense of characters, so that for instance $\chi\omega^{-1}(a) = 0$ if $a \in p\mathbb{Z}_p$.

Proof. By Corollary 11.2.7 we have

$$\frac{\partial \zeta_p}{\partial x}(\chi, s, x) = -\frac{s}{p^v} \sum_{0 \leqslant j < p^v} \frac{\chi(x+j)}{p^v \omega_v((x+j)/p^v)} \zeta_p(s+1, (x+j)/p^v)$$

$$= -\frac{s}{p^v} \sum_{0 \leqslant j < p^v} \chi\omega^{-1}(x+j)\zeta_p(s+1, (x+j)/p^v) ,$$

proving the result. $\qquad\qquad\qquad\qquad\qquad\qquad\qquad\qquad\qquad\quad\square$

Proposition 11.2.17. *For fixed $x \in \mathbb{Z}_p$ the function $\zeta_p(\chi, s, x)$ is a p-adic meromorphic function on $|s| < R_p$, which is analytic, except when $\chi = \chi_0$, in which case it has a simple pole at $s = 1$ with residue $1 - 1/p$.*

Proof. By Corollary 11.2.14 and Proposition 11.2.8, around $s = 1$ we have $\zeta_p(\chi, s, x) = a_{-1}/(s-1) + O(1)$, where

$$a_{-1} = \frac{1}{p^v} \sum_{0 \leqslant j < p^v} \chi(x+j) .$$

The result follows since the sum giving a_{-1} vanishes if $\chi \neq \chi_0$, and otherwise is equal to $p^v(1 - 1/p)$. $\qquad\qquad\qquad\qquad\qquad\qquad\qquad\quad\square$

Corollary 11.2.18. *We have*

$$\frac{\partial \zeta_p}{\partial x}(\chi\omega, 0, x) = -\left(1 - \frac{1}{p}\right)\delta(\chi) ,$$

where here and elsewhere $\delta(\chi) = 0$ if $\chi \neq \chi_0$ and $\delta(\chi) = 1$ if $\chi = \chi_0$.

Proof. By analyticity and Proposition 11.2.16 we have

$$\frac{\partial \zeta_p}{\partial x}(\chi\omega, 0, x) = \lim_{s \to 0} \frac{\partial \zeta_p}{\partial x}(\chi\omega, s, x) = \lim_{s \to 0}(-s\zeta_p(\chi, s+1, x)) = -\left(1 - \frac{1}{p}\right)\delta(\chi)$$

by the above proposition. $\qquad\qquad\qquad\qquad\qquad\qquad\qquad\qquad\qquad\quad\square$

Once again, as in the complex case and as for $\zeta_p(s, x)$ (see the remarks following Proposition 11.2.8), it follows from this corollary that around $s = 1$ we have more precisely

$$\zeta_p(\chi, s, x) = \frac{(1 - 1/p)\delta(\chi)}{s - 1} - \psi_p(\chi, x) + O(s - 1) ,$$

where $\psi_p(\chi, x) = (d/dx)(\mathrm{Log}\Gamma_p(\chi, x))$ and $\mathrm{Log}\Gamma_p(\chi, x) = \frac{\partial \zeta_p}{\partial s}(\chi\omega, 0, x)$; see Proposition 11.5.15.

Proposition 11.2.19. (1) *For any $k \in \mathbb{Z}_{\geqslant 1}$ and $x \in \mathbb{Z}_{\geqslant 0}$ we have*

$$\zeta_p(\chi\omega^k, 1 - k, x) = -\frac{B_k(\chi)}{k} - \sum_{0 \leqslant r < x} \chi(r) r^{k-1} .$$

(2) *In particular, if $\chi \neq \chi_0$ and $x \in \mathbb{Z}_{\geqslant 0}$ we have*

$$\zeta_p(\chi\omega, 0, x) = -\frac{1}{p^v} \sum_{0 \leqslant r < p^v} \chi(r) r - \sum_{0 \leqslant r < x} \chi(r) ,$$

and if, in addition, $\chi \neq \chi_0$ is even we have

$$\zeta_p(\chi\omega, 0, x) = - \sum_{0 \leqslant r < x} \chi(r) .$$

(3) *For all $x \in \mathbb{Z}_p$ we have*

$$\zeta_p(\omega, 0, x) = -\left(x - \left\lceil \frac{x}{p} \right\rceil \right) ,$$

where $\lceil x/p \rceil$ is defined as the p-adic limit of $\lceil x_n/p \rceil$ as $x_n \to x$ in $\mathbb{Z}_{\geqslant 0}$.

Proof. (1) and (2). By Corollary 11.2.14 and Proposition 11.2.6 we have

$$
\begin{aligned}
\zeta_p(\chi\omega^k, 1 - k, x) &= \frac{1}{p^v} \sum_{0 \leqslant j < p^v} \chi\omega^k(x + j) \zeta_p(1 - k, (x + j)/p^v) \\
&= -\frac{1}{kp^v} \sum_{0 \leqslant j < p^v} \chi\omega^k(x + j)\omega_v((x + j)/p^v)^{-k} B_k((x + j)/p^v) \\
&= -\frac{p^{v(k-1)}}{k} \sum_{0 \leqslant j < p^v} \chi(x + j) B_k((x + j)/p^v) .
\end{aligned}
$$

Thus, by Corollary 9.4.6 we have

$$\zeta_p(\chi\omega^k, 1 - k, x) = -\frac{B_k(\chi)}{k} - \sum_{0 \leqslant r < x} \chi(r) r^{k-1} ,$$

proving (1). Statement (2) is an immediate consequence of (1), of the explicit formula for $B_1(\chi)$, of the formula $\sum_{0 \leqslant r < p^v} \chi(r) = 0$ since χ is a nontrivial character, and of the formula $B_1(\chi) = -\chi(0)/2 = 0$ when χ is even. Note that we can also use Proposition 11.2.20 (1) below.

(3). By (1), for $x \in \mathbb{Z}_{\geqslant 0}$ we have

$$\zeta_p(\omega, 0, x) = -B_1(\chi_0) - \sum_{0 \leqslant r < x} \chi_0(r) \ .$$

By Proposition 9.4.9 we have $B_1(\chi_0) = -\chi_0(0)/2 = 0$ (recall that χ_0 is the trivial character modulo p^v for some $v \geqslant 1$). Furthermore, we have

$$\sum_{0 \leqslant r < x} \chi_0(r) = \sum_{0 \leqslant r < x, \ p \nmid r} 1 = \sum_{0 \leqslant r < x} 1 - \sum_{0 \leqslant s < x/p} 1 = x - \lceil x/p \rceil \ ,$$

so (3) follows for $x \in \mathbb{Z}_{\geqslant 0}$, hence for $x \in \mathbb{Z}_p$ by continuity. $\qquad\square$

Note that by continuity, if $k \in \mathbb{Z}_{\geqslant 1}$ and $x \in \mathbb{Z}_p$ we have more generally

$$\zeta_p(\chi\omega^k, 1 - k, x) = -\frac{B_k(\chi)}{k} - \lim_{\substack{n \to x \\ n \in \mathbb{Z}_{\geqslant 0}}} \sum_{0 \leqslant r < x} \chi(r) r^{k-1} \ .$$

Proposition 11.2.20. *Let $x \in \mathbb{Z}_p$ and $|s| < R_p$.*

(1) *We have the functional equation*

$$\zeta_p(\chi, s, x + 1) - \zeta_p(\chi, s, x) = -\chi\omega^{-1}(x)\langle x \rangle^{-s} \ ,$$

where the right-hand side is interpreted to be equal to 0 for $x = 0$.

(2) *We have the reflection formula*

$$\zeta_p(\chi, s, 1 - x) = \chi(-1)\zeta_p(\chi, s, x) \ .$$

(3) *Set $L_p(\chi, s) = \zeta_p(\chi, s, 0)$. If χ is an odd character we have $L_p(\chi, s) = 0$, and more generally if $n \in \mathbb{Z}_{\geqslant 0}$ we have*

$$\zeta_p(\chi, s, n) = \chi(-1)\zeta_p(\chi, s, 1 - n) = L_p(\chi, s) - \sum_{0 \leqslant a < n} \chi\omega^{-1}(a)\langle a \rangle^{-s} \ .$$

(4) *If $p \nmid N$ we have the distribution formula*

$$\sum_{0 \leqslant i < N} \zeta_p\left(\chi, s, x + \frac{i}{N}\right) = \omega\chi^{-1}(N)\langle N \rangle^s \zeta_p(\chi, s, Nx) \ .$$

Note that $L_p(\chi, s)$ is a p-adic *L*-function, whose properties we will study in Section 11.3.2 in more detail, and for general Dirichlet characters χ.

Proof. (1). By definition we have

$$\zeta_p(\chi, s, x + 1) = \frac{1}{s - 1} \lim_{r \to \infty} \frac{1}{p^r} \sum_{0 \leqslant j < p^r} \chi(x + 1 + j)\langle x + 1 + j \rangle^{1-s}$$

$$= \frac{1}{s - 1} \lim_{r \to \infty} \frac{1}{p^r} \sum_{1 \leqslant j \leqslant p^r} \chi(x + j)\langle x + j \rangle^{1-s} \ ,$$

so that for $x \neq 0$ we have

$$\zeta_p(\chi, s, x+1) - \zeta_p(\chi, s, x) = \frac{\chi(x)}{s-1} \lim_{r \to \infty} \frac{1}{p^r}(\langle x+p^r \rangle^{1-s} - \langle x \rangle^{1-s})$$

$$= \frac{\chi(x)}{s-1}\frac{d}{dx}\langle x \rangle^{1-s} = -\chi\omega^{-1}(x)\langle x \rangle^{-s}$$

by Lemma 11.2.3. We could of course also have deduced this result from the corresponding one for $x \in C\mathbb{Z}_p$ thanks to Corollary 11.2.14. On the other hand, since $\chi(0) = \chi(p^r) = 0$, we evidently have $\zeta_p(\chi, s, 1) - \zeta_p(\chi, s, 0) = 0$.

(2). Here, it is slightly preferable to use Corollary 11.2.14. Setting $i = p^v - 1 - j$ we have

$$\zeta_p(\chi, s, 1-x) = p^{-v}\sum_{0 \leqslant j < p^v} \chi(1-x+j)\zeta_p(s, (1-x+j)/p^v)$$

$$= p^{-v}\sum_{0 \leqslant i < p^v} \chi(p^v - i - x)\zeta_p(s, 1 - (x+i)/p^v)$$

$$= \chi(-1)p^{-v}\sum_{0 \leqslant i < p^v} \chi(x+i)\zeta_p(s, (x+i)/p^v) = \chi(-1)\zeta_p(\chi, s, x) \,,$$

using Theorem 11.2.9 (4).

(3). By definition of Dirichlet characters we have $\chi\omega^{-1}(0) = 0$. It follows from (1) that $\zeta_p(\chi, s, 1) = \zeta_p(\chi, s, 0)$ for any character χ, and from (2) that $\zeta_p(\chi, s, 1) = -\zeta_p(\chi, s, 0)$ if $\chi(-1) = -1$, whence $L_p(\chi, s) = \zeta_p(\chi, s, 0) = 0$ in that case. The formula for $\zeta_p(\chi, s, n)$ then follows by induction from (1), and that for $\zeta_p(\chi, s, 1-n)$ follows from (2).

(4). By Corollary 11.2.14 we have

$$\sum_{0 \leqslant i < N} \zeta_p(\chi, s, x+i/N) = p^{-v}\sum_{0 \leqslant i < N}\sum_{0 \leqslant j < p^v} \chi(x+j+i/N)\zeta_p(s, (x+j+i/N)/p^v) \,.$$

Setting $a = Nj + i$, and using the fact that $p \nmid N$ we thus have

$$\sum_{0 \leqslant i < N} \zeta_p(\chi, s, x+i/N) = p^{-v}\chi^{-1}(N)\sum_{0 \leqslant a < Np^v} \chi(Nx+a)\zeta_p(s, (Nx+a)/(Np^v)) \,.$$

Applying once again Corollary 11.2.14, but now with $M = Np^v$ instead of $M = p^v$, we obtain

$$\sum_{0 \leqslant i < N} \zeta_p(\chi, s, x + i/N) = p^{-v}\chi^{-1}(N)(Np^v)\langle Np^v \rangle^{s-1}\zeta_p(\chi, s, Nx)$$

$$= \omega\chi^{-1}\langle N \rangle^s \zeta_p(\chi, s, Nx) \,,$$

since $\langle Np^v \rangle = \langle N \rangle$. $\qquad\qquad\qquad\qquad\qquad\qquad\qquad\qquad\qquad\qquad\square$

Note that in the proof of (4), the use of Corollary 11.2.14 with $M = Np^v$ means that we have implicitly used the distribution formula for $\zeta_p(s, x)$ with $x \in C\mathbb{Z}_p$.

It is perhaps useful to summarize the three distribution formulas that we have obtained, since they are valid in different ranges:

Proposition 11.2.21. *Let χ be a character modulo p^v, let $N \in \mathbb{Z}_{\geqslant 1}$ and $x \in \mathbb{Q}_p$.*

(1) *If $Nx \in C\mathbb{Z}_p$ we have*

$$\sum_{0 \leqslant j < N} \zeta_p\left(s, x + \frac{j}{N}\right) = \omega_v(N)\langle N \rangle^s \zeta_p(s, Nx) .$$

(2) *If $Nx \in \mathbb{Z}_p$ and $p^v \mid N$ we have*

$$\sum_{0 \leqslant j < N} \chi(Nx + j)\zeta_p\left(s, x + \frac{j}{N}\right) = \omega_v(N)\langle N \rangle^s \zeta_p(\chi, s, Nx) .$$

(3) *If $x \in \mathbb{Z}_p$ and $p \nmid N$ we have*

$$\sum_{0 \leqslant j < N} \zeta_p\left(\chi, s, x + \frac{j}{N}\right) = \omega \chi^{-1}(N)\langle N \rangle^s \zeta_p(\chi, s, Nx) .$$

The p-adic Raabe formula (Proposition 11.2.10) is also valid for $x \in \mathbb{Z}_p$, without change:

Proposition 11.2.22. *Let χ be a character modulo p^v. For $|s| < R_p$ and $x \in \mathbb{Z}_p$ we have*

$$\int_{\mathbb{Z}_p} \zeta_p(\chi, s, x + t)\, dt = s\zeta_p(\chi, s, x) + (x - 1)\frac{\partial \zeta_p}{\partial x}(\chi, s, x)$$

$$= s\left(\zeta_p(\chi, s, x) - (x - 1)\zeta_p(\chi\omega^{-1}, s + 1, x)\right) .$$

In particular,

$$\int_{\mathbb{Z}_p} \zeta_p(\chi, s, t)\, dt = s(L_p(\chi, s) + L_p(\chi\omega^{-1}, s + 1)) .$$

Proof. By definition we have

$$\int_{\mathbb{Z}_p} \zeta_p(\chi, s, x + t)\, dt = \lim_{r \to \infty} \frac{S(r)}{p^r} \quad \text{with} \quad S(r) = \sum_{0 \leqslant i < p^r} \zeta_p(\chi, s, x + i) .$$

Let $r \geqslant v$. Applying Corollary 11.2.14 with $M = p^r$ we obtain

$$S(r) = p^{-r} \sum_{0 \leqslant i < p^r} \sum_{0 \leqslant j < p^r} \chi(x + i + j) \zeta_p(s, (x + i + j)/p^r).$$

Set $n = i + j \bmod p^r$, in other words the unique $n \equiv i + j \pmod{p^r}$ such that $0 \leqslant n < p^r$. We can have only $i + j = n$ or $i + j = n + p^r$, and the number of pairs (i, j) such that $i + j = n$ is equal to $n + 1$, while the number of pairs such that $i + j = n + p^r$ is equal to $p^r - (n + 1)$. It follows that

$$S(r) = p^{-r} \sum_{0 \leqslant n < p^r} \chi(x + n) \bigg((n + 1) \zeta_p(s, (x + n)/p^r)$$

$$+ (p^r - (n + 1)) \zeta_p(s, (x + n)/p^r + 1) \bigg)$$

$$= p^{-r} \sum_{0 \leqslant n < p^r} \chi(x + n) \bigg(p^r \zeta_p(s, (x + n)/p^r)$$

$$- (p^r - (n + 1)) \omega_v((x + n)/p^r)^{-1} \langle x + n \rangle^{-s} \bigg)$$

$$= p^r \zeta_p(\chi, s, x) - \sum_{0 \leqslant n < p^r} (p^r - (n + 1)) \chi \omega^{-1}(x + n) \langle x + n \rangle^{-s},$$

using the functional equation of $\zeta(s, x)$ (Theorem 11.2.9 (3)). Now it is clear that

$$(p^r - (n + 1)) \chi \omega^{-1}(x + n) \langle x + n \rangle^{-s}$$
$$= (p^r + x - 1) \chi \omega^{-1}(x + n) \langle x + n \rangle^{-s} - \chi(x + n) \langle x + n \rangle^{1-s},$$

so that by definition

$$\lim_{r \to \infty} p^{-r} (p^r - (n + 1)) \chi \omega^{-1}(x + n) \langle x + n \rangle^{-s}$$
$$= (x - 1) s \zeta_p(\chi \omega^{-1}, s + 1, x) - (s - 1) \zeta_p(\chi, s, x).$$

It follows that

$$\int_{\mathbb{Z}_p} \zeta_p(\chi, s, x + t) \, dt = \lim_{r \to \infty} \frac{S(r)}{p^r} = -(x - 1) s \zeta_p(\chi \omega^{-1}, s + 1, x) + s \zeta_p(\chi, s, x),$$

and we conclude by Proposition 11.2.16 and the definition of $L_p(\chi, s)$ given in Proposition 11.2.20. □

As mentioned, the first formula of the present proposition is identical to the corresponding one for $x \in C\mathbb{Z}_p$, and note that we have not used Raabe's formula in that range to prove it.

We also have the following power series expansion in x of $\zeta_p(\chi, s, x)$, which should be compared with Corollary 9.6.3.

Proposition 11.2.23. *Let χ be a character modulo p^v for some $v \geqslant 1$. For $x \in p^v \mathbb{Z}_p$ we have the power series expansion*

$$\zeta_p(\chi, s, x) = \sum_{k \geqslant 0} \binom{1-s}{k} L_p(\chi\omega^{-k}, s+k) x^k \ ,$$

where $L_p(\chi, s) = \zeta_p(\chi, s, 0)$ is the p-adic L-function.

Proof. Although this result involves p-adic L-functions which we will study in much more detail below, taking simply $L_p(\chi, s) = \zeta_p(\chi, s, 0)$ as a definition is enough for the proof. Indeed, since $x \in p^v \mathbb{Z}_p$ and χ is defined modulo p^v, we have by definition

$$\zeta_p(\chi, s, x) = \frac{1}{s-1} \int_{\mathbb{Z}_p} \chi(x+t)\langle x+t \rangle^{1-s} \, dt$$

$$= \frac{1}{s-1} \int_{\mathbb{Z}_p^*} \chi(t)\langle t \rangle^{1-s} (1 + x/t)^{1-s} \, dt$$

$$= \frac{1}{s-1} \int_{\mathbb{Z}_p^*} \sum_{k \geqslant 0} \binom{1-s}{k} \chi\omega^{-k}(t)\langle t \rangle^{1-s-k} x^k \, dt \ .$$

Since $x \in p\mathbb{Z}_p$ the series is normally convergent, so that we can integrate term by term and obtain

$$\zeta_p(\chi, s, x) = \sum_{k \geqslant 0} \binom{1-s}{k} x^k \frac{1}{s-1} \int_{\mathbb{Z}_p} \chi\omega^{-k}(t)\langle t \rangle^{1-s-k} \, dt \ ,$$

proving the result since

$$L_p(\chi\omega^{-k}, s+k) = \zeta_p(\chi\omega^{-k}, s+k, 0) = \frac{1}{s-1} \int_{\mathbb{Z}_p} \chi\omega^{-k}(t)\langle t \rangle^{1-s-k} \, dt \ .$$

$\square$

Remark. In effect, we have shown that for all $x \in p\mathbb{Z}_p$ the power series on the right-hand side is equal to

$$\frac{1}{s-1} \int_{\mathbb{Z}_p} \chi(t)\langle x+t \rangle^{1-s} \, dt \ ,$$

and this is equal to $\zeta_p(\chi, s, x)$ only when $x \in p^v \mathbb{Z}_p$. Because of this, one could think of using the latter integral as the definition of ζ_p, but it is easily seen that this would lead to a function with very few interesting properties.

11.3 p-adic L-Functions

11.3.1 Dirichlet Characters in the p-adic Context

Before introducing p-adic L-functions, we need some simple but essential preliminaries. First of all, since we will be handling Dirichlet characters χ that have values equal to 0 or roots of unity, we must give them p-adic values. For this, we choose an arbitrary but *fixed* embedding of the algebraic closure $\overline{\mathbb{Q}}$ of $\mathbb{Q}$ into $\mathbb{C}_p$. We can thus consider $\chi(a)$ as an element of $\mathbb{C}_p$. Thus, *in the p-adic context*, when we say that an algebraic number α is p-integral, it means that $|\alpha| \leqslant 1$ (or $v_p(\alpha) \geqslant 0$, or again $\alpha \in \mathcal{Z}_p$, where $\mathcal{Z}_p$ is the ring of p-adic integers of $\mathbb{C}_p$; see Definition 9.5.1) *with respect to this embedding*. Recall also that we write $\alpha \equiv \beta \pmod{\gamma \mathcal{Z}_p}$ or simply $\pmod{\gamma}$ to mean that $(\alpha - \beta)/\gamma \in \mathcal{Z}_p$.

The following definition and lemma emphasizes this:

Definition 11.3.1. *Let p be a prime number and α an algebraic number. We will say that α is p-integral if for every prime ideal $\mathfrak{p}$ above p in the number field $\mathbb{Q}(\alpha)$ we have $v_{\mathfrak{p}}(\alpha) \geqslant 0$.*

Lemma 11.3.2. *Let p be a prime number and α an algebraic number. The following conditions are equivalent:*

(1) *α is p-integral.*
(2) *For any embedding σ of $\overline{\mathbb{Q}}$ into $\mathbb{C}_p$ we have $|\sigma(\alpha)| \leqslant 1$; in other words, $\sigma(\alpha)$ is p-integral as a p-adic number.*
(3) *If we fix an embedding of $\overline{\mathbb{Q}}$ into $\mathbb{C}_p$, then all the conjugates of α are p-integral as p-adic numbers.*

Proof. Clear and left to the reader (Exercise 7). □

Next, let χ_1 and χ_2 be two primitive Dirichlet characters, hence with values in $\overline{\mathbb{Q}}$ (considered as a subfield of $\mathbb{C}$ or of $\mathbb{C}_p$; it does not matter here). We define the character $\chi_1\chi_2$ to be the *primitive* character equivalent to the character $\chi_1(a)\chi_2(a)$. It is clear that the conductor of $\chi_1\chi_2$ divides the LCM of the conductors of χ_1 and χ_2. In addition, we have the following:

Lemma 11.3.3. *If either $\chi_1(a) \neq 0$ or $\chi_2(a) \neq 0$ we have $\chi_1\chi_2(a) = \chi_1(a)\chi_2(a)$.*

Proof. If $\chi_1(a) \neq 0$ and $\chi_2(a) \neq 0$ we have by definition $(\chi_1\chi_2)(a) = \chi_1(a)\chi_2(a)$. If exactly one of them is nonzero, say $\chi_1(a) \neq 0$ and $\chi_2(a) = 0$, then since χ_2 is primitive we have

$$0 = \chi_2(a) = ((\chi_1\chi_2)\chi_1^{-1})(a) = (\chi_1\chi_2)(a)\chi_1^{-1}(a) ,$$

so that $(\chi_1\chi_2)(a) = 0 = \chi_1(a)\chi_2(a)$, as claimed. □

11.3.2 Definition and Basic Properties of *p*-adic *L*-Functions

Since the Hurwitz zeta function is the building block of Dirichlet *L*-functions it is now easy to define *p*-adic *L*-functions. This is essentially due to Kubota–Leopoldt, and I loosely follow the presentation given in Washington's book [Was]. Note, however, that the modern way of giving the definitions and proofs is through the use of *p*-adic measures, but to stay in the spirit of this book (and of the author!) I have avoided doing so. See for instance the paper of Colmez [Colm] for an introduction to the subject using *p*-adic measures.

By Proposition 10.2.5 we know that if χ is a (not necessarily primitive) character modulo f then as a complex function we have

$$L(\chi, s) = f^{-s} \sum_{1 \leqslant a \leqslant f} \chi(a) \zeta(s, a/f) \,.$$

This leads to the following.

Definition 11.3.4. *Let χ be a primitive character of conductor f. For $s \in \mathbb{C}_p$ such that $|s| < R_p$ and $s \neq 1$, we define*

$$L_p(\chi, s) = \frac{\langle f \rangle^{1-s}}{f} \sum_{0 \leqslant a < f} \chi(a) \zeta_p \left(s, \frac{a}{f} \right) \,.$$

We define $L_p(\chi, 1) = \lim_{s \to 1} L_p(\chi, s)$ when the limit exists. In particular, if χ is the trivial character χ_0 we set $\zeta_p(s) = L_p(\chi_0, s) = \zeta_p(s, 0)$, and call this function the Kubota–Leopoldt p-adic zeta function.

Remarks. (1) It is important to note that this definition uses the function $\zeta_p(s, x)$ both for $x \in C\mathbb{Z}_p$ and for $x \in \mathbb{Z}_p$: indeed, when $p \nmid f$ then $a/f \in \mathbb{Z}_p$, so the function that occurs is the function $\zeta_p(\chi_0, s, x)$ defined in Definition 11.2.12. On the other hand, when $p \mid f$ then $q_p \mid f$ (since the conductor of a character cannot be congruent to 2 modulo 4). Furthermore, $\chi(a) \neq 0$ only when $p \nmid a$, so in that case $a/f \in C\mathbb{Z}_p$ and the function that occurs is the initial function $\zeta_p(s, x)$. The above uniform formula is an additional reason to use the same notation for $\zeta_p(s, x)$ when $x \in \mathbb{Z}_p$ and $x \in C\mathbb{Z}_p$.

(2) Note that we sum from $a = 0$ to $f - 1$ instead of from 1 to f in the complex case, where it is essential since $\zeta(s, 0)$ is not defined. Here it makes no difference since we can have $\chi(0) = \chi(f) \neq 0$ only for $\chi = \chi_0$, and by Proposition 11.2.20 we have

$$\zeta_p(\chi, s, 1) = \zeta_p(\chi, s, 0) - \chi\omega^{-1}(0) = \zeta_p(\chi, s, 0)$$

when $\chi \neq \omega$, and in particular when $\chi = \chi_0$. It makes the computations slightly more elegant.

(3) When $f = p^v$ with $v \geqslant v_p(q_p)$, it is clear from Corollary 11.2.14 applied to $M = f$ that $L_p(\chi, s) = \zeta_p(\chi, s, 0)$, so that the above definition indeed generalizes to arbitrary characters the definition that we have already given in Proposition 11.2.20. Since $\zeta_p(\chi, s, x)$ has a Volkenborn integral definition, for future reference we note the following result.

Proposition 11.3.5. *If χ is defined modulo p^v for some $v \geqslant 1$ we have*

$$L_p(\chi, s) = \frac{1}{s-1} \int_{\mathbb{Z}_p} \chi(t) \langle t \rangle^{1-s} \, dt \, .$$

To state the next proposition, it is useful to introduce the following notation.

Definition 11.3.6. (1) *Let $m \in \mathbb{Z}_{>0}$. We define $\chi_{0,m}$ to be the trivial character modulo 1 when $p \nmid m$, and to be the trivial character modulo p when $p \mid m$. In other words, $\chi_{0,m}(a) = 1$ when $p \nmid a$ or when $p \mid a$ but $p \nmid m$, and $\chi_{0,m}(a) = 0$ when $p \mid a$ and $p \mid m$.*
(2) *If $I \subset \mathbb{Z}$, we set*

$$\sum_{\substack{a \in I \\ p \nmid a}}^{(p)} g(a) = \sum_{a \in I} g(a) \quad \text{and similarly} \quad \prod_{\substack{a \in I \\ p \nmid a}}^{(p)} g(a) = \prod_{a \in I} g(a) \, .$$

In particular, if $p \mid m$ we have

$$\sum_{0 \leqslant a < m}^{(p)} g(a) = \sum_{0 \leqslant a < m} \chi_{0,m}(a) g(a) \, .$$

Note that the condition in (2) is $p \nmid a$, and not $p \nmid g(a)$. In certain circumstances it will be essential to have the condition $p \nmid g(a)$ instead, and in that case it will be written explicitly. Note also the following.

Lemma 11.3.7. *Let χ be a nontrivial primitive character of conductor f, and let m be a common multiple of f and p. Then*

$$\sum_{0 \leqslant a < m}^{(p)} \chi(a) = 0 \, .$$

Proof. By multiplicativity we have

$$\sum_{0 \leqslant a < m}^{(p)} \chi(a) = \sum_{0 \leqslant a < m} \chi(a) - \chi(p) \sum_{0 \leqslant b < m/p} \chi(b) \, .$$

Since χ is nontrivial and $f \mid m$ the first sum is zero. If $p \mid f$ we have $\chi(p) = 0$. On the other hand, if $p \nmid f$ we have $fp \mid m$, in other words $f \mid m/p$, so the second sum is zero. $\qquad\square$

Proposition 11.3.8. *Let χ be a primitive character of conductor f, let $m \in \mathbb{Z}_{>0}$ be a multiple of f, and let $s \in \mathbb{C}_p$ be such that $|s| < R_p$ and $s \neq 1$.*

(1) *We have*

$$L_p(\chi, s) = \frac{\langle m \rangle^{1-s}}{m} \sum_{0 \leqslant a < m} \chi_{0,m}(a) \chi(a) \zeta_p\left(s, \frac{a}{m}\right) .$$

(2) *If, in addition, $q_p \mid m$ we have*

$$L_p(\chi, s) = \frac{1}{s-1} \sum_{0 \leqslant a < m}^{(p)} \chi(a) \langle a \rangle^{1-s} \sum_{j \geqslant 0} \binom{1-s}{j} \frac{m^{j-1}}{a^j} B_j .$$

(3) *If $\chi \neq \chi_0$ then $L_p(\chi, 1)$ does indeed exist and is given by the formula*

$$L_p(\chi, 1) = \sum_{0 \leqslant a < m}^{(p)} \chi(a) \left(-\frac{\log_p(\langle a \rangle)}{m} + \sum_{j \geqslant 1}(-1)^j \frac{m^{j-1}}{a^j} \frac{B_j}{j} \right) ,$$

where $m \in \mathbb{Z}_{>0}$ is any common multiple of f and q_p.

Proof. (1). Writing $a = kf + r$ we have

$$\frac{\langle m \rangle^{1-s}}{m} \sum_{0 \leqslant a < m} \chi_{0,m}(a) \chi(a) \zeta_p\left(s, \frac{a}{m}\right)$$

$$= \frac{\langle m \rangle^{1-s}}{m} \sum_{0 \leqslant r < f} \chi(r) \sum_{0 \leqslant k < m/f} \chi_{0,m}(kf + r) \zeta_p\left(s, \frac{r}{m} + \frac{k}{m/f}\right)$$

$$= \frac{\langle m \rangle^{-s}}{\omega_v(m)} \sum_{0 \leqslant r < f} \chi(r) \omega_v(m/f) \langle m/f \rangle^s \zeta_p\left(s, \frac{r}{f}\right)$$

$$= \frac{\langle f \rangle^{-s}}{\omega_v(f)} \zeta_p\left(s, \frac{r}{f}\right) = L_p(\chi, s) ,$$

using all three distribution formulas for $\zeta_p(s, x)$ (Proposition 11.2.21), proving (1). Statement (2) then follows from Definition 11.2.5, and the proof of (3) is immediate by letting $s \to 1$ and is left to the reader (Exercise 10). $\square$

Remarks. (1) If m is a common multiple of f and q_p we thus have

$$L_p(\chi, s) = \frac{\langle m \rangle^{1-s}}{m} \sum_{0 \leqslant a < m}^{(p)} \chi(a) \zeta_p\left(s, \frac{a}{m}\right) ,$$

and the function ζ_p that occurs is now only the one defined on $C\mathbb{Z}_p$, which has the simplest properties. This is usually the definition given in the literature, and we will of course also use it to study $L_p(\chi, s)$.

(2) The formula of (3) gives a convergent series for $L_p(\chi, 1)$. We will see in Theorem 11.5.37 that in fact there exists a "closed" formula for $L_p(\chi, 1)$ completely analogous to the one in the complex case (Proposition 10.3.5), but it is not clear whether it is better for computation.

Proposition 11.3.9. *Keep the above assumptions.*

(1) *The function $L_p(\chi, s)$ is a p-adic analytic function for $|s| < R_p$, except when $\chi = \chi_0$, in which case the function $\zeta_p(s) = L_p(\chi_0, s)$ has a simple pole at $s = 1$ with residue $1 - 1/p$.*

(2) *For $k \in \mathbb{Z}_{\geqslant 1}$ we have*

$$L_p(\chi, 1 - k) = -(1 - (\chi\omega^{-k})(p)p^{k-1})\frac{B_k(\chi\omega^{-k})}{k}$$
$$= (1 - (\chi\omega^{-k})(p)p^{k-1})L(\chi\omega^{-k}, 1 - k),$$

where $\chi\omega^{-k}$ is defined as above.

(3) *If χ is an odd character the function $L_p(\chi, s)$ is identically equal to zero.*

Proof. (1). By definition $L_p(\chi, s)$ is a p-adic meromorphic function with a possible simple pole at $s = 1$ with residue $\sum_{0 \leqslant a < f} \chi(a) \operatorname{Res}_{s=1} \zeta_p(s, a/f)$, and by Propositions 11.2.8 and 11.2.17, the quantity $\operatorname{Res}_{s=1} \zeta_p(a/f)$ is independent of a, so that the residue vanishes unless χ is the trivial character, in which case $f = 1$ and the result follows from Proposition 11.2.17.

(2). For ease of notation write χ_k instead of $\chi\omega^{-k}$. Let m be a common multiple of f and q_p. By Proposition 11.3.8 and Lemma 11.3.3, for $k \geqslant 1$ we have

$$-kL_p(\chi, 1 - k) = m^{k-1} \sum_{0 \leqslant a < m}^{(p)} \chi(a)\omega(a)^{-k} B_k(a/m)$$

$$= m^{k-1} \sum_{0 \leqslant a < m}^{(p)} \chi_k(a) B_k(a/m)$$

$$= m^{k-1} \sum_{0 \leqslant a < m} \chi_k(a) B_k(a/m) - \chi_k(p)m^{k-1} \sum_{0 \leqslant b < m/p} \chi_k(b) B_k(b/(m/p)).$$

Since χ_k is defined modulo a divisor of the LCM of q_p and f that divides m, by Lemma 9.4.7 the first sum is equal to $B_k(\chi_k)$. For the second term, let F_k be the conductor of χ_k, so that in particular $F_k \mid \operatorname{lcm}(f, q_p) \mid m$. If $p \mid F_k$ then $\chi_k(p) = 0$, so the second term is equal to 0. Otherwise, $p \nmid F_k$, so that $F_k \mid m/p$, and once again by Lemma 9.4.7 we have

$$(m/p)^{k-1} \sum_{0 \leqslant b < m/p} \chi_k(b) B_k(b/(m/p)) = B_k(\chi_k).$$

Thus in both cases the second term is equal to $\chi_k(p)p^{k-1}B_k(\chi_k)$, so that

$$-kL_p(\chi, 1-k) = (1 - \chi_k(p)p^{k-1})B_k(\chi_k) \,,$$

proving the first equality of (2). From Corollary 10.2.3 we deduce that

$$L_p(\chi, 1-k) = (1 - (\chi\omega^{-k})(p)p^{k-1})L(\chi\omega^{-k}, 1-k) + S \,,$$

with $S = \chi\omega^{-k}(0)\delta_{k,1}(1 - \chi\omega^{-k}(p)p^{k-1})$. But $\chi\omega^{-k}(0)\delta_{k,1} \neq 0$ only if $k = 1$ and $\chi = \omega$, and in that case $1 - \chi\omega^{-k}(p)p^{k-1} = 1 - 1 = 0$, so that we always have $S = 0$, proving the second equality of (2).

Finally, (3) follows immediately from $\zeta_p(s, 1-x) = \zeta_p(s, x)$, valid for all $x \in \mathbb{Q}_p$.   □

Remarks. (1) The factor $1 - (\chi\omega^{-k})(p)p^{k-1}$ occurring in the formula for $L_p(\chi, 1-k)$ is the inverse of the formal Euler factor at p for the complex function $L(\chi\omega^{-k}, s)$ at $s = 1-k$. Thus $L_p(\chi, s)$ is a p-adic interpolation of the values at negative integers of $(1 - \chi\omega^{s-1}(p)p^{-s})L(\chi, s)$. Note that the values at negative integers of the function $(1 - \chi(p)p^{-s})L(\chi, s)$ (which is exactly the function $L(\chi, s)$ with the Euler p-factor omitted) cannot be p-adically interpolated, in other words that the presence of the Teichmüller character ω is essential.

(2) The Kubota–Leopoldt p-adic zeta function $\zeta_p(s) = L_p(\chi_0, s)$ has a simple pole at $s = 1$ with residue $1 - 1/p$. Furthermore, since $\omega^{-k}(p)$ is equal to 0 if $\phi(q_p) \nmid k$ and to 1 otherwise, we have $\zeta_p(1-k) = -B_k(\omega^{-k})/k$ if $\phi(q_p) \nmid k$ and $\zeta_p(1-k) = -(1 - p^{k-1})B_k/k$ otherwise.

11.3.3 p-adic L-Functions at Positive Integers

The definition of the p-adic L-function is essentially based on its values at negative integers. We now study what happens for *positive* integers. Recall that we have already seen in Proposition 11.2.6 that for $x \in C\mathbb{Z}_p$ and $k \in \mathbb{Z}_{\geqslant 1}$ we have

$$\zeta_p(k+1, x) = \frac{\omega_v(x)^k}{k} \int_{\mathbb{Z}_p} \frac{dt}{(x+t)^k} \cdot$$

Proposition 11.3.10. *Let χ be a primitive character modulo f.*

(1) *For $k \in \mathbb{Z} \setminus \{0\}$ we have*

$$L_p(\chi, k+1) = \frac{1}{k} \lim_{r \to \infty} \frac{1}{fp^r} \sum_{0 \leqslant n < fp^r}^{(p)} \frac{\chi\omega^k(n)}{n^k} \cdot$$

(2) *We have*

$$\lim_{s \to 1} \left(L_p(\chi, s) - \frac{(1 - 1/p)\delta(\chi)}{s - 1} \right) = - \lim_{r \to \infty} \frac{1}{fp^r} \sum_{0 \leqslant n < fp^r}^{(p)} \chi(n) \log_p(n) \,.$$

Of course by continuity the limit on the left-hand side of (2) is equal to $L_p(\chi, 1)$ when $\chi \neq \chi_0$.

Proof. (1). Indeed, by what we have just recalled, for $p \nmid a$ we have

$$\zeta_p\left(k+1, \frac{a}{m}\right) = \frac{\omega(a)^k \langle m \rangle^k}{k} \lim_{r \to \infty} \frac{1}{p^r} \sum_{0 \leqslant j < p^r} \frac{1}{(mj+a)^k} \, .$$

Thus, if we set $n = mj + a$, as j ranges from 0 to $p^r - 1$ and a from 0 to $m - 1$ not divisible by p, n ranges from 0 to $mp^r - 1$ not divisible by p, so that

$$L_p(\chi, k+1) = \frac{1}{mk} \lim_{r \to \infty} \frac{1}{p^r} \sum_{0 \leqslant n < mp^r}^{(p)} \frac{\chi \omega^k(n)}{n^k} \, .$$

(1) follows by choosing for instance $m = fp^2$ and replacing r by $r - 2$.

(2). By the uniformity estimate given in Proposition 11.2.4 and the proof that we have just given, it is clear that the result of (1) can be strengthened to

$$L_p(\chi, k+1) = \frac{1}{k} \frac{1}{fp^r} \sum_{0 \leqslant n < fp^r}^{(p)} \frac{\chi \omega^k(n)}{n^k} + O(p^{r-4}) \, .$$

Since $L_p(\chi, s)$ is analytic, hence continuous at $s = 1$ when $\chi \neq \chi_0$, we thus have

$$\lim_{s \to 1} \left(L_p(\chi, s) - \frac{(1 - 1/p)\delta(\chi)}{s - 1} \right) = \lim_{t \to \infty} L_p(\chi, 1 + \phi(p^{t+1})) - \frac{\delta(\chi)}{p^{t+1}}$$

$$= \lim_{t \to \infty} -\frac{\delta(\chi)}{p^{t+1}} + \frac{1}{(p-1)p^t} \lim_{r \to \infty} \frac{1}{fp^r} \sum_{0 \leqslant n < fp^r}^{(p)} \chi(n) n^{-(p-1)p^t} \, .$$

By the uniformity estimates given above and in Corollary 4.2.13 we can interchange the two limits. Furthermore, since

$$\sum_{0 \leqslant n < fp^r}^{(p)} \chi(n) = f(p-1)p^{r-1}\delta(\chi) \, ,$$

this quantity divided by $fp^r(p-1)p^t$ is equal to $\delta(\chi)/p^{t+1}$, so that

$$\lim_{s \to 1} \left(L_p(\chi, s) - \frac{(1 - 1/p)\delta(\chi)}{s - 1} \right) = \lim_{r \to \infty} \frac{1}{fp^r} \sum_{0 \leqslant n < fp^r}^{(p)} \chi(n) \lim_{t \to \infty} \frac{n^{-(p-1)p^t} - 1}{(p-1)p^t}$$

$$= \lim_{r \to \infty} \frac{1}{fp^r} \sum_{0 \leqslant n < fp^r}^{(p)} \chi(n) \frac{\log_p(n^{-(p-1)})}{p - 1} \, ,$$

proving (2). □

Note that to prove (2) for $\chi \neq \chi_0$, we could also have used the formula for $L_p(\chi, 1)$ in terms of B_j given in Proposition 11.3.9 and the Volkenborn integral representation of the B_j given by Lemma 11.1.7. In that case, we would have to justify the interchange of integration and summation, which can easily be done.

The above result shows that there is indeed some relation between the values of *p*-adic and ordinary *L*-functions also at *positive* integers.

Corollary 11.3.11. *Let $k \in \mathbb{Z} \setminus \{0\}$. If χ is a primitive character modulo a power of p we have*

$$L_p(\chi, k+1) = \frac{1}{k} \int_{\mathbb{Z}_p^*} \frac{\chi \omega^k(t)}{t^k} \, dt$$

and

$$\lim_{s \to 1} \left(L_p(\chi, s) - \frac{(1 - 1/p)\delta(\chi)}{s - 1} \right) = - \int_{\mathbb{Z}_p^*} \chi(t) \log_p(t) \, dt \,.$$

In particular,

$$L_p(\omega^{-k}, k+1) = \frac{1}{k} \int_{\mathbb{Z}_p^*} \frac{1}{t^k} \, dt \,.$$

Proof. Clear. Note that the integrals are over $\mathbb{Z}_p^*$. □

A more useful result is the following.

Proposition 11.3.12. *Let χ be a primitive character modulo f.*

(1) *For all $k \in \mathbb{Z} \setminus \{0\}$ we have*

$$L_p(\chi, k+1) = \lim_{r \to \infty} \frac{B_{\phi(p^r)-k}(\chi \omega^k)}{k} \,.$$

In particular,

$$\lim_{r \to \infty} B_{\phi(p^r)-k} = k L_p(\omega^{-k}, k+1) \,.$$

(2) *We have*

$$\lim_{s \to 1} \left(L_p(\chi, s) - \frac{(1 - 1/p)\delta(\chi)}{s - 1} \right) = - \lim_{r \to \infty} \frac{B_{\phi(p^r)}(\chi) - (1 - 1/p)\delta(\chi)}{\phi(p^r)} \,.$$

Proof. Since $L_p(\chi, s)$ is a continuous function of $s \neq 1$, for all $k \in \mathbb{Z}$ we have

$$L_p(\chi, k+1) = \lim_{r \to \infty} L_p(\chi, k + 1 - \phi(p^r)) \,.$$

Since $\omega^{\phi(q_p)}$ is the trivial character and $\phi(p^r)$ is even for $r \geqslant 2$, for r large enough we have by definition

$$L_p(\chi, k+1 - \phi(p^r)) = -(1 - (\chi \omega^k)(p)) p^{\phi(p^r)-k} \frac{B_{\phi(p^r)-k}(\chi \omega^k)}{\phi(p^r) - k} \,.$$

Taking the limit as $r \to \infty$ and distinguishing cases gives the result. □

Definition 11.3.13. *For $k \in \mathbb{Z} \setminus \{0\}$ we define the p-adic χ-Bernoulli numbers and χ-Euler constant by*

$$B_{k,p}(\chi) = \lim_{r \to \infty} B_{\phi(p^r)+k}(\chi) = -k L_p(\chi \omega^k, 1-k) , \quad and$$

$$\gamma_p(\chi) = -\lim_{r \to \infty} \frac{B_{\phi(p^r)}(\chi) - (1-1/p)\delta(\chi)}{\phi(p^r)}$$

$$= \lim_{s \to 1} \left(L_p(\chi, s) - \frac{(1-1/p)\delta(\chi)}{s-1} \right) .$$

In addition, we set $B_{k,p} = B_{k,p}(\chi_0)$ and $\gamma_p = \gamma_p(\chi_0)$.

Note that γ_p is the p-adic analogue of Euler's constant, and that when $\chi \neq \chi_0$ we evidently have $\gamma_p(\chi) = L_p(\chi, 1)$, so that the notation γ_p is really useful only for $\chi = \chi_0$.

Proposition 11.3.14. *Assume that the conductor of χ is a power of p (which is true in particular when $\chi = \chi_0$). Then for $k \in \mathbb{Z} \setminus \{0\}$ we have*

$$B_{k,p}(\chi) = \lim_{r \to \infty} \frac{1}{p^r} \sum_{0 \leqslant n < p^r}^{(p)} \chi(n) n^k = \int_{\mathbb{Z}_p^*} \chi(t) t^k \, dt \quad and$$

$$\gamma_p(\chi) = -\lim_{r \to \infty} \frac{1}{p^r} \sum_{0 \leqslant n < p^r}^{(p)} \chi(n) \log_p(n) = -\int_{\mathbb{Z}_p^*} \chi(t) \log_p(t) \, dt .$$

Proof. This is a restatement of Corollary 11.3.11.   □

From the definition it is immediate to deduce the following results.

Proposition 11.3.15. (1) *If $\chi(-1) = (-1)^{k-1}$ we have $B_{k,p}(\chi) = 0$, and if $\chi(-1) = -1$ we have $\gamma_p(\chi) = 0$.*
(2) *If $k \geqslant 1$ we have $B_{k,p}(\chi) = (1 - \chi(p)p^{k-1})B_k(\chi)$.*
(3) *As usual let m be a common multiple of f and q_p, and set $H_n(\chi) = \sum_{1 \leqslant a \leqslant m}^{(p)} \chi(a)/a^n$. If $k \geqslant 1$ and $\chi(-1) = (-1)^k$ we have*

$$B_{-k,p}(\chi) = k L_p(\chi \omega^{-k}, k+1) = \sum_{j \geqslant 0} (-1)^j \binom{k+j-1}{k-1} m^{j-1} B_j H_{k+j}(\chi) ,$$

and

$$\gamma_p(\chi) = \lim_{s \to 1} \left(L_p(\chi, s) - \frac{(1-1/p)\delta(\chi)}{s-1} \right)$$

$$= -\frac{1}{m} \sum_{0 \leqslant a < m}^{(p)} \chi(a) \log_p(\langle a \rangle) + \sum_{j \geqslant 1} \frac{(-1)^j}{j} m^{j-1} B_j H_j(\chi) .$$

(4) *For all $k \neq 0$ we have $v_p(B_{k,p}(\chi)) \geqslant -1$.*

(5) If χ is p-adically tame (see Definition 11.3.17 below) then $\gamma_p(\chi)$ is p-integral, and in all cases $v_p(\gamma_p(\chi)) \geqslant -1$.

Note that we will give stronger integrality statements later in Corollary 11.4.8.

Proof. All the statements except the last two are clear from the definitions and Proposition 11.3.9. By Lemma 9.5.11 we know that $v_p(B_k(\chi)) \geqslant -1$, and since $B_{k,p}(\chi) = \lim_{r\to\infty} B_{\phi(p^r)+k}(\chi)$ we also have $v_p(B_{k,p}(\chi)) \geqslant -1$. This also follows from (1), (2), and (3). For (4), since $\gamma_p(\chi) = 0$ if χ is odd, we may assume that χ is even. Since $q_p \mid m$ we have $v_p(m^{j-1}/j) \geqslant 1$ for all $j \geqslant 2$, so by the ordinary Clausen–von Staudt theorem $v_p(m^{j-1}B_j/j) \geqslant 0$ for $j \geqslant 2$, and for $j = 1$ we have $m^{j-1}B_j = -1/2$, which has nonnegative valuation if $p \neq 2$. Since χ is an even character, for $p = 2$ we have

$$H_1(\chi) = \sum_{1\leqslant a\leqslant m}^{(p)} \frac{\chi(a)}{a} = \sum_{1\leqslant a\leqslant m/2}^{(p)} \chi(a)\left(\frac{1}{a} + \frac{1}{m-a}\right) = m\sum_{1\leqslant a\leqslant m/2}^{(p)} \frac{\chi(a)}{a(m-a)} ,$$

so $v_p(H_1(\chi)) \geqslant v_p(m) \geqslant v_p(q_p) = 2$, proving that the valuation of the sum is nonnegative in all cases. Finally, the first sum $(1/m)\sum_{0\leqslant a<m}^{(p)} \chi(a)\log_p(\langle a\rangle)$ will be studied in Theorem 11.3.19 below, which tells us that its valuation is also nonnegative if χ is p-adically tame, and that otherwise it is greater than or equal to -1. □

For future reference, note the following corollary.

Corollary 11.3.16. *Let $k \geqslant 2$ be an even integer.*

(1) *We have*

$$B_{-k,p} \equiv \frac{1}{p} \sum_{1\leqslant a\leqslant p-1} \frac{1}{a^k} \pmod{p^v \mathbb{Z}_p} ,$$

where $v = 1$ if $5 \leqslant p \leqslant k+3$, and $v = 2$ for $p \geqslant k+5$.

(2) *If $p \geqslant k+3$ we have*

$$B_{-k,p} \equiv \frac{k}{k+1} B_{p-1-k} \pmod{p\mathbb{Z}_p} .$$

Proof. Immediate consequence of the proposition and of the Kummer congruences, and left to the reader; see Exercise 50. □

For example for $p \geqslant 7$ we have $B_{-2,p} \equiv (1/p)\sum_{1\leqslant a\leqslant p-1} 1/a^2 \pmod{p^2\mathbb{Z}_p}$, and for $p \geqslant 5$ we have $B_{-2,p} \equiv (2/3)B_{p-3} \pmod{p\mathbb{Z}_p}$. The corresponding congruences for $p \leqslant 5$ can be read off from the table that we give below.

Proposition 11.3.15 gives a practical way of computing the constants $B_{-k,p}(\chi)$ and $\gamma_p(\chi)$, since the definition as a limit of Bernoulli numbers is much slower. For the convenience of the reader, we give a small table for

$\chi = \chi_0$, where as usual the p-adic digits are written from right to left, and the digits from 10 to 18 are coded with the letters A to H.

p	γ_p	$B_{-2,p}$	$B_{-4,p}$
2	$\cdots$110110001100111	$\cdots$00000101000110.1	$\cdots$00111101111100.1
3	$\cdots$112010222121220	$\cdots$01001000002212.2	$\cdots$11210011021012.2
5	$\cdots$321122143203010	$\cdots$214004103314334	$\cdots$00341201131120.4
7	$\cdots$025121026026425	$\cdots$113431404032362	$\cdots$362564350404462
11	$\cdots$9317447545512A1	$\cdots$8682761505A028A	$\cdots$4349913A6604674
13	$\cdots$1893BC946787040	$\cdots$087B14A2BC94ACC	$\cdots$78C4809C3055B95
17	$\cdots$132AE449B942425	$\cdots$60294D387222D3E	$\cdots$539496A1G54488A
19	$\cdots$90489H87B72FHD2	$\cdots$G7GIF9767A0HGDA	$\cdots$F47AB7GDEB7E956

p	$B_{-6,p}$	$B_{-8,p}$	$B_{-10,p}$
2	$\cdots$10111110100010.1	$\cdots$00101010111000.1	$\cdots$10110110111110.1
3	$\cdots$00010112021000.2	$\cdots$22111112220122.2	$\cdots$01101110202202.2
5	$\cdots$241123000012322	$\cdots$20240200211300.4	$\cdots$330344030340240
7	$\cdots$54261355252232.6	$\cdots$033431442506531	$\cdots$506040436364625
11	$\cdots$8A7A967A8664645	$\cdots$625244199481503	$\cdots$17A273A506351A.A
13	$\cdots$0578730584B3284	$\cdots$B4AC3B114A10797	$\cdots$A3C140B38800A3A
17	$\cdots$AE16BFA8D998D2A	$\cdots$2G0ABEGEC44B3E4	$\cdots$EC1E4BCEE3E3G49
19	$\cdots$6EB027DEB2099B1	$\cdots$12ED0C4C01GE318	$\cdots$D958H1DE7004BG4

A Small Table of γ_p and $B_{-2k,p}$

11.3.4 χ-Power Sums Involving p-adic Logarithms

In the rest of this chapter, we will usually consider only primitive characters, although it is not difficult to generalize.

The aim of this subsection is to prove a technical result that will be seen to have several interesting arithmetic applications, essentially in the next subsection. Recall once again that we denote by $\mathcal{Z}_p$ the ring of p-adic integers of $\mathbb{C}_p$. We begin with the following definition.

Definition 11.3.17. *Let χ be a primitive character modulo f, and denote by $o(\chi)$ the order of χ, which divides $\phi(f)$. We say that χ is p-adically wild if χ is nontrivial, p is odd, and if both f and $o(\chi)$ are powers of p, and that χ is p-adically tame otherwise.*

Remarks. (1) This terminology is not completely standard. Properly speaking, we should speak of totally wild and nontotally wild characters, but the above is simpler. In the literature these are sometimes called characters of the second and first kind, respectively, a terminology that is probably even worse.

(2) By Corollary 2.1.35, we know that if χ is p-adically wild then $f = p^v$ and $o(\chi) = p^{v-1}$ for some $v \geqslant 2$.

Lemma 11.3.18. *Let χ be a primitive character modulo p^v for some odd prime p and some $v \geqslant v_p(q_p)$, let g be a primitive root modulo p^v, and let $n \in \mathbb{Z}_{\geqslant 0}$. Then*

$$\sum_{0 \leqslant k < \phi(p^v)} \chi(g)^k k^n \equiv 0 \pmod{p^{v-1-\delta} \mathcal{Z}_p},$$

where $\delta = 0$ if χ is p-adically tame, and $\delta = 1$ if χ is p-adically wild.

Proof. The result being trivial if $v = 1$, we may assume that $v \geqslant 2$. To simplify notation set $\zeta = \chi(g)$, which is a root of unity of order $o(\chi)$, and write $k = p^{v-1}a + b$ with $0 \leqslant a < p-1$ and $0 \leqslant b < p^{v-1}$. If S denotes our sum, we have

$$S \equiv \sum_{0 \leqslant a < p-1} \zeta^{p^{v-1}a} \sum_{0 \leqslant b < p^{v-1}} \zeta^b b^n \pmod{p^{v-1}\mathcal{Z}_p}.$$

If $\zeta^{p^{v-1}} \neq 1$, or equivalently, if $o(\chi) \neq p^{v-1}$, the first sum vanishes, so that $S \equiv 0 \pmod{p^{v-1}\mathcal{Z}_p}$ when χ is tame, as claimed. Thus, assume that $\zeta^{p^{v-1}} = 1$, in other words that χ is wild, so that

$$S \equiv (p-1) \sum_{0 \leqslant b < p^{v-1}} \zeta^b b^n \pmod{p^{v-1}\mathcal{Z}_p}.$$

Here we write $b = p^{v-2}c + d$ with $0 \leqslant c < p$ and $0 \leqslant d < p^{v-2}$, so that

$$S \equiv (p-1) \sum_{0 \leqslant c < p} \zeta^{p^{v-2}c} \sum_{0 \leqslant d < p^{v-2}} \zeta^d d^n \equiv 0 \pmod{p^{v-2}\mathcal{Z}_p},$$

since $\zeta^{p^{v-2}}$ is a primitive pth root of unity, so that the first sum vanishes, proving the lemma. $\qquad\square$

Theorem 11.3.19. *Let χ be a primitive character of conductor f, and let m be the least common multiple of f and q_p.*

(1) *If χ is p-adically tame then for any $n \geqslant 1$ we have*

$$\sideset{}{^{(p)}}\sum_{0 \leqslant a < m} \chi(a) \frac{\log_p(\langle a \rangle)^n}{n!} \equiv 0 \pmod{(q_p^{n-1}/n!)m\mathcal{Z}_p}$$

(note that $q_p^{n-1}/n! \in p\mathbb{Z}_p$ for $n \geqslant 2$).
(2) *If χ is p-adically wild then*

$$\sideset{}{^{(p)}}\sum_{0 \leqslant a < m} \chi(a) \frac{\log_p(\langle a \rangle)^n}{n!} \equiv 0 \pmod{(q_p^{n-1}/n!)(m/p)\mathcal{Z}_p}.$$

Proof. If χ is the trivial character we have $m = q_p$ and $\log_p(\langle a \rangle) \equiv 0$ (mod q_p); hence the result is trivial in that case, so we may assume that χ is nontrivial.

Assume first that f is not a power of p, and write $f = p^v f_2$ and $m = p^w m_2$ with $p \nmid f_2 m_2$, $w \geqslant \max(v, v_p(q_p))$, and $f_2 \mid m_2$. By Proposition 2.1.34 there exist two primitive characters χ_1 modulo p^v and χ_2 modulo f_2 such that $\chi = \chi_1 \chi_2$. Writing $a = p^w r_2 + r_1$ with $0 \leqslant r_1 < p^w$, $p \nmid r_1$, and $0 \leqslant r_2 < m_2$, we have

$$\sum_{0 \leqslant a < m}^{(p)} \chi(a) \frac{\log_p(\langle a \rangle)^n}{n!} = \sum_{0 \leqslant r_1 < p^w}^{(p)} \chi_1(r_1) T(r_1) ,$$

where

$$T(r_1) = \sum_{0 \leqslant r_2 < m_2} \chi_2(p^w r_2 + r_1) \frac{(\log_p(\langle r_1 \rangle) + \log_p(1 + p^w r_2 / r_1))^n}{n!}$$

$$= \sum_{0 \leqslant j \leqslant n} \frac{\log_p(\langle r_1 \rangle)^j}{j!} \sum_{0 \leqslant r_2 < m_2} \chi_2(p^w r_2 + r_1) \frac{\log_p(1 + p^w r_2 / r_1)^{n-j}}{(n-j)!} .$$

Since $w \geqslant v_p(q_p)$, it follows that $v_p(\log_p(1 + p^w r_2 / r_1)) \geqslant w$, so $v_p(\log_p(1 + p^w r_2 / r_1)^{n-j} / (n-j)!) \geqslant w(n-j) - v_p((n-j)!)$. Since $v_p(\langle r_1 \rangle) \geqslant v_p(q_p)$, for $j < n$ we thus have

$$\frac{\log_p(\langle r_1 \rangle)^j}{j!} \sum_{0 \leqslant r_2 < m_2} \chi_2(p^w r_2 + r_1) \frac{\log_p(1 + p^w r_2 / r_1)^{n-j}}{(n-j)!} \equiv 0 \ (\text{mod } p^{v_j} \mathcal{Z}_p) ,$$

where

$$v_j \geqslant j v_p(q_p) - v_p(j!) + w(n-j) - v_p((n-j)!)$$
$$\geqslant w + (n-1)w - j(w - v_p(q_p)) - v_p(n!) + v_p\left(\binom{n}{j} \right)$$
$$\geqslant w + (n-1)w - (n-1)(w - v_p(q_p)) - v_p(n!)$$
$$\geqslant w + (n-1)v_p(q_p) - v_p(n!) ,$$

giving the desired congruence for the terms with $j < n$. For $j = n$, since $p \nmid m_2$ the map $r_2 \mapsto p^w r_2 + r_1$ is a bijection of $\mathbb{Z}/m_2\mathbb{Z}$ onto itself, hence

$$\sum_{0 \leqslant r_2 < m_2} \chi_2(p^w r_2 + r_1) = \sum_{0 \leqslant r_2 < m_2} \chi_2(r_2) = 0$$

since by assumption χ_2 is nontrivial, else f would be equal to a power of p, so the terms with $j = n$ do not contribute, proving the result when f is not a power of p.

Assume now that $f = p^v$ for some $v \geqslant 1$ with $p \geqslant 3$. In that case $m = f$, and the group $(\mathbb{Z}/f\mathbb{Z})^*$ is cyclic. Let g be a primitive root modulo p^v, so that the class of g modulo f generates $(\mathbb{Z}/f\mathbb{Z})^*$. If $p \nmid a$ we have $a \equiv g^k \pmod{f}$ for some k defined uniquely modulo $\phi(f) = (p-1)p^{v-1}$, hence $\omega(a) = \omega(g)^k$, so that

$$\langle a \rangle = \frac{a}{\omega(a)} = \frac{g^k(1 + p^v u_k)}{\omega(g)^k} = \langle g \rangle^k (1 + p^v u_k)$$

for some p-adic integer u_k. It follows that

$$\log_p(\langle a \rangle) \equiv k \log_p(\langle g \rangle) \pmod{p^v} \,.$$

Since $v_p(\log_p(\langle x \rangle)) \geqslant 1$ for all x, an immediate p-adic argument (see Exercise 3 of Chapter 4) shows that for all $n \geqslant 0$ we have

$$\frac{\log_p(\langle a \rangle)^n}{n!} \equiv \frac{k^n \log_p(\langle g \rangle)^n}{n!} \pmod{(p^{n-1}/n!)p^v} \,.$$

Thus

$$\sum_{0 \leqslant a < m}^{(p)} \chi(a) \frac{\log_p(\langle a \rangle)^n}{n!} \equiv \frac{\log_p(\langle g \rangle)^n}{n!} \sum_{0 \leqslant k < \phi(f)} \chi(g)^k k^n \pmod{(p^{n-1}/n!)m} \,.$$

Note that since g is a primitive root, the order of χ as a character is equal to that of $\chi(g)$ as a root of unity, and since χ is nontrivial, that $\chi(g) \neq 1$. Applying Lemma 11.3.18, we see that if $o(\chi) \neq p^{v-1}$ we have

$$\sum_{0 \leqslant k < \phi(f)} \chi(g)^k k^n \equiv 0 \pmod{p^{v-1}\mathbb{Z}_p} \,,$$

otherwise the congruence is only modulo $p^{v-2}\mathbb{Z}_p$. Since $v_p(\log_p(\langle g \rangle)^n/n!) \geqslant v_p(p^n/n!)$, this proves the theorem for $f = p^v$ with $p \geqslant 3$.

Assume finally that $f = 2^v$ with $v \geqslant 2$, and set $p = 2$. For $v = 2$ the result is immediate since the only primitive character modulo f is $\left(\frac{-4}{\cdot}\right)$, so assume that $v \geqslant 3$. We again have $m = f$, and if $p \nmid a$ we can write in a unique way $a \equiv \left(\frac{-4}{a}\right)5^k \pmod{f}$ for some k defined uniquely modulo 2^{v-2}. Since by definition of ω, for $p = 2$ we have $\omega(a) = \left(\frac{-4}{a}\right)$, it follows that $\langle a \rangle = a/\omega(a) \equiv 5^k \pmod{f}$. The same reasoning as in the case $p > 2$ shows that

$$\sum_{0 \leqslant a < m}^{(p)} \chi(a) \frac{\log_p(\langle a \rangle)^n}{n!} \equiv (1 + \chi(-1)) \frac{\log_p(5)^n}{n!} U_n \pmod{(q_p^{n-1}/n!)m} \,, \text{ with}$$

$$U_n = \sum_{0 \leqslant k < 2^{v-2}} \chi(5)^k k^n \,,$$

the factor $1 + \chi(-1)$ coming from the two possible values of $\left(\frac{-4}{a}\right)$. By Lemma 2.1.35 the order of χ is equal to 2^{v-2}. Furthermore, we clearly have

$$U_n \equiv \sum_{0 \leqslant k < 2^{v-3}} \chi(5)^k k^n (1 + \chi(5)^{2^{v-3}}) \pmod{2^{v-3}} \,,$$

and since $\chi(5)$ is a primitive 2^{v-2}th root of unity we have $\chi(5)^{2^{v-3}} = -1$, so that $U_n \equiv 0 \pmod{2^{v-3}}$. Since $q_p = 4$ and $\log_p(5) \equiv 0 \pmod{q_p}$, it follows that

$$\sum_{0 \leqslant a < m}^{(p)} \chi(a) \frac{\log_p(\langle a \rangle)^n}{n!} \equiv 0 \pmod{(q_p^{n-1}/n!)m}\,,$$

as claimed.   □

Corollary 11.3.20. *Let χ be a primitive character of conductor f, let m be the least common multiple of f and q_p, and for simplicity of notation set*

$$T_n(\chi) = \sum_{0 \leqslant a < m}^{(p)} \chi(a) \frac{\log_p(\langle a \rangle)^n}{n!}\,.$$

(1) *For $n \geqslant 2$ we have $T_n(\chi) \equiv 0 \pmod{pm\mathbb{Z}_p}$, except when $n = 3$, $p = 3$, and χ is 3-adically wild, or when $n = 2$ and χ is p-adically wild, in which case the congruence is only modulo $m\mathbb{Z}_p$.*

(2) *For $n = 1$ we have*

$$T_1(\chi) \equiv \begin{cases} 0 & (\mathrm{mod}\ m\mathbb{Z}_p) & \textit{if } \chi \textit{ is } p\textit{-adically tame,} \\[2mm] \dfrac{m}{1 - \chi(1+p)} & (\mathrm{mod}\ m\mathbb{Z}_p) & \textit{if } \chi \textit{ is } p\textit{-adically wild.} \end{cases}$$

(3) *In the special case $n = 1$ and $p = 2$, if χ is odd we have $T_1(\chi) \equiv 0$ $(\mathrm{mod}\ 2m\mathbb{Z}_p)$ except if $f = 4$, in which case $T_1(\chi) \equiv m \pmod{2m\mathbb{Z}_p}$, while if χ is even we have*

$$T_1(\chi) \equiv \begin{cases} 0 & (\mathrm{mod}\ 2m\mathbb{Z}_p) & \textit{if } f \textit{ is not a power of } 2\,, \\[2mm] \dfrac{2m}{1 - \chi(5)} & (\mathrm{mod}\ 2m\mathbb{Z}_p) & \textit{if } f = 2^v \textit{ with } v \geqslant 3\,, \\[2mm] m & (\mathrm{mod}\ 2m\mathbb{Z}_p) & \textit{if } f = 1\,. \end{cases}$$

Proof. (1). As already remarked in the theorem, we have $q_p^{n-1}/n! \in p\mathbb{Z}_p$ for $n \geqslant 2$, so (1) follows when χ is p-adically tame. If on the contrary χ is wild, hence p odd, the theorem says that the congruence is true modulo $(p^{n-2}/n!)m\mathbb{Z}_p$. For $n = 2$ we have $v_p(p^{n-2}/n!) = 0$ since $p \neq 2$, while for $n \geqslant 3$ we have $v_p(p^{n-2}/n!) \geqslant 1$, with the exception of $p = 3$ and $n = 3$, in which case $v_p(p^{n-2}/n!) = 0$, proving (1).

(2). When χ is tame the result is a special case of the theorem, so we may assume that χ is wild, so that p is odd, $m = p^v$, and χ has order p^{v-1}. We use the same reasoning as for Theorem 9.5.5 (3). We can write $a \equiv a_1 a_2$ $(\mathrm{mod}\ m)$, with $a_1 = a^{p^{v-1}}$ and $a_2 \equiv 1 \pmod{p}$, and since χ has order p^{v-1} we have $\chi(a) = \chi(a_2)$. Note that when $p \nmid a$ we have $\log_p(\langle a \rangle) \equiv \log_p(\langle a_1 a_2 \rangle)$ $(\mathrm{mod}\ p^v)$. Since $a_2 \equiv (1+p)^x \pmod{p^v}$ for a unique x modulo p^v and since χ is a nontrivial character we have

$$\sideset{}{^{(p)}}\sum_{0\leqslant a<m}\chi(a)\log_p(\langle a\rangle)\equiv\sideset{}{^{(p)}}\sum_{\substack{a_1\bmod p\\x\bmod p^{v-1}}}\chi((1+p)^x)(\log_p(\langle a_1\rangle)+x\log_p(\langle 1+p\rangle))$$

$$\equiv\sideset{}{^{(p)}}\sum_{a_1\bmod p}\log_p(\langle a_1\rangle)\sum_{x\bmod p^{v-1}}\chi(1+p)^x$$

$$+\log_p(1+p)\sideset{}{^{(p)}}\sum_{a_1\bmod p}\sum_{x\bmod p^{v-1}}x\chi(1+p)^x\ (\bmod\ p^v\mathcal{Z}_p)\,,$$

using the fact that $\langle 1+p\rangle=1+p$. Since χ has exact order p^{v-1} and $v\geqslant 2$, it follows that $\chi(1+p)$ is a primitive p^{v-1}th root of unity, and in particular is different from 1, so that $\sum_{x\bmod p^{v-1}}\chi(1+p)^x=0$. Furthermore, by computing the derivative of a geometric series, we immediately find that

$$\sum_{0\leqslant x<p^{v-1}}x\chi(1+p)^x=\frac{p^{v-1}}{\chi(1+p)-1}\,.$$

Since there are $p-1$ terms in the sum over $a_1\bmod p$, and since $\log_p(1+p)\equiv p$ $(\bmod\ p^2\mathcal{Z}_p)$ and $v_p(\chi(1+p)-1)<1$, (2) follows.

(3). If $f=1$ or $f=4$, we have $m=4$, $\langle 1\rangle=1$, and $\langle 3\rangle=-3$, so our sum is equal to $\chi(-1)\log_2(-3)$, which is easily seen to be congruent to 4 modulo $8\mathbb{Z}_2$. We may therefore assume that $f\neq 1$ and $f\neq 4$, in other words χ nontrivial and $\chi\neq\left(\frac{-4}{\cdot}\right)$. Since $4\mid m$, we can write

$$\sideset{}{^{(p)}}\sum_{0\leqslant a<m}\chi(a)\log_p(a)=\sum_{\substack{0\leqslant a<m\\a\equiv 1\,(\bmod\,4)}}(\chi(a)\log_p(a)+\chi(m-a)\log_p(m-a))$$

$$=(1+\chi(-1))\sum_{\substack{0\leqslant a<m\\a\equiv 1\,(\bmod\,4)}}\chi(a)\log_p(a)$$

$$+\chi(-1)\sum_{\substack{0\leqslant a<m\\a\equiv 1\,(\bmod\,4)}}\chi(a)\log_p(1-m/a)\,.$$

Since $4\mid m$ and $2\nmid a$, by expanding the logarithm we see that $\log_p(1-m/a)\equiv -m/a\equiv m\ (\bmod\ 2m\mathcal{Z}_p)$, so that

$$\sum_{\substack{0\leqslant a<m\\a\equiv 1\,(\bmod\,4)}}\chi(a)\log_p(1-m/a)\equiv m\sum_{\substack{0\leqslant a<m\\a\equiv 1\,(\bmod\,4)}}\chi(a)\ (\bmod\ 2m\mathcal{Z}_p)\,.$$

However, we can write

$$\sum_{\substack{0\leqslant a<m \\ a\equiv 1 \ (\mathrm{mod}\ 4)}} \chi(a) = \frac{1}{2}\sum_{0\leqslant a<m}^{(p)}\left(1+\left(\frac{-4}{a}\right)\right)\chi(a)$$

$$= \frac{1}{2}\left(\sum_{0\leqslant a<m}^{(p)}\chi(a) + \sum_{0\leqslant a<m}^{(p)}\chi_1(a)\right),$$

where $\chi_1(a) = \left(\frac{-4}{a}\right)\chi(a)$. Since m is a common multiple of 4 and the conductors of χ and χ_1, and since by assumption χ is both nontrivial and different from $\left(\frac{-4}{\cdot}\right)$, it follows from Lemma 11.3.7 that both sums above vanish. We thus have

$$\sum_{0\leqslant a<m}^{(p)}\chi(a)\log_p(a) \equiv (1+\chi(-1))\sum_{\substack{0\leqslant a<m \\ a\equiv 1 \ (\mathrm{mod}\ 4)}}\chi(a)\log_p(a) \ (\mathrm{mod}\ 2m\mathcal{Z}_p)\,.$$

This proves the theorem when χ is an odd character, and also when $8 \nmid m$ (in other words, $8 \nmid f$), since $1+\chi(-1) \equiv 0 \ (\mathrm{mod}\ 2)$ and $\log_p(a) \equiv 0 \ (\mathrm{mod}\ 4)$.

Thus, assume now that $8 \mid f$, so that $m = f$, and that χ is an even character. Since $f/2 \equiv 0 \ (\mathrm{mod}\ 4)$, by Corollary 2.1.30 we have

$$\sum_{\substack{0\leqslant a<f \\ a\equiv 1 \ (\mathrm{mod}\ 4)}}\chi(a)\log_p(a) = \sum_{\substack{0\leqslant a<f/2 \\ a\equiv 1 \ (\mathrm{mod}\ 4)}}\chi(a)(\log_p(a) - \log_p(a+f/2))$$

$$= -\sum_{\substack{0\leqslant a<f/2 \\ a\equiv 1 \ (\mathrm{mod}\ 4)}}\chi(a)\log_p(1+f/(2a))\,.$$

Since $8 \mid f$ we check that $\log_p(1+f/(2a)) \equiv f/(2a) \equiv f/2 \ (\mathrm{mod}\ f)$ (note that we only need the congruence modulo f and not $2f$ here), so that

$$\sum_{\substack{0\leqslant a<f \\ a\equiv 1 \ (\mathrm{mod}\ 4)}}\chi(a)\log_p(a) \equiv (f/2)\sum_{\substack{0\leqslant a<f/2 \\ a\equiv 1 \ (\mathrm{mod}\ 4)}}\chi(a) \ (\mathrm{mod}\ f\mathcal{Z}_p)\,.$$

Using once again the characters $\left(\frac{-4}{\cdot}\right)$ and $\chi_1 = \left(\frac{-4}{\cdot}\right)\chi$, and since we have $\sum_{0\leqslant a<f/2}\chi(a) = 0$ for an even nontrivial character χ, we obtain

$$\sum_{0\leqslant a<m}^{(p)}\chi(a)\log_p(a) \equiv (f/2)\sum_{0\leqslant a<f/2}\chi_1(a) \ (\mathrm{mod}\ 2f\mathcal{Z}_p)\,.$$

When f is not a power of 2, by Corollary 9.5.10 this last sum is divisible by 4, proving the theorem in that case. On the other hand, when $f = 2^v$ with $v \geqslant 3$, by the same corollary we have

$$\sum_{0\leqslant a<f/2}\chi_1(a) \equiv 4/(1-\chi(5)) \ (\mathrm{mod}\ 4\mathcal{Z}_p)\,,$$

finishing the proof of the theorem. $\square$

11.3.5 The Function $L_p(\chi, s)$ Around $s = 1$

Since $L_p(\chi, s)$ is a p-adic holomorphic function in the disk of radius $R_p > 1$ when $\chi \neq 1$, we can look at its expansion around 1. It happens that the simple p-adic properties of its coefficients gives important arithmetic information on Bernoulli numbers and other quantities, which are more precise than those obtained in Chapter 9. The result, which easily follows from Theorem 11.2.11 and Corollary 11.3.20, is the following (we may of course assume that χ is an even character, otherwise $L_p(\chi, s)$ is identically zero).

Theorem 11.3.21. *Let χ be an even primitive character of conductor f. The Taylor series expansion of $L_p(\chi, s)$ around $s = 1$ has the form*

$$L_p(\chi, s) = \frac{a_{-1}}{s - 1} + a_0 + a_1(s - 1) + a_2(s - 1)^2 + \cdots ,$$

where $a_{-1} = 0$ if χ is not the trivial character and $a_{-1} = 1 - 1/p$ if χ is the trivial character, and where the coefficients a_j satisfy the following:

(1) *For $j \geqslant 2$ we have $p \mid a_j$ (in other words $|a_j/p| \leqslant 1$), except if $j = 2$, $p = 3$, and χ is 3-adically wild, in which case we have only $|a_2| \leqslant 1$.*

(2) *For $j = 1$ we have $p \mid a_1$, except if p is odd and either χ is p-adically wild, or if χ is the trivial character and $p = 3$, in which cases we have only $|a_1| \leqslant 1$ (and more precisely $a_1 \equiv 2 \pmod{3\mathbb{Z}_3}$ when χ is the trivial character and $p = 3$).*

(3) *For $j = 0$ we have $|a_0| \leqslant 1$, except if p is odd and χ is p-adically wild, in which case $|pa_0| \leqslant 1$, and more precisely*

$$a_0 \equiv \frac{1}{\chi(1 + p) - 1} \pmod{\mathcal{Z}_p} .$$

(4) *In addition, if $p = 2$ then*

$$a_0 \equiv \begin{cases} 0 & \pmod{2\mathcal{Z}_p} & \text{if } f \text{ is not a power of } 2 , \\ \dfrac{2}{\chi(5) - 1} & \pmod{2\mathcal{Z}_p} & \text{if } f = 2^v \text{ with } v \geqslant 3 , \\ 1 & \pmod{2\mathcal{Z}_p} & \text{if } f = 1 . \end{cases}$$

Note that if χ is nontrivial we have $a_0 = L_p(\chi, 1)$ (hence $a_0 = \gamma_p(\chi)$ when the conductor of χ is a power of p), which we will compute in Section 11.5.6, while if χ is trivial, so that $L_p(\chi, s) = \zeta_p(s)$ is the Kubota–Leopoldt p-adic zeta function, we have by definition $a_0 = \gamma_p$, the p-adic Euler constant.

Proof. Choose $m = \text{lcm}(f, q_p)$, and for simplicity of notation, set

$$S_{-n}(\chi) = \sum_{0 \leqslant a < m}^{(p)} \frac{\chi(a)}{a^n} \quad \text{and} \quad T_n(\chi) = \sum_{0 \leqslant a < m}^{(p)} \chi(a) \frac{\log_p(\langle a \rangle)^n}{n!} .$$

By Proposition 11.3.8 we have

$$L_p(\chi, s) = \sum_{0 \leqslant a < m}^{(p)} \chi(a) \frac{\langle m \rangle^{1-s}}{m} \zeta_p\left(s, \frac{a}{m}\right),$$

so Theorem 11.2.11 (3) tells us that

$$L_p(\chi, s) = \frac{a_{-1}}{s-1} + a_0 + a_1(s-1) + a_2(s-1)^2 + \cdots,$$

with

$$a_{-1} = \frac{1}{m} \sum_{0 \leqslant a < m}^{(p)} \chi(a) = \frac{S_0(\chi)}{m},$$

$$a_0 \equiv \frac{S_{-1}(\chi)}{2} + \frac{m}{12} S_{-2}(\chi) - \frac{T_1(\chi)}{m} \pmod{q_p \mathbb{Z}_p},$$

$$a_1 \equiv \frac{T_2(\chi)}{m} - \frac{1}{2} \sum_{0 \leqslant a < m}^{(p)} \chi(a) \frac{\log_p(\langle a \rangle)}{a} + \frac{m}{12} S_{-2}(\chi) \pmod{q_p \mathbb{Z}_p},$$

$$a_j \equiv (-1)^{j+1} \frac{T_{j+1}(\chi)}{m} \pmod{q_p \mathbb{Z}_p} \quad \text{for } j \geqslant 2.$$

We have already shown that $a_{-1} = 1 - 1/p$ if χ is trivial, and $a_{-1} = 0$ otherwise.

(1). By Corollary 11.3.20, for $j \geqslant 2$ we have $T_{j+1}(\chi) \equiv 0 \pmod{pm\mathbb{Z}_p}$, except if $j = 2$, $p = 3$, and χ is 3-adically wild, in which case we have only $T_{j+1}(\chi) \equiv 0 \pmod{m\mathbb{Z}_p}$, so (1) follows.

(2). We have similarly $T_2(\chi) \equiv 0 \pmod{pm\mathbb{Z}_p}$, except if χ is p-adically wild, in which case we have only $T_2(\chi) \equiv 0 \pmod{m\mathbb{Z}_p}$. Furthermore, $m/12 \in p\mathbb{Z}_p$ for $p \geqslant 5$, while for $p = 2$ and $p = 3$ we have $m/12 \in \mathbb{Z}_p$, and $a^2 \equiv 1$ $\pmod{p}$ when $p \nmid a$, so that $S_{-2}(\chi) \equiv \sum_{0 \leqslant a < m}^{(p)} \chi(a) \pmod{p}$, and this last sum is equal to 0, except when χ is the trivial character, in which case it is equal to $m(1 - 1/p)$. This is even for $p = 2$ since $4 \mid m$, but is congruent to 2 $\pmod 3$ for $p = 3$ and $m = \text{lcm}(f, q_p) = \text{lcm}(1, 3) = 3$. Finally, since $\log_p(\langle a \rangle) \equiv 0 \pmod{q_p}$, it follows trivially that

$$\frac{1}{2} \sum_{0 \leqslant a < m}^{(p)} \chi(a) \frac{\log_p(\langle a \rangle)}{a} \equiv 0 \pmod{p\mathbb{Z}_p},$$

proving (2).

(3). We have seen in the proof of (2) that $(m/12)S_{-2}(\chi) \in p\mathbb{Z}_p$, except when $p = 3$ and χ is the trivial character, in which case $(m/12)S_{-2}(\chi) \in \mathbb{Z}_p$. Furthermore, using the symmetry $a \mapsto m - a$ and the fact that χ is an even character, it is clear that $S_{-1}(\chi) \equiv 0 \pmod{m\mathbb{Z}_p}$, and since $q_p \mid m$, we have $S_{-1}(\chi)/2 \equiv 0 \pmod{p\mathbb{Z}_p}$, so that $S_{-1}(\chi)/2 + (m/12)S_{-2}(\chi) \in p\mathbb{Z}_p$ except in the special case mentioned above. Furthermore, by Corollary 11.3.20 we have $T_1(\chi) \equiv 0 \pmod{m\mathbb{Z}_p}$ except if χ is p-adically wild, in which case the congruence is modulo $(m/p)\mathcal{Z}_p$, and the more precise congruence follows from Corollary 11.3.20.

(4). By the proof of (3), we already know that for $p = 2$ we have $S_{-1}(\chi)/2 + (m/12)S_{-2}(\chi) \in p\mathcal{Z}_p$, so the result follows from Corollary 11.3.20 (3). $\qquad\square$

Corollary 11.3.22. *Let* χ *be an even primitive character modulo* f, *and assume that* χ *is p-adically tame. Define*

$$M_p(\chi, s) = \begin{cases} L_p(\chi, s) & \text{if } \chi \text{ is nontrivial}, \\ L_p(\chi, s) - \dfrac{1 - 1/p}{s - 1} & \text{if } \chi \text{ is trivial and } p \neq 3, \\ L_p(\chi, s) - \dfrac{1 - 1/p}{s - 1} + (s - 1) & \text{if } \chi \text{ is trivial and } p = 3. \end{cases}$$

The Taylor series expansion of $M_p(\chi, s)$ *around* $s = 1$ *has the form* $\sum_{j \geqslant 0} b_j(s - 1)^j$ *with* $p \mid b_j$ *for all* $j \geqslant 1$, *and* $|b_0| = |a_0| \leqslant 1$.

Proof. Clear. $\qquad\square$

11.4 Applications of p-adic L-Functions

11.4.1 Integrality and Parity of L-Function Values

The existence and basic properties seen above for $L_p(\chi, s)$ (not including the value at $s = 1$), especially Theorem 11.3.21, imply in a simple way several integrality results on values of ordinary L-functions, congruences on Bernoulli numbers and additional nontrivial and important results. Here are a few examples.

Proposition 11.4.1. *Let* χ *be an even primitive character of conductor* f *and let* $k \in \mathbb{Z}$ *be arbitrary, not necessarily positive.*

(1) *If* χ *is nontrivial and p-adically tame then* $L_p(\chi, 1 - k)$ *is p-integral and*

$$L_p(\chi, 1 - k) \equiv L_p(\chi, 1) \pmod{p\mathcal{Z}_p}.$$

(2) *If* χ *is nontrivial and p-adically wild then* $v_p(L_p(\chi, 1 - k)) = -1/\phi(f)$, *and more precisely*

$$L_p(\chi, 1 - k) \equiv \frac{1}{\chi(1 + p) - 1} \pmod{\mathcal{Z}_p}.$$

(3) *If* $k \neq 0$ *then* $\zeta_p(1 - k) + (1 - 1/p)/k$ *is p-integral and all these quantities are congruent modulo* $p\mathbb{Z}_p$, *except for* $p = 3$, *where the congruent quantities are the* $\zeta_p(1 - k) + (1 - 1/p)/k - k$.

(4) *For $p = 2$ we have*

$$L_2(\chi, 1 - k) \equiv \begin{cases} 0 & (\mathrm{mod}\ 2\mathbb{Z}_2) & \text{if } f \text{ is not a power of } 2, \\ \dfrac{2}{\chi(5) - 1} & (\mathrm{mod}\ 2\mathbb{Z}_2) & \text{if } f = 2^v \text{ with } v \geqslant 3, \\ 1 - \dfrac{1}{2k} & (\mathrm{mod}\ 2\mathbb{Z}_2) & \text{if } f = 1 \text{ and } k \neq 0. \end{cases}$$

Proof. Since $L_p(\chi, s)$ is an analytic function for $|s| < R_p$ and $R_p > 1$, it follows that the radius of convergence of its Taylor series around $s = 1$ is greater than or equal to R_p. Thus, by Theorem 11.3.21, if s is p-integral (and in particular if $s \in \mathbb{Z}$) we have $L_p(\chi, s) \equiv a_0 \pmod{p}$, therefore (1) and (2) follow from the same theorem, (3) is proved similarly, and (4) follows in the same way from Theorem 11.3.21 (4).   □

An important consequence of this proposition is that, apart from some well-understood exceptions, the values at negative integers of L-functions of Dirichlet characters are twice algebraic integers.

Corollary 11.4.2. *Let $k \in \mathbb{Z}_{\geqslant 1}$, let χ be a nontrivial primitive character such that $\chi(-1) = (-1)^k$, and denote by f its conductor and by $u = o(\chi) \mid \phi(f)$ its order.*

(1) *Let p be a prime. The algebraic number $L(\chi, 1 - k)$ is p-integral (see Definition 11.3.1), except possibly when p is odd, $f = p^v$ for some $v \geqslant 1$, and $u = p^{v-1}(p - 1)/\gcd(p - 1, k)$.*

(2) *If f is not a power of 2 the algebraic number $L(\chi, 1 - k)/2$ is 2-integral, except possibly if $k = 1$ and f is an odd prime power, in which case, in general, only $L(\chi, 1 - k)$ is 2-integral.*

(3) *In particular, assume that f is not a power of 2. Then $L(\chi, 1 - k)/2$ is an algebraic integer (hence is in $\mathbb{Z}[\zeta_u]$), except possibly when $f = p^v$ for some odd prime p with $v \geqslant 1$, and either $u = p^{v-1}(p - 1)/\gcd(p - 1, k)$ or $k = 1$.*

(4) *If $f = p^v$ and $u = p^{v-1}(p - 1)/\gcd(p - 1, k)$ for some odd prime p and some $v \geqslant 1$, there exists a unique prime ideal $\mathfrak{p}$ of $\mathbb{Q}(\chi) = \mathbb{Q}(\zeta_u)$ above p such that $v_{\mathfrak{q}}(L(\chi, 1 - k)) \geqslant 0$ for all prime ideals $\mathfrak{q} \neq \mathfrak{p}$, and such that*

$$v_{\mathfrak{p}}\left(L(\chi, 1 - k) + \frac{1 - 1/p}{k}\right) \geqslant 0 \quad \text{if } v = 1,$$

$$v_{\mathfrak{p}}\left(L(\chi, 1 - k) - \frac{1}{\chi(1 + p) - 1}\right) \geqslant 0 \quad \text{if } v \geqslant 2.$$

(5) *If $f = 2^v$ with $v \geqslant 3$, we have*

$$L(\chi, 1 - k) - \frac{2}{\chi(5) - 1} \in 2\mathbb{Z}[\zeta_u].$$

(6) If $\chi = \left(\frac{-4}{\cdot}\right)$, for all odd k we have $L(\chi, 1 - k) + k/2 - 1 \in 4\mathbb{Z}$.

(7) If $\chi = \left(\frac{-8}{\cdot}\right)$, for all odd k we have $L(\chi, 1 - k) - 1 \in 4\mathbb{Z}$.

(8) If $\chi = \left(\frac{8}{\cdot}\right)$, for all even k we have $L(\chi, 1 - k) + 1 \in 4\mathbb{Z}$.

Remarks. (1) It is absolutely necessary to assume $k \geqslant 1$, since for $k \leqslant 0$ the value of $L(\chi, 1 - k)$ is usually not algebraic.

(2) If $\chi(-1) = (-1)^{k-1}$ we have $L(\chi, 1 - k) = 0$, so we may indeed assume that $\chi(-1) = (-1)^k$.

(3) Let χ be a not necessarily primitive character modulo m. Then if we denote by f is its conductor and by χ_f the primitive character modulo f equivalent to χ we have

$$L(\chi, 1 - k) = \prod_{p \mid m, \, p \nmid f} (1 - \chi_f(p)p^{k-1})L(\chi_f, 1 - k) \,.$$

It follows that the integrality and parity results given in the corollary are still valid for χ, the restrictions being on the conductor f and *not* on m.

(4) Since $L(\chi, 1-k) = -B_k(\chi)/k$ for a nontrivial character, the above results can be restated as results on $B_k(\chi)/k$. The reader is invited to compare with the corresponding results for $B_k(\chi)$ itself given in Theorem 9.5.13, which are weaker.

(5) The corresponding statements for the trivial character χ will be given, in a slightly stronger form, as Proposition 11.4.4 below.

Proof. Since this corollary is a result on ordinary and not p-adic L-functions, the notion of p-integrality is a little different (see Definition 11.3.1). More precisely, by Corollary 10.2.3 we know that $L(\chi, 1 - k) = -B_k(\chi)/k$ is an algebraic number (belonging to $\mathbb{Q}(\zeta_u)$), and since $B_k(\chi)$ is a rational linear combination of values of χ, it follows that the conjugates of $L(\chi, 1 - k)$ in $\overline{\mathbb{Q}}$ are the $L(\chi^j, 1 - k)$ for j modulo u and coprime to u. Note that χ^j has the same conductor and the same order as χ itself. By Lemma 11.3.2 it follows that if q is any prime number then $L(\chi, 1 - k)$ is q-integral as an algebraic number if and only if $L(\chi^j, 1 - k)$ is q-integral in $\mathbb{C}_q$ for all j coprime to u. This being said, we can now give the (straightforward) proof proper.

(1), (2), and (3). Let p be any prime number. By Proposition 11.3.9 (2), for $k \in \mathbb{Z}_{\geqslant 1}$ we have

$$L_p(\chi\omega^k, 1 - k) = (1 - \chi(p)p^{k-1})L(\chi, 1 - k) \,.$$

Assume first that $k \geqslant 2$. Since $\chi(p)$ is either 0 or a root of unity it is p-integral, so that $1 - \chi(p)p^{k-1}$ is a p-adic unit, congruent to 1 modulo p, and hence $L(\chi, 1 - k) \equiv L_p(\chi\omega^k, 1 - k) \pmod{p\mathbb{Z}_p}$ as elements of $\mathbb{C}_p$.

Assume first that p is odd. Since the conductor of ω^k divides p, writing $\chi = (\chi\omega^k)\omega^{-k}$ it follows that f divides the LCM of the conductor of $\chi\omega^k$ and of p, so that if f is not a power of p, the conductor of $\chi\omega^k$ is also not a power of p (and $\chi\omega^k$ is nontrivial), so it is p-adically tame. Thus, assume that f is a power of p, and write $f = p^v$ for some $v \geqslant 1$. Note that the order of ω^{-k} is equal to $(p-1)/\gcd(p-1,k)$. Thus, if the order of $\chi\omega^k$ is a power of p, the order of $\chi = (\chi\omega^k)\omega^{-k}$ will be equal to $p^w(p-1)/\gcd(p-1,k)$ for a certain integer $w \leqslant v-1$, and since by Corollary 2.1.35 this must be divisible by p^{v-1}, we must have $w = v-1$, hence $u = p^{v-1}(p-1)/\gcd(p-1,k)$. It follows that if this equality does *not* hold, then $\chi\omega^k$ is p-adically tame.

If $p = 2$ then by definition $\chi\omega^k$ is p-adically tame unless it is the trivial character, in other words if $\chi = \omega^{-k}$. However, for $p = 2$ we have $\omega = \left(\frac{-4}{\cdot}\right)$, and k is odd, else χ is trivial, so $\chi\omega^k$ is p-adically tame when $\chi \neq \left(\frac{-4}{\cdot}\right)$. Statement (1) thus follows from Proposition 11.4.1 (1), since the order and conductor of χ^j for j coprime to u are the same as those of χ. For (2) the same reasoning holds using Proposition 11.4.1 (3), since the condition $\chi\omega^{-k}$ nontrivial and different from $\left(\frac{8}{\cdot}\right)$ means that χ is (nontrivial and) different from $\left(\frac{-4}{\cdot}\right)$, $\left(\frac{8}{\cdot}\right)$, and $\left(\frac{-8}{\cdot}\right)$.

For $k = 1$, we must reason differently since the local Euler factor $1 - \chi(p)p^{k-1} = 1 - \chi(p)$ may vanish. In this case, by Corollary 10.3.2 we have $L(\chi, 0) = -B_1(\chi)/f = -S_1(\chi)/f$. It follows from Corollary 9.5.7 (1) that $L(\chi, 0)$ is an algebraic integer, except possibly if $f = q^v$ with q an odd prime such that $u = q^{v-1}(q-1)$, or if $f = 2^2 = 4$. In these cases $L(\chi, 0) = -S_1(\chi)/q^v$ is evidently p-integral for $p \neq q$, proving (1) for $k = 1$. Applying Corollary 9.5.7 (2) proves (2) when f is not an odd prime power, since the only additional condition is that f is not a power of 2. When f is an odd prime power the formula $L(\chi, 0) = -S_1(\chi)/f$ shows that $L(\chi, 0)$ is 2-integral, proving (2) in all cases.

Assume now that $f = p$. If $\mathfrak{q}$ is a prime ideal that is above a prime $q \neq p$ then $v_\mathfrak{q}(L(\chi, 1-k)) \geqslant 0$ since $L(\chi, 1-k)$ is q-integral by what we have already shown. Since $\chi(p) = 0$ we have $L_p(\chi\omega^k, 1-k) = L(\chi, 1-k)$ for all $k \in \mathbb{Z}_{\geqslant 1}$, so it is not necessary to consider separately the case $k = 1$. By Proposition 11.4.1 (1), if $\chi \neq \omega^{-k}$ then $L_p(\chi, 1-k)$ is p-integral. It follows from Lemma 11.3.2 that $L(\chi, 1-k)$ is p-integral as an algebraic number as soon as $\chi^j \neq \omega^{-k}$ for all j coprime to u. It is clear that the set of such χ^j is equal to the set of characters of exact order u. On the other hand, ω^{-k} is of order $(p-1)/\gcd(p-1,k)$. It follows that $L(\chi, 1-k)$ will be p-integral, hence an algebraic integer, as soon as $\gcd(p-1,k) \neq (p-1)/u$, proving (1) in all cases.

It is clear that (3) follows immediately from (1) and (2) since an algebraic number is an algebraic integer if and only if it is p-integral for all primes p.

(4). Assume first that $v = 1$, so that $f = p$ and $\gcd(p-1,k) = (p-1)/u$. There exists a unique a modulo u and coprime to u such that $\chi^a = \omega^{-k}$. If follows that, as elements of $\mathbb{C}_p$, the $L(\chi^j, 1-k)$ are p-integral, except for $j = a$, in which case $L(\chi^j, 1-k) + (1-1/p)/k$ is p-integral by Proposition 11.4.1 (2), and since the $L(\chi^j, 1-k)$ are the conjugates of $L(\chi, 1-k)$, this is exactly the statement given in (4).

Assume now that $f = p^v$ for $v \geqslant 2$. Once again we have $\chi(p) = 0$, so that for all j coprime to $u = o(\chi)$ we have $L(\chi^j, 1 - k) = L_p(\chi^j \omega^k, 1 - k)$. Write $\chi = \chi_1 \chi_2$ with $\chi_1 = \chi^{p^{v-1}}$. It is clear that χ_1 is defined modulo p, and that χ_2 is a primitive character of order p^{v-1}. As in the case $f = p$, since $u = p^{v-1}(p-1)/\gcd(p-1,k)$ we have $o(\chi_1) = u_1$, say, where $u_1 = u/p^{v-1} = (p-1)/\gcd(p-1,k)$. It follows that there exists a unique a modulo u_1 and coprime to u_1 such that $\chi_1^a = \omega^{-k}$. Since $v \geqslant 2$, the character $\chi^j \omega^k$ is never trivial, and $\chi^j \omega^k = \chi_1^j \omega^k \chi_2^j$ is p-adically wild if and only if $\chi_1^j = \omega^{-k} = \chi_1^a$, hence if and only if $j \equiv a \pmod{u_1}$. It follows from Proposition 11.4.1 that, as an element of $\mathbb{C}_p$, $L(\chi^j, 1 - k) = L_p(\omega^k \chi_1^j \chi_2^j, 1 - k)$ is p-integral for $j \not\equiv a$ $\pmod{u_1}$, and that for $j \equiv a \pmod{u_1}$,

$$L(\chi^j, 1 - k) \equiv \frac{1}{\chi_2(1+p)^j - 1} \equiv \frac{1}{\chi(1+p)^j - 1} \pmod{\mathcal{Z}_p}.$$

Since the $L(\chi^j, 1-k)$ are the conjugates of $L(\chi, 1-k)$ and the $1/(\chi(1+p)^j - 1)$ are the corresponding conjugates of $1/(\chi(1+p) - 1)$, this proves (4). Note that we use here implicitly the fact that the prime ideals above p are totally ramified in the extension $\mathbb{Q}(\zeta_u)/\mathbb{Q}(\zeta_{u_1})$.

(5). If $f = 2^v$ with $v \geqslant 3$ the proof is similar, now using Proposition 11.4.1 (4). In this case we have necessarily $u = 2^{v-1}$, so the prime 2 is totally ramified in $\mathbb{Q}(\zeta_u)/\mathbb{Q}$, and in particular, the prime $\mathfrak{p}$ above 2 is unique. The details are left to the reader, as are the proofs of statements (6), (7), and (8) (Exercise 11). $\qquad\square$

Corollary 11.4.3. *Let $k \in \mathbb{Z}_{\geqslant 1}$, let D be the discriminant of a quadratic field, and assume that $\operatorname{sign}(D) = (-1)^k$.*

(1) *Assume that $D \neq -4$, $D \neq \pm 8$, and that either $D \neq (-1)^{(p-1)/2}p$ for some odd prime p, or that $D = (-1)^{(p-1)/2}p$, $k \not\equiv (p-1)/2 \pmod{p-1}$, and $k > 1$. Then $L((\frac{D}{\cdot}), 1 - k) \in 2\mathbb{Z}$.*

(2) *If $D = (-1)^{(p-1)/2}p$ for some odd prime p and if $k \equiv (p-1)/2 \pmod{p-1}$ and $k > 1$, then $L((\frac{D}{\cdot}), 1-k)/2$ is q-integral for all $q \neq p$ and $L((\frac{D}{\cdot}), 1 - k) + (1 - 1/p)/k$ is p-integral.*

(3) *If $D = -p$ with p an odd prime and $k = 1$, then $L((\frac{D}{\cdot}), 1 - k) - 1 \in 2\mathbb{Z}$, except that $L((\frac{-3}{\cdot}), 1 - k) = 1/3$.*

(4) *We have $L((\frac{-4}{\cdot}), 1 - k) + k/2 - 1 \in 4\mathbb{Z}$, $L((\frac{-8}{\cdot}), 1 - k) - 1 \in 4\mathbb{Z}$, and $L((\frac{8}{\cdot}), 1 - k) + 1 \in 4\mathbb{Z}$.*

Proof. Immediate from the preceding corollary since the absolute value of the discriminant of a quadratic field is not divisible by the square of an odd prime and is a power of 2 only for $|D| = 4$ and $|D| = 8$; see Exercise 12. $\qquad\square$

Note that by Exercise 41 of Chapter 10 we have $L((\frac{-4}{\cdot}), -2k) = E_{2k}/2$, where the E_{2k} are the Euler numbers (see Definition 9.1.8); hence for instance the above corollary says that E_{2k} is an odd integer such that $E_{2k} \equiv (-1)^k$ (mod 4); see Exercise 16 of Chapter 9 for a direct proof.

Remark. As we have mentioned above, the results of Corollary 11.4.2 are stronger than those obtained in Chapter 9 using more "elementary" methods, such as the general Clausen–von Staudt congruence (Theorem 9.5.13) and also the Kummer congruences (Theorem 9.5.24), which we will give below in a stronger form. The reader can check that the *only* p-adic result that we needed to prove is that the power series expansion of $L_p(\chi, s)$ around $s = 1$ has a radius of convergence R_p that is strictly greater than 1.

11.4.2 Bernoulli Numbers and Regular Primes

In Corollary 11.4.2 we could have included the case $\chi = 1$, which corresponds to the Kummer congruences (Theorem 9.5.24), but in view of its importance we treat it separately, since it gives a stronger statement that includes the case $(p-1) \mid k$.

Proposition 11.4.4. *For any* $k \geqslant 2$ *even, set*

$$
z_p(k) = \begin{cases}
(p^{k-1} - 1)\dfrac{B_k}{k} & \text{if } (p-1) \nmid k \,, \\[2mm]
(p^{k-1} - 1)\dfrac{B_k}{k} + \dfrac{1 - 1/p}{k} & \text{if } (p-1) \mid k \text{ and } p \neq 3 \,, \\[2mm]
(p^{k-1} - 1)\dfrac{B_k}{k} + \dfrac{1 - 1/p}{k} - k & \text{if } p = 3 \,.
\end{cases}
$$

Then $z_p(k)$ *is* p*-integral, and if* $k' \equiv k$ *(mod* $\phi(p^e)$*) we have* $z_p(k') \equiv z_p(k)$ *(mod* p^e*).*

Proof. Since $k' \equiv k$ (mod $p-1$) we have $\omega^{k'} = \omega^k$, so by Corollary 11.3.22, since $p \mid b_i$ for $i \geqslant 1$ we have

$$
M_p(\omega^k, 1 - k) = \sum_{j \geqslant 0} b_j(-k)^j \equiv \sum_{j \geqslant 0} b_j(-k')^j = M_p(\omega^{k'}, 1 - k') \text{ (mod } p^e) \,.
$$

On the other hand, since k is even, by Proposition 11.3.9 (2) we have $L_p(\omega^k, 1 - k) = (p^{k-1} - 1)B_k/k$, hence $M_p(\omega^k, 1 - k) = z_p(k)$, proving the proposition. $\qquad\square$

The following corollary generalizes Corollary 9.5.25 (essentially the Kummer congruences) to the case $(p-1) \mid k$.

Corollary 11.4.5. *Let* k *and* k' *be even and such that* $(p-1) \mid k$ *and* $(p-1) \mid k'$*, and assume that* $\min(k - 2 - v_p(k), k' - 2 - v_p(k')) \geqslant e$*. Then if* $k' \equiv k$ *(mod* $\phi(p^e)$*) we have*

$$\frac{B_{k'} - (1 - 1/p)}{k'} \equiv \frac{B_k - (1 - 1/p)}{k} \pmod{p^e} \quad for\ p \neq 3\ and$$

$$\frac{B_{k'} - (1 - 1/p)}{k'} + k' \equiv \frac{B_k - (1 - 1/p)}{k} + k \pmod{p^e} \quad for\ p = 3\ .$$

Proof. Left to the reader (Exercise 25). □

Corollary 11.4.6. *Let $k \geqslant 2$ be even.*

(1) *If $(k, p) \neq (2, 2)$, $(2, 3)$, and $(4, 2)$ we have*

$$\frac{B_k}{k} \equiv B_1(\omega^{k-1}) \pmod{p} \ if\ (p - 1) \nmid k\ ,$$

$$\frac{B_k - (1 - 1/p)}{k} \equiv B_1(\omega^{k-1}) - (1 - 1/p) \pmod{p} \ if\ (p-1) \mid k\ and\ p \neq 3\ ,$$

$$\frac{B_k - (1 - 1/p)}{k} \equiv B_1(\omega^{k-1}) - k + 1/p \pmod{p} \ if\ p = 3\ .$$

(2) *The right-hand side of each of the above expressions is always p-integral, and the left-hand side is p-integral except if $(k, p) = (2, 2)$.*

Proof. By Proposition 11.3.9 (2) we have

$$L_p(\omega^k, 0) = -(1 - \omega^{k-1}(p))B_1(\omega^{k-1}) = -B_1(\omega^{k-1})$$

since $k - 1$ is odd, so ω^{k-1} is a nontrivial character (recall that ω has order exactly equal to $\phi(q_p)$, which is even). It follows that

$$M_p(\omega^k, 0) = \begin{cases} -B_1(\omega^{k-1}) & if\ (p-1) \nmid k\ , \\ -B_1(\omega^{k-1}) + 1 - 1/p & if\ (p-1) \mid k\ and\ p \neq 3\ , \\ -B_1(\omega^{k-1}) - 1/p & if\ p = 3\ . \end{cases}$$

By Corollary 11.3.22 we have $M_p(\omega^k, s) \equiv a_0 \pmod{p}$ for all $s \in \mathbb{Z}_p$, hence

$$z_p(k) = M_p(\omega^k, 1 - k) \equiv a_0 \equiv M_p(\omega^k, 0) \pmod{p}\ .$$

This is immediately seen to imply the congruences given in the corollary for pairs (k, p) such that $p^{k-2}B_k/k$ is p-integral. If $(p - 1) \nmid k$ this is true by Adams's Proposition 9.5.23. On the other hand, if $(p - 1) \mid k$ then $v_p(B_k) = -1$ hence $v_p(p^{k-2}B_k/k) = k - 3 - v_p(k)$, and it is immediate that when $(p-1) \mid k$ this is greater than or equal to 0 if and only if $(k, p) \neq (2, 2), (4, 2)$, and $(2, 3)$, proving (1), and (2) is immediate since a_0 is p-integral. □

Note that the integrality statement includes the result of Adams (Proposition 9.5.23) and is stronger than the Clausen–von Staudt Theorem 9.5.14. In fact, as examples of Proposition 11.4.4 and Corollary 11.4.5, we give the following additional congruences:

Corollary 11.4.7. *Let $k \geqslant 2$ be an even integer.*

(1) *We have $B_k \equiv k + 1/2 \pmod{2^{2+v_2(k)}}$, so in particular $B_k \equiv k + 1/2 \pmod 4$ and $B_k \equiv 1/2 \pmod 2$.*

(2) *If $3 \nmid k$ we have $B_k \equiv -1/3 \pmod 3$, and if $3 \mid k$ we have $B_k \equiv 2/3 \pmod 9$.*

(3) *If $4 \mid k$ we have $B_k \equiv 4/5 \pmod 5$.*

(4) *If $12 \mid k$ we have $B_k \equiv 12/13 \pmod{13}$.*

Proof. Left to the reader (Exercise 25). $\qquad\square$

Corollary 11.4.8. *If $k \neq 0$ is even then $B_{k,p}/k$ is p-integral if $(p-1) \nmid k$, and $(B_{k,p} - (1 - 1/p))/k$ is p-integral if $(p-1) \mid k$ (including for $p = 3$ and for $(k,p) = (2,2)$). In addition, γ_p is always p-integral.*

Proof. Immediate from the above results and the definitions of $B_{k,p}$ and γ_p, and left to the reader (Exercise 35). $\qquad\square$

Corollary 11.4.9. *Assume that $p \geqslant 5$. Then $p \mid h_p^-$ if and only if $v_p(B_k) \geqslant 1$ for some even k such that $2 \leqslant k \leqslant p - 3$.*

Proof. Indeed, since the odd characters of $(\mathbb{Z}/p\mathbb{Z})^* \simeq \mathrm{Gal}(\mathbb{Q}(\zeta_p)/\mathbb{Q})$ are the ω^{k-1} for k even with $2 \leqslant k \leqslant p - 1$, by Proposition 10.5.26 we have

$$v_p(h_p^-) = 1 + v_p(B_1(\omega^{p-2})) + \sum_{2 \leqslant k \leqslant p-3, \; k \text{ even}} v_p(B_1(\omega^{k-1})) \,.$$

By Corollary 11.4.6, $B_1(\omega^{p-2}) + 1/p$ is p-integral (so that $1 + v_p(B_1(\omega^{p-2})) = 0$) and $B_1(\omega^{k-1})$ is p-integral for $2 \leqslant k \leqslant p - 3$. Thus $p \mid h_p^-$ if and only if $v_p(B_1(\omega^{k-1})) \geqslant 1$ for some even $k \leqslant p - 3$, hence by the corollary if and only if $v_p(B_k/k) \geqslant 1$, so if and only if $v_p(B_k) \geqslant 1$ since $0 < k < p$. $\qquad\square$

Since it can be shown that $p \mid h_p^+$ implies that $p \mid h_p^-$ (see for instance [Was]), this implies the following theorem due to Kummer:

Theorem 11.4.10. *A prime $p \geqslant 3$ is irregular if and only if it divides the numerator of some B_k for an even k such that $2 \leqslant k \leqslant p - 3$.*

11.4.3 Strengthening of the Almkvist–Meurman Theorem

Thanks to Corollary 11.4.6 we can give a refinement of the Almkvist–Meurman Theorem 9.5.29. We essentially follow the same method as the one used in Section 9.5.5, so we begin by proving the following generalization of Hermite's Lemma 9.5.28, due to Carlitz.

Proposition 11.4.11 (Carlitz). *Let p be a prime number and let $n \geqslant 1$ be an integer. We have the congruence*

$$p + (p-1) \sum_{1 \leqslant j \leqslant (n-1)/(p-1)} \binom{n}{(p-1)j} \equiv 0 \pmod{p^{v_p(n)+1}} \,,$$

or equivalently,

$$\left(1 - \frac{1}{p}\right) \sum_{\substack{1 \leqslant m \leqslant n-1 \\ (p-1) \mid m}} \binom{n}{m} \equiv -1 \pmod{p^{v_p(n)} \mathbb{Z}_p} \,.$$

Proof. For $p = 2$ the left-hand side is equal to 2^n, and since $n - 1 \geqslant v_2(n)$ for all $n \geqslant 1$ the result is clear, so we may assume that $p \geqslant 3$. By the recurrence formula for Bernoulli numbers (Proposition 9.1.3) we have

$$1 - \frac{n}{2} + n \sum_{2 \leqslant k \leqslant n-1} \binom{n-1}{k-1} \frac{B_k}{k} = 0 \,.$$

By Adams's Proposition 9.5.23, B_k/k is p-integral if $(p-1) \nmid k$, and by Corollary 11.4.6, since $p \geqslant 3$ we have $B_k/k \equiv (1 - 1/p)/k \pmod{\mathbb{Z}_p}$ when $(p-1) \mid k$. It follows that

$$0 \equiv 1 + n \left(1 - \frac{1}{p}\right) \sum_{\substack{2 \leqslant k \leqslant n-1 \\ (p-1) \mid k}} \frac{1}{k} \binom{n-1}{k-1}$$

$$\equiv 1 + \left(1 - \frac{1}{p}\right) \sum_{\substack{2 \leqslant k \leqslant n-1 \\ (p-1) \mid k}} \binom{n}{k} \pmod{p^{v_p(n)} \mathbb{Z}_p} \,,$$

proving the proposition after multiplying by p. $\qquad\square$

Theorem 11.4.12. *For $n \geqslant 0$, $k \in \mathbb{Z} \setminus \{0\}$, and $h \in \mathbb{Z}$ set $b_n(h,k) = k^n(B_n(h/k) - B_n)$. We have $b_0(h,k) = 0$, $b_1(h,k) = h$, $b_2(h,k) = h(h-k)$, and for $n > 2$,*

$$b_n(h,k) \equiv \begin{cases} 0 & \pmod{n/d(n,k)} \,, \\ h^n & \pmod{kd(n,k)/2^{v_2(kd(n,k))}} \,, \\ h^n - \dfrac{nh^{n-1}k}{2} + \dfrac{n(n-1)h^{n-2}k^2}{12} & \pmod{2^{v_2(kd(n,k))}} \,, \end{cases}$$

where for simplicity we set $d(n,k) = \gcd(n, k^\infty)$ (recall that $\gcd(n, k^\infty) = \prod_{p \mid k} p^{v_p(n)}$).

Proof. As in the proof of Theorem 9.5.29, for $n > 2$ we have

$$\frac{b_n(h,k)}{nk} = \frac{h^n}{nk} - \frac{h^{n-1}}{2} + \sum_{2 \leqslant m \leqslant n-1} \binom{n-1}{m-1} \frac{B_m}{m} h^{n-m} k^{m-1} .$$

Fix some prime number p, and assume first that $p \mid k$. By the Clausen–von Staudt Theorem 9.5.14 we have $v_p(B_m k^{m-1}/m) \geqslant m - 2 - v_p(m) \geqslant 0$ if either $p \geqslant 3$ and $m \geqslant 2$, or if $p = 2$ and $m \geqslant 3$. It follows that for $p \geqslant 3$ we have

$$\frac{b_n(h,k)}{nk} - \frac{h^n}{nk} \in \mathbb{Z}_p ,$$

while for $p = 2$ we have

$$\frac{b_n(h,k)}{nk} - \frac{h^n}{nk} + \frac{h^{n-1}}{2} - \frac{(n-1)h^{n-2}k}{12} \in \mathbb{Z}_2 ,$$

proving the last two congruences.

Assume now that $p \nmid k$ and $p \mid h$. We can then write instead

$$\binom{n-1}{m-1} \frac{B_m}{m} h^{n-m} = \frac{1}{n} \binom{n}{n-m} B_m h^{n-m} = \binom{n-1}{n-m-1} B_m \frac{h^{n-m}}{n-m} ,$$

and once again by the Clausen–von Staudt theorem we have $v_p(B_m h^{n-m}/(n-m)) \geqslant n - m - 1 - v_p(n-m) \geqslant 0$ when $m \leqslant n-1$, so the sum is p-integral, and since $h^{n-1}/2$ and $h^n/(nk)$ are also p-integral (since $n \geqslant v_p(n)$ and $p \nmid k$), it follows that $b_n(h,k) \in p^{v_p(n)}\mathbb{Z}$ as claimed. Finally, assume that $p \nmid k$ and $p \nmid h$, hence that $p \mid n$, since otherwise there is nothing to prove. Consider first the case $p > 2$. As in the proof of Carlitz's result we have

$$\frac{b_n(h,k)}{n} \equiv \frac{h^n}{n} - \frac{kh^{n-1}}{2} + h^n \left(1 - \frac{1}{p}\right) \sum_{\substack{2 \leqslant m \leqslant n-1 \\ \mathrm{lcm}(2,p-1) \mid m}} \binom{n-1}{m-1} \frac{1}{m} \frac{k^m}{h^m} \pmod{\mathbb{Z}_p} .$$

Since $(p-1) \mid m$ we may apply Lemma 2.1.22 to $s = k^{p-1} \equiv 1 \pmod{p}$ and deduce that $k^m \equiv 1 \pmod{p^{v_p(m)+1}}$, hence that $k^m/m \equiv 1/m \pmod{p\mathbb{Z}_p}$, and similarly for h. Since $p > 2$, it follows from Carlitz's result that

$$\frac{b_n(h,k)}{n} \equiv \frac{h^n}{n} + \frac{h^n}{n} \left(1 - \frac{1}{p}\right) \sum_{\substack{1 \leqslant m \leqslant n-1 \\ (p-1) \mid m}} \binom{n}{m} \equiv 0 \pmod{\mathbb{Z}_p} ,$$

so $b_n(h,k) \in n\mathbb{Z}_p$ as claimed, and since $p \nmid k$ this is equivalent to $b_n(h,k) \in p^{v_p(nk)}\mathbb{Z}$. Consider now the case $p = 2$, so that n is even and k and h are odd. Using Corollary 11.4.6 and taking into account the given exception for $(m,p) = (2,2)$, since $n \geqslant 4$ we obtain

$$\frac{b_n(h,k)}{n} \equiv \frac{h^n}{n} - \frac{kh^{n-1}}{2} + \frac{(n-1)k^2h^{n-2}}{12} + \frac{1}{2} \sum_{\substack{3 \leqslant m \leqslant n-1 \\ 2|m}} \binom{n-1}{m-1} \frac{1}{m} h^{n-m} k^m$$

$$\equiv \frac{h^n}{n} - \frac{kh^{n-1}}{2} - \frac{(n-1)k^2h^{n-2}}{6}$$

$$+ \frac{h^n}{2} \sum_{\substack{1 \leqslant m \leqslant n-1 \\ 2|m}} \binom{n-1}{m-1} \frac{1}{m} \frac{k^m}{h^m} \quad (\mathrm{mod}\ \mathbb{Z}_2) \ .$$

Since $2 \mid m$ we may apply Lemma 2.1.22 to $s = k^2 \equiv 1 \ (\mathrm{mod}\ 4)$; hence $k^m = (k^2)^{m/2} \equiv 1 \ (\mathrm{mod}\ p^{v_p(m)+1})$, and similarly $h^m \equiv 1 \ (\mathrm{mod}\ p^{v_p(m)+1})$, so that $h^{-m}k^m/m \equiv 1/m \ (\mathrm{mod}\ 2\mathbb{Z}_2)$. Replacing in the above formula and using the fact that k and h are odd and n is even gives

$$\frac{b_n(h,k)}{n} \equiv \frac{h^n}{n} + \frac{h^n}{2n} \sum_{\substack{1 \leqslant m \leqslant n-1 \\ 2|m}} \binom{n}{m} \quad (\mathrm{mod}\ \mathbb{Z}_2) \ .$$

Since n is even this last sum is equal to $2^{n-1} - 2$, so that

$$\frac{b_n(h,k)}{n} \equiv \frac{h^n 2^{n-2}}{n} \equiv 0 \ (\mathrm{mod}\ \mathbb{Z}_2)$$

since $n - 2 - v_2(n) \geqslant 0$ for $n \geqslant 4$, finishing the proof of the theorem. $\qquad \square$

Remark. From the proof, it is immediate to see that in the last two congruences (but not in the first) we can replace $\gcd(n, k^\infty)$ by $\gcd(nh, k^\infty)$.

11.5 p-adic Log Gamma Functions

In Section 11.2 we have defined and studied the main properties of the functions $\zeta_p(s,x)$ and $\zeta_p(\chi, s, x)$ which are the basic building blocks of all the functions that we study in this chapter. As first and essential application, we then studied p-adic L-functions, which are finite linear combinations of the functions $\zeta_p(s,x)$. Now that we have these tools, we can study p-adic log gamma functions, which are constructed in a way very similar to the complex case. Most of their properties are immediate consequences of the corresponding ones for ζ_p, with two notable exceptions: the value of the ψ functions at rational numbers, and the Gross–Koblitz formula.

We have seen that the p-adic Hurwitz zeta function $\zeta_p(s,x)$ has two closely related but distinct definitions: one for $x \in C\mathbb{Z}_p$, and one for $x \in \mathbb{Z}_p$, and in the latter case we can even introduce a character χ of conductor a power of p. Correspondingly, we have two closely related but distinct *log gamma* functions: one for $x \in C\mathbb{Z}_p$, introduced by J. Diamond, the other for $x \in \mathbb{Z}_p$, basically introduced by Y. Morita. One of the remarkable facts is that one can take the exponential of Morita's log gamma function, thus leading to a p-adic gamma function, but only for $x \in \mathbb{Z}_p$. As we have done for ζ_p, we study these functions in turn, starting with Diamond's.

11.5.1 Diamond's p-adic Log Gamma Function

For more details on this function, see [Dia1] and [Dia2].

Recall that in the complex case we defined $\log(\Gamma(x)/\sqrt{2\pi}) = \zeta'(0, x)$. We thus give a similar definition in the p-adic case.

Definition 11.5.1. *For $x \in C\mathbb{Z}_p$ we define Diamond's log gamma function* $\mathrm{Log}\Gamma_p(x)$ *by the formula*

$$\mathrm{Log}\Gamma_p(x) = \omega_v(x)\frac{\partial\zeta_p}{\partial s}(0, x) .$$

Since $x \in C\mathbb{Z}_p$ the function $\zeta_p(s, x)$ is analytic for $|s| < R_p$ except for a simple pole at $s = 1$, so the definition makes sense. The notation $\mathrm{Log}\Gamma_p$ is due to the author, and is simply suggestive of log gamma. The normalization factor $\omega_v(x)$ will be seen to be essential.

Although the function $\mathrm{Log}\Gamma_p(x)$ is a priori defined only for $x \in C\mathbb{Z}_p$, we are going to see that it can be extended to all $x \in \mathbb{C}_p$ such that $|x| > 1$.

Proposition 11.5.2. *Assume as above that $x \in C\mathbb{Z}_p$.*

(1) We have the functional equation

$$\mathrm{Log}\Gamma_p(x + 1) = \mathrm{Log}\Gamma_p(x) + \log_p(x) .$$

(2) We have the Laurent series expansion

$$\mathrm{Log}\Gamma_p(x) = \left(x - \frac{1}{2}\right)\log_p(x) - x + \sum_{k\geqslant 1}\frac{B_{2k}}{2k(2k - 1)}x^{1-2k} ,$$

where the right-hand side converges for $|x| > 1$, and more generally if $|u| < |x|$ (and $x \in C\mathbb{Z}_p$), we have

$$\mathrm{Log}\Gamma_p(x + u) = \left(x + u - \frac{1}{2}\right)\log_p(x) - x + \sum_{j\geqslant 2}\frac{(-1)^j}{j(j - 1)}B_j(u)x^{1-j} .$$

(3) We have the reflection formula

$$\mathrm{Log}\Gamma_p(1 - x) + \mathrm{Log}\Gamma_p(x) = 0 .$$

(4) *If* $N \in \mathbb{Z}_{>0}$ *is such that* $Nx \in C\mathbb{Z}_p$ *(in particular if* $p \nmid N$*) we have the distribution formula*

$$\sum_{0 \leqslant j < N} \mathrm{Log}\Gamma_p\left(x + \frac{j}{N}\right) = \mathrm{Log}\Gamma_p(Nx) - \left(Nx - \frac{1}{2}\right)\log_p(N) .$$

Corollary 11.5.3. *For* $x \in C\mathbb{Z}_p$ *define* $\psi_p(x) = (d/dx)(\mathrm{Log}\Gamma_p(x))$. *We have* $\psi_p(x+1) = \psi_p(x) + 1/x$, $\psi_p(1-x) = \psi_p(x)$, *the expansion*

$$\psi_p(x) = \log_p(x) + \sum_{j \geqslant 1}(-1)^{j-1}\frac{B_j}{j}x^{-j} ,$$

and for $Nx \in C\mathbb{Z}_p$ *the distribution formula*

$$\sum_{0 \leqslant j < N} \psi_p\left(x + \frac{j}{N}\right) = N\,\psi_p(Nx) - N\log_p(N) .$$

Proof. The proofs of the proposition and its corollary follow immediately from the corresponding properties of $\zeta_p(s,x)$ and are left to the reader (Exercise 20). $\qquad\square$

In the complex case, we have set $x = 1/N$ in the distribution formula for the ordinary gamma function and deduced the formula

$$\sum_{1 \leqslant k \leqslant N} \mathrm{Log}\Gamma\left(\frac{k}{N}\right) = -\frac{1}{2}\log(N) + \frac{N-1}{2}\log(2\pi)$$

(see Proposition 9.6.33). Here this is not possible since we must have $Nx \in C\mathbb{Z}_p$. The corresponding results are the following:

Proposition 11.5.4. *Let* χ *be a primitive character of conductor* f, *and let* $N \in \mathbb{Z}_{\geqslant 1}$ *be a common multiple of* f *and* q_p. *Recall from Proposition 11.3.9 that* $L_p(\chi\omega, 0) = -(1 - \chi(p))B_1(\chi)$.

(1) *We have*

$$\sum_{0 \leqslant k < N}^{(p)} \chi(k)\,\mathrm{Log}\Gamma_p\left(\frac{k}{N}\right) = L_p'(\chi\omega, 0) + L_p(\chi\omega, 0)\log_p(\langle N \rangle) .$$

(2) *We have*

$$\sum_{0 \leqslant k < N}^{(p)} \chi(k)\,\psi_p\left(\frac{k}{N}\right) = -\left(1 - \frac{1}{p}\right)N\log_p(\langle N \rangle)\delta(\chi) - N\gamma_p(\chi) ,$$

where as usual $\delta(\chi) = 0$ *if* $\chi \neq \chi_0$ *and* $\delta(\chi_0) = 1$, *and where we recall that* $\gamma_p(\chi) = L_p(\chi, 1)$ *if* $\chi \neq \chi_0$, *and* $\gamma_p(\chi_0) = \gamma_p$.

Proof. Recall that by definition of the p-adic L-function, for any N divisible by q_p and by f we have

$$L_p(\chi, s) = \sum_{0 \leqslant k < N}^{(p)} \chi(k)\zeta_p(s, k, N) = \omega_v(N)^{-1}\langle N \rangle^{-s} \sum_{0 \leqslant k < N}^{(p)} \chi(k)\zeta_p(s, k/N) \ .$$

The important point, which follows from Proposition 11.3.8, is that this is independent of the choice of $N \equiv 0 \pmod{q_p}$.

If we differentiate this formula with respect to s, set $s = 0$, and replace $\zeta_p'(0, k/N)$ by $\omega_v^{-1}(k/N)\operatorname{Log}\Gamma_p(k/N)$, we obtain (1) after replacing χ by $\chi\omega$.

The proof of (2) is similar, but around $s = 1$. By Proposition 11.5.6, which we will prove presently, around $s = 1$ we have $\zeta_p(s, x) = 1/(s-1) - \psi_p(x) + O(s-1)$. When χ is a nontrivial character, the formula follows since $\sum_{1 \leqslant k < N}^{(p)} \chi(k) = 0$, and when $\chi = \chi_0$ is the trivial character then

$$L_p(\chi, s) = \frac{1 - 1/p}{s - 1} + \gamma_p + O(s-1) \ ,$$

so the result again follows since

$$\omega_v(N)^{-1}\langle N \rangle^{-s} = \frac{1}{N}(1 - (s-1)\log_p(\langle N \rangle) + O(s-1)^2)$$

and since $\sum_{1 \leqslant k < N}^{(p)} \chi_0(k) = N(1 - 1/p)$. $\qquad\square$

Remarks. (1) By the reflection formula $\operatorname{Log}\Gamma_p(1-x) + \operatorname{Log}\Gamma_p(x) = 0$ (or by the above result applied to $\chi = \chi_0$, so that $\chi\omega$ is an odd character), we have

$$\sum_{1 \leqslant k \leqslant N}^{(p)} \operatorname{Log}\Gamma_p(k/N) = 0 \ .$$

(2) In Proposition 11.5.17 below we will give a more general result.

Corollary 11.5.5. *Let χ be a primitive character of conductor f, and let $N \in \mathbb{Z}_{\geqslant 1}$ be a common multiple of f and q_p. We have*

$$L_p'(\chi, 0) = (1 - \chi\omega^{-1}(p))B_1(\chi\omega^{-1})\log_p(N) + \sum_{0 \leqslant k < N}^{(p)} \chi\omega^{-1}(k)\operatorname{Log}\Gamma_p\left(\frac{k}{N}\right) \ .$$

In particular,

$$\zeta_p'(0) = \sum_{0 \leqslant k < q_p}^{(p)} \omega^{-1}(k)\operatorname{Log}\Gamma_p\left(\frac{k}{q_p}\right) \ ,$$

where we recall that $\zeta_p(s) = L_p(\chi_0, s)$ is the Kubota–Leopoldt zeta function.

Proof. Clear.   □

I do not know whether there is a more explicit expression for $\zeta_p'(0)$ (recall that in the complex case $\zeta'(0) = -\log(2\pi)/2$). However, when $\chi\omega^{-1}(p) = 1$ we will prove that $L_p'(\chi, 0)$ is a $\overline{\mathbb{Q}}$-linear combination of p-adic logarithms of algebraic numbers; see Proposition 11.7.10.

Proposition 11.5.6. (1) *For all $k \in \mathbb{Z}_{\geq 1}$ we have*

$$\zeta_p(k+1, x) = (-1)^{k-1} w_v(x)^k \frac{\psi_p^{(k)}(x)}{k!}.$$

(2) *Around $s = 1$ we have*

$$\zeta_p(s, x) = \frac{1}{s-1} - \psi_p(x) + O(s-1).$$

Proof. (1) immediately follows by comparing the expansions

$$\zeta_p(k+1, x) = \frac{\langle x \rangle^{-k}}{k} \sum_{j \geq 0} (-1)^j \binom{k+j-1}{k-1} B_j x^{-j} \quad \text{and}$$

$$(-1)^{k-1} x^k \frac{\psi_p^{(k)}(x)}{(k-1)!} = \sum_{j \geq 0} (-1)^j \binom{k+j-1}{k-1} B_j x^{-j}.$$

For (2) we note that for $j \geq 1$, in the neighborhood of $s = 0$ we have $\binom{s}{j} = ((-1)^{j-1}/j)s + O(s^2)$, hence by definition

$$\sum_{j \geq 0} \binom{s}{j} B_j x^{-j} = 1 + s \sum_{j \geq 1} (-1)^{j-1} \frac{B_j}{j} x^{-j} + O(s^2)$$

$$= 1 + s(\psi_p(x) - \log_p(x)) + O(s^2).$$

Since $\langle x \rangle^s = 1 + s\log_p(\langle x \rangle) + O(s^2))$ and $\log_p(\langle x \rangle) = \log_p(x)$, (2) follows by changing s into $1 - s$ and dividing by $s - 1$, and (3) is an immediate consequence of the definitions.   □

See also Exercise 21.

Remarks. The following remarks show that $\mathrm{Log}\Gamma_p(x)$ shares very similar properties with $\log(\Gamma(x)/\sqrt{2\pi})$.

(1) The Laurent series expansion of $\mathrm{Log}\Gamma_p(x)$ for $x \in C\mathbb{Z}_p$ is *identical* to the Euler–MacLaurin asymptotic expansion of $\log(\Gamma(x)/\sqrt{2\pi})$ as $x \to \infty$; see Section 9.2.5. In addition, since it converges for $|x| > 1$, it can be taken as a new *definition* of the function $\mathrm{Log}\Gamma_p(x)$, now valid for all $x \in \mathbb{C}_p$ such that $|x| > 1$, while the initial one was valid only for $x \in C\mathbb{Z}_p$.

(2) The functional equation is identical to that of $\log(\Gamma(x)/\sqrt{2\pi})$, while the reflection formula is similar (the function $\pi/\sin(\pi x)$ does not occur).

(3) The distribution formula is identical to that of the function $\log(\Gamma(s)/\sqrt{2\pi})$, as can be seen from Proposition 9.6.33.

(4) We would like to define an exponential of $\mathrm{Log}\Gamma_p(x)$ so as to have a *p*-adic gamma function defined for $x \in C\mathbb{Z}_p$. This is not possible in general since $\mathrm{Log}\Gamma_p(x)$ is not the domain of convergence of $\exp_p$ (we will in fact see in Proposition 11.5.10 that we usually have $v_p(\mathrm{Log}\Gamma_p(x)) < 0$), and the possible extensions of $\exp_p$ to $\mathbb{C}_p$ are not canonical. However, see Exercise 18 for a partial answer.

Recall that we have defined $\zeta_p(s, x)$ using a Volkenborn integral, although afterward we have worked only with the infinite series. We can of course recover the Volkenborn integrals for the functions $\mathrm{Log}\Gamma_p$ and ψ_p as follows.

Proposition 11.5.7. *For $|x| > 1$ we have*

$$\mathrm{Log}\Gamma_p(x) = \int_{\mathbb{Z}_p} ((x+t)\log_p(x+t) - (x+t))\, dt \;,$$

$$\psi_p(x) = \int_{\mathbb{Z}_p} \log_p(x+t)\, dt \;,$$

$$\psi_p^{(k)}(x) = (-1)^{k-1}(k-1)! \int_{\mathbb{Z}_p} \frac{dt}{(x+t)^k}$$

for $k \geqslant 1$.

Proof. From the formal power series expansion for $\log(1+T)$ we deduce that $(1+T)\log(1+T) - T = \sum_{j \geqslant 1}(-1)^{j-1}T^{j+1}/(j(j+1))$. Thus, since $|x| > 1$ we have for $n \in \mathbb{Z}$,

$$(n+x)\log_p(n+x) = (n+x)\log_p(x) + x(1 + n/x)\log_p(1 + n/x)$$

$$= (n+x)\log_p(x) + n + \sum_{j \geqslant 1}(-1)^{j-1}\frac{n^{j+1}}{j(j+1)x^j} \;.$$

It follows from the Euler–MacLaurin formula that

$$\frac{1}{p^r}\sum_{0 \leqslant n < p^r}(n+x)\log_p(n+x) = x\log_p(x) + \frac{p^r - 1}{2}(\log_p(x) + 1)$$

$$+ \sum_{j \geqslant 1}(-1)^{j-1}\frac{B_{j+2}(p^r) - B_{j+2}(0)}{p^r(j+2)}\frac{1}{j(j+1)x^j} \;.$$

By absolute and uniform convergence it is immediate to see that we can take the limit term by term in this expression. Since $\lim_{r \to \infty}(B_{j+2}(p^r) - B_{j+2}(0))/p^r = B'_{j+2}(0) = (j+2)B_{j+1}$, we obtain

$$\lim_{r\to\infty} \frac{1}{p^r} \sum_{0\leqslant n < p^r} (n+x)\log_p(n+x)$$

$$= (x-1/2)\log_p(x) - 1/2 + \sum_{j\geqslant 1}(-1)^{j-1}\frac{B_{j+1}}{j(j+1)x^j}$$

$$= x - 1/2 + \mathrm{Log}\Gamma_p(x)$$

by Proposition 11.5.2 (2). The first formula then follows from Lemma 11.1.7. The formulas for ψ_p and its derivatives are proved similarly, or by differentiating the formula for $\mathrm{Log}\Gamma_p(x)$, although in that case one must justify the exchange of limit and differentiation. $\square$

Proposition 11.5.8. *For $|x| > 1$ we have*

$$\lim_{n\to 0}\frac{B_n(x) - x^n}{nx^n} = \psi_p(x) - \log_p(x)\,,$$

where n tends p-adically to 0 in $\mathbb{Z}_{>0}$.

Proof. By definition

$$\frac{B_n(x) - x^n}{nx^n} = \frac{1}{n}\sum_{1\leqslant j\leqslant n}\binom{n}{j}B_j x^{-j} = \sum_{1\leqslant j\leqslant n}\binom{n-1}{j-1}\frac{B_j}{j}x^{-j}\,.$$

If we let n tend p-adically to 0, $n \in \mathbb{Z}_{>0}$, then $\binom{n-1}{j-1}$ tends to $\binom{-1}{j-1} = (-1)^{j-1}$, and since $|x| > 1$, by normal convergence we have

$$\lim_{n\to 0}\frac{B_n(x) - x^n}{nx^n} = \sum_{j\geqslant 1}(-1)^{j-1}\frac{B_j}{j}x^{-j}\,,$$

so we conclude by the expansion of $\psi_p(x)$ given by Corollary 11.5.3. $\square$

Raabe's formula for the function $\mathrm{Log}\Gamma_p$ is as follows.

Proposition 11.5.9. *If $x \in C\mathbb{Z}_p$ we have*

$$\int_{\mathbb{Z}_p}\mathrm{Log}\Gamma_p(x+t)\,dt = (x-1)\,\psi_p(x) - x + \frac{1}{2}\,.$$

Proof. It is not difficult to show that we can differentiate with respect to s under the integral sign in Proposition 11.2.10; hence after setting $s = 0$ we obtain

$$\int_{\mathbb{Z}_p}\omega_v^{-1}(x+t)\mathrm{Log}\Gamma_p(x+t) = (x-1)\omega_v(x)^{-1}\psi_p(x) + \zeta_p(0,x)\,.$$

Since $x \in C\mathbb{Z}_p$ and $t \in \mathbb{Z}_p$ we have $\omega_v(x+t) = \omega_v(x)$, and by Proposition 11.2.6 we have $\zeta_p(0,x) = -\omega_v(x)^{-1}(x-1/2)$, so the result follows. $\square$

Proposition 11.5.10. *For all* $a \in \mathbb{Z}_p^*$ *we have*

$$\mathrm{Log}\Gamma_p\left(\frac{a}{q_p}\right) \equiv \begin{cases} -\omega(a)/p & (\mathrm{mod}\ p\mathbb{Z}_p) & \textit{if } p \geqslant 5, \\ 2\omega(a)/3 & (\mathrm{mod}\ p\mathbb{Z}_p) & \textit{if } p = 3, \\ -\omega(a)/4 + a/2\ (\mathrm{mod}\ q_p\mathbb{Z}_p) & \textit{if } p = 2. \end{cases}$$

In particular, we have $v_p(\mathrm{Log}\Gamma_p(a/q_p)) = -v_p(q_p)$.

Proof. Assume first that $p \geqslant 3$. By Proposition 11.5.2 we have

$$\mathrm{Log}\Gamma_p\left(\frac{a}{p}\right) = \frac{1}{p}(a\log_p(a) - a) - \frac{1}{2}\log_p(a) + \sum_{k \geqslant 1} \frac{B_{2k}}{2k(2k-1)} \frac{p^{2k-1}}{a^{2k-1}}.$$

By Proposition 4.4.46 we have $a\log_p(a) - a \equiv -\omega(a)\ (\mathrm{mod}\ p^2\mathbb{Z}_p)$, and $\log_p(a) = \log_p(\langle a \rangle) \equiv 0\ (\mathrm{mod}\ p\mathbb{Z}_p)$. Furthermore, since $v_p(B_{2k}) \geqslant -1$, it is immediate to check that $v_p(B_{2k}p^{2k-1}/(2k(2k-1))) \geqslant 1$ for $k \geqslant 2$. Thus

$$\mathrm{Log}\Gamma_p\left(\frac{a}{p}\right) \equiv -\frac{\omega(a)}{p} + \frac{p}{12a}\ (\mathrm{mod}\ p\mathbb{Z}_p).$$

Since $v_p(a) = 0$, this gives the result for $p \geqslant 5$, and for $p = 3$ we have $p/(12a) = 1/(4a) \equiv a \equiv \omega(a)\ (\mathrm{mod}\ 3\mathbb{Z}_3)$.

Assume now that $p = 2$. A similar computation to the proof of Proposition 4.4.46 shows that $(a/4 - 1/2)\log_p(a) - a/4 \equiv -\omega(a)/4\ (\mathrm{mod}\ 4\mathbb{Z}_2)$ (Exercise 8). Furthermore, it is easy to check that $v_2(B_{2k}p^{2k-1}/(2k(2k-1))) \geqslant 3$ for $k \geqslant 3$. Thus

$$\mathrm{Log}\Gamma_p\left(\frac{a}{4}\right) \equiv -\frac{\omega(a)}{4} + \frac{p}{12a} - \frac{p^3}{360a^3}\ (\mathrm{mod}\ 4\mathbb{Z}_2).$$

Since $a \in 1 + 2\mathbb{Z}_2$ we have

$$\frac{p}{12a} - \frac{p^3}{360a^3} \equiv \frac{1}{6a} - \frac{1}{45a^3} \equiv \frac{a}{6} - a \equiv -\frac{5a}{6} \equiv \frac{3a}{6} \equiv \frac{a}{2}\ (\mathrm{mod}\ 4\mathbb{Z}_2),$$

finishing the proof. $\qquad\qquad\qquad\qquad\qquad\qquad\qquad\qquad\qquad\qquad\square$

11.5.2 Morita's p-adic Log Gamma Function

In complete similarity with Definition 11.5.1 we set the following (see [Morit1] and [Morit2]).

Definition 11.5.11. *Let* χ *be a character modulo* p^v *for some* $v \geqslant v_p(q_p)$. *For* $x \in \mathbb{Z}_p$ *we define Morita's log gamma function* $\mathrm{Log}\Gamma_p(\chi, x)$ *by the formula*

$$\mathrm{Log}\Gamma_p(\chi, x) = \frac{\partial \zeta_p}{\partial s}(\chi\omega, 0, x),$$

where $\zeta_p(\chi, s, x)$ is given by Definition 11.2.12, and we write for simplicity $\mathrm{Log}\Gamma_p(x)$ instead of $\mathrm{Log}\Gamma_p(\chi_0, x)$, where we recall that χ_0 is the trivial character modulo p^v for any $v \geqslant 1$ (and not for $v = 0$). We define the p-adic ψ function by $\psi_p(\chi, x) = (d/dx)(\mathrm{Log}\Gamma_p(\chi, x))$ and write $\psi_p(x) = \psi_p(\chi_0, x) = (d/dx)(\mathrm{Log}\Gamma_p(x))$.

Note that we ask that $q_p \mid p^v$, so that even if we want to take for χ a trivial character χ_0, it must be the trivial character modulo a multiple of q_p, so in particular we ask that $\chi_0(p) = 0$. By the remarks made before and after Definition 11.2.12, it is clear that the definition then does not depend on the value of v such that p^v is a multiple of the conductor of χ and of q_p.

Once again, since $\zeta_p(\chi, s, x)$ is analytic for $|s| < R_p$ this definition makes sense. Furthermore, Corollary 11.2.14 gives us an immediate link between this function and Diamond's log gamma function:

Proposition 11.5.12. *Let M be such that $p^v \mid M$. For $x \in \mathbb{Z}_p$ we have*

$$\mathrm{Log}\Gamma_p(\chi, x) = \sum_{0 \leqslant j < M} \chi(x + j)\,\mathrm{Log}\Gamma_p\left(\frac{x + j}{M}\right)$$

$$+ \log_p(M) \sum_{0 \leqslant j < M} \chi(x + j)\left(\frac{x + j}{M} - \frac{1}{2}\right).$$

In particular, we have

$$\mathrm{Log}\Gamma_p(\chi, x) = \sum_{0 \leqslant j < p^v} \chi(x + j)\,\mathrm{Log}\Gamma_p\left(\frac{x + j}{p^v}\right).$$

Proof. Immediate from the definition and Corollary 11.2.14, since by Proposition 11.2.6 we have $\zeta_p(0, x) = -\omega_v(x)^{-1}(x - 1/2)$ when $x \in C\mathbb{Z}_p$. $\square$

Thus, as for the function $\zeta_p(\chi, s, x)$ for $x \in \mathbb{Z}_p$ we can find properties of $\mathrm{Log}\Gamma_p(\chi, x)$ either directly from the definition or from the above proposition. For instance, the analogue of Proposition 11.5.2 is the following.

Proposition 11.5.13. *Let χ be a character modulo p^v and let $x \in \mathbb{Z}_p$.*

(1) *We have the functional equation*

$$\mathrm{Log}\Gamma_p(\chi, x + 1) = \mathrm{Log}\Gamma_p(\chi, x) + \chi(x)\log_p(x)\,,$$

where $\chi(x)\log_p(x)$ is to be interpreted as 0 for $x = 0$.
(2) *We have the reflection formula*

$$\mathrm{Log}\Gamma_p(\chi, 1 - x) + \chi(-1)\,\mathrm{Log}\Gamma_p(\chi, x) = 0\,.$$

(3) *We have*

$$\mathrm{Log}\Gamma_p(\chi, x) = L_p'(\chi\omega, 0) + \lim_{\substack{n \to x \\ n \in \mathbb{Z}_{\geqslant 0}}} \sum_{0 \leqslant a < n} \chi(a) \log_p(a) \,,$$

and in particular if $n \in \mathbb{Z}_{\geqslant 0}$ we have

$$\mathrm{Log}\Gamma_p(\chi, n) = -\chi(-1)\,\mathrm{Log}\Gamma_p(\chi, 1-n) = L_p'(\chi\omega, 0) + \sum_{0 \leqslant a < n} \chi(a) \log_p(a) \,.$$

(4) *If $N \in \mathbb{Z}_{>0}$ is such that $p \nmid N$ we have the distribution formula*

$$\sum_{0 \leqslant j < N} \mathrm{Log}\Gamma_p\left(\chi, x + \frac{j}{N}\right) = \chi^{-1}(N)\left(\mathrm{Log}\Gamma_p(\chi, Nx) + \log_p(N)\zeta_p(\chi\omega, 0, Nx)\right) \,,$$

where $\zeta_p(\chi\omega, 0, Nx)$ is given in Proposition 11.2.19.

Corollary 11.5.14. *Let χ be a character modulo p^v and let $x \in \mathbb{Z}_p$.*

(1) *We have the functional equation*

$$\psi_p(\chi, x+1) = \psi_p(\chi, x) + \frac{\chi(x)}{x} \,.$$

(2) *We have the reflection formula*

$$\psi_p(\chi, 1-x) = \chi(-1)\psi_p(\chi, x) \,.$$

(3) *We have*

$$\psi_p(\chi, x) = -\gamma_p(\chi) + \lim_{\substack{n \to x \\ n \in \mathbb{Z}_{\geqslant 0}}} \sum_{0 \leqslant a < n} \frac{\chi(a)}{a} \,,$$

and in particular if $n \in \mathbb{Z}_{\geqslant 0}$ we have

$$\psi_p(\chi, n) = \chi(-1)\psi_p(\chi, 1-n) = -\gamma_p(\chi) + \sum_{0 \leqslant a < n} \frac{\chi(a)}{a} \,.$$

(4) *If $N \in \mathbb{Z}_{>0}$ is such that $p \nmid N$ we have the distribution formula*

$$\sum_{0 \leqslant j < N} \psi_p\left(\chi, x + \frac{j}{N}\right) = N\chi^{-1}(N)\psi_p(\chi, Nx) - \left(1 - \frac{1}{p}\right) N \log_p(N)\delta(\chi) \,.$$

In particular,

$$\sum_{0 \leqslant j < N} \psi_p\left(\chi, \frac{j}{N}\right) = -N\chi^{-1}(N)\gamma_p(\chi) - \left(1 - \frac{1}{p}\right) N \log_p(N)\delta(\chi) \,.$$

Proposition 11.5.15. *Let χ be a character modulo p^v.*

(1) *For all $k \in \mathbb{Z}_{\geqslant 1}$ we have*

$$\zeta_p(\chi, k+1, x) = (-1)^{k-1}\frac{\psi_p^{(k)}(\chi\omega^k, x)}{k!} .$$

(2) *Around $s = 1$ we have*

$$\zeta_p(\chi, s, x) = \frac{(1 - 1/p)\delta(\chi)}{s - 1} - \psi_p(\chi, x) + O(s - 1) .$$

Proof. The proofs of these results are immediate from Proposition 11.2.20, Corollary 11.2.15, and Proposition 11.5.6, and left to the reader (Exercise 15). $\square$

Corollary 11.5.16. *For all $k \in \mathbb{Z}_{\geqslant 1}$ we have*

$$\psi_p^{(k)}(\chi, 0) = (-1)^{k-1}k!L_p(\chi\omega^{-k}, k+1) = (-1)^{k-1}(k-1)!B_{-k,p}(\chi) ,$$

and for $k = 0$ we have $\psi_p(\chi, 0) = -\gamma_p(\chi)$, in other words $\psi_p(\chi, 0) = -L_p(\chi, 1)$ when $\chi \neq \chi_0$ and $\psi_p(0) = -\gamma_p$.

Proof. Clear from the above proposition since $L_p(\chi, s) = \zeta_p(\chi, s, 0)$ if χ has p-power conductor. $\square$

Proposition 11.5.17. *Let χ be a character modulo p^v, let $x \in \mathbb{Z}_p$, and let $N \in \mathbb{Z}_{\geqslant 1}$ be such that $p^v \mid N$.*

(1) *We have the distribution formula*

$$\sum_{0 \leqslant k < N} \chi(x + k) \operatorname{Log}\Gamma_p\left(\frac{x + k}{N}\right) = \operatorname{Log}\Gamma_p(\chi, x) + \log_p(N)\zeta_p(\chi\omega, 0, x) ,$$

where $\zeta_p(\chi\omega, 0, x)$ is given in Proposition 11.2.19 (note that the log gamma function on the left-hand side is Diamond's).

(2) *We have the distribution formula*

$$\sum_{0 \leqslant k < N} \chi(x + k)\psi_p\left(\frac{x + k}{N}\right) = N\psi_p(\chi, x) - \left(1 - \frac{1}{p}\right)N\log_p(\langle N \rangle)\delta(\chi) .$$

(3) *In particular, if χ is an even character then for $x \in \mathbb{Z}_p$ we have*

$$\sum_{0 \leqslant k < N} \chi(x + k) \operatorname{Log}\Gamma_p\left(\frac{x + k}{N}\right) = \sum_{0 \leqslant r < x} \chi(r)\log_p(\langle r/N \rangle) ,$$

where the right-hand side is interpreted to be extended by continuity to $x \in \mathbb{Z}_p$ if $x \notin \mathbb{Z}_{\geqslant 0}$.

(4) *In particular, for $x \in \mathbb{Z}_p$ we have*

$$\sum_{\substack{0 \leqslant k < N \\ p \nmid (x+k)}} \text{Log}\Gamma_p\left(\frac{x+k}{N}\right) = \text{Log}\Gamma_p(x) - \left(x - \left\lceil \frac{x}{p} \right\rceil\right) \log_p(\langle N \rangle) ,$$

where as above on the left-hand side $\text{Log}\Gamma_p((x + k)/N)$ *is Diamond's p-adic log gamma function.*

Note that this proposition is a generalization of Proposition 11.5.4.

Proof. By Corollary 11.2.14 we have

$$\zeta_p(\chi, s, x) = \frac{\langle N \rangle^{1-s}}{N} \sum_{0 \leqslant k < N} \chi(x + k) \zeta_p\left(\frac{x+k}{N}\right) .$$

Differentiating this equality with respect to s, setting $s = 0$, replacing χ by $\chi\omega$, and using the definitions proves (1), and (2) follows by differentiating with respect to x and using Corollary 11.2.18. If χ is an even character, we know by Proposition 11.2.19 that when $x \in \mathbb{Z}_{\geqslant 0}$ we have $\zeta_p(\chi\omega, 0, x) = -\sum_{0 \leqslant r < x} \chi(r)$. On the other hand, since $\chi\omega$ is an odd character, by Proposition 11.5.13 we have $\text{Log}\Gamma_p(\chi, x) = \sum_{0 \leqslant r < x} \chi(r) \log_p(r)$, so (3) follows by continuity, and (4) is also a special case of (1) using the value of $\zeta_p(\omega, 0, x)$ given by Proposition 11.2.19. □

Proposition 11.5.18. *For all $x \in \mathbb{Z}_p$ we have the Volkenborn integral representations*

$$\text{Log}\Gamma_p(\chi, x) = \int_{\mathbb{Z}_p} \chi(x + t)((x + t) \log_p(x + t) - (x + t)) \, dt ,$$

$$\psi_p(\chi, x) = \int_{\mathbb{Z}_p} \chi(x + t) \log_p(x + t) \, dt ,$$

$$\psi_p^{(k)}(\chi, x) = (-1)^{k-1}(k - 1)! \int_{\mathbb{Z}_p} \frac{\chi(x + t)}{(x + t)^k} \, dt$$

for $k \geqslant 1$.

Proof. Recall that by definition we have

$$\zeta_p(\chi, s, x) = \frac{1}{s - 1} \int_{\mathbb{Z}_p} \chi(x + t) \langle x + t \rangle^{1-s} \, dt .$$

Using the uniformity estimate given in Proposition 11.2.4 it is easy to show that we can differentiate with respect to s or to x under the integral sign. Differentiating with respect to s, setting $s = 0$, and replacing χ by $\chi\omega$ gives the formula for $\text{Log}\Gamma_p(\chi, x)$, and the others are obtained from that one by differentiating with respect to x. The details are left to the reader (Exercise 15). □

Remark. The reader should compare the above Volkenborn integrals with those for $\mathrm{Log}\Gamma_p$ and ψ_p for $x \in C\mathbb{Z}_p$ (Proposition 11.5.7).

Proposition 11.5.19. *For $x \in p^v\mathbb{Z}_p$ we have the convergent power series expansion*

$$\mathrm{Log}\Gamma_p(\chi, x) = L'_p(\chi\omega, 0) - \gamma_p(\chi)x + \sum_{k \geqslant 2}(-1)^k \frac{B_{1-k,p}(\chi)}{k(k-1)}x^k$$

$$= L'_p(\chi\omega, 0) - \gamma_p(\chi)x + \sum_{k \geqslant 2}(-1)^k \frac{L_p(\chi\omega^{1-k}, k)}{k}x^k \ .$$

Furthermore, the radius of convergence of this power series in $\mathbb{C}_p$ is equal to 1.

Remarks. (1) If χ is an even character, and in particular if $\chi = \chi_0$, we have $L_p(\chi\omega, s) = 0$ hence in particular $\mathrm{Log}\Gamma(\chi, x) = L'_p(\chi\omega, 0) = 0$.
(2) If $\chi \neq \chi_0$ the second formula can be written as

$$\mathrm{Log}\Gamma_p(\chi, x) = L'_p(\chi\omega, 0) + \sum_{k \geqslant 1}(-1)^k \frac{L_p(\chi\omega^{1-k}, k)}{k}x^k \ .$$

(3) By Corollary 11.5.16, since by Proposition 11.5.13 we have $\mathrm{Log}\Gamma_p(\chi, 0) = L'_p(\chi\omega, 0)$, the right-hand side of the above formulas is the Taylor expansion of the left-hand side. We must show, however, that the left-hand side is indeed equal to the sum of its Taylor expansion for $x \in p^v\mathbb{Z}_p$, and for this we will reason directly without using Corollary 11.5.16. Indeed, we will prove it only for $x \in p^v\mathbb{Z}_p$ as stated in the proposition, and in fact it is *false* in general for $x \in p\mathbb{Z}_p \setminus p^v\mathbb{Z}_p$.

Proof. The proof is essentially the same as that of Proposition 11.2.23, and in fact the result can be deduced from that proposition. Nonetheless, I prefer to redo it here. By Proposition 11.5.18 we have

$$\mathrm{Log}\Gamma_p(\chi, x) = \int_{\mathbb{Z}_p} \chi(x + t)((x + t)\log_p(x + t) - (x + t))\, dt \ .$$

Since we assume that $x \in p^v\mathbb{Z}_p$ we have $\chi(x + t) = \chi(t)$ (this is the only but essential place where we use the assumption that $x \in p^v\mathbb{Z}_p$, and not only $x \in p\mathbb{Z}_p$), and since $\chi(t) = 0$ if $t \notin \mathbb{Z}_p^*$ we have

$$\mathrm{Log}\Gamma_p(\chi, x) = \int_{\mathbb{Z}_p^*} \chi(t)f(x, t)\, dt$$

with

$$f(x, t) = (x + t)\log_p(t)\, dt + (x + t)\log_p(1 + x/t) - (x + t) = \sum_{k \geqslant 0} a_k(t)x^k \ ,$$

where, after a small computation we find that

$$a_0(t) = t\log_p(t) - t, \quad a_1(t) = \log_p(t), \quad \text{and} \quad a_k(t) = \frac{(-1)^k}{k(k-1)}t^{1-k} \text{ for } k \geqslant 2 .$$

Since $|x| < 1$ and $|t| = 1$ the series is normally convergent; hence as in the complex case we can integrate term by term, in other words exchange the sum and the integral, so we obtain $\text{Log}\Gamma_p(\chi, x) = \sum_{k\geqslant 0} b_k x^k$, where

$$b_k = \int_{\mathbb{Z}_p} \chi(t)a_k(t)\, dt ,$$

and where we have replaced the integral over $\mathbb{Z}_p^*$ by the integral over $\mathbb{Z}_p$ since once again $\chi(t) = 0$ if $t \notin \mathbb{Z}_p^*$.

By Proposition 11.3.5 (or Proposition 11.3.14) we have

$$\int_{\mathbb{Z}_p} \chi(t)t^{1-k}\, , dt = (k-1)L_p(\chi\omega^{1-k}, k) ,$$

so for $k \geqslant 2$ we have $b_k = (-1)^k L_p(\chi\omega^{1-k}, k)/k$, and by Proposition 11.3.14 we have

$$b_1 = \int_{\mathbb{Z}_p} \chi(t)\log_p(t)\, dt = -\gamma_p(\chi) .$$

Finally, we have

$$b_0 = \int_{\mathbb{Z}_p} \chi(t)(t\log_p(t) - t)\, dt ,$$

but we note simply that $b_0 = \text{Log}\Gamma_p(\chi, 0)$, hence that $b_0 = L_p'(\chi\omega, 0)$ by Proposition 11.5.13. The statement concerning the radius of convergence follows from the fact that we always have $v_p(B_{k,p}(\chi)) \geqslant -1$ (see Proposition 11.3.15).   □

Remark. Since the radius of convergence of the power series for $\text{Log}\Gamma_p(\chi, x)$ is equal to 1, it is reasonable to ask whether its sum is equal to $\text{Log}\Gamma_p(\chi, x)$ for all $x \in p\mathbb{Z}_p$, and not only for $x \in p^v\mathbb{Z}_p$. As mentioned above, it is easy to check on numerical examples that in fact the result is false in general; see Exercise 16. In fact, similarly to what we have already remarked after Proposition 11.2.23, we have shown that for $x \in p\mathbb{Z}_p$ the power series sums to

$$\int_{\mathbb{Z}_p} \chi(t)((x+t)\log_p(x+t) - (x+t))\, dt ,$$

which is equal to $\text{Log}\Gamma_p(\chi, x)$ only for $x \in p^v\mathbb{Z}_p$.

To use the power series given in Proposition 11.5.19, we need lower bounds for the p-adic valuations of the coefficients. A sufficient result is as follows, which for simplicity we give only in the case $\chi = \chi_0$, for which we only need the coefficients of x^{2k+1}.

Lemma 11.5.20. *Set*

$$v(k,p) = v_p(B_{-2k,p}p^{2k+1}/(2k(2k+1))) \,.$$

Then $v(k,p) \geqslant 4$ except for the following values: $v(1,2) = 1$, $v(1,3) = 1$, $v(1,p) = 3$ for $p \geqslant 5$, $v(2,2) = 2$, and $v(2,5) = 3$.

Proof. Assume first that $(p-1) \nmid 2k$. By Corollary 11.4.8, $B_{-2k,p}/(2k)$ is p-integral, so that $v(k,p) \geqslant 2k+1 - v_p(2k+1)$. It is immediate to check that for $k \geqslant 2$ we have $2k+1 - v_p(2k+1) \geqslant 4$. On the other hand, for $k=1$ we have $v(k,p) \geqslant 3 - v_p(3) = 3$ if $p \geqslant 5$, and in fact it is clear that $v(k,p) = 3$ for $p \geqslant 5$. Assume now that $(p-1) \mid 2k$. Again by Corollary 11.4.8 we have $v_p(B_{-2k,p}) = -1$, hence $v(k,p) = 2k - \max(v_p(2k), v_p(2k+1))$. Once again we check that for $k \geqslant 3$ this is always greater than or equal to 5, that for $k=2$ it is greater or equal to 4 unless $p=2$ or $p=5$, and on the other hand, for $k=1$ the only possible values of p are $p=2$ and $p=3$. The lemma follows by an explicit computation of the special cases. $\qquad\square$

Analogously to Proposition 11.5.8, we have the following.

Proposition 11.5.21. *For $x \in p^v \mathbb{Z}_p$ we have*

$$\lim_{r \to \infty} \frac{B_{\phi(p^r)}(\chi, x) - (1 - 1/p)\delta(\chi)}{\phi(p^r)} = \psi_p(\chi, x) \,.$$

If, in addition, $\chi = \chi_0$ is the trivial character modulo 1, the above is true for all $x \in \mathbb{Z}_p$.

Note that in all the other results of this chapter, the trivial character is always assumed to be defined modulo p^v for $v \geqslant 1$. Here, exceptionally, we accept the trivial character modulo 1.

Proof. As mentioned after the proof of Proposition 11.3.10 (2), there are two ways to prove this kind of formula: the most natural one is to use Volkenborn integrals, the second being to use power series expansions. In both cases one must justify an exchange of limits. In the proof of Proposition 11.3.10 (2) we used Volkenborn integrals, so for a change we use here power series expansions. By definition we have

$$\frac{B_n(\chi, x) - B_0(\chi)x^n - (1 - 1/p)\delta(\chi)}{n} = \frac{B_n(\chi) - (1 - 1/p)\delta(\chi)}{n} + S, \text{ with}$$

$$S = \frac{1}{n}\sum_{j=1}^{n-1}\binom{n}{n-j}B_{n-j}(\chi)x^j = \sum_{j=1}^{n-1}\binom{n-1}{j}\frac{B_{n-j}(\chi)}{n-j}x^j \,.$$

If we replace n by $\phi(p^r)$ and make $r \to \infty$, by Definition 11.3.13 we have

$$(B_n(\chi) - (1 - 1/p)\delta(\chi))/n \to -\gamma_p(\chi) \,,$$

and for $j \geqslant 1$ we have $B_{n-j}(\chi)/(n-j) \to B_{-j,p}(\chi)/(-j)$. Furthermore, $\binom{n-1}{j}$ tends to $\binom{-1}{j} = (-1)^j$. Since $x \in p^v\mathbb{Z}_p \subset p\mathbb{Z}_p$, by normal convergence we can take the limit inside the sum, and since $x^n/n \to 0$ we deduce that

$$\lim_{r \to \infty} \frac{B_{\phi(p^r)}(\chi, x) - (1 - 1/p)\delta(\chi)}{\phi(p^r)} = -\gamma_p(\chi) + \sum_{j \geqslant 1} (-1)^{j-1} \frac{B_{-j,p}(\chi)}{j} x^j$$

$$= \psi_p(\chi, x)$$

by Proposition 11.5.19, proving the result when $x \in p^v\mathbb{Z}_p$.

Now assume that $\chi = \chi_0$, where χ_0 is the trivial character modulo 1. Thus we know that the result is true for $x \in p\mathbb{Z}_p$, so now let $x \in \mathbb{Z}_p \setminus p\mathbb{Z}_p$. There exists a unique $r \in [1, p-1]$ such that $y = x - r \in p\mathbb{Z}_p$. By Proposition 9.1.3 we have $B_n(x) = B_n(y) + n\sum_{y \leqslant m < x} m^{n-1}$, so that

$$\frac{B_n(x) - (1 - 1/p)}{n} = \frac{B_n(y) - (1 - 1/p)}{n} + \sum_{y \leqslant m < x} m^{n-1} .$$

If we replace n by $\phi(p^r)$ and make $r \to \infty$, we have $m^{n-1} = m^{-1}m^n \to m^{-1}$ when $p \nmid m$, and $m^{n-1} \to 0$ when $p \mid m$. Since by definition of r the only $m \in [y, x[$ such that $p \mid m$ is $m = y$, it follows from the first part of the proof that

$$\lim_{r \to \infty} \frac{B_{\phi(p^r)}(x) - (1 - 1/p)}{\phi(p^r)} = \psi_p(y) + \sum_{y < m < x} \frac{1}{m} = \psi_p(x)$$

by the functional equation given in Corollary 11.5.14, finishing the proof. $\quad\square$

Note that since $\psi_p(\chi, 0) = -\gamma_p(\chi)$, this proposition can be considered as a generalization of Definition 11.3.13, which we used in the proof.

Raabe's formula is the following.

Proposition 11.5.22. *Let χ be a character modulo p^v. If $x \in \mathbb{Z}_p$ we have*

$$\int_{\mathbb{Z}_p} \mathrm{Log}\Gamma_p(\chi, x + t)\, dt = (x - 1)\psi_p(\chi, x) + \zeta_p(\chi\omega, 0, x) ,$$

where $\zeta_p(\chi\omega, 0, x)$ is given in Proposition 11.2.19. In particular, we have

$$\int_{\mathbb{Z}_p} \mathrm{Log}\Gamma_p(x + t)\, dt = (x - 1)\psi_p(x) - x + \left\lceil \frac{x}{p} \right\rceil .$$

Proof. As for Proposition 11.5.9, this follows immediately by differentiation with respect to s of Raabe's formula for $\zeta_p(\chi\omega, s, x)$ (Proposition 11.2.22), and is left to the reader. The special case $\chi = \chi_0$ then follows from Proposition 11.2.19. $\quad\square$

It is of course immediate to deduce properties of $\psi_p(\chi, x)$ from the above properties of $\mathrm{Log}\Gamma_p(\chi, x)$.

The analogue of Proposition 11.5.10 is the following.

Proposition 11.5.23. *Let χ be a character modulo q_p and let $x \in \mathbb{Z}_p$. We have*

$$\mathrm{Log}\Gamma_p(\chi, x) \equiv \begin{cases} 0 & (\mathrm{mod}\ p\mathbb{Z}_p) & \text{if } \chi \neq \omega^{-1} \text{ and } p \geqslant 3, \\ 1 + (-1)^{x(x-1)/2} & (\mathrm{mod}\ q_p\mathbb{Z}_p) & \text{if } \chi \neq \omega^{-1} \text{ and } p = 2, \\ -(1 - 1/p) & (\mathrm{mod}\ p\mathbb{Z}_p) & \text{if } \chi = \omega^{-1} \text{ and } p \geqslant 5, \\ 4/3 & (\mathrm{mod}\ p\mathbb{Z}_p) & \text{if } \chi = \omega^{-1} \text{ and } p = 3, \\ -(1/2 + (-1)^{x(x-1)/2}) & (\mathrm{mod}\ q_p\mathbb{Z}_p) & \text{if } \chi = \omega^{-1} \text{ and } p = 2. \end{cases}$$

Proof. Assume first that $p \geqslant 3$. By Propositions 11.5.12 and 11.5.10 we have

$$\mathrm{Log}\Gamma_p(\chi, x) \equiv c(p) \sum_{0 \leqslant j < p} \chi\omega(x + j) \pmod{p\mathbb{Z}_p},$$

with $c(p) = -1/p$ for $p \geqslant 5$ and $c(p) = 2/3$ for $p = 3$. Note that the fact that $v_p(c(p)) < 0$ is irrelevant for the validity of this congruence. Since $\chi\omega$ is defined modulo p we have $\sum_{0 \leqslant j < p} \chi\omega(x + j) = \sum_{0 \leqslant j < p} \chi\omega(j)$, and this is equal to 0 if $\chi\omega \neq \chi_0$, and otherwise is equal to $p - 1$, proving the lemma for $p \geqslant 3$. Assume now that $p = 2$. By the same propositions we now have

$$\mathrm{Log}\Gamma_p(\chi, x) \equiv -\frac{1}{4} \sum_{0 \leqslant j < 4} \chi\omega(x + j) + \frac{1}{2} \sum_{0 \leqslant j < 4} \chi(x + j)(x + j) \pmod{4\mathbb{Z}_2}.$$

Here we can have only $\chi = \chi_0$ or $\chi = \omega = \omega^{-1}$. If $\chi = \chi_0$ the first sum vanishes as usual, while if $\chi = \omega^{-1}$ it is equal to 2. We find by inspection that the second sum is equal to $2(1 + (-1)^{x(x-1)/2})$ if $\chi = \chi_0$, and to $-2(-1)^{x(x-1)/2}$ if $\chi = \omega^{-1}$, proving the proposition. $\square$

The remarkable fact about this proposition is the first statement, in other words, the fact that $\mathrm{Log}\Gamma_p(\chi, x) \equiv 0 \pmod{p\mathbb{Z}_p}$ if $\chi \neq \omega^{-1}$ and $p \geqslant 3$. Indeed, this implies that the exponential of this function makes sense, and will have analogous functional properties. There are, however, two related obstructions to this construction. First, although we could set $\Gamma_p(\chi, x) = \exp_p(\mathrm{Log}\Gamma_p(\chi, x))$ (when $\chi \neq \omega^{-1}$ and $p \geqslant 3$), this is not really canonical since we could just as well set $\Gamma_p(\chi, x) = \zeta_{p-1}^{f(x)} \exp_p(\mathrm{Log}\Gamma_p(\chi, x))$ for any reasonable integral-valued function f, since $\log_p(\zeta_{p-1}) = 0$. Second, consider for instance the functional equation for $\mathrm{Log}\Gamma_p(\chi, x + 1)$. Taking exponentials gives $\Gamma_p(\chi, x+1) = \exp_p(\chi(x) \log_p(x))\Gamma_p(\chi, x)$, and once again we would like to say that $\exp_p(\chi(x) \log_p(x)) = x^{\chi(x)}$, but this does not make much sense, except when $\chi(x) = \pm 1$. Thus, it is preferable to define the gamma function from scratch and study its properties directly. This will be done later, in Section 11.6.

By Proposition 11.5.19, we know that when the conductor of χ is a power of p we have $L'_p(\chi, 0) = \mathrm{Log}\Gamma_p(\chi\omega^{-1}, 0)$. It is easy to generalize this to a character of arbitrary conductor.

Proposition 11.5.24. *Let χ be a primitive character of conductor f, and denote by f_1 the conductor of the character $\chi_1 = \chi\omega^{-1}$.*

(1) *If $p \nmid f$ then $f_1 = q_p f$ and*

$$L_p'(\chi, 0) = B_1(\chi_1) \log_p(f) + \omega^{-1}(f) \sum_{0 \leqslant k < f} \chi(k) \operatorname{Log}\Gamma_p\left(\omega^{-1}, \frac{k}{f}\right).$$

(2) *If $p \nmid f_1$ then $f = q_p f_1$ and*

$$L_p'(\chi, 0) = (1 - \chi_1(p))B_1(\chi_1) \log_p(f) + \sum_{0 \leqslant k < f_1} \chi_1(k) \operatorname{Log}\Gamma_p\left(\chi_0, \frac{k}{f_1}\right).$$

(3) *If $p \mid f$ and $p \mid f_1$ then $f_1 = f$ and $L_p'(\chi, 0)$ is given by Corollary 11.5.5 with $N = f$.*

Proof. Immediate consequence of Corollary 11.5.5 and Proposition 11.5.12 and left to the reader (Exercise 17). $\qquad\square$

11.5.3 Computation of some p-adic Logarithms

In view of applications to special values of the p-adic ψ functions at rational numbers, we need to prove some formulas involving expressions of the form $\log_p(1 - \zeta_m)$ for an mth root of unity ζ_m, which are interesting in their own right. I heartily thank P. Colmez for the alternative proofs that are briefly indicated as exercises, which use formal power series, well known in p-adic analysis at least since Serre and Iwasawa. I refer to Washington's book [Was] and to the course notes of Colmez available on his web site for much more on this.

Recall from Definition 4.5.9 that we have defined

$$\mathcal{D} = \{z \in \mathbb{C}_p, \ |z - 1| \geqslant 1\} = \{z \in \mathbb{C}_p, \ v_p(z - 1) \leqslant 0\},$$

and that Lemma 4.5.11 tells us that the condition $|z - 1| \geqslant 1$ is equivalent either to $|z/(1 - z)| \leqslant 1$ or to $|z^{p^N}/(1 - z^{p^N})| \leqslant 1$ for all $N \geqslant 0$.

Theorem 11.5.25. *For all $z \in \mathcal{D}$ we have*

$$\lim_{N \to \infty} \frac{1}{z^{p^N} - 1} \sum_{0 \leqslant a < p^N}^{(p)} \frac{z^a}{a} = \log_p(1 - z) - \frac{1}{p}\log_p(1 - z^p).$$

Proof. Set

$$S_N(z) = \frac{1}{z^{p^N} - 1} \sum_{0 \leqslant a < p^N}^{(p)} \frac{z^a}{a}.$$

It is clear that this is a rational function all of whose poles ζ are p^Nth roots of unity, hence satisfy $|\zeta - 1| < 1$ by Proposition 3.5.5. Thus, if we show that

$S_N(z)$ converges, and uniformly on $\mathcal{D}$, by definition this will show that the limit exists and is Krasner analytic on $\mathcal{D}$ (see Definition 4.5.9). Thanks to the ultrametric property, this is equivalent to showing that $S_{N+1}(z) - S_N(z)$ tends uniformly to 0, which is easily done: setting $a = p^N q + r$ with $0 \leqslant q < p$ and $0 \leqslant r < p^N$ we have

$$\sideset{}{^{(p)}}\sum_{0 \leqslant a < p^{N+1}} \frac{z^a}{a} = \sum_{0 \leqslant q < p} z^{p^N q} \sideset{}{^{(p)}}\sum_{0 \leqslant r < p^N} \frac{z^r}{r + p^N q} .$$

Now

$$\sideset{}{^{(p)}}\sum_{0 \leqslant r < p^N} \frac{z^r}{r + p^N q} = \sideset{}{^{(p)}}\sum_{0 \leqslant r < p^N} \frac{z^r}{r} (1 + p^N q/r)^{-1} = (z^{p^N} - 1)(S_N(z) + O(p^N)) ,$$

where $f(N) = O(p^N)$ means that $|f(N)/p^N|$ remains bounded, and here uniformly in z. Since $\sum_{0 \leqslant q < p} z^{p^N q} = (z^{p^{N+1}} - 1)/(z^{p^N} - 1)$, we deduce that $S_{N+1}(z) = S_N(z) + O(p^N)$, with a uniform O constant, so that $S_{N+1}(z) - S_N(z)$ tends uniformly to 0 on $\mathcal{D}$, as claimed. Note that in this computation we have implicitly used Lemma 4.5.11, which tells us that $|z^{p^N} - 1| \geqslant 1$ when $z \in \mathcal{D}$.

We have thus shown that the left-hand side exists and is a Krasner analytic function on $\mathcal{D}$. Let us now consider the right-hand side. Set

$$U(z) = \frac{(1 - z)^p - 1 + z^p}{1 - z^p} .$$

For $p > 2$ we have

$$(1 - z)^p - 1 + z^p = \sum_{1 \leqslant n \leqslant p-1} \binom{p}{n} (-z)^n ,$$

while for $p = 2$ we have $(1 - z)^p - 1 + z^p = -2z + 2z^2$. Since all the binomial coefficients have a p-adic valuation equal to 1 (or directly for $p = 2$), an easy computation using Lemma 4.5.11 shows that $v_p(U(z)) \geqslant 1$ (see Exercise 26). We thus have

$$p \log_p(1 - z) - \log_p(1 - z^p) = \log_p \left(\frac{(1 - z)^p}{1 - z^p} \right)$$

$$= \log_p(1 + U(z)) = \sum_{k \geqslant 1} (-1)^{k-1} \frac{U(z)^k}{k} ,$$

and since $v_p(U(z)) \geqslant 1$ for all $z \in \mathcal{D}$, this series converges uniformly in $\mathcal{D}$. Since $U(z)$ is a rational function with poles at the pth roots of unity, hence outside $\mathcal{D}$, it follows that the right-hand side of the formula of the theorem is also a Krasner analytic function on $\mathcal{D}$.

Thanks to the crucial Lemma 4.5.10, to prove the theorem it is sufficient to prove that the two sides are equal on a nonempty open subset S of $\mathcal{D}$. We choose S to be the "open" unit ball $S = \{z \in \mathbb{C}_p, |z| < 1\}$, which is clearly a nonempty open subset of $\mathcal{D}$. In the set S the formula is trivial since $\sum_{a \geqslant 1} z^a/a$ converges to $-\log_p(1-z)$, so that $\sum_{a \geqslant 1}^{(p)} z^a/a$ converges to the right-hand side, proving the theorem. $\square$

For an important alternative proof of this theorem, see Exercise 29.

As an example, we give the following:

Corollary 11.5.26. *We have*

$$\lim_{N \to \infty} \sum_{0 \leqslant a < p^N}^{(p)} \frac{(-1)^{a-1}}{a} = 2 \left(1 - \frac{1}{p} \right) \log_p(2) \ .$$

Proof. Clear from the above theorem for $p \geqslant 3$, and immediate for $p = 2$ by grouping terms for a and $p^N - a$. $\square$

See also Exercise 27.

We now want to generalize Theorem 11.5.25 to the case $|z - 1| < 1$, in other words, $v_p(z-1) > 0$. The formulas are here slightly different. We begin with the following auxiliary result.

Theorem 11.5.27. *Let $r \in \mathbb{Z}_{\geqslant 0}$ be coprime to p, and let $z \in \mathbb{C}_p$ be such that $v_p(z - 1) > 0$.*

(1) *We have*

$$\lim_{N \to \infty} \sum_{1 \leqslant a \leqslant p^N} \frac{z^{ra} - 1}{a} \left\lceil \frac{ra}{p^N} \right\rceil = r \log_p \left(\frac{z^r - 1}{r(z-1)} \right) \ ,$$

where of course $(z^r - 1)/(z - 1) = \sum_{0 \leqslant i < r} z^i = r$ for $z = 1$.

(2) *We have*

$$\lim_{N \to \infty} \sum_{0 \leqslant a < p^N}^{(p)} \frac{z^{ra}}{a} \left\lceil \frac{ra}{p^N} \right\rceil = r \log_p \left(\frac{z^r - 1}{z - 1} \right) - \frac{r}{p} \log_p \left(\frac{z^{pr} - 1}{z^p - 1} \right) \ ,$$

and in particular

$$\lim_{N \to \infty} \sum_{0 \leqslant a < p^N}^{(p)} \frac{1}{a} \left\lceil \frac{ra}{p^N} \right\rceil = \left(1 - \frac{1}{p} \right) r \log_p(r) \ .$$

Proof. (1). For any positive real number u it is clear that $\lceil u \rceil = \sum_{0 \leqslant m < u} 1$. It follows that

$$\sum_{1 \leqslant a \leqslant p^N} \frac{z^{ra}}{a} \left\lfloor \frac{ra}{p^N} \right\rfloor = \sum_{1 \leqslant a \leqslant p^N} \frac{z^{ra}}{a} \sum_{0 \leqslant m < ra/p^N} 1$$

$$= \sum_{0 \leqslant m < r} \sum_{p^N m/r < a \leqslant p^N} \frac{z^{ra}}{a} = \sum_{0 \leqslant m < r} \left(S(p^N) - S(p^N m/r) \right),$$

where for any positive real number M we set

$$S(M) = \sum_{1 \leqslant a \leqslant M} \frac{z^{ra}}{a} .$$

Note that in the above we do not exclude indexes divisible by p. Set $\pi = z^r - 1$, so that $v_p(\pi) > 0$, since $v_p(z - 1) > 0$, and hence $v_p(z) = v_p(1 + (z - 1)) = 0$ by assumption. Thus

$$S(M) = \sum_{1 \leqslant a \leqslant M} \frac{z^{ra}}{a} = \sum_{1 \leqslant a \leqslant M} \frac{(\pi + 1)^a}{a} = \sum_{1 \leqslant a \leqslant M} \frac{1}{a} \sum_{0 \leqslant k \leqslant a} \binom{a}{k} \pi^k$$

$$= \sum_{1 \leqslant a \leqslant M} \frac{1}{a} + \sum_{1 \leqslant k \leqslant M} \frac{\pi^k}{k} \sum_{1 \leqslant a \leqslant M} \binom{a-1}{k-1} .$$

By the recurrence formula for binomial coefficients and the fact that $a \leqslant M$ is equivalent to $a \leqslant \lfloor M \rfloor$, we have

$$\sum_{1 \leqslant a \leqslant M} \binom{a-1}{k-1} = \binom{\lfloor M \rfloor}{k} ,$$

so that, since we do not need to include the condition $k \leqslant M$,

$$S(M) = \sum_{1 \leqslant a \leqslant M} \frac{1}{a} + \sum_{k \geqslant 1} \frac{\pi^k}{k} \binom{\lfloor M \rfloor}{k} ,$$

so that

$$S(p^N) - S(p^N m/r) = \sum_{p^N m/r < a \leqslant p^N} \frac{1}{a} + \sum_{k \geqslant 1} \frac{\pi^k}{k} \left(\binom{p^N}{k} - \binom{\lfloor p^N m/r \rfloor}{k} \right) .$$

By Lemma 4.2.8 we have $v_p(\binom{p^N}{k}) \geqslant N - v_p(k)$, so that

$$v_p((\pi^k/k) \binom{p^N}{k}) \geqslant N + k v_p(\pi) - 2 v_p(k) .$$

Set

$$B = \max_{k \geqslant 1} (2 \log(k) / \log(p) - k v_p(\pi)) ,$$

which exists since $v_p(\pi) > 0$. Thus $\min_{k \geqslant 1} (k v_p(\pi) - 2 v_p(k)) \geqslant -B$, so that $v_p((\pi^k/k) \binom{p^N}{k}) \geqslant N - B$. Furthermore, we have the following lemma.

Lemma 11.5.28. *Let a and b in $\mathbb{Z}_p$, and for $k \in \mathbb{Z}_{\geqslant 0}$ set*

$$d_k = \binom{p^N a + b}{k} - \binom{b}{k} .$$

Then $v_p(d_k) \geqslant N - \log(k)/\log(p)$.

Proof. Recall the binomial identity

$$\sum_{0 \leqslant j \leqslant k} \binom{p^N a}{j} \binom{b}{k-j} = \binom{p^N a + b}{k} ,$$

which is proved by expanding the product of the binomial expansions of $(1 + X)^b$ with $(1 + X)^{p^N a}$. It follows that

$$d_k = \sum_{1 \leqslant j \leqslant k} \binom{p^N a}{j} \binom{b}{k-j} ,$$

so by Lemma 4.2.8

$$v_p(d_k) \geqslant N - \max_{1 \leqslant j \leqslant k} v_p(j) \geqslant N - \max_{1 \leqslant j \leqslant k} \log(j)/\log(p) \geqslant N - \log(k)/\log(p) ,$$

as claimed. $\qquad\qquad\qquad\qquad\qquad\qquad\qquad\qquad\qquad\qquad\qquad\qquad\square$

Resuming the proof of the theorem, we thus have

$$v_p((\pi^k/k)d_k) \geqslant N + k v_p(\pi) - v_p(k) - \log(k)/\log(p) \geqslant N - B ,$$

with the same B as above. Putting everything together, it follows that if we set

$$\left\lfloor \frac{p^N m}{r} \right\rfloor = \frac{p^N m - a_m}{r}$$

with $0 \leqslant a_m < r$, then noting that m/r and a_m/r are in $\mathbb{Z}_p$, we have

$$S(p^N) - S(p^N m/r) = \sum_{p^N m/r < a \leqslant p^N} \frac{1}{a} - \sum_{k \geqslant 1} \frac{\pi^k}{k} \binom{-a_m/r}{k} + O(p^{N-B}) ,$$

where B depends on z, but not on N. Thus

$$\sum_{1 \leqslant a \leqslant p^N} \frac{z^{ra}}{a} \left\lceil \frac{ra}{p^N} \right\rceil = \sum_{0 \leqslant m < r} (S(p^N) - S(p^N m/r))$$

$$= \sum_{0 \leqslant m < r} \sum_{p^N m/r < a \leqslant p^N} \frac{1}{a} - \sum_{k \geqslant 1} \frac{\pi^k}{k} \sum_{0 \leqslant m < r} \binom{-a_m/r}{k} + O(p^{N-B}) .$$

It is clear that the a_m form a permutation of $[0, r-1]$: indeed, a_m is the representative in $[0, r-1]$ of $p^N m \bmod r$, and since p^N is coprime to r the map $m \mapsto p^N m \bmod r$ is a bijection from $\mathbb{Z}/r\mathbb{Z}$ onto itself. Thus, if X is a formal variable, we have

$$\sum_{k \geqslant 1} X^k \sum_{0 \leqslant m < r} \binom{-a_m/r}{k} = \sum_{0 \leqslant j < r} \sum_{k \geqslant 1} \binom{-j/r}{k} X^k = \sum_{0 \leqslant j < r} ((1+X)^{-j/r} - 1)$$

$$= \frac{(1+X)^{-1} - 1}{(1+X)^{-1/r} - 1} - r = -r - X\left(\frac{1}{(1+X)((1+X)^{-1/r} - 1)}\right)$$

$$= -r + X\left(\frac{1}{1+X} - \frac{(1+X)^{-1/r-1}}{(1+X)^{-1/r} - 1}\right).$$

Dividing by X and integrating formally, it follows after a small computation that

$$\sum_{k \geqslant 1} \frac{X^k}{k} \sum_{0 \leqslant m < r} \binom{-a_m/r}{k} = r \log\left(\frac{r((1+X)^{1/r} - 1)}{X}\right).$$

Since $v_p(\pi) > 0$ and $p \nmid r$ we may replace X by $\pi = z^r - 1$ in the above formula. Note that $(1+\pi)^{1/r} \equiv 1 \pmod{\pi}$. Since $p \nmid r$, by Proposition 4.3.2 we know that if η is an rth root of unity different from 1 then $v_p(\eta - 1) = 0$, and in particular $\eta \not\equiv 1 \pmod{\pi}$, so that $(1+\pi)^{1/r} = z$. Thus

$$\sum_{1 \leqslant a \leqslant p^N} \frac{z^{ra}}{a}\left[\frac{ra}{p^N}\right] = S_0 + r \log_p\left(\frac{z^r - 1}{r(z - 1)}\right) + O(p^{N-B}),$$

with

$$S_0 = \sum_{0 \leqslant m < r} \sum_{p^N m/r < a \leqslant p^N} \frac{1}{a} = \sum_{1 \leqslant a \leqslant p^N} \frac{1}{a}\left[\frac{ra}{p^N}\right],$$

proving (1).

(2). We have

$$\sum_{\substack{1 \leqslant a \leqslant p^N \\ p \mid a}} \frac{z^{ra} - 1}{a}\left[\frac{ra}{p^N}\right] = \sum_{1 \leqslant b \leqslant p^{N-1}} \frac{z^{rpb} - 1}{pb}\left[\frac{rb}{p^{N-1}}\right],$$

and by (1), as $N \to \infty$ this tends to $(r/p) \log_p((z^{pr} - 1)/(r(z^p - 1)))$, so it follows that

$$\lim_{N \to \infty} \sideset{}{^{(p)}}\sum_{0 \leqslant a < p^N} \frac{z^{ra} - 1}{a}\left[\frac{ra}{p^N}\right] = r \log_p\left(\frac{z^r - 1}{r(z - 1)}\right) - \frac{r}{p} \log_p\left(\frac{z^{pr} - 1}{r(z^p - 1)}\right).$$

As in the proof of (1), we have

$$\sum_{1 \leqslant a \leqslant p^N} \frac{1}{a}\left[\frac{ra}{p^N}\right] = \sum_{0 \leqslant m < r} (H(p^N) - H(mp^N/r)),$$

where $H(M) = \sum_{1 \leqslant n \leqslant M} 1/n$ is the harmonic sum, so that

$$\sum_{1 \leqslant a \leqslant p^N}^{(p)} \frac{1}{a} \left\lceil \frac{ra}{p^N} \right\rceil = \sum_{0 \leqslant m < r} (H_p(p^N) - H_p(mp^N/r)) ,$$

where $H_p(M) = \sum_{1 \leqslant n \leqslant M}^{(p)} 1/n$ is the harmonic sum with the indexes divisible by p removed. To compute the limit as $N \to \infty$, we note that by definition of ψ_p we have

$$H_p(p^N) - H_p(mp^N/r) = \psi_p(p^N + 1) - \psi_p(\lfloor mp^N/r \rfloor + 1) .$$

Since ψ_p is continuous at 0, by Corollary 11.5.14 we have $\lim_{N \to \infty} \psi_p(p^N + 1) = \psi_p(1) = -\gamma_p$. Furthermore, as above we have $\lfloor mp^N/r \rfloor = (mp^N - a_m)/r$, so that $\lim_{N \to \infty} \psi_p(\lfloor mp^N/r \rfloor + 1) = \psi_p(-a_m/r + 1)$. Since the a_m form a permutation of $[0, r-1]$ it follows that

$$\lim_{N \to \infty} \sum_{0 \leqslant m < r} (H_p(p^N) - H_p(mp^N/r)) = -r\gamma_p - \sum_{0 \leqslant j < r} \psi_p(1 - j/r)$$

$$= -r\gamma_p - \sum_{1 \leqslant k \leqslant r} \psi_p(k/r) .$$

Applying the distribution formula for the function ψ_p (Corollary 11.5.14 once again) we obtain the formula

$$\lim_{N \to \infty} \sum_{0 \leqslant a < p^N}^{(p)} \frac{1}{a} \left\lceil \frac{ra}{p^N} \right\rceil = \left(1 - \frac{1}{p}\right) r \log_p(r) ,$$

proving (2) after adding to the formula obtained above for $\sum_{0 \leqslant a < p^N}^{(p)} ((z^{ra} - 1)/a) \lceil ra/p^N \rceil$. $\qquad \square$

Before proving the main theorem, we also need the following easy result.

Lemma 11.5.29. *Let* $z \in \mathbb{C}_p$ *be such that* $v_p(z - 1) > 0$. *There exists a constant* B *depending on* z, *but not on* N, *such that*

$$\sum_{0 \leqslant a < p^N}^{(p)} \frac{z^a}{a} = O(p^{N-B}) .$$

Proof. As in the computation of $S(M)$ made above, we have

$$\sum_{1 \leqslant a \leqslant p^N} \frac{z^a}{a} = \sum_{1 \leqslant a \leqslant p^N} \frac{1}{a} + \sum_{k \geqslant 1} \frac{(z-1)^k}{k} \binom{p^N}{k} ,$$

and using this formula with z replaced by z^p and N by $N - 1$, and dividing by p and subtracting gives

$$\sum_{0 \leqslant a < p^N}^{(p)} \frac{z^a}{a} = \sum_{0 \leqslant a < p^N}^{(p)} \frac{1}{a} + \sum_{k \geqslant 1} \frac{(z-1)^k}{k} \binom{p^N}{k} - \frac{1}{p} \sum_{k \geqslant 1} \frac{(z^p - 1)^k}{k} \binom{p^{N-1}}{k} .$$

As in the proof above, since $v_p(\binom{p^N}{k}) = N - v_p(k)$ and $v_p(z-1) > 0$, both infinite sums on k are $O(p^{N-B})$ for a suitable constant B independent of N, but dependent on z, so that to prove the lemma we must simply show that $\sum_{0 \leqslant a < p^N}^{(p)} 1/a$ tends to 0. But this is easily done directly since

$$\sum_{0 \leqslant a < p^N}^{(p)} \frac{1}{a} = \sum_{0 \leqslant a < p^N/2}^{(p)} \left(\frac{1}{a} + \frac{1}{p^N - a} \right) = p^N \sum_{0 \leqslant a < p^N/2}^{(p)} \frac{1}{a(p^N - a)} = O(p^N) ,$$

proving the lemma.   □

See Exercise 31 for a precise result.

We can now prove the theorem that we are after, which generalizes Theorem 11.5.25 to the case $v_p(z-1) > 0$:

Theorem 11.5.30. *Let $r \in \mathbb{Z}_{\geqslant 0}$ be such that $p \nmid r$, let $z \in \mathbb{C}_p$ be such that $v_p(z-1) > 0$, and set*

$$S_N(z) = \frac{1}{p^N} \sum_{0 \leqslant a < p^N}^{(p)} z^a \log_p(a) .$$

(1) *As $N \to \infty$ the sequence $S_N(z)$ tends to a limit, which we denote by $S(z)$.*

(2) *If $z^p \neq 1$ we have*

$$S(z^r) - S(z) = \log_p \left(\frac{z^r - 1}{z - 1} \right) - \frac{1}{p} \log_p \left(\frac{z^{pr} - 1}{z^p - 1} \right)$$
$$- r \log_p(z) \log_p(r) \frac{z^{pr} - z^r}{(z^r - 1)(z^{pr} - 1)} ,$$

and if $z^p = 1$ but $z \neq 1$ we have

$$S(z^r) - S(z) = \log_p \left(\frac{z^r - 1}{z - 1} \right) .$$

(3) *We have $S(1) = -\gamma_p$, and if as usual ζ_{p^v} denotes a primitive p^vth root of unity with $v \geqslant 1$, we have*

$$S(\zeta_{p^v}^r) = \begin{cases} \log_p(1 - \zeta_{p^v}^r) - \dfrac{1}{p} \log_p(1 - \zeta_{p^v}^{rp}) & \text{for } v \geqslant 2 , \\ \log_p(1 - \zeta_p^r) + \dfrac{\gamma_p}{p-1} & \text{for } v = 1 . \end{cases}$$

Proof. (1). By the ultrametric property it is enough to show that $S_{N+1}(z) - S_N(z)$ tends to 0. Writing $a = p^N q + b$ with $0 \leqslant q < p$ and $0 \leqslant b < p^N$ we have

$$
\begin{aligned}
S_{N+1}(z) &= \frac{1}{p^{N+1}} \sum_{0 \leqslant q < p} z^{p^N q} \sum_{0 \leqslant b < p^N}^{(p)} z^b \log_p(b + p^N q) \\
&= S_N(z)\frac{1}{p} \sum_{0 \leqslant q < p} z^{p^N q} + \frac{1}{p} \sum_{0 \leqslant q < p} \sum_{0 \leqslant b < p^N}^{(p)} q \frac{z^b}{b} + O(p^{N-2}) \\
&= S_N(z)\frac{1}{p} \sum_{0 \leqslant q < p} (1 + O(p^{N-B})) + O(p^{N-B}) + O(p^{N-2})
\end{aligned}
$$

by Lemma 11.5.29 and the fact that $z^{p^N} = (1 + (z-1))^{p^N} = 1 + O(p^{N-B})$ as we have shown and used above, proving (1).

(2). Consider the expression

$$
E_N = - \sum_{0 \leqslant a < p^N}^{(p)} z^{ra} \log_p \left(\frac{p^N}{ra} \left[\frac{ra}{p^N} \right] - 1 \right) .
$$

On the one hand, we have

$$
E_N = - \sum_{0 \leqslant a < p^N}^{(p)} z^{ra} \log_p \left(p^N \left[\frac{ra}{p^N} \right] - ra \right)
$$

$$
+ \log_p(r) \sum_{0 \leqslant a < p^N}^{(p)} z^{ra} + \sum_{0 \leqslant a < p^N}^{(p)} z^{ra} \log_p(a) .
$$

If we set $b = p^N \lceil ra/p^N \rceil - ra$ then $0 \leqslant b < p^N$ and $p \nmid b$, since $p \nmid r$, so the map $a \mapsto b$ is a bijection from the integers in $[0, p^N[$ coprime to p to itself. Thus

$$
E_N = p^N (S_N(z^r) - S_N(z)) + \log_p(r)(z^{rp^N} - 1)\frac{z^{pr} - z^r}{(z^r - 1)(z^{pr} - 1)} ,
$$

and with an evident interpretation when some denominator vanishes. More precisely, since $p \nmid r$ and $v(z-1) > 0$, by Proposition 3.5.5 we cannot have $z^r = 1$, except if $z = 1$, which is excluded. Thus we have $z^{pr} = 1$ if and only if $z = \zeta_p$ is a primitive pth root of unity. In that case $z^{pr} - z^r = -(z^r - 1)$, and $(z^{rp^N} - 1)/(z^{pr} - 1) = \sum_{0 \leqslant j < p^{N-1}} z^{prj} = p^{N-1}$, so that $E_N = p^N (S_N(z^r) - S_N(z)) - p^{N-1} \log_p(r)$.

If $z^p \neq 1$, then since $v_p(z-1) > 0$ the p-adic logarithm of z is defined by the usual power series in $z - 1$, and by the property of the p-adic exponential, for N sufficiently large (but in general not for all N) we have

$$
z^{rp^N} = \exp_p(rp^N \log_p(z)) = 1 + rp^N \log_p(z) + O(p^{2N-1}) ,
$$

and since we have shown above that $S_N(z)$ tends to a limit $S(z)$ we thus have

$$\lim_{N\to\infty}\frac{E_N}{p^N} = \begin{cases} S(z^r) - S(z) + r\log_p(z)\log_p(r)\dfrac{z^{pr}-z^r}{(z^r-1)(z^{pr}-1)} & \text{if } z^p\neq 1, \\[2ex] S(z^r) - S(z) - \dfrac{\log_p(r)}{p} & \text{if } z^p = 1. \end{cases}$$

On the other hand, again since $p\nmid r$ we can write

$$E_N = \sum_{0\leqslant a<p^N}^{(p)} z^{ra}\frac{p^N}{ra}\left[\frac{ra}{p^N}\right] + O(p^{2N-1}),$$

so by Theorem 11.5.27 we have

$$\lim_{N\to\infty}\frac{E_N}{p^N} = \log_p\left(\frac{z^r-1}{z-1}\right) - \frac{1}{p}\log_p\left(\frac{z^{pr}-1}{z^p-1}\right).$$

Comparing the two expressions that we have obtained proves (2).

(3). By definition we have $S(1) = \int_{\mathbb{Z}_p^*}\log_p(t)\,dt$, so that $S(1) = -\gamma_p$ by Proposition 11.3.14. This also follows from the fact that, also by definition, $S(1) = \psi_p(0) = -\gamma_p$.

Assume $v\geqslant 1$ and set $\zeta = \zeta_{p^v}$, so that $v_p(\zeta-1) > 0$ by Proposition 3.5.5. We have $\log_p(\zeta) = 0$, hence $L(\zeta,r) = \delta_{v,1}\log_p(r)/p$, where we set $\delta_{v,1} = 1$ for $v = 1$ and $\delta_{v,1} = 0$ for $v\geqslant 2$. We are going to sum the formula for $S(\zeta^r) - S(\zeta)$ over all values of r coprime to p such that $0\leqslant r < p^v$. For $p\nmid a$ we have

$$\sum_{0\leqslant r<p^v}^{(p)}\zeta^{ra} = \sum_{0\leqslant r<p^v}\zeta^{ra} - \sum_{0\leqslant s<p^{v-1}}\zeta^{psa} = -\delta_{v,1}.$$

Thus, by Proposition 11.3.14 we have

$$\sum_{0\leqslant r<p^v}^{(p)} S(\zeta^r) = -\delta_{v,1}\lim_{N\to\infty}\frac{1}{p^N}\sum_{0\leqslant a<p^N}^{(p)}\log_p(a) = -\delta_{v,1}\int_{\mathbb{Z}_p^*}\log_p(t)\,dt = \delta_{v,1}\gamma_p.$$

Equivalently, we can write

$$\sum_{0\leqslant r<p^v}^{(p)} S(\zeta^r) = -\delta_{v,1}\lim_{N\to\infty}\frac{1}{p^N}\sum_{0\leqslant a<p^N}^{(p)}\log_p(a)$$

$$= -\delta_{v,1}\lim_{N\to\infty}\frac{\mathrm{Log}\Gamma_p(p^N)}{p^N} = -\delta_{v,1}\psi_p(0) = \delta_{v,1}\gamma_p.$$

Note that by Proposition 3.5.4 we have

$$\sum_{0\leqslant r<p^v}^{(p)}\log_p(1-\zeta^r) = \log_p(\Phi_{p^v}(1)) = \log_p(p) = 0,$$

and similarly with ζ replaced by ζ^p when $v \geqslant 2$. We thus obtain

$$\delta_{v,1}\gamma_p - \phi(p^v)S(\zeta) = -\phi(p^v)\log_p(1-\zeta) + \phi(p^v)\frac{1-\delta_{v,1}}{p}\log_p(1-\zeta^p)\,,$$

since when $v = 1$ the term $(-1/p)\log_p((\zeta^{pr}-1)/(\zeta^p-1)) = (-1/p)\log_p(r)$ cancels with $L(\zeta,r)$. It follows that

$$S(\zeta) = \log_p(1-\zeta) - \frac{1-\delta_{v,1}}{p}\log_p(1-\zeta^p) + \frac{\delta_{v,1}}{\phi(p^v)}\gamma_p\,,$$

proving (3).   □

Remark. I do not know whether $S(z)$ can be evaluated in closed form for other values of z than those given in (3).

11.5.4 Computation of Limits of some Logarithmic Sums

The goal of this section is the proof of the technical Corollary 11.5.32, which will be needed to compute the values at rational numbers both for the function ψ_p and for the function $\psi_p(\chi)$, but apart from that it is essentially independent of p-adic gamma and L-functions.

Theorem 11.5.31. *Let* $m \geqslant 1$, *and denote as usual by* ζ_m *a primitive mth root of unity. For r and u in $\mathbb{Z}$ we set*

$$F_N(u) = \frac{1}{p^N m}\sum_{r\leqslant a<p^N m+r}^{(p)} \zeta_m^{ua}\log_p(a)\,.$$

Assume that $0 \leqslant r < m$. *Then* $F(u) = \lim_{N\to\infty} F_N(u)$ *exists and is given by the following formulas:*

$$F(u) = \begin{cases} L(0) - \gamma_p & \text{when } m \mid u\,, \\ L(u) + \log_p(1-\zeta_m^u) + \dfrac{\gamma_p}{p-1} & \text{when } m \mid up \text{ and } m \nmid u\,, \\ L(u) + \log_p(1-\zeta_m^u) - \dfrac{1}{p}\log_p(1-\zeta_m^{up}) & \text{when } m \nmid up\,, \end{cases}$$

where we have set

$$L(u) = \sum_{0\leqslant a<r}^{(p)} \frac{\zeta_m^{ua}}{a}\,.$$

Note that for simplicity of notation, we have not indicated the dependence in r and m.

Proof. Since $r \geqslant 0$, setting $a = p^N m + b$ and using the fact that $\zeta_m^m = 1$, we have

$$\sum_{p^N m \leqslant a < p^N m+r}^{(p)} \zeta_m^{ua} \log_p(a) = \sum_{0 \leqslant b < r}^{(p)} \zeta_m^{ub} (\log_p(b) + p^N m/b + O(p^{2N})) \, ,$$

in other words

$$\sum_{p^N m \leqslant a < p^N m+r}^{(p)} \zeta_m^{ua} \log_p(a) = \sum_{0 \leqslant b < r}^{(p)} \zeta_m^{ub} \log_p(b) + p^N m L(u) + O(p^{2N}) \, ,$$

using the notation $L(u)$ of the theorem. Thus

$$F_N(u) = L(u) + \frac{1}{p^N m} \sum_{0 \leqslant b < p^N m}^{(p)} \zeta_m^{ub} \log_p(b) \, .$$

As in the proof of Theorem 11.5.30, it is easily checked, either from the Volkenborn integral representation of γ_p (Proposition 11.3.14) or from the elementary properties of Morita's p-adic log gamma function, that

$$\lim_{N \to \infty} \frac{1}{p^N m} \sum_{0 \leqslant b < p^N m}^{(p)} \log_p(b) = -\gamma_p \, ,$$

so that $F(0) = L(0) - \gamma_p$. By the functional equation for ψ_p we know that $L(0) = \psi_p(r) - \psi_p(0)$, so that $F(0) = \psi_p(r)$, but it is better to leave the expression as it is.

We may thus assume that $m \nmid u$. In the sum occurring in the above expression for $F_N(u)$ we set $b = p^N c + d$, so that $0 \leqslant c < m$, $0 \leqslant d < p^N$, and $p \nmid d$. We consider two cases.

Case 1. $(m/p^{v_p(m)}) \nmid u$. We have

$$\sum_{0 \leqslant b < p^N m}^{(p)} \zeta_m^{ub} \log_p(b) = \sum_{0 \leqslant d < p^N}^{(p)} \zeta_m^{ud} \sum_{0 \leqslant c < m} \zeta_m^{up^N c} (\log_p(d) + p^N c/d + O(p^{2N}))$$

$$= \sum_{0 \leqslant d < p^N}^{(p)} \zeta_m^{ud} \log_p(d) \sum_{0 \leqslant c < m} \zeta_m^{up^N c}$$

$$+ p^N \sum_{0 \leqslant d < p^N}^{(p)} \frac{\zeta_m^{ud}}{d} \sum_{0 \leqslant c < m} c \zeta_m^{up^N c} + O(p^{2N}) \, .$$

Since $(m/p^{v_p(m)}) \nmid u$, for all N we have $m \nmid up^N$, in other words $\zeta_m^{up^N} \neq 1$, so that $\sum_{0 \leqslant c < m} \zeta_m^{up^N c} = 0$, and an immediate calculation gives

$$\sum_{0 \leqslant c < m} c \zeta_m^{up^N c} = \frac{m}{\zeta_m^{up^N} - 1} \, .$$

Thus

$$F_N(u) = L(u) + \frac{1}{\zeta_m^{up^N} - 1} \sum_{0 \leqslant d < p^N}^{(p)} \frac{\zeta_m^{ud}}{d} \, .$$

Since ζ_m^u is not a p^nth root of unity for any n it follows from Proposition 3.5.5 that $|1 - \zeta_m^u| = 1$, so by Theorem 11.5.25, $F(u)$ exists and we have

$$F(u) = L(u) + \log_p(1 - \zeta_m^u) - \frac{1}{p}\log_p(1 - \zeta_m^{up}) \,.$$

Case 2. $(m/p^{v_p(m)}) \mid u$. Note that this case can occur only when $p \mid m$, since otherwise $m \mid u$, which has been excluded. The condition implies that for N sufficiently large (more precisely for $N \geqslant v_p(m)$) we have $\zeta_m^{up^N} = 1$. For simplicity, write $v = v_p(m) - v_p(u)$ and

$$a = \frac{u}{m/p^v} = \frac{u}{p^{v_p(u)}(m/p^{v_p(m)})} \,,$$

so that $p \nmid a$. Note that $v > 0$, since otherwise $m \mid u$. For $N \geqslant v_p(m)$ we thus have

$$\sideset{}{^{(p)}}\sum_{0 \leqslant b < p^N m} \zeta_m^{ub} \log_p(b)$$

$$= m \sideset{}{^{(p)}}\sum_{0 \leqslant d < p^N} \zeta_m^{ud} \log_p(d) + p^N \frac{m(m-1)}{2} \sideset{}{^{(p)}}\sum_{0 \leqslant d < p^N} \frac{\zeta_m^{ud}}{d} + O(p^{2N})$$

$$= m \sideset{}{^{(p)}}\sum_{0 \leqslant d < p^N} \zeta_{p^v}^{ad} \log_p(d) + p^N \frac{m(m-1)}{2} \sideset{}{^{(p)}}\sum_{0 \leqslant d < p^N} \frac{\zeta_{p^v}^{ad}}{d} + O(p^{2N}) \,.$$

Now by Lemma 11.5.29 we have

$$\sideset{}{^{(p)}}\sum_{0 \leqslant d < p^N} \frac{\zeta_{p^v}^{ad}}{d} = O(p^{N-B})$$

for some constant B (which can in fact be taken equal to 0 here, but we do not need this), and by Theorem 11.5.30, since $\zeta_{p^v}^a = \zeta_m^u$ we have

$$\lim_{N \to \infty} \frac{1}{p^N} \sideset{}{^{(p)}}\sum_{0 \leqslant d < p^N} \zeta_{p^v}^{ad} \log_p(d) = \log_p(1 - \zeta_m^u) - \frac{1}{p}\log_p(1 - \zeta_m^{up})$$

for $v \geqslant 2$, and

$$\lim_{N \to \infty} \frac{1}{p^N} \sideset{}{^{(p)}}\sum_{0 \leqslant d < p^N} \zeta_{p^v}^{ad} \log_p(d) = \log_p(1 - \zeta_m^u) + \frac{\gamma_p}{p - 1}$$

for $v = 1$. It follows that $F(u)$ exists, and since $v = 1$ is equivalent to $m \mid up$ we have

$$F(u) = L(u) + \begin{cases} \log_p(1 - \zeta_m^u) - \dfrac{1}{p}\log_p(1 - \zeta_m^{up}) & \text{if } m \nmid up \,, \\[2mm] \log_p(1 - \zeta_m^u) + \dfrac{\gamma_p}{p - 1} & \text{if } m \mid up \,, \end{cases}$$

proving the theorem. $\qquad\qquad\qquad\qquad\qquad\qquad\qquad\qquad\qquad\qquad\qquad\quad \square$

Corollary 11.5.32. *Keep the same notation and assumptions, in particular that $0 \leqslant r < m$, and assume in addition that $p \nmid \gcd(r,m)$. Then*

$$\sum_{u \bmod m} \zeta_m^{-ur} F(u) = -\left(1 + \frac{\delta_{m,p}}{p-1}\right) \gamma_p$$

$$+ \sum_{1 \leqslant u \leqslant m-1} \zeta_m^{-ur} \log_p(1 - \zeta_m^u) - \frac{1}{p} \sum_{\substack{1 \leqslant u \leqslant m-1 \\ m \nmid up}} \zeta_m^{-ur} \log_p(1 - \zeta_m^{up}),$$

where $\delta_{m,p} = 1$ when $p \mid m$, and $\delta_{m,p} = 0$ when $p \nmid m$.

Note that when $p \nmid m$ the condition $m \nmid up$ is unnecessary.

Proof. First we have

$$\sum_{u \bmod m} \zeta_m^{-ur} L(u) = \sum_{0 \leqslant a < r}^{(p)} \frac{1}{a} \sum_{u \bmod m} \zeta_m^{-u(r-a)}.$$

Since $1 \leqslant r - a \leqslant r < m$ we have $m \nmid (r-a)$, so that the inner sum is equal to 0 and the whole sum vanishes.

If $p \nmid m$ the case $m \mid up$ and $m \nmid u$ cannot occur, so we directly obtain the result. Thus assume that $p \mid m$. The condition $m \mid up$ is equivalent to $u = k(m/p)$, where k is defined modulo p, so by Theorem 11.5.31, since the terms involving $L(u)$ cancel we have

$$\sum_{u \bmod m} \zeta_m^{-ur} F(u) = \gamma_p \left(-1 + \frac{1}{p-1} \sum_{\substack{k \bmod p \\ p \nmid k}} \zeta_p^{-kr}\right)$$

$$+ \sum_{1 \leqslant u \leqslant m-1} \zeta_m^{-ur} \log_p(1 - \zeta_m^u) - \frac{1}{p} \sum_{\substack{1 \leqslant u \leqslant m-1 \\ m \nmid up}} \zeta_m^{-ur} \log_p(1 - \zeta_m^{up}),$$

giving the desired result since $\sum_{k \bmod p,\, p \nmid k} \zeta_p^{-kr} = -1$, because we have assumed that $p \nmid \gcd(r,m)$, so that $p \nmid r$ in the present case. $\square$

11.5.5 Explicit Formulas for $\psi_p(r/m)$ and $\psi_p(\chi, r/m)$

Theorem 11.5.33. *Let $m \in \mathbb{Z}_{\geqslant 1}$ be such that $q_p \mid m$, denote as usual by ζ_m a primitive mth root of unity in $\overline{\mathbb{Q}} \subset \mathbb{C}_p$, and let $r \in \mathbb{Z}$ be such that $p \nmid r$ and $0 \leqslant r < m$. We have the explicit formula*

$$\psi_p\left(\frac{r}{m}\right) = -\log_p(m) - \frac{p}{p-1}\gamma_p$$

$$+ \sum_{1 \leqslant a \leqslant m-1} \zeta_m^{-ar} \log_p(1 - \zeta_m^a) - \frac{1}{p} \sum_{\substack{1 \leqslant a \leqslant m-1 \\ m \nmid ap}} \zeta_m^{-ar} \log_p(1 - \zeta_m^{ap}).$$

Proof. By Proposition 11.5.7 we have

$$\psi_p(r/m) = \lim_{N\to\infty} \frac{1}{p^N} \sum_{0\leqslant a < p^N} \log_p(a + r/m)$$

$$= -\log_p(m) + \lim_{N\to\infty} \frac{1}{p^N} \sum_{\substack{r\leqslant b < p^N m + r \\ b \equiv r \;(\mathrm{mod}\; m)}} \log_p(b) \,.$$

Since $p \mid m$ and $p \nmid r$ we have $p \nmid b$, so in this last sum we may replace the symbol $\sum$ by $\sum^{(p)}$. In addition, since $\sum_{u \bmod m} \zeta_m^{u(b-r)}$ is equal to 0 if $b \not\equiv r$ (mod m) and to m otherwise, we have

$$\psi_p(r/m) = -\log_p(m) + \lim_{N\to\infty} \frac{1}{p^N m} \sum_{r\leqslant b < p^N m + r}^{(p)} \sum_{u \bmod m} \zeta_m^{u(b-r)} \log_p(b)$$

$$= -\log_p(m) + \sum_{u \bmod m} \zeta_m^{-ur} F(u) \,,$$

where $F(u)$ is as in Theorem 11.5.31. The result now follows from Corollary 11.5.32.   □

Note that by using the functional equation $\psi_p(x + 1) = \psi_p(x) + 1/x$ (Corollary 11.5.3), the theorem also gives an explicit formula for $\psi_p(r/m)$ for any $r \in \mathbb{Z}$.

We now prove the analogous result for Morita's ψ function $\psi_p(\chi, x)$. Recall that we write again $\psi_p(x)$ instead of $\psi(\chi_0, x)$, but now with $x \in \mathbb{Z}_p$. It is now preferable to separate the cases $\chi = \chi_0$ and $\chi \neq \chi_0$. The result for $\chi = \chi_0$ is the following.

Theorem 11.5.34. *Let $m \in \mathbb{Z}_{\geqslant 1}$ be such that $p \nmid m$, denote as usual by ζ_m a primitive mth root of unity in $\overline{\mathbb{Q}} \subset \mathbb{C}_p$, and let $r \in \mathbb{Z}$ be such that $0 \leqslant r < m$. We have the explicit formula*

$$\psi_p\left(\frac{r}{m}\right) = -\left(1 - \frac{1}{p}\right)\log_p(m) - \gamma_p$$

$$+ \sum_{1\leqslant a \leqslant m-1} \zeta_m^{-ar}\left(\log_p(1 - \zeta_m^a) - \frac{1}{p}\log_p(1 - \zeta_m^{ap})\right) \,.$$

Proof. By Proposition 11.5.18 we have

$$\psi_p(r/m) = \lim_{N\to\infty} \frac{1}{p^N} \sum_{\substack{0\leqslant a < p^N \\ v_p(a+r/m)=0}} \log_p(a + r/m) \,.$$

Since $p \nmid m$, the condition $v_p(a + r/m) = 0$ means that $p \nmid am + r$, and also that $a \not\equiv -rm^{-1}$ (mod p). The number of such a satisfying $0 \leqslant a < p^N$ is equal to $p^N(1 - 1/p)$, so

$$\psi_p(r/m) = -(1 - 1/p)\log_p(m) + \lim_{N\to\infty} \frac{1}{p^N} \sum_{\substack{r\leqslant b < p^N m+r \\ p\nmid b,\ b\equiv r \pmod m}} \log_p(b) .$$

Since $\sum_{u\bmod m} \zeta_m^{u(b-r)}$ is equal to 0 if $b \not\equiv r \pmod m$ and to m otherwise, we have

$$\psi_p(r/m) = -(1 - 1/p)\log_p(m) + \lim_{N\to\infty} \frac{1}{p^N m} \sum_{r\leqslant b < p^N m+r}^{(p)} \sum_{u\bmod m} \zeta_m^{u(b-r)} \log_p(b)$$

$$= -(1 - 1/p)\log_p(m) + \sum_{u\bmod m} \zeta_m^{-ur} F(u) ,$$

where $F(u)$ is as in Theorem 11.5.31; in other words,

$$F(u) = \lim_{N\to\infty} \frac{1}{p^N m} \sum_{r\leqslant b < p^N m+r}^{(p)} \zeta_m^{ub} \log_p(b) .$$

Note that the use of the notation $\sum^{(p)}$ is justified since the remaining condition is $p \nmid b$. The result now follows from Corollary 11.5.32.   □

The result for $\chi \neq \chi_0$ is the following.

Theorem 11.5.35. *Let χ be a primitive character modulo p^v for some $v \geqslant 1$, let $m \in \mathbb{Z}_{\geqslant 1}$ be such that $p \nmid m$, and let $r \in \mathbb{Z}$ be such that $0 \leqslant r < m$. We have the explicit formula*

$$\psi_p\left(\chi, \frac{r}{m}\right) = \frac{\chi^{-1}(-m)\tau(\chi)}{p^v} \sum_{\substack{x\bmod p^v \\ u\bmod m}} \chi^{-1}(x)\zeta_m^{-ur} \log_p\left(1 - \zeta_{p^v}^x \zeta_m^u\right) .$$

Proof. Since the proof is very similar, but using simple properties of Gauss sums, we leave it as an excellent but long exercise for the reader (Exercise 28).   □

Remarks. (1) As for the function $\psi_p(x)$, using the functional equation for the function $\psi_p(\chi, x)$ (Corollary 11.5.14 (1)), these theorems also give explicit formulas for $\psi_p(\chi, r/m)$ for any $r \in \mathbb{Z}$.

(2) These theorems can also be proved directly from the corresponding theorem for ψ_p (Theorem 11.5.33) together with Proposition 11.5.12, but since all these theorems rely on Corollary 11.5.32 there is not much point in doing so, except to check the correctness of the formulas.

11.5.6 Application to the Value of $L_p(\chi, 1)$

We have seen in Proposition 10.3.5 that when χ is an even character there exists an explicit expression for the complex value $L(\chi, 1)$ in terms of the values of χ and logarithms. By analogy, it seems reasonable to expect that there is a similar formula for $L_p(\chi, 1)$ in the p-adic case. This is indeed the case, and it is an easy consequence of the theorems proved in the preceding section. First note the following.

Proposition 11.5.36. *Let χ be a nontrivial primitive character of conductor f. We have*

$$L_p(\chi, 1) = -\frac{1}{f} \sum_{0 \leqslant r < f} \chi(r) \psi_p\left(\frac{r}{f}\right) .$$

Proof. Clear from the definition and Propositions 11.5.6 and 11.5.15. Note that, as in the definition of $L_p(\chi, s)$, we use here the function $\psi_p(x)$ for $x \in C\mathbb{Z}_p$ when $q_p \mid f$, and for $x \in \mathbb{Z}_p$ when $p \nmid f$.   □

Theorem 11.5.37. *Let χ be a nontrivial even primitive character of conductor f, let $\zeta = \zeta_f$ be a primitive fth root of unity, and as usual let $\tau(\chi) = \sum_{1 \leqslant a < f} \chi(a) \zeta^a \in \mathbb{C}_p$ be the Gauss sum. Then*

$$L_p(\chi, 1) = -\left(1 - \frac{\chi(p)}{p}\right) \frac{\tau(\chi)}{f} \sum_{1 \leqslant r < f} \chi^{-1}(r) \log_p(1 - \zeta^r) .$$

Note that as usual χ^{-1} is taken in the sense of the group of characters, so that $\chi(r) = \chi^{-1}(r) = 0$ if $\gcd(r, f) > 1$.

Note also that if $\chi \neq \chi_0$ and if χ is defined modulo a power of p then $\psi_p(\chi, 0) = -\gamma_p(\chi) = -L_p(\chi, 1)$, and since χ is even, Theorem 11.5.35 applied to $r = 0$ and $m = 1$ exactly gives the desired formula. The proof that we now give is essentially the same, generalized to characters of arbitrary conductor.

Proof. We separate the cases $q_p \mid f$ and $p \nmid f$. Assume first that $q_p \mid f$. By the above proposition and Theorem 11.5.33, and since $\sum_{0 \leqslant r < f} \chi(r) = 0$, we have $L_p(\chi, 1) = -(S_1 - S_p/p)/f$, where

$$S_1 = \sum_{0 \leqslant r < f} \chi(r) \sum_{1 \leqslant a < f} \zeta_f^{-ar} \log_p(1 - \zeta_f^a) \quad \text{and}$$

$$S_p = \sum_{0 \leqslant r < f} \chi(r) \sum_{\substack{1 \leqslant a < f \\ (f/p) \nmid a}} \zeta_f^{-ar} \log_p(1 - \zeta_f^{ap}) .$$

By Corollary 2.1.42, since χ is an even primitive character we have

$$S_1 = \sum_{1 \leqslant a < f} \log_p(1 - \zeta_f^a) \sum_{0 \leqslant r < f} \chi(r) \zeta_f^{-ar} = \tau(\chi) \sum_{1 \leqslant a < f} \chi^{-1}(a) \log_p(1 - \zeta_f^a) .$$

Similarly, we have

$$S_p = \tau(\chi) \sum_{\substack{1 \leqslant a < f \\ (f/p) \nmid a}} \chi^{-1}(a) \log_p(1 - \zeta_f^{ap}) .$$

If $f = p$ we cannot have $(f/p) \nmid a$, so $S_p = 0$. On the other hand, if $f/p > 1$ then $\gcd(f, f/p) > 1$, so that $\chi^{-1}(f/p) = 0$. It follows that

$$S_p = \tau(\chi) \sum_{1 \leqslant a < f} \chi^{-1}(a) \log_p(1 - \zeta_f^{ap}) .$$

Since the function $\log_p(1 - \zeta_f^{ap})$ is periodic of period f/p and since χ^{-1} is primitive, it follows from Corollary 2.1.33 that $S_p = 0$ in all cases, proving the theorem when $q_p \mid f$, since $\chi(p) = 0$ in that case.

Assume now that $p \nmid f$. By the above proposition, Theorem 11.5.34, and since $\sum_{0 \leqslant r < f} \chi(r) = 0$, we again have $L_p(\chi, 1) = -(S_1 - S_p/p)/f$, where S_1 is as before, and

$$S_p = \sum_{0 \leqslant r < f} \chi(r) \sum_{1 \leqslant a < f} \zeta_f^{-ar} \log_p(1 - \zeta_f^{ap}) .$$

As above, we have

$$S_p = \tau(\chi) \sum_{1 \leqslant a < f} \chi^{-1}(a) \log_p(1 - \zeta_f^{ap}) ,$$

so that by setting $b = ap$ and noting that multiplication by p is a bijection of $(\mathbb{Z}/f\mathbb{Z})^*$ onto itself, we have

$$S_p = \tau(\chi) \sum_{1 \leqslant b < f} \chi^{-1}(bp^{-1}) \log_p(1 - \zeta_f^b) = \chi(p)\tau(\chi) \sum_{1 \leqslant a < f} \chi^{-1}(a) \log_p(1 - \zeta_f^a) ,$$

proving the result in this case. $\qquad\square$

As already mentioned, although the formula of the above theorem expresses $L_p(\chi, 1)$ as a finite linear combination of values of $\log_p(1 - \zeta^r)$, it is not clear that this last formula is better in practice for computing $L_p(\chi, 1)$ than the convergent series given by Proposition 11.3.8 (3).

In Section 10.5.5 we have also seen the fundamental result that $L(\chi, 1) \neq 0$. This is also true in the p-adic context; in other words, we have $L_p(\chi, 1) \neq 0$ when χ is a nontrivial even character. However, the proof involves much deeper arguments: as in the complex case we consider the product of the $L_p(\chi, 1)$ over all nontrivial even characters χ of given conductor, which is a sort of p-adic Dedekind zeta function. However, the crucial part of the proof appeals to the deep theorem à la Baker on linear forms in p-adic logarithms; see [Was], Section 5.5, for details.

Corollary 11.5.38. *Let $D > 1$ be a fundamental discriminant, let ε_D be a fundamental unit of $\mathbb{Q}(\sqrt{D})$, and denote by χ_D the Legendre–Kronecker symbol $\chi_D(n) = \left(\frac{D}{n}\right)$. We have*

$$L_p(\chi_D, 1) = 2 \left(1 - \frac{\chi_D(p)}{p}\right) \frac{\log_p(\varepsilon_D)}{\sqrt{D}}$$

for a suitable choice of the p-adic square root $\sqrt{D}$ corresponding to the choice of the embedding of ε_D in $\mathbb{Q}_p(\sqrt{D})$.

Note that by Exercise 34 of Chapter 4, we know that $\log_p(\varepsilon_D)/\sqrt{D} \in \mathbb{Z}_p$.

Proof. By the above theorem and the basic property of the p-adic logarithm we have

$$L_p(\chi_D, 1) = -\left(1 - \frac{\chi_D(p)}{p}\right) \frac{\tau(\chi_D)}{D} \sum_{1 \leqslant r < D} \left(\frac{D}{r}\right) \log_p(1 - \zeta_D^r)$$

$$= -\left(1 - \frac{\chi_D(p)}{p}\right) \frac{\tau(\chi_D)}{D} \log_p(E_D) \,,$$

where

$$E_D = \prod_{1 \leqslant r < D} (1 - \zeta_D^r)^{\chi_D(r)} \,.$$

Thus $E_D \in \overline{\mathbb{Q}}$ and has nothing more to do with p-adic numbers. By Corollary 10.3.6 we have $E_D = \varepsilon_D^{-2}$ for *some* fundamental unit ε_D of $K = \mathbb{Q}(\sqrt{D})$, depending on the choice of primitive Dth root of unity ζ_D. Changing ε_D into $-\varepsilon_D$ does not change $\log_p(\varepsilon_D)$, but changing ε_D into ε_D^{-1} changes $\log_p(\varepsilon_D)$ into its opposite. Since χ_D is an even primitive character such that $\chi_D^{-1} = \chi_D$, by Corollary 2.1.47 we have $\tau(\chi_D)^2 = D$, so the corollary follows.   □

11.6 Morita's p-adic Gamma Function

11.6.1 Introduction

As we have seen in Section 11.5 we can naturally define two log gamma functions, one defined for $x \in C\mathbb{Z}_p$, and one defined for $x \in \mathbb{Z}_p$. In the latter case, we can even introduce a character χ, and we have seen that when for instance $\chi \neq \omega$ and $p \geqslant 3$, we can in fact take the p-adic exponential of this log gamma function. However, although this exponential is well defined, it is not canonical since we could multiply it by any root of unity without changing its logarithm. It is thus preferable to give a new definition from scratch. Three comments about this:

(1) When taking the logarithm of the formulas involving products, we will evidently recover the formulas that we have already given for the function $\mathrm{Log}\Gamma_p(x)$. What will be new in the formulas is the precise root of unity that occurs, which is not always easy to compute.
(2) It is not completely clear how to define the gamma function for a general character χ, but only, for instance, for a real character. Thus we will in fact restrict to the case that χ is the trivial character χ_0 modulo q_p. This

means in practice that χ will not appear in the formulas, but instead we will have to exclude arguments divisible by p, which we denote as usual by $\sum^{(p)}$ (see Definition 11.3.6).

(3) We will prove a very important formula for products of p-adic gamma functions, the Gross–Koblitz formula, from which therefore we can also deduce a formula for the function $\mathrm{Log}\Gamma_p(x)$, but it seems that it is impossible to prove it directly at the level of $\mathrm{Log}\Gamma_p(x)$.

11.6.2 Definitions and Basic Results

As we have seen in Section 9.6, there are many possible definitions of $\Gamma(s)$. Among those, we are going to adapt to the p-adic case the formula

$$\Gamma(s) = \lim_{n \to \infty} \frac{n^{s-1} n!}{s(s+1)\cdots(s+n-1)} ,$$

which will lead to Morita's gamma function.

Recall that in Definition 11.3.6 we have introduced the notation $\sum^{(p)}$ and $\prod^{(p)}$ to indicate that we exclude indexes that are divisible by p. In addition, we set the following:

Definition 11.6.1. (1) *For $s \in \mathbb{Z}_p$ we write $s = \sum_{j \geqslant 0} a_j(s) p^j$, where the $a_j(s)$ are uniquely defined by the inequality $0 \leqslant a_j(s) < p$.*

(2) *For $s \in \mathbb{Z}_p$ we write $s \backslash p$ instead of $(s - a_0(s))/p = \sum_{j \geqslant 1} a_j(s) p^{j-1}$, so that $s \backslash p \in \mathbb{Z}_p$ and*

$$s - s\backslash p = a_0(s) + (p-1) \sum_{j \geqslant 1} a_j(s) p^{j-1} = \sum_{j \geqslant 0} (a_j(s) - a_{j+1}(s)) p^j .$$

To show the usefulness of these definitions, we begin with the following easy result.

Proposition 11.6.2. *Let u be a p-adic unit.*

(1) *For all $s \in \mathbb{Z}_p$ the quantity $u^{[s-s\backslash p]}$ defined by*

$$u^{[s-s\backslash p]} = \lim_{m \to s} \prod_{1 \leqslant k \leqslant m}^{(p)} u = \lim_{m \to s} u^{m - m\backslash p}$$

is well defined, where m tends p-adically to s in $\mathbb{Z}_{\geqslant 0}$.

(2) *More precisely, we have*

$$u^{[s-s\backslash p]} = \begin{cases} u^{a_0(s)} \exp_p((s\backslash p) \log_p(u^{p-1})) & \text{when } p > 2 , \\ u^{a_0(s)+a_1(s)} \exp_p(((s\backslash p)\backslash p) \log_p(u^2)) & \text{when } p = 2 , \end{cases}$$

so that

$$u^{[s-s\backslash p]} \equiv \begin{cases} u^{a_0(s)} \pmod{p\mathbb{Z}_p} & \text{when } p > 2 , \\ u^{a_0(s)+a_1(s)} \pmod{8\mathbb{Z}_2} & \text{when } p = 2 . \end{cases}$$

(3) *The function* $u^{[s-s\backslash p]}$ *is differentiable on* $\mathbb{Z}_p$ *and we have*

$$\frac{d}{ds}\left(u^{[s-s\backslash p]}\right) = \left(1 - \frac{1}{p}\right)\log_p(u)u^{[s-s\backslash p]} .$$

Proof. The equality of the last two quantities is clear. Let m_i be a sequence of elements of $\mathbb{Z}_{\geqslant 0}$ tending to s. For i sufficiently large we will have $m_i \equiv a_0(s)$ (mod p), so that $m_i\backslash p = (m_i - a_0(s))/p$ and

$$m_i - m_i\backslash p = a_0(s) + \frac{m_i - a_0(s)}{p}(p-1) .$$

Define $n_i = (m_i - a_0(s))/p \in \mathbb{Z}$. Since u is a p-adic unit, $u^{p-1} \equiv 1$ (mod $p\mathbb{Z}_p$), so by Corollary 4.2.18, for $p > 2$ we have

$$u^{(p-1)n_i} = \exp_p(n_i \log_p(u^{p-1})) .$$

Since $\exp_p$ and $\log_p$ are p-adically continuous inside their domains of convergence and since n_i tends to $(s - a_0(s))/p \in \mathbb{Z}_p$, it follows that $u^{(p-1)n_i}$ converges to $\exp_p(((s - a_0(s))/p)\log_p(u^{p-1}))$, so that finally $u^{m_i - m_i\backslash p}$ converges to

$$u^{a_0(s)}\exp_p(((s - a_0(s))/p)\log_p(u^{p-1})) .$$

Once again, since $p > 2$ we have $v_p(\log_p(u^{p-1})) > 0$ and $\exp_p(((s - a_0(s))/p)\log_p(u^{p-1})) \equiv 1$ (mod $p\mathbb{Z}_p$), proving the proposition for $p > 2$.

For $p = 2$, for i sufficiently large we will have $m_i \equiv a_0(s)+2a_1(s)$ (mod 4). Set $n_i = (m_i - a_0(s) - 2a_1(s))/4 \in \mathbb{Z}$, so that $m_i = a_0(s) + 2a_1(s) + 4n_i$ and $m_i\backslash p = a_1(s) + 2n_i$. Thus

$$m_i - m_i\backslash p = a_0(s) + a_1(s) + 2n_i ,$$

and since u is a 2-adic unit, $u^2 \equiv 1$ (mod $8\mathbb{Z}_2$), so that by Corollary 4.2.18 we have

$$u^{2n_i} = \exp_2(n_i \log_2(u^2)) .$$

The rest of the reasoning is exactly as in the case $p > 2$.

To compute the derivative of $u^{[s-s\backslash p]}$ we note that $a_0(t)$ and $a_1(t)$ are ultimately constant as $t \to s$, and equal to $a_0(s)$ and $a_1(s)$ respectively. Thus $u^{a_0(t)}$ is ultimately constant, and also $t\backslash p = (t - a_0(t))/p = (t - a_0(s))/p$. The formula immediately follows.   □

Warning. The notation $u^{[s-s\backslash p]}$ should not be taken too literally, and in fact this is why we write it in this way and not simply as $u^{s-s\backslash p}$, since it is *not* true in general that $u^{[s-s\backslash p]} = u^s u^{-s\backslash p}$, except when $s \in \mathbb{Z}$. For instance, if $s = -a/(p-1)$ with $0 \leqslant a \leqslant p-1$, then $s = a\sum_{j \geqslant 0} p^j$ and $s\backslash p = a\sum_{j \geqslant 0} p^j = s$, so that literally we should have $u^{s-s\backslash p} = u^0 = 1$, while in fact as we shall see in the following corollary, $u^{[s-s\backslash p]} = \omega(u)^a$, where $\omega(u)$ is the Teichmüller character of u. As a special case, note for instance that by Proposition 11.6.2, in $\mathbb{Q}_5$ we have

$$3^{[1/2-(1/2)\backslash 5]} = 3^3 \exp_5(-(1/2)\log_5(3^4)) = 3^3(-3^2) = -3 \ ,$$

although $1/2 - (1/2)\backslash 5 = 1$. The discrepancy comes from the fact that $\exp_5(-(1/2)\log_5(3^4)) = (3^4)^{-1/2}$ is congruent to 1 modulo 5, hence is equal to -9 and not to 9.

As an example, note the following corollary.

Corollary 11.6.3. *Assume that $p > 2$, let $s = a/(p-1)$ with $a \in \mathbb{Z}$, and let u be a p-adic unit. Then*

$$u^{[s-1-(s-1)\backslash p]} = u^{a\backslash p}\omega(u)^{-a} \quad and \quad u^{[(-s)-(-s)\backslash p]} = u^{-a\backslash p}\omega(u)^a \ ,$$

where $\omega(u)$ is the Teichmüller character of u.

Proof. Since $p > 2$, the proposition shows that $u^{[s-1-(s-1)\backslash p]} = u^{a_0(s-1)}E$ with

$$E = \exp_p(((s-1-a_0(s-1))/p)\log_p(u^{p-1})) \ .$$

If $s = a/(p-1)$ then it is easily checked that $a_0(s-1) = p-1-a-p((p-1-a)\backslash p)$, so that $(s-1-a_0(s-1))/p = (a+1-p)/(p-1) + (p-1-a)\backslash p$. Thus

$$E = u^{(p-1)((p-1-a)\backslash p)}\exp_p(-(p-1-a)\log_p(u^{p-1})/(p-1))$$
$$= u^{(p-1)((p-1-a)\backslash p)}(\omega(u)/u)^{p-1-a}$$

by Proposition 4.3.4, and since $\omega(u)^{p-1} = 1$ we obtain

$$u^{[s-1-(s-1)\backslash p]} = u^{-(p-1-a)\backslash p}\omega(u)^{-a} \ .$$

Since it is easily checked that $-(p-1-a)\backslash p = a\backslash p$, the first formula of the corollary follows, as does the second since it is immediately checked that $((s-1)-(s-1)\backslash p) + ((-s)-(-s)\backslash p) = 0$; see Exercise 37. $\qquad\square$

We are now ready to define Morita's p-adic gamma function. The following proposition gives the congruences necessary to use p-adic interpolation.

Proposition 11.6.4. *For any a and N in $\mathbb{Z}_{\geqslant 1}$ and $m \in \mathbb{Z}$ we have*

$$\prod_{m \leqslant k < m+p^N a}^{(p)} k \equiv (-1)^{p^N a} \pmod{p^N} \ ,$$

except for $(p, N) = (2, 2)$, in which case the left-hand side is congruent to $(-1)^a$ modulo p^N.

Proof. Note that the special case $a = 1$, $N = 1$, and $m = 0$ is Wilson's theorem. We prove this proposition in the same way. Assume first that $a = 1$. Let $G = (\mathbb{Z}/(p^N\mathbb{Z}))^*$. The integers k such that $m \leqslant k < m + p^N$ and $p \nmid k$ form a complete set of representatives of G in $\mathbb{Z}$. Thus the left-hand side is congruent modulo p^N to $\prod_{g \in G} g$. When $g^2 \neq 1$, we pair g with g^{-1} in the product. Thus

$$\prod_{g \in G} g = \prod_{\substack{g \in G \\ g^2 = 1}} g .$$

However, if $g = \overline{k}$ modulo p^N, then $g^2 = 1$ in G means that $k^2 \equiv 1 \pmod{p^N}$. If $p \neq 2$, p cannot divide both $k-1$ and $k+1$; hence we have $k \equiv \pm 1 \pmod{p^N}$, so that

$$\prod_{g \in G} g \equiv -1 \equiv (-1)^{p^N} \pmod{p^N} .$$

Assume now that $p = 2$, and consider three cases. When $N = 1$ the result is trivial since the proposition says that both sides are odd. When $N = 2$ the congruence $k^2 \equiv 1 \pmod 4$ means that $k \equiv \pm 1 \pmod 4$, so as in the case $p > 2$ we have $\prod_{g \in G} g \equiv -1 \pmod 4$. Finally, when $N \geqslant 3$ then 4 cannot divide both $k - 1$ and $k + 1$, so the congruence $k^2 \equiv 1 \pmod{2^N}$ is equivalent to $k \equiv \pm 1 \pmod{2^{N-1}}$; in other words, $k \equiv 1, 1 + 2^{N-1}, -1,$ or $-1 + 2^{N-1}$ modulo 2^N, which are distinct since $N \geqslant 3$. Thus in that case

$$\prod_{g \in G} g \equiv 1 - 2^{2N-2} \equiv 1 \equiv (-1)^{p^N} \pmod{p^N} ,$$

proving the result in the case $a = 1$. The case of general a is immediate by induction on a. □

Definition 11.6.5. *Let $s \in \mathbb{Z}_p$. We define*

$$\Gamma_p(s) = \lim_{m \to s} (-1)^m \prod_{0 \leqslant k < m}^{(p)} k = \lim_{m \to s - 1} (-1)^{m+1} \frac{m!}{p^{m \backslash p}(m \backslash p)!} ,$$

where as above, the limits are for m tending to s and $s - 1$ respectively, p-adically in $\mathbb{Z}_{\geqslant 0}$.

We are going to see that the limit does exist, and that in a suitable sense this definition generalizes the usual gamma function.

Proposition 11.6.6. *The above definition makes sense (in other words the limit always exists) for all $s \in \mathbb{Z}_p$, and $\Gamma_p(s)$ is a p-adic unit. Furthermore, for all s and t in $\mathbb{Z}_p$ we have $\Gamma_p(s) \equiv \Gamma_p(t) \pmod{p^{v_p(s-t)}\mathbb{Z}_p}$, except when $p = 2$ and $v_p(s - t) = 2$, in which case $\Gamma_p(s) \equiv -\Gamma_p(t) \pmod{p^{v_p(s-t)}\mathbb{Z}_p}$.*

Proof. Set $u_m = (-1)^m \prod_{0 \leqslant k < m}^{(p)} k$, and let m_i be any sequence of positive integers tending to s as $i \to \infty$. This means that m_i is a Cauchy sequence, in other words because of the ultrametric inequality, that $m_{i+1} - m_i$ tends to 0 p-adically as $i \to \infty$. We must show that u_{m_i} is also a Cauchy sequence. Since u_m is a p-adic unit, this is clearly equivalent again by the ultrametric inequality to the fact that $u_{m_{i+1}}/u_{m_i}$ tends to 1 p-adically. Let $N \geqslant 3$ (to avoid the special case $(p, N) = (2, 2)$). Since $m_{i+1} - m_i$ tends to 0, there exists i_0 such that for $i \geqslant i_0$ we have $v_p(m_{i+1} - m_i) \geqslant N$, in other words $m_{i+1} - m_i = p^N a$ for some $a \in \mathbb{Z}$. Assume for instance $a \geqslant 0$ (otherwise we compute $u_{m_i}/u_{m_{i+1}}$ instead). Then

$$\frac{u_{m_{i+1}}}{u_{m_i}} = (-1)^{p^N a} \prod_{m_i \leqslant k < m_i + p^N a}^{(p)} k \equiv 1 \pmod{p^N}$$

by the above proposition, so that u_{m_i} is indeed a Cauchy sequence. Since by definition u_m is a p-adic unit, it follows that so is $\Gamma_p(s)$. For the last statement, let m_i and n_i be sequences of nonnegative integers tending to s and t respectively. In particular, $v_p(m_i - n_i) = v_p(s - t)$ for i sufficiently large. Thus for such i, by the above proposition if $(p, v_p(s - t)) \neq (2, 2)$ we have $u_{m_i} \equiv u_{n_i} \pmod{p^{v_p(s-t)}}$, so by definition of Γ_p we have $\Gamma_p(s) \equiv \Gamma_p(t) \pmod{p^{v_p(s-t)}}$, while if $(p, v_p(s-t)) = (2, 2)$ the above proposition says that we must include a minus sign. $\square$

11.6.3 Main Properties of the p-adic Gamma Function

Lemma 11.6.7. *Let n_i and m_i be two sequences of elements of $\mathbb{Z}$ such that $n_i \leqslant m_i$, converging p-adically respectively to s and t. Then*

$$\lim_{i \to \infty} (-1)^{m_i - n_i} \prod_{n_i \leqslant k < m_i}^{(p)} k = \frac{\Gamma_p(t)}{\Gamma_p(s)} \ .$$

Proof. If we had $n_i \geqslant 0$ for all i, the result would be immediate from the definition. Thus, we must handle this difficulty. For this, choose a sequence of exponents N_i such that $N_i \to \infty$ and $n_i + p^{N_i} \geqslant 0$ for all i, and for simplicity of notation set $P_i = p^{N_i}$. We may of course assume that $N_i \geqslant 3$ when $p = 2$. By Proposition 11.6.4 we have

$$\prod_{n_i \leqslant k < n_i + P_i}^{(p)} k \equiv (-1)^{P_i} \pmod{P_i} \ ,$$

and similarly with n_i replaced by m_i since $n_i \leqslant m_i$. It follows that

$$\frac{\prod_{n_i + P_i \leqslant k < m_i + P_i}^{(p)} k}{\prod_{n_i \leqslant k < m_i}^{(p)} k} = \frac{\prod_{n_i \leqslant k < m_i + P_i}^{(p)} k}{\prod_{n_i \leqslant k < n_i + P_i}^{(p)} k \prod_{n_i \leqslant k < m_i}^{(p)} k}$$

$$= \frac{\prod_{m_i \leqslant k < m_i + P_i}^{(p)} k}{\prod_{n_i \leqslant k < n_i + P_i}^{(p)} k} \equiv 1 \pmod{P_i} \ ,$$

so that

$$\prod_{n_i \leqslant k < m_i}^{(p)} k \equiv \prod_{n_i + P_i \leqslant k < m_i + P_i}^{(p)} k \pmod{P_i} .$$

We can now directly apply the definition and deduce that $(-1)^{n_i - m_i}$ times the right-hand side tends to $\Gamma_p(t)/\Gamma_p(s)$, proving the lemma. □

The main advantage of this lemma is the possibility of choosing sequences m_i and n_i in $\mathbb{Z}$, not necessarily in $\mathbb{Z}_{\geqslant 0}$.

Corollary 11.6.8. (1) *The function* $\Gamma_p(s)$ *is continuous on* $\mathbb{Z}_p$; *more precisely, it satisfies*

$$|\Gamma_p(s) - \Gamma_p(t)| \leqslant |s - t| ,$$

except for $p = 2$, *for which this inequality is valid only for* $v_p(s - t) \geqslant 3$.
(2) *We have* $\Gamma_p(s + 1) = -s\Gamma_p(s)$ *if* $v_p(s) = 0$, *and* $\Gamma_p(s + 1) = -\Gamma_p(s)$ *if* $v_p(s) > 0$.
(3) *More generally, if* $m \in \mathbb{Z}_{\geqslant 0}$ *we have*

$$\frac{\Gamma_p(s + m)}{\Gamma_p(s)} = (-1)^m \prod_{\substack{0 \leqslant k < m \\ v_p(s+k)=0}} (s + k) .$$

(4) *When* $m \in \mathbb{Z}_{\geqslant 0}$ *we have*

$$\Gamma_p(m + 1) = (-1)^{m+1} \frac{m!}{p^{m \backslash p}(m \backslash p)!} \quad and \quad \Gamma_p(-m) = (-p)^{(m \backslash p)} \frac{(m \backslash p)!}{m!} .$$

In particular, $\Gamma_p(0) = 1$, $\Gamma_p(1) = -1$, $\Gamma_p(2) = 1$, *and* $\Gamma_p(-1) = 1$.

Proof. (1) is a restatement of the second part of Proposition 11.6.6. For (2), let $n_i \to s$, $n_i \in \mathbb{Z}_{\geqslant 0}$, so that $m_i = n_i + 1 \to s + 1$. Applying the lemma to n_i and m_i we deduce that

$$\frac{\Gamma_p(s + 1)}{\Gamma_p(s)} = - \lim_{i \to \infty} \begin{cases} n_i & \text{if } p \nmid n_i , \\ 1 & \text{if } p \mid n_i , \end{cases}$$

proving (2), and (3) follows from (2) by induction on m or directly from the lemma. Note that in (3) we cannot use the expression $\prod_{1 \leqslant k \leqslant m}^{(p)}(s + k)$ since this would mean $p \nmid k$ and *not* $p \nmid (s + k)$.

Since from the definition it is clear that $\Gamma_p(0) = 1$ and $\Gamma_p(1) = -1$, the formulas of (4) follow from (3) by choosing $s = 1$ and $s = -m$ respectively. □

Corollary 11.6.9. *If* n *and* m *are in* $\mathbb{Z}$ *with* $n \leqslant m$ *and* $s \in \mathbb{Z}_p$, *we have*

$$\prod_{\substack{n \leqslant k < m \\ v_p(s+k)=0}} (s + k) = (-1)^{m-n} \frac{\Gamma_p(s + m)}{\Gamma_p(s + n)} .$$

Proof. Clear from (3) of the above corollary.   □

The following lemma generalizes Lemma 11.6.7:

Lemma 11.6.10. *Let n_i and m_i be two sequences of elements of $\mathbb{Z}$ such that $n_i \leqslant m_i$, converging p-adically respectively to s and t, and let $a \in \mathbb{Z}_p$. Then*

$$\lim_{i \to \infty} (-1)^{m_i - n_i} \prod_{\substack{n_i \leqslant k < m_i \\ v_p(k+a)=0}} (k + a) = \frac{\Gamma_p(t + a)}{\Gamma_p(s + a)} \,.$$

Proof. By the above corollary we have

$$\prod_{\substack{n_i \leqslant k < m_i \\ v_p(k+a)=0}} (k + a) = (-1)^{m_i - n_i} \frac{\Gamma_p(a + m_i)}{\Gamma_p(a + n_i)} \,,$$

so the result follows by the continuity of the function Γ_p proved above.   □

For future reference we also give the following result.

Corollary 11.6.11. *Let $f \geqslant 1$, set $q = p^f$, and for all $k \in \mathbb{Z}_{\geqslant 0}$ write $k = qm + r$ with $0 \leqslant r < q$. Then*

$$\prod_{0 \leqslant i < f} \Gamma_p(-\lfloor k/p^i \rfloor) = (-p)^{m(q-1)/(p-1)+v_p(r!)} \frac{m!}{k!} = (-p)^{v_p(k!)-v_p(m!)} \frac{m!}{k!} \,.$$

Proof. Set $k_i = \lfloor k/p^i \rfloor$. By Corollary 11.6.8 (4), we have $\Gamma_p(-k_i) = (-p)^{k_{i+1}} \dfrac{k_{i+1}!}{k_i!}$. The factorials give a telescoping product, so that

$$\prod_{0 \leqslant i < f} \Gamma_p(-k_i) = (-p)^{\sum_{1 \leqslant i \leqslant f} k_i} \frac{k_f!}{k_0!} \,.$$

Since $k = qm + r = p^f m + r$ we have $k_i = p^{f-i}m + \lfloor r/p^i \rfloor$; hence using Lemma 4.2.8 and summing a geometric series we have $\sum_{1 \leqslant i \leqslant f} k_i = m(q-1)/(p-1) + v_p(r!)$. Since $k_f = m$ and $k_0 = k$ we obtain the first formula, and the second follows from an immediate computation using Lemma 4.2.8.   □

The second formula of Corollary 11.6.8 (4) is a special case of the following reflection formula.

Proposition 11.6.12. *For all $s \in \mathbb{Z}_p$ we have*

$$\Gamma_p(s)\Gamma_p(1 - s) = (-1)^{[s-1-(s-1)\backslash p]+1} = \begin{cases} (-1)^{a_0(s-1)+1} & \text{if } p > 2 \,, \\ (-1)^{a_0(s-1)+a_1(s-1)+1} & \text{if } p = 2 \,. \end{cases}$$

Proof. Let $m_i \to s$ as $i \to \infty$ with $m_i \geq 0$. Then $1 - m_i \to 1 - s$. It follows from Lemma 11.6.7 that

$$\frac{\Gamma_p(0)}{\Gamma_p(1-s)} = \lim_{i \to \infty} (-1)^{m_i+1} \prod_{1-m_i \leq k < 0}^{(p)} k = (-1)^{m_i - (m_i-1)\backslash p}(-1)^{m_i} \prod_{0 \leq k \leq m_i - 1}^{(p)} k \,,$$

and this converges to $(-1)^{[s-1-(s-1)\backslash p]+1}\Gamma_p(s)$, proving the first formula. The others follow from Proposition 11.6.2, since both sides are equal to ± 1 and congruent modulo p for $p > 2$, hence equal, and congruent modulo 8 for $p = 2$, hence equal. □

Corollary 11.6.13. (1) *For $p \neq 2$ we have $\Gamma_p(1/2)^2 = (-1)^{(p+1)/2}$, so that $\Gamma_p(1/2) = \pm 1$ when $p \equiv 3 \pmod 4$, and $\Gamma_p(1/2) = \pm i$ when $p \equiv 1 \pmod 4$, where i is one of the square roots of -1 in $\mathbb{Z}_p$.*
(2) *For $p \neq 2$ we have*

$$\Gamma_p(1/2) \equiv (-1)^{(p+1)/2}((p-1)/2)! \pmod{p\mathbb{Z}_p} \,.$$

Proof. Note that when p is odd we have $-1/2 = \sum_{j \geq 0}((p-1)/2)p^j$. It follows that $a_0(-1/2) = (p-1)/2$, so that by the above proposition $\Gamma_p(1/2)^2 = (-1)^{(p+1)/2}$, proving (1). For (2) we have

$$\Gamma_p(1/2) \equiv \Gamma_p((p+1)/2) \equiv ((p-1)/2)! \pmod{p\mathbb{Z}_p}$$

by Proposition 11.6.6 and Corollary 11.6.8. □

Remarks. (1) Statement (1) determines $\Gamma_p(1/2)$ up to sign, and statement (2) makes the sign unambiguous.
(2) When $p \equiv 1 \pmod 4$ it is reasonable to define $\Gamma_p(1/2)$ as a canonical square root i of -1, in other words the square root such that $i \equiv -((p-1)/2)! \pmod{p\mathbb{Z}_p}$. On the other hand, when $p \equiv 3 \pmod 4$ we have $\Gamma_p(1/2) = \pm 1 \equiv ((p-1)/2)! \pmod{p\mathbb{Z}_p}$, and the sign can be given by a number of equivalent formulas; see Exercise 43.

We also have a duplication formula, and more generally a distribution formula.

Theorem 11.6.14. *Let p be a prime number and $n \geq 1$ such that $p \nmid n$. Then for all $s \in \mathbb{Z}_p$ we have the distribution formula*

$$\prod_{0 \leq j < N} \Gamma_p\left(s + \frac{j}{N}\right) = c_{p,N} \frac{\Gamma_p(Ns)}{N^{[Ns-1-(Ns-1)\backslash p]}} \,,$$

where

$$c_{p,N} = \begin{cases} \left(\dfrac{-p}{N}\right) & \text{if } N \text{ is odd,} \\ (-1)^{N/2+1}\left(\dfrac{(-1)^{N/2+1}N/2}{p}\right)\Gamma_p(1/2) & \text{if } N \text{ is even (hence } p \neq 2) \,, \end{cases}$$

where we recall that $\Gamma_p(1/2)^2 = (-1)^{(p+1)/2}$. In particular, for $p \neq 2$ we have

$$\Gamma_p(s)\Gamma_p(s + 1/2) = \Gamma_p(1/2)\frac{\Gamma_p(2s)}{2^{[2s-1-(2s-1)\backslash p]}} \ .$$

Note that $N^{[Ns-1-(Ns-1)\backslash p]}$ must be computed as explained in Definition 11.6.1 and Proposition 11.6.2, and *not* in a naïve manner.

Proof. Let m_i be a sequence of positive integers tending to s. By Lemma 11.6.10 applied to $n_i = 0$, m_i, and $a = j/N \in \mathbb{Z}_p$, we have

$$\lim_{i \to \infty} (-1)^{m_i} \prod_{\substack{0 \leqslant k < m_i \\ v_p(k+j/N)=0}} (k + j/N) = \frac{\Gamma_p(s + j/N)}{\Gamma_p(j/N)} \ .$$

Thus

$$\prod_{0 \leqslant j < N} \Gamma_p(s + j/N) = c_{p,N} \lim_{i \to \infty} u_i \ ,$$

where

$$c_{p,N} = \prod_{0 \leqslant j < N} \Gamma_p(j/N) = \prod_{1 \leqslant j \leqslant N-1} \Gamma_p(j/N)$$

and

$$u_i = (-1)^{Nm_i} \prod_{0 \leqslant j < N} \prod_{\substack{0 \leqslant k < m_i \\ v_p(k+j/N)=0}} (k + j/N)$$

$$= (-1)^{Nm_i} \prod_{\substack{0 \leqslant r < Nm_i \\ v_p(r/N)=0}} (r/N) = (-1)^{Nm_i} \prod_{0 \leqslant r < Nm_i}^{(p)} (r/N)$$

$$= (1/N)^{Nm_i - 1 - (Nm_i - 1)\backslash p} (-1)^{Nm_i} \prod_{0 \leqslant r < Nm_i}^{(p)} r \ ,$$

and since $Nm_i - 1$ tends to $Ns - 1$, u_i clearly converges to

$$\frac{\Gamma_p(Ns)}{N^{[Ns-1-(Ns-1)\backslash p]}} \ .$$

It remains to compute $c_{p,N}$. We consider two cases.

Case $p > 2$.

Let m be an inverse of $-N$ modulo p. Changing m into $m + p$ if necessary, we can assume that m is even. Then $-j/N \equiv jm \pmod{p}$, so $a_0(-j/N) = jm - p((jm)\backslash p)$. Applying the reflection formula to $s = 1 - j/N$, we deduce that

$$\Gamma_p(1 - j/N)\Gamma_p(j/N) = (-1)^{jm+1-p((jm)\backslash p)} = -(-1)^{(jm)\backslash p}$$

since m is even and p is odd. Letting as usual $(N-1) \bmod 2$ equal 0 if $N - 1$ is even and 1 if $N - 1$ is odd, it follows that

$$c_{p,N} = (-1)^{d_1}\Gamma_p(1/2)^{(N-1)\bmod 2} ,$$

where

$$d_1 = \left\lfloor \frac{N-1}{2} \right\rfloor + \sum_{1 \leqslant j \leqslant (N-1)/2} \left\lfloor \frac{jm}{p} \right\rfloor .$$

Write $mN + 1 = rp$, so that r is odd. From the equality

$$\frac{jm}{p} + \frac{j}{Np} = \frac{jr}{N} ,$$

we deduce that for $1 \leqslant j < N/2$ we have

$$\frac{jm}{p} < \frac{jr}{N} < \frac{jm+1/2}{p} .$$

Clearly the open interval $]jm/p, (jm+1/2)/p[$ does not contain any integers, so that $\lfloor jm/p \rfloor = \lfloor jr/N \rfloor$. It follows that

$$d_1 = \left\lfloor \frac{N-1}{2} \right\rfloor + \sum_{1 \leqslant j \leqslant (N-1)/2} \left\lfloor \frac{jr}{N} \right\rfloor ,$$

and since m is even, r is such that $rp \equiv 1 \pmod{2N}$.

Since $\Gamma_p(1/2)^2 = (-1)^{(p+1)/2}$, it follows that $c_{p,N} = \Gamma_p(1/2)^{N-1}c_1$, with

$$c_1 = (-1)^{\lfloor (N-1)/2 \rfloor (p-1)/2}(-1)^{S(r,N)} ,$$

and $S(r,N) = \sum_{1 \leqslant j \leqslant (N-1)/2} \lfloor jr/N \rfloor$. In Corollary 2.2.14 we have shown that

$$(-1)^{S(r,N)} = \begin{cases} (-1)^{(N-1)(r-1)/4} \left(\dfrac{N}{r} \right) & \text{if } N \text{ is odd,} \\[3mm] (-1)^{(N-2)(r-1)/4} \left(\dfrac{2N}{r} \right) & \text{if } N \text{ is even.} \end{cases}$$

Since $rp \equiv 1 \pmod{2N}$, the properties of the Kronecker–Jacobi symbol thus imply that we can replace r by p in the above formulas (this would not be true if we only had $rp \equiv 1 \pmod{N}$). It is easily seen that for $p > 2$ the formulas of the theorem follow from this and the quadratic reciprocity law.

Case $p = 2$.

This case is simpler. We can choose $m = -N$ as the inverse of $-N$ modulo 4. Thus if we write $jm - 4(jms) = r_j$, we have

$$a_0(-j/N)+a_1(-j/N) \equiv r_j(r_j+1)/2 \equiv jm(jm+1)/2 \equiv j(jm+1)/2 \pmod 2 .$$

Here N is odd, so applying the reflection formula, we obtain in a manner similar to the preceding case $c_{p,N} = (-1)^{d_1}$ with

$$d_1 = (N-1)/2 + \sum_{1 \leqslant j \leqslant (N-1)/2} j(jm+1)/2 \,.$$

Since we may choose $m = -N$, we compute explicitly that

$$d_1 = (N-1)/2 + ((N^2 - 1)/8)(-N^2/3 + 1)/2 \,.$$

Now in $\mathbb{Z}_2$ we have $(-N^2/3 + 1)/2 \equiv 1/3 \equiv 1 \pmod{2\mathbb{Z}_2}$, hence $d_1 \equiv (N-1)/2 + (N^2-1)/8 \pmod 2$. It follows that $c_{p,N} = \left(\frac{-4}{N}\right)\left(\frac{N}{2}\right) = \left(\frac{-2}{N}\right)$, finishing the proof. $\qquad\square$

11.6.4 Mahler–Dwork Expansions Linked to $\Gamma_p(x)$

The definition of the p-adic gamma function as given is totally unsuitable for practical computation. In this section we give the Mahler expansions of functions closely related to $\Gamma_p(x)$, which in particular will allow us to compute $\Gamma_p(x)$ numerically much more efficiently. It also allows the extension of $\Gamma_p(x)$ to a nonempty ball of $\mathbb{C}_p$. I am indebted to F. Rodriguez-Villegas for making available a GP script implementing part of this.

Proposition 11.6.15. *As in Section 4.2.3, let u_k be the sequence of rational numbers defined formally by $\exp(X + X^p/p) = \sum_{k \geqslant 0} u_k X^k$. Then we have the following convergent expansions for $x \in \mathbb{Z}_p$:*

(1)

$$\sum_{k \geqslant 0} (-1)^k k! u_k \binom{x}{k} = \begin{cases} \Gamma_p(x) & \text{if } |x| < 1 \,, \\ 0 & \text{if } |x| = 1 \,. \end{cases}$$

(2) *More generally, if $0 \leqslant r \leqslant p-1$ then*

$$\sum_{k \geqslant r} (-1)^{k+1} k! u_{k-r} \binom{x}{k} = \begin{cases} \Gamma_p(x+1) & \text{if } x - r \in p\mathbb{Z}_p \,, \\ 0 & \text{if } x - r \notin p\mathbb{Z}_p \,. \end{cases}$$

(3) *If $0 \leqslant r \leqslant p-1$ then*

$$\Gamma_p(px - r) = \sum_{k \geqslant 0} \binom{x+k-1}{k} k! u_{pk+r} p^k \,.$$

(4)

$$\Gamma_p(x) = \sum_{k \geqslant 0} (-1)^{k-1} k! t_k \binom{x-1}{k}, \quad \text{where} \quad t_k = \sum_{\max(0,k-p+1) \leqslant j \leqslant k} u_j \,.$$

Proof. (1). Let $f(x)$ be the function defined as $f(x) = \Gamma_p(x)$ for $|x| < 1$ and $f(x) = 0$ for $|x| = 1$. Since by Corollary 11.6.8 the function $\Gamma_p(x)$ is continuous and since the p-adic topology is totally disconnected, the function

$f(x)$ is *continuous* on $|x| \leqslant 1$ (this would of course be trivially false over the complex numbers). We may thus apply Mahler's Theorem 4.2.26, which tells us that $f(x) = \sum_{k \geqslant 0} a_k \binom{x}{k}$ with

$$a_k = \sum_{m=0}^{k} (-1)^{k-m} \binom{k}{m} f(m)$$

tending to 0 as k tends to infinity. By Corollary 11.6.8 we thus have

$$a_k = \sum_{0 \leqslant m \leqslant k,\ p \mid m} (-1)^{k-m} \binom{k}{m} \Gamma_p(m)$$

$$= \sum_{0 \leqslant n \leqslant k/p} (-1)^{k-pn} \frac{k!}{(pn)!(k-pn)!} (-1)^{pn} \frac{(pn)!}{p^n n!}$$

$$= (-1)^k k! \sum_{0 \leqslant n \leqslant k/p} \frac{1}{p^n (k-pn)! n!} = (-1)^k k! u_k$$

by Corollary 4.2.23, proving (1). (2) follows by applying (1) to $x - r$ and using Corollary 11.6.8 (note that $\Gamma_p(x+1) = -\Gamma_p(x)$ when $x \in p\mathbb{Z}_p$, so (1) is indeed the special case $r = 0$ of (2)). Formulas (3) and (4) are proved in a manner similar to (1) and are left to the reader (Exercise 46). □

Remarks. (1) In all of the above formulas we can of course replace $k!\binom{a}{k}$ by $a(a-1)\cdots(a-k+1)$.

(2) The fact that the expansions converge follows from Mahler's Theorem 4.2.26 and the continuity of the function Γ_p. However, by Corollary 4.2.23 we know in fact that for example $v_p(k!u_k)$ grows to infinity with k approximately like $(k/p)(1 - 1/p)$, so the convergence of all of the above series is quite fast and completely controlled.

(3) In particular, the above proposition gives efficient methods to compute $\Gamma_p(x)$ for all $x \in \mathbb{Z}_p$.

(4) We will see below that these Mahler expansions imply that $\Gamma_p(x)$ has a power series expansion that converges for $|x| < p^{-(2p-1)/(p(p-1))}$, so that this allows us to extend the definition of $\Gamma_p(x)$ to such x. We can then use the formulas that we have seen, such as functional equation, reflection formula, and distribution formula, to extend to other elements of $\mathbb{C}_p$. It is not clear whether this is useful.

Corollary 11.6.16. (1) *We have*

$$\sum_{k \geqslant 0} k^m k! u_k = \begin{cases} 0 & \text{if } 0 \leqslant m \leqslant p-2, \\ -1 & \text{if } m = p-1. \end{cases}$$

(2) *For $p \neq 2$ we have*

$$\Gamma_p\left(\frac{1}{2}\right) = \sum_{k \geqslant (p-1)/2} \frac{(2k)!}{2^{2k}k!} u_{k-(p-1)/2} \cdot$$

Proof. Left to the reader (Exercise 47).   □

For the proof of the Gross–Koblitz formula below we need a more general statement than Proposition 11.6.15 (3). Recall from Chapter 4 that there exists $\pi \in \mathbb{Q}_p(\zeta_p)$ such that $\pi^{p-1} = -p$, and that for all $f \geqslant 1$ we define coefficients $d_{k,f} \in \mathbb{Q}_p(\zeta_p)$ by the formal power series expansion $\exp(\pi(X - X^q)) = \sum_{k \geqslant 0} d_{k,f} X^k$, where $q = p^f$.

Proposition 11.6.17. *Let r be fixed such that $0 \leqslant r < q$. We have the Mahler expansion*

$$\prod_{0 \leqslant i < f} \Gamma_p(-(\lfloor r/p^i \rfloor + p^{f-i}x)) = \pi^{-s_p(r)} \sum_{k \geqslant 0} \frac{d_{qk+r,f}}{\pi^k} k! \binom{x}{k},$$

where $s_p(r)$ is the sum of the digits of r in base p.

Proof. As usual, if we set

$$g(x) = \prod_{0 \leqslant i < f} \Gamma_p(-(\lfloor r/p^i \rfloor + p^{f-i}x)),$$

then g is continuous on $\mathbb{Z}_p$, so by Mahler's theorem $g(x) = \sum_{k \geqslant 0} c_k \binom{x}{k}$ with $c_k = \sum_{0 \leqslant m \leqslant k} (-1)^{k-m} g(m) \binom{k}{m}$. By Corollary 11.6.11 we have $g(m) = (-p)^{m(q-1)/(p-1)+v_p(r!)} m!/(qm+r)!$, so that

$$c_k = (-1)^k (-p)^{v_p(r!)} k! \sum_{0 \leqslant m \leqslant k} \frac{(-1)^m (-p)^{m(q-1)/(p-1)}}{(k-m)!(qm+r)!}$$

$$= (-p)^{k(q-1)/(p-1)+v_p(r!)} k! \sum_{0 \leqslant m \leqslant k} \frac{(-1)^m (-p)^{-m(q-1)/(p-1)}}{m!(qk+r-qm)!}$$

after changing m into $k - m$. Now changing X into X/π in the definition of $d_{k,f}$ and using $\pi^{p-1} = -p$ we obtain

$$\exp\left(X - \frac{X^q}{(-p)^{(q-1)/(p-1)}}\right) = \sum_{k \geqslant 0} \frac{d_{k,f}}{\pi^k} X^k.$$

Using the power series expansion of $\exp(X)$ we thus obtain

$$\frac{d_{k,f}}{\pi^k} = \sum_{0 \leqslant m \leqslant k/q} \frac{(-1)^m (-p)^{-m(q-1)/(p-1)}}{m!} \frac{1}{(k-qm)!},$$

so thanks to the expression obtained above for c_k we have

$$c_k = \frac{(-p)^{k(q-1)/(p-1)+v_p(r!)} k! d_{qk+r,f}}{\pi^{qk+r}}$$

$$= \pi^{-k-r}(-p)^{v_p(r!)} k! d_{qk+r,f} = \pi^{-s_p(r)-k} k! d_{qk+r,f}$$

using once again $\pi^{p-1} = -p$ and the formula for $v_p(r!)$, proving the result.
□

11.6.5 Power Series Expansions Linked to $\Gamma_p(x)$

In this subsection we study the power series expansion of $\Gamma_p(x)$ and some consequences. There are essentially two methods to compute this power series. The first one is to transform the explicit Mahler series expansion (Proposition 11.6.15) into a power series, using Proposition 4.2.28. The second is to compute the exponential of the explicit power series of $\text{Log}\Gamma(x)$ (Proposition 11.5.19). We will see that this second method gives stronger results than the first, but it is still interesting to see what can be obtained from the first.

Theorem 11.6.18. (1) *For $p > 2$ the function $\Gamma_p(x)$ has a power series expansion around $x = 0$ with a radius of convergence at least equal to $p^{-(2p-1)/(p(p-1))}$.*

(2) *More precisely, if we write $\Gamma_p(x) = \sum_{k \geq 0} g_k x^k$ with $g_0 = 1$ then*

$$v_p(g_k) \geq -\frac{2p-1}{p(p-1)} k .$$

(3) *For $n \geq 1$ we have the identity $\sum_{0 \leq k \leq n}(-1)^k g_k g_{n-k} = 0$, and in particular $g_2 = g_1^2/2$, or equivalently, $\Gamma_p''(0) = (\Gamma_p'(0))^2$.*

Proof. (1) and (2). In the proof of Proposition 11.6.15 we have seen that

$$\Gamma_p(-px) = \sum_{k \geq 0} a_{0,k} \binom{x}{k} ,$$

where by Corollary 4.2.23 we have $v_p(a_{0,k}) \geq (1 - 1/p)(k - s_p(k)/p)$. Applying Proposition 4.2.28 with $\alpha = (1 - 1/p)$, $\alpha' = -(1 - 1/p)/p$, and $\alpha'' = 0$ (which can be applied only when $(1 - 1/p) > 1/(p - 1)$, hence when $p \geq 3$), we deduce that $\Gamma_p(-px)$ is equal to the sum of a power series with radius of convergence greater than or equal to $R = p^{\alpha - 1/(p-1)} = p^{(p^2-3p+1)/(p(p-1))}$, hence the radius of convergence of $\Gamma_p(x)$ itself is greater than or equal to $R/p = p^{-(2p-1)/(p(p-1))}$, proving (1). Now note that $\alpha' + 1/(p - 1) = (2p - 1)/(p^2(p - 1)) \geq 0$. Since by definition $\Gamma_p(-px) = \sum_{k \geq 0}(-p)^k g_k x^k$, Proposition 4.2.28 tells us also that $v_p((-p)^k g_k) \geq ((p^2 - 3p + 1)/(p(p - 1)))k$, proving (2).

(3). Let $x \in p\mathbb{Z}_p$ when $p \geq 3$, $x \in 4\mathbb{Z}_2$ when $p = 2$. Using the reflection formula (Proposition 11.6.12) we check that whether $p \geq 3$ or not we have

$\Gamma_p(x)\Gamma_p(1-x) = -1$ since $a_0(x-1) = p-1$ if $p \geqslant 3$ and $a_0(x-1)+a_1(x-1) = 2$ if $p = 2$. Since $x \in p\mathbb{Z}_p$ we have $\Gamma_p(1-x) = -\Gamma_p(-x)$ by Corollary 11.6.8, so we deduce that $\Gamma_p(x)\Gamma_p(-x) = 1$. Expanding the product of the two power series gives $\sum_{n \geqslant 0} G_n x^n = 1$ with $G_n = \sum_{0 \leqslant k \leqslant n} (-1)^k g_k g_{n-k} = 0$. Since the radius of convergence is not zero we deduce from Corollary 4.2.4 that $G_n = 0$ for $n \geqslant 1$. $\qquad\square$

It is not difficult to show that for $p > 2$, computing the exponential of the power series for $\mathrm{Log}\Gamma_p(x)$ essentially gives the same result as Theorem 11.6.18, which uses Mahler expansions. On the other hand, for $p = 2$ we could not use Mahler expansions, but using the power series for $\mathrm{Log}\Gamma_p(x)$ we can give (a lower bound for) the radius of convergence.

Proposition 11.6.19. (1) *Let* $x \in 2\mathbb{Z}_2$. *We have with evident notation*

$$\exp_2(\log_2(\Gamma_2(x))) = (-1)^{x(x-2)/8}\Gamma_2(x) \ .$$

(2) *The function* $\Gamma_2(x)$ *has a power series expansion around* $x = 0$ *with a radius of convergence greater than or equal to* $1/2$. *More precisely, if we write* $\Gamma_2(x) = \sum_{k \geqslant 0} g_k x^k$ *with* $g_0 = 1$ *then* $v_2(g_k) \geqslant -k$.

(3) *For* $f \geqslant 3$ *the function* $\Gamma_2(2^f x)$ *has a Mahler expansion* $\Gamma_2(2^f x) = \sum_{k \geqslant 0} a_{f,k}\binom{x}{k}$ *with* $v_2(a_{f,k}) \geqslant fk - s_2(k)$, *and the same is true for the function* $\Gamma_2(-2^f x)$.

Note that if $p > 2$ and $x \in p\mathbb{Z}_p$, it immediately follows from Proposition 11.6.6 that $\exp_p(\log_p(\Gamma_p(x))) = \Gamma_p(x)$.

Proof. (1). By the easy Exercise 41, we know that $\Gamma_2(2n) \equiv (-1)^{n(n-1)/2}$ (mod 4). Taking $n \to x/2 \in \mathbb{Z}_2$ implies that $\Gamma_2(x) \equiv (-1)^{x(x-2)/8}$ (mod 4). Setting $y = (-1)^{x(x-2)/8}\Gamma_2(x) \equiv 1$ (mod 4), it follows from Proposition 4.2.10 (5) that $\exp_2(\log_2(y)) = y$, proving (1).

(2) and (3). Recall from Proposition 11.5.19 that if we set $d_{k,p} = B_{-2k,p}p^{2k+1}/(2k(2k+1))$ (see Definition 11.3.13) then

$$\log_p(\Gamma_p(px)) = \mathrm{Log}\Gamma_p(px) = -p\gamma_p x - \sum_{k \geqslant 1} d_{k,p} x^{2k+1} \ ,$$

and that Lemma 11.5.20 gives us information on $v_p(d_{k,p}) = v(k,p)$. We easily compute that $\gamma_2 \equiv 1$ (mod 2), so that $v_2(2\gamma_2) \geqslant 1$, and $v_p(d_{k,2}) = v(k,2) \geqslant 1$ for all $k \geqslant 1$. Thus all the coefficients of the formal power series $\exp(\log(\Gamma_p(pX)))$ (which is the formal product of $\exp(-p\gamma_p X)$ and the $\exp_p(-d_{k,p}X^{2k+1})$ for $k \geqslant 1$) are p-integral, and (2) follows. For (3) we note that by (1), if $f \geqslant 3$ we have $\exp_2(\log_2(\Gamma_2(2^f x))) = \Gamma_2(2^f x)$, so that statement (3) follows from (2) and Exercise 17 of Chapter 4. $\qquad\square$

Note that (3) is *not* true for $f = 1$ or $f = 2$, and in fact one can show that $v_2(a_{1,k}) \sim k/2$ and $v_2(a_{2,k}) = k$, and similarly for the coefficients of $\Gamma_2(-2^f x)$; see Exercise 36.

Corollary 11.6.20. *For $f \geqslant 1$ set*

$$a_{f,k} = \sum_{m=0}^{k} (-1)^{k-m} \binom{k}{m} 1 \cdot 3 \cdots (2^f m - 1) \quad and$$

$$b_{f,k} = \sum_{m=0}^{k} (-1)^{k-m+2^{f-1}m} \binom{k}{m} \frac{1}{1 \cdot 3 \cdots (2^f m - 1)} \,,$$

which are the Mahler coefficients of $\Gamma_2(2^f x)$ and $\Gamma_2(-2^f x)$, respectively. For $f \geqslant 3$ we have $v_2(a_{f,k}) \geqslant fk - s_2(k)$ and $v_2(b_{f,k}) \geqslant fk - s_2(k)$.

Proof. Follows immediately from the above corollary, the explicit formula for the Mahler coefficients (Theorem 4.2.26), and the values on $\mathbb{Z}$ of the function $\Gamma_2(s)$ (Corollary 11.6.8). □

Note that I do not know of a direct combinatorial proof of the above corollary, and that numerically it seems that we have the slightly stronger inequality $v_2(a_{f,k}) \geqslant fk - s_2(k)/2$ (and similarly for $b_{f,k}$).

11.6.6 The Jacobsthal–Kazandzidis Congruence

Proposition 11.6.21. *Set*

$$D_p(x,y) = \frac{\log_p(\Gamma_p(x+y)) - \log_p(\Gamma_p(x)) - \log_p(\Gamma_p(y))}{xy(x+y)} \,.$$

If x and y are in $p\mathbb{Z}_p$ we have the following congruences:

(1) *For $p \geqslant 5$,*

$$D_p(x,y) \equiv -\frac{1}{2p} \sum_{1 \leqslant a \leqslant p-1} \frac{1}{a^2} \equiv -\frac{B_{p-3}}{3} \pmod{p\mathbb{Z}_p} \,.$$

(2) *For $p = 3$,*

$$D_p(x,y) \equiv \frac{5}{3} \pmod{p\mathbb{Z}_p} \,.$$

(3) *For $p = 2$,*

$$D_p(x,y) \equiv \frac{3}{4} + \frac{7}{8}(x^2 + xy + y^2) \pmod{p\mathbb{Z}_p} \,.$$

Proof. By Proposition 11.5.19 we have

$$D_p(x,y) = -\sum_{k \geqslant 1} \frac{d_{k,p}}{p^3} \frac{(x+y)^{2k+1} - x^{2k+1} - y^{2k+1}}{p^{2k-2}xy(x+y)}$$

with $d_{k,p}$ as above. Since the numerator of

$$A_{2k-2}(x,y) = \frac{(x+y)^{2k+1} - x^{2k+1} - y^{2k+1}}{xy(x+y)}$$

vanishes for $x = 0$, $y = 0$, and $y = -x$, it follows that $A_{2k-2}(x,y)$ is a polynomial, which is clearly homogeneous of degree $2k - 2$, so that $A_{2k-2}(x,y) \in p^{2k-2}\mathbb{Z}_p$ when x and y are in $p\mathbb{Z}_p$. By Lemma 11.5.20 we thus have

$$D_p(x,y) \equiv -3\frac{d_{1,p}}{p^3} - 5\frac{d_{2,p}}{p^3}\frac{x^2 + xy + y^2}{p^2}$$
$$\equiv -\frac{B_{-2,p}}{2} - (x^2 + xy + y^2)\frac{B_{-4,p}}{4} \pmod{p\mathbb{Z}_p} .$$

Since x and y are in $p\mathbb{Z}_p$ and $v_p(B_{-4,p}) \geqslant -1$, it is clear that for $p \neq 2$ the term involving $B_{-4,p}$ is congruent to 0 modulo p hence may be ignored (this of course follows from Lemma 11.5.20 for $p \neq 5$, but not for $p = 5$). Assume first that $p \geqslant 5$. By Corollary 11.3.16 we have

$$B_{-2,p} \equiv \frac{1}{p} \sum_{1 \leqslant a \leqslant p-1} \frac{1}{a^2} \pmod{p\mathbb{Z}_p} ,$$

proving the first congruence. Furthermore, by the same corollary for $p \geqslant 5$ we have $B_{-2,p} \equiv (2/3)B_{p-3} \pmod{p}$, which proves (1), and for $p = 3$ we have $B_{-2,p} \equiv 2/3 + 2 \pmod{3\mathbb{Z}_3}$, which proves (2). For $p = 2$ we must use the formula involving $B_{-4,p}$. We have $B_{-2,2} \equiv 1/2 + 2 \pmod{2^2\mathbb{Z}_2}$ and we easily compute that $B_{-4,2} \equiv 1/2 + 2^2 \pmod{2^3\mathbb{Z}_2}$, proving (3).   □

This proposition allows us to prove the following congruences, due in a weaker form to Jacobstahl and Kazandzidis.

Corollary 11.6.22 (Jacobstahl, Kazandzidis). *Let m and n be two integers such that $0 \leqslant m \leqslant n$ and let p be a prime number. We have the congruence*

$$\binom{pn}{pm} \equiv K_p(n,m)\binom{n}{m} \pmod{p^4 nm(n-m)\binom{n}{m}\mathbb{Z}_p} ,$$

where

$$K_p(n,m) = \begin{cases} 1 - (B_{p-3}/3)p^3 mn(n-m) & \text{if } p \geqslant 5, \\ 1 + 45nm(n-m) & \text{if } p = 3, \\ (-1)^{m(n-m)}P(n,m) & \text{if } p = 2, \end{cases}$$

with $P(n,m) = 1 + 6nm(n-m) - 4nm(n-m)(n^2 - nm + m^2) + 2(nm(n-m))^2$.

Proof. By Corollary 11.6.8 (4), we know that for $n \in \mathbb{Z}_{\geqslant 0}$ we have $\Gamma_p(pn) = (-1)^n(pn)!/(p^n n!)$, hence

$$\frac{\Gamma_p(pn)}{\Gamma_p(pm)\Gamma_p(p(n-m))} = \frac{(pn)!m!(n-m)!}{(pm)!(p(n-m))!n!} = \frac{\binom{pn}{pm}}{\binom{n}{m}} .$$

Assume first that $p \geqslant 3$. By Corollary 11.6.8 (1), if $x \in p\mathbb{Z}_p$ we have $\Gamma_p(x) \equiv 1 \pmod{p\mathbb{Z}_p}$, hence $|\Gamma_p(x) - 1| \leqslant 1/p < r_p = |p|^{1/(p-1)}$ since $p \geqslant 3$. It follows from Proposition 4.2.10 (5) and (4) that for all $x \in p\mathbb{Z}_p$ we have $\exp_p(\log_p(\Gamma_p(x))) = \Gamma_p(x)$, hence that if x and y are in $p\mathbb{Z}_p$ we have

$$\exp_p(xy(x+y)D_p(x,y)) = \frac{\Gamma_p(x+y)}{\Gamma_p(x)\Gamma_p(y)} ,$$

so in particular

$$\frac{\binom{pn}{pm}}{\binom{n}{m}} = \exp_p(p^3 nm(n-m)D_p(pm, p(n-m))) .$$

On the other hand, since we are inside the disk of convergence of the p-adic exponential it follows from the above proposition that if $p \geqslant 5$ and x and y are in $p\mathbb{Z}_p$ we have

$$\exp_p(xy(x+y)D_p(x,y)) \equiv 1 - \frac{B_{p-3}}{3}xy(x+y) \pmod{pxy(x+y)\mathbb{Z}_p} ,$$

proving the result in that case. Similarly, for $p = 3$ we have

$$\exp_p(xy(x+y)D_p(x,y)) \equiv 1 + \frac{5}{3}xy(x+y)$$

$$\equiv 1 + 45\frac{xy(x+y)}{27} \pmod{pxy(x+y)\mathbb{Z}_p} ,$$

and the result again follows.

Assume now that $p = 2$. By Proposition 11.6.19, if x and y are in $2\mathbb{Z}_2$ we have

$$\exp_p(xy(x+y)D_p(x,y)) = (-1)^{xy/4}\frac{\Gamma_p(x+y)}{\Gamma_p(x)\Gamma_p(y)} .$$

Furthermore, since $v_2(xy(x+y)) \geqslant 4$ when x and y are in $2\mathbb{Z}_2$, a similar computation to that above gives

$$\exp_p(xy(x+y)D_p(x,y)) \equiv Q(x,y) \pmod{pxy(x+y)\mathbb{Z}_p} ,$$

where

$$Q(x,y) = 1 + 6\frac{xy(x+y)}{8} + 28\frac{xy(x+y)(x^2+xy+y^2)}{32} + 18\frac{(xy(x+y))^2}{64} .$$

Since $P(n,m) \equiv Q(2m, 2(n-m)) \pmod{p^4 nm(n-m)}$ it follows that

$$\frac{\binom{2n}{2m}}{\binom{n}{m}} = (-1)^{m(n-m)}\exp_p(p^3 nm(n-m)D_p(pm, p(n-m)))$$

$$\equiv (-1)^{m(n-m)}P(n,m) \pmod{p^4 nm(n-m)\mathbb{Z}_p} ,$$

as claimed.   □

Remarks. (1) In the same way we could find more complicated congruences
modulo higher powers of p, but there is no point in doing so.
(2) If $p \geqslant 5$ we have in particular

$$\binom{pn}{pm} \equiv \binom{n}{m} \pmod{p^3 nm(n-m)}\binom{n}{m}\mathbb{Z}_p ,$$

and this congruence is valid modulo $p^4 nm(n-m)\binom{n}{m}\mathbb{Z}_p$ if and only if
p divides the numerator of B_{p-3}. Such a prime is sometimes called a
Wolstenholme prime; see Exercise 30 of Chapter 2. By Exercise 50, p is
such a prime if and only if p^3 divides the numerator of $\sum_{1 \leqslant a \leqslant p-1} 1/a$, if
and only if p^2 divides the numerator of $\sum_{1 \leqslant a \leqslant p-1} 1/a^2$. The only known
Wolstenholme primes are $p = 16843$ and $p = 2124679$, and there are
no others up to 10^9. As usual, however, on probabilistic grounds there
should exist infinitely many; see also Exercise 51.
(3) The congruences are of course valid only in $\mathbb{Z}_p$ and not in $\mathbb{Z}$, since for
instance $\binom{pn}{pm}/\binom{n}{m}$ is not necessarily an integer.

11.7 The Gross–Koblitz Formula and Applications

We have seen that Morita's p-adic gamma function (as well as Diamond's log
gamma function) satisfies essentially all the usual properties of the complex
gamma function. A remarkable fact is that it satisfies additional beautiful
finite identities due to B. Gross and N. Koblitz, whose equivalent for the classical gamma function is the Chowla–Selberg formula (Proposition 10.5.11).
The Gross–Koblitz formula was proved initially using results of N. Katz dealing with crystalline cohomology, but it has recently been proved using much
more elementary methods by A. Robert in [Rob2], and we reproduce his
proof.

11.7.1 Statement and Proof of the Gross–Koblitz Formula

Let $q = p^f$ be a prime power, and as in Section 4.4.8 let $\pi \in \mathcal{K} = \mathbb{Q}_p(\zeta_p)$ be
such that $\pi^{p-1} = -p$. We let $\mathfrak{p}$ be the prime ideal above p in $\mathbb{Q}(\zeta_p)$, so that π is
a generator of $\mathfrak{p}\mathbb{Z}_{\mathcal{K}}$ and $v_{\mathfrak{p}}(p) = p-1$. Recall from Proposition 4.4.40 that we
have defined coefficients $d_{k,f}$ by the formal power series expansion $\exp(\pi(X - X^q)) = \sum_{k \geqslant 0} d_{k,f} X^k$, and we have seen their relation to the function Γ_p
in Proposition 11.6.17. There are two crucial ingredients in Robert's proof.
The first is an identity involving the $d_{k,f}$, and the other is the lower bound
for $v_p(d_{k,f})$ which we have given in Proposition 4.4.40. Since f is fixed, for
notational simplicity we write d_k instead of $d_{k,f}$.
   For $r \in \mathbb{Z}_{\geqslant 0}$ we define $G_r(x)$ by the Mahler expansion

$$G_r(x) = \sum_{k \geqslant 0} \frac{d_{qk+r}}{\pi^k} k! \binom{x}{k} = \sum_{k \geqslant 0} \frac{d_{qk+r}}{\pi^k}(x)_k$$

with the usual notation $(x)_k = k!\binom{x}{k} = x(x-1)\cdots(x-k+1)$. Robert's crucial identity is the following:

Proposition 11.7.1. *For all r and N in $\mathbb{Z}_{\geqslant 0}$ we have*

$$(1-q) \sum_{0 \leqslant k < N} d_{(q-1)k+r} = G_r\left(\frac{r}{1-q}\right) - G_{(q-1)N+r}\left(\frac{r}{1-q} - N\right).$$

Proof. We first prove the proposition for $N = 1$. For this, we transform the term $G_{q-1+r}(x-1)$ (where we will set $x = r/(1-q)$ later) as follows. First note that by differentiating the defining formula $\exp(\pi(X - X^q)) = \sum_{k \geqslant 0} d_k X^k$ we get

$$\pi(1 - qX^{q-1}) \sum_{k \geqslant 0} d_k X^k = \sum_{k \geqslant 0}(k+1)d_{k+1}X^k,$$

whence the recurrence $(k+1)d_{k+1} = \pi(d_k - qd_{k-(q-1)})$ for $k \geqslant q-1$, so that

$$d_{k-1} = \frac{k}{\pi}d_k + qd_{k-q} \quad \text{for } k \geqslant q.$$

We thus have

$$G_{q-1+r}(x-1) = \sum_{k \geqslant 0} \frac{d_{q(k+1)-1+r}}{\pi^k}(x-1)_k$$

$$= \sum_{k \geqslant 0} \left(\frac{q(k+1)+r}{\pi}d_{q(k+1)+r} + qd_{qk+r}\right) \frac{(x-1)_k}{\pi^k}$$

$$= \sum_{k \geqslant 1} \frac{qk+r}{\pi}d_{qk+r}\frac{(x-1)_{k-1}}{\pi^{k-1}} + q\sum_{k \geqslant 0} d_{qk+r}\frac{(x-1)_k}{\pi^k}.$$

Since for $k \geqslant 1$ we have $(x)_k = x(x-1)_{k-1}$ and $(x-1)_k = (x-k)(x-1)_{k-1}$, it follows that $G_r(x) - G_{q-1+r}(x-1)$ is equal to

$$(1-q)d_r + \sum_{k \geqslant 1} d_{qk+r}\frac{(x-1)_{k-1}}{\pi^k}(x-(qk+r) - q(x-k))$$

$$= (1-q)d_r + \sum_{k \geqslant 1} d_{qk+r}\frac{(x-1)_{k-1}}{\pi^k}(x(1-q) - r),$$

so that

$$G_r\left(\frac{r}{1-q}\right) - G_{q-1+r}\left(\frac{r}{1-q} - 1\right) = (1-q)d_r,$$

proving the proposition for $N = 1$.

Applying the case $N = 1$ with r replaced by $(q-1)k+r$ we deduce that

$$G_{(q-1)k+r}\left(\frac{r}{1-q} - k\right) - G_{(q-1)(k+1)+r}\left(\frac{r}{1-q} - k - 1\right) = (1-q)d_{(q-1)k+r} ,$$

and summing this from $k = 0$ to $N - 1$ proves the proposition in general. $\qquad\Box$

Corollary 11.7.2. *For $r \in \mathbb{Z}_{\geqslant 0}$ we have*

$$(1 - q) \sum_{k \geqslant 0} d_{(q-1)k+r} = G_r\left(\frac{r}{1-q}\right) .$$

Proof. Note that since by Proposition 4.4.40 the valuation of d_k tends to infinity with k, we already know that the series on the left-hand side converges. In any case, thanks to the proposition, to prove that it converges and that its sum is as given is equivalent to showing that $G_{(q-1)N+r}(r/(1 - q) - N)$ tends to 0 as $N \to \infty$. Now if $x \in \mathbb{Z}_p$ we have $\binom{x}{k} \in \mathbb{Z}_p$, so by definition of $G_r(x)$, since $v_p(\pi) = 1/(p-1)$ we have

$$v_p(G_r(x)) \geqslant \min_{k \geqslant 0}(v_p(d_{qk+r}) - k/(p-1) + v_p(k!)) .$$

By Proposition 4.4.40 and Lemma 4.2.8 we have

$$v_p(d_{qk+r}) - k/(p-1) + v_p(k!) \geqslant \frac{(qk+r)(p-1)}{p^{f+1}} - s_p(k)/(p-1) ,$$

where as usual $s_p(k)$ is the sum of the base-p digits of k. Since $s_p(k)$ grows only logarithmically with k it follows that $qk(p-1)/p^{f+1} - s_p(k)/(p-1)$ tends to infinity with k, hence that there exists a constant c_p depending only on p such that $qk(p-1)/p^{f+1} - s_p(k)/(p-1) \geqslant c_p$ for all $k \geqslant 0$. Thus for all $x \in \mathbb{Z}_p$ we have $v_p(G_r(x)) \geqslant c_p + r(p-1)/p^{f+1}$, and since $r/(1-q) - N \in \mathbb{Z}_p$ it follows that

$$v_p(G_{(q-1)N+r}(r/(1-q) - N)) \geqslant c_p + ((q-1)N + r)(p-1)/p^{f+1} ,$$

which tends to infinity with N, proving the corollary. $\qquad\Box$

Now recall from Section 4.4.8 that since the power series $D_{\pi,f}(X) = \exp(\pi(X - X^q))$ has a radius of convergence strictly greater than 1, we can define $D_{\pi,f}(a)$ for all $a \in \mathbb{C}_p$ such that $|a| \leqslant 1$, and in particular if $a^q = a$ we have seen that $D_{\pi,f}(a)$ is a pth root of unity given by the formula $D_{\pi,f}(a) = \zeta_\pi^{\mathrm{Tr}_{\mathcal{L}/\mathcal{K}}(a)}$, where $\mathcal{L} = \mathcal{K}(\zeta_{q-1})$ is the unique unramified extension of degree f of $\mathcal{K}$ (see Corollary 4.4.27), and ζ_π is the unique pth root of unity congruent to $1 + \pi$ modulo $\mathfrak{p}^2$.

Definition 11.7.3. *For any* $r \in \mathbb{Z}_{\geqslant 0}$ *we define the Gauss sum*

$$\tau_q(r) = \sum_{a \in \mathcal{L},\, a^{q-1}=1} a^{-r} \zeta_\pi^{\mathrm{Tr}_{\mathcal{L}/\mathcal{K}}(a)} \, .$$

Note that there are $q-1$ terms in the sum, and that evidently $\tau_q(r) \in \mathcal{L}$.

Proposition 11.7.4. *For any* $r \in \mathbb{Z}$ *such that* $0 \leqslant r < q-1$ *we have*

$$\tau_q(r) = (q-1) \sum_{m \geqslant 0} d_{(q-1)m+r} = -G_r\left(\frac{r}{1-q}\right) \, .$$

Proof. Indeed

$$\tau_q(r) = \sum_{a^{q-1}=1} a^{-r} D_{\pi,f}(a) = \sum_{a^{q-1}=1} a^{-r} \sum_{k \geqslant 0} d_k a^k$$

$$= \sum_{k \geqslant 0} d_k \sum_{a^{q-1}=1} a^{k-r} \, .$$

Since $\mathcal{L}$ contains all $(q-1)$st roots of unity the inner sum is a geometric series whose sum is equal to 0 if $(q-1) \nmid (k-r)$ and is equal to $q-1$ otherwise, so writing $k = (q-1)m + r$ we obtain $\tau_q(r) = (q-1) \sum_{m \geqslant 0} d_{(q-1)m+r}$. Note that this is where we need $0 \leqslant r < q-1$, since otherwise this last sum would not begin at $m = 0$. The second formula follows from this combined with the crucial Corollary 11.7.2. $\qquad \square$

The Gross–Koblitz formula is now immediate.

Theorem 11.7.5 (Gross–Koblitz). *If* $0 \leqslant r < q-1$ *then*

$$\tau_q(r) = -\pi^{s_p(r)} \prod_{0 \leqslant i < f} \Gamma_p\left(\frac{r^{(i)}}{q-1}\right) \, ,$$

where $0 \leqslant r^{(i)} < q-1$ *have base-p expansions obtained by a cyclic permutation from that of the f base-p digits of r.*

Proof. By definition of $G_r(x)$, Proposition 11.6.17 tells us that

$$\prod_{0 \leqslant i < f} \Gamma_p(-(\lfloor r/p^i \rfloor + p^{f-i} x)) = \pi^{-s_p(r)} G_r(x) \, .$$

Setting $x = r/(1-q)$, it follows that $\tau_q(r) = -\pi^{s_p(r)} \prod_{0 \leqslant i < f} \Gamma_p(x_i/(q-1))$, where $x_i = -(q-1)\lfloor r/p^i \rfloor + p^{f-i} r$. If we let $r = \sum_{0 \leqslant j < f} r_j p^j$ be the p-adic expansion of r with $0 \leqslant r_j \leqslant p-1$, we have

$$p^{f-i} r = \sum_{0 \leqslant j < i} r_j p^{f-(i-j)} + q \sum_{i \leqslant j < f} r_j p^{j-i} = \sum_{0 \leqslant j < i} r_j p^{f-(i-j)} + q \lfloor r/p^i \rfloor \, ,$$

hence

$$x_i = \sum_{0 \leqslant j < i} r_j p^{f-(i-j)} + \sum_{i \leqslant j < f} r_j p^{j-i} = r^{(i)} \,,$$

proving the theorem.   □

Note that it is essential to take the f coefficients in the base-p expansion of r, including the possible leading zeros. For instance, if $f = 2$ and $r = 1$ then $r^{(0)} = 1$ and $r^{(1)} = p$. The following corollary makes this totally unambiguous.

Corollary 11.7.6. *Recall that we define $s(r) = s_p(r \bmod (q-1))$ and that $\{x\}$ denotes the fractional part of x. For all $r \in \mathbb{Z}$ we have*

$$\tau_q(r) = -\pi^{s(r)} \prod_{0 \leqslant i < f} \Gamma_p \left(\left\{ \frac{p^{f-i} r}{q-1} \right\} \right) \,.$$

Proof. Since both sides are now periodic in r of period $q-1$, it is sufficient to prove the result when $0 \leqslant r < q-1$. In that case we have $p^{f-i}r/(q-1) = (r + r/(q-1))/p^i$ and since $0 \leqslant r/(q-1) < 1$ we deduce that $\lfloor p^{f-i}r/(q-1) \rfloor = \lfloor r/p^i \rfloor$. Since $r^{(i)} = (1-q)\lfloor r/p^i \rfloor + p^{f-i}r$ the corollary follows.   □

Note that since the fractional part of $p^f r/(q-1)$ is equal to that of $r/(q-1)$, the product can be indifferently from 0 to $f-1$ as above, or from 1 to f, and we may also replace p^{f-i} by p^i if desired.

The case $f = 1$, hence $q = p$, is especially interesting (the proof would have been only slightly simpler if we had restricted to this case).

Corollary 11.7.7. *Let $m/n \in \mathbb{Q}$ be such that $n \mid (p-1)$.*

(1) *If $0 \leqslant m/n < 1$ we have the formula*

$$\Gamma_p(m/n) = -\pi^{-r} \tau_p(r) \,,$$

where $r = m(p-1)/n$.

(2) *$\Gamma_p(m/n)$ is an algebraic number, more precisely*

$$\Gamma_p(m/n) \in \mathbb{Q}(\zeta_{np}, (-p)^{1/n}) \,,$$

where as usual ζ_N denotes a primitive Nth root of unity.

Proof. (1) is evidently the special case $f = 1$ of the theorem. Since $\pi = (-p)^{1/(p-1)}$ we have $\pi^{-r} = (-p)^{-m/n} \in \mathbb{Q}((-p)^{1/n})$. Furthermore, ζ_π is a pth root of unity; in other words, ζ_π can be considered as an element of $\mathbb{Q}(\zeta_p)$. Finally, if $a \in \mathbb{Z}_p$ is such that $a^{p-1} = 1$, then once again a can be considered as an element of $\mathbb{Q}(\zeta_{p-1})$, hence a^r as an element of $\mathbb{Q}(\zeta_n)$, proving (2) for $0 \leqslant m/n < 1$, and the general case follows from the functional equation giving $\Gamma_p(x+1)$ in terms of $\Gamma_p(x)$.   □

Note that this theorem is in marked contrast to the complex case, where the values of the complex gamma function at nonintegral rational arguments are believed to be transcendental.

Recall from Proposition 4.3.4 and Exercise 19 of Chapter 4 that the set G_p of $a \in \mathbb{Z}_p$ such that $a^p = a$ can be given a natural finite field structure thanks to the Teichmüller map ω extended by 0 outside p-adic units, which is canonically isomorphic to $\mathbb{F}_p$. Thus if we set $\chi(a) = a^{-r}$ and $\psi(a) = \zeta_\pi^a$ then χ is a multiplicative character and ψ is a nontrivial additive character on G_p, hence $\tau_p(r) = \tau(\chi, \psi)$ in the sense of Gauss sums associated with finite fields, which we have studied in Section 2.5.2. A character such as ψ (which depends on the choice of π) is called a *Dwork character*. For instance, the translation of Corollary 2.5.17 (1) in our context is the following:

Proposition 11.7.8. *If $r + s \not\equiv 0 \pmod{p-1}$ then*

$$\frac{\tau_p(r)\tau_p(s)}{\tau_p(r+s)} = \sum_{\substack{a^{p-1}=1 \\ a \neq 1}} a^{-r}\omega(1-a)^{-s} .$$

Examples. We have already seen in Corollary 11.6.13 as a consequence of the reflection formula that $\Gamma_p(1/2) = \pm 1$ when $p \equiv 3 \pmod 4$ and $\Gamma_p(1/2) = \pm i$ when $p \equiv 1 \pmod 4$, and in particular is algebraic. Let us consider a less immediate example.

Proposition 11.7.9. *We have*

$$\Gamma_5(1/4) = \sqrt{-2+i} ,$$

where i is the square root of -1 congruent to 3 modulo 5, and the outer square root is chosen congruent to 1 modulo 5.

Proof. By the theorem and the above proposition, we have

$$(\Gamma_5(1/4))^2 = (-\pi^{-1}\tau_p(1))^2 = \pi^{-2}\tau_p(2) \sum_{\substack{a^{p-1}=1 \\ a \neq 1}} a^{-1}\omega(1-a)^{-1} .$$

The Jacobi sum on the right is equal to

$$-\omega(2)^{-1} - i\omega(1-i)^{-1} + i\omega(1+i)^{-1} = -\omega(3) - i\omega(2) + i\omega(-1)$$
$$= -i - i(-i) - i = -1 - 2i$$

(look at Proposition 4.3.4 to see how easy it is to compute Teichmüller values). On the other hand, since $\chi(a) = a^{-2}$ is a character of order 2 it follows from Corollary 2.5.17 that $\tau_p(2) = \pm\sqrt{5}$ for some sign $\pm$ depending on the choice of π, and since $\pi^2 = \sqrt{-5} = \pm i\sqrt{5}$, it follows that $\Gamma_5(1/4)^2 = \pm(-2+i)$, and the signs are immediately determined by looking modulo 5. $\qquad\square$

See Exercise 48 for other examples.

11.7.2 Application to $L_p'(\chi, 0)$

As a first immediate application of the Gross–Koblitz formula, we have the following result; see [Fer-Gre].

Proposition 11.7.10 (Ferrero–Greenberg). *Let χ be a primitive even character, let $\chi_1 = \chi\omega^{-1}$, and denote by f_1 the conductor of χ_1. Assume that $\chi_1(p) = 1$, so that in particular $p \nmid f_1$, let u be the order of p in $(\mathbb{Z}/f_1\mathbb{Z})^*$, let $q = p^u$, and finally let $c_1, \ldots, c_g$ be a system of representatives in $\mathbb{Z}$ of $(\mathbb{Z}/f_1\mathbb{Z})^*/\langle p \rangle$, where $\langle p \rangle$ is as usual the subgroup of $(\mathbb{Z}/f_1\mathbb{Z})^*$ generated by p. We have $L_p(\chi, 0) = 0$ and the formula*

$$L_p'(\chi, 0) = \sum_{i=1}^{g} \chi_1(c_i) \log_p(\tau_q(c_i(q-1)/f_1)) \ .$$

In particular, $L_p'(\chi, 0)$ is a $\mathbb{Q}(\chi_1)$-linear combination of p-adic logarithms of algebraic numbers.

Proof. Since $\chi_1(p) = 1$, by Propositions 11.3.9 and 11.5.24 we have $L_p(\chi, 0) = 0$ and

$$L_p'(\chi, 0) = \sum_{0 \leqslant k < f_1} \chi_1(k) \operatorname{Log}\Gamma_p \left(\chi_0, \frac{k}{f_1} \right) \ .$$

For $k \in \mathbb{Z}$ we have $k \equiv c_i p^j \pmod{f_1}$ for a unique i and j modulo u, and for $0 \leqslant k < f_1$ we thus have $k/f_1 = \{c_i p^j / f_1\}$. Once again, since $\chi_1(p) = 1$ we have $\chi_1(k) = \chi_1(c_i)$; hence, using $\operatorname{Log}\Gamma_p(\chi_0, x) = \log_p(\Gamma_p(x))$ we have

$$L_p'(\chi, 0) = \sum_{i=1}^{g} \chi_1(c_i) \sum_{j \bmod u} \log_p \left(\Gamma_p \left(\left\{ \frac{c_i p^j}{f_1} \right\} \right) \right) = \sum_{i=1}^{g} \chi_1(c_i) \log_p(P_i) \ ,$$

where

$$P_i = \prod_{j \bmod u} \log_p \left(\Gamma_p \left(\left\{ \frac{(c_i(q-1)/f_1)p^j}{q-1} \right\} \right) \right)$$

$$= \log_p \left(-\tau_q(c_i(q-1)/f_1)\pi^{-s(c_i(q-1)/f_1)} \right)$$

using the Gross–Koblitz formula (Corollary 11.7.6). The result follows since $\pi^{p-1} = -p$, so that $\log_p(\pi) = 0$. $\qquad\square$

Remark. Since $L_p(\chi, 0) = (1 - \chi_1(p))L(\chi_1, 0)$ and since $\chi_1 = \chi\omega^{-1}$ is an odd character, it follows from the functional equation *in the complex case* and the nonvanishing of $L(\chi_1, 1)$ (Theorem 10.5.29) that we always have $L(\chi_1, 0) \neq 0$. Thus we have $L_p(\chi, 0) = 0$ if and only if $\chi_1(p) = 1$, which is exactly the condition of the proposition. Thanks to the above result and the deep results of Baker already mentioned in the context of the nonvanishing of $L_p(\chi, 1)$, Ferrero–Greenberg then prove that in the present situation we have $L_p'(\chi, 0) \neq 0$, so that $L_p(\chi, s)$ cannot have a multiple zero at $s = 0$. This has important arithmetic consequences mentioned in their paper.

11.7.3 Application to the Stickelberger Congruence

It is interesting to note that the Gross–Koblitz formula gives an immediate proof of Stickelberger's Theorem 3.6.6.

Proposition 11.7.11. *If $r \in \mathbb{Z}$ is such that $0 \leqslant r < q - 1$ then*

$$\tau_q(r) \equiv -\pi^{s_p(r)} \frac{(-p)^{v_p(r!)}}{r!} \equiv -\frac{\pi^r}{r!} \pmod{\mathfrak{p}^{s_p(r)+p-1}\mathbb{Z}_K} .$$

Proof. We have seen that

$$r^{(i)} = (1 - q)\left\lfloor \frac{r}{p^i} \right\rfloor + p^{f-i}r \equiv \left\lfloor \frac{r}{p^i} \right\rfloor \pmod{p}$$

since $0 \leqslant i < f$ and $p \mid q$. Since $q - 1 \equiv -1 \pmod{p}$ it follows from Proposition 11.6.6 that

$$\Gamma_p(r^{(i)}/(q - 1)) \equiv \Gamma_p(-\lfloor r/p^i \rfloor) \pmod{p\mathbb{Z}_p}$$

(note that the special case $(p, f) = (2, 2)$ does not need to be treated separately since $-1 \equiv 1 \pmod{2}$). Since $0 \leqslant r < q - 1$ it follows from the Gross–Koblitz formula and Corollary 11.6.11 that

$$\pi^{-s_p(r)}\tau_q(r) \equiv - \prod_{0 \leqslant i < f} \Gamma_p(-\lfloor r/p^i \rfloor) \equiv -\frac{(-p)^{v_p(r!)}}{r!} \pmod{p\mathbb{Z}_p} ,$$

and the result follows since $\pi^{p-1} = -p$ and $v_p(r!) = (r - s_p(r))/(p - 1)$. On the other hand, if $(p, f) = (2, 2)$ a direct computation immediately shows that $-\tau_q(r)/\pi^{s_p(r)} = 1$ for $0 \leqslant r < q - 1 = 3$, which agrees with the congruence except for $r = 2$, where the sign must be changed. $\square$

The above result is stated in a $\mathfrak{p}$-adic context. It is easy to see that it implies the usual form of Stickelberger's theorem. Indeed, we have the following:

Lemma 11.7.12. *If π is the unique element of $\mathbb{Q}_p(\zeta_p)$ such that $\pi^{p-1} = -p$ and $\pi/(\zeta_p - 1) \equiv 1 \pmod{\mathfrak{p}\mathbb{Z}_K}$, then with the notation of Section 3.6.2 we have $\tau(\omega_{\mathfrak{P}}^{-r}, \psi_1) = \tau_q(r)$.*

Proof. By definition $\omega_{\mathfrak{P}}(x)$ is the unique $(q - 1)$st root of unity congruent to x modulo $\mathfrak{P}$. Furthermore, $\mathrm{Tr}_{(\mathbb{Z}_L/\mathfrak{P})/(\mathbb{Z}_K/\mathfrak{p})}(x)$ depends only on the class of x modulo $\mathfrak{P}$. It follows that we can write

$$\tau(\omega_{\mathfrak{P}}^{-r}, \psi_1) = \sum_{a \in L, \, a^{q-1}=1} a^{-r} \zeta_p^{\mathrm{Tr}_{(\mathbb{Z}_L/\mathfrak{P})/(\mathbb{Z}_K/\mathfrak{p})}(a)} \, .$$

Since both sides are algebraic numbers, we can replace the sum by a sum in the completion $\mathcal{L}$ of L. If we choose the unique $\pi \in \mathcal{K}$ such that $\pi^{p-1} = -p$ and $\pi/(\zeta_p-1) \equiv 1 \pmod{\mathfrak{p}\mathbb{Z}_\mathcal{K}}$ then since $\zeta_p = 1+(\zeta_p-1)$, by definition we will have the equality $\zeta_p = \zeta_\pi$ in $\mathcal{K}$. Finally, note that since $\mathcal{L}/\mathcal{K}$ is an unramified extension of $\mathfrak{p}$-adic fields and since $a^{q-1} = 1$, it follows from Corollary 4.4.29 that $\mathrm{Tr}_{(\mathbb{Z}_L/\mathfrak{P})/(\mathbb{Z}_K/\mathfrak{p})}(a) = \mathrm{Tr}_{\mathcal{L}/\mathcal{K}}(a)$, proving the lemma. $\qquad \square$

Remark. In the above proof we have in fact considered $\mathbb{F}_q$ as $\mathbb{Z}_\mathcal{L}/\mathfrak{P}\mathbb{Z}_\mathcal{L}$. In that context the character $\omega_{\mathfrak{P}}$ is essentially the Teichmüller character studied in Chapter 4.

We can now give an alternative proof of Stickelberger's Theorem 3.6.6, which we restate here in a slightly different form, using Lemma 3.6.7.

Theorem 11.7.13. *With the above notation we have*

$$\frac{\tau(\omega_{\mathfrak{P}}^{-r}, \psi_1)}{(\zeta_p - 1)^{s(r)}} \equiv -\frac{(-p)^{v_p(r!)}}{r!} \pmod{\mathfrak{P}} \, .$$

Proof. As usual by periodicity we may assume that $0 \leqslant r < q-1$. By the above lemma and Proposition 11.7.11 we have (in the $\mathfrak{p}$-adic context)

$$
\begin{aligned}
\frac{\tau(\omega_{\mathfrak{P}}^{-r}, \psi_1)}{(\zeta_p - 1)^{s_p(r)}} &= \frac{\tau_q(r)}{\pi^{s_p(r)}} \left(\frac{\zeta_p - 1}{\pi}\right)^{s_p(r)} \\
&\equiv -\frac{(-p)^{v_p(r!)}}{r!} \left(\frac{\zeta_p - 1}{\pi}\right)^{s_p(r)} \pmod{\mathfrak{p}^{p-1}\mathbb{Z}_\mathcal{K}} \, .
\end{aligned}
$$

Since by definition of π we have $(\zeta_p - 1)/\pi \equiv 1 \pmod{\mathfrak{p}\mathbb{Z}_\mathcal{K}}$ we deduce that

$$\frac{\tau(\omega_{\mathfrak{P}}^{-r}, \psi_1)}{(\zeta_p - 1)^{s_p(r)}} \equiv -\frac{(-p)^{v_p(r!)}}{r!} \pmod{\mathfrak{p}\mathbb{Z}_\mathcal{K}} \, .$$

Since $\tau(\omega_{\mathfrak{P}}^{-r}, \psi_1) \in L$ and $\mathfrak{P}/\mathfrak{p}$ is unramified, the result follows. $\qquad \square$

Remark. Note that since by Exercise 35 we have $v_\pi((\zeta_p - 1)/\pi - 1) = 1$ for $p \geqslant 3$, the exponent $p-1$ reduces to 1, so we lose on the power of $\mathfrak{p}$. However, by the same exercise, for $p \geqslant 3$ we can deduce the more precise congruence

$$\frac{\tau(\omega_{\mathfrak{P}}^{-r}, \psi_1)}{(\zeta_p - 1)^{s(r)}} \equiv -\frac{(-p)^{v_p(r!)}}{r!} \left(1 + \frac{\pi}{2} s_p(r)\right) \pmod{\mathfrak{P}^2} \, .$$

11.7.4 Application to the Hasse–Davenport Product Relation

We have seen in Section 3.7.2 that (at least for $p \geqslant 3$) the Hasse–Davenport (HD) product relation is a rather easy consequence of Stickelberger's theorem together with the distribution relations for the functions $s(r)$ and $t(r)$ proved in Section 3.7.1. In this section we are going to see that it can be proved directly and painlessly from the Gross–Koblitz formula together with the distribution formula for the p-adic gamma function without using any algebraic number theory. We will in fact obtain some additional information. We begin with the following.

Lemma 11.7.14. *As usual let* $q = p^f$, $m \mid q - 1$, *and* $d = (q - 1)/m$. *For integers* r *and* i *such that* $0 \leqslant r < q - 1$ *and* $0 \leqslant i < f$ *set*

$$e(r, i) = \left\{ \frac{p^{f-i}r}{q-1} \right\} = \frac{r^{(i)}}{q-1} \,,$$

and for general integers r *and* i *set* $e(r, i) = e(r \bmod (q-1), i \bmod f)$. *Finally, for* $b \in \mathbb{Z}$ *set*

$$g(m, b, i) = \begin{cases} m^{r_i} \exp_p(-e(mb, i+1) \log_p(m^{p-1})) & \text{if } p > 2 \,, \\ m^{r_i + r_{i+1}} \exp_p(-e(mb, i+2) \log_p(m^2)) & \text{if } p = 2 \,, \end{cases}$$

where $mb \bmod (q-1) = \sum_{0 \leqslant i < f} r_i p^i$ *is the usual base-p expansion of* $mb \bmod (q-1)$. *Then for all* $b \in \mathbb{Z}$ *we have*

$$\prod_{0 \leqslant a < m} \Gamma_p(e(ad + b, i)) = c_{p,m} g(m, b, i) \Gamma_p(e(mb, i)) \,,$$

where $c_{p,m}$ *is given by Theorem 11.6.14.*

Proof. Denote the left-hand side by $P(b)$. It is clear that $P(b)$ is periodic in b of period dividing d, hence we may assume that $0 \leqslant b < d$. Writing $p^{f-i}(ad + b)/(q - 1) = p^{f-i}a/m + x$ with $x = p^{f-i}b/(q - 1)$ it is clear that the argument of the gamma function depends only on the value of a modulo m and not on a itself, so we may replace the product from 0 to $m - 1$ by the product for a modulo m. Since p is coprime to $q - 1 = p^f - 1$ hence to m, the map $z \mapsto p^{f-i}z$ is a bijection of $\mathbb{Z}/m\mathbb{Z}$ onto itself, so we can remove the factor p^{f-i}. Similarly the map $a \mapsto a + k$ is also a bijection, so we may replace x by $y = x - k/m$, where $k = \lfloor mx \rfloor$. Since $0 \leqslant x - k/m < 1/m$ we have

$$P(b) = \prod_{0 \leqslant a < m} \Gamma_p(a/m + y) = c_{p,m} m^{-(my-1-(my-1)\backslash p)} \Gamma_p(my) \,,$$

using the distribution formula for the p-adic gamma function (Theorem 11.6.14). By definition if we set $r = mb$ we have $0 \leqslant r < q - 1$, so that

$$my = mx - k = \{mx\} = \{p^{f-i}mb/(q-1)\} = (mb)^{(i)}/(q-1) = e(mb,i) \ .$$

An immediate calculation then shows that

$$(my - 1)\backslash p = (mb)^{(i+1)}/(q-1) - 1 = e(mb,i+1) - 1 \ ,$$

and the lemma follows from this and Proposition 11.6.2.   □

Corollary 11.7.15. (1) *With the same notation, for all $b \in \mathbb{Z}$ we have*

$$\prod_{0 \leqslant a < m} \tau_q(ad + b) = -\omega(m)^{s(mb)} \tau_q(mb) \prod_{0 \leqslant a < m} \tau_q(ad) \ ,$$

where $\omega(m)$ is the Teichmüller character (see Proposition 4.3.4), here by convention taken equal to 1 if $p = 2$.
(2) *For all $n \in \mathbb{Z}$, with n even if $p = 2$, define $(-p)^{n/2} = \pi^{n(p-1)/2}$. Then*

$$\prod_{0 \leqslant a < m} \tau_q(ad) = (-1)^m c_{p,m}^f (-p)^{(m-1)f/2} \ .$$

(3) *We have*

$$\prod_{0 \leqslant a < m,\ a \neq m/2} \tau_q(ad) = \begin{cases} -\left(\dfrac{p}{m}\right)^f q^{(m-1)/2} & \text{if } m \text{ is odd,} \\[2ex] -\left(\dfrac{(-1)^{m/2+1}m/2}{p}\right)^f q^{(m-2)/2} & \text{if } m \text{ is even.} \end{cases}$$

Proof. (1). Either directly or by Lemma 3.6.7 it is clear that we have $\sum_{0 \leqslant i < f} e(mb,i) = s(mb)/(p-1)$. Thus, if we denote by LHS the left hand side of (1), then by the above lemma and the Gross–Koblitz formula (more precisely Corollary 11.7.6) we have

$$\text{LHS} = (-1)^m c_{p,m}^f \pi^{S(b)} h(m,b) \prod_{0 \leqslant i < f} \Gamma_p(e(mb,i)) \ ,$$

with $S(b) = \sum_{0 \leqslant a < m} s(ad + b)$ and

$$h(m,b) = \begin{cases} m^{s(mb)} \exp_p(-s(mb) \log_p(m^{p-1})/(p-1)) & \text{if } p \geqslant 3 \ , \\ m^{2s(mb)} \exp_p(-s(mb) \log_p(m^2)) = 1 & \text{if } p = 2 \ . \end{cases}$$

Thus using again Corollary 11.7.6 together with the basic formula for the Teichmüller character $\omega(x)$ (Proposition 4.3.4, except that we set $\omega(m) = 1$ for $p = 2$), we deduce that for all p we have

$$\text{LHS} = (-1)^{m-1} c_{p,m}^f \omega(m)^{s(mb)} \pi^{S(b)-s(mb)} \tau_q(mb)$$
$$= (-1)^{m-1} c_{p,m}^f \omega(m)^{s(mb)} \pi^{S(0)} \tau_q(mb) \ ,$$

using the distribution formula for the function s (Proposition 3.7.1). In particular, since $\tau_q(0) = -1$ we have

$$\prod_{0 \leqslant a < m} \tau_q(ad) = (-1)^m c_{p,m}^f \pi^{S(0)} ,$$

proving (1), and (2) also follows by using the value $S(0) = (p-1)(m-1)f/2$ given by Proposition 3.7.1 and the formula $\pi^{p-1} = -p$. Applying (2) with $m = 2$ we obtain $-\tau_q((q-1)/2) = c_{p,2}^f(-p)^{f/2}$, so (3) follows after a short calculation by considering separately m odd and m even and using the formula for $c_{p,m}$ given by Theorem 11.6.14. □

Remark. Since $\omega(m)$ is a $(p-1)$st root of unity and since $mb \equiv s(mb)$ (mod $(p-1)$), we may replace $\omega(m)^{s(mb)}$ by $\omega(m)^{mb}$. Indeed, thanks to this remark it is clear that we immediately deduce the HD product relation (Theorem 3.7.3), together with an additional result:

Theorem 11.7.16. *Let ρ be a multiplicative character of exact order m dividing $(q-1)$, and let ψ be any nontrivial additive character on $\mathbb{F}_q$.*

(1) *For any multiplicative character χ on $\mathbb{F}_q$ we have*

$$\prod_{0 \leqslant a < m} \tau(\chi\rho^a, \psi) = -\chi^{-m}(m)\tau(\chi^m, \psi) \prod_{0 \leqslant a < m} \tau(\rho^a, \psi) .$$

(2) *In addition, if $\psi = \psi_1$ in the notation of Proposition 2.5.4, the product on the right-hand side is given by*

$$\prod_{0 \leqslant a < m} \tau(\rho^a, \psi_1) = k(p, f, m)q^{(m-1)/2} ,$$

where $k(p, f, m)$ is a fourth root of unity given by

$$k(p, f, m) = \begin{cases} -\left(\dfrac{p}{m}\right)^f & \text{if } m \text{ is odd,} \\ (-1)^f\left(\dfrac{(-1)^{m/2+1}m/2}{p}\right)^f\left(\dfrac{-1}{p}\right)^{f/2} & \text{if } m \text{ is even,} \end{cases}$$

where $(-1)^{f/2}$ is to be understood as i^f when f is odd.

(3) *Equivalently, we have*

$$J_{m-2}(\rho, \dots, \rho^{m-2}) = \left(\frac{p}{m}\right)^f q^{(m-3)/2} \qquad \text{if } m \text{ is odd,}$$

$$J_{m-1}(\rho, \dots, \rho^{m-1}) = \left(\frac{(-1)^{m/2+1}m/2}{p}\right)^f q^{(m-2)/2} \qquad \text{if } m \text{ is even.}$$

Proof. (1) clearly follows from the above corollary, Lemma 11.7.12, and the fact that $\omega_{\mathfrak{P}}(m)$ can be identified with the Teichmüller character $\omega(m)$ in $\mathbb{Q}_p$, since they are both $(p-1)$st roots of unity congruent to m modulo $\mathfrak{P}$. (2) is also immediate by inspection when m is odd. When m is even, by (3) of the above corollary we have

$$\prod_{0\leqslant a<m,\ a\neq m/2} \tau(\rho^a,\psi_1) = -\left(\frac{(-1)^{m/2+1}m/2}{p}\right)^f q^{(m-2)/2}\ ,$$

and on the other hand, by Corollary 3.7.6 we have

$$\tau(\rho^{m/2},\psi_1) = (-1)^{f-1}\left(\frac{-1}{p}\right)^{f/2} q^{1/2}\ ,$$

with the interpretation of $(-1)^{f/2}$ given in the theorem. Finally, (3) follows immediately from (2) and Proposition 2.5.14.   □

Remarks. (1) When m is even it is necessary to remove the term $a = m/2$ in the p-adic product, and then put it back in the complex product. Indeed, to take the specific example of $p = 5$, $f = 1$, and $m = 2$, the identification of the p-adic and complex products gives the "identity"

$$\Gamma_5(1/2)\pi^2 = -5^{1/2}\ ,$$

where π is the fourth root of -5 in $\mathbb{Q}_5(\zeta_5)$ such that $\pi \equiv \zeta_5 - 1 \pmod{\pi^2}$, and $5^{1/2}$ is the positive square root in $\mathbb{R}$. Although the square of this identity gives the correct equality $5 = 5$ (since $\Gamma_5(1/2)^2 = -1$), the identity does not seem to mean much in itself.

(2) If we want the product for $\psi = \psi_b$ with $b \in \mathbb{F}_q^*$ in the notation introduced in Proposition 2.5.4, it is clear from Lemma 2.5.6 that the right-hand side of the formula stays the same when m is odd, and is multiplied by $\rho^{m/2}(b) = \chi_q(b)$ when m is even, hence is unchanged if b is a square in $\mathbb{F}_q$ and multiplied by -1 otherwise.

(3) See also Exercise 55 for another result.

11.8 Exercises for Chapter 11

1. Prove Proposition 11.1.6 and the formulas in the examples following it.

2. (E. Friedman.) Let D be a subset of $\mathbb{C}_p$ closed under translation by $\mathbb{Z}_p$, for instance $D = \mathbb{Z}_p$ or $D = C\mathbb{Z}_p$. Let f be a function from D to $\mathbb{C}_p$, and assume that $\Delta(f)(x) = f(x+1) - f(x)$ is strongly differentiable on D. Finally, let h be a strongly differentiable function from D to $\mathbb{C}_p$, and for $x \in \mathbb{Z}_{\geqslant 0}$ set $H(x) = \sum_{0\leqslant a<x} h(a)$.

(a) Show that $H(x)$ is a continuous p-adic function on $\mathbb{Z}_{\geqslant 0}$, so that we can define $H(x)$ for any $x \in \mathbb{Z}_p$ by $H(x) = \lim_{n\to x,\ n\in\mathbb{Z}_{\geqslant 0}} H(n)$, and show that the resulting function H is strictly differentiable on $\mathbb{Z}_p$.

(b) Show that for any $x \in D$ the Volkenborn integral $\int_{\mathbb{Z}_p} h(t) f(x+t)\, dt$ exists and that we have

$$\int_{\mathbb{Z}_p} h(t) f(x+t)\, dt = f(x) \int_{\mathbb{Z}_p} h(t)\, dt - \int_{\mathbb{Z}_p} H(t+1)\Delta(f)(x+t)\, dt \ .$$

(c) Assume that $f(x) = \int_{\mathbb{Z}_p} g(x+t)\, dt$. Deduce that under suitable sufficient conditions on g we have

$$\int_{\mathbb{Z}_p} h(t) f(x+t)\, dt = f(x) \int_{\mathbb{Z}_p} h(t)\, dt - \int_{\mathbb{Z}_p} H(t+1) g'(x+t)\, dt$$

(this is essentially "integration by parts" for Volkenborn integrals).

(d) In particular, show that under these conditions we have

$$\int_{\mathbb{Z}_p} f(x+t)\, dt = f(x) + (x-1) f'(x) - \int_{\mathbb{Z}_p} (x+t) g'(x+t)\, dt \ .$$

(e) Deduce from this alternate proofs of the Raabe formulas seen in the text.

3. (Applications of the preceding exercise.)

(a) Set $b_j = \int_{\mathbb{Z}_p} B_j(t)\, dt$, and let $F(T)$ be the formal power series $F(T) = \sum_{j \geq 0} b_j T^j / j!$. Show that $F(T) = (T/(\exp(T)-1))^2$, and by computing the derivative of $T/(\exp(T)-1)$, deduce that $b_j = -(j B_{j-1} + (j-1) B_j)$.

(b) Using the preceding exercise, show that

$$\int_{\mathbb{Z}_p} t^k B_n(x+t)\, dt = \frac{B_{k+1}}{k+1} n B_{n-1}(x) + B_k B_n(x)$$

$$- \frac{n}{k+1} \sum_{j=-1}^{k} (-1)^{k-j} \binom{k+1}{j+1} B_{k-j}(x) B_{n+j}(x) \ .$$

In particular, generalizing (a) we have

$$\int_{\mathbb{Z}_p} B_j(x+t)\, dt = -(j(1-x) B_{j-1}(x) + (j-1) B_j(x)) \ .$$

(c) Deduce the following reciprocity formula for Bernoulli numbers:

$$m! \sum_{j=0}^{m} \frac{B_{m-j}}{(m-j)!} \frac{B_{n+j+1}}{(j+1)!} + n! \sum_{j=0}^{n} \frac{B_{n-j}}{(n-j)!} \frac{B_{m+j+1}}{(j+1)!} = -B_{m+n} \ .$$

See Exercise 5 of Chapter 9 for another proof.

(d) Using once again the preceding exercise, show that for $x \in C\mathbb{Z}_p$ we have

$$\int_{\mathbb{Z}_p} t^k \zeta_p(s, x+t)\, dt = -\frac{B_{k+1}}{k+1} \omega_v^{-1}(x) s \zeta_p(s+1, x) + B_k \zeta_p(s, x)$$

$$+ \frac{1}{k+1} \sum_{j=-1}^{k} (-1)^{k-j} \binom{k+1}{j+1} (s-j-1) B_{k-j}(x) \omega_v(x)^j \zeta_p(s-j, x) \ .$$

Note that this is the p-adic analogue of the formula proved in Exercise 45 (b) of Chapter 9.

(e) Deduce that for $x \in C\mathbb{Z}_p$ we have

$$\int_{\mathbb{Z}_p} t^k \operatorname{Log}\Gamma_p(x+t)\,dt = \frac{(-1)^k B_{k+1}(x) + B_{k+1}}{k+1}\,\psi_p(x) + B_k \operatorname{Log}\Gamma_p(x)$$

$$+ \frac{1}{k+1}\sum_{j=0}^{k}(-1)^{k-j}\binom{k+1}{j+1}B_{k-j}(x)\omega_v(x)^{j+1}\left(\zeta_p(-j,x) - (j+1)\frac{\partial\zeta_p}{\partial s}(-j,x)\right)$$

(recall that $\zeta_p(0,x) = -\omega_v^{-1}(x)(x-1/2)$ and $\frac{\partial\zeta_p}{\partial s}(0,x) = \omega_v^{-1}(x)\operatorname{Log}\Gamma_p(x)$).
This is the analogue of the formula of Exercise 45 (c) of Chapter 9.

4. Show that under suitable sufficient conditions on the function f, we have the following general Volkenborn integral evaluations, which are p-adic forms of the Euler–MacLaurin summation formula:

$$\int_{\mathbb{Z}_p} f(x+t)\,dt = \sum_{j\geqslant 0}\frac{B_j}{j!}f^{(j)}(x)$$

$$\text{and}\quad \int_{\mathbb{Z}_p^*} f(x+t)\,dt = \sum_{j\geqslant 0}(1-p^{j-1})\frac{B_j}{j!}f^{(j)}(x)\,.$$

5. The aim of this exercise is to give a definition and properties of $\zeta_p(s,x)$ for $p=2$ and $v_p(x) = -1$. All the proofs will need to use continuity and the fact that $\mathbb{Z}$ is dense in $\mathbb{Z}_p$, and *not* analyticity and power series. Thus, assuming that $p=2$ we set

$$\zeta_p(s,x) = \frac{1}{s-1}\int_{\mathbb{Z}_p}\varepsilon(t/x,s)\langle x+t\rangle^{1-s}\,dt\,,$$

where for simplicity of notation, for $y \notin \mathbb{Z}_2$ we have set $\varepsilon(y,s) = \omega(1+y)^{1-s}$, so that $\varepsilon(y,s) = 1$ unless $v_p(y) = -1$ and $s \in 2\mathbb{Z}_2$.

(a) Prove that $\zeta_p(s,x)$ is a continuous function of s on $\mathbb{Z}_p \setminus \{1\}$, and that the Laurent series expansion of Definition 11.2.5 is still valid for $s \in \mathbb{Z}_p \setminus \{1\}$.

(b) Prove that the results of Theorem 11.2.9 are still valid with the following modifications: in (2), the right-hand side must be multiplied by $\varepsilon(u/x,s)$, (3) must be replaced by

$$\varepsilon(1/x,s)\zeta_p(s,x+1) - \zeta_p(s,x) = -\frac{\langle x\rangle^{1-s}}{x} = -\omega_v(x)^{-1}\langle x\rangle^{-s}$$

$$\text{and}\quad \zeta_p(s,1-x) = \varepsilon(1/x,s)\zeta_p(s,x)\,,$$

and (4) by

$$\sum_{0\leqslant j<m}\varepsilon\left(\frac{j}{mx},s\right)\zeta_p\left(s,x+\frac{j}{m}\right) = m\langle m\rangle^{s-1}\zeta_p(s,mx)$$

$$= \omega_v(m)\langle m\rangle^s\zeta_p(s,mx)\,.$$

6. Continuing the preceding exercise, assume that $p=2$, $v_p(x) = -1$, and let now $s \in \mathbb{C}_p$. Using in particular Lemma 4.2.8, show that the series of Proposition 11.2.4 for $\zeta_p(s,x)$ converges if $|1-s| \leqslant 1/2$, and that there exist values of s such that $|1-s| = 1$ for which it does not converge. Does it converge for any value of s such that $1/2 < |1-s| < 1$?

7. Prove Lemma 11.3.2.

8. Show that if $a \in \mathbb{Z}_2^*$ we have $(a/4 - 1/2)\log_p(a) - a/4 \equiv -\omega(a)/4 \pmod{4\mathbb{Z}_2}$.

9.

(a) By introducing a primitive p^N th root of unity for p odd and proceeding similarly for $p = 2$, prove directly that if χ is a primitive character of conductor a power of p then

$$\sum_{0 \leqslant n < p^N}^{(p)} \chi(n) \langle n \rangle^{-s}$$

tends to 0 p-adically as $N \to \infty$, and in fact give an upper bound for its p-adic absolute value.

(b) Deduce directly that $\zeta_p(\chi, s, x)$ is a continuous function of $x \in \mathbb{Z}_p$.

10. Prove Proposition 11.3.8 (3).

11. Prove Corollary 11.4.2 (6), (7), and (8).

12. Prove Corollary 11.4.3.

13.

(a) Prove that if $|u| < |x|$ and $x \in C\mathbb{Z}_p$ we have

$$\text{Log}\Gamma_p(x + u) = \text{Log}\Gamma_p(x) + \psi_p(x)u + \sum_{k \geqslant 2}(-1)^k w_v(x)^{1-k} \zeta_p(k, x)\frac{u^k}{k}.$$

(b) Let $\zeta_p(s)$ be the Kubota–Leopoldt p-adic zeta function. For $a \in 2\mathbb{Z}$, compute $\sum_{k \geqslant 1} \zeta_2(k+1)a^k$ in terms of the function $\psi_p(x)$.

14. Prove that if $k \in \mathbb{Z}_{\geqslant 0}$ we have

$$\sum_{0 \leqslant a < p}^{(p)} \psi_p\left(k + \frac{a}{p}\right) = -p\gamma_p + \sum_{0 \leqslant m < pk}^{(p)} \frac{1}{m}.$$

15. Prove Proposition 11.5.13, Corollary 11.5.14, and Proposition 11.5.15, and fill in the details of the proof of Proposition 11.5.18.

16. As mentioned in the text, the power series for $\text{Log}\Gamma_p(\chi, x)$ given by Proposition 11.5.19 is in general not valid for $x \in p\mathbb{Z}_p \setminus p^v\mathbb{Z}_p$. Show this on a numerical example as follows. Let χ be one of the two even primitive characters modulo 9. Prove that $\text{Log}\Gamma_p(\chi, 3) \in \rho\mathbb{Z}_3$, where ρ is a suitable primitive cube root of unity. On the other hand, evaluate the power series expansion at $x = 3$ and show that it definitely is not in $\rho\mathbb{Z}_3$ by computing it for instance modulo 3^2.

17. Let χ be a primitive character of conductor f, let $\chi_1 = \chi\omega^{-1}$, and let f_1 be the conductor of χ_1.

(a) Show that if $p \nmid f$ then $f_1 = q_p f$, that if $p \nmid f_1$ then $f = q_p f_1$, and that if $p \mid f$ and $p \mid f_1$ then $f_1 = f$.

(b) Using Corollary 11.5.5 and Proposition 11.5.12, prove Proposition 11.5.24.

18. Although we cannot define a canonical exponential of the function $\text{Log}\Gamma_p(x)$ for $x \in C\mathbb{Z}_p$, in particular having nice properties, show the following.

(a) For $a \in \mathbb{C}_p$ such that $|a| \leqslant 1$, prove that there exists a canonical exponential $\Gamma_{L,p}(x, a)$ of the function $\text{Log}\Gamma_p(x + a) - \text{Log}\Gamma_p(x)$, defined for $x \in C\mathbb{Z}_p$. Note that if $b \in \mathbb{Z}_p$, we have the trivial cocycle relation $\Gamma_{L,p}(x, a + b) = \Gamma_{L,p}(x + b, a)\Gamma_{L,p}(x, b)$.

(b) Prove that under suitable conditions we have the following relations:

$$\Gamma_{L,p}(x, a+1) = \langle x + a \rangle \Gamma_{L,p}(x, a) ,$$
$$\Gamma_{L,p}(x+1, a) = (1 + a/x) \Gamma_{L,p}(x, a) ,$$
$$\Gamma_{L,p}(1 - x, -a) = 1/\Gamma_{L,p}(x, a) ,$$
$$\prod_{0 \leqslant j < m} \Gamma_{L,p}(x + j/m, a) = \Gamma_{L,p}(mx, ma) \langle m \rangle^{-ma} .$$

19. Consider Gauss's hypergeometric series

$$F(a, b, c; x) = \sum_{n \geqslant 0} \frac{a(a+1) \cdots (a+n-1)b(b+1) \cdots (b+n-1)}{c(c+1) \cdots (c+n-1)} \frac{x^n}{n!}$$

introduced in Exercise 115 of Chapter 9, but now in the p-adic domain. We Assume that all the variables are in $\mathbb{Q}_p$, that a and b are in $\mathbb{Z}_p$, and that $v_p(c) < 0$.

(a) Compute the radius of convergence of the series. Deduce in particular that $f(a, b, c) = F(a, b, c; 1)$ is well defined, and that the contiguity relation proved in the above-mentioned exercise is still valid in the p-adic domain of convergence.

(b) Set $\phi_n(a, c) = \binom{-a}{n} / \binom{-c}{n}$. Prove that for $n \geqslant 1$,

$$\phi_n(a, c) = \frac{a}{a+1-c} (\phi_n(a+1, c) - \phi_{n-1}(a+1, c)) ,$$

and deduce that $f(a, 1, c) = (c - 1)/(c - a - 1)$ (this is of course a special case of Gauss's evaluation, but now in the p-adic domain).

(c) Compute $f(a, b, c)$ when a or b is in $\mathbb{Z}$, inside the domain of convergence.

(d) Deduce that we have the following analogue of Gauss's evaluation in the p-adic domain, for a and b in $\mathbb{Z}_p$ and $c \in \mathbb{Q}_p$ with $v_p(c) < 0$:

$$\log_p(f(a, b, c)) = \text{Log}\Gamma_p(c) + \text{Log}\Gamma_p(c - a - b) - \text{Log}\Gamma_p(c - a) - \text{Log}\Gamma_p(c - b) ,$$

where $\text{Log}\Gamma_p$ is Diamond's log gamma function.

(e) Under the same assumptions on a, b, and c, let $(a_n)_{n \geqslant 0}$ be a sequence of positive integers tending p-adically to a as $n \to \infty$. Show that in fact

$$f(a, b, c) = \lim_{n \to \infty} \frac{(c-1)(c-2) \cdots (c - a_n)}{(c - b - 1)(c - b - 2) \cdots (c - b - a_n)} .$$

(Hint: you will first have to prove that the right-hand side is a Cauchy sequence, which is not completely trivial, which can be done for instance using Exercise 37 of Chapter 9.)

(f) Under the same assumptions, prove that we also have

$$f(a, b, c) = w(1 + ab/c) \frac{\Gamma_{L,p}(c - a, a)}{\Gamma_{L,p}(c - a - b, a)} = w(1 + ab/c) \frac{\Gamma_{L,p}(c - b, b)}{\Gamma_{L,p}(c - a - b, b)} ,$$

where $\Gamma_{L,p}(x, a)$ is the function introduced in Exercise 18. By what should this formula be replaced if a, b, or c is in $\mathbb{C}_p$ instead? (I do not know the answer.)

(g) Let c be a parameter and let $k \in \mathbb{Z}_{\geqslant 0}$. Using the above results, prove that the following identity is valid in the complex domain for $\Re(c) > k+1$, and in the p-adic domain for $v_p(c) < 0$:

$$\sum_{n \geqslant 0} \binom{n}{k} \frac{n!}{c(c+1)\cdots(c+n)} = \frac{k!}{(c-1)(c-2)\cdots(c-k-1)} \,.$$

(Hint: apply (c) of Exercise 115 of Chapter 9 and (c) of the present exercise to $a = b = k+1$, with c replaced by $k+1+c$.) Note that we have already proved this result differently in Exercise 39 of Chapter 4.

(h) Set

$$g(c;x) = \sum_{n \geqslant 0} \frac{n!}{c(c+1)\cdots(c+n)} x^n \,.$$

Deduce that if $v_p(c) < 0$ and if $x \in \mathbb{C}_p$ with $|x| \leqslant 1$ then $g(1-c,x) = -g(c, 1-x)$. For which values of (c, x) is this formula true in the complex domain? (Warning: it is not what you may expect.)

20. Prove Proposition 11.5.2, and Corollary 11.5.3.

21. The following exercises are sequels to Exercises 88 and 89 of Chapter 9. Their goal is to prove in the p-adic case the results of the above-mentioned exercises, of which we keep the notation. I thank F. Beukers for telling me of the problems.

(a) Using Corollary 11.5.3 and Exercise 38 of Chapter 4, prove that for $|x| > 1$ in $\mathbb{C}_p$ we have

$$\psi_p'(x) = \sum_{n \geqslant 0} \frac{1}{n+1} \begin{bmatrix} n \\ x \end{bmatrix} \,.$$

(b) For instance, deduce that in $\mathbb{Q}_2$ we have

$$\sum_{n \geqslant 0} \frac{1}{n+1} \begin{bmatrix} n \\ 1/2 \end{bmatrix} = 0 \,.$$

(c) Deduce from Proposition 11.5.6 that

$$\zeta_p(2,x) = \omega_v(x) \sum_{n \geqslant 0} \frac{1}{n+1} \begin{bmatrix} n \\ x \end{bmatrix} \,.$$

(d) Similarly, show that

$$\zeta_p(3,x) = \omega_v(x)^2 \sum_{n \geqslant 0} \frac{H_n}{n+1} \begin{bmatrix} n \\ x \end{bmatrix} \,,$$

where as usual $H_n = \sum_{1 \leqslant j \leqslant n} 1/j$ is the harmonic sum.

(e) It is clear that if we set $p = 1$ and $x = 1$ in the formula for $\zeta_p(2, x)$ we obtain $\sum_{n \geqslant 0} 1/(n+1)^2$, which is the usual real sum giving $\zeta(2)$. Show that the same is true for $\zeta_p(3, x)$; in other words, prove that in $\mathbb{R}$ we have the identity

$$\sum_{n \geqslant 0} \frac{H_n}{(n+1)^2} = \zeta(3) \,.$$

(Hint: by expanding in powers of t, prove that the left-hand side is equal to $\int_0^1 \log^2(1-t)/(2t)\, dt$, and then change t into $1-t$ and expand again.)

The reader will note that many p-adic formulas reduce to real formulas in this way. For instance, it is easy to see that the convergent expansion of the p-adic Hurwitz zeta function (Proposition 11.2.4) reduces to the nonconvergent but asymptotic expansion of the ordinary Hurwitz zeta function $\zeta(s,x)$ for $x \in \mathbb{Z}_{\geqslant 1}$ if we set $p = 1$.

22. Show that the formula of Exercise 88 (e) of Chapter 9 is still valid in the p-adic case, as usual with $|x| > 1$ and with $\psi'(x)$ replaced by $\psi'_p(x)$.

23.

(a) Using Exercise 89 (d) of Chapter 9, show that for $x \in \mathbb{C}_p$ with $|x| > 1$ we have the absolutely convergent series

$$T_p(x) = \sum_{n \geqslant 0} \frac{1}{n+x} \begin{bmatrix} n \\ x \end{bmatrix} = -\sum_{j \geqslant 0} \frac{2(2^j - 1)B_j}{x^{j+1}} \ .$$

(b) Deduce from Corollary 11.5.3 that for $|x| > 1$ in $\mathbb{C}_p$ we have

$$\psi'_p(x/2) - 2\psi'_p(x) = \sum_{n \geqslant 0} \frac{1}{n+x} \begin{bmatrix} n \\ x \end{bmatrix} = T_p(x) \ ,$$

which is the exact analogue of the formula in the complex case.

(c) Assume $x \in \mathbb{Q}_p$ is such that $v_p(x) < 0$. By Exercise 39 of Chapter 4, we know that

$$T_p(x) = -\sum_{n \geqslant 0} \begin{bmatrix} n \\ x \end{bmatrix} \begin{bmatrix} n \\ 1-x \end{bmatrix} = \sum_{n \geqslant 0} \frac{1}{n+x} \begin{bmatrix} n \\ x \end{bmatrix} \ .$$

Using Proposition 11.5.6, prove the following formulas:

(i) We have

$$T_p(x) = \omega_v(x)^{-1}(\omega_v(2)\zeta_p(2, x/2) - 2\zeta_p(2, x)) \ .$$

(ii) If $v_p(x) \leqslant -v_p(q_p)$ we also have

$$T_p(x) = \omega_v(x)^{-1}(2\zeta_p(2, x) - \omega_v(2)\zeta_p(2, (x+1)/2))$$
$$= \frac{\omega_v(x/2)^{-1}}{2}(\zeta_p(2, x/2) - \zeta_p(2, (x+1)/2)) \ .$$

(iii) If $p = 2$ and $v_p(x) = -1$, using the extended definition of $\zeta_p(s, x)$ given in Exercise 5, we also have

$$T_p(x) = \frac{\omega_v(x/2)^{-1}}{2}(\zeta_p(2, x/2) + \zeta_p(2, (x+1)/2)) \ .$$

(d) In particular, prove that

$T_2(1/2) = 8\zeta_2(2)$, $T_2(1/4) = 16L_2(\chi_8, 2)$, $T_2(1/6) = 40\zeta_2(2)$,

$T_3(1/3) = (27/2)\zeta_3(2)$, $T_3(1/6) = 36L_3(\chi_{12}, 2)$,

$T_2(1/12) = 8(8L_2(\chi_8, 2) + 9L_2(\chi_{24}, 2))$, $T_3(1/12) = 72(L_3(\chi_8, 2) + L_3(\chi_{24}, 2))$,

where $\chi_D(n) = \left(\frac{D}{n}\right)$.

24.

(a) Using similar methods to those of the preceding exercises, prove that, in perfect analogy with Exercise 90 of Chapter 9, we have

$$\sum_{n \geqslant 0} \frac{1}{n+1} \begin{bmatrix} n \\ x \end{bmatrix} \begin{bmatrix} n \\ 1-x \end{bmatrix} = \psi_p''(x) \ .$$

(b) Deduce for instance that for all $k \in \mathbb{Z}$ such that $-k/2 \notin \mathbb{Z}_{\geqslant 0}$ we have

$$\sum_{n \geqslant 0} \frac{1}{n+k/2} \begin{bmatrix} n \\ m+1/2 \end{bmatrix}^2 = a_{k,m} \zeta_2(3) + b_{k,m}$$

for some rational numbers $a_{k,m}$ and $b_{k,m}$ (for instance $a_{1,0} = 32$, $a_{2,0} = -16$, and $b_{1,0} = b_{2,0} = 0$). You may also want to compute these numbers explicitly, in analogy with Exercise 88 (e) of Chapter 9.

25. Prove Corollaries 11.4.5 and 11.4.7 (a similar congruence exists modulo 563; see Exercise 39).

26. With the notation of Theorem 11.5.25, show that when $z \in \mathcal{D}$ we have $v_p(U(z)) \geqslant 1$.

27. Prove the following formula, analogous to Corollary 11.5.26:

$$\lim_{N \to \infty} (-1)^{p(p-1)N/2} \sum_{\substack{0 \leqslant a < p^N \\ 2 \nmid a}}^{(p)} \frac{(-1)^{(a-1)/2}}{a} = \frac{1}{2} \left(1 - \frac{1}{p}\right) \log_p(2) \ .$$

28. Let χ be a nontrivial character modulo p^v with $v \geqslant 1$, and assume that χ is *primitive*.

(a) Let f be any function from $\mathbb{Z} \setminus p\mathbb{Z}$ to $\mathbb{C}_p$, and let χ be a character modulo p^v for some $v \geqslant 1$, let $I \subset \mathbb{Z}$, and set

$$S(\chi, z) = \sum_{a \in I} \chi(a) z^a f(a) \ .$$

For $\chi \neq \chi_0$ show that

$$S(\chi, z) = \frac{\chi(-1)\tau(\chi)}{p^v} \sum_{x \bmod p^v} \chi^{-1}(x) S(\chi_0, \zeta_{p^v}^x z) \ .$$

(b) In addition, show that $|z - 1| \geqslant 1$ (in other words $z \in \mathcal{D}$) implies that $\zeta_{p^v}^x z \in \mathcal{D}$ and that $|z - 1| < 1$ implies that $|\zeta_{p^v}^x z - 1| < 1$.

(c) Using (1) and Proposition 2.1.44, deduce from Theorem 11.5.25 that if $z \in \mathcal{D}$ we have

$$\lim_{N \to \infty} \frac{1}{z^{p^N} - 1} \sum_{0 \leqslant a < p^N} \chi(a) \frac{z^a}{a} = \frac{\chi(-1)\tau(\chi)}{p^v} \sum_{r \bmod p^v} \chi^{-1}(r) \log_p(1 - \zeta_{p^v}^r z) \ .$$

(d) Using a similar method, prove a corresponding result with the character χ for Theorem 11.5.27 (2), Lemma 11.5.29, and Theorem 11.5.30.

(e) Show that the corresponding result for Theorem 11.5.31 is the following. Assume that $p \nmid m$ and the same assumptions as the theorem, and let $F_N(\chi, u)$ be the corresponding sum with a character. Prove that $F(\chi, u) = \lim_{N \to \infty} F_N(\chi, u)$ exists and that

$$F(\chi, u) = L(\chi, u) + \frac{\chi(-1)\tau(\chi)}{p^v} \sum_{r \bmod p^v} \chi^{-1}(x) \log_p \left(1 - \zeta_{p^v}^x \zeta_m^u\right) ,$$

where

$$L(\chi, u) = \sum_{0 \leqslant a < r} \chi(a) \frac{\zeta_m^{ua}}{a} .$$

(Note: although the result is uniform, it is necessary to distinguish several cases here.)

(f) Deduce finally a proof of Theorem 11.5.35, giving the value of $\psi_p(\chi, r/m)$ for $p \nmid m$ and $0 \leqslant r < m$.

29. The aim of this long exercise and the next is to give alternative proofs of Theorems 11.5.25 and 11.5.27 using formal power series, which is an important tool in p-adic analysis. For any $c \in \mathbb{C}_p$, denote by I_c the formal integration operator from $\mathbb{Z}_p[[T]]$ to $\mathbb{C}_p[[T]]$ sending $\sum_{n \geqslant 0} a_n T^n$ to $c + \sum_{n \geqslant 1}(a_{n-1}/n)T^n$, by U the linear form defined on $\mathbb{C}_p[T]$ by $U(G) = G(0) - (1/p) \sum_{\eta^p = 1} G(\eta - 1)$, and let $V = U \circ I_c$ as a linear form on $\mathbb{C}_p[T]$, which does not depend on c.

(a) Show that $U((1 + T)^n) = 0$ if $p \mid n$, $U((1 + T)^n) = 1$ if $p \nmid n$, $V((1 + T)^n) = 0$ if $p \mid (n + 1)$, and $V((1 + T)^n) = 1/(n + 1)$ if $p \nmid (n + 1)$.
(b) Show that $v_p(V(T^m)) \geqslant (n + 1)/(p - 1) - 1 - v_p(n + 1)$, and that V can be naturally extended to a continuous linear form on $\mathbb{Z}_p[[T]]$.
(c) Show that $v_p(V(((1 + T)^{p^N} - 1)f(T))) \geqslant N$ for any $f \in \mathbb{Z}_p[[T]]$.
(d) Set $f(T) = z/(z(1 + T) - 1) - 1/(1 + T)$. Show that $f \in \mathbb{Z}_p[[T]]$ and compute $V(f)$ in terms of p-adic logarithms. (Warning: you must justify the switch from formal power series to p-adic power series.)
(e) Show that

$$\frac{1}{z^{p^N} - 1} \sum_{0 \leqslant a < p^N} z^a (1 + T)^{a-1} = \left(1 + \frac{z^{p^N}}{z^{p^N} - 1}\left((1 + T)^{p^N} - 1\right)\right) f(T) ,$$

and assuming that $z \in \mathcal{D}$ and applying the form V to both sides, deduce Theorem 11.5.25.

30. Denote by $\mathfrak{P}$ the ideal of $\mathbb{Z}_p[[T]]$ generated by T and p, so that $\sum_{n \geqslant 0} a_n T^n \in \mathfrak{P}^N$ if and only if $v_p(a_n) \geqslant N - n$ for $0 \leqslant n \leqslant N$, and in addition $v_p(a_n) \geqslant 0$ for $n > N$.

(a) Let $H \in \mathbb{Z}_p[[T]]$ be such that for some $u \in \mathbb{C}_p$ such that $v_p(u) \geqslant -1$ and some $k \in \mathbb{Z}_{\geqslant 0}$ we have $(1 + uT)^k H(T) \in \mathfrak{P}^N$ for some $N \geqslant ((p - 1)/\log(p)) - 1$. Show that $v_p(V(H)) \geqslant (N + 1)/(p - 1) - 1 - \log(N + 1)/\log(p)$.
(b) Let $r \in \mathbb{Z}_{\geqslant 0}$ be such that $p \nmid r$ be fixed, let $\zeta = \zeta_{p^v}$, and set

$$S_N(T) = \sum_{0 \leqslant a < p^N} \zeta^a \left\lceil \frac{ar}{p^N} \right\rceil (1 + T)^{a-1} .$$

By imitating the proof given in the preceding exercise, but using (a) and $S_N(T)$ instead, prove Theorem 11.5.27 (2) for $z = \zeta_{p^v}$.

31.

(a) Prove that

$$\lim_{N\to\infty}\frac{1}{p^{2N}}\sum_{0\leqslant a<p^N}^{(p)}\frac{1}{a}=-\frac{B_{-2,p}}{2}\,.$$

(b) Let $\mathrm{Li}_2(x)=\sum_{n\geqslant1}x^n/n^2$ be the dilogarithm function, defined for $|x|<1$ (see Exercise 22 of Chapter 4 for another occurrence of this function). Let z be such that $v_p(z-1)>0$. By expanding in powers of $z-1$ and using (a), prove that

$$\lim_{N\to\infty}\frac{1}{p^N}\sum_{0\leqslant a<p^N}^{(p)}\frac{z^a}{a}=-\mathrm{Li}_2(1-z)+\frac{1}{p^2}\mathrm{Li}_2(1-z^p)\,.$$

32. Let $m\geqslant1$ be such that $q_p\mid m$, and let $r\in\mathbb{Z}$ be such that $p\nmid r$. Using Theorem 11.5.33, show that

$$\int_{\mathbb{Z}_p}\log_p(r+mt)\,dt=\lim_{N\to\infty}\frac{1}{p^N}\sum_{0\leqslant a<p^N}\log_p(r+ma)=-\frac{p}{p-1}\gamma_p$$

$$+\sum_{1\leqslant a\leqslant m-1}\zeta_m^{-ar}\log_p(1-\zeta_m^a)-\frac{1}{p}\sum_{\substack{1\leqslant a\leqslant m-1\\m\nmid ap}}\zeta_m^{-ar}\log_p(1-\zeta_m^{ap})\,.$$

33. Show that for $k\in\mathbb{Z}_{\geqslant1}$ we have

$$\lim_{N\to\infty}\frac{1}{p^N}\sum_{0\leqslant a<p^N}^{(p)}a^k\log_p(a)=kL_p'(\omega^k,1-k)-L_p(\omega^k,1-k)\,.$$

Note that $L_p(\omega^k,1-k)=-(1-p^{k-1})B_k/k$, and that for $k=0$ the above limit is equal to $-\gamma_p$.

34. Show that Theorem 11.5.33 implies the formula of Proposition 11.5.4 (2). Similarly, show that Theorem 11.5.34 implies the special case $s=0$ of the distribution formula for the function ψ_p given in Corollary 11.5.14.

35. Prove Corollary 11.4.8.

36. With the notation of Corollary 11.6.20, prove directly the following results, valid for all $k\geqslant0$:

$$v_2(a_{1,k})=\lfloor(k+1)/2\rfloor+v_2(k-1)\,,$$
$$v_2(b_{1,k})=\lfloor(k+1)/2\rfloor\,,$$
$$v_2(a_{2,k})=v_2(b_{2,k})=k\,.$$

37.

(a) Prove the second formula of Corollary 11.6.3 by showing that $((s-1)-(s-1)\backslash p)+((-s)-(-s)\backslash p)=0$.

(b) Let $a\in\mathbb{Z}_p$. Show that

$$-(-a)\backslash p=\begin{cases}1+(a\backslash p)&\text{if }v_p(a)=0\,,\\(a\backslash p)&\text{if }v_p(a)>0\,.\end{cases}$$

38. Prove directly (i.e., without using L-functions) that $\Gamma'_p(0) \equiv -((p-1)! + 1)/p$ (mod $p\mathbb{Z}_p$).

39. Prove that $v_p(\gamma_p) \geqslant 1$ if and only if $p = 3$ or $(p-1)! \equiv -1$ (mod p^2) if and only if $v_p(B_{p-1} - (1 - 1/p)) \geqslant 1$. Such primes are called *Wilson primes*, and the only known primes satisfying this congruence are $p = 5$, 13, and 563, and there are no others up to $5 \cdot 10^8$, but as usual there should be infinitely many.

40. Assume that $p \geqslant 5$.

   (a) Show that
   $$(p-1)! \equiv -1 + p\gamma_p - \frac{p^2}{2}\gamma_p^2 \pmod{p^3\mathbb{Z}_p} .$$

   (b) Show that
   $$\gamma_p \equiv -\frac{B_{p-1} - (1 - 1/p)}{p - 1} \pmod{p\mathbb{Z}_p} .$$

   (c) Deduce that
   $$(p-1)! \equiv p(B_{p-1} - 1) \pmod{p^2\mathbb{Z}_p} .$$

   (d) Show directly or from the preceding congruence that
   $$(p-1)! \equiv p - 2 - \sum_{1 \leqslant r \leqslant p-1} \langle r \rangle \equiv -p + \sum_{1 \leqslant r \leqslant p-1} r^{p-1} \pmod{p^2\mathbb{Z}_p} .$$

   (e) Find similar congruences modulo $p^3\mathbb{Z}_p$, this time involving $B_{p(p-1)}$.

41. Prove that
$$\Gamma_2(2n) = \frac{(2n)!}{2^n n!} = 1 \cdot 3 \cdots (2n - 1) \equiv (-1)^{n(n-1)/2} \pmod{4} .$$

42. Using a method similar to that of the proof of Corollary 11.6.22 and the duplication formula for Morita's p-adic gamma function, prove that for $p \geqslant 5$ we have the congruence
$$\binom{p-1}{(p-1)/2} \equiv (-1)^{(p-1)/2} 4^{p-1} \left(1 + \frac{p^3}{12} B_{p-3}\right) \pmod{p^4} .$$

43. Let $p \equiv 3$ (mod 4), so that by Corollary 11.6.13 we have $\Gamma_p(1/2) = \pm 1$. Prove the following results:

   (a) If k is the number of integers a such that $1 \leqslant a \leqslant (p-1)/2$ and $\left(\frac{a}{p}\right) = -1$, then $\Gamma_p(1/2) = (-1)^k$.

   (b) If $h(-p)$ denotes the class number of the imaginary quadratic field $\mathbb{Q}(\sqrt{-p})$, which is always odd, then for $p > 3$ we have $\Gamma_p(1/2) = (-1)^{(h(-p)+1)/2}$.

44.

   (a) Using Proposition 11.6.6, compute a lower bound for $v_p(a_{r,k})$ tending to infinity with k, where $a_{r,k} = (-p)^k k! u_{pk+r}$ as in Corollary 4.2.23 is the sequence that enters in Proposition 11.6.15 (2).

   (b) Show that $\Gamma_p(x)$ is an infinitely differentiable function of x.

   (c) (Hard.) Deduce the much better lower bound for $v_p(a_{r,k})$ given in Corollary 4.2.23.

45. Write $\Gamma_p(x) = 1 + \sum_{k \geqslant 1} g_k x^k$. Using the Mahler expansion of $\Gamma_p(px)$ given in Proposition 11.6.15, prove that we have the following values, where the numbers are written in base p (hence with infinitely many digits on the left):

(a) For $p = 3$, $g_1 = \ldots 2000101010$, $g_2 = \ldots 0101121200$, $g_3 = \ldots 2121122200.12$, $g_4 = \ldots 0100221121.2$, $g_5 = \ldots 2220221002.2$, $g_6 = \ldots 2222110111.1222$, $g_7 = \ldots 122200201.012$, $g_8 = \ldots 1111111222.201$, $g_9 = 201202021.0010211$, $g_{10} = \ldots 2200011012.111111$, $g_{11} = \ldots 212210120.1212$, $g_{12} = \ldots 122122100.102022112$.

(b) For $p = 5$, $g_1 = \ldots 22301241440$, $g_2 = \ldots 2041302300$, $g_3 = \ldots 3341243101$.

Check these computations using Proposition 11.3.15 and the table following it, which is a more efficient way of computing these numbers.

46. By considering the Mahler expansions of the functions $\Gamma_p(-px-r)$ and $\Gamma_p(x+1)$ respectively, prove Proposition 11.6.15 (3) and (4).

47. Prove Corollary 11.6.16, and compute the result of (1) also for $m = p$ and $m = p + 1$ (in this last case it is a fourth-degree polynomial in p for $p \geqslant 3$).

48.

(a) Using a method similar to that of Proposition 11.7.9, prove that $\Gamma_7(1/3)^3 = 1 + 3j$, where j is the cube root of 1 congruent to 4 modulo 7.

(b) Using the general Gross–Koblitz formula, show that

$$\Gamma_3(1/8)\Gamma_3(3/8) = -(1 + \sqrt{-2}) ,$$

where $\sqrt{-2}$ is the square root of -2 in $\mathbb{Q}_3$ congruent to 1 modulo 3.

49. Prove that

$$\sum_{\substack{1 \leqslant a \leqslant 9 \\ \gcd(a,20)=1}} \binom{20}{a} \mathrm{Log}\Gamma_5\left(\frac{a}{20}\right) = \frac{1}{2} \log_5(2 - i) ,$$

where $i^2 = -1$ is such that $i \equiv 3 \pmod{5\mathbb{Z}_5}$. Note that this is a formula for *Diamond's* log gamma function, not Morita's.

50.

(a) Using Proposition 11.3.15 (3) and Exercise 30 of Chapter 2 prove Corollary 11.3.16 (1), and give the corresponding congruences for $p = 2$ and $p = 3$.

(b) Using the Kummer congruences (Corollary 9.5.25) prove Corollary 11.3.16 (2).

(c) Strengthening Wolstenholme's congruence given in the above-mentioned exercise, show that if $p \geqslant 5$ is prime we have

$$\frac{1}{p^2} \sum_{1 \leqslant a \leqslant p-1} \frac{1}{a} \equiv -\frac{1}{3} B_{p-3} \pmod{p\mathbb{Z}_p} \text{ and } \frac{1}{p} \sum_{1 \leqslant a \leqslant p-1} \frac{1}{a^2} \equiv \frac{2}{3} B_{p-3} \pmod{p\mathbb{Z}_p} .$$

(d) Show directly that

$$\sum_{1 \leqslant a \leqslant p-1} \frac{1}{a^{2k-1}} \equiv -kp \sum_{1 \leqslant a \leqslant p-1} \frac{1}{a^{2k}} \pmod{p^2\mathbb{Z}_p} .$$

(e) Generalizing (c), show that if $k \geqslant 1$ and $p \geqslant 2k + 3$ is prime we have

$$\frac{1}{p^2} \sum_{1 \leqslant a \leqslant p-1} \frac{1}{a^{2k-1}} \equiv -\frac{k(2k-1)}{(2k+1)} B_{p-1-2k} \pmod{p\mathbb{Z}_p}$$

and $\frac{1}{p} \sum_{1 \leqslant a \leqslant p-1} \frac{1}{a^{2k}} \equiv \frac{2k}{2k+1} B_{p-1-2k} \pmod{p\mathbb{Z}_p} .$

(f) Prove similar results modulo $p^2\mathbb{Z}_p$.

51. Recall from the remarks following Corollary 11.6.22 that a prime p is called a Wolstenholme prime if it divides the numerator of B_{p-3}, or equivalently by the preceding exercise, if p^3 divides the numerator of $\sum_{1\leqslant a\leqslant p-1} 1/a$, or if p^2 divides the numerator of $\sum_{1\leqslant a\leqslant p-1} 1/a^2$. Prove that $p > 7$ is a Wolstenholme prime if and only if it divides the numerator of $\sum_{p/6<a<p/4} 1/a^3$ (see Exercise 11 of Chapter 9).

52. Prove Lemma 3.6.7 by using Morita's p-adic gamma function.

53. In the proof of Theorem 11.7.16 (2) for m even it was necessary to use Corollary 3.7.6 for the character $\rho^{m/2}$, which itself is a consequence of the HD lifting relation. Prove directly the necessary identity.

54. Recall that in Section 3.7.2 we proved the HD product relation for $p \geqslant 3$ using algebraic number theory, essentially the Stickelberger congruence, while in Section 11.7.4 we proved it directly from the distribution formula for Morita's p-adic gamma function together with the Gross–Koblitz formula. The goal of this exercise is to prove it for $p = 2$ using the methods of Section 3.7.2. Thus, we let $q = 2^f$ and use the notation of Section 3.7.2. For $0 \leqslant r < q-1$ define $w(r) = r!/(-2)^{v_p(r!)}$, $u(r) = (-1)^{r(r-1)/2}$ if $0 \leqslant r < q/2$, $u(r) = (-1)^{r(r+1)/2}$ if $q/2 \leqslant r < q-1$, and extend $u(r)$ and $w(r)$ by periodicity of period $q-1$.

(a) Using Lemma 11.7.12, the Gross–Koblitz formula, and Proposition 11.6.6, prove that
$$\tau(\omega_{\mathfrak{P}}^{-r}, \psi_1)/(-2)^{s(r)} \equiv -u(r)/w(r) \pmod{\mathfrak{P}^2}.$$
Also, try to prove this directly without appealing to the Gross–Koblitz formula.

(b) Let $m \mid (q-1)$, $d = (q-1)/m$, and $b \in \mathbb{Z}$. Prove that the functions $f(r) = u(r)$ and $f(r) = w(r)$ both satisfy the distribution relation
$$\frac{\prod_{0\leqslant a<m} f(da+b)}{f(mb)\prod_{0\leqslant a<m} f(da)} \equiv (-1)^{(\lfloor 2b/d\rfloor+(b\bmod d))(m-1)/2} \pmod 4,$$
where $b \bmod d$ is the least nonnegative residue of b modulo d.

(c) Using the same proof as for Theorem 3.7.3, prove the HD product relation for $p = 2$.

55. Let $L(x) = \exp(\Lambda(x))$ be the multiplicative arithmetic function defined by $L(x) = \ell$ if $x = \ell^k$ with $k \geqslant 1$ for some prime ℓ, and $L(x) = 1$ otherwise. With the same notation as in Theorem 11.7.16, compute
$$\prod_{1\leqslant a\leqslant m-1,\ \gcd(a,m)=1} \tau(\rho^a, \psi_1)$$
in terms of the function $L(x)$.

Part IV

Modern Tools

12. Applications of Linear Forms in Logarithms

By Yann Bugeaud, Guillaume Hanrot,
and Maurice Mignotte

12.1 Introduction

A linear form in logarithms of algebraic numbers is an expression of the form

$$\beta_1 \log \alpha_1 + \cdots + \beta_n \log \alpha_n,$$

where the α's and the β's denote complex algebraic numbers, and log denotes any determination of the logarithm.

12.1.1 Lower Bounds

The first lower bound for such a sum was obtained in 1935 by Gel'fond [Gel] for the case $n = 2$ of two logarithms. A giant step was made in 1966 by A. Baker [Bak1], who was able to deal with such a form for arbitrary n. Subsequently, many papers were published on this problem, by Baker, Fel'dman, etc., and this field is often called Baker's theory. We give only some references of works after 1990, see [Bak-Wus, BBGMS, BMS2, Lau, Lau-Mig-Nes, Mat, Wald1], and much more information can be found in the book by Waldschmidt [Wald2]. It is important to note that there are essentially two kinds of results: general estimates valid for any n as in [Bak-Wus, Mat, Wald1], and specific results for two logarithms as in [Lau, Lau-Mig-Nes], or for three logarithms as in [BBGMS, BMS2], which are crucial for the complete resolution of Diophantine equations.

Here we will consider only the case in which the β's are in $\mathbb{Z}$, and they will be denoted by $b_1, \ldots, b_n$. This is the only case which has applications to Diophantine equations. Also, in this chapter log will always denote the principal determination of the complex logarithm.

It will be sufficient for us to give the general lower bound for linear forms in logarithms due to Matveev [Mat]. Let $\mathbb{L}$ be a number field of degree D, let $\alpha_1, \ldots, \alpha_n$ be nonzero elements of $\mathbb{L}$, and let $b_1, \ldots, b_n$ be integers. Set

$$B = \max\{|b_1|, \ldots, |b_n|\},$$

and

$$\Lambda^* = \alpha_1^{b_1} \cdots \alpha_n^{b_n} - 1.$$

We wish to bound $|\Lambda^*|$ from below, assuming that it is nonzero. Since $\log(1 + x)$ is asymptotic to x as $|x|$ tends to 0, our problem consists in bounding from below the "linear form in logarithms"

$$\Lambda = b_1 \log \alpha_1 + \cdots + b_n \log \alpha_n + b_{n+1} \log(-1) ,$$

where $b_{n+1} = 0$ if $\mathbb{L}$ is real, and $|b_{n+1}| \leqslant nB$ otherwise. Although Λ and Λ^* are closely linked (in particular one vanishes if and only if the other does), it is useful to keep both, and this dual notation will be used in the sequel without further explanation.

Recall that we define the *absolute logarithmic height* of an algebraic number α as follows.

Definition 12.1.1. *Let $\mathbb{L}$ be a number field of degree D, let $\alpha \in \mathbb{L}^*$ be of degree $d \mid D$, and let $\sum_{0 \leqslant k \leqslant d} a_k X^k$ be its minimal primitive polynomial in $\mathbb{Z}[X]$ with $a_d \neq 0$. We define the absolute logarithmic height $\mathrm{h}(\alpha)$ of α by one of the two equivalent formulas*

$$\mathrm{h}(\alpha) = \frac{1}{d} \left(\log(|a_d|) + \sum_{1 \leqslant i \leqslant d} \max(\log(|\alpha_i|), 0) \right)$$

$$= \frac{1}{D} \sum_{v \in \mathcal{P}(\mathbb{L})} \max(\log(|\alpha|_v), 0) ,$$

where the α_i are the conjugates of α and $\mathcal{P}(\mathbb{L})$ denotes the set of places of $\mathbb{L}$ (see Definition 4.1.12).

It is immediate to check that these formulas are equivalent, hence do not depend on the number field $\mathbb{L}$ containing α, and that if $\alpha = n/d \in \mathbb{Q}$ with $\gcd(n, d) = 1$ then $\mathrm{h}(\alpha) = \max(\log(|n|), \log(|d|))$, so this generalizes to algebraic numbers the usual notion of height that we have already used in Section 8.1.5.

Let $A_1, \ldots, A_n$ be real numbers such that

$$A_j \geqslant \mathrm{h}'(\alpha_j) := \max\{D \, \mathrm{h}(\alpha_j), |\log \alpha_j|, 0.16\}, \quad 1 \leqslant j \leqslant n.$$

We call h' the modified height (with respect to the field $\mathbb{L}$). With this notation, the main result of Matveev [Mat] implies the following estimate.

Theorem 12.1.2. *Assume that Λ^* (defined above) is nonzero. We then have*

$$\log|\Lambda^*| > -3 \cdot 30^{n+4} (n+1)^{5.5} D^2 A_1 \cdots A_n (1 + \log D) (1 + \log nB).$$

Furthermore, if $\mathbb{L}$ is real, we have

$$\log|\Lambda^*| > -1.4 \cdot 30^{n+3} n^{4.5} D^2 A_1 \cdots A_n (1 + \log D) (1 + \log B).$$

Remark. In several applications to Diophantine equations we need a better estimate in terms of B, as in the following result ([Wald1], [Wald2] Theorem. 9.1): with the above notation and assuming that the algebraic numbers $\alpha_1, \ldots, \alpha_n$ are multiplicatively independent and that $b_n \neq 0$, there exists a positive effective constant $C(n)$, which depends only on n, such that

$$\log |\Lambda| > -C(n) \cdot D^2 (\log D) A_1 \cdots A_n \log B',$$

where

$$B' = \max_{1 \leqslant j < n} \left\{ \frac{|b_n|}{A_j} + \frac{|b_j|}{A_n} \right\}.$$

Compared to Theorem 12.1.2, this gives an improvement especially when α_n has a large height and $|b_n|$ is small (see Section 12.10 for the application to Thue equations).

It is important to compare the above lower bound for $|\Lambda|$ with the elementary lower bound obtained by a Liouville-type argument, which gives

$$\log |\Lambda| > -D\big(1 + |b_1| \, \mathrm{h}(\alpha_1) + |b_2| \, \mathrm{h}(\alpha_2) + \cdots + |b_n| \, \mathrm{h}(\alpha_n)\big).$$

In this estimate the dependence with respect to D and to each $\mathrm{h}(\alpha_j)$ is better than in the theorem, *but* in the theorem the dependence on B is logarithmic, while it is linear in this elementary estimate. This makes all the difference, and this elementary estimate has no applications to Diophantine equations. Actually, for applications we do not need a lower bound that is logarithmic in terms of B: in many cases (but not in all; see Catalan's equation in Section 12.9) a result like the following would be sufficient: for any $\varepsilon > 0$ there exists a positive constant C_ε such that

$$\log |\Lambda| > -\varepsilon B \quad \text{for } B > C_\varepsilon,$$

where C_ε does not depend of B, but on $\alpha_1, \ldots, \alpha_n$ and their logarithms. In practice, the best results for two or three logarithms depend on a term in $\log^2 B$.

12.1.2 Applications to Diophantine Equations and Problems

For applications to Diophantine problems, the strategy is the following. In a first step, various and often ad hoc algebraic manipulations associate to a "large" solution of the equation a "very small" value of a certain linear form in logarithms, which means that we have an upper bound for the values of this linear form corresponding to a solution of the equation. Comparing this upper bound with the lower bound provided by Theorem 12.1.2, we get an absolute upper bound M for the absolute values of the unknowns of our equation.

At this point, there are two main cases in which we can go from a bound for the unknowns to a complete solution of the equation.

The first case is that M is not too large. Then, using various methods including sieves, we find the complete list of solutions below M. For this to be possible, it is crucial to get a "reasonably small" value for M. Actually, its size is directly related to the size of the "numerical constant" that appears in Theorem 12.1.2, in other words the factor $1.4 \cdot 30^{n+3} n^{4.5}$ occurring in the second estimate.

Many celebrated Diophantine equations lead to estimates of linear forms in two or three logarithms. In these cases, Theorem 12.1.2 gives numerical constants around 10^{12} and 10^{14}, respectively. Both are usually too large for practical applications. Fortunately, an alternative approach, developed by Waldschmidt, Laurent, and Mignotte, among others, yields much better numerical constants with, however, a worse dependence on B (or in B'), namely with the factor $\log B$ replaced by its square. Despite this worse dependence on B, the results from [Lau, Lau-Mig-Nes, BBGMS, BMS2] are, for many practical applications, better than Theorem 12.1.2.

A second very important special case is that the linear form constructed above has the following property: *only the coefficients b_i are unknown*. From the bound $|b_i| \leqslant M$ obtained using the type of argument outlined above, one can derive a much smaller bound. Indeed, effective Diophantine approximation techniques (continued fractions, LLL algorithm) can be used to obtain a good lower bound for

$$\min_{\substack{|b_i| \leqslant M \\ (b_i)_{1 \leqslant i \leqslant n} \neq 0}} \left| \sum_{i=1}^{n} b_i x_i \right| ,$$

which can be used in place of the Baker-type estimate. Comparing this new bound with the upper bound, we get a value M' with $|b_i| \leqslant M'$; we can repeat this process until no new improvement is obtained. The technicalities of this method have already been worked out in Section 2.3.5, and the method applied in Section 8.7.

12.1.3 A List of Applications

The list of applications given in this chapter may look unusual to experts. We want to begin with results for which the reduction

$$\text{Diophantine problem} \longrightarrow \text{linear form in logarithms}$$

is almost obvious. The chosen examples correspond to such reductions in increasing order of difficulty. This is the reason why the most important example of Thue equations is near the end of our list. A more classical presentation can be found in the book by Baker [Bak2], but our presentation has some similarities with that in the book by Shorey and Tijdeman [Sho-Tij].

12.2 A Lower Bound for $|2^m - 3^n|$

One of the simplest applications of linear forms in logarithms is to prove that $|2^m - 3^n|$ tends to infinity with $m + n$, and, in addition, to give an explicit lower bound for this quantity.

Let $n \geqslant 2$ be an integer and define m and m' by the conditions

$$2^{m'} < 3^n < 2^{m'+1} \quad \text{and} \quad |3^n - 2^m| = \min\{3^n - 2^{m'}, 2^{m'+1} - 3^n\}.$$

Then

$$|2^m - 3^n| < 2^m, \quad (m-1)\log 2 < n \log 3 < (m+1)\log 2,$$

and the problem of finding a lower bound for $|2^m - 3^n|$ clearly reduces to this special case.

Thus, consider the "obvious" linear form

$$\Lambda^* = 3^n 2^{-m} - 1.$$

Applying Matveev's theorem we get

$$\log |\Lambda^*| > -c_0 (1 + \log m),$$

where it is easy to verify that we can take $c_0 = 5.87 \times 10^8$.

This implies the following estimate.

Theorem 12.2.1. *Let m and n be any strictly positive integers, then*

$$|2^m - 3^n| > 2^m (em)^{-5.87 \times 10^8}.$$

More generally, if S denotes a finite set of prime numbers and if $(x_j)_{j \geqslant 1}$ is the increasing sequence of all integers whose prime divisors belong to S, then it follows from Theorem 12.1.2 that

$$|x_{j+1} - x_j| \geqslant x_j (\log x_j)^{-c},$$

where the constant c can be explicitly computed in terms of the prime numbers in S (see [Sho-Tij]).

Theorem 12.2.1 enables us to find the list of all powers of 3 that increased by 5 give a power of 2.

Corollary 12.2.2. *The only integer solutions to the Diophantine equation*

$$2^m - 3^n = 5$$

are $(m, n) = (3, 1)$ and $(m, n) = (5, 3)$, which correspond respectively to $8 - 5 = 3$ and to $32 - 27 = 5$.

Proof. Applying the above theorem to this equation we get

$$5 > 2^m \cdot (em)^{-5.87 \times 10^8},$$

which implies

$$\log 5 > m \log 2 - 5.87 \times 10^8 \left(1 + \log m\right),$$

so that $m < 2.1 \times 10^{10}$ and $n < m \log 2 / \log 3 < 1.4 \times 10^{10}$.

Moreover, the relation $2^m - 3^n = 5$ implies

$$\left| m - n \frac{\log 3}{\log 2} \right| < \frac{5}{\log 2} 3^{-n}.$$

This identity has a first consequence: since $(5/\log 2)\, 3^{-n} < \frac{1}{2n}$ for $n \geqslant 4$, we see that if (m, n) is a solution to our problem with $n \geqslant 4$ then m/n is a convergent of the continued fraction expansion of $\xi := \log 3 / \log 2$.

But we can also notice the following: for $n < N = 1.4 \times 10^{10}$, the smallest value of $|m - n\xi|$ is obtained for the largest convergent of the continued fraction expansion of ξ with denominator less than N. The computation of this expansion shows that

$$\frac{5}{\log 2} 3^{-n} > \left| m - n \frac{\log 3}{\log 2} \right| > 10^{-11}, \qquad \text{for } 0 < n < 1.4 \times 10^{10}.$$

Comparing those two estimates, we see that if (m, n) is a solution of our problem then $n \leqslant 24$. A trivial verification in the range $1 \leqslant n \leqslant 24$ (or a less trivial verification using the fact that for $n \geqslant 4$, m/n must be a convergent of ξ, hence either $19/12$ or $8/5$) completes the proof.   □

Notice that we are in effect solving the Diophantine inequality $0 \leqslant 2^m - 3^n \leqslant 5$ in all steps except the final verification.

More generally, we have the following result (see [Ben3]):

Theorem 12.2.3 (Bennett). *For given nonzero integers a, b, and c the equation $a^m - b^n = c$ has at most two integer solutions.*

To conclude this section, we indicate how a similar, generalized strategy can be used to solve the equation $0 \leqslant u - v \leqslant X$, where u and v are integers with all their prime factors in a given set $\{p_1, \ldots, p_r\}$. Write $u = \prod_{i=1}^{r} p_i^{u_i}$, $v = \prod_{i=1}^{r} p_i^{v_i}$. If we restrict to primitive solutions, i.e., with $(u, v) = 1$ (non-primitive solutions can be easily enumerated as multiples of primitive ones), we can assume that for all i, at least one of u_i, v_i is zero.

We have

$$\Lambda^* := \prod_{i=1}^{r} p_i^{u_i - v_i} - 1 \leqslant X v^{-1},$$

which implies that

$$0 \leqslant \Lambda := \sum_{i=1}^{r}(u_i - v_i)\log p_i \leqslant X v^{-1} .$$

Put $m_i = u_i - v_i$, and $M = \max_i |m_i|$. Assume further that the p_i are sorted in increasing order, i.e., $p_1 < p_2 < \cdots < p_r$.

We need to rewrite these upper bounds in terms of M rather than v. Either $M = \max_i |v_i|$, in which case we have $v \geqslant p_1^M$, or $M = \max_i |u_i|$, in which case we have $v \geqslant u - X \geqslant u/2 \geqslant p_1^M/2$ if $u \geqslant 2X$; hence $X v^{-1} \leqslant 2X \exp(-M \log p_1)$.

Comparing this upper bound with the lower bound coming from Matveev's theorem, we deduce

$$M \log p_1 \leqslant \log(2X) + 1.4 \cdot 30^{r+3} r^{4.5} \left(\prod_{i=1}^{r} \log p_i \right)(1 + \log M).$$

We obtain an upper bound on M by using, for instance, the following easy lemma, due in this form to Pethő and de Weger:

Lemma 12.2.4. *Let B be a nonnegative integer such that*

$$\alpha \log B + \beta \geqslant \gamma B .$$

If $\alpha \geqslant e\gamma$, we have

$$B \leqslant \frac{2}{\gamma}\left(\alpha \log \frac{\alpha}{\gamma} + \beta \right) .$$

Proof. Exercise for the reader. $\square$

Once combined with the reduction technique described in Section 2.3, this leads to the following:

Algorithm 12.2.5 (Find all $0 \leqslant u - v \leqslant X$ with given prime factors) Let $p_1 < \cdots < p_r$ be primes and X a nonnegative real number. This algorithm outputs all the solutions of the Diophantine equation $0 \leqslant u - v \leqslant X$, with u, v integers having all their prime factors in the set $\{p_1, \ldots, p_r\}$.

1. [Compute constants for Baker's bound] Compute the constants involved in the upper and lower bounds above:

$$\lambda_1 = \log p_1, \quad \lambda_3 = 1.4 \cdot 30^{r+3} r^{4.5} \left(\prod_{i=1}^{r} \log p_i \right), \quad \lambda_2 = \lambda_3 + \log(2X) .$$

2. [Derive a large upper bound on M] Compute the bound

$$M = \frac{2}{\lambda_3}\left(\lambda_1 \log \frac{\lambda_1}{\lambda_3} + \lambda_2 \right) .$$

3. [Reduction step, initialization] Let C be a nonnegative real number somewhat larger than $(rM)^r$, say $100(rM)^r$.

4. [Reduction step] Apply the LLL algorithm to the lattice generated by the columns of the following matrix:

$$\begin{pmatrix} 1 & 0 & \cdots & 0 & 0 \\ 0 & 1 & \cdots & 0 & 0 \\ \vdots & \vdots & \ddots & \vdots & \vdots \\ 0 & 0 & \cdots & 1 & 0 \\ \lfloor C \log p_1 \rfloor & \lfloor C \log p_2 \rfloor & \cdots & \lfloor C \log p_{r-1} \rfloor & \lfloor C \log p_r \rfloor \end{pmatrix}$$

Let l_0 be the L^2-norm of the first vector of the reduced basis.

5. [Reduction step, new upper bound] If $l_0 < 2^{(r-1)/2}\sqrt{(r^2/4 + (r-1))M}$, put $C \leftarrow 100C$ and go to Step 4. Otherwise, set

$$M' = -\frac{1}{\log p_1} \log\left(\frac{\sqrt{2^{1-r}l_0^2 - (r-1)M} - rM/2}{2CX} \right).$$

If $\lfloor M' \rfloor < M$, set $M \leftarrow \lfloor M' \rfloor$ and go to Step 3.

6. [Final enumeration] Set $\mathcal{S}_0 = \emptyset$. For all $2r$-tuples $((u_i)_{1 \leqslant i \leqslant r}, (v_i)_{1 \leqslant i \leqslant r})$ with $u_i v_i = 0$ and $u_i, v_i \leqslant M$, set $u = \prod_{i=1}^r u_i, v = \prod_{i=1}^r v_i$, check whether $0 \leqslant u - v \leqslant X$, and if this is the case, add (u, v) to the set $\mathcal{S}_0$ of the solutions.

7. [Small solutions] Look for solutions with $u \leqslant 2C$ and add them to $\mathcal{S}_0$.

8. [Nonprimitive solutions] Set $\mathcal{S} = \emptyset$. For all solutions (u, v) in $\mathcal{S}_0$ and all primes p of $\{p_1, \ldots, p_r\}$, do the following: $(u_0, v_0) \leftarrow (u, v)$; while $|u_0 - v_0| \leqslant C$, add (u_0, v_0) to $\mathcal{S}$ and do $(u_0, v_0) \leftarrow (pu_0, pv_0)$.

9. [Terminate] Return $\mathcal{S}$.

Remarks. (1) The constant 100 involved in the choice of C is somewhat arbitrary and should be tuned somehow.

(2) The algorithm given above is somewhat naïve, especially regarding the final enumeration.

(3) This method can again be generalized to solve the equation $A + B = C$ when A, B, C have all their prime factors in the set $\{p_1, \ldots, p_r\}$, using non-Archimedean arguments (see Section 12.12). This approach has been used, in particular, to obtain "worst cases" for the abc conjecture (Conjecture 14.6.4). See [DeW1] for extensive computations on this problem.

12.3 Lower Bounds for the Trace of α^n

Let α be a nonzero algebraic integer of degree $d > 1$ that is not a root of unity. Assume that its conjugates $\alpha = \alpha_1, \ldots, \alpha_d$ (in the field $\mathbb{C}$) satisfy

$$|\alpha_1| \geqslant |\alpha_2| > |\alpha_3| \geqslant \cdots \geqslant |\alpha_d| .$$

Note that Kronecker's theorem (Corollary 3.3.10) implies that $|\alpha| > 1$, and that the trace of α^n is by definition

$$\mathrm{Trace}(\alpha^n) = \alpha_1^n + \alpha_2^n + \cdots + \alpha_d^n.$$

The purpose of this section is to show how one derives from Theorem 12.1.2 a lower bound for $\mathrm{Trace}(\alpha^n)$.

In the (trivial) case $|\alpha_1| > |\alpha_2|$ we have

$$|\alpha_1|^n - (d-1)|\alpha_2|^n \leqslant |\mathrm{Trace}(\alpha^n)| \leqslant |\alpha_1|^n + (d-1)|\alpha_2|^n ,$$

so $|\mathrm{Trace}(\alpha^n)| \approx |\alpha_1|^n$. But suppose now that $|\alpha_1| = |\alpha_2|$ and that we are not in the "degenerate" case $\alpha_2 = -\alpha_1$. Then α_2 is the complex conjugate of α_1. Set $\alpha = \rho e^{i\varphi}$, with $\rho > 0$, and observe that

$$\alpha_1^n + \alpha_2^n = \rho^n(e^{in\varphi} + e^{-in\varphi}) = 2\rho^n \cos(n\varphi).$$

This elementary formula shows that obtaining a lower bound for $|\alpha_1^n + \alpha_2^n|$ is exactly equivalent to obtaining a lower bound for

$$\Lambda_1 := ni\varphi - ki\pi ,$$

where $i = \sqrt{-1}$ and k is an integer. Our problem thus becomes a Diophantine approximation problem. More precisely, Λ_1 is a *linear form in the logarithms of algebraic numbers with integer coefficients*. Indeed,

$$i\varphi = \log(\alpha/|\alpha|), \quad i\pi = \log(-1) ,$$

where $\alpha/|\alpha|$ and -1 are algebraic numbers. Applying Theorem 12.1.2, it is easy to see that

$$\log |\Lambda_1| \geqslant -c_1 \log n ,$$

where c_1 is a positive real number that depends only on α. We immediately conclude that in the present situation,

$$|\mathrm{Trace}(\alpha^n)| \geqslant 0.5 |\alpha|^n n^{-c_1}, \quad \text{for } n > c_2 ,$$

for some positive *constants* c_1 and c_2 that depend only on α. Here and in the sequel all the constants c_i are positive, effective, and indeed easy to compute explicitly, and we will indicate their dependence in terms of the parameters.

More generally, we can prove the following result.

Theorem 12.3.1. *Assume that a and α are nonzero algebraic numbers with α nonreal. Then*

$$|a\alpha^n + \bar{a}\bar{\alpha}^n| > |\alpha|^n n^{-c_3},$$

for $n > c_4$, where c_3 and c_4 depend only on a and α.

It is clear that such a result has applications to linear recurrent sequences; for this question, see [Sho-Tij] for more details.

12.4 Pure Powers in Binary Recurrent Sequences

Consider first the example of Fibonacci and Lucas numbers. Recall that these two sequences (F_n) and (L_n) are defined respectively by

$$F_n = \frac{\alpha^n - \beta^n}{\sqrt{5}}, \ L_n = \alpha^n + \beta^n, \quad \text{where} \quad \alpha = \frac{1+\sqrt{5}}{2} \text{ and } \beta = \frac{1-\sqrt{5}}{2} \,,$$

so that for $n \geqslant 0$,

$$(F_n) = 0, 1, 1, 2, 3, 5, 8, 13, 21, 34, 55, 89, 144, 233, \dots \text{ and}$$
$$(L_n) = 2, 1, 3, 4, 7, 11, 18, 29, 47, 76, 123, 199, 322, \dots \,.$$

Suppose that $F_n = y^p$ is a pure power. Then, we have

$$\alpha^n - \sqrt{5} y^p = O(\alpha^{-n}) \,,$$

so that

$$\Lambda_2 := n \log \alpha - p \log y - \log \sqrt{5} = O(\alpha^{-2n}) = O(y^{-2p}) \,.$$

There exist integers k and r such that $n = kp + r$ with $|r| \leqslant p/2$, so that we have

$$\Lambda_2 = p \log \left(\frac{\alpha^k}{y} \right) + r \log \alpha - \log \sqrt{5} \,,$$

which is a linear form in three logarithms. If we apply the above theorem of Matveev, we get

$$\log |\Lambda_2| \geqslant -c_5 \log y \log p.$$

Comparing both estimates of $|\Lambda_2|$, we see that the exponent p is bounded. More precisely, Matveev's theorem above implies $p < 3 \times 10^{13}$, but a special estimate for linear forms in three logarithms proved in [BMS1] implies the sharper upper bound $p < 2 \times 10^8$, a range that is suitable for computer calculations. For Lucas numbers, a similar study leads to a linear form in two logarithms—a much better situation—and an application of [Lau-Mig-Nes] leads to $p < 300$ if $L_n = y^p$. In the case of Fibonacci numbers, the application of the "modular method" (see Chapter 15) allows us to prove that $r = \pm 1$, for any prime p in the above range. Thus

$$\Lambda_2 = p \log \left(\frac{\alpha^k}{\log y} \right) + \log \left(\frac{\alpha^{\pm 1}}{\sqrt{5}} \right)$$

is a linear form in two logarithms (!), and an application of [Lau-Mig-Nes] now leads to $p < 733$.

Coming back to a more general situation we have the following result.

Theorem 12.4.1. *Suppose that* (u_n) *is a sequence of integers of the form*

$$u_n = a\alpha^n + O(|\alpha|^{\theta n}), \quad \text{with} \quad 0 < \theta < 1,$$

where a and α are nonzero algebraic numbers, with $|\alpha| > 1$ and θ fixed, and that $u_n - a\alpha^n \neq 0$ for all n. The equation

$$u_n = y^p, \quad u_n \notin \{0, \pm 1\},$$

implies that $p < c_6$, where the upper bound c_6 depends only on a, α, θ, and on the implicit constant in the above O.

This result evidently applies in particular to nondegenerate linear binary recurrent sequences with real "roots" and, more generally, to linear recurrent sequences with exactly one "dominant root," and moreover for which this dominant root is simple. See again [Sho-Tij] for more details.

Note that it has been recently proved in [BMS1] that the only perfect powers in the Fibonacci and Lucas sequences are exactly the powers that appear in the previous list, in other words 0, 1, 8, and 144 for the Fibonacci numbers, and 1 and 4 for the Lucas numbers.

12.5 Greatest Prime Factors of Terms of Some Recurrent Sequences

As in the previous section consider a sequence of nonzero integers u_n such that

$$u_n = a\alpha^n + O(|\alpha|^{\theta n}), \quad 0 < \theta < 1, \quad u_n \neq a\alpha^n,$$

where a and α are nonzero algebraic numbers with $|\alpha| > 1$, and where θ is fixed. Let $p_1 = 2$, $p_2 = 3$, $p_3 = 5$, ... be the sequence of prime numbers in increasing order. Assume that the largest prime factor of u_n is equal to p_k, in other words that

$$u_n = p_1^{r_1} \cdots p_k^{r_k},$$

with $r_k > 0$. Set

$$\Lambda_3 = n \log \alpha + \log a - r_1 \log p_1 - \cdots - r_k \log p_k.$$

The definition of u_n implies that

$$\log |\Lambda_3| \leqslant -c_7 n,$$

where c_7 is a positive constant that depends only on a, α, θ, and on the implicit constant in the above O.

In the other direction, Theorem 12.1.2 implies that

$$\log |\Lambda_3| \geqslant -c_8 (\log p_k)^k \log n,$$

where c_8 is a constant depending only on a and α. Comparing both estimates and using the estimate $p_k \sim k \log k$, which is equivalent to the prime number theorem, we obtain the following result.

Theorem 12.5.1. *Let (u_n) be a sequence of nonzero integers of the form*

$$u_n = a\alpha^n + O(|\alpha|^{\theta n}), \quad 0 < \theta < 1,$$

where a and α are nonzero algebraic numbers, with $|\alpha| > 1$ and θ fixed, and assume that $u_n - a\alpha^n \neq 0$ for all n. Let $p_1 = 2$, $p_2 = 3$, $p_3 = 5$, ... be the sequence of prime numbers, and suppose that the largest prime factor of u_n is equal to p_k. Then

$$k \geqslant c_9 \log n / \log \log \log n,$$

where c_9 is a positive constant that depends only on a, α, θ, and on the implicit constant in the above O.

12.6 Greatest Prime Factors of Values of Integer Polynomials

Let $f(X)$ be an irreducible polynomial in $\mathbb{Z}[X]$ of degree greater than or equal to 2, and let x be a strictly positive integer. Using Baker's theory, it is possible to give a lower bound for the greatest prime factor of $f(x)$. Take for instance $f(X) = X(X - 1)$ and, with the same notation as in the previous section, write

$$x(x - 1) = p_1^{r_1} \cdots p_k^{r_k},$$

with $r_k > 0$. Then, for suitable ε_i in $\{\pm 1\}$, we get

$$|p_1^{\varepsilon_1 r_1} \cdots p_k^{\varepsilon_k r_k} - 1| \leqslant \frac{1}{x - 1}.$$

Since $p_j \sim j \log j$ we deduce from Theorem 12.1.2 that there exists an absolute positive constant c_{10} such that

$$\log x \leqslant c_{10}^{k \log \log k} \log \log x.$$

This implies that

$$p_k \sim k \log k \gg \log \log x \frac{\log \log \log x}{\log \log \log \log x}.$$

A similar result holds for any irreducible polynomial in $\mathbb{Z}[X]$ of degree greater than or equal to 2.

12.7 The Diophantine Equation $ax^n - by^n = c$

In this section we consider the exponential Diophantine equation

$$ax^n - by^n = c,$$

where a, b, and c are nonzero fixed integers, with a and b strictly positive, and where x, y, and n are unknowns. If for some exponent n there exists a solution (x, y) with $|y| > 1$ then

$$\Lambda_4 = \log |a/b| - n \log |x/y| = O(|y|^{-n}) .$$

In the other direction, Theorem 12.1.2 above implies that

$$\log |\Lambda_4| \geqslant -c_{11} \log |y| \cdot \log n .$$

Comparing both estimates we see that

$$n < c_{12} ,$$

where c_{12} depends on a, b, and c.

We give an explicit version of this result, established in [Mig]. Its proof does not depend on Theorem 12.1.2, but on [Lau-Mig-Nes]. Actually, Theorem 12.1.2 does not include all known refinements. In particular, as first noticed by Shorey, it can be considerably improved when $\alpha_1, \ldots, \alpha_n$ are real numbers all very close to 1: roughly speaking, the product of the A_i's can then be replaced by their sum. This is precisely what is used in the proof of the next result.

Theorem 12.7.1. *Assume that the exponential Diophantine inequality*

$$|ax^n - by^n| \leqslant c, \qquad \text{with } a, b, c \in \mathbb{Z}_{>0} \text{ and } a \neq b,$$

has a solution in strictly positive integers x and y with $\max\{x, y\} > 1$. Then

$$n \leqslant \max \left\{ 3 \log(1.5 \, |c/b|), \, 7400 \, \frac{\log A}{\log\left(1 + (\log A)/\log |a/b|\right)} \right\} ,$$

where $A = \max\{a, b, 3\}$.

It is remarkable that when $|c|$ is very small and the ratio $|a/b|$ is very close to 1, the upper bound given by Theorem 12.7.1 is absolute; in other words, it does not depend on a, b, and c. For instance, if there exist strictly positive integers $n \geqslant 3$, b, c, x, and y with $\max\{x, y\} > 1$ and $b \geqslant c^2$, satisfying

$$|(b + c)x^n - by^n| \leqslant c ,$$

then n is less than some *absolute* constant.

Using Padé approximations, Bennett and de Weger [Ben-deW] improved the previous result, and ultimately Bennett [Ben1] obtained the following definitive result for $c = \pm 1$ (the case $n = 3$ is essentially Skolem's Theorem 6.4.30).

Theorem 12.7.2 (Bennett). *If $n \geqslant 3$ the equation*

$$|ax^n - by^n| = 1, \qquad with \ a, \ b \in \mathbb{Z}_{>0} \ ,$$

has at most one solution in strictly positive integers x and y.

For instance, we thus know the complete list of solutions to the parametric family of equations $(b+1)x^n - by^n = \pm 1$.

Note that such equations are called *Thue equations*, and will be considered in more detail below.

12.8 Simultaneous Pell Equations

Consider the so-called simultaneous Pell equations

$$x^2 - ay^2 = 1, \qquad x^2 - bz^2 = 1,$$

where $a \geqslant 2$ and $b \geqslant 2$ are distinct squarefree integers, in the integer unknowns x, y, and z.

12.8.1 General Strategy

Let ε and η be the fundamental units of norm 1 of the real quadratic orders $\mathbb{Z}[\sqrt{a}]$ and $\mathbb{Z}[\sqrt{b}]$, respectively. Changing if necessary x, y, and z into their opposites we then have the relations

$$x + y\sqrt{a} = \varepsilon^m \quad \text{and} \quad x + z\sqrt{b} = \eta^n \ ,$$

for suitable nonnegative integers m and n. In particular,

$$2x = \varepsilon^m + \varepsilon^{-m} = \eta^n + \eta^{-n} \ .$$

Consider the linear form

$$\Lambda_5 = m \log \varepsilon - n \log \eta \ ,$$

which satisfies

$$\Lambda_5 = O(\varepsilon^{-m} + \eta^{-n}).$$

It follows from Theorem 12.1.2 that m and n are bounded (we leave the proof as an exercise for the reader). Consequently, the above system of two simultaneous Pell equations has only a finite number of solutions (x, y, z).

As an example of a problem that leads to a system of simultaneous Pell equations consider the following. A Diophantine m-tuple is a set of m integers such that the product of any two of them, increased by 1, is a perfect square. A famous example is the quadruple $\{1, 3, 8, 120\}$ found by Fermat. In 1969,

Baker and Davenport [Bak-Dav] proved that the set $\{1, 3, 8, 120\}$ cannot be extended to a Diophantine quintuple. Actually, they established a stronger result: the only strictly positive integer t such that $\{1, 3, 8, t\}$ is a Diophantine quadruple is 120. It is an easy exercise to see that this problem reduces to the study of the simultaneous equations

$$3x^2 - 2 = y^2 \quad \text{and} \quad 8x^2 - 7 = z^2 .$$

They have only the "trivial" solution $(x, y, z) = (\pm 1, \pm 1, \pm 1)$ and the solution $(x, y, z) = (\pm 11, \pm 19, \pm 31)$, which corresponds to $t = 120$. This was the first Diophantine problem completely solved by Baker's theory and it involved computations with 1000 decimal digits, something very new at that time.

Another problem that can be reduced to a system of two simultaneous Pell equations is the so-called cannonball equation $6y^2 = x(x+1)(2x+1)$. Indeed, by studying the factors of x, $x + 1$, $2x + 1$ modulo squares and elementary congruence arguments, we find that either $x = u^2$, $x + 1 = 2v^2$, $2x + 1 = 3w^2$ or $x = 6u^2$, $x + 1 = v^2$, $2x + 1 = w^2$ for some integers (u, v, w). Both cases lead to a system of simultaneous Pell equations. Note that we have already solved this problem by completely elementary methods in Section 6.8.2, and that it can also be solved using the methods of Section 8.7, which also rely on Baker-type arguments, involving linear forms in elliptic logarithms instead of ordinary logarithms.

12.8.2 An Example in Detail

We now study a particular example in detail. Consider the simultaneous Pell equations

$$x^2 - 2y^2 = x^2 - 3z^2 = 1.$$

It is easy to see that if $x > 1$ is a solution then there exist strictly positive integers m (with m even) and n such that

$$u_m = v_n, \quad \text{where } u_j = (\alpha^j + (-\alpha)^{-j})/2 \text{ and } v_k = (\beta^k + \beta^{-k})/2 ,$$

with

$$\alpha = 1 + \sqrt{2}, \quad \beta = 2 + \sqrt{3} .$$

If we assume $m, n \neq 0$, we obtain

$$|\alpha^m - \beta^n| = |\alpha^{-m} - \beta^{-n}| \leqslant 1 ,$$

so that in fact

$$|\alpha^{-m} - \beta^{-n}| \leqslant \alpha^{-m}\beta^{-n} .$$

From this, we deduce that

$$|\alpha^m \beta^{-n} - 1| \leqslant \alpha^{-m}\beta^{-2n} .$$

If $n \neq 0$, we have $|\alpha^m - \beta^n| \leqslant \alpha^m/2$, so that the upper bound can be replaced by $4\alpha^{-3m}$, and we can see that $m \geqslant n$. Theorem 12.1.2 (with $D = 4$) shows that

$$\log|\alpha^m \beta^{-n} - 1| \geqslant -1.37 \times 10^{12}(1 + \log m) .$$

Comparing this with the upper bound, we obtain $m < 1.63 \times 10^{13}$.

To obtain a more practical bound on m, we need to work with $\Lambda = m \log \alpha - n \log \beta$ rather than with Λ^*. An upper bound for Λ can be obtained, for instance, by means of the following elementary lemma whose proof is left to the reader:

Lemma 12.8.1. *If $x \in \mathbb{C}$ is such that $|x - 1| \leqslant 1/2$, then $|\log x| \leqslant 2|x - 1| \log 2$.*

We obtain

$$|m \log \alpha - n \log \beta| \leqslant 8\alpha^{-3m} \log 2$$

as soon as $4\alpha^{-3m} < 1/2$, i.e., $m \geqslant 1$.

Thus, as in the case of the equation $2^n - 3^m = 5$, we have to consider the continued fraction expansion of some quotient of two logarithms, here of $\xi = \log \alpha / \log \beta$. The computation of this continued fraction expansion shows that

$$\left| n - m \frac{\log \alpha}{\log \beta} \right| > 3.91 \times 10^{-14} \quad \text{when } m \leqslant 1.63 \times 10^{13}.$$

We thus deduce that $1 \leqslant m \leqslant 12$. Finally, a trivial computation shows that the only solution for $u_m = v_n$ is $(m, n) = (3, 2)$. However, this does not correspond to a solution to our problem since m is odd.

To conclude, we have thus proved that the only strictly positive solution to the equations

$$|x^2 - 2y^2| = x^2 - 3z^2 = 1$$

is $(x, y, z) = (7, 5, 4)$.

12.8.3 A General Algorithm

We can formalize this procedure as an algorithm. First, we need a technical lemma, which will give us both a bound on $\Lambda^* = \gamma \epsilon^n \delta^{-1} \eta^{-m}$, to which we want to apply Matveev's theorem, and on $\Lambda = \log \gamma \delta^{-1} + n \log \epsilon - m \log \eta$, which we shall use in the reduction process.

Lemma 12.8.2. *Let $\varepsilon > 1, \eta > 1, \gamma$, and δ be real numbers. Under the assumptions*

$$|\gamma \varepsilon^n - \delta \eta^m| \leqslant C \cdot \max(\varepsilon^{-n}, \eta^{-m}) , \quad n \neq 0 , \quad \text{and}$$

$$m > M_0 := \frac{1}{\log \eta} \max\left(\log \frac{C|\gamma|}{|\delta|}, \frac{1}{2} \log \left\{ C \max \left(\frac{1}{2|\delta|}, \frac{1}{|\gamma|} \right) \right\} \right) ,$$

we have

$$|\log \gamma \delta^{-1} + n \log \varepsilon - m \log \eta| \leqslant 2C(\log 2) \max(\delta^{-1}, 2\gamma^{-1})\eta^{-2m} .$$

Proof. Since $n, m > 0$, we have $|\gamma\varepsilon^n - \delta\eta^m| \leqslant C$, hence $|\varepsilon^n| \leqslant C + |\delta\gamma^{-1}\eta^m| \leqslant 2|\delta\gamma^{-1}\eta^m|$ for $m > \log(C|\gamma\delta^{-1}|)/\log\eta$. As a consequence, we have

$$\left| \frac{\gamma\varepsilon^n}{\delta\eta^m} - 1 \right| \leqslant C \max(\delta^{-1}, 2\gamma^{-1})\eta^{-2m} \ .$$

Since $m > M_0$, the left-hand side is at most $1/2$ and we can apply Lemma 12.8.1. □

Two cases can occur in the reduction step: either we are in the homogeneous case (as above), and we can use elementary properties of continued fractions; or we are in the inhomogeneous case (the general case in which the right-hand side is nontrivial) and we can also use continued fractions as in the following lemma.

Lemma 12.8.3 (Baker–Davenport). *Let x_0, x_1, x_2 be real numbers, b_1 an integer with $|b_1| \leqslant B$, and Q a nonnegative integer. Then*

$$d(x_0 + b_1 x_1, \mathbb{Z}) \geqslant \frac{1}{Q}\{d(Qx_0, \mathbb{Z}) - Bd(Qx_1, \mathbb{Z})\} \ .$$

Proof. This follows from the chain of inequalities

$$\begin{aligned} Qd(x_0 + b_1 x_1, \mathbb{Z}) &\geqslant d(Qx_0 + Qb_1 x_1, \mathbb{Z}) \\ &\geqslant d(Qx_0, \mathbb{Z}) - d(Qb_1 x_1, \mathbb{Z}) \\ &\geqslant d(Qx_0, \mathbb{Z}) - Bd(Qx_1, \mathbb{Z}) \ . \end{aligned}$$

□

This lemma should be applied with Q the denominator of a convergent of the continued fraction expansion of x_1, with $Q \approx \kappa B$, since in that case, we expect $Bd(Qx_1, \mathbb{Z}) \approx \kappa^{-1}$. We obtain the following general algorithm:

Algorithm 12.8.4 (Simultaneous Pell equations) Let a_0, a_1, b_0, b_1 be integers such that $\sqrt{a_0/b_0}$ is not a rational number. This algorithm gives the list of solutions of the system $x^2 - a_0 y^2 = a_1, x^2 - b_0 z^2 = b_1$.

1. [Algebraic precomputations] Compute fundamental units ε of $\mathbb{Q}(\sqrt{a_0})$ and η of $\mathbb{Q}(\sqrt{b_0})$; we shall assume that they are chosen such that $\varepsilon, \eta > 1$. Compute two sets S_0 and S_1 of inequivalent solutions in algebraic integers of the norm equations $N_{\mathbb{Q}(\sqrt{a_0})/\mathbb{Q}}(\gamma) = a_1$, $N_{\mathbb{Q}(\sqrt{b_0})/\mathbb{Q}}(\delta) = b_1$. The following steps should be executed for all $(\gamma, \delta) \in S_0 \times S_1$.

2. [Compute Baker's bound] Compute the constants M_0,

$$\lambda_0 = a_1|\gamma|^{-1} + b_1|\delta^{-1}|, \quad \lambda_1 = 2.94 \times 10^{10} h'(\varepsilon)h'(\eta)h'(\delta/\gamma),$$
$$\lambda_2' = C\max(\delta^{-1}, 2\gamma^{-1}), \quad \lambda_2 = \lambda_2' + \lambda_1, \quad \lambda_3 = 2\log\eta, \quad \text{and}$$
$$M = \frac{2}{\lambda_3}\left(\lambda_1\log\frac{\lambda_1}{\lambda_3} + \lambda_2\right).$$

3. [Reduction, choice of the parameter] If $\delta = \gamma$, go to Step 5. Otherwise, choose C somewhat larger than M, say $C = 100M$.

4. [Reduction, inhomogeneous case] Compute the largest convergent p/q of the real number $\log\eta/\log\varepsilon$ for which $q \leqslant C$. If

$$D := d(q\log\gamma\delta^{-1}/\log\varepsilon, \mathbb{Z}) - Md(q\log\eta/\log\varepsilon, \mathbb{Z}) < 0,$$

set $M \leftarrow 100M$ and go to Step 4. Otherwise, compute

$$M' = -\frac{\log(D/2q\lambda_2'\log 2)}{2\log\eta}.$$

If $\lfloor M' \rfloor < M$, set $M \leftarrow \lfloor M' \rfloor$ and go to Step 3; otherwise, go to Step 6.

5. [Reduction, homogeneous case] Compute the largest convergent p/q of the real number $\log\eta/\log\varepsilon$ with $q \leqslant M$,

$$D = |q\log\eta - p\log\varepsilon|, \quad \text{and} \quad M' = -\frac{\log\{D/(2q\lambda_2'\log 2)\}}{2\log\eta}.$$

If $M' < M$, set $M \leftarrow M'$ and go to Step 5.

6. [Enumeration] Initialize $\mathcal{S}$ to $\emptyset$. For each $m < \max(M_0, M)$, compute $x = (\delta\eta^m + a_1\delta^{-1}\eta^{-m})/2$; put $y = \sqrt{(x^2 - a_1)/a_0}$, $z = \sqrt{(x^2 - b_1)/b_0}$. If y and z are rational integers, add (x, y, z) to $\mathcal{S}$.

7. [Terminate] Return $\mathcal{S}$.

Remarks. (1) In practice, the case $\delta = \gamma$ should be treated via forms in 2 logarithms rather than via Matveev's bound.

(2) The enumeration step should be optimized, especially if $\max(M_0, M)$ is large. For instance, one could use congruence conditions on m.

12.9 Catalan's Equation

Catalan's problem [Cat] posed in 1844 is the following: do there exist consecutive positive integers other than 8 and 9 that are both pure powers? This corresponds to the exponential Diophantine equation

$$x^m - y^n = 1.$$

Although this problem was solved completely in the negative by Mihăilescu in 2002 (see Chapter 16), it is still interesting to study the application of Baker's theory to this problem (even if Baker's theory is not used in Mihăilescu's latest proof, which we give in Chapter 16). In 1976, Tijdeman [Tij] proved that Catalan's problem is a "finite" problem, and we will sketch his proof.

Theorem 12.9.1 (Tijdeman). *Let x, y, $m \geqslant 2$, and $n \geqslant 2$ be strictly positive integers such that $x^m - y^n = 1$. There exists an effectively computable, absolute constant C such that $\max\{x, y, m, n\} < C$.*

Proof. By the results of Lebesgue and Ko Chao (Proposition 6.7.12 and Theorem 6.11.8) we may assume that m and n are odd. We consider the equation

$$x^m - y^n = \varepsilon \, ,$$

where $\varepsilon = \pm 1$ and x, y, m, n are strictly positive integers with $n > m > 2$.

Since

$$(c^n - 1)/(c - 1) = n + (c - 1) \sum_{1 \leqslant k \leqslant n-1} \binom{n}{k+1}(c-1)^{k-1}$$

it follows that $\gcd((c^n - 1)/(c - 1), c - 1) \mid n$. We thus have the relations

$$x - \varepsilon = u^n/m^* \qquad \text{and} \qquad y + \varepsilon = v^m/n^* \, ,$$

where u and v are integers, $|u|$, $|v| > 1$, and where m^* (respectively n^*) is a divisor of m (respectively of n).

It follows from the assumption $n > m$ that $x > y$. Throughout the proof, the constants implied by $\ll$ and $\gg$ are absolute.

Consider now the following two linear forms in logarithms:

$$\Lambda_6 := n \log(yu^{-m}) + m \log m^*$$

and

$$\Lambda_7 := mn \log(u/v) - m \log m^* + n \log n^* \, .$$

Since

$$\left| \frac{y^n}{u^{mn}(m^*)^{-m}} - 1 \right| = \left| \frac{x^m - \varepsilon}{(x - \varepsilon)^m} - 1 \right| \ll \frac{m}{x} \, ,$$

we note that $|\Lambda_6| \ll m/x$, so that using Theorem 12.1.2 and the upper bound $y \leqslant 2u^m$, we obtain

$$\log x \ll m(\log m)(\log n)(\log u) \, ,$$

hence,

$$n \ll m(\log m)(\log n). \tag{9.1}$$

Furthermore, it follows from

$$\left| \frac{u^{mn}}{(m^*)^m} \frac{(n^*)^n}{v^{mn}} - 1 \right| = \left| \frac{(x - \varepsilon)^m}{(y + \varepsilon)^n} - 1 \right| \ll \frac{n}{y}$$

that $|\Lambda_7| \ll n/y$. Thus, we deduce from $u \leqslant 3v$ and Theorem 12.1.2 that

$$\log y \ll (\log mn)\,(\log m)\,(\log n)\,(\log v)\,,$$

hence that

$$m \ll (\log m)\,(\log n)^2\,. \tag{9.2}$$

Combining equations (9.1) and (9.2) we find that n is bounded by an absolute constant. It then follows from (9.2) that m is also bounded by an absolute constant. The fact that x and y are also bounded follows from general results on superelliptic Diophantine equations given below in Section 12.11.   □

12.10 Thue Equations

12.10.1 The Main Theorem

Let $\mathbb{K}$ be an algebraic number field of degree d, let $\alpha_1,\dots,\alpha_n$ be $n \geqslant 3$ distinct algebraic integers in $\mathbb{K}$, and let a and m be nonzero integers. We have the following result (we refer to [Bug-Gyo] for a general totally explicit statement, and to [Bug-Gyo, Sho-Tij] for an extensive list of bibliographic references).

Theorem 12.10.1. *With the above notation, the equation*

$$a(x - \alpha_1 y)\cdots(x - \alpha_n y) = m$$

has only finitely many solutions in integers x and y, and all of these can be effectively determined. Moreover, the bound for $\max\{|x|, |y|\}$ is polynomial in $|m|$.

Proof. Let x and y be integers satisfying the above equation. For the sake of simplicity assume that $a = 1$. Without loss of generality, we may assume that x and y are very large, and that x/y is very close to α_1. More precisely, setting $X := \max\{|x|, |y|\}$, we assume that

$$|x - \alpha_1 y| \ll X^{-n+1} \quad \text{and} \quad |x - \alpha_i y| \gg X, \quad \text{for } i = 2,\dots,n. \tag{10.1}$$

Here and throughout the proof, the constants implied by $\ll$ and $\gg$ are effectively computable, and depend only on $\alpha_1,\dots,\alpha_n$ and $\mathbb{K}$. Some of them will be made explicit in the section devoted to algorithmic aspects. Set

$$\beta_i = x - \alpha_i y \quad (1 \leqslant i \leqslant 3)\,,$$

and note that

$$(\alpha_1 - \alpha_2)\beta_3 + (\alpha_2 - \alpha_3)\beta_1 + (\alpha_3 - \alpha_1)\beta_2 = 0\,.$$

Consider now the very small "linear form"

$$\Lambda_8^* := \frac{\alpha_2 - \alpha_3}{\alpha_3 - \alpha_1} \cdot \frac{x - \alpha_1 y}{x - \alpha_2 y} = \frac{x - \alpha_3 y}{x - \alpha_2 y} \cdot \frac{\alpha_2 - \alpha_1}{\alpha_3 - \alpha_1} - 1 .$$

Let r be the rank of the group of units of $\mathbb{K} := \mathbb{Q}(\alpha_1)$ and let $\varepsilon_{1,1}, \ldots, \varepsilon_{1,r}$ be a system of fundamental units of $\mathbb{K}$. Denote by $\varepsilon_{2,1}, \ldots, \varepsilon_{2,r}$ and $\varepsilon_{3,1}, \ldots, \varepsilon_{3,r}$ the conjugates of $\varepsilon_{1,1}, \ldots, \varepsilon_{1,r}$ in $\mathbb{Q}(\alpha_2)$ and $\mathbb{Q}(\alpha_3)$, respectively, which all belong to the Galois closure $\mathbb{L}$ of $\mathbb{K}$.

There exist an algebraic integer γ_1 in $\mathbb{K}$ of norm at most $|m|$, and integers $b_1, \ldots, b_r$ such that $x - \alpha_1 y = \gamma_1 \varepsilon_{1,1}^{b_1} \cdots \varepsilon_{1,r}^{b_r}$. We thus have

$$\Lambda_8^* = \left(\frac{\varepsilon_{3,1}}{\varepsilon_{2,1}}\right)^{b_1} \cdots \left(\frac{\varepsilon_{3,r}}{\varepsilon_{2,r}}\right)^{b_r} \frac{\gamma_3(\alpha_2 - \alpha_1)}{\gamma_2(\alpha_3 - \alpha_1)} - 1 ,$$

where γ_j denotes the conjugate of γ_1 in $\mathbb{Q}(\alpha_j)$, for $j = 2, 3$. We then note that

$$B := \max\{|b_1|, \ldots, |b_r|\} \ll \mathrm{h}(x - \alpha_1 y) \ll \log X . \qquad (10.2)$$

Write $\nu := \gamma_3(\alpha_2 - \alpha_1)/(\gamma_2(\alpha_3 - \alpha_1))$ and note that $\mathrm{h}'(\nu) \ll \log M$, where $M := |m|$. Theorem 12.1.2 gives

$$\log \Lambda_8^* \gg -(\log M)(\log B) ,$$

while our assumptions (10.1) imply that

$$\log \Lambda_8^* \ll - \log X .$$

Combining (10.2) with both estimates, we obtain

$$\log X \ll (\log M)(\log \log X) .$$

We conclude that X is bounded, but this is not sufficient to obtain a bound that is polynomial in M.

For establishing the last statement of the theorem it is crucial to apply the improvement quoted just after Theorem 12.1.2, in other words to replace B by B'. This gives

$$\log \Lambda_8^* \gg -\mathrm{h}'(\nu) \log\left(\frac{B}{\mathrm{h}'(\nu)}\right) ,$$

so that we obtain

$$\left(\frac{\log X}{\mathrm{h}'(\nu)}\right) \ll \log\left(\frac{\log X}{\mathrm{h}'(\nu)}\right) .$$

From this, we deduce that

$$\log X \ll \mathrm{h}'(\nu) \ll \log M ,$$

and that there exists a (very small) numerical constant τ such that

$$\max\{|x|, |y|\} \leqslant M^{1/\tau}. \tag{10.3}$$

Thus we have proved not only that $|x|$ and $|y|$ are effectively bounded, but also that the bound is polynomial in M. □

As a by-product of an explicit version of Theorem 12.10.1, we obtain an effective improvement of Liouville's inequality. Let α be an algebraic number of degree $n \geqslant 3$. As proved by Liouville, there exists a positive constant $c_{13}(\alpha)$ such that

$$\left|\alpha - \frac{p}{q}\right| \geqslant \frac{c_{13}(\alpha)}{q^n}, \tag{10.4}$$

for any rational number p/q. More than one century afterward, Roth established that for any $\varepsilon > 0$, there exists a positive constant $c_{14}(\alpha, \varepsilon)$ such that

$$\left|\alpha - \frac{p}{q}\right| \geqslant \frac{c_{14}(\alpha, \varepsilon)}{q^{2+\varepsilon}},$$

for any rational number p/q. However, Roth's proof does not yield an explicit value for the constant $c_{14}(\alpha, \varepsilon)$. A challenging open problem in Diophantine approximation is to establish an effective version of Roth's theorem. At present, the best that we can do, thanks to (10.3), is to improve only slightly on Liouville's statement, as we now explain.

Let us apply Theorem 12.10.1 with the α_i's equal to the complex conjugates of α, with $\alpha_1 = \alpha$. Let x and y be integers such that $x \neq \alpha_i y$ for any $i = 1, \ldots, n$, and again set $X := \max\{|x|, |y|\}$. We then obtain from (10.3) that

$$|x - \alpha y| = \frac{|m|}{|x - \alpha_2 y| \cdots |x - \alpha_n y|} \gg X^{-(n-1)+\tau}.$$

We thus obtain an estimate of the shape (10.4), with the exponent n replaced by $n-\tau$, where τ is strictly positive, but very small. Notice that an alternative proof of this result, independent of Baker's theory, has been given by Bombieri [Bom].

12.10.2 Algorithmic Aspects

We are in a situation where obtaining an algorithm is possible, since in the linear form $\log \Lambda_8$, only the coefficients b_i are unknown. We need, however, to make explicit the constants in the estimate

$$\log \Lambda_8^* \ll -B, \quad \log \Lambda_8 \ll -B,$$

in order to be able to compute an explicit upper bound for B, and use the reduction process. This is the purpose of the following lemma. We also derive an explicit version of (10.1), which may be needed for efficient final enumeration.

In this subsection we restrict to the irreducible case that the left-hand side of the equation is a monic irreducible polynomial in $\mathbb{Z}[X, Y]$.

Before stating the lemma, note that the constants derived depend on two choices, the choice of a conjugate α_1 of a root of $P(X, 1)$ (in practice, a real conjugate; see the end of the lemma) and of an algebraic integer $\gamma_1 \in \mathbb{Q}(\alpha_1)$ of norm m. From an algorithmic point of view, we will have to loop over all possible choices.

Lemma 12.10.2. *Let (x, y) be a solution of the Thue equation of Theorem 12.10.1, and assume that $(x - \alpha_1 y)/\gamma_1$ is a unit of $\mathbb{Q}(\alpha_1)$ and $|x - \alpha_1 y| = \min_{1 \leqslant i \leqslant n} |x - \alpha_i y|$. Let $M = (m_{ij})$ be the inverse of the matrix $(\log|\varepsilon_{i+1,j}|)_{\substack{1 \leqslant i \leqslant r \\ 1 \leqslant j \leqslant r}}$, and define*

$$\lambda_0 = \frac{2^{n-1}|m|}{|g'(\alpha_1)|}, \quad \lambda_1 = 2\left|\frac{\alpha_2 - \alpha_3}{(\alpha_3 - \alpha_1)(\alpha_1 - \alpha_2)}\right|, \quad \lambda_2 = \min_{1 \leqslant i \leqslant r} \frac{n}{\left|\sum_{j=1}^r m_{ij}\right|},$$

$$\lambda_3 = \log \lambda_1 + \lambda_2 \max_{1 \leqslant i \leqslant r} \left(\log(3/2) \sum_{j=1}^r |m_{ij}| + \left|\sum_{j=1}^r m_{ij} \log\left|\frac{\alpha_1 - \alpha_{j+1}}{\gamma_{j+1}}\right|\right|\right),$$

$$Y_0 = \max_{1 \leqslant i \leqslant n-1} \left(\frac{2\lambda_0}{|\alpha_1 - \alpha_{i+1}|}\right)^{1/n}.$$

Then, if $|y| > Y_0$, we have

$$|x - \alpha_1 y| \leqslant \lambda_0 |y|^{1-n}$$

and

$$\log|\Lambda_8^*| \leqslant -n \log|y| + \log \lambda_1 \leqslant -\lambda_2 B + \lambda_3.$$

Proof. For $j \neq 1$ we have

$$2|x - \alpha_j y| \geqslant |x - \alpha_j y| + |x - \alpha_1 y| \geqslant |y||\alpha_1 - \alpha_j|. \tag{10.5}$$

Thus,

$$|x - \alpha_1 y| = \frac{|m|}{\prod_{2 \leqslant i \leqslant n} |x - \alpha_i y|}$$

$$\leqslant |m||y|^{-n+1} \frac{2^{n-1}}{\prod_{2 \leqslant i \leqslant n} |\alpha_i - \alpha_1|} = |m||y|^{-n+1} \frac{2^{n-1}}{|g'(\alpha_1)|},$$

which proves the first assertion.

Further, (10.5) shows that for $|y| \geqslant Y_0$,

$$|\Lambda_8^*| = \left|\frac{\alpha_2 - \alpha_3}{\alpha_3 - \alpha_1} \cdot \frac{x - \alpha_1 y}{x - \alpha_2 y}\right| \leqslant 2\left|\frac{\alpha_2 - \alpha_3}{(\alpha_3 - \alpha_1)(\alpha_1 - \alpha_2)}\right| |y|^{-n}.$$

We now need to compare B and $\log|y|$. For this, write

$$\log\left|\frac{x - \alpha_{i+1}y}{\gamma_{i+1}}\right| = \sum_{j=1}^{r} b_j \log|\varepsilon_{i+1,j}|, \quad 1 \leqslant i \leqslant r.$$

From this and the definition of M, we deduce that

$$b_i = \sum_{j=1}^{r} m_{ij} \log\left|\frac{x - \alpha_{j+1}y}{\gamma_{j+1}}\right|.$$

Hence,

$$|b_j| \leqslant \left|\sum_{j=1}^{r} m_{ij}\right| \log|y| + \left|\sum_{j=1}^{r} m_{ij} \log\left|\frac{\alpha_1 - \alpha_{i+1}}{\gamma_{i+1}}\right|\right| + \left|\sum_{j=1}^{r} m_{ij} \log\left|\frac{x - \alpha_{i+1}y}{y(\alpha_1 - \alpha_{i+1})}\right|\right|.$$

Finally, note that

$$\left|\frac{x - \alpha_{i+1}y}{y(\alpha_1 - \alpha_{i+1})}\right| = \left|1 - \frac{x - \alpha_1 y}{y(\alpha_1 - \alpha_{i+1})}\right| \leqslant 3/2, \quad \text{for } |y| \geqslant Y_0,$$

finishing the proof. $\qquad\square$

A comparison with Matveev's bound (note that Λ_8^* is nonzero under our assumptions) then yields an explicit upper bound for B, which can then be used to initiate the reduction process.

Finally, note the following additional consequences of the lemma:

– For

$$|y| > Y_1 := \max\left(Y_0, \left(\frac{\lambda_0}{\min_{s+1 \leqslant k \leqslant s+t} |\mathrm{Im}\,(\alpha_k)|}\right)^{1/n}\right),$$

we have $\alpha_1 \in \mathbb{R}$. This can be used to solve very efficiently totally imaginary Thue equations, and otherwise to restrict the set of roots over which one has to loop. Hence, we can assume that the field $\mathbb{K}$ has at least one real embedding, so that the only roots of unity are ± 1.

– For $|y| > Y_2 := (2\lambda_0)^{1/(n-2)}$, x/y is a convergent of the continued fraction expansion of x/y; this can be used for an efficient final enumeration of "medium-sized" solutions, as in the case of simultaneous Pell equations.

This leads to the following algorithm:

Algorithm 12.10.3 (Solve a Thue equation) Let P be an irreducible monic polynomial with rational integer coefficients, of degree $d \geqslant 3$. This algorithm computes all the solutions of the Thue equation $Y^n P(X/Y) = m$.

1. [Algebraic precomputations] Compute the roots $\alpha_1, \ldots, \alpha_n$ of P (ordered so that the first s are real, and $\alpha_{s+t+i} = \overline{\alpha_{s+i}}$, $1 \leqslant i \leqslant t$), a system of fundamental units $\varepsilon_{1,1}, \ldots, \varepsilon_{1,r}$ of $\mathbb{Q}(\alpha_1)$, and a set Γ of nonassociate solutions of the norm equation $N_{\mathbb{Q}(\alpha_1)/\mathbb{Q}}(u) = m$. Set $Y = 0$.

   The following Steps 2–7 should be done for all values of $(\alpha, \gamma) \in \{\alpha_1, \ldots, \alpha_s\} \times \Gamma$. In the following, the roots should be reordered so that the root under consideration is α_1.

2. [Computation of various constants] Compute $\lambda_0, \lambda_1, \lambda_2, \lambda_3, Y_0, Y_1, Y_2, Y = \max(Y, Y_0, Y_1, Y_2)$.

3. [Computation of Baker's bound] Compute the corresponding Matveev's constant

$$\kappa = 3 \cdot 30^{r+4}(r+1)^{5.5}(n(n-1))^2 h'(\varepsilon_1) \cdots h'(\varepsilon_r) \cdot \log(2\pi)(1 + \log(n(n-1))) \,,$$

and deduce Baker's bound

$$B = \frac{2}{\lambda_2}\left(\kappa \log \frac{\kappa}{\lambda_2} + \lambda_3 + \kappa(1 + \log(r+1))\right) \,.$$

4. [Reduction, preparation] Let $\ell = \max(2, r)$. Compute with sufficient precision approximations of the vector $(\delta_i)_{1 \leqslant i \leqslant \ell}$ defined by

$$\delta_i = \begin{cases} \log|\varepsilon_{2,i}/\varepsilon_{3,i}| & \text{when } r > 1\,, \\ \operatorname{Arg}(\varepsilon_{2,1}/\varepsilon_{3,1}) & \text{when } r = 1 \text{ and } i = 1\,, \\ 2\pi & \text{when } r = 1 \text{ and } i = 2\,, \end{cases}$$

and θ defined by

$$\theta = \begin{cases} \log|\gamma_3(\alpha_2 - \alpha_1)/(\gamma_2(\alpha_3 - \alpha_1))| & \text{when } r > 1\,, \\ \operatorname{Arg}(\gamma_3(\alpha_2 - \alpha_1)/(\gamma_2(\alpha_3 - \alpha_1))) & \text{when } r = 1\,. \end{cases}$$

5. [Reduction, choice of the parameter] Set C somewhat larger than B^r, say $C = 100B^r$. If δ_i and θ are not computed with enough precision to determine exactly $\lfloor C\delta_i \rfloor$, increase the working precision and go back to Step 4.

6. [Reduction step] Define

$$G_0 = \begin{pmatrix} 1 & 0 & \cdots & 0 & 0 \\ 0 & 1 & \cdots & 0 & 0 \\ \vdots & \vdots & \ddots & \vdots & \vdots \\ 0 & 0 & \cdots & 1 & 0 \\ \lfloor C\delta_1 \rfloor & \lfloor C\delta_2 \rfloor & \cdots & \lfloor C\delta_{\ell-1} \rfloor & \lfloor C\delta_\ell \rfloor \end{pmatrix} \,, \quad v_0 = \begin{pmatrix} 0 \\ 0 \\ \vdots \\ 0 \\ \lfloor B\theta \rfloor \end{pmatrix} \,.$$

   Let G be the matrix of the LLL reduced basis of the lattice generated by the columns of G_0, and $v = G^{-1}v_0$.

Let i be the smallest index such that $v_i \notin \mathbb{Z}$. Set $d_0 = 2^{(1-r)/2} d(v_i, \mathbb{Z}) \|g_1\|_2$, where g_1 is the first column vector of G. If $d_0 < \sqrt{(r^2/4 + (r-1))B}$, set $C \leftarrow 100C$ and go to Step 5. Otherwise, set

$$d_0' = \sqrt{d_0^2 - (r-1)B} - rB/2 \quad \text{and} \quad B' = \frac{1}{\lambda_2}\left(\lambda_3 + \log\frac{2C\log 2}{d_0'}\right).$$

If $\lfloor B' \rfloor < B$, let $B \leftarrow \lfloor B' \rfloor$ and go to Step 5; otherwise, set

$$Y' \leftarrow \max\left(Y', \left(\frac{2\lambda_1 C\log 2}{d_0'}\right)^{1/n}\right).$$

7. [Medium solutions] Compute the convergents p/q of $\alpha_1, \ldots, \alpha_s$ with $Y \leqslant q \leqslant Y'$; if $F(p, q) = m$, add (p, q) to $\mathcal{S}$. If $F(-p, -q) = m$, add $(-p, -q)$ to $\mathcal{S}$.
8. [Small solutions] For all y with $|y| \leqslant Y$, find the integer roots of the polynomial $F(X, y) - m$ (e.g., by computing roots modulo p and Hensel lifting), and add the corresponding solutions (x, y) to $\mathcal{S}$.
9. [Terminate] Return $\mathcal{S}$.

Remarks. (1) This algorithm follows rather closely the presentation by Tzanakis and de Weger [Tza-Weg]. It can be optimized in many ways, but should perform quite well as it stands for equations of small degree. A few phenomena can be encountered in small degree (especially in degree 3) that induce a failure of the algorithm: there can be a relation between the δ_i (in which case this relation should be taken into account to eliminate one of the variables); or similarly $\log|\gamma_3(\alpha_2 - \alpha_1)/(\gamma_2(\alpha_3 - \alpha_1))|$ is 0, in which case the homogeneous version of the reduction should be used.

(2) As in the other algorithms, for the solution to be rigorous, the value of the integers $\lfloor B\delta_i \rfloor$ should be known exactly. This implies either a careful error analysis in the computations or the use of an interval arithmetic package.

(3) Various ideas can be used to take into account the fact that the choice of the second and third conjugates to build the linear form is somewhat arbitrary. For instance, Bilu and Hanrot [Bil-Han] have shown how one can use all the conjugates to build a set of $r - 1$ simultaneously small linear forms in r indeterminates, which, by linear algebra, can be used to build a linear form (the coefficients of which are no longer a priori logarithms of algebraic numbers) in two variables that is very small. The last two parts (reduction, enumeration) become then very similar to the case of simultaneous Pell equations.

12.11 Other Classical Diophantine Equations

We begin by stating a special case of Theorem 12.10.1.

Theorem 12.11.1. *Let $F(X,Y)$ be an irreducible, homogeneous, binary integral form of degree $n \geqslant 3$, and let b be a nonzero integer. The equation*

$$F(x,y) = b, \quad \text{with } x,\, y \text{ in } \mathbb{Z}\,, \tag{11.1}$$

has only finitely many solutions, and all of them can be effectively determined.

An ineffective version of this result goes back to Thue, and equation (11.1) is commonly called a *Thue equation*. Baker's theory also yields an effective version of a result of Siegel from the 1930s (we refer to [Sho-Tij, Bug] for bibliographical references and for a more general statement; see also Theorem 8.1.2):

Theorem 12.11.2. *Let $f(X)$ be an irreducible polynomial in $\mathbb{Z}[X]$ of degree $n \geqslant 3$ (respectively $\geqslant 2$), and let $m \geqslant 2$ (respectively $m \geqslant 3$) be an integer. The equation*

$$f(x) = y^m, \quad \text{with } x,\, y \text{ in } \mathbb{Z}\,,$$

has only finitely many solutions, and all of them can be effectively determined.

As already shown by Tijdeman's theorem, the theory of linear forms in logarithms is sufficiently powerful to deal with Diophantine equations in which the exponents are unknown. A general result, due to Schinzel and Tijdeman, is the following.

Theorem 12.11.3. *Let $f(X)$ be an irreducible polynomial in $\mathbb{Z}[X]$ of degree $n \geqslant 3$. The equation*

$$f(x) = y^z, \quad \text{with } x,\, y,\, z \text{ in } \mathbb{Z}, \text{ where } z \geqslant 2 \text{ and } |y| \geqslant 2\,, \tag{11.2}$$

has only finitely many solutions, and all of them can be effectively determined.

Proof. We prove this theorem, following Brindza, Evertse, and Győry [Bri-Eve-Gyo]. For the sake of simplicity, we will assume that $f(X)$ is monic. Our goal is to bound z in terms of $f(X)$, and then to conclude by applying Theorem 12.11.2.

Let D be the discriminant of $f(X)$, and H the maximum of the absolute values of its coefficients. Throughout the proof, the numerical constants implied by $\ll$ depend only on $f(X)$.

Let (x, y, z) be a solution of (11.2). First note that if $|x| \leqslant H + 2$ then we have $2^z \leqslant (2H + 3)^n$ since the roots $\alpha_1, \dots, \alpha_n$ of $f(X)$ are bounded in absolute value by $H + 1$. We can therefore assume that $|x| > H + 2$, which implies that $|x - \alpha_j| \geqslant 1$ and $|x - \alpha_j| \leqslant |y|^z$ for any $j = 1, \dots, n$.

In the number field $\mathbb{K} = \mathbb{Q}(\alpha_1)$ the greatest common divisor of the ideals $(x - \alpha_1)\mathbb{Z}_{\mathbb{K}}$ and $(x - \alpha_2) \cdots (x - \alpha_n)\mathbb{Z}_{\mathbb{K}}$ divides the ideal $f'(\alpha_1)\mathbb{Z}_{\mathbb{K}}$. Since $\mathcal{N}_{\mathbb{K}/\mathbb{Q}}\big(f'(\alpha_1)\big) = \pm D$ there exist integral ideals $\mathfrak{a}$, $\mathfrak{b}$, and $\mathfrak{c}$ of $\mathbb{K}$ such that

$$\mathcal{N}_{\mathbb{K}/\mathbb{Q}}(\mathfrak{a}) \leqslant |D|, \quad \mathcal{N}_{\mathbb{K}/\mathbb{Q}}(\mathfrak{b}) \leqslant |D|, \quad \text{and} \quad \mathfrak{a}\,(x - \alpha_1)\mathbb{Z}_{\mathbb{K}} = \mathfrak{b}\,\mathfrak{c}^m.$$

Let h be the class number of $\mathbb{K}$. The last equation implies that

$$\alpha \left(x - \alpha_1\right)^h = \varepsilon \beta \gamma^z , \qquad (11.3)$$

where α, β, and γ are, respectively, generators of the principal ideals $\mathfrak{a}^h$, $\mathfrak{b}^h$, $\mathfrak{c}^h$, and ε is a unit of $\mathbb{K}$. In particular, we may assume that

$$\max\{\mathrm{h}(\alpha), \mathrm{h}(\beta)\} \ll 1 ,$$

where $\mathrm{h}(\xi)$ denotes the absolute logarithmic height of the algebraic number ξ as given by Definition 12.1.1.

For $i = 1, \ldots, n$, let ϕ_i be the $\mathbb{Q}$-automorphism defined on $\mathbb{K}$ by $\phi_i(\alpha_1) = \alpha_i$, and let $\{\eta_1, \ldots, \eta_r\}$ be a system of fundamental units in $\mathbb{K}$. Equation (11.3) gives

$$(x - \alpha_i)^h = \left(\phi_i(\eta_1)\right)^{k_{i,1}} \cdots \left(\phi_i(\eta_r)\right)^{k_{i,r}} \phi_i(\beta/\alpha) \phi_i(\gamma)^z , \qquad i = 1, \ldots, n ,$$

where the $k_{i,j}$ are integers such that $|k_{i,j}| < z$. Since $\max_{1 \leqslant i \leqslant t} |x - \alpha_i|^h \leqslant |y|^{zh}$, we obtain

$$\mathrm{h}(\gamma) \ll \log |y| .$$

If necessary after reordering the roots of $f(X)$, we may assume that

$$\min_{1 \leqslant i \neq j \leqslant n} \left| \frac{\alpha_i - \alpha_j}{x - \alpha_i} \right| = \left| \frac{\alpha_1 - \alpha_2}{x - \alpha_1} \right| .$$

Recalling that

$$\prod_{1 \leqslant i \neq j \leqslant n} \left| \frac{\alpha_i - \alpha_j}{x - \alpha_i} \right| = \frac{|D|}{|y|^{z(n-1)}} ,$$

we obtain

$$\left| \frac{\alpha_1 - \alpha_2}{x - \alpha_1} \right| \leqslant \frac{|D|}{|y|^{z/n}} .$$

In addition, we may assume that $|y|^{z/(2n)} > h\,|D|$, since otherwise our theorem is proved. Since $|x - \alpha_2| \leqslant |x - \alpha_1|$, we obtain

$$\Lambda_9^* := \left| \left(\frac{x - \alpha_2}{x - \alpha_1} \right)^h - 1 \right| \leqslant \left| \frac{x - \alpha_2}{x - \alpha_1} - 1 \right| \cdot h \leqslant \frac{|D|h}{|y|^{z/n}} < |y|^{-z/(2n)} . \qquad (11.4)$$

If $\Lambda_9^* = 0$, then $(\alpha_1 - \alpha_2)/(x - \alpha_1)$ is an algebraic integer and (11.2) gives the upper bound $z \leqslant n^2 \log_2 |D|$.

If $\Lambda_9^* \neq 0$, we apply Theorem 12.1.2, and we obtain

$$\log \Lambda_9^* \gg - \log |y| \, \log z . \qquad (11.5)$$

Comparing (11.4) and (11.5), we deduce that

$$z \ll \log z .$$

This proves that z is bounded. It then remains only to apply Theorem 12.11.2 to conclude the proof of our theorem. $\square$

12.12   A Few Words on the Non-Archimedean Case

Shortly after the publication of the first papers by Baker on the theory of linear forms in logarithms, a few papers on non-Archimedean analogues appeared. The problem becomes the following. Let p be a given prime number. Let $\alpha_1, \ldots, \alpha_n$ be algebraic numbers whose norms are not divisible by p, and let $b_1, \ldots, b_n$ be integers. We wish to bound from above the p-adic valuation of

$$\alpha_1^{b_1} \cdots \alpha_n^{b_n} - 1.$$

It turns out that the presently available estimates are of comparable quality to those provided in the Archimedean case by Theorem 12.1.2, apart from a single point: the dependence on the prime p. Namely, a supplementary factor p^D appears, which is essentially due to the fact that the p-adic exponential function has a bounded disk of convergence (Proposition 4.2.10).

An important application of linear forms in p-adic logarithms concerns the families of Diophantine equations discussed in the previous section. Instead of looking only for *integer* solutions, we can now deal with rational solutions whose denominators are divisible only by prime numbers from a given finite set.

We end this section with a concrete Diophantine equation solved in [Bug-Mig]. It shows that, sometimes, the use of p-adic logarithmic forms yields better results than the use of Archimedean ones. Consider the equation

$$\frac{10^n - 1}{10 - 1} = y^q , \tag{12.1}$$

corresponding to the search for perfect powers written in base ten with only the digit 1. Rewriting this equation in the form

$$9y^q 10^{-n} - 1 = -10^{-n},$$

we obtain an upper bound for q by using estimates for linear forms in three logarithms. However, we can also consider this equation as

$$10^n = 9y^q + 1 ,$$

and consider the 5-adic valuation v_5 of both sides. On the one hand, it is trivially equal to n. On the other hand, the 5-adic analogue of Baker's theory allows us to bound $v_5(9y^q + 1)$ from above: this is not greater than $c(\log y)(\log q)$, with a reasonably small constant c, since it is derived from a linear form in only two logarithms. Observing further that $n \geqslant q\,(\log y)/(\log 10)$, we immediately obtain that

$$q \log y \leqslant c\,(\log 10)(\log y)(\log q) ,$$

and hence an upper bound for q. This upper bound is much smaller than the one that can be derived from estimates for three Archimedean logarithms. In practice, we get $q \leqslant 2063$: since q may obviously be assumed to be prime, we have replaced our equation (12.1) in three unknowns by about three hundred equations in two unknowns. Various methods are then used to prove that (12.1) has no solution with $n \geqslant 2$ and $q \geqslant 2$.

Remark. Linear forms in non-Archimedean logarithms can be used to extend some of the algorithms presented above to obtain not only integral solutions, but also S-integral solutions, i.e., solutions with denominators having prime factors in a prescribed finite set.

13. Rational Points on Higher-Genus Curves

By Sylvain Duquesne

13.1 Introduction

In Chapter 8 we have seen that elliptic curves and associated tools allow us to solve many Diophantine problems, essentially those coming from cubic or hyperelliptic quartic equations. The goal of the present chapter is to give some idea of the methods that are used for more general equations. For instance, Diophantus himself poses a problem (Problem 17 of book VI of the Arabic manuscript of *Arithmetica* [Ses]) that is equivalent to finding a nontrivial rational point on the curve defined by the equation

$$y^2 = x^6 + x^2 + 1 \,.$$

This is a curve of genus 2, whereas elliptic curves are curves of genus 1, and it is the only example of a curve of genus greater than or equal to 2 considered by Diophantus. In this chapter we will be interested in curves of genus greater than or equal to 2.

Even if curves of higher genus appear to be simply a generalization of elliptic curves, the Diophantine problems and the necessary tools to solve them are quite different. Indeed, the two main Diophantine questions on elliptic curves are the structure of the set of rational solutions and the determination of the set of integral points, and the crucial tool to solve both of these problems is the group structure on elliptic curves, a tool that is not available in the higher-genus case. In higher genus, a deep theorem of Faltings proves Mordell's conjecture stating that the set of rational points is finite, so that the set of integral points is also finite and can be easily deduced from the set of rational points. Thus the main Diophantine problem in higher genus is the determination of the set of rational points. This question is far from being solved in all generality, however, so that it may be interesting to be able to find the set of integral points even if we are unable to find the set of rational points. For this purpose it is sometimes possible to use a Diophantine approximation method due to de Weger (see [DeW2]), but we will not consider this subject here.

We will restrict to smooth projective algebraic curves over some field K (in other words, a projective algebraic variety of dimension 1 with no singular points), together with a point defined over K. As in the case of genus 1, i.e., elliptic curves, this rational point will be denoted by $\mathcal{O}$.

In genus 2, but not in higher genus, the situation is similar to the situation in genus 1: thanks to the Riemann–Roch theorem it can be shown that there exists a plane model of the curve with affine part described by an equation of the form

$$y^2 + h(x)y = f(x) , \tag{13.1}$$

where h and f are polynomials defined over K, with $\deg(f) = 5$ or 6 and $\deg(h) \leqslant 3$. Since this curve must be smooth, the partial derivatives $2y + h(x)$ and $h'(x)y - f'(x)$ must not vanish simultaneously on the curve defined by equation 13.1. Such a curve is called a hyperelliptic curve (of genus 2).

In the case of higher genus we can consider similar hyperelliptic equations, the only change being that $\deg(g) = 2g + 1$ or $2g + 2$ and $\deg(h) \leqslant g + 1$, but not all curves are hyperelliptic if $g \geqslant 3$.

Remarks. (1) When the characteristic of K is different from 2 we can choose h equal to zero by completing the square. Since we are mainly interested in curves defined over number fields, we will always make this assumption. Most of the definitions and basic algorithms can be generalized to fields of characteristic 2.

(2) If there is a K-rational point on the curve with $y = 0$ (assuming $h(x) = 0$) we may assume that $\deg(f) = 2g + 1$ by sending this point to infinity. In this case we may also assume that f is monic by a suitable change of variables, and the homogenized equation has exactly one singularity at infinity. Without loss of generality, we can assume that this point at infinity is $\mathcal{O}$.

For simplicity, in the following we will consider only hyperelliptic equations, although most theoretical results (but usually not practical results) remain true for nonhyperelliptic ones. There are two reasons for this restriction. The first one is that most Diophantine problems that have been solved by the methods that we will explain are genus-2 curves. The second reason is that nonhyperelliptic curves are more complicated to use in practice and very few tools have been developed for such curves.

13.2 The Jacobian

Recall that the main tool that we used for solving Diophantine problems on elliptic curves was the group law, and that it is not available for higher-genus curves, so that the situation seems to be considerably more difficult. As a first step it is thus natural to define a new algebraic object having a group structure and related to such curves, and this is the Jacobian variety.

In the following we let $\mathcal{C}$ be a hyperelliptic curve of genus g defined over K by an equation of the form

$$y^2 = f(x) \quad \text{with} \quad \deg(f) = 2g + 1 \text{ or } 2g + 2 .$$

13.2.1 Functions on Curves

Definition 13.2.1. *The function field* $\overline{K}(\mathcal{C})$ *is the field of fractions of*
$$\overline{K}[\mathcal{C}] = \overline{K}[x, y]/(y^2 - f(x)) .$$

It is clear that any element of this field can be written in the form $(a(x) + b(x)y)/(c(x) + d(x)y)$, where a, b, c, and d are polynomials in x defined over $\overline{K}$. We will use this description of $\overline{K}(\mathcal{C})$ in the following.

Definition 13.2.2. *Let* $n \in \overline{K}[\mathcal{C}]^*$ *be a nonzero polynomial function and let* P *be a point in* $\mathcal{C}(\overline{K})$. *We say that* n *has a zero at* P *if* $n(P) = 0$.

Let $n(x, y) = a(x) + b(x)y$ be a polynomial function in $\overline{K}(\mathcal{C})$ and let $P = (x_0, y_0)$ be a zero of n. Let r be the largest integer such that $(x - x_0)^r$ divides $n(x, y)$, so that we can write
$$n(x, y) = (x - x_0)^r (\alpha(x) + \beta(x)y) .$$

Let s be the largest integer such that $(x - x_0)^s$ divides $\alpha^2(x) - \beta^2(x)f(x)$.

(1) If $y_0 \neq 0$ we define the *order* $\mathrm{ord}_P(n)$ of P to be $r + s$, and we say that P is a zero of n of order $\mathrm{ord}_P(n)$.
(2) If $y_0 = 0$, we define the order of P to be $2r + s$.
(3) If the degree of f is odd and P is the point at infinity, we define the order of P to be $-\max(2 \deg(a), \deg(f) + 2 \deg(b))$.
(4) If the degree of f is even and P is one of the two points on the nonsingular curve that lie over the point at infinity, we define the order of P to be $-\max(\deg(a), \deg(f)/2 + \deg(b))$.

Example 13.2.3. Let $\mathcal{C}$ be the hyperelliptic curve of genus 2 defined over $\mathbb{Q}$ by the equation
$$y^2 = x^5 + 4 .$$

Let n_1 be the function $n_1(x, y) = x - 2$. This function has a zero if the x-coordinate of the point is 2, so that the points $(2, 6)$ and $(2, -6)$ are zeros of n_1, and their order is 1. The coefficient of y in n_1 is equal to zero, so that its degree is $-\infty$; hence the order of the point at infinity $\mathcal{O}$ is -2.

Example 13.2.4. Let $\mathcal{C}$ be the hyperelliptic curve of genus 1 (in other words the elliptic curve) defined over $\mathbb{Q}$ by the equation
$$y^2 = x^3 + 1 .$$

Let n_2 be the function $n_2(x, y) = x + 1 - y$ (which is the equation of a line). The zeros of this function are the points $(2, 3)$, $(0, 1)$, and $(-1, 0)$, each with order 1. Here the degree of f is equal to 3 and the degree of the coefficient of y is equal to 0, so that the point at infinity has order -3. This of course corresponds to the group law on the elliptic curve.

We can now define the order of a zero or a pole of a function of $\overline{K}(\mathcal{C})$.

Definition 13.2.5. *Let* $m \in \overline{K}(\mathcal{C})^*$ *be a nonzero function and let* P *be a point on* $\mathcal{C}\left(\overline{K}\right)$. *Write* $m = n/d$, *where* n *and* d *are polynomial functions, and set* $\operatorname{ord}_P(m) = \operatorname{ord}_P(n) - \operatorname{ord}_P(d)$. *If* $\operatorname{ord}_P(m)$ *is strictly positive (respectively strictly negative), we say that* m *has a zero (respectively a pole) at* P *of order* $|\operatorname{ord}_P(m)|$.

Theorem 13.2.6. *Let* m *be a function in* $\overline{K}(\mathcal{C})^*$. *Counting orders,* m *has as many zeros as poles; in other words,*

$$\sum_{P \in \mathcal{C}(\overline{K})} \operatorname{ord}_P(m) = 0 \ .$$

13.2.2 Divisors

We have seen that the group structure on the set of points of an elliptic curve is a powerful tool for solving many problems concerning elliptic curves, and in particular Diophantine problems that can be reduced to elliptic curves. The set of points of a curve of higher genus does not have a natural group structure, so we are going to embed this set into a larger one that does have a natural group structure by introducing the free abelian group generated by these points.

Definition 13.2.7. *Let* $\mathcal{C}$ *be a smooth projective algebraic curve defined over* K. *The* divisor group $\operatorname{Div}_{\overline{K}}$ *of* $\mathcal{C}$ *is the free abelian group over the points of* $\mathcal{C}\left(\overline{K}\right)$.

An element D of $\operatorname{Div}_{\overline{K}}(\mathcal{C})$ is called a *divisor* and is thus of the form

$$D = \sum_{P \in \mathcal{C}(\overline{K})} n_P P \ ,$$

where the integer n_P is called the order of D at P and is zero for almost all points P on the curve.

Definition 13.2.8. (1) *Let* $D = \sum_{P \in \mathcal{C}(\overline{K})} n_P P \in \operatorname{Div}_{\overline{K}}(\mathcal{C})$. *We define* $\deg(D) = \sum_{P \in \mathcal{C}(\overline{K})} n_P$.
(2) *We say that* D *is* effective *if* $n_P \geqslant 0$ *for all* P.
(3) *Let* m *be a function on* $\mathcal{C}$. *We define the* divisor *of* m *by*

$$\operatorname{div}(m) = \sum_{P \in \mathcal{C}(\overline{K})} \operatorname{ord}_P(m) P \in \operatorname{Div}_{\overline{K}}(\mathcal{C}) \ .$$

Such divisors are called principal divisors, *and the set of principal divisors is denoted by* $\operatorname{Pr}_{\overline{K}}(\mathcal{C})$.

Example 13.2.9. Let us come back to Examples 13.2.3 and 13.2.4. The divisors of the functions n_1 and n_2 are

$$\mathrm{div}(n_1) = (2,6) + (2,-6) - 2\mathcal{O} \,,$$
$$\mathrm{div}(n_2) = (2,3) + (0,1) + (-1,0) - 3\mathcal{O} \,.$$

Denote by $\mathrm{Div}^0_{\overline{K}}(\mathcal{C})$ the group of all divisors of degree 0. It follows from Theorem 13.2.6 that the set of principal divisors is a subgroup of $\mathrm{Div}^0_{\overline{K}}(\mathcal{C})$. We can thus set the following.

Definition 13.2.10. *The quotient group* $\mathrm{Div}^0_{\overline{K}}(\mathcal{C})/\mathrm{Pr}_{\overline{K}}(\mathcal{C})$ *is called the Picard group of* $\mathcal{C}$.

We will not define the *Jacobian variety* $J(\mathcal{C})$ of $\mathcal{C}$, which is an Abelian variety functorially associated with $\mathcal{C}$, but simply note that the group $J_{\overline{K}}(\mathcal{C})$ of $\overline{K}$-rational points on this variety, which is the only structure that we will use, is naturally isomorphic to the Picard group defined above.

13.2.3 Rational Divisors

Definition 13.2.11. *The set of* K-*rational divisors, denoted by* $\mathrm{Div}_K(\mathcal{C})$, *is defined by*

$$\mathrm{Div}_K(\mathcal{C}) = (\mathrm{Div}_{\overline{K}}(\mathcal{C}))^{\mathrm{Gal}(\overline{K}/K)} \,.$$

This definition means that a divisor D is rational over K if it is invariant under the Galois action of $\mathrm{Gal}(\overline{K}/K)$. In other words, if P is a point such that the order of D at P is nonzero, then D has the same order at all the conjugates of P.

Example 13.2.12. Let $\mathcal{C}$ be the curve of genus 2 defined over $\mathbb{Q}$ by the equation

$$y^2 = x^5 + x - 3 \,.$$

The point $(1, i)$ is of course not a $\mathbb{Q}$-rational point, but the divisor $D = (1, i) + (1, -i)$ is a $\mathbb{Q}$-rational divisor of degree 2.

The group of K-rational elements on the Jacobian, $J_K(\mathcal{C})$, is the group of classes of K-rational divisors modulo functions of $K(\mathcal{C})$ or, equivalently,

$$J_K(\mathcal{C}) = (J_{\overline{K}}(\mathcal{C}))^{\mathrm{Gal}(\overline{K}/K)} \,.$$

All these definitions are of course also valid for elliptic curves and in this case the curve is isomorphic to its Jacobian. This indicates that the Jacobian is the correct generalization of elliptic curves if we are interested in the group structure. In fact we have the following theorem, which generalizes the Mordell–Weil theorem for elliptic curves.

Theorem 13.2.13 (Weil). *If K is a number field, $J_K(\mathcal{C})$ is a finitely gen-erated abelian group.*

In the following we will assume that K is a number field. The structure of $J_K(\mathcal{C})$ can be computed analogously to the computation of the Mordell–Weil group for elliptic curves that we have studied in Chapter 8, in other words by using descent methods. It is evidently more difficult, and we will not explain this computation here. The interested reader is referred to [Sch] or to [Sto].

We have already mentioned that in genus 1, the Jacobian is isomorphic to the curve. This isomorphism is given by the map $P \mapsto P - \mathcal{O}$ from $\mathcal{C}(K)$ to $J_K(\mathcal{C})$; in other words, an element of the Jacobian is represented by a point on the curve. In higher genus g, the following theorem states that the situation is analogous since a rational element of the Jacobian can be represented by a g-tuple of points on the curve stable by $\text{Gal}\left(\overline{K}/K\right)$.

Theorem 13.2.14. *Fix a K-rational divisor D_0 of degree g in $\text{Div}(\mathcal{C})$ (for instance $g\mathcal{O}$ if the degree of f is $2g + 1$). Every K-rational element of the Jacobian can be represented by a divisor of degree 0 of the form $D - D_0$, where D is an effective K-rational divisor of degree g.*

Such a D is not unique. In some cases, additional conditions can make it unique. For instance, if $\deg(f) = 2g+1$ a K-rational element of the Jacobian can be uniquely represented by a divisor of degree 0 of the form

$$P_1 + \cdots + P_r - r\mathcal{O}$$

with $r \leqslant g$ and such that $P_i \neq \mathcal{O}$ and a point and its image under the hyperelliptic involution (the map that sends a point (x, y) to $(x, -y)$ in our case) do not both occur in this divisor. Such a representative is called a *reduced* divisor. If the condition on r is omitted, it is called a *semireduced* divisor. We will now describe the group law explicitly and explain how to compute on Jacobians of curves of higher genus.

13.2.4 The Group Law: Cantor's Algorithm

The construction of the Jacobian that we presented in Section 13.2.2 is anal-ogous to the definition of the class group of an algebraic number field as the quotient of the group of fractional ideals modulo principal ideals. In [Can], Cantor uses this analogy to provide an efficient algorithm to compute on the Jacobian of a hyperelliptic curve given by an equation of the form

$$y^2 = f(x)\,, \quad \text{with} \quad \deg(f) = 2g + 1\,. \tag{13.2}$$

The first point is to represent divisors as polynomials.

Theorem 13.2.15 (Mumford's Representation). *Let $\mathcal{C}$ be a hyperellip-tic curve of genus g defined over K as in (13.2). Any semireduced K-rational* *divisor D can be represented by a pair of polynomials a and b in $K[x]$. If $D = P_1 + \cdots + P_r - r\mathcal{O}$ with $P_i = (x_i, y_i) \in \mathcal{C}(\overline{K})$, a and b are defined by the following conditions:*

(1) $a(x) = \prod_{i=1}^{r}(x - x_i)$,
(2) $\deg(b) < \deg(a)$,
(3) *a divides* $b^2 - f$.

This means that for each point $P_i = (x_i, y_i)$, x_i is a root of a with multiplicity $\mathrm{ord}_{P_i}(D)$. The last conditions ensure that $b(x_i) = y_i$ with the correct multiplicity. In this representation, the neutral element of the Jacobian is represented by $a = 1$ and $b = 0$. For reduced divisors, we have the additional condition that the degree of a must be less than or equal to g, and such a representation becomes unique.

This representation of divisors is more natural than the representation with a g-tuple of points. In fact, a K-rational element is now represented by polynomials with coefficients in K, whereas the points of the previous description were defined over $\overline{K}$. Moreover, Cantor noticed that this representation together with $c = (b^2 - f)/a$ is analogous to the representation of the quadratic forms $aX^2 + bXY + cY^2$ with discriminant f. Thus, to add two elements of the Jacobian represented by pairs of polynomials (a_1, b_1) and (a_2, b_2), he proceeds as for the classical composition of quadratic forms (see Chapter 5 of [Coh0]). In this way, he obtains a semireduced representative for the sum that can then be reduced. We now describe these two steps.

Composition. Let d denote the GCD of a_1, a_2, and $b_1 + b_2$, and let s_1, s_2, and $s_3 \in K[x]$ obtained by Euclid's extended algorithm be such that

$$d = s_1 a_1 + s_2 a_2 + s_3(b_1 + b_2).$$

The sum of the elements represented by (a_1, b_1) and (a_2, b_2) is the class of a semireduced divisor represented by the polynomials

$$a = \frac{a_1 a_2}{d^2},$$

$$b = \frac{s_1 a_1 b_2 + s_2 a_2 b_1 + s_3(b_1 b_2 + f)}{d} \bmod a.$$

Reduction. If $\deg(a) > g$ we can decrease it by replacing a by $(f - b^2)/a$, and then replacing b by $-b \bmod a$. These operations must be repeated until $\deg(a) \leqslant g$, so that after dividing a by its leading coefficient to make it monic, the pair of polynomials (a, b) represents the unique reduced divisor in the class of the sum of the two initial divisors. Note that all computations take place in K, and this would not necessarily be the case if the elements of the Jacobian were represented by g-tuples of points in $\mathcal{C}(\overline{K})$.

13.2.5 The Group Law: The Geometric Point of View

Here we want to generalize to the higher-genus case the definition of the group law given for elliptic curves, where we used lines to define the group law. For simplicity, we restrict our study to the genus-2 case and mention the general case at the end.

In genus 2 we have seen in Theorem 13.2.14 that K-rational elements on the Jacobian can be represented by pairs of conjugate or K-rational points. Thus, to add two such pairs of points $\{P_1, Q_1\}$ and $\{P_2, Q_2\}$, we must find a polynomial going through these four points. We of course choose a degree-three polynomial m in $K[x]$, so that the line used for elliptic curves is now replaced by a cubic. This cubic generically meets the hyperelliptic curve in two new points R_3 and S_3 whose opposites are denoted by P_3 and Q_3. The function $y - m(x)$ vanishes at P_1, Q_1, P_2, Q_2, R_3, and S_3, so that the divisor $P_1 + Q_1 + P_2 + Q_2 + R_3 + S_3 - 3D_0$ is the divisor of the function $y - m(x)$. This means that this divisor is the neutral element of the Jacobian. Thus in the Jacobian we have

$$P_3 + Q_3 - D_0 = P_1 + Q_1 - D_0 + P_2 + Q_2 - D_0 .$$

This can be seen in the following picture.

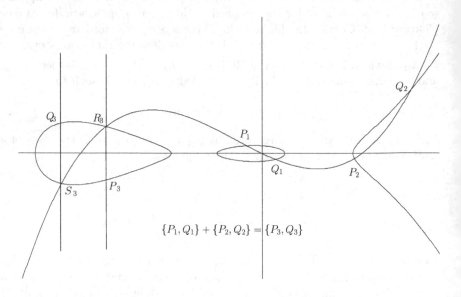

$$\{P_1, Q_1\} + \{P_2, Q_2\} = \{P_3, Q_3\}$$

Remark. This picture does not really correspond to the general situation over $\mathbb{R}$ since points occurring in $\mathbb{R}$-rational divisors are not necessary defined over $\mathbb{R}$ but usually only over $\mathbb{C}$, so that they cannot be represented in our picture. Nonetheless, the method is still valid, and moreover it is also valid for other base fields.

In practice, after computing m, we must simply solve the equation $m(x)^2 = f(x)$. The degree of this equation is equal to 6, and four roots are already known, so that the last two are easy to find. Since m and f are defined over K and since the four known roots are either rational or pairwise conjugate, the two new roots are also rational or conjugate, so they represent a K-rational divisor.

Even if this method for computing the group law is slower than Cantor's algorithm, it has the advantage of working for every curve of genus 2 and of helping to compute in the formal group. Thus, in [Fly] Flynn used it to compute rational points on hyperelliptic curves, which is the goal of this chapter.

Remark. In higher genus the situation is quite similar. We must find two polynomials p and q such that the function $yq(x) - p(x)$ passes through the $2g$ points defining the two elements of the Jacobian that we want to add, and we must ensure that the intersection of $yq(x) - p(x) = 0$ with the curve gives exactly $3g$ points generically. It is an easy exercise to show that we must have $\deg(q) = \lfloor (g-1)/2 \rfloor$ and $\deg(p) = \lfloor 3g/2 \rfloor$, except when g is odd and $\deg(f) = 2g + 2$, in which case we must take $\deg(p) = (3g+1)/2$ and the quotient of the leading coefficients of p and q must be equal to a square root of the leading coefficient of f. The principle is then the same as in genus 2.

13.3 Rational Points on Hyperelliptic Curves

We now focus on Diophantine problems, and particularly on the determination of the rational points on the curve.

Mordell's conjecture, proved in 1983 by Faltings [Fal], states that a curve of genus greater than or equal to 2 defined over a number field K has only finitely many K-rational points. Unfortunately, Faltings's proof is not effective, in that it does not provide a bound for the heights of the K-rational points, and it does not give a bound on the number of such points. However, before Faltings's proof, several papers had appeared that proved effective versions of Mordell's conjecture, but under some restrictive conditions. The best known is the Chabauty–Coleman proof of Mordell's conjecture under a condition on the rank of the Jacobian, but there is also a method due to Dem′yanenko [Dem2], and generalized by Manin [Man]. We begin withn the latter.

13.3.1 The Method of Dem′yanenko–Manin

The idea of this method is to study the decomposition, up to isogeny, of the Jacobian as an Abelian variety. Assume for instance that the Jacobian of a curve $\mathcal{C}$ splits as a product of elliptic curves such that one of these elliptic curves has rank 0 (so that it has finitely many rational points that are easy to compute, see Section 8.1.3). In that case, it is clear that there are finitely many rational points on C itself, which can be found in the preimage of the rational points of the elliptic curve of rank 0. More generally we have the following.

Theorem 13.3.1 (Dem'yanenko–Manin). *Let C be a curve defined over a number field K. Assume that A is a K-simple Abelian variety such that A^m occurs in the decomposition of the Jacobian of C up to isogeny over K and that*

$$m > \frac{\mathrm{rk}(A(K))}{\mathrm{rk}(\mathrm{End}_K(A))} ,$$

where as usual rk *denotes the rank. Then $C(K)$ is finite and can be determined explicitly.*

In practice, we will often use the following corollary.

Corollary 13.3.2. *Let C be a nonsingular projective curve defined over $\mathbb{Q}$. Let E be an elliptic curve defined over $\mathbb{Q}$ such that there exist l independent morphisms from C to E defined over $\mathbb{Q}$. If $l > \mathrm{rk}(E(\mathbb{Q}))$ then $C(\mathbb{Q})$ is finite and can be determined explicitly.*

This method is used in [Dem2], [Sil4], and [Kul] to solve some examples or families of examples of curves. We now describe in detail how this corollary can be made explicit in the first nontrivial case, namely when the elliptic curve E has rank 1 and $l = 2$.

Thus, assume that E has rank 1 over $\mathbb{Q}$ and that we have determined both a generator G of the free part of $E(\mathbb{Q})$ and the torsion subgroup $E(\mathbb{Q})_t$. Assume also we have two independent morphisms ϕ_1 and ϕ_2 from C to E of the same degree d. For each point P in $C(\mathbb{Q})$, there exist two integers n_1 and n_2 and two points T_1 and T_2 in $E(\mathbb{Q})_t$ such that $\phi_i(P) = n_i \cdot G + T_i$ for $i = 1$, 2, so that $\widehat{h}(\phi_i(P)) = n_i^2 \widehat{h}(G)$ for $i = 1$, 2, hence

$$n_2^2 - n_1^2 \;=\; \frac{\widehat{h}(\phi_2(P)) - \widehat{h}(\phi_1(P))}{\widehat{h}(G)} , \tag{13.3}$$

where as usual $\widehat{h}$ denotes the canonical height (see Theorem 8.1.17).

We now bound $n_2^2 - n_1^2$. We have seen in Theorem 8.1.18 that there exist two explicit bounds B_1 and B_2 such that for all P in $E(\mathbb{Q})$, we have

$$-B_1 \leqslant \widehat{h}(P) - h(P) \leqslant B_2 .$$

Moreover, $h(\phi_i(P)) = dh(P) + O(1)$, so that there exists a constant B_3 such that for all P in $E(\mathbb{Q})$, we have

$$|h(\phi_2(P)) - h(\phi_1(P))| \leqslant B_3 .$$

We deduce from (13.3) that

$$|n_2^2 - n_1^2| \leqslant \frac{B_1 + B_2 + B_3}{\widehat{h}(G)} \; .$$

Thus, if $n_1 \neq \pm n_2$, it is immediate to obtain a bound for n_1 and n_2, namely

$$\max(|n_1|, |n_2|) \leqslant \frac{1}{2} \left(\frac{B_1 + B_2 + B_3}{\widehat{h}(G)} + 1 \right) \; .$$

If $n_1 = \pm n_2$ then $\phi_1(P) \pm \phi_2(P)$ is a torsion point and it is easy to find all the possibilities for such a P. We thus obtain a bound on the cardinality of $\mathcal{C}(\mathbb{Q})$, and if, in addition, ϕ_i^{-1} can be computed, we can find exactly all the rational points on $\mathcal{C}$.

Example 13.3.3. Let $\mathcal{C}$ be the Fermat quartic curve defined over $\mathbb{Q}$ by

$$\mathcal{C}: \quad x^4 + y^4 = 2 \; ,$$

and let E be the elliptic curve defined over $\mathbb{Q}$ by

$$E: \quad y^2 = x^3 - 2x \; .$$

The point $T = (0,0)$ is the only nontrivial torsion point on $E(\mathbb{Q})$, $E(\mathbb{Q})$ has rank 1, and its free part is generated by $G = (-1,1)$. Let ϕ_1 and ϕ_2 be the morphisms

$$
\begin{array}{llll}
\phi_1: & \mathcal{C}(\mathbb{Q}) & \to & E(\mathbb{Q}) \, , \\
& (x,y) & \mapsto & (-x^2, xy^2) \, ,
\end{array}
\qquad
\begin{array}{llll}
\phi_2: & \mathcal{C}(\mathbb{Q}) & \to & E(\mathbb{Q}) \, , \\
& (x,y) & \mapsto & (-y^2, x^2 y) \, .
\end{array}
$$

These two morphisms have degree 2, and Silverman proves in a more general context in [Sil4] that they are independent. The bounds B_1 and B_2 are easy to compute. Moreover, we have trivially that $h(\phi_1(P)) = 2h(P)$ for all $P \in \mathcal{C}(\mathbb{Q})$, and using the equation defining $\mathcal{C}$ we can prove that

$$2h(P) - \frac{\log(2)}{2} \leqslant h(\phi_2(P)) \leqslant 2h(P) + \frac{\log(2)}{2} \; ,$$

so that

$$B_3 = \frac{\log(2)}{2} \leqslant 0.35, \quad B_1 \leqslant 4.85, \quad \text{and} \quad B_2 \leqslant 4.43 \; .$$

We deduce that if $n_1 \neq \pm n_2$ then

$$\max(|n_1|, |n_2|) \leqslant \frac{1}{2} \left(\frac{B_1 + B_2 + B_3}{\widehat{h}(G)} + 1 \right) \leqslant 8.42 \; .$$

If $n_1 = \pm n_2$ we have $\phi_1(P) \pm \phi_2(P) = (0,0)$, so that $x = y$, and we find the trivial rational points on $\mathcal{C}(\mathbb{Q})$. The map ϕ_1^{-1} is given by

$$\phi_1^{-1}: \quad E(\mathbb{Q}) \;\;\rightarrow\;\; C(\mathbb{Q})\,,$$

$$(x, y) \;\;\mapsto\;\; \left(\sqrt{-x}, \sqrt{\frac{y}{\sqrt{-x}}}\right),$$

if the square roots that occur are defined over $\mathbb{Q}$. Thus, we must simply check, for n less than or equal to 8, whether the points $\pm nG$ and $\pm nG + T$ have a rational preimage under ϕ_1. This happens only for the point G, and we again find the trivial solutions. Finally, we have thus shown in this way that the only rational solutions to the equation $x^4 + y^4 = 2$ are the trivial ones. This can also be proved using descent methods; see Exercise 24 of Chapter 6. See Section 6.6.3 and in particular Theorem 6.6.13 for the general case of the equation $x^4 + y^4 = c$.

13.3.2 The Method of Chabauty–Coleman

Chabauty-like methods are the best-nknown methods for solving Mordell's conjecture in an effective way. They are based on a theorem due to Chabauty [Cha] preceding Faltings's work by more than 40 years.

Theorem 13.3.4 (Chabauty). *Let C be an algebraic curve defined over a number field K, and let $J_K(C)$ be the group of K-rational points on the Jacobian of C. If the rank of $J_K(C)$ is strictly less than the genus of C there are only finitely many K-rational points on this curve.*

To prove this, Chabauty used p-adic integration, and Coleman noticed in 1985 [Col] that it is possible to deduce from Chabauty's proof a bound on the number of K-rational points on the curve. For this, he obtained an upper bound on the number of zeros of an integral of the first kind on $C(K_\nu)$, where ν is a place of K above a prime p of good reduction for C, and K_ν is the completion of K at ν. We will give Coleman's theorem only for curves defined over $\mathbb{Q}$ since this is the most common case.

Theorem 13.3.5 (Coleman). *Let C be a curve of genus g defined over $\mathbb{Q}$ such that the rank of its Jacobian is less than or equal to $g - 1$. Let p be a prime number such that $p > 2g$ and such that C has good reduction at p. Then*

$$|C(\mathbb{Q})| \leqslant |C(\mathbb{F}_p)| + 2g - 2\,.$$

In [McC] McCallum has generalized Coleman's proof of the above result to prove the second case of Fermat's last theorem for regular primes. Of course this is superseded by the work of Wiles, but it shows the power of the method.

Example 13.3.6. Grant gave the first nontrivial example of a Diophantine equation solved by this method [Gran]. Let C be the curve of genus 2 defined over $\mathbb{Q}$ by the equation

$$y^2 = x(x-1)(x-2)(x-5)(x-6) \ .$$

It can be shown that the Jacobian of $\mathcal{C}$ has rank 1, so that Chabauty's condition is satisfied. The curve has good reduction at 7, so that $|\mathcal{C}(\mathbb{Q})| \leqslant |\mathcal{C}(\mathbb{F}_7)| + 2$. We check that $|\mathcal{C}(\mathbb{F}_7)| = 8$, so that there are at most 10 points on $\mathcal{C}(\mathbb{Q})$. Since it is easy to find 10 points on $\mathcal{C}(\mathbb{Q})$ this bound is sharp, so we have shown that

$$\mathcal{C}(\mathbb{Q}) = \{\mathcal{O}, (0,0), (1,0), (2,0), (5,0), (6,0), (3,\pm 6), (10,\pm 120)\} \ .$$

Grant also proves that $\mathcal{C}$ does not cover an elliptic curve of rank 0 over $\mathbb{Q}$, so that this example is not trivial (in fact, $J(\mathcal{C})$ is absolutely simple; in other words, $\mathcal{C}$ does not cover any elliptic curve).

Unfortunately, even under the restriction on the rank of the Jacobian, Chabauty–Coleman-type theorems are not wholly satisfactory since they provide a bound on the *number* of rational points and not on the *height* of such rational points. Thus, if the bound obtained in Theorem 13.3.5 is not sharp, we cannot say exactly which are the rational points on the curve. In fact, Coleman's bound is often not sharp. This situation is similar to the one that we have already met in applications of Strassmann's theorem; see Section 4.5.4.

Following this idea, Flynn studies in more detail the Jacobian of each curve instead of proving a theorem valid for all curves. This leads to an explicit method for finding rational points on curves satisfying Chabauty's condition that provides a better bound than Coleman's theorem, so that we can hope that we can now obtain a sharp bound. This method is explained in full detail in [Fly], but we will not describe it here since it involves advanced tools on Jacobians, so we will give only the main ideas. Afterward, we will describe the elliptic Chabauty method, also due to Flynn [Fly-Wet1]. It has the advantage of being very similar, but uses only the classical tools on elliptic curves already described in the rest of this book. In addition, it also allows us to solve certain Diophantine problems for which Chabauty's condition is not satisfied.

13.3.3 Explicit Chabauty According to Flynn

As described above, This method provides a better bound than Coleman's, and in contrast to Coleman's, this bound is often sharp. Unfortunately, at present it is available only for curves of genus 2 (for which the rank of the Jacobian is 1), since it involves an intricate study of the group law on the Jacobian and of the formal group law. Moreover, it requires that the structure of $J_K(\mathcal{C})$ have been computed. Thus, in what follows we assume that we have computed the torsion subgroup and a generator $\mathcal{G}$ of the free part of $J_K(\mathcal{C})$.

The method is described in detail in [Fly] and [Cas-Fly], and uses the representation of elements of the Jacobian by pairs of points on the curve that are conjugate over some quadratic extension of K. It can be split into six main steps.

(1) The first step is just the remark that finding K-rational points on C is equivalent to finding elements on the Jacobian represented by a pair $\{P, P\}$. Indeed a point P defined over a quadratic extension of K is K-rational if and only if it is equal to its conjugate.

(2) We choose a prime p such that the curve has good reduction at p, we let m be the order of $\mathcal{G}$ in the Jacobian reduced modulo p, and we set $\mathcal{E} = m \cdot \mathcal{G}$. It is easy to prove that every element of the Jacobian can be written uniquely in the form

$$\mathcal{A} + n \cdot \mathcal{E}, \text{ with } n \in \mathbb{Z} \text{ and } \mathcal{A} \text{ is in a finite set } U .$$

(3) Using the formal group law on the Jacobian (see Section 7.3.5 for the case of elliptic curves), for each $\mathcal{A}$ in the finite set U we can express the coordinates representing an element of the form $\mathcal{A} + n \cdot \mathcal{E}$ as a power series in n. The fact that $\mathcal{E}$ is in the kernel of the reduction modulo p ensures that these power series are convergent in $\mathbb{Z}_p$.

(4) There exists a trivial relation between the coordinates representing the element $\mathcal{A} + n \cdot \mathcal{E}$ that expresses that the two points representing $\mathcal{A} + n \cdot \mathcal{E}$ are equal. Flynn deduces that for each $\mathcal{A} \in U$, if an element of the Jacobian of the form $\mathcal{A} + n \cdot \mathcal{E}$ is represented by two equal points, n must be a zero of a power series that converges in $\mathbb{Z}_p$.

(5) We use Strassmann's theorem described in Section 4.5 to bound the number of such zeros.

(6) The sum of these bounds for all $\mathcal{A}$ in the finite set U gives a bound on the number of elements of the Jacobian represented by a pair of the form (P, P) and hence a bound on the number of rational points on the curve itself.

Remark. As in Coleman's method it can happen that the bound is not sharp, but this happens much less frequently. Moreover, if the bound is not sharp, we can try another prime p of good reduction. Finally, even if the bound is not sharp after several attempts, we obtain a great deal of local information on the missing points, and this can be used to find exactly all the rational points as in the complete example that we will give in Section 13.3.6.

Example 13.3.7. Let C be the curve of genus 2 defined over $\mathbb{Q}$ by the equation

$$y^2 = 2x(x^2 - 2x - 2)(-x^2 + 1) .$$

The Jacobian of this curve has rank 1, so that Chabauty's condition is satisfied, and in [Cas-Fly] Flynn gives a detailed proof based on this method that the only rational points on this curve are

$$\mathcal{C}(\mathbb{Q}) = \left\{ \mathcal{O}, (0,0), (1,0), (-1,0), \left(-\frac{1}{2}, \frac{3}{4}\right), \left(-\frac{1}{2}, -\frac{3}{4}\right) \right\} .$$

Since these six points cannot be equal over $\mathbb{F}_p$ if $p \geqslant 5$, we will always have $|\mathcal{C}(\mathbb{F}_p)| = 6$, so that Coleman's bound will always be equal to 8, which is not sharp. In particular, this example shows why Coleman's bound is often not sharp.

This method provides a powerful tool for finding rational points, but as already mentioned, it can unfortunately be used only for curves of genus 2 whose Jacobian has rank 1. In higher genus, there are no theoretical obstructions but the arithmetic on the Jacobian and the formal group law are much more complicated. In recent years, mathematicians have preferred to focus on curves of genus 2 but whose Jacobians have higher rank, in other words when Chabauty's condition is not satisfied.

13.3.4 When Chabauty Fails

We have already seen that a trivial instance of Dem'yanenko's method allows us to find rational points on a curve $\mathcal{C}$ by covering them by rational points on an elliptic curve of rank 0. We are now going to see how to generalize this. The idea remains the same, but we will also treat the case in which the elliptic curves have nonzero rank.

The first step is to find a covering collection for $\mathcal{C}$ over the base field K. A covering collection is a set $\{D_i \to \mathcal{C}\}$ of K-rational covers of $\mathcal{C}$ in a single $\overline{K}$-isomorphism class such that every point in $\mathcal{C}(K)$ is the image of a point in some $D_i(K)$. Given a covering collection, the question of determining or of bounding the set of rational points on $\mathcal{C}$ can be reduced to determining or bounding the set of rational points on each of the covers. There are several techniques to find such covering collections, which all use the same idea: we first find an Abelian variety A that maps to the Jacobian of $\mathcal{C}$ under an isogeny ϕ. The pullbacks under ϕ of a suitably chosen set of embeddings of the curve in its Jacobian then give a covering collection on curves lying over A.

Example 13.3.8. In his PhD dissertation [Wet], Wetherell solves, thanks to this kind of technique, what seems to be the only curve considered by Diophantus that has genus strictly greater than 1 (Problem 17 of Book VI of the Arabic manuscript of *Arithmetica* [Ses]). This curve of genus 2 is given by the equation

$$\mathcal{C}: \quad y^2 = x^6 + x^2 + 1 .$$

It covers two elliptic curves with the maps $(x,y) \mapsto (x^2, y)$ and $(x,y) \mapsto (1/x^2, y/x^3)$. Let A denote the product of these elliptic curves. Both elliptic curves have rank 1 over $\mathbb{Q}$, so that the Jacobian of $\mathcal{C}$, which is isogenous to A, has rank 2, and hence Chabauty's condition is not satisfied. Wetherell uses pullbacks of the isogeny from A to $J(\mathcal{C})$ to find a pair of genus-3 curves D_1 and D_2 whose rational points cover the rational points on $\mathcal{C}$ (up to the hyperelliptic involution). The Jacobians of these genus-3 curves have rank 1 and 0, so that the rational points on these curves can be found using Chabauty's methods. Finally, one finds that the finite rational points on $\mathcal{C}$ are $(0, \pm 1)$ and $(\pm 1/2, \pm 9/8)$.

Example 13.3.9. In [Fly-Wet2], Flynn and Wetherell use another method, whose principle has been used for several other Diophantine equations, to solve the Diophantine equation

$$x^4 + y^4 = 17$$

proposed by Serre. They use the map introduced for 2-descent on the Jacobian that generalizes the fundamental 2-descent map for elliptic curves studied in Sections 8.2 and 8.3. Thanks to this map, they obtain a covering collection for the curve defined over $\mathbb{Q}$ by the equation

$$y^2 = (9x^2 - 28x + 18)(x^2 + 12x + 2)(x^2 - 2) \,,$$

whose rational points are sufficient to solve Serre's equation.

There is a more classical approach using resultants to obtain covering collections. It is used by Bruin to solve some generalized Fermat equations, see [Bru1] and Chapter 14, and in [Duq1] to solve a hyperelliptic curve of genus 4 whose Jacobian has rank 4. This method consists in factoring the polynomial f defining the curve over some number field so that both factors must be squares but have lower degrees. We will illustrate this method in the complete example given in Section 13.3.6.

In most cases, such covering methods give rise to a new Diophantine problem on elliptic curves that is very similar to the problem of finding rational points on curves of higher genus. This new problem can be solved, under certain conditions, by the method called elliptic curve Chabauty. We will describe this method in detail for two reasons. The first one is that it allows us to solve many Diophantine problems that do not satisfy Chabauty's condition. The second one is that, as explained above, it is very similar to the explicit Chabauty method developed by Flynn and much easier to describe and to understand.

13.3.5 Elliptic Curve Chabauty

We want to solve the following problem: given an elliptic curve E defined over a number field $K = \mathbb{Q}(\alpha)$ of degree d over $\mathbb{Q}$, find all the points in $E(K)$ having their x-coordinate in $\mathbb{Q}$.

It follows from Faltings's theorem that the number of such points is finite and this also follows from Chabauty's theorem if the Mordell–Weil rank of $E(K)$ is strictly less than d.

As in the explicit method developed by Flynn for curves of genus 2, we will be able to conclude only under this condition. For this, we will use the formal group law on elliptic curves to construct power series with coefficients in $\mathbb{Z}_p$ whose zeros correspond to the points that we want (in other words, having a $\mathbb{Q}$-rational x-coordinate). The number of these zeros can then be bounded thanks to Strassmann's theorem or to the Weierstrass preparation theorem seen in Chapter 4. Once again, we will obtain an upper bound only for the number of points and not for the heights of the points, so that the method fails if the bound obtained is not sharp.

The first step of the method is evidently the determination of the Mordell–Weil group $E(K)$. This can be done using descent methods generalizing those explained for the case of $\mathbb{Q}$ in Chapter 8; see [Sim2].

Thus, in what follows we will assume that we know the torsion subgroup and a set of generators of $E(K)$. It is clearly not necessary to treat the trivial case of rank 0. We thus write

$$E(K) = \langle P_1, \dots, P_r \rangle \oplus E(K)_t \ .$$

The second step is to transform the condition of $\mathbb{Q}$-rationality into a condition of vanishing of power series. To do this, choose a prime number p such that α is p-integral, and let $\widetilde{\alpha}$ denote the image of α in $\mathbb{Z}_K/p\mathbb{Z}_K$. Assume that p is chosen so that the following conditions hold:

(1) $[\mathbb{Q}_p(\alpha) : \mathbb{Q}_p] = [\mathbb{Q}(\alpha) : \mathbb{Q}] = d$,
(2) $\mathbb{Q}(\alpha)/\mathbb{Q}$ is unramified at p ,
(3) $|\alpha|_p = 1$,
(4) The residue field of $\mathbb{Q}_p(\alpha)$ is $\mathbb{F}_p(\widetilde{\alpha})$,
(5) The curve E has good reduction at p ,
(6) The coefficients of the equation defining E are p-integral.

Remark. The first condition is always satisfied if d is 2 or 3. In higher degrees it is much more difficult to satisfy. However, in [Fly-Wet2] Flynn and Wetherell prove that we can remove this condition. Conditions (2), (5), and (6) are not difficult to satisfy, and conditions (3) and (4) ensure that α (or $\widetilde{\alpha}$) is a generator for all fields and rings that we will consider.

Denote by $\widetilde{E}$ the reduction of E modulo p. Thanks to condition (5), $\widetilde{E}$ is an elliptic curve defined over $\mathbb{F}_p(\widetilde{\alpha})$. Let us now define for all integers $i \leqslant r$,

$\widetilde{P}_i$,   the reduction of P_i modulo p,
m_i ,   the order of $\widetilde{P}_i$ in $\widetilde{E}(\mathbb{F}_p(\widetilde{\alpha}))$,
Q_i ,   the multiple $m_i \cdot P_i$ of P_i in $E(\mathbb{Q}(\alpha))$,
t_i ,   the t-coordinate of Q_i ($t_i = -x_i/y_i$ if $Q_i = (x_i, y_i)$).

We now want to write any point in $E(\mathbb{Q}(\alpha))$ in terms of the Q_i instead of the P_i. To do this, we define a finite set $\mathcal{U}$ in the following way:

$$\mathcal{U} = \{T + n_1 \cdot P_1 + \cdots + n_r \cdot P_r : T \in E(\mathbb{Q}(\alpha))_t, \ -m_i/2 < n_i \leqslant m_i/2\} \ ,$$

so that every point P in $E(\mathbb{Q}(\alpha))$ can be written uniquely in the form

$$P = U + n_1 \cdot Q_1 + \cdots + n_r \cdot Q_r \ , \tag{13.4}$$

where U lies in the finite set $\mathcal{U}$ and $n_i \in \mathbb{Z}$. In order to express the x-coordinate of a point in this form as a power series, we use the formal group law described in Section 7.3.5. The curves that we deal with are given by equations of the form

$$y^2 = g_3 x^3 + g_2 x^2 + g_1 x + g_0 \ .$$

Thus, let us first give the explicit formulas that we need for such curves. These formulas can be easily deduced from the formulas for the formal group law given in 7.3.5. We set $t = -x/y$. The inverse of the x-coordinate is in $\mathbb{Z}[g_0, g_1, g_2, g_3][[t]]$ and the beginning of its expansion is

$$\frac{1}{x} = g_3 \left(t^2 + g_2 t^4 + \left(g_1 g_3 + g_2^2 \right) t^6 + O\left(t^8 \right) \right) \ . \tag{13.5}$$

Moreover, we can express the x-coordinate of the sum of a point (x_0, y_0) and a point (x, y) as a power series ψ in t with coefficients in $\mathbb{Z}[g_0, g_1, g_2, g_3, x_0, y_0]$:

$$\begin{aligned} \psi(t) = x_0 + 2y_0 t &+ \left(3g_3 x_0^2 + 2g_2 x_0 + g_1 \right) t^2 \\ &+ (4g_3 x_0 y_0 + 2g_2 y_0) t^3 + O\left(t^4 \right) \ . \end{aligned} \tag{13.6}$$

We will also use the formal logarithm and the formal exponential. The beginning of their expansions are given by

$$L(t) = t + \frac{1}{3} g_2 t^3 + \frac{1}{5} \left(g_2^2 + 2g_1 g_3 \right) t^5 + O\left(t^7 \right) \ , \tag{13.7}$$

$$E(t) = t - \frac{2}{3!} g_2 t^3 + \frac{8}{5!} \left(2g_2^2 - 6g_1 g_3 \right) t^5 + O\left(t^7 \right) \ . \tag{13.8}$$

Remark. Note that as usual the coefficients of $E(t)$ are not in the ring $\mathbb{Z}[g_0, g_1, g_2, g_3]$, but the denominator of the kth coefficient divides $k!$.

We can now use this formal group structure. The t-coordinate of $n_1 \cdot Q_1 + \cdots + n_r \cdot Q_r$ can be expressed as a formal power series in the r variables $n_1, \ldots, n_r$. Indeed we have

$$t\text{-coord}(n_1 \cdot Q_1 + \cdots + n_r \cdot Q_r) = E(n_1 L(t_1) + \cdots + n_r L(t_r)) \ .$$

The sixth condition on p, the above remark, and the fact that each Q_i is in the kernel of reduction modulo p (so that $|t_i|_p \leqslant p^{-1}$) ensure that this power series has coefficients in $\mathbb{Z}_p[\alpha]$ converging to zero in $\mathbb{Z}_p[\alpha]$.

To describe all points in $E(\mathbb{Q}(\alpha))$ as in (13.4), it is now necessary to add such a linear combination of the Q_i with an element U of the finite set $\mathcal{U}$. If $U = (x_0, y_0)$, we use formula (13.6) to express the x-coordinate of $U + n_1 \cdot Q_1 + \cdots + n_r \cdot Q_r$ as a power series in $\mathbb{Z}_p[\alpha][[n_1, \ldots, n_r]]$. If U is the point at infinity, we use formula (13.5) to express the inverse of the x-coordinate of $U + n_1 \cdot Q_1 + \cdots + n_r \cdot Q_r$ as a power series in $\mathbb{Z}_p[\alpha][[n_1, \ldots, n_r]]$. In any case, let us call θ_U the resulting power series. We can split θ_U into its components

$$\theta_U = \theta_U^{(0)} + \theta_U^{(1)} \alpha + \cdots + \theta_U^{(d-1)} \alpha^{d-1} \,,$$

where each $\theta_U^{(i)}$ is a power series in $\mathbb{Z}_p[[n_1, \ldots, n_r]]$ whose coefficients converge to zero in $\mathbb{Z}_p$. Finally, a point $P = U + n_1 \cdot Q_1 + \cdots + n_r \cdot Q_r$ has its x-coordinate in $\mathbb{Q}$ if and only if

$$\theta_U^{(1)}(n_1, \ldots, n_r) = \cdots = \theta_U^{(d-1)}(n_1, \ldots, n_r) = 0 \,.$$

The strategy is now clear: for each U in the finite set $\mathcal{U}$, we compute these $d - 1$ power series in r variables. Then we use Strassmann's theorem given in Section 4.5 or variants in several variables (see [Sug, Duq1]) to obtain a bound on the number of zeros of such a system of power series. This bound is also a bound on the number of points with a $\mathbb{Q}$-rational x-coordinate of the form $U + n_1 \cdot Q_1 + \cdots + n_r \cdot Q_r$. Doing this for all the elements of the finite set $\mathcal{U}$ give a bound for the elliptic Chabauty problem.

Remarks. (1) By this method, we obtain a system of $d - 1$ power series in r variables, so that the Chabauty-like restriction on the rank of $E(\mathbb{Q}(\alpha))$ (namely $r < d$) is crucial for the success of this method.
(2) This method can easily be adapted if $\mathbb{Q}$ is replaced by a number field.

13.3.6 A Complete Example

In this section, we will apply and illustrate the elliptic curve Chabauty method to prove the following theorem [Duq2].

Theorem 13.3.10. *Let $\mathcal{C}$ be the curve of genus 2 defined over $\mathbb{Q}$ by the equation*

$$\mathcal{C}: \quad y^2 = (x^2 + 1)(x^2 + 3)(x^2 + 7) \,.$$

The rational points on this curve are

$$\mathcal{C}(\mathbb{Q}) = \{\infty^+, \infty^-, (1, \pm 8), (-1, \pm 8)\} \,,$$

where ∞^+ and ∞^- are the points on the nonsingular curve that lie over the point at infinity on $\mathcal{C}$.

This curve was introduced by Flynn and Wetherell in [Fly-Wet1] together with about fifty semirandom curves of the same type (where the polynomial f is even) in order to test their method for such curves. It is the only one for which their method fails. Thus we use another method by changing the problem of finding rational points on a hyperelliptic curve into an elliptic Chabauty problem.

Since the polynomial $x^2 + 1$ factors in the number field $\mathbb{Q}(i)$ we are going to apply the resultant method as follows. If (x, y) is a point on $\mathcal{C}(\mathbb{Q})$ there exist y_1, y_2, and α in $\mathbb{Q}(i)$ such that we have simultaneously

$$\alpha y_1^2 = (x^2 + 3)(x + i) \,,$$
$$\alpha y_2^2 = (x^2 + 7)(x - i) \,.$$

These equations are those of elliptic curves defined over $\mathbb{Q}(i)$ with a point (x, y_i) having its x-coordinate in $\mathbb{Q}$. Therefore, if for each α we are able to solve the corresponding elliptic Chabauty problem on one of these two curves, then we are able to prove Theorem 13.3.10.

The first step is to enumerate the possible values for α. The resultant of $(x^2 + 3)(x + i)$ and $(x^2 + 7)(x - i)$ is equal to $-2^7 3i$, so that if $y \neq 0, \infty$, we can assume, without loss of generality, that α is a square-free $\{2, 3\}$-unit in $\mathbb{Q}(i)$. The $\{2, 3\}$-units of $\mathbb{Q}(i)$ are generated by i, $1 + i$, and 3, so that

$$\alpha \in \{1, i, 1 + i, 3, 1 - i, 3i, 3(1 + i), 3(1 - i)\}.$$

We can reduce this set by standard local arguments. Indeed, if x is in $\mathbb{Q}$, $\alpha(x^2 + 3)(x + i)$ and $\alpha(x^2 + 7)(x - i)$ can be simultaneously squares modulo 9 in $\mathbb{Q}(i)$ only if $\alpha = 1 + i$ or $1 - i$. We will treat only the case $\alpha = 1 - i$ (in fact the other case can be deduced from this one by an easy change of variables). Let E_1 and E_2 be the elliptic curves defined over $Q(i)$ by the equations

$$E_1 : \quad y^2 = (1 - i)(x^2 + 3)(x + i) \,.$$
$$E_2 : \quad y^2 = (1 - i)(x^2 + 7)(x - i) \,.$$

If $(x, y) \in \mathcal{C}(\mathbb{Q})$, there exist a point on $E_1(\mathbb{Q}(i))$ and a point on $E_2(\mathbb{Q}(i))$ having the same x-coordinate x (which is in $\mathbb{Q}$). Thus if the elliptic Chabauty method succeeds either for E_1 or for E_2, Theorem 13.3.10 is proved. In fact, the rank of $E_1(\mathbb{Q}(i))$ is equal to 2, so that the method cannot be applied since the Chabauty-like condition is not satisfied. On the other hand, the rank of $E_2(\mathbb{Q}(i))$ is equal to 1, so hopefully the method can be applied. The torsion subgroup is equal to $\{\mathcal{O}, (i, 0)\}$, and $G = (4i - 3, 12)$ is a generator for the free part of $E_2(\mathbb{Q}(i))$; see [Sim2].

The smallest prime number p satisfying the six conditions given in Section 13.3.5 is $p = 11$. The reduction $\widetilde{G}$ of G modulo 11 on $\widetilde{E_2}$ has order 5, so that we set

$$m = 5 \, ,$$
$$Q = 5 \cdot G \, ,$$
$$t = \text{the } t\text{-coordinate of } Q \ (t = -x/y \text{ if } Q = (x, y)) \, ,$$
$$\mathcal{U} = \{\mathcal{O}, (i, 0), \pm G, \pm 2 \cdot G, \pm G + (i, 0), \pm 2 \cdot G + (i, 0)\} \, .$$

Therefore every point P in $E_2(\mathbb{Q}(i))$ can be written in the form $P = U + nQ$ with U in the finite set $\mathcal{U}$ and n in $\mathbb{Z}$. In fact, $\mathcal{U}$ can be easily reduced in two ways.

First, since Q is in the kernel of the reduction modulo 11, $\tilde{P} = \tilde{U}$. Thus if the x-coordinate of P is in $\mathbb{Q}$, the x-coordinate of $\tilde{U}$ is in $\mathbb{F}_{11}$. Hence we can eliminate the points $(i, 0)$, $\pm G$ and $\pm 2 \cdot G$ from $\mathcal{U}$.

Second, since n is in $\mathbb{Z}$, if we know the values of n such that the x-coordinate of $U + nG$ is in $\mathbb{Q}$, we know the values of n such that the x-coordinate of $-U + nG$ is in $\mathbb{Q}$. This explains why $\mathcal{U}$ has not been defined by

$$\{T + n_1 \cdot P_1 + \cdots + n_r \cdot P_r : \ T \in E(\mathbb{Q}(\alpha))_t, \ 0 \leqslant n_i \leqslant m_i - 1\} \, .$$

Finally, we choose

$$\mathcal{U} = \{\mathcal{O}, G + (i, 0), 2 \cdot G + (i, 0)\} \, .$$

We will now compute the beginning of the expansions of the power series involved in the method. In fact, working modulo 11^5 will be sufficient for our purpose. Thanks to the standard estimate $|k!|_p \geqslant p^{-(k-1)/(p-1)}$ and the remark given after equation 13.8, the terms in $O(t^7)$ can be ignored. Let us first compute the t-coordinate of Q $(\text{mod } 11^5)$,

$$t = 11(10763 + 7311i) \ (\text{mod } 11^5) \, ,$$

then its formal logarithm thanks to (13.7),

$$L(t) = 11(1446 + 5496i) \ (\text{mod } 11^5) \, .$$

We can now compute the t-coordinate of nQ modulo 11^5 thanks to the formal exponential (13.8)

$$t\text{-coord}(nQ) = 11(1446 + 5496i)n + 11^3(77 + 15i)n^3 \ (\text{mod } 11^5) \, .$$

Thanks to the expansions (13.5) and (13.6), we deduce that

$$\theta_{\mathcal{O}}(n) = 11^2(574 + 17i)n^2 + 11^4(8 + 7i)n^4 \ (\text{mod } 11^5) \, ,$$
$$\theta_{G+(i,0)}(n) = -1 + 11(45 + 3073i)n + 11^2(382 + 1318i)n^2$$
$$+ 11^3(68 + 16i)n^3 + 11^4(10 + 5i)n^4 \ (\text{mod } 11^5) \, ,$$
$$\theta_{2 \cdot G+(i,0)}(n) = 154608 + 11 \cdot 10541i + 11(1484 + 10609i)n$$
$$+ 11^2(445 + 247i)n^2 + 11^3(115 + 93i)n^3 + 11^4(7 + 8i)n^4 \ (\text{mod } 11^5) \, .$$

Thus, when $U + nQ$ has its x-coordinate in $\mathbb{Q}$ the following power series $\theta_U^{(1)}(n)$ must vanish:

$$\theta_{\mathcal{O}}^{(1)}(n) = 11^2 \cdot 17n^2 + 11^4 \cdot 7n^4 \pmod{11^5},$$

$$\theta_{G+(i,0)}^{(1)}(n) = 11 \cdot 3073n + 11^2 \cdot 1318n^2 + 11^3 \cdot 16n^3 + 11^4 \cdot 5n^4 \pmod{11^5},$$

$$\theta_{2 \cdot G+(i,0)}^{(1)}(n) = 11 \cdot 10541 + 11 \cdot 10609n$$
$$+ 11^2 \cdot 247n^2 + 11^3 \cdot 93n^3 + 11^4 \cdot 8n^4 \pmod{11^5}.$$

We now use Strassmann's theorem to bound the number of zeros of these power series.

- If $U = \mathcal{O}$, we assume that $n \neq 0$ and apply Strassmann's theorem to the power series

$$\frac{\theta_{\mathcal{O}}^{(1)}(n)}{n^2} = 11^2 \cdot 17 + 11^4 \cdot 7n^2 \pmod{11^5},$$

  which proves that this power series has no 11-adic solution. Thus $\theta_{\mathcal{O}}^{(1)}$ has only one zero. We already know that $\mathcal{O}$ is $\mathbb{Q}$-rational, so that the bound is sharp in this case.
- If $U = G + (i,0)$, we deduce that the power series $\theta_{G+(i,0)}^{(1)}$ has at most one zero in $\mathbb{Z}_{11}$. Again this bound is sharp since for $n = 0$, $G + (i,0) + nQ = (-1, -4i)$ has its x-coordinate in $\mathbb{Q}$.
- If $U = 2 \cdot G + (i,0)$, the bound given by Strassmann's theorem is again equal to 1 but we do not know any point of the form $2 \cdot G + (i,0) + nQ$ having its x-coordinate in $\mathbb{Q}$. Thus the bound is not sharp and the method fails.

Finally, the elliptic Chabauty method only allows us to prove that the points on $E_2(\mathbb{Q}(i))$ having their x-coordinate in $\mathbb{Q}$ are $\mathcal{O}$, $(-1, \pm 4i)$, and at most one point P_0 (and its opposite) such that

$$P_0 = 2 \cdot (4i - 3, 12) + (i,0) + nQ \text{ with } n \in \mathbb{Z}.$$

We could of course try other prime numbers p, but the method also fails for $p = 19$ and $p = 23$.

Thus the elliptic Chabauty method does not allow us to prove Theorem 13.3.10 directly. However, we get some local information. Indeed, Let x_0 denote the x-coordinate of P_0. Since Q is in the kernel of the reduction modulo 11, x_0 is equal to the x-coordinate of $2 \cdot (4i - 3, 12) + (i,0)$ modulo 11. It follows that $x_0 \equiv 3 \pmod{11}$. It is now easy to check that there is no point in $\mathcal{C}(\mathbb{Q})$ whose x-coordinate is congruent to 3 modulo 11. Thus, even if the elliptic Chabauty method used alone fails here (and this is not often the case), we still have proved Theorem 13.3.10.

14. The Super-Fermat Equation

This chapter gives a detailed survey of the work done on the super-Fermat equation by many authors, assuming without proof the most difficult results. The parametrizations given for the elliptic case were initially obtained (with a few errors and omissions) by F. Beukers and D. Zagier, and completed by J. Edwards for the most difficult and interesting icosahedral case. Although I have included this chapter in the part dealing with "modern methods," most of its contents is the treatment of the elliptic cases not including the icosahedral case. This is not at all modern, but is exactly the type of reasoning done using simple algebraic number theory that we have employed many times in Chapter 6. Sections 14.2, 14.3, and 14.4 should therefore not be studied directly (it would probably be rather boring to do so), but considered instead as exercises that the reader is invited to solve by himself without looking at the completely detailed solutions given in these sections. On the other hand, the solution to the icosahedral case, due to Beukers and Edwards, uses classical invariant theory, but in a very original manner linked to the modern theory of Grothendieck dessins d'enfants, and the results on the hyperbolic case use modern methods for finding rational points on curves of higher genus (Chapter 13), and the modular method of Ribet–Wiles (Chapter 15).

14.1 Preliminary Reductions

The general super-Fermat equation is the equation $Ax^p + By^q + Cz^r = 0$ for given nonzero integers A, B, C and integral exponents p, q, and r greater than or equal to 2 (otherwise the equation would have little interest). The number of integers less than or equal to some large X of the form Ax^p is $O(X^{1/p})$, and similarly for By^q and Cz^r. Thus, to be able to obtain 0 as a sum of such quantities by something other than pure accident, it is reasonable to believe that we must have $X \leqslant O(X^{1/p+1/q+1/r})$, in other words $1/p + 1/q + 1/r \geqslant 1$. Thus, we *expect* (of course we have no proof) that when $1/p + 1/q + 1/r < 1$ (the so-called *hyperbolic* case), we will have only finitely many solutions. On the other hand, when $1/p + 1/q + 1/r > 1$ (the so-called *elliptic* or *spherical* case), we expect an infinity of solutions. Finally, as we have seen in Sections 6.4.2, 6.4.3, 6.4.4, 6.4.5, and 6.5, in the intermediate case $1/p + 1/q + 1/r = 1$ (the so-called *parabolic*) we can have infinitely many or finitely many solutions, depending on A, B, C.

This heuristic reasoning is almost correct, but not quite. Indeed, I claim that for many triples (p, q, r) it is easy to construct an infinite number of "nontrivial" solutions. Assume for instance that $A = B = 1$ and $C = -1$ and that (p, q, r) are pairwise coprime. Let a and b be integers strictly greater than 1, and set $c = a + b$. Multiplying this equation by $a^{uqr}b^{vpr}c^{wpq}$ for some integers u, v, and w we obtain

$$a^{uqr+1}b^{vpr}c^{wpq} + a^{uqr}b^{vpr+1}c^{wpq} = a^{uqr}b^{vpr}c^{wpq+1} \ .$$

This is a "nontrivial" solution to our equation if we choose $u \equiv (qr)^{-1}$ (mod p), $v \equiv (pr)^{-1}$ (mod q), and $w \equiv (pq)^{-1}$ (mod r). Therefore it is necessary to add a further condition to exclude this type of solution, and the natural choice is to ask that x, y, and z be pairwise coprime. With that additional restriction, our heuristic reasoning is correct.

A second reduction can be made most of the time. Assume that two among p, q, and r are coprime. Without loss of generality, assume for example that $\gcd(p, q) = 1$. There exist unique positive integers u and v such that $up - vq = 1$ and $1 \leqslant u \leqslant p$, $1 \leqslant v \leqslant q$. Multiplying our equation by $A^{vq}B^{pq-up}$ gives the equation $x_1^p + y_1^q + C_1 z^r = 0$ with $x_1 = (AB)^u x$, $y_1 = A^v B^{p-v} y$ and $C_1 = A^{vq} B^{pq-up} C$. We may thus in that case assume that $A = B = 1$. Note, however, that the coprimality of the solutions may be destroyed by this transformation.

In this chapter, we will in fact often consider the case $A = B = 1$ and $C = \pm 1$. It is easy to see that we can then reduce to the case $C = -1$: indeed, if $C = 1$ and if $x^p + y^q + z^r = 0$, then if p, q, and r are all three even it is clear by positivity that there are no nontrivial solutions, or else at least one of them, say r, is odd, and then the equation can be written $x^p + y^q - (-z)^r = 0$, thus with $C = -1$. Therefore we will consider mainly the equations $x^p + y^q = z^r$, as usual with $\gcd(x, y) = 1$.

Finally, given a triple (p, q, r) up to permutation, if we want to fix the right-hand side, say z^r, then we must consider the four equations $-x^p - y^q = z^r$, $x^p - y^q = z^r$, $-x^p + y^q = z^r$, and $x^p + y^q = z^r$. If p (respectively q, respectively r) is odd, we may change x into $-x$ (respectively y into $-y$, respectively z into $-z$). Then it is easily seen by examination of cases that we can reduce to the examination of a smaller number of equations. More precisely, if at least two of p, q, and r are odd, it is sufficient to consider the equation $x^p + y^q = z^r$; if exactly one is odd, we must in addition consider the equation $x^p - y^q = z^r$ if p or r is odd, and the equation $-x^p + y^q = z^r$ if q is odd. Finally, if p, q, and r are even, we must consider the three equations $x^p + y^q = z^r$, $x^p - y^q = z^r$, and $-x^p + y^q = z^r$, except if $p = q$, in which case it is enough to consider the first two.

We will begin by considering the elliptic case. Up to permutation of (p, q, r), this corresponds to the cases $(p, q, r) = (2, 2, r)$ for $r \geqslant 2$, $(2, 3, 3)$,

$(2, 3, 4)$, $(2, 3, 5)$, which for reasons that will be seen below can be called the dihedral, tetrahedral, octahedral, and icosahedral cases respectively.

14.2 The Dihedral Cases $(2, 2, r)$

This case is the simplest. We must consider the two equations $x^2 - y^2 = z^r$ and $x^2 + y^2 = z^r$.

14.2.1 The Equation $x^2 - y^2 = z^r$

We set $a = x + y$, $b = x - y$, so that $ab = z^r$. Since x and y are coprime, there are two cases. The first is $x \not\equiv y \pmod 2$, in which case a and b are coprime and z is odd, hence $a = \pm s_1^r$, $b = \pm t_1^r$ for coprime odd integers s_1 and t_1, so that $x = \pm(s_1^r + t_1^r)/2$, $y = \pm(s_1^r - t_1^r)/2$, and $z = s_1 t_1$ (and also $-s_1 t_1$ if r is even). If we insist on not having denominators, we set $s = (s_1 + t_1)/2$, $t = (s_1 - t_1)/2$, which are coprime integers of opposite parity, hence we obtain the parametrization

$$(x, y, z) = (\pm((s+t)^r + (s-t)^r)/2, \pm((s+t)^r - (s-t)^r)/2, s^2 - t^2)$$

(and also $z = t^2 - s^2$ if r is even). Note that here we can insist either that the $\pm$ signs be the same (this is how they have been obtained), or that they be independent, since a change of t into $-t$ changes only y into $-y$.

The second case is $x \equiv y \equiv 1 \pmod 2$, so that a and b are even but $a/2$ and $b/2$ are coprime of opposite parity. Changing y into $-y$ if necessary, we may therefore assume that $a/2$ is even and $b/2$ is odd. Since $(a/2)(b/2) = 2^{r-2}(z/2)^r$, we have $a = \pm 2^{r-1}s^r$, $b = \pm 2t^r$ for coprime integers s and t with t odd if $r \geqslant 3$, $t \not\equiv s \pmod 2$ if $r = 2$, so that we obtain

$$(x, y, z) = (\pm(2^{r-2}s^r + t^r), \pm(2^{r-2}s^r - t^r), 2st)$$

(and also $z = -2st$ if r is even).

We thus obtain the following special cases, where we always assume that s and t are coprime, plus indicated additional conditions modulo 2. Often the additional sign of x, y, or z when r is even can be absorbed by changing s into $-s$, or t into $-t$, or by exchanging s and t.

$r = 2$: $(x, y, z) = (\pm(s^2 + t^2), 2ts, (s - t)(s + t))$, where $s \not\equiv t \pmod 2$, up to exchange of y and z.

$r = 3$: $(x, y, z) = (s(s^2 + 3t^2), t(3s^2 + t^2), (s - t)(s + t))$, where $s \not\equiv t$ $\pmod 2$, or $(x, y, z) = (\pm(2s^3 + t^3), 2s^3 - t^3, 2ts)$, where $2 \nmid t$.

$r = 4$: $(x, y, z) = (\pm(s^4 + 6t^2s^2 + t^4), 4ts(s^2 + t^2), (s - t)(s + t))$, where $s \not\equiv t$ $\pmod 2$, or $(x, y, z) = (\pm(2s^2 - 2st + t^2)(2s^2 + 2st + t^2), \pm(2s^2 - t^2)(2s^2 + t^2), 2ts)$, where $2 \nmid t$.

$r = 5$: $(x, y, z) = (s(s^4 + 10t^2s^2 + 5t^4), t(5s^4 + 10t^2s^2 + t^4), (s - t)(s + t))$, where $s \not\equiv t \pmod 2$, or $(x, y, z) = (\pm(8s^5 + t^5), 8s^5 - t^5, 2ts)$, where $2 \nmid t$.

14.2.2 The Equation $x^2 + y^2 = z^r$

Here we set $a = x + iy$, $b = x - iy$ so that $ab = z^r$. If we had $x \equiv y \equiv 1$ (mod 2), we would have $z^r \equiv 2$ (mod 8), which is impossible since $r \geqslant 2$. Thus since x and y are coprime, x and y have opposite parity and a and b are coprime in the principal ideal domain $\mathbb{Z}[i]$. It follows that there exist $\alpha = s + it \in \mathbb{Z}[i]$ and some $v = 0$, 1, 2, or 3 such that $x + iy = i^v \alpha^r$, hence $x - iy = i^{-v} \overline{\alpha}^r$, $z = \alpha \overline{\alpha}$ (and also $z = -\alpha \overline{\alpha}$ if r is even). Clearly multiplication by i^v corresponds to changing signs of x and/or y and exchange of x and y, so that up to exchange of x and y we obtain the parametrization

$$\begin{cases} x = \pm \sum_{0 \leqslant k \leqslant \lfloor r/2 \rfloor} (-1)^k \binom{r}{2k} t^{2k} s^{r-2k} , \\ y = \pm \sum_{0 \leqslant k \leqslant \lfloor (r-1)/2 \rfloor} (-1)^k \binom{r}{2k+1} t^{2k+1} s^{r-2k-1} , \\ z = s^2 + t^2 \quad \text{(and also } -(s^2 + t^2) \text{ if } r \text{ is even)} . \end{cases}$$

Furthermore, the condition $\gcd(x, y) = 1$ of course implies that s and t are coprime, and since $r \geqslant 2$, if s and t were both odd we would have $\alpha^r \equiv (1+i)^r \equiv 0$ (mod $2\mathbb{Z}[i]$), so that x and y would both be even. It follows that in addition s and t have opposite parity. Conversely, it is easy to see that if this is the case then x and y are coprime.

We thus obtain the following special cases, where we assume that s and t are coprime of opposite parity. Again the additional sign of x, y, or z when r is even, or the exchange of x and y can be absorbed by changing s into $-s$, or t into $-t$, or by exchanging s and t, or a combination.

$r = 2$: $(x, y, z) = (2ts, s^2 - t^2, \pm(s^2 + t^2))$, up to exchange of x and y.

$r = 3$: $(x, y, z) = (s(s^2 - 3t^2), t(3s^2 - t^2), s^2 + t^2)$.

$r = 4$: $(x, y, z) = (\pm(s^2 - 2st - t^2)(s^2 + 2st - t^2), 4ts(s-t)(s+t), \pm(s^2 + t^2))$, up to exchange of x and y.

$r = 5$: $(x, y, z) = (s(s^4 - 10t^2 s^2 + 5t^4), t(5s^4 - 10t^2 s^2 + t^4), s^2 + t^2)$.

14.2.3 The Equations $x^2 + 3y^2 = z^3$ and $x^2 + 3y^2 = 4z^3$

As additional examples of dihedral equations we prove the following results, which we will need elsewhere.

Proposition 14.2.1. (1) *The equation $x^2 + 3y^2 = z^3$ in nonzero integers x, y, and z with x and y coprime can be parametrized by*

$$(x, y, z) = (s(s - 3t)(s + 3t), 3t(s - t)(s + t), s^2 + 3t^2) ,$$

where s and t denote coprime integers of opposite parity such that $3 \nmid t$.

(2) *The equation $x^2 + 3y^2 = 4z^3$ in nonzero integers x, y, and z with x and y coprime has the two disjoint parametrizations*

$$(x, y, z) = ((s+t)(s-2t)(2s-t), 3st(s-t), s^2 - st + t^2) ,$$
$$(x, y, z) = (\pm(s^3 + 3s^2 t - 6st^2 + t^3), s^3 - 3s^2 t + t^3, s^2 - st + t^2) ,$$

where in both cases s and t are coprime integers such that $3 \nmid s+t$. The first parametrization corresponds to the case $6 \mid y$, and the second to the case that y is coprime to 6.

Proof. For (1) we set $x_1 = x + 3y$ and the equation becomes $x_1^2 - 3x_1(2y) + 3(2y)^2 = z^3$. Thanks to Proposition 6.4.16 of Chapter 6 we know that this equation has three disjoint parametrizations. Among these, only the first gives an even value for the second variable, so that $x_1 = s^3 + 3s^2 t - 6st^2 + t^3$, $y = 3st(s-t)/2$, and $z = s^2 - st + t^2$. If s is even we set $S = t - s/2$, $T = s/2$, if t is even we set $S = s - t/2$, $T = t/2$, and if s and t are both odd we set $S = (s+t)/2$ and $T = (s-t)/2$. In all three cases we check that up to sign we obtain the given parametrizations and the conditions at the primes 2 and 3.

For (2) we note that x and y are both odd, so we set $x_1 = (x+3y)/2$, and the equation is $x_1^2 - 3x_1 y + 3y^2 = z^3$. By Proposition 6.4.16 once again we obtain three parametrizations, but it is immediate that (up to the sign of x, which does not matter) the last two are interchanged by exchanging s and t, so we have only the two parametrizations given above. Note that because of this sign change in the interchange of the last two parametrizations we have to add a $\pm$ sign for the parametrization of x. $\square$

Note that by looking modulo 8 it is clear that the equation $x^2 + 3y^2 = 2z^3$ is impossible in coprime x, y.

14.3 The Tetrahedral Case $(2,3,3)$

14.3.1 The Equation $x^3 + y^3 = z^2$

Thanks to the reductions made above, for $(p, q, r) = (2, 3, 3)$ it is sufficient to consider the single equation $x^3 + y^3 = z^2$. We will imitate what we did in the case of FLT, by factoring $x^3 + y^3$ in $\mathbb{Z}[\zeta]$, where ζ is a primitive cube root of unity. Thus, we write

$$(x + y)(x + \zeta y)(x + \zeta^2 y) = z^2 .$$

Case 1: $3 \nmid z$. If $\pi \in \mathbb{Z}[\zeta]$ is a prime element that divides two distinct factors on the left, then $\pi \mid 1 - \zeta$, hence $\pi = 1 - \zeta$, which is excluded since $3 \nmid z$. Thus the factors are coprime in $\mathbb{Z}[\zeta]$, and each one is equal to a unit multiplied by a square. If we had factored directly in $\mathbb{Z}$, we would have obtained that $x + y = \pm a^2$ for $a \in \mathbb{Z}$, and since the cofactor $x^2 - xy + y^2$ is always positive and $z^2 > 0$, we necessarily have $x + y = a^2$. Thus our equation implies that $x + y = a^2$ with $a \in \mathbb{Z}$ and $x + \zeta y = (-\zeta)^k \alpha^2$ for some integer k, and conversely this implies also $x + \zeta^2 y = (-\zeta^2)^k \overline{\alpha}^2$, hence $z = \pm a\alpha\overline{\alpha}$. In addition, since $\zeta = \zeta^4$ is a square, we may write $(-\zeta)^k \alpha^2 = (-1)^k (\alpha \zeta^{2k})^2$. Finally, our equation is thus equivalent to the equations $x + y = a^2$, $x + \zeta y = \varepsilon \alpha^2$, $z = \pm a\alpha\overline{\alpha}$, where $a \in \mathbb{Z}$, $\alpha \in \mathbb{Z}[\zeta]$, and $\varepsilon = \pm 1$.

If we set $\beta = \zeta^2 \alpha$ then

$$\beta^2 + \overline{\beta}^2 = \zeta\alpha^2 + \zeta^2\overline{\alpha}^2 = \varepsilon(\zeta(x + \zeta y) + \zeta^2(x + \zeta^2 y)) = \varepsilon(-x - y) = -\varepsilon a^2 \ .$$

Conversely, if $\beta \in \mathbb{Z}[\zeta]$ satisfies $\beta^2 + \overline{\beta}^2 = -\varepsilon a^2$ and if we set $\alpha = \zeta\beta$, then one checks that

$$\frac{\varepsilon\overline{\alpha}^2 - a^2}{\varepsilon\alpha^2 - a^2} = -\zeta^2 = \frac{\overline{\zeta} - 1}{\zeta - 1} \ ,$$

so that $y = (\varepsilon\alpha^2 - a^2)/(\zeta - 1) \in \mathbb{Q}$. However, $3 \nmid a$, so that $a^2 \equiv 1 \pmod 3$, and also $(1 - \zeta) \nmid \beta$, so that $\alpha^2 \equiv \pm\varepsilon \pmod{1 - \zeta}$. However, if $\alpha^2 \equiv -\varepsilon \pmod{1 - \zeta}$, we would have $-a^2 \equiv \varepsilon(-\varepsilon)(\zeta + \zeta^2) \equiv 1 \pmod{1 - \zeta}$, which is absurd. Thus $\alpha^2 \equiv \varepsilon \pmod{1 - \zeta}$, so in fact $y \in \mathbb{Z}$. Thus our equation is now equivalent to the single simpler equation $a^2 = -\varepsilon(\beta^2 + \overline{\beta}^2)$. If we write $\beta = u + v\zeta$ with u and v in $\mathbb{Z}$, this gives finally the equation

$$a^2 = \varepsilon(v^2 + 2uv - 2u^2) \ .$$

Note that the condition $\gcd(x, y) = 1$ implies that a and β are coprime in $\mathbb{Z}[\zeta]$, hence that $\gcd(u, v) = 1$. Note also for future reference that this implies that $u + v$ and a are coprime (easy exercise left to the reader). Also, since we are in the case $3 \nmid z$, we have $3 \nmid a$ hence $3 \nmid u + v$. Thus

$$1 \equiv a^2 \equiv \varepsilon((u + v)^2 - 3u^2) \equiv \varepsilon(u + v)^2 \equiv \varepsilon \pmod 3 \ ,$$

so that we must have $\varepsilon = 1$.

We have thus reduced our problem to the solution of a Diophantine equation of degree 2, for which an algorithmic solution is always possible.

We can do one more important reduction. It is clear that exchanging x and y is equivalent to changing β into $\overline{\beta}$, or in other words the pair (u, v) into the pair $(u - v, -v)$. Note that $v \equiv a \pmod 2$. Thus, if a is odd, v is odd, so either u or $u - v$ is odd. If a is even, then v is even, so that u is odd since it is coprime to v. Thus in all cases we may assume, possibly after exchanging x and y, that $u + v \not\equiv a \pmod 2$.

We write $3u^2 = (u + v)^2 - a^2 = (u + v - a)(u + v + a)$. Since $3 \nmid a$ and $3 \nmid (u + v)$, if necessary by changing β into $-\beta$ (or a into $-a$) we may assume that $3 \mid u + v - a$, and then $3 \nmid u + v + a$. Since $u + v$ and a are coprime, and since we have reduced above to the case that they do not have the same parity, it follows that $u + v - a$ and $u + v + a$ are coprime. Thus $u + v - a = 3\varepsilon_1 s_1^2$, $u + v + a = \varepsilon_1 t_1^2$, so $u = \varepsilon_2 s_1 t_1$ with s_1 and t_1 coprime and odd, and with $\varepsilon_1 = \pm 1$ and $\varepsilon_2 = \pm 1$. If we change simultaneously u, v, and a into their opposites, we may assume that $\varepsilon_1 = 1$. Changing s_1 into $-s_1$ we may also assume that $\varepsilon_2 = 1$. Finally, we set $s = s_1$, $t = (t_1 - s_1)/2$, which are coprime with s odd, which gives $u = s(s + 2t)$, $v = s^2 + 2t^2$, $a = -s^2 + 2st + 2t^2)$. The condition $3 \nmid u + v$ (or $3 \mid a$) is equivalent to $3 \nmid (s^2 + st + t^2)$, hence (since s and t are coprime) to $s \not\equiv t \pmod 3$. Replacing everywhere gives the first parametrization

$$
\begin{cases}
x = s(s + 2t)(s^2 - 2ts + 4t^2) \,, \\
y = -4t(s - t)(s^2 + ts + t^2) \,, \\
z = \pm(s^2 - 2ts - 2t^2)(s^4 + 2ts^3 + 6t^2 s^2 - 4t^3 s + 4t^4) \,,
\end{cases}
$$

where s is odd and $s \not\equiv t \pmod 3$, up to exchange of x and y.

Note that if we had $s \equiv t \pmod 3$ we would have $3 \mid \gcd(x, y)$, contrary to our assumption. Note also that if we had not done the reduction equivalent to exchanging x and y, we would have obtained a second parametrization, which would have been equivalent to the first one where x and y are exchanged.

Case 2: $3 \mid z$. In this case $x + y$, $x + \zeta y$, and $x + \zeta^2 y$ are all three divisible by $1 - \zeta$, and their quotient by $1 - \zeta$ are pairwise coprime. Thus

$$
\frac{x + y}{3} \frac{x + \zeta y}{1 - \zeta} \frac{x + \zeta^2 y}{1 - \zeta^2} = (z/3)^2
$$

with the three factors on the left pairwise coprime, so as above, our equation is equivalent to $x + y = 3a^2$, $x + \zeta y = \varepsilon(1 - \zeta)\alpha^2$, $z = \pm 3a\alpha\overline{\alpha}$, with $\varepsilon = \pm 1$. We note that since $3 \nmid xy$ (otherwise $3 \mid \gcd(x, y)$) then $v_{\mathfrak{p}}(x + \zeta y) = v_{\mathfrak{p}}(x + y + (\zeta - 1)y) = 1$, where $\mathfrak{p} = (1 - \zeta)\mathbb{Z}[\zeta]$, so that α is coprime to $1 - \zeta$.

We have $(1 - \zeta)y = 3a^2 - \varepsilon(1 - \zeta)\alpha^2$, hence $y = (1 - \zeta^2)a^2 - \varepsilon\alpha^2$, and since $y \in \mathbb{Q}$ we obtain $\alpha^2 - \overline{\alpha}^2 = \varepsilon a^2(\zeta - \zeta^2)$. Conversely, if this is satisfied for some $\alpha \in \mathbb{Z}[\zeta]$, then we can take $y = (1 - \zeta^2)a^2 - \varepsilon\alpha^2 \in \mathbb{Z}$. Thus as before, our equation is equivalent to the single simpler equation $a^2(\zeta - \zeta^2) = \varepsilon(\alpha^2 - \overline{\alpha}^2)$. If we write $\alpha = u + v\zeta$ with u and v in $\mathbb{Z}$, this finally gives the equation

$$
a^2 = \varepsilon v(2u - v) \,.
$$

We have already mentioned that α is coprime to $1 - \zeta$, which is equivalent to $3 \nmid u + v$. In addition, the condition $\gcd(x, y) = 1$ implies that a and α are coprime in $\mathbb{Z}[\zeta]$, so that $\gcd(u, v) = 1$. Thus the GCD of v and $2u - v$ is equal to 1 if v is odd, and to 2 if v is even.

It is easily seen that exchanging x and y is here equivalent to simultaneously changing α into $\overline{\alpha}$ and ε into $-\varepsilon$. Thus, we may assume that $\varepsilon = 1$, so we have two possibilities according to the parity of v.

– If v is odd, then $v = \varepsilon_1 s_1^2$, $2u - v = \varepsilon_1 t_1^2$, so that $a = \varepsilon_2 s_1 t_1$ with s_1 and t_1 odd and $\varepsilon_1 = \pm 1$. Changing α into $-\alpha$, we may assume that $\varepsilon_1 = 1$, and changing s_1 into $-s_1$ that $\varepsilon_2 = 1$. We set $s = (s_1 + t_1)/2$, $t = (s_1 - t_1)/2$, which are coprime integers of opposite parity, so that $v = (s + t)^2$, $u = s^2 + t^2$, $a = s^2 - t^2$. The condition $3 \nmid u + v$ is again equivalent to $s \not\equiv t$ (mod 3). Replacing everywhere gives the second parametrization, where the sign of z can be absorbed by exchanging s and t:

$$\begin{cases} x = s^4 - 4ts^3 - 6t^2s^2 - 4t^3s + t^4 \,, \\ y = 2(s^4 + 2ts^3 + 2t^3s + t^4) \,, \\ z = 3(s - t)(s + t)(s^4 + 2s^3t + 6s^2t^2 + 2st^3 + t^4) \,, \end{cases}$$

where $s \not\equiv t$ (mod 2) and $s \not\equiv t$ (mod 3), up to exchange of x and y.

- If v is even, then $v = 2\varepsilon_1 s^2$, $2u - v = 2\varepsilon_1 t^2$, so $u = \varepsilon_1(s^2 + t^2)$ and $a = 2\varepsilon_2 st$, where s and t are coprime integers of opposite parity. As before, we may reduce to the case $\varepsilon_1 = \varepsilon_2 = 1$. The condition $3 \nmid u + v$ is now equivalent to $3 \nmid t$. Replacing everywhere gives the third and final parametrization, where the sign of z can be absorbed by changing s into $-s$:

$$\begin{cases} x = -3s^4 + 6t^2s^2 + t^4 \,, \\ y = 3s^4 + 6t^2s^2 - t^4 \,, \\ z = 6st(3s^4 + t^4) \,, \end{cases}$$

where $s \not\equiv t$ (mod 2) and $3 \nmid t$, up to exchange of x and y.

We have thus proved the following theorem.

Theorem 14.3.1. *The equation $x^3 + y^3 = z^2$ in integers x, y, z with $\gcd(x, y) = 1$ can be parametrized by one of the above three parametrizations, up to exchange of x and y, where s and t denote coprime integers satisfying the given congruences modulo 2 and 3. In addition, these parametrizations are disjoint, in that any solution to our equation belongs to a single parametrization (up to exchange of x and y).*

14.3.2 The Equation $x^3 + y^3 = 2z^2$

This equation is very similar to the preceding one, and will be needed in the octahedral case. Thus we give only a brief sketch. We can factor $x^3 + y^3$ as usual, and we use the fact that 2 is inert in $\mathbb{Z}[\zeta]$. As usual we distinguish two cases.

Case 1: $3 \nmid z$. Using the same technique as above, it is easily seen that our equation is equivalent to the equations $x + y = 2a^2$, $z = \pm a\beta\overline{\beta}$, and $2a^2 = -(v^2 + 2uv - 2u^2)$, where $\beta = u + v\zeta$, and u and v are coprime. Thus $v = 2w$ must be even; hence u is odd, so we obtain $(u - w - a)(u - w + a) = 3w^2$. It follows that $a^2 = (u - w)^2 - 3w^2 \equiv (1 - w)^2 - 3w^2 \equiv 1$ (mod 2), so a is odd. Since $3 \mid a$ and $3 \nmid u + v \equiv u - w$ (mod 3), we may assume that

$3 \mid (u - w - a)$. Thus, we have two different cases (where as usual we can get rid of the signs):

Case 1.1: $2 \nmid w$. Here $u - w - a = 3s_1^2$, $u - w + a = t_1^2$ with s_1 and t_1 odd and coprime; hence setting $s = s_1$, $t = (t_1 - s_1)/2$ coprime with s odd and with $s \not\equiv t \pmod 3$, we obtain $w = s(s+2t)$, $a = -s^2 + 2st + 2t^2$, $u = 3s^2 + 4st + 2t^2$. Replacing gives the first parametrization, where the exchange of x and y can be absorbed by the exchange of s and t:

$$\begin{cases} x = -(s^2 + 4ts - 2t^2)(3s^2 + 4ts + 2t^2) \,, \\ y = (s^2 + 2t^2)(5s^2 + 8ts + 2t^2) \,, \\ z = \pm(s^2 - 2ts - 2t^2)(7s^4 + 20ts^3 + 24t^2s^2 + 8t^3s + 4t^4) \,, \end{cases}$$

where s is odd and $s \not\equiv t \pmod 3$.

Case 1.2: $2 \mid w$. Here $u - w - a = 6s^2$, $u - w + a = 2t^2$, $w = 2st$; hence $a = t^2 - 3s^2$, $u = 3s^2 + 2st + t^2$, and s and t are coprime integers of opposite parity with $3 \nmid t$. Replacing gives the second parametrization, where the exchange of x and y can be absorbed by changing s into $-s$:

$$\begin{cases} x = (3s^2 - 6ts + t^2)(3s^2 + 2ts + t^2) \,, \\ y = (3s^2 - 2ts + t^2)(3s^2 + 6ts + t^2) \,, \\ z = \pm(3s^2 - t^2)(9s^4 + 18t^2s^2 + t^4) \,, \end{cases}$$

where $s \not\equiv t \pmod 2$ and $3 \nmid t$.

Case 2: $3 \mid z$. Using the same technique as above, it is easily seen that our equation is equivalent to the equations $x + y = 6a^2$, $z = \pm 3a\alpha\overline{\alpha}$, $y = 2(1 - \zeta^2)a^2 - \varepsilon\alpha^2$, and $2a^2 = \varepsilon v(2u - v)$, where $\alpha = u + v\zeta$ and u and v are coprime, and α is coprime to $1 - \zeta$. Thus $v = 2w$ must be even, and hence u is odd, so we deduce that a is even and $\varepsilon w(u - w)/2 = (a/2)^2$. Since exchanging x and y is equivalent to changing α into $\overline{\alpha}$ and ε into $-\varepsilon$, we may assume that $\varepsilon = 1$. Once again we have two cases, where as usual we can get rid of the signs.

Case 2.1: $2 \nmid w$. Here $w = s^2$, $u - w = 2t^2$, $a = 2st$, so $u = s^2 + 2t^2$, $v = 2s^2$, where s and t are coprime with s odd. Replacing, we obtain the third parametrization:

$$\begin{cases} x = -3s^4 + 12t^2s^2 + 4t^4 \,, \\ y = 3s^4 + 12t^2s^2 - 4t^4 \,, \\ z = 6ts(3s^4 + 4t^4) \,, \end{cases}$$

where s is odd and $3 \nmid t$, up to exchange of x and y.

Case 2.2: $2 \mid w$. Here $w = 2s^2$, $u - w = t^2$, $a = 2st$, so $u = 2s^2 + t^2$, $v = 4s^2$, where s and t are coprime with t odd. Replacing, we obtain the fourth and final parametrization:

$$\begin{cases} x = -12s^4 + 12t^2s^2 + t^4\,, \\ y = 12s^4 + 12t^2s^2 - t^4\,, \\ z = 6ts(12s^4 + t^4)\,, \end{cases}$$

where t is odd and $3 \nmid t$, up to exchange of x and y.

We have thus proved the following theorem.

Theorem 14.3.2. *The equation $x^3 + y^3 = 2z^2$ in integers x, y, z with $\gcd(x,y) = 1$ can be parametrized by one of the above four parametrizations, up to exchange of x and y, where s and t denote coprime integers with the indicated congruence conditions modulo 2 and 3. In addition, these parametrizations are disjoint, in that any solution to our equation belongs to a single parametrization (up to exchange of x and y).*

14.3.3 The Equation $x^3 - 2y^3 = z^2$

We will also need this equation in the octahedral case. Note first that z is necessarily odd, since otherwise x is even, so y is also even, a contradiction. Similarly, it is easy to check that the congruence $x^3 - 2y^3 \equiv 0 \pmod 9$ implies that $x \equiv y \equiv 0 \pmod 3$, which is impossible. Thus we must have $3 \nmid z$, i.e., the "second case" does not occur.

We now work in the number field $K = \mathbb{Q}(\theta)$, where $\theta^3 = 2$, whose ring of integers is $\mathbb{Z}[\theta]$ and is a principal ideal domain. Note also that 3 is totally ramified in $\mathbb{Z}[\theta]$. Our equation is a *norm equation* of the type $\mathcal{N}(\alpha) = z^2$, for $\alpha = x - y\theta \in \mathbb{Z}[\theta]$. We factor our equation as $(x - y\theta)(x^2 + xy\theta + y^2\theta^2) = z^2$. Since $3 \nmid z$, as usual it is easily seen that the two factors on the left are coprime in $\mathbb{Z}[\theta]$, so that $x - y\theta = \pm\varepsilon^k\beta^2$ for $\varepsilon = \theta - 1$ the fundamental unit, and some $\beta \in \mathbb{Z}[\theta]$. We may of course assume that $k = 0$ or 1. Taking norms and using the fact that $\mathcal{N}(\varepsilon) = 1$ gives $z^2 = \pm\mathcal{N}(\beta)^2$, so that the sign must be $+$, and then $z = \pm\mathcal{N}(\beta)$. The only condition is thus that the coefficient of θ^2 in $\varepsilon^k\beta^2$ be equal to 0. Writing $\beta = u + v\theta + w\theta^2$, we thus have two cases.

Case 1: $k = 0$. Here we obtain the equations $v^2 + 2uw = 0$, $x = u^2 + 4vw$, $y = -2(w^2 + uv)$, $z = \pm(u^3 + 2v^3 + 4w^3 - 6uvw)$. Thus $v = 2v_1$ is even, so u is odd. Since x and y are coprime, so are u and w. Thus the equation $uw = -2v_1^2$ implies that $u = \varepsilon_1 s^2$, $w = -\varepsilon_1 2t^2$, $v = \varepsilon_2 2st$ for some ε_1 and ε_2 equal to ±1, with s and t coprime and s odd. As usual, changing if necessary β into $-\beta$, and s into $-s$, we may assume that $\varepsilon_1 = \varepsilon_2 = 1$. Replacing gives the first parametrization:

$$\begin{cases} x = s(s^3 - 16t^3)\,, \\ y = -4t(s^3 + 2t^3)\,, \\ z = \pm(s^6 + 40t^3s^3 - 32t^6)\,, \end{cases}$$

where s is odd and $s \not\equiv t \pmod 3$.

Case 2: $k = 1$. Here we obtain the equations $(2v - 2w)u - v^2 + 2w^2 = 0$, $x = -u^2 + 4wu + 2v^2 - 4wv$, $y = -u^2 + 2vu - 4wv + 2w^2$. The first equation can be written $(u - w)^2 + w^2 = (v - u)^2$. Since $\gcd(u, v, w) = 1$, the solution to the Pythagorean triple equation gives the parametrizations $u - w = 2st$, $w = s^2 - t^2$, $v - u = \varepsilon_1(s^2 + t^2)$ or $w = 2st$, $u - w = s^2 - t^2$, $v - u = \varepsilon_1(s^2 + t^2)$ for $\varepsilon_1 = \pm 1$, where s and t are coprime integers of opposite parity. Since we can change β into $-\beta$, we may assume that $\varepsilon_1 = 1$. Replacing gives the following two further parametrizations:

$$\begin{cases} x = 3s^4 + 12ts^3 + 6t^2 s^2 + 4t^3 s + 3t^4 , \\ y = -3s^4 + 6t^2 s^2 + 8t^3 s + t^4 , \\ z = \pm(9s^6 + 18ts^5 + 45t^2 s^4 + 60t^3 s^3 + 15t^4 s^2 - 6t^5 s - 5t^6) , \end{cases}$$

where $s \not\equiv t \pmod 2$ and $3 \nmid t$, and

$$\begin{cases} x = 7s^4 + 4ts^3 + 6t^2 s^2 - 4t^3 s - t^4 , \\ y = 3s^4 - 8ts^3 - 6t^2 s^2 - t^4 , \\ z = \pm(17s^6 + 30ts^5 - 15t^2 s^4 + 20t^3 s^3 + 15t^4 s^2 + 6t^5 s - t^6) , \end{cases}$$

where $s \not\equiv t \pmod 2$ and $s \not\equiv t \pmod 3$.

We have thus proved the following theorem:

Theorem 14.3.3. *The equation $x^3 - 2y^3 = z^2$ in integers x, y, z with $\gcd(x, y) = 1$ can be parametrized by one of the above three parametrizations, where s and t denote coprime integers with the indicated congruence conditions modulo 2 and 3. In addition, these parametrizations are disjoint, in that any solution to our equation belongs to a single parametrization.*

14.4 The Octahedral Case $(2, 3, 4)$

According to the reductions made above, this case reduces to the two equations $x^2 \pm y^4 = z^3$. We consider both separately.

14.4.1 The Equation $x^2 - y^4 = z^3$

Factoring gives $(x - y^2)(x + y^2) = z^3$. Since x and y are coprime, either $x - y^2$ and $x + y^2$ are coprime, or x and y are odd and $(x - y^2)/2$ and $(x + y^2)/2$ are coprime.

Case 1: $2 \nmid z$. Here $x - y^2$ and $x + y^2$ are coprime, so that $x - y^2 = a^3$, $x + y^2 = b^3$, $z = ab$ (the possible sign can be removed by changing the sign of a). This is equivalent to $x = y^2 + a^3$, $z = ab$, and $2y^2 + a^3 = b^3$. Changing variable names, we are thus reduced to the equation $x^3 + y^3 = 2z^2$ with x

and y odd, which we have studied above. Note that the exchange of x and y in this latter equation is equivalent to the exchange of b with $-a$, hence to the exchange of x with $-x$ in our initial equation. Thus after replacing we obtain the following four different parametrizations of our equation, where in each case s and t are coprime integers satisfying the indicated additional congruence conditions modulo 2 and 3:

$$\begin{cases} x = 4s(s + 2t)(s^2 + ts + t^2)(s^4 + 4ts^3 + 16t^2s^2 + 24t^3s + 12t^4) \\ \quad \times (19s^4 - 4ts^3 + 8t^3s + 4t^4) \,, \\ y = \pm(s^2 - 2ts - 2t^2)(7s^4 + 20ts^3 + 24t^2s^2 + 8t^3s + 4t^4) \,, \\ z = (s^2 + 2t^2)(s^2 + 4ts - 2t^2)(3s^2 + 4ts + 2t^2)(5s^2 + 8ts + 2t^2) \,, \end{cases}$$

where s is odd and $s \not\equiv t \pmod 3$.

Note that changing t into $-s-t$ changes x into $-x$, hence we do not need to put a $\pm$ sign in front of x. The three other parametrizations are

$$\begin{cases} x = 4ts(3s^2 + t^2)(3s^4 - 2t^2s^2 + 3t^4)(81s^4 - 6t^2s^2 + t^4) \,, \\ y = \pm(3s^2 - t^2)(9s^4 + 18t^2s^2 + t^4) \,, \\ z = -(3s^2 - 6ts + t^2)(3s^2 - 2ts + t^2)(3s^2 + 2ts + t^2)(3s^2 + 6ts + t^2) \,, \end{cases}$$

where $s \not\equiv t \pmod 2$ and $3 \nmid t$,

$$\begin{cases} x = \pm(3s^4 - 4t^4)(9s^8 + 408t^4s^4 + 16t^8) \,, \\ y = 6ts(3s^4 + 4t^4) \,, \\ z = (3s^4 - 12t^2s^2 - 4t^4)(3s^4 + 12t^2s^2 - 4t^4) \,, \end{cases}$$

where s is odd and $3 \nmid t$, and

$$\begin{cases} x = \pm(12s^4 - t^4)(144s^8 + 408t^4s^4 + t^8) \,, \\ y = 6ts(12s^4 + t^4) \,, \\ z = (12s^4 - 12t^2s^2 - t^4)(12s^4 + 12t^2s^2 - t^4) \,, \end{cases}$$

where t is odd and $3 \nmid t$.

Note that the exchange of x and y in the parametrizations of $x^3 + y^3 = 2z^3$ corresponds only to the exchange of x and $-x$ in the present ones.

Case 2: $2 \mid z$. Here we must have $2 \mid ((x - y^2)/2)(x + y^2)/2$, so that changing x into $-x$ if necessary, we may assume that $4 \mid x - y^2$. It follows that $x - y^2 = 4a^3$, $x + y^2 = 2b^3$, $z = 2ab$. This is equivalent to $x = y^2 + 4a^3$, $z = 2ab$, and $y^2 + 2a^3 = b^3$, with y odd. We are thus reduced to the equation $x^3 - 2y^3 = z^2$, which we have studied above. We thus obtain three parametrizations, which after replacing gives the following three additional parametrizations of our equation, for a total of seven:

$$\begin{cases} x = \pm(s^6 - 176t^3s^3 - 32t^6)(s^6 + 32t^6) \,, \\ y = \pm(s^6 + 40t^3s^3 - 32t^6) \,, \\ z = -8ts(s^3 - 16t^3)(s^3 + 2t^3) \,, \end{cases}$$

where s is odd and $s \not\equiv t \pmod 3$,

$$\begin{cases} x = \pm(-27s^{12} + 324ts^{11} + 1782t^2s^{10} + 3564t^3s^9 + 3267t^4s^8 \\ \quad + 2376t^5s^7 + 2772t^6s^6 + 3960t^7s^5 + 4059t^8s^4 \\ \quad + 2420t^9s^3 + 726t^{10}s^2 + 156t^{11}s + 29t^{12}) \,, \\ y = \pm(9s^6 + 18ts^5 + 45t^2s^4 + 60t^3s^3 + 15t^4s^2 - 6t^5s - 5t^6) \,, \\ z = -2(3s^4 - 6t^2s^2 - 8t^3s - t^4)(3s^4 + 12ts^3 + 6t^2s^2 + 4t^3s + 3t^4) \,, \end{cases}$$

where $s \not\equiv t \pmod 2$ and $3 \nmid t$, and

$$\begin{cases} x = \pm(397s^{12} + 156ts^{11} + 2046t^2s^{10} + 1188t^3s^9 - 1485t^4s^8 - 2376t^5s^7 \\ \quad - 924t^6s^6 - 792t^7s^5 + 99t^8s^4 + 44t^9s^3 - 66t^{10}s^2 - 12t^{11}s - 3t^{12}) \,, \\ y = \pm(17s^6 + 30ts^5 - 15t^2s^4 + 20t^3s^3 + 15t^4s^2 + 6t^5s - t^6) \,, \\ z = 2(3s^4 - 8ts^3 - 6t^2s^2 - t^4)(7s^4 + 4ts^3 + 6t^2s^2 - 4t^3s - t^4) \,, \end{cases}$$

where $s \not\equiv t \pmod 2$ and $s \not\equiv t \pmod 3$.

Remark. We could have used the parametrizations of the dihedral equation $x^2 - y^2 = z^3$, but it would not have been really simpler. The same is true for the next equation.

We have thus proved the following theorem:

Theorem 14.4.1. *The equation $x^2 - y^4 = z^3$ in integers x, y, z with $\gcd(x,y) = 1$ can be parametrized by one of the above seven parametrizations, where s and t denote coprime integers with the indicated congruence conditions modulo 2 and 3. In addition, these parametrizations are disjoint, in that any solution to our equation belongs to a single parametrization.*

14.4.2 The Equation $x^2 + y^4 = z^3$

We note that here we cannot have x and y both odd, since otherwise $z^3 \equiv 2$ (mod 8), absurd. We work in $\mathbb{Z}[i]$ and factor the equation as $(x + iy^2)(x - iy^2) = z^3$. Since x and y are coprime and not both odd, $x + iy^2$ and $x - iy^2$ are coprime in $\mathbb{Z}[i]$. Thus there exists $\alpha \in \mathbb{Z}[i]$ such that $x + iy^2 = \alpha^3$, so that $x - iy^2 = \overline{\alpha}^3$ and $z = \alpha\overline{\alpha}$, where the possible power of i can be absorbed in α. We write $\alpha = u + iv$, so that $z = u^2 + v^2$, $x = u^3 - 3uv^2$, and $y^2 = 3u^2v - v^3$; hence we must solve this equation. Note that since x and y are coprime, we have $\gcd(u,v) = 1$ and u and v have opposite parity. We write $y^2 = v(3u^2 - v^2)$ and consider two cases.

Case 1: $3 \nmid v$. Then v and $3u^2 - v^2$ are coprime, so $v = \varepsilon a^2$, $3u^2 - v^2 = \varepsilon b^2$, $y = \pm ab$ with $\varepsilon = \pm 1$, and then a and b are coprime, b is odd, and $3 \nmid ab$. We note that $3u^2 - v^2 \equiv -(u^2 + v^2) \equiv -1 \pmod 4$ since u and v have opposite parity; hence we must have $\varepsilon = -1$, so the equations to be solved are $v = -a^2$ and $3u^2 = v^2 - b^2$. Since $3 \nmid v$ and $3 \nmid b$, changing if necessary b into $-b$, we may assume that $3 \mid v - b$, so the second equation is $u^2 = ((v - b)/3)(v + b)$. Note that v and b are coprime. I claim that v is odd. Indeed, otherwise a is even, so that $4 \mid v = -a^2$; hence $v^2 - b^2 \equiv 7 \pmod 8$, while $3u^2 \equiv 3 \pmod 8$, a contradiction. Thus v is indeed odd, so u is even and $v - b$ and $v + b$ are even with $(v - b)/2$ and $(v + b)/2$ coprime. Thus we can write $v - b = 6\varepsilon_1 c^2$, $v + b = 2\varepsilon_1 d^2$, $u = 2cd$ (where the sign of u can be removed by changing c into $-c$) with c and d coprime, and $3 \nmid d$. Thus $v = \varepsilon_1(3c^2 + d^2)$, $b = \varepsilon_1(d^2 - 3c^2)$, and since $v = -a^2$ we have $\varepsilon_1 = -1$. The last remaining equation to be solved is the second-degree equation $d^2 + 3c^2 = a^2$. Corollary 6.3.15 gives us a priori the two parametrizations $d = \pm(s^2 - 3t^2)$, $c = 2st$, $a = \pm(s^2 + 3t^2)$ with coprime integers s and t of opposite parity such that $3 \nmid s$, and $d = \pm(s^2 + 4st + t^2)$, $c = s^2 - t^2$, $a = \pm 2(s^2 + st + t^2)$, with coprime integers s and t of opposite parity such that $s \not\equiv t \pmod 3$. However, since $v = -a^2$ is odd, a is odd so this second parametrization is impossible. Thus there remains only the first one, so replacing everywhere gives the first parametrization

$$\begin{cases} x = 4ts(s^2 - 3t^2)(s^4 + 6t^2s^2 + 81t^4)(3s^4 + 2t^2s^2 + 3t^4)\,, \\ y = \pm(s^2 + 3t^2)(s^4 - 18t^2s^2 + 9t^4)\,, \\ z = (s^4 - 2t^2s^2 + 9t^4)(s^4 + 30t^2s^2 + 9t^4)\,, \end{cases}$$

where $s \not\equiv t \pmod 2$ and $3 \nmid s$.

Case 2: $3 \mid v$. Set $w = v/3$. Then $3 \nmid u$, and w and $u^2 - 3w^2$ are coprime, so $v = \varepsilon 3a^2$, $u^2 - 3w^2 = \varepsilon b^2$, $y = \pm 3ab$ with $\varepsilon = \pm 1$, and then a and b are coprime and b is odd. Since u and v (hence w) have opposite parity, we have $u^2 - 3w^2 \equiv u^2 + w^2 \equiv 1 \pmod 4$; hence we must have $\varepsilon = 1$, so the equations to be solved are $w = a^2$ and $u^2 - 3w^2 = b^2$. Corollary 6.3.15 tells us that there exist coprime integers c and d of opposite parity such that either $u = c^2 + 3d^2$, $w = 2cd$, $b = c^2 - 3d^2$ with $3 \nmid c$, or $u = 2(c^2 + cd + d^2)$, $w = c^2 - d^2$, $b = c^2 + 4cd + d^2$ with $c \not\equiv d \pmod 3$, where the signs can be absorbed as usual by changing either x into $-x$ or b into $-b$. Thus in the first case the final equation to be solved is $2cd = a^2$, so that there exist coprime s and t with $3 \nmid s$ such that either $c = 2s^2$, $d = t^2$, $a = \pm 2st$ and t odd, or $c = s^2$, $d = 2t^2$, $a = \pm 2st$ and s odd. Replacing everywhere gives the second and third parametrizations:

$$\begin{cases} x = \pm(4s^4 + 3t^4)(16s^8 - 408t^4s^4 + 9t^8)\,, \\ y = 6ts(4s^4 - 3t^4)\,, \\ z = 16s^8 + 168t^4s^4 + 9t^8\,, \end{cases}$$

where t is odd and $3 \nmid s$, and

$$\begin{cases} x = \pm(s^4 + 12t^4)(s^8 - 408t^4s^4 + 144t^8)\,, \\ y = 6ts(s^4 - 12t^4)\,, \\ z = s^8 + 168t^4s^4 + 144t^8\,, \end{cases}$$

where s is odd and $3 \nmid s$.

In the second case the final equation to be solved is $c^2 - d^2 = a^2$ with c and d of opposite parity, hence with a odd, so that by the solution to the Pythagorean equation there exist coprime integers s and t of opposite parity such that $c = s^2 + t^2$, $d = 2st$, $a = s^2 - t^2$ with $s \not\equiv t \pmod 3$. Replacing everywhere gives the fourth and final parametrization:

$$\begin{cases} x = \pm 2(s^4 + 2ts^3 + 6t^2s^2 + 2t^3s + t^4)(23s^8 - 16ts^7 - 172t^2s^6 - 112t^3s^5 \\ \qquad - 22t^4s^4 - 112t^5s^3 - 172t^6s^2 - 16t^7s + 23t^8)\,, \\ y = 3(s - t)(s + t)(s^4 + 8ts^3 + 6t^2s^2 + 8t^3s + t^4)\,, \\ z = 13s^8 + 16ts^7 + 28t^2s^6 + 112t^3s^5 + 238t^4s^4 \\ \qquad + 112t^5s^3 + 28t^6s^2 + 16t^7s + 13t^8\,, \end{cases}$$

where $s \not\equiv t \pmod 2$ and $s \not\equiv t \pmod 3$.

We have thus proved the following theorem:

Theorem 14.4.2. *The equation $x^2 + y^4 = z^3$ in integers x, y, z with $\gcd(x, y) = 1$ can be parametrized by one of the above four parametrizations, where s and t denote coprime integers with the indicated congruence conditions modulo 2 and 3. In addition, these parametrizations are disjoint, in that any solution to our equation belongs to a single parametrization.*

14.5 Invariants, Covariants, and Dessins d'Enfants

There is a completely different way of attacking the super-Fermat equation in the elliptic case, which is based on geometrical methods. This is an alternative way for the tetrahedral and octahedral cases, but is the only known way of solving the icosahedral case. The reason is that in the tetrahedral case $(2, 3, 3)$ and the octahedral case $(2, 3, 4)$, we can *factor* the equation (possibly in some number field), hence reduce to a simpler equation, and we treated these cases with complete success. On the other hand, in the icosahedral case $(2, 3, 5)$, it is not possible to factor the equation. Thus another approach is needed, which will be given by the considerations of the present section.

14.5.1 Dessins d'Enfants, Klein Forms, and Covariants

The present subsection will serve as a motivation for the results that will be given without proof below, and we refer to [Edw] for details. For the moment we ignore all rationality questions and we look for one-variable polynomials P, Q, and R with complex coefficients satisfying $P^3 + Q^k = R^2$ for $k = 3, 4$, and 5. We could try to solve this by indeterminate coefficients, but there is no guarantee that we would succeed. However, we can use a very important theorem due to G. Belyi, which tells us (in our special case) that for any graph inscribed in the Riemann sphere (a "dessin d'enfant," the name coined by A. Grothendieck), there exists a rational function ϕ from the sphere to $\mathbb{P}_1$ such that the zeros of ϕ have order equal to the number of edges meeting at the vertices V of the graph, the poles of ϕ have order equal to the number of vertices along the faces F of the graph, and finally the values where $\phi = -1$ have order 2, one for each edge E of the graph, and the coefficients of ϕ may be chosen in a number field.

We apply this to the five platonic solids, and we index the polynomials according to their degrees.

- For the tetrahedron, we have $\phi = P_4^3/Q_4^3$ and $\phi + 1 = R_6^2/Q_4^3$, so that $P_4^3 + Q_4^3 = R_6^2$.
- For the cube, we have $\phi = P_8^3/Q_6^4$ and $\phi + 1 = R_{12}^2/Q_6^4$, so that $P_8^3 + Q_6^4 = R_{12}^2$.
- For the octahedron, we have $\phi = P_6^4/Q_8^3$ and $\phi + 1 = R_{12}^2/Q_8^3$, so that $P_6^4 + Q_8^3 = R_{12}^2$. This is exactly the same equation as for the cube (coming from the fact that the cube and the octahedron are dual), so we do not need to consider the cube.
- For the dodecahedron, we have $\phi = P_{20}^3/Q_{12}^5$ and $\phi + 1 = R_{30}^2/Q_{12}^5$, so that $P_{20}^3 + Q_{12}^5 = R_{30}^2$.
- For the icosahedron, we have $\phi = P_{12}^5/Q_{20}^3$ and $\phi + 1 = R_{30}^2/Q_{20}^3$, so that $P_{12}^5 + Q_{20}^3 = R_{30}^2$. This is exactly the same equation as for the dodecahedron (coming from the fact that the dodecahedron and the icosahedron are dual), so we do not need to consider the dodecahedron.

This geometric interpretation explains the origin of the tetrahedral, octahedral, and icosahedral terminology, which is always used in relation to finite subgroups of $\mathrm{PSL}_2(\mathbb{C})$.

Almost a century before Belyi, Klein had already shown the existence of the Belyi functions ϕ in the case of platonic solids. More precisely, he proved the following:

Theorem 14.5.1. *Let G be the vertices of a regular tetrahedron, octahedron, or icosahedron inscribed in the Riemann sphere, let N be the north pole of the sphere, and for $g \in G$ let $(\alpha_g : \beta_g) \in \mathbb{P}_1(\mathbb{C})$ be the point obtained by stereographic projection from N (if $g = N$, choose the point at infinity $(1 : 0)$).*

Let $k = |G|$ be the number of vertices (4, 6, or 12 respectively), let r be the number of edges meeting at each vertex (3, 4, or 5 respectively), and set

$$f_G(s,t) = \prod_{g \in G} (\beta_g s - \alpha_g t),$$

$$h_G(s,t) = \frac{1}{k^2(k-1)^2} \left(\frac{\partial^2 f_G}{\partial s^2} \frac{\partial^2 f_G}{\partial t^2} - \left(\frac{\partial^2 f_G}{\partial s \partial t} \right)^2 \right),$$

$$j_G(s,t) = \frac{1}{2k(k-2)} \left(\frac{\partial f_G}{\partial s} \frac{\partial h_G}{\partial t} - \frac{\partial f_G}{\partial t} \frac{\partial h_G}{\partial s} \right).$$

Then after a suitable rotation of the sphere there exists a constant $u_G \in \mathbb{C}$ such that

$$j_G^2 + h_G^3 + f_G^r / u_G = 0.$$

Although for the moment the polynomials are with coefficients in $\mathbb{C}$, this is exactly what we need for solving the $(2,3,r)$ equation in the elliptic case. To make this clearer, we look at all three cases. Consider first the regular tetrahedron. Up to rescaling and rotation we can choose $f_G(s,t) = t(s^3 - t^3)$ (the factor t corresponds to the north pole N, and the roots of $s^3 - t^3$ to the cube roots of unity, i.e., to the face of the tetrahedron opposite to N). A short computation shows that $j_G^2 + h_G^3 + f_G^3/64 = 0$, so that $u_G = 64$. Similarly, consider the regular octahedron. Clearly we can choose $f_G(s,t) = st(s^4 - t^4)$ (draw a picture!), and a short computation shows that $j_G^2 + h_G^3 + f_G^4/432 = 0$, so that $u_G = 432$. Finally, consider the regular icosahedron. Here the geometry is slightly more complicated, but after a little work it can be seen that we may choose $f_G(s,t) = st(s^{10} - 11s^5t^5 - t^{10})$ (Exercise 1), and a short computation shows that $j_G^2 + h_G^3 + f_G^5/1728 = 0$, so that $u_G = 1728$.

Starting from these basic solutions, if we apply an element of $\mathrm{GL}_2(\mathbb{C})$ we obtain a new relation of the same type (this is in fact the meaning of the word covariant), hence as many as we want. The basic problem is now to obtain polynomials with coefficients in $\mathbb{Q}$, or even in $\mathbb{Z}$, and to separate equivalent parametrizations under $\mathrm{GL}_2(\mathbb{Z})$. This can be done using a suitable reduction theory; see [Edw].

14.5.2 The Icosahedral Case $(2, 3, 5)$

It can be checked that up to signs, all the parametrizations that we have given in the preceding sections correspond to special cases of Klein's theorem: let us introduce a convenient shorthand, copied from [Edw]. We simply write $f = [a_k, \ldots, a_0]$ as an abbreviation for

$$f(s,t) = \sum_{0 \leqslant i \leqslant k} \binom{k}{i} a_i s^i t^{k-i}.$$

The inclusion of the binomial coefficient is natural and simplifies the formulas. Starting from f we define h and j as in the theorem, and since we now want arithmetic solutions, we will impose $u_G = \pm 1$, so that the parametrizations of $x^2 + y^3 \pm z^r = 0$ will be $x = \pm j$, $y = h$, and $z = \pm f$ for any sign in x and any sign in z if $r = 4$.

So that the reader can relate to what we have done in the cases $r = 3$ and $r = 4$, we give in abbreviated form the results that we have obtained, in the same order.

For the equation $x^2 + y^3 - z^3 = 0$ the three parametrizations are $f_1 = [1, 0, 0, 2, 0]$, $f_2 = [2, 1, 0, 1, 2]$, and $f_3 = [3, 0, 1, 0, -1]$.

For the equation $x^2 + y^3 - z^4 = 0$ the seven parametrizations are $f_1 = [7, 1, -2, -4, -4, -4, -8]$, $f_2 = [27, 0, 3, 0, -1, 0, -1]$, $f_3 = [0, 3, 0, 0, 0, 4, 0]$, $f_4 = [0, 12, 0, 0, 0, 1, 0]$, $f_5 = [1, 0, 0, 2, 0, 0, -32]$, $f_6 = [9, 3, 3, 3, 1, -1, -5]$, and $f_7 = [17, 5, -1, 1, 1, 1, -1]$.

For the equation $x^2 + y^3 + z^4 = 0$ the four parametrizations are $f_1 = [1, 0, -1, 0, -3, 0, 27]$, $f_2 = [0, 4, 0, 0, 0, -3, 0]$, $f_3 = [0, 1, 0, 0, 0, -12, 0]$, and $f_4 = [3, 4, 1, 0, -1, -4, -3]$.

We can now give without proof Edwards's result on the $(2, 3, 5)$ equation.

Theorem 14.5.2. *Up to changing x into $-x$ there are exactly 27 distinct parametrizations of $x^2 + y^3 + z^5 = 0$ given by*

$f_1 = [0, 1, 0, 0, 0, 0, -144/7, 0, 0, 0, 0, -20736, 0]$,

$f_2 = [-1, 0, 0, -2, 0, 0, 80/7, 0, 0, 640, 0, 0, -102400]$,

$f_3 = [-1, 0, -1, 0, 3, 0, 45/7, 0, 135, 0, -2025, 0, -91125]$,

$f_4 = [1, 0, -1, 0, -3, 0, 45/7, 0, -135, 0, -2025, 0, 91125]$,

$f_5 = [-1, 1, 1, 1, -1, 5, -25/7, -35, -65, -215, 1025, -7975, -57025]$,

$f_6 = [3, 1, -2, 0, -4, -4, 24/7, 16, -80, -48, -928, -2176, 27072]$,

$f_7 = [-10, 1, 4, 7, 2, 5, 80/7, -5, -50, -215, -100, -625, -10150]$,

$f_8 = [-19, -5, -8, -2, 8, 8, 80/7, 16, 64, 64, -256, -640, -5632]$,

$f_9 = [-7, -22, -13, -6, -3, -6, -207/7, -54, -63, -54, 27, 1242, 4293]$,

$f_{10} = [-25, 0, 0, -10, 0, 0, 80/7, 0, 0, 128, 0, 0, -4096]$,

$f_{11} = [6, -31, -32, -24, -16, -8, -144/7, -64, -128, -192, -256, 256, 3072]$,

$f_{12} = [-64, -32, -32, -32, -16, 8, 248/7, 64, 124, 262, 374, 122, -2353]$,

$f_{13} = [-64, -64, -32, -16, -16, -32, -424/7, -76, -68, -28, 134, 859, 2207]$,

$f_{14} = [-25, -50, -25, -10, -5, -10, -235/7, -50, -49, -34, 31, 614, 1763]$,

$f_{15} = [55, 29, -7, -3, -9, -15, -81/7, 9, -9, -27, -135, -459, 567]$,

$f_{16} = [-81, -27, -27, -27, -9, 9, 171/7, 33, 63, 141, 149, -67, -1657]$,

$f_{17} = [-125, 0, -25, 0, 15, 0, 45/7, 0, 27, 0, -81, 0, -729]$,

$f_{18} = [125, 0, -25, 0, -15, 0, 45/7, 0, -27, 0, -81, 0, 729]$,

$f_{19} = [-162, -27, 0, 27, 18, 9, 108/7, 15, 6, -51, -88, -93, -710]$,

$f_{20} = [0, 81, 0, 0, 0, 0, -144/7, 0, 0, 0, 0, -256, 0]$,

$f_{21} = [-185, -12, 31, 44, 27, 20, 157/7, 12, -17, -76, -105, -148, -701]$,

$f_{22} = [100, 125, 50, 15, 0, -15, -270/7, -45, -36, -27, -54, -297, -648]$,

$f_{23} = [192, 32, -32, 0, -16, -8, 24/7, 8, -20, -6, -58, -68, 423]$,

$f_{24} = [-395, -153, -92, -26, 24, 40, 304/7, 48, 64, 64, 0, -128, -512]$,

$f_{25} = [-537, -205, -133, -123, -89, -41, 45/7, 41, 71, 123, 187, 205, -57]$,

$f_{26} = [359, 141, -1, -21, -33, -39, -207/7, -9, -9, -27, -81, -189, -81]$,

$f_{27} = [295, -17, -55, -25, -25, -5, 31/7, -5, -25, -25, -55, -17, 295]$.

For instance, one of the simplest parametrizations, given by f_{20}, is explicitly

$$\begin{cases} x = \pm(81s^{10} + 256t^{10}) \\ \quad \times (6561s^{20} - 6088608t^5 s^{15} - 207484416t^{10}s^{10} + 19243008t^{15}s^5 + 65536t^{20}) \,, \\ y = -6561s^{20} - 2659392t^5 s^{15} - 10243584t^{10}s^{10} + 8404992t^{15}s^5 - 65536t^{20} \,, \\ z = 12st(81s^{10} - 1584t^5 s^5 - 256t^{10}) \,. \end{cases}$$

14.6 The Parabolic and Hyperbolic Cases

14.6.1 The Parabolic Case

We now consider the parabolic case $1/p + 1/q + 1/r = 1$. Up to permutation of (p, q, r), this corresponds to the three cases $(p, q, r) = (2, 3, 6)$, $(2, 4, 4)$, and $(3, 3, 3)$. The result is then simply as follows.

Proposition 14.6.1. *In the parabolic case $1/p + 1/q + 1/r = 1$, the equation $x^p + y^q = z^r$ has no solutions in nonzero coprime integers, except that the equation $x^3 + y^6 = z^2$ has the solutions $(x, y, z) = (2, \pm 1, \pm 3)$, and the equation $x^3 + y^2 = z^6$ has the solutions $(x, y, z) = (-2, \pm 3, \pm 1)$.*

Proof. The $(3, 3, 3)$ case is FLT for exponent 3, which has been proved in Section 6.9, the case of general coefficients having been studied in Sections 6.4.2, 6.4.3, 6.4.4, and 6.4.5. The $(2, 4, 4)$ case corresponds to the equations $x^4 \pm y^4 = z^2$ which have been solved in Proposition 6.5.3, the case of general coefficients having been treated in Sections 6.5.1, 6.5.2, and 6.5.3. We are left with the $(2, 3, 6)$ case which, thanks to the reductions made above, reduces to the equations $x^3 \pm y^6 = z^2$, the case of general coefficients having been treated in Section 6.5.5. As a special case of Proposition 6.5.9 (with variable names changed), it is clear that if we set $X = x/y^2$, $Y = z/y^3$ our equations are equivalent to finding rational points on the curves $Y^2 = X^3 \pm 1$. This is done by the 2-descent technique, explained in Chapter 8, which is here very easy to apply since the 2-torsion point $(\mp 1, 0)$ has rational coordinates. We have treated this example in Proposition 8.2.14, but we can of course be lazy and use Cremona's `mwrank` program, which tells us that both curves have rank 0. The curve $Y^2 = X^3 - 1$ has only the point $(1, 0)$ as torsion point (in addition to the point at infinity). On the other hand, the curve $Y^2 = X^3 + 1$ has a torsion subgroup of order 6, and apart from the point at infinity the torsion points are $(-1, 0)$, $(0, \pm 1)$, and $(2, \pm 3)$. The points $(-1, 0)$ and $(0, \pm 1)$ correspond to $z = 0$ and $x = 0$, respectively, while the points $(2, \pm 3)$ give $x = 2y^2$, $z = \pm 3y^3$. Since x and y are coprime, we must have $y = \pm 1$, hence $x = 2$ and $z = \pm 3$, proving the proposition. $\square$

Thus in the parabolic case there are only finitely many nonzero coprime solutions. As already mentioned, this is because we consider the super-Fermat equation only with coefficients ± 1, and in the case of general coefficients, the equation may have finitely or infinitely many coprime solutions, depending on whether the rank of the corresponding elliptic curve is zero or not, see the sections mentioned above. For instance, the equations $x^2 + y^4 = 2z^4$ and $x^4 + 8y^4 = z^2$ have infinitely many coprime solutions, see Exercise 13 of Chapter 8.

14.6.2 General Results in the Hyperbolic Case

We finally consider what is by far the most difficult case, the hyperbolic case $1/p + 1/q + 1/r < 1$. Proving all that is known would require a book in itself, so we will give only a survey with few proofs. Note that when we talk of solutions to our equations, we always mean integral nonzero coprime solutions.

First, there is a beautiful theorem of Darmon and Granville [Dar-Gra] as follows.

Theorem 14.6.2. *For fixed p, q, and r such that $1/p + 1/q + 1/r < 1$ and fixed nonzero integers A, B, and C, there exist only finitely many solutions to the equation $Ax^p + By^q + Cz^r = 0$ in integers x, y, and z with x and y coprime.*

To prove this theorem, Darmon and Granville succeed in reducing it to Faltings's famous theorem on the finiteness of the number of rational points on a curve of genus greater than or equal to 2 (Mordell's conjecture), which is not a trivial task since $Ax^p + By^q + Cz^r = 0$ does not a priori represent a curve.

Second, we recall the very important *abc conjecture* of Masser–Oesterlé, which implies many results or other conjectures in number theory (for instance, Elkies has proved that it implies Faltings's result above: *abc* implies Mordell; see [Elk1]). There are several possible statements of this conjecture, but the following is sufficient.

Definition 14.6.3. *For a nonzero natural integer N we define the* radical $\operatorname{rad}(N)$ *of N as the product of the prime numbers dividing N, i.e., $\operatorname{rad}(N) = \prod_{p \mid N} p$.*

The *abc* conjecture is then as follows.

Conjecture 14.6.4. *Let $\varepsilon > 0$. If a, b, and c are three nonzero pairwise coprime integers such that $a + b + c = 0$ then*

$$\max(|a|, |b|, |c|) = O_\varepsilon(\operatorname{rad}(abc)^{1+\varepsilon}) \ .$$

We then have the following result:

Proposition 14.6.5. *The abc conjecture implies that the total number of nonzero coprime solutions to $x^p \pm y^q \pm z^r = 0$ with $1/p + 1/q + 1/r < 1$ is finite, even allowing p, q, and r to vary. Here, if $x = \pm 1$ (respectively $y = \pm 1$, respectively $z = \pm 1$), we identify solutions having the same value of x^p (respectively y^q, respectively z^r).*

Proof. Order p, q, and r such that $p \leqslant q \leqslant r$. Then the hyperbolic cases correspond to the triples $(2, 3, r)$ for $r \geqslant 7$, $(2, 4, r)$ for $r \geqslant 5$, $(2, q, r)$ for $r \geqslant q \geqslant 5$, $(3, 3, r)$ for $r \geqslant 4$, $(3, q, r)$ for $r \geqslant q \geqslant 4$, or (p, q, r) with $r \geqslant q \geqslant p \geqslant 4$. In all these cases one checks immediately that $1/p + 1/q + 1/r \leqslant 41/42$, attained for $(p, q, r) = (2, 3, 7)$. We apply the *abc* conjecture to $a = x^p$, $b = \pm y^q$, and $c = \pm z^r$, and we choose $\varepsilon = 1/42$. Note that $\operatorname{rad}(abc) = \operatorname{rad}(xyz) \leqslant xyz$. If we set $M = \max(|x^p|, |y^q|, |z^r|)$, we thus have

$$M = O((xyz)^{1+\varepsilon}) = O(M^{(1/p+1/q+1/r)(1+\varepsilon)})$$
$$= O(M^{(41/42)(43/42)}) = O(M^{1763/1764}) \ ,$$

which is impossible if M is sufficiently large. Thus M is bounded, hence so are x, y, z, p, q, and r (except in the special case $\min(|x|, |y|, |z|) = 1$), proving the proposition. $\qquad \square$

A stronger statement is given in Exercise 2.

Remark. As already mentioned in Chapter 1, it has been proved by P. Mihăilescu in 2002 that Catalan's conjecture is true, i.e., that $x^p \pm y^q = 1$ is possible only if $x^p = 9$ and $y^q = 8$, and I refer the reader to Section 6.11 and Chapter 16 for a detailed description of the proof (see also [Bilu] and [Boe-Mis]). Thus the special case mentioned in the proposition occurs only (up to ordering of p, q, and r) for $p = 2$, $q = 3$, $r \geqslant 7$, with $(\pm 3)^p - 2^q = 1^r$ (and also $(-1)^r$ when r is even).

A computer search gives the following 10 essentially different solutions (where as above the first one is counted only once, and we also count once solutions differing only by sign changes):

$$1^r + 2^3 = (\pm 3)^2 \quad (\text{for } r \geqslant 7, \text{ with also } (-1)^r \text{ for } r \text{ even}) ,$$
$$(\pm 3)^4 + (-2)^5 = (\pm 7)^2 ,$$
$$2^9 + (-7)^3 = (\pm 13)^2 ,$$
$$2^7 + 17^3 = (\pm 71)^2 ,$$
$$3^5 + (\pm 11)^4 = (\pm 122)^2 ,$$
$$15613^3 - (\pm 33)^8 = (\pm 1549034)^2 ,$$
$$65^7 + (-1414)^3 = (\pm 2213459)^2 ,$$
$$113^7 + (-9262)^3 = (\pm 15312283)^2 ,$$
$$17^7 + 76271^3 = (\pm 21063928)^2 ,$$
$$(\pm 43)^8 + 96222^3 = (\pm 30042907)^2 .$$

These solutions can easily be found in a few seconds by a systematic search on a PC. A search for several weeks has not revealed any additional solutions. There may be no more, and on probabilistic grounds one would expect at most two or three more. Note also that the number of solutions found decreases with $\chi = 1/p + 1/q + 1/r - 1$, as can be expected: counting the first one when possible, we have five solutions for $\chi = -1/42$, three solutions for $\chi = -1/24$, two solutions for $\chi = -1/20$, one solution for $\chi = -1/18$, and no solution for other χ (apart from the first when applicable).

14.6.3 The Equations $x^4 \pm y^4 = z^3$

We now study a few hyperbolic equations. In each case, we proceed as follows. We reduce the equation to finding integral or rational points on curves. We then use general methods to find this set of points. When the curve is an elliptic curve, we use Cremona's `mwrank` program, which does all the work for us, or the methods explained in Chapter 8. When the curve is a curve of higher genus, or an elliptic curve of nonzero rank, the problem becomes more difficult and we will mention only the known results. We begin with the simple equations $x^4 \pm y^4 = z^3$.

Proposition 14.6.6. *The equations $x^4 \pm y^4 = z^3$ have no solution in nonzero coprime integers x, y, z.*

Proof. Write our equation as $x^4 + \eta y^4 = z^3$ with $\eta = \pm 1$. Although we could use the solution of the elliptic equation $X^2 + \eta y^4 = z^3$ given in Section 14.4.2, it is much simpler to use only the solution to the dihedral equation $X^2 + \eta Y^2 = Z^3$ seen above. We consider two cases.

Case 1: $\eta = 1$, or $\eta = -1$ and $2 \nmid z$. In this case we deduce from Section 14.2.2 that there exist coprime integers s and t of opposite parity such that

$x^2 = s(s^2 - 3\eta t^2)$, $y^2 = t(3s^2 - \eta t^2)$, and $z = s^2 + \eta t^2$, since the second parametrization for $\eta = -1$ is excluded by the condition $2 \nmid z$. Since s and t are coprime, one at least is not divisible by 3. If necessary by exchanging x and y, and changing z into $-z$ if $\eta = -1$, we may assume that $3 \nmid s$. It follows that s is coprime to $s^2 - 3\eta t^2$, and so there exist integers a and b and a sign $\varepsilon_1 = \pm 1$ such that $s = \varepsilon_1 a^2$ and $s^2 - 3\eta t^2 = \varepsilon_1 b^2$. Since $3 \nmid s$ we have $s^2 - 3\eta t^2 \equiv 1 \pmod 3$, therefore $\varepsilon_1 = 1$, and we obtain the equations $s = a^2$, $s^2 - 3\eta t^2 = b^2$, and $b^2 = a^4 - 3\eta t^2$. Since s and t have opposite parity one is even and the other is odd. Clearly $2 \mid s$ would imply $4 \mid s$ (since $s = a^2$) hence $s^2 - 3\eta t^2 \equiv \pm 3 \pmod 8$, in contradiction to $s^2 - 3\eta t^2 = b^2$. Thus $2 \mid t$ and $2 \nmid s$. I now claim that $3 \mid t$. Indeed, otherwise t would be coprime to $3s^2 - \eta t^2$, so we would have similarly $t = \varepsilon_2 c^2$ and $3s^2 - \eta t^2 = \varepsilon_2 d^2$. Thus $4 \mid t$, so $3s^2 - \eta t^2 \equiv 3 \pmod 8$, which is impossible for $\varepsilon_2 d^2$. Thus $t = 3t_1$ for some integer t_1, and we deduce that $(y/3)^2 = t_1(s^2 - 3\eta t_1^2)$. The factors are now coprime, so $t_1 = \pm e^2$ for some integer e. We thus obtain the equation $b^2 = a^4 - 27\eta e^4$. This is a hyperelliptic quartic equation, which is therefore easy to reduce to an elliptic curve using Corollary 7.2.2. In fact, we have already done the work in Corollary 6.5.8, and so we are reduced to the computation of the rank of the elliptic curves $Y^2 = X^3 - 27\eta X$, which is easily done using 2-descent or `mwrank`. Since we find that both ranks are equal to 0 this case follows. Note that if we had found a nonzero rank it would not have implied that our equation has nonzero solutions since we dropped some conditions along the way.

Case 2: $\eta = -1$ and $2 \mid z$. In this case we deduce from Section 14.2.2 that there exist coprime integers s and t and some sign $\varepsilon = \pm 1$ such that $2 \nmid t$ and $x^2 = \varepsilon(2s^3 + t^3)$, $y^2 = 2s^3 - t^3$, and $z = 2st$. Since $x^2 y^2 = -\varepsilon(t^6 - 4s^6)$ and t is odd, we have $t^6 - 4s^6 \equiv 1 \pmod 4$, so we deduce that $\varepsilon = -1$, and x and y are odd. We can thus write $((y-x)/2)((y+x)/2) = (y^2 - x^2)/4 = s^3$, and since x and y are coprime so are $(y-x)/2$ and $(y+x)/2$. Thus there exist integers a and b such that $y - x = 2a^3$ and $y + x = 2b^3$; therefore $y = a^3 + b^3$, $x = b^3 - a^3$. But then $(-t)^3 = (x^2 + y^2)/2 = a^6 + b^6$, which has no nonzero solution since it is a special case of Fermat's equation $A^3 + B^3 = C^3$ with $A = a^2$, $B = b^2$, and $C = -t$, proving the proposition. □

14.6.4 The Equation $x^4 + y^4 = z^5$

Proposition 14.6.7. *The equation $x^4 + y^4 = z^5$ has no solution in nonzero coprime integers x, y, z.*

Proof. Once again we use the solution to the dihedral equation. Our equation is thus equivalent to $x^2 = s(s^4 - 10t^2 s^2 + 5t^4)$, $y^2 = t(5s^4 - 10t^2 s^2 + t^4)$, and $z = s^2 + t^2$, where s and t are coprime integers of opposite parity. I claim that $5 \mid st$. Indeed, if s and t are not divisible by 5, then the factors in the expressions for x^2 and y^2 are coprime; hence in particular there exist integers u and v such that $s^4 - 10t^2 s^2 + 5t^4 = \pm u^2$ and $5s^4 - 10t^2 s^2 + t^4 = \pm v^2$.

Then if t is even s is odd, and the second equation gives a contradiction modulo 8, and similarly if t is odd then s is even and now the first equation gives a contradiction modulo 8, proving my claim. Thus $5 \mid st$, and if necessary exchanging x and y, hence s and t, we may assume that $5 \mid s$. Writing $s = 5s_1$, we thus have in particular $(x/5)^2 = s_1(125s_1^4 - 50t^2s_1^2 + t^4)$, hence $125s_1^4 - 50t^2s_1^2 + t^4 = \pm u^2$ for some integer u. If t is even then s_1 is odd, and this gives a contradiction modulo 8. Thus $s = 5s_1$ is even, and since t is coprime to 5 the equation for y^2 gives $5s^4 - 10t^2s^2 + t^4 = \pm v^2$, and since s is even and t is odd, again looking modulo 8 we see that the sign must be $+$, so we finally obtain the hyperelliptic quartic equation $V^2 = T^4 - 10T^2 + 5$, with $V = v/s^2$ and $T = t/s$. Corollary 7.2.2 tells us that if we set $X = 2(T^2 - V - 5)$ and $Y = 4T(T^2 - V - 5)$ this is a birational transformation whose inverse is $T = Y/(2X)$ and $V = Y^2/(4X^2) - X/2 - 5$, and which transforms our genus-1 curve into the Weierstrass equation in minimal form $Y^2 = X(X^2 + 20X + 80)$. This curve has a rational point of order 2, so the 2-descent method of Section 8.2.4 is easily applicable and shows that our curve has rank 0 (or we can be lazy and use `mwrank`). The only nontrivial torsion point has $X = 0$, so $V = 5 - T^2$, but replacing in the quartic we obtain the contradiction $25 = 5$, proving the proposition.   □

Remark. The equation $x^4 - y^4 = z^5$ leads to elliptic curves of nonzero rank, and I do not know whether it can be treated by similar methods (although the nonexistence of nontrivial solutions follows from the $(2, 4, 5)$ case treated by Bruin; see below).

14.6.5 The Equation $x^6 - y^4 = z^2$

Proposition 14.6.8. *The equation $x^6 - y^4 = z^2$ has no solution in nonzero coprime integers x, y, z.*

Proof. Once again, we use the solution to the dihedral equation $X^2 + Y^2 = Z^3$. We deduce that $x^6 - y^4 = z^2$ is equivalent to $x^2 = s^2 + t^2$, $y^2 = s(s^2 - 3t^2)$, $z = t(3s^2 - t^2)$, where s and t are coprime with $s \not\equiv t \pmod 2$. The first equation is equivalent to $s = 2uv$, $t = u^2 - v^2$, $x = \pm(u^2 + v^2)$, up to exchange of s and t, where u and v are coprime integers of opposite parity. We consider both cases.

Case 1: $2 \mid s$. Set $a = u + v$, $b = u - v$, which are coprime and both odd. Then $s = (a^2 - b^2)/2$ and $t = ab$, so the last equation to be solved can be written $8y^2 = (a^2 - b^2)(a^4 - 14a^2b^2 + b^4)$. Since b is odd, we can set $Y = y/b^3$, $X = a^2/b^2$, and we obtain the elliptic curve $8Y^2 = (X - 1)(X^2 - 14X + 1)$, which can be given in reduced Weierstrass form as $y^2 = (x + 2)(x^2 - 2x - 11)$. In any case, the `mwrank` program tells us that (outside the point at infinity) the only rational point has $y = 0$, which does not correspond to a solution of our equation.

Case 2: $2 \nmid s$. Here $s = u^2 - v^2$, $t = 2uv$, so that the last equation to be solved can be written $y^2 = (u^2 - v^2)(u^4 - 14u^2v^2 + v^4)$. We cannot have $v = 0$, since otherwise $t = 0$ hence $z = 0$, which is impossible. Thus, we can set $Y = y/v^3$, $X = u^2/v^2$ and we obtain the elliptic curve $Y^2 = (X - 1)(X^2 - 14X + 1)$, which can be given in reduced Weierstrass form as $y^2 = (x+4)(x^2 - 4x - 44)$. In any case the `mwrank` program again tells us that the only rational point has $y = 0$, which again does not correspond to a solution.   □

14.6.6 The Equation $x^4 - y^6 = z^2$

Proposition 14.6.9. *The equation $x^4 - y^6 = z^2$ has no solution in nonzero coprime integers x, y, z.*

Proof. Once again, we use the solution to the dihedral equation. We find that our equation is equivalent to $x^2 = s(s^2 + 3t^2)$, $z = t(3s^2 + t^2)$, $y^2 = s^2 - t^2$ with s and t coprime of opposite parity, or to $x^2 = \pm(2s^3 + t^3)$, $z = 2s^3 - t^3$, $y^2 = 2ts$ with s and t coprime and t odd. We consider both cases separately.

Case 1: $2 \nmid y$. This corresponds to the first parametrization. Since $y^2 + t^2 = s^2$ and y is odd, there exist coprime u and v of opposite parity such that $y = u^2 - v^2$, $t = 2uv$, and $s = \pm(u^2 + v^2)$. Since $x^2 > 0$, we have $s > 0$, so the $\pm$ is $+$. The last equation to be solved is thus is $x^2 = (u^2 + v^2)(u^4 + 14u^2v^2 + v^4)$. Note for future reference that since u and v have opposite parity, x is odd. We have $v \neq 0$, since otherwise $t = 0$ hence $z = 0$, which is impossible. Thus we set $Y = z/v^3$, $X = u^2/v^2$ and obtain the elliptic curve $Y^2 = (X + 1)(X^2 + 14X + 1)$, which can be given in reduced Weierstrass form as $y^2 = (x - 4)(x^2 + 4x - 44)$. However, here the `mwrank` program tells us that this is a curve of rank 1, so we must proceed differently. Note that when we set $X = u^2/v^2$, we implicitly forget the information that X is a square. In order to keep it, we must return to the equation $x^2 = (u^2 + v^2)(u^4 + 14u^2v^2 + v^4)$. First, we set $a = u + v$, $b = u - v$, which are coprime and both odd. We obtain $2x^2 = (a^2 + b^2)(a^4 - a^2b^2 + b^4)$. Since $a^4 - a^2b^2 + b^4 = (a^2 + b^2)^2 - 3a^2b^2$, it follows that the only possible common prime divisor of $a^2 + b^2$ and $a^4 - a^2b^2 + b^4$ is $p = 3$. But this is impossible, since $a^2 + b^2 \equiv 0 \pmod 3$ if and only if $a \equiv b \equiv 0 \pmod 3$, which is excluded since a and b are coprime. Thus the factors are coprime, and by positivity we obtain that there exist integers c and d such that $a^2 + b^2 = 2c^2$, $a^4 - a^2b^2 + b^4 = d^2$, and $x = \pm cd$, and c and d are both odd since x is odd.

We consider only the second equation. Setting $D = d/b^2$, $A = a/b$ we obtain the hyperelliptic quartic curve $D^2 = A^4 - A^2 + 1$ of genus 1. Corollary 7.2.2 tells us that if we set $X = 2A^2 - 2D - 1$, $Y = 2A(2A^2 - D - 1)$ this is a birational transformation whose inverse is $A = Y/(2X)$, $D = Y^2/(4X^2) - (X+1)/2$ and that transforms our genus-1 curve into the Weierstrass equation $Y^2 = X(X^2 + 2X - 3)$, and now the `mwrank` program tells us that the rank is zero, but there are eight rational torsion points on the curve: not counting the point at infinity, they are $(-3, 0)$, $(-1, \pm2)$, $(0, 0)$, $(1, 0)$, and $(3, \pm6)$.

Because of our birational transformation, we cannot have $Y = 0$. It is easy to check that the other four points, $(X, Y) = (-1, \pm 2)$ and $(3, \pm 6)$, correspond to the four points $(A, D) = (\pm 1, \pm 1)$. Since $A = a/b$ and a and b are coprime, we must have therefore $a = \pm 1$ and $b = \pm 1$, hence $a = \pm b$. Now recall that $a = u + v$ and $b = u - v$. It follows that $a = \pm b$ is equivalent to $uv = 0$, hence to $t = 0$, which is impossible since this implies $z = 0$, so that there are no solutions in this case as claimed.

Case 2: $2 \mid y$. This corresponds to the second parametrization $x^2 = \pm(2s^3 + t^3)$, $z = 2s^3 - t^3$, $y^2 = 2ts$ with s and t coprime and t odd. Thus there exist coprime a and b with a odd and $\varepsilon = \pm 1$ such that $t = \varepsilon a^2$, $s = 2\varepsilon b^2$, and $y = \pm 2ab$. The last equation to be solved is thus $x^2 = \pm \varepsilon(a^6 + 16b^6)$. Since a is odd, we have $\varepsilon = \pm$ hence the equation is $x^2 = a^6 + 16b^6$. We have $b \neq 0$, since otherwise $y = 0$, so that setting $Y = x/b^3$ and $X = a^2/b^2$ we obtain the elliptic curve $Y^2 = X^3 + 16$, whose minimal Weierstrass equation is $y^2 + y = x^3$, and `mwrank` tells us that the only rational points outside the point at infinity of this curve are those with $x = 0$, hence $X = 0$, hence $a = 0$, which is impossible since a is odd.   □

14.6.7 The Equation $x^6 + y^4 = z^2$

Proposition 14.6.10. *The equation $x^6 + y^4 = z^2$ has no solution in nonzero coprime integers x, y, z.*

Proof. Once again, we use the solution to the dihedral equation. We find that our equation is equivalent to $z = s(s^2 + 3t^2)$, $y^2 = t(3s^2 + t^2)$, $x^2 = s^2 - t^2$ with s and t coprime of opposite parity, or to $z = \pm(2s^3 + t^3)$, $y^2 = 2s^3 - t^3$, $x^2 = 2ts$ with s and t coprime and t odd. We consider both cases separately.

Case 1: $2 \nmid x$. This corresponds to the first parametrization. Since $x^2 + t^2 = s^2$ and x is odd, there exist coprime u and v of opposite parity such that $x = u^2 - v^2$, $t = 2uv$, $s = \pm(u^2 + v^2)$. The last equation to be solved is thus $y^2 = 2uv(3u^4 + 10u^2v^2 + 3v^4)$. We set $a = u + v$, $b = u - v$, which are both odd, and this gives $2y^2 = (a^2 - b^2)(a^4 + a^2b^2 + b^4)$. We set $Y = y/b^3$, $X = a^2/b^2$ and we obtain the elliptic curve $2Y^2 = (X - 1)(X^2 + X + 1) = X^3 - 1$, whose reduced Weierstrass equation is $y^2 = x^3 - 8$. Once again `mwrank` tells us that this equation has no solutions with $y \neq 0$.

Case 2: $2 \mid x$. This corresponds to the second parametrization. Since t is odd, s is even, and since s and t are coprime the equation $x^2 = 2ts$ gives $t = \varepsilon u^2$, $s = 2\varepsilon v^2$, and $x = \pm 2uv$ with u and v coprime, u odd, and $\varepsilon = \pm 1$. The last equation to be solved is thus $y^2 = \varepsilon(16v^6 - u^6)$. Since u is odd, y is odd, so that $y^2 \equiv 1 \pmod 8$, while $16v^6 - u^6 \equiv -1 \pmod 8$. Thus $\varepsilon = -1$, and we have $y^2 = u^6 - 16v^6$. We have $v \neq 0$, since otherwise $s = 0$, hence $z = 0$, which is impossible. Thus setting $Y = y/v^3$ and $X = u^2/v^2$ gives the elliptic curve $Y^2 = X^3 - 16$ in reduced Weierstrass form, and once again `mwrank` tells us that this curve has no solutions with $y \neq 0$.   □

Putting together the three equations above, we obtain the following:

Corollary 14.6.11. *The equations $\pm x^6 \pm y^4 = z^2$ have no solutions in nonzero coprime integers x, y, z.*

14.6.8 Further Results

The reason for which it has not been difficult to treat the $(4,4,3)$ cases, one of the $(4,4,5)$ cases, and the $(2,4,6)$ cases is that we have always been able to reduce to curves of genus 1 with only a finite number of rational points. In only one case, we had a curve of genus 1 with infinitely many rational points, but we were able to bypass it by using additional information given by the elliptic parametrizations. Unfortunately, in other hyperbolic cases, when reducing to finding rational points on curves, some of these curves will have infinitely many rational points, and some will be of genus greater than or equal to 2, and our knowledge of algorithmic methods for finding all rational points on such curves is much smaller. One of the only general methods, due to Chabauty, unfortunately works only in certain cases; see Chapter 13. In other cases, such as FLT itself, one can also use the method of Ribet–Wiles for finding all the solutions. Thus, we give a brief survey of known results. For equations with fixed small exponents (p, q, r), one method is to find covering curves for the solutions. These curves may be of genus 1, as we have seen in the $(4,4,3)$, $(4,4,5)$, and $(2,4,6)$ examples, but are in general of higher genus. We summarize below the known results, including the highest genus that is necessary and the name of the authors. Recall that we consider only nonzero coprime solutions.

Equation	Solutions	Genus	Author(s)
$\pm x^6 \pm y^4 = z^2$	none	1	Bruin
$x^2 + y^4 = z^5$	none	2	Bruin
$x^2 - y^4 = z^5$	$(\pm 7, \pm 3, -2)$, $(\pm 122, \pm 11, 3)$	2	Bruin
$x^2 + y^8 = z^3$	$(\pm 1549034, \pm 33, 15613)$	2	Bruin
$x^2 - y^8 = z^3$	$(\pm 3, \pm 1, 2)$, $(\pm 30042907, \pm 43, 96222)$	2	Bruin
$x^2 + y^3 = z^7$	$(\pm 3, -2, 1)$, $(\pm 71, -17, 2)$, $(\pm 2213459, 1414, 65)$, $(\pm 15312283, 9262, 113)$, $(\pm 21063928, -76271, 17)$	3	Poonen–Schaefer–Stoll
$x^2 + y^3 = z^9$	$(\pm 3, -2, 1)$, $(\pm 13, 7, 2)$	3	Bruin

The results of N. Bruin can be found in [Bru1], [Bru2], [Bru3], and [Bru4]. The recent $(2, 3, 7)$ result (see [Poo-Sch-Sto]) deserves special mention. Using

Galois representation techniques and level lowering à la Ribet–Wiles (see Chapter 15), the authors show that solutions come from rational points on twists of the modular curve $X(7)$ that come from a finite list of elliptic curves, and this leads to finding the rational points satisfying congruence conditions modulo 2 and 3 on precisely 10 curves of genus 3 defined over $\mathbb{Q}$, which over $\mathbb{C}$ are all isomorphic to the so-called Klein quartic curve, whose projective equation is $x^3y + y^3z + z^3x = 0$.

Using known techniques it is possible to find the rational points satisfying the congruence conditions on 9 of the 10 curves, leading to the given solutions. To prove that the tenth curve does not have any rational point is more difficult, but has been achieved by the authors. It is interesting to note that the large solutions for the $(2, 3, 7)$ come from extremely small solutions on the twisted Klein curves.

Once again we note that the large number of solutions in this case is (heuristically) due to the fact that $\chi = 1/p + 1/q + 1/r - 1 = -1/42$ is as close to zero as it can be in the hyperbolic case.

All the other results on the super-Fermat equation (including the original Fermat equation itself) have also been proved using Galois representation techniques. We refer to Chapter 15, written by S. Siksek, for a black-box explanation of this method, and we also refer to the excellent papers [Kra2] and [Ben2] for surveys, details, and references. Among the results obtained to date (2007) using this method we cite the following:

Equation	Conditions	Author(s)
$x^n + y^n = z^n$	$3 \leqslant n$	Ribet–Taylor–Wiles
$x^3 + y^3 = z^n$	$17 \leqslant n \leqslant 10000$ or z even	Kraus
$x^5 + y^5 = z^n$	$3 \leqslant n$ and z even	Darmon–Kraus
$x^n + y^n = z^2$	$4 \leqslant n$	Darmon–Merel, Poonen
$x^n + y^n = z^3$	$3 \leqslant n$	Darmon–Merel, Poonen
$x^4 + y^n = z^4$	$2 \leqslant n$	Darmon
$x^2 + y^4 = z^n$	$211 \leqslant n$	Ellenberg, Ramakrishnan
$x^2 - y^4 = z^n$	$5 \leqslant n$	Bennett–Skinner
$x^2 + y^{2n} = z^3$	$11 \leqslant n \leqslant 10^7$ and $n \neq 31$, or $3 \nmid y$	Chen
$x^4 + y^{2n} = z^3$	$2 \leqslant n$	Bennett–Chen
$x^2 + y^{4n} = z^3$	$2 \leqslant n$	Bennett–Chen
$x^{2n} + y^{2n} = z^5$	$2 \leqslant n$	Bennett

14.7 Applications of Mason's Theorem

It is interesting to note that most of the important Diophantine problems that we have met in this book, such as Fermat's last theorem, Catalan's equation, the super-Fermat equation, and others, have a very simple answer if we look at them in the context of *polynomials*, in other words if we look for polynomial, as opposed to rational or integral, solutions. This essentially follows from a single, elementary, result, due to Mason. It should be emphasized that these results have no use whatsoever for the initial Diophantine equations to be solved over $\mathbb{Q}$. Nonetheless I believe that they have a place in this book.

14.7.1 Mason's Theorem

The reader should compare the following with Definition 14.6.3 and Conjecture 14.6.4.

Definition 14.7.1. *For a nonzero polynomial P in one variable, we define* $\mathrm{rad}(P)$ *to be the monic polynomial with no multiple roots having the same roots as P, in other words* $\mathrm{rad}(P) = \prod_{P(\alpha)=0}(X - \alpha) = P/\gcd(P, P')$.

Proposition 14.7.2 (Mason). *Let A, B, C be pairwise coprime polynomials in one variable, not all constant and such that $A + B + C = 0$. Then*

$$\max(\deg(A), \deg(B), \deg(C)) \leqslant \deg(\mathrm{rad}(ABC)) - 1 \,.$$

In other words, the *abc* conjecture is true for polynomials.

Proof. Let $f = A/C$ and $g = B/C$, so that f and g are rational functions such that $f + g + 1 = 0$. Note that g is not constant; otherwise, f would also be constant and A, B, and C would be proportional hence constant since they are pairwise coprime. Differentiating, it follows that $f' = -g'$, so

$$\frac{B}{A} = \frac{g}{f} = -\frac{f'/f}{g'/g} \,.$$

If we write

$$A(X) = a \prod_i (X - \alpha_i)^{a_i}, \quad B(X) = b \prod_j (X - \beta_j)^{b_j}, \quad C(X) = c \prod_k (X - \gamma_k)^{c_k}$$

we have

$$\frac{f'}{f}(X) = \sum_i \frac{a_i}{X - \alpha_i} - \sum_k \frac{c_k}{X - \gamma_k} \quad \text{and} \quad \frac{g'}{g}(X) = \sum_j \frac{b_j}{X - \beta_j} - \sum_k \frac{c_k}{X - \gamma_k} \,.$$

Thus if we multiply f'/f and g'/g by $N = \mathrm{rad}(ABC)$ we obtain polynomials, and the degree of these polynomials is less than or equal to $\deg(N) - 1$. From the equality

$$\frac{B}{A} = -\frac{Nf'/f}{Ng'/g}$$

and the fact that A and B are coprime we deduce that B divides Nf'/f and A divides Ng'/g, hence that $\max(\deg(A), \deg(B)) \leqslant \deg(N) - 1$, so $\deg(C) = \deg(-A - B) \leqslant \deg(N) - 1$, proving the proposition. $\square$

14.7.2 Applications

Corollary 14.7.3. *FLT is true for polynomials in one variable that are not all constant, in other words if f, g, and h are nonzero polynomials, not all constant and such that $f^n + g^n = h^n$ then $n \leqslant 2$.*

Proof. Dividing the equation by $\gcd(f, g)^n$ we may assume that f, g, and h are pairwise coprime. Setting $A = f^n$, $B = g^n$, $C = -h^n$ we have $A + B + C = 0$ and $\operatorname{rad}(ABC) \mid fgh$. Thus by the above proposition we have

$$n \max(\deg(f), \deg(g), \deg(h)) \leqslant \deg(fgh) - 1 = \deg(f) + \deg(g) + \deg(h) - 1 .$$

Adding the corresponding inequalities for f, g, and h we obtain

$$n(\deg(f) + \deg(g) + \deg(h)) \leqslant 3(\deg(f) + \deg(g) + \deg(h)) - 3 ,$$

hence $n < 3$ as claimed. □

Note that since we have a two-parameter coprime integer solution to FLT for $n = 2$, a fortiori there exists a solution with polynomials in one variable, for instance $f = 2x$, $g = x^2 - 1$, and $h = x^2 + 1$.

More generally, a similar proof shows that the super-Fermat equation can have solutions only in the elliptic case:

Corollary 14.7.4. *Let p, q, r be integers such that $2 \leqslant p \leqslant q \leqslant r$, and assume that f, g, and h are pairwise coprime polynomials, not all constant and satisfying the super-Fermat equation $f^p + g^q = h^r$. We are then in the elliptic case, in other words $(p, q, r) = (2, 2, r)$ for some $r \geqslant 2$, $(2, 3, 3)$, $(2, 3, 4)$, or $(2, 3, 5)$.*

Proof. Once again we have $A + B + C = 0$ with $A = f^p$, $B = g^q$, and $C = -h^r$, which are pairwise coprime by assumption, and $\operatorname{rad}(ABC) \mid fgh$. If we denote by a, b, and c respectively the degrees of f, g, and h, the above proposition tells us that $\max(pa, qb, rc) \leqslant a + b + c - 1$. Since $p \leqslant q \leqslant r$ we have $p(a + b + c) \leqslant pa + qb + rc \leqslant 3(a + b + c) - 3$; hence as for FLT we deduce that $p < 3$, so that $p = 2$, and the inequality $pa \leqslant a + b + c - 1$ gives $a \leqslant b + c - 1$. If $q = 2$ we are in the dihedral case $(2, 2, r)$. Otherwise, assume that $q \geqslant 3$. Our basic inequality now gives $2a + qb + rc \leqslant 3a + 3b + 3c - 3$; hence since $q \leqslant r$,

$$q(b + c) \leqslant qb + rc \leqslant a + 3b + 3c - 3 \leqslant 4(b + c) - 4 ,$$

so that $q < 4$, hence $q = 3$. Finally, for $p = 2$ and $q = 3$ the inequality for qb gives $2b \leqslant a + c - 1 \leqslant b + 2c - 2$, so that $b \leqslant 2c - 2$, hence $a \leqslant b + c - 1 \leqslant 3c - 3$. Thus $rc \leqslant a + b + c - 1 \leqslant 6(c - 1)$, hence $r < 6$, so $r = 3$, 4, or 5, proving the corollary. □

Note that we have seen in Sections 14.2, 14.3, 14.4, and 14.5.2, that in all the elliptic cases we have a two-variable parametrization, so in the given cases of the corollary, solutions do indeed exist.

14.8 Exercises for Chapter 14

1. Show that, as claimed in the text, in Theorem 14.5.1 we can choose $f_G(s,t) = st(s^{10} - 11s^5t^5 - t^{10})$ in the case of the regular icosahedron.

2. (M. Stoll.) Assume that the following weaker form of the abc Conjecture 14.6.4 is valid: there exists $\varepsilon < 1/5$ such that for all nonzero pairwise coprime integers a, b, c with $a + b + c = 0$ we have $\max(|a|, |b|, |c|) = O(\mathrm{rad}(abc)^{1+\varepsilon})$.

(a) Prove that there are in total only finitely many solutions to the super-Fermat equations with $1/p + 1/q + 1/r < 5/6 + \delta$ for some $\delta > 0$ depending on ε.

(b) Deduce from the Darmon–Granville Theorem 14.6.2 that there are in total only finitely many solutions in the hyperbolic case.

3. Let p and q be integers such that $p \geqslant 2$ and $q \geqslant 2$. Using Mason's theorem (Proposition 14.7.2), prove that if f and g are nonconstant coprime polynomials then

$$\deg(f^p - g^q) \geqslant (p - 1 - p/q)\deg(f) + 1 \ .$$

The special case $\deg(f^3 - g^2) \geqslant (\deg(f)/2) + 1$ is due to Davenport and is the polynomial analogue of *Hall's conjecture*, which states that if a, b are coprime positive integers different from 1 then for every $\varepsilon > 0$ we have $|a^3 - b^2| \geqslant a^{1/2 - \varepsilon}$ except for finitely many (a, b).

4. Let p, q, and r be strictly positive integers. Show that there do not exist any solutions to the *negative* super-Fermat equation $x^{-p} + y^{-q} = z^{-r}$ with x, y, and z pairwise coprime.

15. The Modular Approach to Diophantine Equations

By Samir Siksek

15.1 Newforms

15.1.1 Introduction and Necessary Software Tools

One of the most powerful tools in the study of Diophantine equations, extensively developed in the past few years, has been the use of special types of elliptic curves *associated* with possible solutions of the Diophantine equation (but not considered as Diophantine equations in themselves), now called Hellegouarch–Frey curves, or simply Frey curves. The three very deep theorems that are necessary to use these tools are on the one hand the Taniyama–Shimura–Weil conjecture, now proved thanks to Wiles and successors (Theorems 8.1.4 and 8.1.5), Ribet's *level-lowering* Theorem 15.2.5, and Mazur's Theorem 15.2.6; see below. The aim of this chapter is to explain these tools so that they can be used by the reader as a black box, in particular with a minimal knowledge of the underlying (beautiful) mathematics. Since the first great success of this method was the complete proof of Fermat's last theorem in 1995, it is not surprising that the method is difficult, and requires more prerequisites than assumed in the rest of this book. However, considering its importance, we have decided to include it as a chapter in the last part of this book. We will see for instance that FLT is the *easiest* case to which the method applies.

Apart from the black box that we will explain in detail, the reader absolutely needs to have at his disposal a number of software tools, which are all available in the `magma` computer algebra system, and some of which are available in `Pari/GP`. First he will need to be able to compute the minimal model, the conductor, and the minimal discriminant of an elliptic curve defined over $\mathbb{Q}$. Given such an elliptic curve E, he will also need to compute the coefficients $a_\ell(E)$ such that $|E(\mathbb{F}_\ell)| = \ell + 1 - a_\ell(E)$ for reasonable values of ℓ at which E has good reduction. All of these functions are available in both the above-mentioned systems. He will also need to be able to compute the list of newforms at a given level, together with the totally real number fields that they generate (see below for the definitions). This is available in `magma` in a package written by W. Stein.

15.1.2 Newforms

The fundamental objects that we will need to use are *normalized newforms* of weight 2 without character on $\Gamma_0(N)$, which we simply abbreviate to newforms of level N. Although a newform is a modular form for a certain subgroup of $\mathrm{SL}_2(\mathbb{Z})$ that is an eigenfunction of important operators called *Hecke operators*, in keeping with the black-box principle we need to know only the following.

– A newform is a q-expansion

$$f = q + \sum_{n \geqslant 2} c_n q^n$$

with no term in q^0 and normalized so that the coefficient of q is equal to 1. The c_n will be called the *Fourier coefficients* of f.
– The field $K = \mathbb{Q}(c_2, c_3, \dots)$ obtained by adjoining to $\mathbb{Q}$ the Fourier coefficients of f is a *finite and totally real* extension of $\mathbb{Q}$, in other words is a totally real number field.
– The Fourier coefficients c_n are algebraic integers, in other words they belong to $\mathbb{Z}_K$.
– Let L be the Galois closure of K. If f is a newform and σ is any element of $\mathrm{Gal}(L/\mathbb{Q})$ then $q + \sum_{n \geqslant 2} \sigma(c_n) q^n$ is again a newform, denoted by $\sigma(f)$ and called a conjugate of f. We will usually identify a newform with all of its conjugates.
– The Ramanujan conjecture, proved by Deligne in the general case, but much easier in the weight 2 case: If ℓ is a prime we have $|c_\ell| \leqslant 2\ell^{1/2}$. Since this is also true for the conjugates of f, we have in fact $|\sigma(c_\ell)| \leqslant 2\ell^{1/2}$ for all σ.
– For a given level N, the number of newforms (up to conjugacy or not) is finite. For the sake of completeness, we give the formula for the number of newforms (the number of newforms up to conjugacy does not have any known closed form). This formula appears in several places; see for instance [Hal-Kra1] and [Mart].

Proposition 15.1.1. *We define five arithmetic functions $A_i(N)$ for $1 \leqslant i \leqslant 5$ by asking that they be multiplicative, and that their values on prime powers p^k be given as follows:*

(1) $A_1(p) = -1$, $A_1(p^k) = 0$ *when* $k \geqslant 2$.
(2) $A_2(p) = p(1-1/p)$, $A_2(p^2) = p^2(1-1/p-1/p^2)$, $A_2(p^k) = p^k(1-1/p)(1-1/p^2)$ *when* $k \geqslant 3$.
(3) $A_3(p) = \left(\frac{-4}{p}\right) - 1$, $A_3(p^2) = -\left(\frac{-4}{p}\right)$, $A_3(p^k) = 0$ *when* $k \geqslant 3$ *for* $p \neq 2$, *while* $A_3(2) = A_3(2^2) = -1$, $A_3(2^3) = 1$, *and* $A_3(2^k) = 0$ *for* $k \geqslant 4$.
(4) $A_4(p) = \left(\frac{-3}{p}\right) - 1$, $A_4(p^2) = -\left(\frac{-3}{p}\right)$, $A_4(p^k) = 0$ *when* $k \geqslant 3$ *for* $p \neq 3$, *while* $A_4(3) = A_4(3^2) = -1$, $A_4(3^3) = 1$, *and* $A_4(3^k) = 0$ *for* $k \geqslant 4$.

(5) $A_5(p^2) = p(1 - 2/p)$, $A_5(p^{2k}) = p^k(1 - 1/p)^2$ when $k \geqslant 2$, and $A_5(p^{2k-1}) = 0$ for $k \geqslant 1$.

The number of newforms of level N (counting conjugate ones as distinct) is equal to

$$A_1 + \frac{A_2}{12} - \frac{A_3}{4} - \frac{A_4}{3} - \frac{A_5}{2}.$$

Corollary 15.1.2. There are no newforms for levels

$$1, 2, 3, 4, 5, 6, 7, 8, 9, 10, 12, 13, 16, 18, 22, 25, 28, 60.$$

For all other levels there are newforms.

Proof. Follows from an immediate (computer-aided) computation from the proposition. $\qquad\square$

Example. If $N = 110$ the formula shows that there are five newforms. In fact there are three newforms for which $K = \mathbb{Q}$, hence alone in their conjugacy class, and one conjugacy class of newforms for which $K = \mathbb{Q}(\sqrt{33})$, which gives the two other conjugate newforms. Explicitly we have

$$f_1 = q - q^2 + q^3 + q^4 - q^5 - q^6 + 5q^7 + \cdots,$$
$$f_2 = q + q^2 + q^3 + q^4 - q^5 + q^6 - q^7 + \cdots,$$
$$f_3 = q + q^2 - q^3 + q^4 + q^5 - q^6 + 3q^7 + \cdots,$$
$$f_4 = q - q^2 + \theta q^3 + q^4 + q^5 - \theta q^6 - \theta q^7 + \cdots,$$
$$f_5 = \sigma(f_4),$$

where $\theta = (-1 + \sqrt{33})/2$ and σ is the nontrivial automorphism of $\mathbb{Q}(\sqrt{33})$.

15.1.3 Rational Newforms and Elliptic Curves

We will say that a newform f is *rational* when the field K associated with f is equal to $\mathbb{Q}$, in other words if all the Fourier coefficients of f are in $\mathbb{Z}$, such as the first three forms in the above example. These will be particularly important for us. We recall the modularity theorem for elliptic curves, proved by Wiles and successors (formerly the Taniyama–Shimura–Weil conjecture).

Theorem 15.1.3 (The Modularity Theorem for Elliptic Curves). *Let $N \geqslant 1$ be an integer. There is a one-to-one correspondence $f \mapsto E_f$ between rational newforms of level N and isogeny classes of elliptic curves E defined over $\mathbb{Q}$ and of conductor equal to N. Under this correspondence, for all primes $\ell \nmid N$ we have $c_\ell = a_\ell(E_f)$, where c_ℓ is the ℓth Fourier coefficient of f and $a_\ell(E_f) = \ell + 1 - |E_f(\mathbb{F}_\ell)|$ as in Section 7.3.4.*

Remarks. (1) The correspondence $f \mapsto E_f$ was found by Shimura, and is not difficult. The fact that this correspondence is surjective, in other words that any elliptic curve over $\mathbb{Q}$ comes from a newform is much deeper, and was first proved by Wiles and Taylor–Wiles in [Wil], [Tay-Wil] for squarefree N, and in complete generality by Breuil, Conrad, Diamond, and Taylor in [BCDT].

(2) Note that with the definition that we have given in Section 7.3.4 we have $c_\ell = a_\ell(E_f)$ for *all* primes ℓ, including those dividing the conductor, but we will not need this.

To take again the example given above at level 110, we see from the theorem that up to isogeny there exists exactly three elliptic curves of conductor 110 defined over $\mathbb{Q}$. An immediate computation shows that f_1, f_2, f_3 correspond respectively to the curves denoted by $110C1$, $110B1$, and $110A1$ in the tables of Cremona [Cre2].

The above theorem, which is among one of the crowning achievements of number theory of the second half of the twentieth century (together with Deligne's proof of the Weil conjectures and of the Ramanujan conjecture, and with Faltings's proof of the Mordell and Shafarevich conjectures), is needed to go back and forth with ease between rational newforms and elliptic curves. However, there is really no need to understand what is going on in detail: we simply remember that with each (isogeny class of) elliptic curve(s) of conductor N is associated a rational newform of level N, and conversely. This is not at all the case with the second essential tool that we need, Ribet's lowering theorem, for which we need to understand a little more what is going on.

15.2 Ribet's Level-Lowering Theorem

We keep the notation c_ℓ for the Fourier coefficients of a newform, and $a_\ell(E) = \ell + 1 - |E(\mathbb{F}_\ell)|$.

15.2.1 Definition of "Arises From"

Definition 15.2.1. *Let E be an elliptic curve over $\mathbb{Q}$ of conductor N, let f be a newform of level N' not necessarily equal to N, and let K be the number field generated by the Fourier coefficients of f. We will say that E arises modulo p from f, and write $E \sim_p f$, if there exists a prime ideal $\mathfrak{p}$ of K above p such that $c_\ell \equiv a_\ell(E) \pmod{\mathfrak{p}}$ for all but finitely many prime numbers ℓ.*

For instance, if $E = E_f$ is the elliptic curve of level N' corresponding to a rational newform f then $c_\ell = a_\ell(E)$ for $\ell \nmid N'$, so that $E \sim_p f$ for all primes p. This will be an uninteresting case. On the other hand, if E is an elliptic curve of conductor N such that $E \sim_p f$ with f a *rational* newform of level N', then by the modularity theorem above we know that f corresponds to an elliptic curve $F = E_f$ defined over $\mathbb{Q}$ of conductor N', and we will also write $E \sim_p F$.

It is not difficult to prove that the above definition implies the following important properties, which makes it more precise.

Proposition 15.2.2. *Assume that $E \sim_p f$. There exists a prime ideal $\mathfrak{p}$ of K above p such that for all prime numbers ℓ we have:*

(1) *If $\ell \nmid pNN'$ then $a_\ell(E) \equiv c_\ell \pmod{\mathfrak{p}}$.*
(2) *If $\ell \| N$ but $\ell \nmid pN'$ then $c_\ell \equiv \pm(\ell + 1) \pmod{\mathfrak{p}}$.*

There is, however, a slight but essential refinement of this proposition due to Kraus–Oesterlé [Kra-Oes], which is the final form of the definition of $\sim_p$ that we will use:

Proposition 15.2.3. *Let E and F be elliptic curves over $\mathbb{Q}$ with respective conductors N and N', and assume that $E \sim_p F$ as defined above. Then for all primes numbers ℓ we have:*

(1) *If $\ell \nmid NN'$ then $a_\ell(E) \equiv a_\ell(F) \pmod{p}$.*
(2) *If $\ell \| N$ but $\ell \nmid N'$ then $a_\ell(F) \equiv \pm(\ell + 1) \pmod{p}$.*

The crucial refinement of this proposition is that we have removed the assumption that $\ell \neq p$. This will be important in applications since p will be an unknown exponent in the equations that we want to solve, and it would be awkward to have conditions depending on p.

Note that the condition $\ell \nmid NN'$ means that the elliptic curves E and F have good reduction modulo ℓ, in other words that their reduction is nonsingular. The condition that $\ell \| N$ and $\ell \nmid N'$ means that F has good reduction, and it is easily shown that this means that E has multiplicative reduction at ℓ, see Section 7.1.4, but we will not need this interpretation.

15.2.2 Ribet's Level-Lowering Theorem

Let E be an elliptic curve defined over $\mathbb{Q}$. For any finite place p of $\mathbb{Q}$ we can find a generalized Weierstrass equation that is integral at p and whose discriminant has minimal p-adic valuation. Since $\mathbb{Q}$ has class number 1, it is possible to glue these local equations and obtain a global integral generalized Weierstrass equation whose discriminant $\Delta_{\min}$ has minimal p-adic valuation for all prime numbers p. This equation is not unique (it is unique up to simple changes of coordinates), but $\Delta_{\min}$ is unique, and it will be called the minimal discriminant of E. On the other hand, E has a conductor N that can be explicitly computed using an algorithm due to Tate (see for example Algorithms 7.5.1 and 7.5.2 of [Coh0], and [Pap]), and N and $\Delta_{\min}$ are related by the fact that $N \mid \Delta_{\min}$ and that N and $\Delta_{\min}$ have the same prime divisors, the primes of bad reduction.

Definition 15.2.4. *Keep the above notation and let p be a prime number. We define N_p by the formula*[1]

$$N_p = N \Big/ \prod_{\substack{q \| N \\ p | v_q(\Delta_{\min})}} q \ ;$$

in other words, N_p is equal to N divided by the product of all prime numbers q such that $v_q(N) = 1$ and $p \mid v_q(\Delta_{\min})$.

We emphasize that the $\Delta_{\min}$ occurring in the definition of N_p must be the minimal discriminant.

We can now state a simplified special case of Ribet's level-lowering theorem that will be sufficient for our applications (see [Rib1] for the full statement).

Theorem 15.2.5 (Ribet's Level-Lowering Theorem). *Let E be an elliptic curve defined over $\mathbb{Q}$ and let $p \geqslant 5$ be a prime number. Assume that there does not exist a p-isogeny (i.e., of degree p) defined over $\mathbb{Q}$ from E to some other elliptic curve, and let N_p be as above. There exists a newform f of level N_p such that $E \sim_p f$.*

As mentioned, Ribet's theorem is much more general than this, but the present statement is sufficient. In addition, in Ribet's general theorem there is a modularity assumption, but since we restrict to the case of elliptic curves this assumption is automatically satisfied thanks to the modularity theorem.

Example. Let E be the elliptic curve with minimal Weierstrass equation

$$y^2 = x^3 - x^2 - 77x + 330 \ ,$$

referenced as $132B1$ in [Cre2]. We compute that the minimal discriminant and the conductor are respectively

---

[1] **Highbrow remark, to be omitted on first reading.** This N_p is not always the same as the Serre conductor. If N_p' denotes the Serre conductor then $N_p' \mid N_p$, and N_p/N_p' is a power of p. More precisely,

$$N_p = \begin{cases} N_p' & \text{if } E \text{ has good reduction at } p \ , \\ pN_p' & \text{if } E \text{ has multiplicative reduction at } p \ , \\ p^2 N_p' & \text{if } E \text{ has additive reduction at } p \ . \end{cases}$$

Ribet's theorem allows us to obtain a newform of level N_p' and weight $k_p \geqslant 2$ (where k_p is the Serre weight). Since we have limited ourselves to weight-2 newforms, it turns out that we obtain a newform of level N_p and not N_p'. To understand why we have chosen to restrict to weight 2, note that later p will be an unknown exponent in some Diophantine equation. Often we will not know whether p is a prime of good reduction. The restriction that we have made allows us to deal with all these cases uniformly, by giving a unique level and weight regardless of whether E has good, multiplicative, or additive reduction at p.

$$\Delta_{\min} = 2^4 \cdot 3^{10} \cdot 11 \quad \text{and} \quad N = 2^2 \cdot 3 \cdot 11 \,.$$

Using [Cre2] we see that the only isogeny that the curve has is a 2-isogeny, so we may apply Ribet's theorem with $p = 5$. We find that $N_p = 2^2 \cdot 11 = 44$, so Ribet's theorem asserts the existence of a newform f at level 44 such that $E \sim_5 f$. The formula given for the number of newforms shows that there is a single one at level 44, necessarily rational, and Cremona's tables show that it corresponds to the elliptic curve $F = 44A1$ with equation

$$y^2 = x^3 + x^2 + 3x - 1 \,,$$

so that $E \sim_5 F$. In order for the reader to understand what is expected from Proposition 15.2.3 we give the values of $a_\ell(E)$ and $a_\ell(F)$ for $\ell \leqslant 37$.

ℓ	2	3	5	7	11	13	17	19	23	29	31	37
$a_\ell(E)$	0	−1	2	2	−1	6	−4	−2	−8	0	0	−6
$a_\ell(F)$	0	1	−3	2	−1	−4	6	8	−3	0	5	−1

15.2.3 Absence of Isogenies

There are a number of technical difficulties that must be solved in order to be able to apply Ribet's theorem in practice. The most important one is the restriction that E should not have any p-isogenies defined over $\mathbb{Q}$ (for simplicity we will say that E has no p-isogenies), in other words that there should be no subgroup of order p of E that is stable under conjugation (see Definition 8.4.1). This is not always easy to check, but there are several results that help us in doing so. We give here two of the most useful.

Theorem 15.2.6 (Mazur [Maz]). *Let E be an elliptic curve defined over $\mathbb{Q}$ of conductor N. Then E does not have any p-isogeny if at least one of the following conditions holds:*

(1) *$p \geqslant 17$ and $j(E) \notin \mathbb{Z}[1/2]$.*
(2) *$p \geqslant 11$ and N is squarefree.*
(3) *$p \geqslant 5$, N is squarefree, and $4 \mid |E_t(\mathbb{Q})|$, this last condition meaning that $E(\mathbb{Q})$ has full 2-torsion.*

Theorem 15.2.7 (Diamond–Kramer [Dia-Kra]). *Let E be an elliptic curve defined over $\mathbb{Q}$ of conductor N. If $v_2(N) = 3$, 5, or 7, then E does not have any p-isogeny for p an odd prime.*

Example. Let E be an elliptic curve defined over $\mathbb{Q}$ of conductor N and minimal discriminant $\Delta_{\min}$. We have the following theorem, conjectured by Brumer–Kramer in [Bru-Kra] and proved by Serre in [Ser3] assuming a hypothesis that has now been removed thanks to the theorems of Ribet and Wiles.

Theorem 15.2.8. *Assume that N is squarefree. If $\Delta_{\min}$ is a pth power for some prime p then $p \leqslant 5$ and E has a rational point of order p.*

Proof. We prove only the statement $p \leqslant 7$ using the tools that we have introduced. Since N is squarefree and $\Delta_{\min}$ is a pth power, the definition shows that $N_p = 1$. Assume first that $p \geqslant 11$. Since N is squarefree, the second condition of Mazur's theorem shows that E does not have any p-isogeny. We can thus apply Ribet's theorem, which tells us that $E \sim_p f$ for a newform f of level 1. Since there are no such newforms, we have a contradiction showing that $p \leqslant 7$. With some extra work Serre shows that E has a rational point of order p. In addition one can prove that there are no curves with N squarefree whose discriminant is a power of 7 (see [Mes-Oes]). $\square$

Remark. If E has no p-isogenies then Ribet's theorem implies that $E \sim_p f$ for some newform f of level N_p. At that level there may be rational newforms, but also nonrational newforms defined over number fields of relatively large degree. In fact, the following proposition shows that the degree is unbounded:

Proposition 15.2.9. *An elliptic curve defined over $\mathbb{Q}$ can arise from a newform whose field of definition K has arbitrarily large degree.*

Proof. Let $p \geqslant 5$ be a prime, set $L = 2^{p+4} + 1$, and let E be the elliptic curve with equation

$$Y^2 = X(X+1)(X - 2^{p+4}) \,.$$

Using Tate's algorithm we easily compute that the minimal discriminant and conductor are given by

$$\Delta_{\min} = 2^{2p} L^2, \quad N = 2\,\mathrm{rad}(L)$$

(see Definition 14.6.3). From Mazur's Theorem 15.2.6 we know that E has no p-isogenies, so we can apply Ribet's theorem, which tells us that $E \sim_p f$ for some newform at level N_p whose field of definition is some number field K. We cannot compute N_p, but since L is odd, $2\|N$, and $v_2(\Delta_{\min}) = 2p$, it follows from the definition of N_p that N_p is *odd*. Thus by Proposition 15.2.2 (2) applied to $\ell = 2$ we deduce that $p \mid \mathcal{N}_{K/\mathbb{Q}}(3 \pm c_2)$, where we denote by c_ℓ the Fourier coefficients of the newform f. However, we know that all the conjugates of c_2 in $\overline{\mathbb{Q}}$ are bounded in absolute value by $2\sqrt{2}$, and that c_2 is an algebraic integer. It follows that $p \leqslant (3 + 2\sqrt{2})^{[K:\mathbb{Q}]} < 6^{[K:\mathbb{Q}]}$, hence that

$$[K : \mathbb{Q}] > \frac{\log(p)}{\log(6)} \,,$$

proving the proposition. $\square$

15.2.4 How to use Ribet's Theorem

The general strategy for applying to a Diophantine equation the tools that we have introduced is the following. We assume that it has a solution, and to such a solution we associate if possible in some way an elliptic curve, called a Hellegouarch–Frey curve, or simply a Frey curve.[2] The key properties that a "Frey curve" E must have are the following:

- The coefficients of E depend on the solution of the Diophantine equation.
- The minimal discriminant $\Delta_{\min}$ of E can be written in the form $\Delta_{\min} = C \cdot D^p$, where D depends on the solution of the Diophantine equation, p is an unknown prime occurring as an exponent in the Diophantine equation, and most importantly C *does not depend on the solution* of the Diophantine equation, but only on the equation itself.
- If ℓ is a prime dividing D then E has multiplicative reduction at ℓ, in other words $v_\ell(N) = 1$, where N is the conductor of E.

The conductor N will be divisible by the primes dividing C and D, but because of the last condition above, the primes dividing D will be removed when computing N_p (see Definition 15.2.4); in other words, N_p is a divisor of N that is divisible only by primes dividing C, hence depending only on the equation. Without knowing the solutions to the Diophantine equation we can thus easily write a finite number of possibilities for N_p depending only on the equation. Using Ribet's theorem we will then be able to list a finite set of newforms f such that $E \sim_p f$.

From then on we have to work more. Knowing the newform gives local information on the elliptic curve E, and since the equation of E has coefficients that depend on the solution to the Diophantine equation, we may obtain useful information about these solutions, including of course the fact that they do not exist.

The rest of this chapter is devoted to giving concrete examples of how Ribet's theorem is used to obtain information about solutions to certain Diophantine equations, and occasionally to solve them.

15.3 Fermat's Last Theorem and Similar Equations

The foremost example, for which in fact this whole machinery was developed and successfully applied, is FLT. As we will see, FLT is in fact one of the *simplest* applications, because no newform exists corresponding to a solution, so we do not need to do the extra work mentioned above.

---

[2] Y. Hellegouarch was the first to have this idea, but G. Frey realized that it would become very fruitful once the modularity theorem was proved.

15.3.1 A Generalization of FLT

Thus, since FLT is too easy (!!!), we will solve a more general problem:

Theorem 15.3.1. *Let $p \geqslant 5$ be a prime. The equation*

$$x^p + 2^r y^p + z^p = 0$$

has no solution with $xyz \neq 0$ and x, y, and z pairwise coprime, except when $r = 1$, for which it has the solutions $(x, y, z) = \pm(1, -1, 1)$.

This theorem is the celebrated theorem of Wiles for $r = 0$, and it was proved by Ribet in [Rib2] for $r \geqslant 2$, and by Darmon–Merel in [Dar-Mer] for $r = 1$.

Proof. Assume that (x, y, z) is a nontrivial primitive solution to our equation (in other words that $xyz \neq 0$, and that x, y, and z are pairwise coprime). It is clear that without loss of generality we may assume that $2^r y^p \equiv 0$ (mod 2) (this is automatic if $r \geqslant 1$, and is obtained by a suitable permutation of x, y, and z if $r = 0$ since at least one must be even), and that $x^p \equiv -1$ (mod 4) (x must be odd; otherwise, x and z will both be even hence not coprime, and we change if necessary (x, y, z) into $(-x, -y, -z)$, since p is odd). We associate to this solution the Frey elliptic curve E with equation

$$Y^2 = X(X - x^p)(X + 2^r y^p) \,.$$

It is easily checked that the invariants are given by

$$c_4 = 16(z^{2p} - 2^r (xy)^p), \quad \mathrm{disc}(E) = 2^{2r+4}(xyz)^{2p}, \quad j(E) = \frac{(z^{2p} - 2^r x^p y^p)^3}{2^{2r-8}(xyz)^{2p}} \,.$$

Note that $\mathrm{disc}(E)$ is not necessarily the minimal discriminant. In fact, using Tate's algorithm we easily compute that the minimal discriminant and the conductor are given by the following formulas:

$$\Delta_{\min} = \begin{cases} 2^{2r+4}(xyz)^{2p} & \text{if } 16 \nmid 2^r y^p \,, \\ 2^{2r-8}(xyz)^{2p} & \text{if } 16 \mid 2^r y^p \,, \end{cases}$$

and

$$N = \begin{cases} 2\,\mathrm{rad}_2(xyz) & \text{if } r \geqslant 5 \text{ or } y \text{ is even} \,, \\ \mathrm{rad}_2(xyz) & \text{if } r = 4 \text{ and } y \text{ is odd} \,, \\ 8\,\mathrm{rad}_2(xyz) & \text{if } r = 2 \text{ or } r = 3 \text{ and } y \text{ is odd} \,, \\ 32\,\mathrm{rad}_2(xyz) & \text{if } r = 1 \text{ and } y \text{ is odd} \,, \end{cases}$$

where for an integer X and a prime q we define $\mathrm{rad}_q(X)$ by the formula

$$\mathrm{rad}_q(X) = \prod_{\substack{\ell \mid X, \ \ell \text{ prime} \\ \ell \neq q}} \ell$$

(see also Definition 14.6.3).

Applying Definition 15.2.4 we find that

$$
N_p = \begin{cases}
2 & \text{if } r = 0 \text{ or } r \geqslant 5 \text{ and } r \not\equiv 4 \pmod{p}, \\
1 & \text{if } r \equiv 4 \pmod{p}, \\
2 & \text{if } 1 \leqslant r \leqslant 3 \text{ and } y \text{ is even}, \\
8 & \text{if } r = 2 \text{ or } r = 3 \text{ and } y \text{ is odd}, \\
32 & \text{if } r = 1 \text{ and } y \text{ is odd}.
\end{cases}
$$

Before applying Ribet's level-lowering theorem we must ensure that E has no p-isogenies. However, note that by construction the Frey curve has full 2-torsion, so that $4 \mid |E(\mathbb{Q})_{\text{tors}}|$. Thus, if N is squarefree Mazur's Theorem 15.2.6 tells us that E does not have p-isogenies (since $p \geqslant 5$ is prime). From the formulas given for N we see that if N is not squarefree then $v_2(N) = 3$ or $v_2(N) = 5$ (since by definition $\text{rad}_2(X)$ is squarefree and odd). We may therefore apply Diamond–Kramer's Theorem 15.2.7, which shows that there are no p-isogenies in this case also.

The hypotheses being satisfied, we may therefore apply Ribet's level-lowering Theorem 15.2.5. This theorem tells us that there exists a newform of level N_p such that $E \sim_p f$. Now by Corollary 15.1.2 there do not exist newforms at levels 1, 2, and 8, so we deduce that $N_p = 32$, hence that $r = 1$ and y is odd. Note in passing that we have already solved FLT (!). In addition, there do exist newforms in level 32, and this is a *good* (if somewhat annoying) thing since there exists the solution $(1, -1, 1)$ for $r = 1$, so of course this case cannot be eliminated. This is of course a special case of a natural-philosophical remark: it is in general easier to show that a Diophantine equation has no solutions at all than to show that the list of solutions is as given.

To deal with the remaining case, we look at newforms at level 32. Proposition 15.1.1 tells us that there is a single newform, which is therefore necessarily rational, and it is easy to compute using [Cre2] that under the modularity theorem it corresponds to the elliptic curve F with equation

$$
Y^2 = X(X + 1)(X + 2),
$$

referred to as 32A2 in [Cre2]. Note that if we take the known solution $(x, y, z) = (-1, 1, -1)$ to our initial Diophantine equation, the corresponding Frey curve is exactly F.

For the moment the only thing that we know is that $E \sim_p F$, in other words that E arises modulo p from F. We must now perform the additional work mentioned above. Here we are helped by quite a special circumstance, which does not often happen: the curve F has complex multiplication.

15.3.2 E Arises from a Curve with Complex Multiplication

In this case we can use the following theorem.

Theorem 15.3.2. *Let E and F be two elliptic curves defined over $\mathbb{Q}$. Assume that F has complex multiplication by some order in an imaginary quadratic field L, and that p is a prime number such that $E \sim_p F$.*

(1) *If $p = 11$ or $p \geqslant 17$ and p splits in L the conductors of E and F are equal.*

(2) *If $p \geqslant 5$, p is inert in L, and E has a $\mathbb{Q}$-rational subgroup of order 2 or 3, then $j(E) \in \mathbb{Z}[1/p]$.*

Remarks. (1) Part (1) was proved by Halberstadt and Kraus in [Hal-Kra1] as a consequence of work of Momose [Mom], and part (2) was proved by Darmon and Merel in [Dar-Mer].

(2) For other Diophantine applications of this theorem see [Dar-Mer], [Kra2], [Ivo1], and [Ivo2].

(3) In part (2) of the theorem we can in fact say a little more:

Proposition 15.3.3. *With the assumptions of Theorem 15.3.2 (2), assume in addition that $p^2 \nmid N$ and that $p \nmid N'$, where N is the conductor of E and N' that of F. Then in fact $j(E) \in \mathbb{Z}$.*

Proof. By (2) we know that $j(E) \in \mathbb{Z}[1/p]$. Assume by contradiction that $j(E) \notin \mathbb{Z}$, so that the denominator of $j(E)$ is divisible by p. Thus $p \mid N$, and since $p^2 \nmid N$ we have $p \| N$. Since $p \nmid N'$ we can apply Proposition 15.2.3 (2) so that $a_p(F) \equiv \pm(p+1) \pmod{p}$. However, since p is inert in L and F has complex multiplication by an order in L, we have $a_p(F) = 0$, which contradicts the congruence. $\qquad\square$

15.3.3 End of the Proof of Theorem 15.3.1

Using the above theorem we can now complete the proof of Theorem 15.3.1. Recall that we have shown that $r = 1$, that y is odd, and that $E \sim_p F$, where F is the curve with equation $Y^2 = X(X+1)(X+2)$. To simplify we will assume that $p \neq 5$ and $p \neq 13$. The proof of the theorem having been achieved by Dénes in [Den] for $p \leqslant 31$ using classical tools from algebraic number theory, this is not an important restriction.

We note that F has complex multiplication by $\mathbb{Z}[i]$: indeed, setting $X_1 = X + 1$ in the equation gives the isomorphic curve $Y^2 = X_1^3 - X_1$, which is the prototypical example of a curve with CM; see Section 7.3.1 (change (X_1, Y) to $(-X_1, iY)$). We thus consider two cases.

Case 1. $p \equiv 1 \pmod{4}$. In this case p splits in $\mathbb{Z}[i]$, so we deduce from Theorem 15.3.2 (1) (which is applicable since we have excluded $p = 5$ and $p = 13$) that the conductor N of E is equal to that of F, hence to 32. It follows from the formula given above for N that $\mathrm{rad}_2(xyz) = 1$, in other words that x, y, and z are powers of 2. But since x, y, and z are odd we thus have x, y, z equal to ± 1, giving the two solutions $(x, y, z) = \pm(1, -1, 1)$.

Case 2. $p \equiv 3 \pmod 4$. In this case p is inert in $\mathbb{Z}[i]$. In addition the Frey curve E clearly has rational 2-torsion (in fact full 2-torsion), so we may apply Theorem 15.3.2 (2), which tells us that $j(E) \in \mathbb{Z}[1/p]$. Furthermore, from the formula for N and the fact that the conductor N' of F is equal to 32, it follows that $p^2 \nmid N$ and $p \nmid N'$, so by Proposition 15.3.3 we have in fact $j(E) \in \mathbb{Z}$. However, since $j(E) = 64(z^{2p} - 2x^p y^p)^3/(xyz)^{2p}$ and x, y, and z are pairwise coprime, it follows that x, y, and z cannot have any odd prime divisors, hence that they are powers of 2, so we conclude again that the only possibilities are $(x, y, z) = \pm(1, -1, 1)$. $\qquad\square$

15.3.4 The Equation $x^2 = y^p + 2^r z^p$ for $p \geqslant 7$ and $r \geqslant 2$

We treat another equation that also leads to curves with CM. Consider the Diophantine equation

$$x^2 = y^p + 2^r z^p \,,$$

where we assume that $p \geqslant 7$ is prime, and as usual $xyz \neq 0$ and x, y, and z pairwise coprime. We first solve this equation for $r \geqslant 2$ (see [Ben-Ski], [Ivo1], and [Sik]).

Theorem 15.3.4. *The only nonzero pairwise coprime solutions to $x^2 = y^p + 2^r z^p$ for $r \geqslant 2$ and $p \geqslant 7$ prime are for $r = 3$, for which $(x, y, z) = (\pm 3, 1, 1)$ is a solution for all p.*

We will prove a slightly weaker version of this theorem in which we assume that $p \neq 7$ when $r \geqslant 6$, $p \neq 7$, 11, and 13 when $r = 2$, and $p \neq 13$ when $r = 3$. The other cases can be treated similarly but using other theorems, or attacked directly using classical algebraic number theory techniques.

Proof. Changing if necessary x into $-x$ we may assume that $x \equiv 3 \pmod 4$. We consider the Frey curve E with equation

$$Y^2 = X(X^2 + 2xX + y^p) \,.$$

Note that in contrast to the Frey curve considered in the previous section, this curve has nontrivial 2-torsion, but not full 2-torsion.

We compute that the invariants are given by

$$c_4 = 16(4x^2 - 3y^p), \quad \text{disc}(E) = 2^{r+6}(y^2 z)^p, \quad j(E) = \frac{(4x^2 - 3y^p)^3}{2^{r-6}(y^2 z)^p} \,.$$

Using Tate's algorithm we easily compute that the minimal discriminant and the conductor are given by the following formulas:

$$\Delta_{\min} = \begin{cases} 2^{r+6}(y^2 z)^p & \text{if } 64 \nmid 2^r z^p \,, \\ 2^{r-6}(y^2 z)^p & \text{if } 64 \mid 2^r z^p \,, \end{cases}$$

and

$$N = \begin{cases} 2\,\mathrm{rad}_2(yz) & \text{if } r \geqslant 7 \text{ or } z \text{ is even}, \\ \mathrm{rad}_2(yz) & \text{if } r = 6 \text{ and } z \text{ is odd}, \\ 8\,\mathrm{rad}_2(yz) & \text{if } r = 4 \text{ or } r = 5 \text{ and } z \text{ is odd}, \\ 32\,\mathrm{rad}_2(yz) & \text{if } r = 3 \text{ and } z \text{ is odd}, \\ 4\,\mathrm{rad}_2(yz) & \text{if } r = 2 \text{ and } z \equiv 3 \ (\mathrm{mod}\ 4), \\ 16\,\mathrm{rad}_2(yz) & \text{if } r = 2 \text{ and } z \equiv 1 \ (\mathrm{mod}\ 4). \end{cases}$$

As usual we must ensure that E has no p-isogenies for $p \geqslant 7$. In the first two cases for N above, N is squarefree, so Mazur's Theorem 15.2.6 (2) applies, at least for $p \neq 7$, which we assume. Note that since E does not have full 2-torsion we cannot apply (3). In the next two cases we have $v_2(N) = 3$ or 5, so we can apply Theorem 15.2.7. In the final two cases we note that from the formula for $j(E)$ and the fact that x, y, and z are pairwise coprime and odd, $j(E) \in \mathbb{Z}[1/2]$ can occur only if y and z are equal to ± 1, which is not possible for $r = 2$, so Mazur's theorem implies that there are no p-isogenies if $p \geqslant 17$. Thus, using the above-mentioned two theorems we have shown the absence of p-isogenies except perhaps for $p = 7$ if $r \geqslant 6$ or z even, or for $r = 2$, z odd, and $p = 7$, 11, and 13, cases that we exclude for simplicity.

This annoying but essential technical step being done, we are ready to apply Ribet's theorem. Using Definition 15.2.4 we find that

$$N_p = \begin{cases} 2 & \text{if } r \geqslant 7 \text{ and } r \not\equiv 6 \ (\mathrm{mod}\ p), \\ 1 & \text{if } r \equiv 6 \ (\mathrm{mod}\ p), \\ 2 & \text{if } 2 \leqslant r \leqslant 5 \text{ and } z \text{ is even}, \\ 8 & \text{if } r = 4 \text{ or } r = 5 \text{ and } z \text{ is odd}, \\ 32 & \text{if } r = 3 \text{ and } z \text{ is odd}, \\ 4 & \text{if } r = 2 \text{ and } z \equiv 3 \ (\mathrm{mod}\ 4), \\ 16 & \text{if } r = 2 \text{ and } z \equiv 1 \ (\mathrm{mod}\ 4). \end{cases}$$

Since there are no newforms in levels 1, 2, 4, 8, and 16, these cases are completely solved, and the only remaining case is $r = 3$ and z odd, which must of course remain since there does exist a solution for $r = 3$. We obtain a newform of level 32, and since there is a single one, it corresponds to the same elliptic curve F with complex multiplication by $\mathbb{Z}[i]$ seen in the preceding section. Once again we consider two cases: if $p \equiv 1 \ (\mathrm{mod}\ 4)$ then p is split in $\mathbb{Q}(i)$, so by Theorem 15.3.2 (1) (which we can apply only if $p \neq 13$) we deduce that $N = 32$, and from the explicit formula for N and the fact that y and z are odd this means that y and z are equal to ± 1, giving as only solutions $(\pm 3, 1, 1)$ (the other three values of (y, z) lead to $x \notin \mathbb{Z}$). If $p \equiv 3 \ (\mathrm{mod}\ 4)$ then p is inert in $\mathbb{Q}(i)$, and E has a rational point of order 2 (even if it does not have full 2-torsion). Thus by Theorem 15.3.2 (2) and Proposition 15.3.3 we deduce that $j(E) \in \mathbb{Z}$, which is impossible unless y and z are equal to ± 1, again giving the known solutions.   □

Remark. I am indebted to A. Kraus for pointing out that the above theorem is false if we assume only that $\gcd(x, y, z) = 1$, as is shown by the example $(x, y, z) = (3 \cdot 2^{(p-3)/2}, 2, 1)$ for $r = p - 3$, see [Ivo1].

15.3.5 The Equation $x^2 = y^p + z^p$ for $p \geqslant 7$

Theorem 15.3.5 (Darmon–Merel [Dar-Mer]**).** *If $p \geqslant 7$ is prime there are no nonzero pairwise coprime solutions to $x^2 = y^p + z^p$.*

We will prove this for $p \neq 13$.

Proof. First note that if z (or y) is even we can write $x^2 = y^p + 2^p(z/2)^p$, which does not have any pairwise coprime solutions for $p \geqslant 7$ by Theorem 15.3.4. We may therefore assume that both y and z are odd, hence that x is even, and since $x^2 \equiv y + z \pmod 4$ we have $y + z \equiv 0 \pmod 4$. Thus, if necessary exchanging y and z and changing x into $-x$ we may assume that $y \equiv 3 \pmod 4$ and $x \equiv 3 \pmod 4$. As in the preceding section, to any possible solution we associate the Frey curve E with equation

$$Y^2 = X(X^2 + 2xX + y^p) \ .$$

We easily compute that the minimal discriminant is given by $\Delta_{\min} = 2^6(y^2 z)^p$, and the conductor is $N = 2^5 \operatorname{rad}(yz)$. By Theorem 15.2.7, E does not have any isogenies of degree p, so we may apply Ribet's level-lowering theorem, which tells us that $E \sim_p f$ for some newform f of level N_p with $N_p = 32$, corresponding once again to the elliptic curve F with complex multiplication by $\mathbb{Z}[i]$. Once again we consider two cases: if $p \equiv 1 \pmod 4$ then p is split in $\mathbb{Q}(i)$; hence by Theorem 15.3.2 (1) (which we can apply only if $p \neq 13$) we deduce that $N = 32$, so that $\operatorname{rad}(yz) = 1$, which means that y and z are equal to ± 1, which is not possible unless $x = 0$, which has been excluded. If $p \equiv 3 \pmod 4$ then p is inert in $\mathbb{Q}(i)$; hence by Theorem 15.3.2 (2), since E has a rational point of order 2, we deduce from Proposition 15.3.3 that $j(E) \in \mathbb{Z}$, which implies y and z equal to ± 1 once again by the explicit formula for $j(E)$ given in the preceding section, again a contradiction. $\square$

To finish this section on the equation $x^2 = y^p + 2^r z^p$, we note that the case $r = 1$ remains unresolved, although it is straightforward to show that in this case there are no pairwise coprime solutions with z *even*, see [Ivo1] and [Ivo2] for these and additional results.

15.4 An Occasional Bound for the Exponent

The above three examples are in fact quite miraculous. In many cases there were no newforms at all, and in the cases in which a newform existed, it was a rational newform corresponding to an elliptic curve with complex multiplication. In general we cannot expect that these phenomena will occur: we will usually find a finite collection of newforms, some rational, and some irrational, which we have to study separately. It is, however, sometimes possible to obtain a bound for the exponent p occurring in the Diophantine equation thanks to the following proposition.

Proposition 15.4.1. *Let $E/\mathbb{Q}$ be an elliptic curve of conductor N, and let t be an integer such that $t \mid |E_t(\mathbb{Q})|$, where we recall that $E_t(\mathbb{Q})$ is the torsion subgroup of $E(\mathbb{Q})$. Let f be a newform of level N' with Fourier coefficients c_n, let K be the totally real number field that they generate, and let ℓ be a prime number such that $\ell^2 \nmid N$ and $\ell \nmid N'$. Finally, define*

$$S_\ell = \left\{ a \in \mathbb{Z}, \quad -2\ell^{1/2} \leqslant a \leqslant 2\ell^{1/2} \text{ and } a \equiv \ell + 1 \pmod{t} \right\},$$

$$B'_\ell(f) = \mathcal{N}_{K/\mathbb{Q}}((\ell+1)^2 - c_\ell^2) \prod_{a \in S_\ell} \mathcal{N}_{K/\mathbb{Q}}(a - c_\ell), \text{ and}$$

$$B_\ell(f) = \begin{cases} \ell B'_\ell(f) & \text{if } f \text{ is not rational}, \\ B'_\ell(f) & \text{if } f \text{ is rational}. \end{cases}$$

Then if $E \sim_p f$ we have $p \mid B_\ell(f)$.

Proof. First note that if ℓ is a prime of good reduction, in other words such that $\ell \nmid N$, then by Proposition 8.1.13, $|E_t(\mathbb{Q})|$ divides $E(\mathbb{F}_\ell) = \ell + 1 - a_\ell(E)$, so if $t \mid |E_t(\mathbb{Q})|$ we have $a_\ell(E) \equiv \ell + 1 \pmod{t}$.

Assume first that $\ell \neq p$. Since $\ell \nmid N'$ and $\ell^2 \nmid N$, by Proposition 15.2.2 either $\ell \nmid pN$, in which case $p \mid \mathcal{N}_{K/\mathbb{Q}}(a_\ell(E) - c_\ell)$, or $\ell \| N$, in which case $p \mid \mathcal{N}_{K/\mathbb{Q}}((\ell+1)^2 - c_\ell^2)$ since $\ell \neq p$. Since by Hasse's theorem $-2\ell^{1/2} \leqslant a_\ell(E) \leqslant 2\ell^{1/2}$, and since in this case ℓ is a prime of good reduction, so that $a_\ell(E) \equiv \ell + 1 \pmod{t}$, it follows that if $\ell \neq p$ we have $p \mid B'_\ell(f)$, hence in particular $p \mid \ell B'_\ell(f)$ in all cases. If, in addition, f is rational, we can use the more precise Proposition 15.2.3 to conclude that $p \mid B'_\ell(f)$ whether or not ℓ is equal to p, proving the proposition. $\square$

Remark. The reader may wonder why we do not simply choose $t = |E_t(\mathbb{Q})|$ in the above proposition, but suppose only that $t \mid |E_t(\mathbb{Q})|$. The reason is that later on we will apply the proposition to Frey curves related to some Diophantine equations. We will often know that our Frey curves have points of order 2 or 3, say, and we can in those cases take $t = 2$ or 3. However, our Frey curves will be elliptic curves whose coefficients depend on some unknown variables appearing in the Diophantine equations. It is therefore not possible, or not convenient, to compute exactly the order of the torsion subgroup.

This proposition enables us to bound p if we can find a prime ℓ such that $B_\ell(f) \neq 0$. This is not always possible (see below), but it is guaranteed that we will succeed in the following cases:

Proposition 15.4.2. *In each of the following cases there are infinitely many ℓ for which $B_\ell(f) \neq 0$:*

(1) *When f is irrational.*
(2) *When f is rational, t is either a prime number or is equal to 4, and for every elliptic curve F isogenous to the elliptic curve corresponding to f we have $t \nmid |F_t(\mathbb{Q})|$.*
(3) *If f is rational and $t = 4$, and if for every elliptic curve F isogenous to the elliptic curve corresponding to f then $F(\mathbb{Q})$ does not have full 2-torsion.*

Note that for (1), if f is not rational it is easy to show that $c_\ell \notin \mathbb{Q}$ for infinitely many ℓ, so that $B_\ell(f) \neq 0$ at least for all such ℓ since $(\ell+1)^2 - c_\ell^2 \neq 0$ (recall that $|c_\ell| \leqslant 2\ell^{1/2}$).

Of course we do not need to know whether we will succeed, we are happy if we find some ℓ for which $B_\ell(f) \neq 0$.

Example. We give an example in which it can easily be shown that $B_\ell(f) = 0$ for all ℓ. Let $m \geqslant 5$ and assume that $L = 2^m - 1$ is a (Mersenne) prime number. Consider the elliptic curve F with equation

$$Y^2 = X(X+1)(X+2^m) \,.$$

Using Tate's algorithm it is easy to show that the conductor N of F is equal to $2L$. By the modularity theorem, we can let f be the newform corresponding to F. If $\ell \neq 2$ and $\ell \neq L$ then ℓ is a prime of good reduction. Thus $|F_t(\mathbb{Q})|$ divides $|F(\mathbb{F}_\ell)|$, so if t is any divisor of it we have $t \mid |F(\mathbb{F}_\ell)| = \ell + 1 - c_\ell$; hence $c_\ell \equiv \ell + 1 \pmod{t}$ and of course $|c_\ell| \leqslant 2\ell^{1/2}$; hence $c_\ell \in S_\ell$, so indeed $B_\ell(f) = 0$ for all $\ell \neq 2$ and L.

15.5 An Example of Serre–Mazur–Kraus

In this section we consider the Diophantine equation

$$x^p + L^r y^p + z^p = 0 \,,$$

where L is a fixed odd prime number (we have treated the case $L = 2$ in Theorem 15.3.1) and $p \geqslant 5$ is prime. As usual we may assume that $xyz \neq 0$ and that x, y, and z are pairwise coprime. We may also assume that $r < p$ (otherwise we can include the extra powers of L in y), and in fact that $0 < r < p$ since the case $r = 0$ is FLT, which has already been treated (!). This equation was studied by Serre in [Ser3] and by Kraus in [Kra3], and the connection with Mazur will become apparent below. Since we will come back to this equation several times, we will call it the SMK equation (with the implicit assumptions that $p \geqslant 5$ is prime, that L is an odd prime, and that $0 < r < p$), and say that (x, y, z) is a nontrivial solution of the SMK equation if x, y, z are nonzero and pairwise coprime.

Let A, B, and C be some permutation of x^p, $L^r y^p$, and z^p chosen such that $B \equiv 0 \pmod 2$ and $A \equiv -1 \pmod 4$, which is always possible by coprimality. Consider the Frey curve E with equation

$$Y^2 = X(X - A)(X + B) \,.$$

One easily computes that the minimal discriminant $\Delta_{\min}$ is given by $\Delta_{\min} = 2^{-8} L^{2r} (xyz)^{2p}$, and that the conductor N is equal to $N = \operatorname{rad}(Lxyz)$. Note that this is much simpler than the formulas that we obtained for $L = 2$. Thus the conductor is squarefree, and E has full 2-torsion, so by Mazur's Theorem 15.2.6 (3), E does not have any p-isogeny.

Since L is an odd prime we see that $v_L(\Delta_{\min}) \equiv 2r \not\equiv 0 \pmod p$ since $0 < r < p$, and $v_2(\Delta_{\min}) \equiv -8 \not\equiv 0 \pmod p$. Thus using Definition 15.2.4 we see that $N_p = 2L$. Applying Ribet's theorem we deduce that E arises modulo p from some newform f of level $2L$.

Now we have the following lemma.

Lemma 15.5.1. *Assume that F is an elliptic curve defined over $\mathbb{Q}$ with conductor $2L$, and assume that F has full 2-torsion. Then L is either a Mersenne or a Fermat prime (in other words L is a prime such that $L = 2^m - 1$ or $L = 2^{2^k} + 1$).*

Proof. Since F has full 2-torsion, up to isomorphism it has an equation of the form $Y^2 = X(X - a)(X + b)$ with a and b in $\mathbb{Z}$. It is easily shown that we can choose a and b such that this equation is minimal at all primes different from 2. Since the conductor is equal to $2L$ and the above model is minimal outside 2 its discriminant has the form $2^u L^v$. On the other hand, it is equal to $16a^2b^2(a + b)^2$, so that a, b, and $a + b$ are products of powers of 2 and L. Since the model is minimal at L it must have bad reduction at L, so that at least one of a, b, and $a + b$ is divisible by L. Since L^2 does not divide the conductor, by Proposition 8.1.6 the curve reduced modulo L cannot have a cusp, which clearly means that at most one (hence exactly one) of a, b, and $a + b$ is divisible by L. Writing $a + b - (a + b) = 0$ we thus obtain a relation of the form

$$\pm 2^\alpha \pm 2^\beta \pm 2^\gamma L^\delta = 0$$

with $\delta \geqslant 1$. Thus $2^\gamma L^\delta = \pm 2^\alpha \pm 2^\beta$. If $\alpha = \beta$ then the right-hand side is either equal to 0 or to a power of 2, which is impossible since it is divisible by the odd prime L. Thus $\alpha \neq \beta$, so assume $\alpha < \beta$, say. Thus $2^\gamma L^\delta = \pm 2^\alpha (2^{\beta - \alpha} \pm 1)$, from which we deduce that $\gamma = \alpha$ and $L^\delta = 2^m \pm 1$ for some m; in other words, $L^\delta \pm 1 = 2^m$ for some sign $\pm$. If the sign is $-$ the left-hand side is divisible by $L - 1$, so $L - 1 \mid 2^m$; hence $L - 1 = 2^k$, so $L = 2^k + 1$ is a Fermat prime (since it is trivially shown that $2^k + 1$ can be a prime only if k is a power of 2). If the sign is $+$, write $\delta = 2^t \varepsilon$ with ε odd. The left-hand side is divisible by $L^{2^t} + 1$, so $L^{2^t} + 1 = 2^k$ for some integer k. However, if $t \geqslant 1$ then since L is odd $L^{2^t} = (L^2)^{2^{t-1}} \equiv 1 \pmod 8$, so that $2^k \equiv 2 \pmod 8$ which is impossible since $L \geqslant 3$. Thus $t = 0$, and hence $L = 2^k - 1$ is a Mersenne prime, proving the lemma.   □

The following theorem stated in [Ser3] is now immediate:

Theorem 15.5.2 (Mazur). *Let L be an odd prime that is neither a Mersenne nor a Fermat prime. There exists a constant C_L such that for any nontrivial solution (x, y, z, p) to the SMK equation we have $p \leqslant C_L$.*

Proof. By the discussion preceding the lemma we know that $E \sim_p f$ for a newform f at level $N_p = 2L$. If f is irrational then by Proposition 15.4.2 (1) we know that there are infinitely many ℓ for which $B_\ell(f) \neq 0$. If f is rational, then since L is not a Mersenne or Fermat prime the lemma implies that none of the elliptic curves F isogenous to the curve corresponding to f under the modularity theorem has full 2-torsion, so it follows from Proposition 15.4.2 (3) that there are infinitely many ℓ for which $B_\ell(f) \neq 0$. In both cases any suitable ℓ gives a bound C_L on the exponent p.   □

In [Kra3] Kraus shows that we can choose the bound

$$C_L = (((L+1)/2)^{1/2} + 1)^{(L+11)/6} .$$

This bound is very large, but in practice we can obtain a much lower bound since for a given newform f we compute several $B_\ell(f)$ and p must divide the greatest common divisor of all of them.

Theorem 15.5.3. *Suppose that $3 \leqslant L \leqslant 100$ is prime. Then the SMK equation does not have any nontrivial solutions, except for $L = 31$, in which case $E \sim_p F$, where F is the curve 62A1 in [Cre2].*

Proof. This follows from the use of Proposition 15.4.1, the method of Kraus (Proposition 15.6.3 below), and the method of predicting exponents (Section 15.7 below). We will see examples of each of these methods applied to this theorem.   □

Note that for $L = 31$ and $r = 1$ there is the evident (but nontrivial!) solution $(x, y, z, p) = (2, -1, -1, 5)$, so this value of L cannot be excluded.

Example. As an illustration, we treat the case $L = 19$:

Proposition 15.5.4. *The above result is true for $L = 19$.*

Proof. From what we have done we already know that $E \sim_p f$ for some newform of level 38. There are two newforms of level 38, and although this cannot be seen purely from the dimension formulas that we have given, both are rational. Their q-expansions are:

$$f_1 = q - q^2 + q^3 + q^4 - q^6 - q^7 + \cdots ,$$
$$f_2 = q + q^2 - q^3 + q^4 - 4q^5 - q^6 + 3q^7 + \cdots .$$

Since E has full 2-torsion we can apply Proposition 15.4.1 with $t = 4$, and we can then compute that $B_3(f_1) = -15$ and $B_5(f_1) = -144$. Since p must divide both and $p \geqslant 5$ we obtain a contradiction, so that necessarily $E \sim_p f_2$. A similar computation gives

$$B_3(f_2) = 15, \quad B_5(f_2) = 240, \quad B_7(f_2) = 1155, \quad B_{11}(f_2) = 3360 .$$

Thus $p = 5$, which solves our equation for all $p \geqslant 7$. We now treat the case $p = 5$. We first note that *all* the $B_\ell(f_2)$ will be divisible by 5, hence that there is no point in pushing these computations any further. The reason for this is the following. Let F be the elliptic curve corresponding to f_2, which is $38B1$ in [Cre2]. Looking at [Cre2] we see that this curve has a rational point of order 5. It follows that $5 \mid |F(\mathbb{F}_\ell)| = \ell + 1 - a_\ell(F)$ for all $\ell \nmid 38$. Thus from the definition of B_ℓ we see that this implies that $5 \mid B_\ell(f_2)$ for all primes $\ell \nmid 38$, so it is impossible to eliminate $p = 5$ in this way.

However, we can turn this to our advantage as follows. Assume that $E \sim_5 f_2$, or equivalently, that $E \sim_5 F$. Then $a_\ell(E) \equiv a_\ell(F) \pmod 5$ for all but finitely many ℓ, so by what we have said above, $5 \mid (\ell + 1 - a_\ell(E))$ for all but finitely many ℓ. We now use an important theorem from algebraic number theory, the Čebotarev density theorem, which implies that E has necessarily a 5-isogeny (we will assume this fact, see [Ser4], IV-6). On the other hand, the conductor $N = \mathrm{rad}(Lxyz)$ of E is squarefree and E has full 2-torsion, so by Theorem 15.2.6 (3) we deduce that E has no 5-isogeny, a contradiction. $\square$

For the analogue of this trick when the newform is irrational see [Kra3], pages 1155–1156.

15.6 The Method of Kraus

For convenience we begin with the following elementary lemma (recall Definition 7.3.15 for the notion of quadratic twist).

Lemma 15.6.1. *Let A, B, and C be nonzero integers such that $A + B + C = 0$, and let E be the elliptic curve with equation*

$$Y^2 = X(X - A)(X + B) .$$

Then any permutation of A, B, and C gives rise to a curve that is either isomorphic to E, or to its quadratic twist by -1, in other words to the curve E' with equation $Y^2 = X(X + A)(X - B)$.

Proof. The exchange of A and B clearly changes E into E'. The exchange of A and C changes E into the curve with equation $Y^2 = X(X + A + B)(X + B)$, which we see is again isomorphic to E' by changing X into $X - B$. The exchange of B and C changes E into the curve $Y^2 = X(X - A)(X - A - B)$, isomorphic to E' by changing X into $X + A$. Conversely, these three transpositions change E' into a curve isomorphic to E. Thus the remaining two nontrivial permutations, which are products of two transpositions, will change E into a curve isomorphic to E. $\qquad\square$

As we have seen, Proposition 15.4.1 is often capable of bounding the exponent p of our Diophantine equation when our (hypothetical Frey) curve arises modulo p from a newform f. There is another important method due to Kraus [Kra1] that can often be used to derive a contradiction, but *for a fixed value of p*. In fact:

– Kraus introduced and used this method to show that the equation

$$a^3 + b^3 = c^p \quad \text{with } abc \neq 0 \text{ and pairwise coprime}$$

has no solutions for $11 \leqslant p \leqslant 10000$.
– A combination of Kraus's method, Proposition 15.4.1, and classical techniques for Diophantine equations, recently led to the complete solution in integers of the equations $y^2 = x^n + t$ for $n \geqslant 3$ and $1 \leqslant t \leqslant 100$, which we already considered at length in Section 6.7 (see [BMS2] and [Sik-Cre]). As an application of the study of exponent-$(p, p, 2)$ super-Fermat equations, it is sometimes possible to determine all the *rational* points on the hyperelliptic curves $y^2 = x^n + t$; see [Ivo2] or [Ivo-Kra].

In this section we adapt the method of Kraus to the Serre–Mazur–Kraus equation considered in the preceding section. Recall that E is the curve $Y^2 = X(X - A)(X + B)$, where A, B, and C is some permutation of x^p, $L^r y^p$, and z^p such that $A \equiv -1 \pmod 4$ and $2 \mid B$. It is somewhat awkward to work with the curve E since there are six possibilities for the triple A, B, C. However, if we let E' be the curve with equation

$$Y^2 = X(X - x^p)(X + z^p) ,$$

we see from the above lemma that E and E' are either isomorphic or quadratic twists of each other by -1. If we write $\delta = (z/x)^p$, it is clear that E' is the quadratic twist by x^p of the curve E_δ with equation

$$Y^2 = X(X - 1)(X + \delta) .$$

Thus by Proposition 7.3.16 it follows that if $\ell \nmid x$ then $a_\ell(E) = \pm a_\ell(E_\delta)$, where the sign $\pm$ is a product of a Legendre symbol $\left(\frac{\pm 1}{\ell}\right)$ with the Legendre symbol $\left(\frac{x^p}{\ell}\right)$. From this remark together with Proposition 15.2.2 and the fact that the conductor N of E is equal to $\mathrm{rad}(Lxyz)$, we deduce the following.

Lemma 15.6.2. *Keep the above notation, assume that $E \sim_p f$ for some newform f of level $2L$, and let K be the number field generated by the Fourier coefficients c_n of f. Let ℓ be a prime different from 2, L, and p. Then*

(1) *If $\ell \nmid xyz$ then $p \mid \mathcal{N}_{K/\mathbb{Q}}(a_\ell(E_\delta)^2 - c_\ell^2)$.*
(2) *If $\ell \mid xyz$ then $p \mid \mathcal{N}_{K/\mathbb{Q}}((\ell+1)^2 - c_\ell^2)$.*

The following proposition is now immediate.

Proposition 15.6.3. *Let $p \geqslant 5$ be a fixed prime number, and let E be as above. Assume that for every newform f of level $2L$ there exists a positive integer n satisfying the following conditions, where as usual c_ℓ denotes the ℓth Fourier coefficient of f:*

(1) *$\ell = np + 1$ is prime.*
(2) *$\ell \neq L$.*
(3) *$p \nmid \mathcal{N}_{K/\mathbb{Q}}((\ell+1)^2 - c_\ell^2)$.*
(4) *For all $\delta \in \mathbb{F}_\ell$ such that $\delta^n = 1$ and $\delta \neq -1$ we have $p \nmid \mathcal{N}_{K/\mathbb{Q}}(a_\ell(E_\delta)^2 - c_\ell^2)$.*

Then the SMK equation for exponent p does not have any nontrivial solution.

Proof. If ℓ satisfies (1) and (2) it is a prime different from 2, L, and p, so the above lemma is applicable. Because of (3) and the lemma we have $\ell \nmid xyz$. But then the reduction modulo ℓ of $\delta = (x/z)^p$ is well defined and is in $\mathbb{F}_\ell^*$, and evidently satisfies $\delta^n = 1$. In addition, we cannot have $\delta \equiv -1 \pmod{\ell}$, since otherwise $x^p \equiv -z^p \pmod{\ell}$, so that by the SMK equation we would have $\ell \mid L^r y^p$, which is impossible since $\ell \neq L$ and $\ell \nmid y$. Thus the lemma and (4) give a contradiction. $\qquad\square$

As an example, we have the following theorem.

Theorem 15.6.4. *Assume that $L = 31$. Then the SMK equation does not have any nontrivial solution for $11 \leqslant p \leqslant 10^6$.*

Proof. Thanks to Ribet's theorem we have seen that $E \sim_p f$ for a newform f of level $2L = 62$. The number of newforms is equal to 3, but one is rational and the other two are conjugate over $\mathbb{Q}(\sqrt{3})$. If we set $\theta = \sqrt{3}$, the forms are

$$f_1 = q + q^2 + q^4 - 2q^5 + \cdots,$$
$$f_2 = q - q^2 + (1+\theta)q^3 + q^4 - 2\theta q^5 - (1+\theta)q^6 + 2q^7 + \cdots.$$

Since the conductor N of E is equal to $\mathrm{rad}(Lxyz)$ hence squarefree, if we choose $\ell \neq 2$ or 31 we may apply Proposition 15.4.1 with $t = 4$, which tells us that $p \mid B_\ell(f)$ for all such ℓ. We compute that $B_3(f_2) = -792$, $B_5(f_2) = 184320$, $B_7(f_2) = 14515200$, and since the only primes p dividing the first two are $p = 2$ and $p = 3$, we deduce that $E \sim_p f_2$ is not possible, so if E exists we must have $E \sim_p f_1$. On the other hand, it is easy to see that the same method cannot work for f_1, in other words that $B_\ell(f_1) = 0$ for all ℓ. However, we proceed as follows. Under the modularity theorem the rational newform f_1 corresponds to an elliptic curve F that is the curve $62A1$ in [Cre2], with minimal generalized Weierstrass equation

$$Y^2 + XY + Y = X^3 - X^2 - X + 1 .$$

It is now immediate to write a small program that for a *given* prime p looks for a prime ℓ satisfying the conditions of Proposition 15.6.3, and the program easily finds such a prime for $11 \leqslant p \leqslant 10^6$, the total time being less than 20 minutes. On the other hand, the program fails to find an ℓ for $p = 5$ (which is a good thing since solutions exist), and fails also for $p = 7$, proving the theorem. We will treat the case $p = 7$ later (see Corollary 15.7.5). $\qquad\square$

Remark. Using a variant of the modular method that they call the symplectic method, Halberstadt and Kraus [Hal-Kra1] show that for every prime number $p \geq 7$ with $p \equiv 3 \pmod 4$, the equation $x^p + 31y^p + z^p = 0$ has no nontrivial solutions.

15.7 "Predicting Exponents of Constants"

The title of this section is in quotes because it is rather vague. For various Diophantine equations the modular approach explained in this chapter is very effective at predicting exponents of terms with constant base. For instance, this method is central to the recent determination of all perfect powers in the Fibonacci and Lucas sequences [BMS1].

15.7.1 The Diophantine Equation $x^2 - 2 = y^p$

We will illustrate this method by studying the Diophantine equation

$$x^2 - 2 = y^p, \quad p \geqslant 5 \text{ prime} ,$$

which is a special case of the equations studied in Section 6.7.

We will explain shortly what is the exponent that we would like to predict. We give two motivations for studying this equation.

- The more general equation $x^2 - 2^m = y^p$ has been solved in Theorem 15.3.4 for $m \geqslant 2$. For $m = 0$ it is due to Ko Chao and has been proved as Theorem 6.11.8, but can now be proved using modular techniques as a consequence of Theorem 15.3.1. Thus there remains the case $m = 1$.
- The equation $x^2 - 2 = y^p$ is now considered to be one of the most difficult exponential equations. This section presents a partial attempt at solving this equation, due to Bugeaud, Mignotte, and Siksek.

Similarly to the case $m \geqslant 2$, we associate to any solution of the equation the Frey curve E with equation

$$Y^2 = X(X^2 + 2xX + 2) .$$

We easily compute that

$$\Delta_{\min} = 2^8 y^p, \quad N = 2^7 \operatorname{rad}(y), \quad \text{and} \quad N_p = 128 .$$

Thanks to Theorem 15.2.7 we know that E does not have any p-isogenies. It follows from Ribet's theorem that E arises from a newform of level 128. Proposition 15.1.1 tells us that there are four newforms of level 128, and it is easily shown that they are all rational. Thus they correspond under the modularity theorem to the four elliptic curves $F_1 = 128A1$, $F_2 = 128B_1$, $F_3 = 128C_1$, $F_4 = 128D_1$, so $E \sim_p F_i$ for some i. Note that the Diophantine equation has the universal solutions $(x, y) = (\pm 1, -1)$ valid for all $p \geqslant 3$. It follows that any attempt to show that p is bounded by some result similar to Proposition 15.4.1 will fail, and so will any method mimicking Kraus's method. However, we can still use the modular approach to derive some nontrivial information on the Diophantine equation.

The classical line of attack for an equation of the type $x^2 + t = y^p$, which we have used at length in Section 6.7, is to factor it in the quadratic field $\mathbb{Q}(\sqrt{-t})$. As we did in that section, since $\mathbb{Q}(\sqrt{2})$ has class number 1 and a fundamental unit is $1 + \sqrt{2}$, we deduce that there exist U, V in $\mathbb{Z}$ such that

$$x + \sqrt{2} = (1 + \sqrt{2})^r (U + V\sqrt{2})^p ,$$

where in addition we may assume that $-(p-1)/2 < r \leqslant (p-1)/2$ by including all pth powers of the fundamental unit in the second factor. We deduce that

$$\frac{1}{2\sqrt{2}} \left((1 + \sqrt{2})^r (U + V\sqrt{2})^p - (1 - \sqrt{2})^r (U - V\sqrt{2})^p \right) = 1 .$$

Thus to solve our equation for any particular exponent p we must solve p Thue equations of the above form, one for each value of r. As p gets large the coefficients of these equations become very unpleasant, making it difficult to solve the Thue equations. This has already been noted in Section 6.7. However, based on a short search, we believe that the only solutions are the universal solutions $(x, y) = (\pm 1, -1)$ mentioned above. Since $x + \sqrt{2} = (1 + \sqrt{2})^r (U + V\sqrt{2})^p$, we suspect that the only values of r that can give rise to solutions are $r = \pm 1$. Indeed, we prove this using the modular approach together with a result proved by classical means.

Proposition 15.7.1. *With the above notation we have $r = \pm 1$.*

Proof. Let F be one of the four elliptic curves F_i above, and assume that $E \sim_p F$. Let ℓ be a prime number satisfying the following conditions:

(1) $\ell = np + 1$ for some integer n.
(2) $\ell \equiv \pm 1 \pmod 8$.
(3) $a_\ell(F) \not\equiv \pm(\ell + 1) \pmod p$.
(4) $(1 + \theta)^n \not\equiv 1 \pmod \ell$, where θ is a square root of 2 in $\mathbb{F}_\ell$, which exists by
(2).

Since $N_p = 128$ and $N = 2^7 \operatorname{rad}(y)$, if $\ell \mid y$ we would have $\ell \nmid N_p$ and
$\ell \| N$. It follows from Proposition 15.2.3 that $a_\ell(F) \equiv \pm(\ell + 1) \pmod p$,
contradicting condition (c). Thus $\ell \nmid y$, so that $\overline{y}^p \in \mu_n(\mathbb{F}_\ell)$, where $\overline{y}$ is the
class of y in $\mathbb{F}_\ell$, and

$$\mu_n(\mathbb{F}_\ell) = \left\{\delta \in \mathbb{F}_\ell / \ \delta^n = 1\right\}.$$

Set

$$\mathcal{X}'_\ell = \left\{\delta \in \mathbb{F}_\ell / \ \delta^2 - 2 \in \mu_n(\mathbb{F}_\ell)\right\},$$

so that $\overline{x} \in \mathcal{X}'_\ell$. Since $|\mu_n(\mathbb{F}_\ell)| = n$, we have $|\mathcal{X}'_\ell| \leqslant 2n$. We want to refine $\mathcal{X}'_\ell$
so as to have better information on the value of x modulo ℓ. For $\delta \in \mathcal{X}'_\ell$ let
E_δ be the elliptic curve over $\mathbb{F}_\ell$ with equation

$$Y^2 = X(X^2 + 2\delta X + 2).$$

We let

$$\mathcal{X}_\ell = \left\{\delta \in \mathcal{X}'_\ell / \ a_\ell(E_\delta) \equiv a_\ell(F) \pmod p\right\}.$$

Since $\ell \nmid y$, it follows from Proposition 15.2.3 that $\overline{x} \in \mathcal{X}_\ell$, and we can hope
that the set $\mathcal{X}_\ell$ is much smaller than $\mathcal{X}'_\ell$. We want to obtain information on
r from the fact that $\overline{x} \in \mathcal{X}_\ell$. Note first that from the formula linking $x + \sqrt{2}$
to r, U, and V there exists $\delta \in \mathcal{X}_\ell$ such that

$$\delta + \theta = (1 + \theta)^r (U + V\theta)^p \quad \text{in } \mathbb{F}_\ell.$$

Since $U^2 - 2V^2 = \pm y$ and $\ell \nmid y$, it follows that $U + V\theta \neq 0$ in $\mathbb{F}_\ell$.

To obtain information about r we use the discrete logarithm in $\mathbb{F}_\ell$. Let g
be a fixed primitive root modulo ℓ. The discrete logarithm with respect to g is
the isomorphism from $(\mathbb{F}_\ell^*, \times)$ to $(\mathbb{Z}/(\ell - 1)\mathbb{Z}, +)$ given by $g^k \mapsto k \bmod \ell - 1$.
Let Φ be the group homomorphism from $(\mathbb{F}_\ell^*, \times)$ to $(\mathbb{Z}/p\mathbb{Z}, +)$ obtained by
composing the discrete logarithm with the natural projection from $\mathbb{Z}/(\ell - 1)\mathbb{Z}$
to $\mathbb{Z}/p\mathbb{Z}$. Applying Φ to the fundamental identity obtained above we deduce
that

$$\Phi(\delta + \theta) \equiv r\Phi(1 + \theta) \pmod p.$$

By the fourth assumption that we have made at the beginning of this proof
we have $\Phi(1 + \theta) \not\equiv 0 \pmod p$, since otherwise $1 + \theta = g^{kp}$ for some k, so
that $(1 + \theta)^n = g^{kpn} = g^{k(\ell - 1)} = 1$. Thus we deduce that

$$r \bmod p \in \mathcal{R}_\ell(F), \text{ where } \mathcal{R}_\ell(F) = \left\{\frac{\Phi(\delta + \theta)}{\Phi(1 + \theta)}, \ \delta \in \mathcal{X}_\ell\right\}.$$

Since $-(p-1)/2 < r \leqslant (p-1)/2$, to show that $r = \pm 1$ it is sufficient to show that $r \equiv \pm 1 \pmod{p}$. Therefore if we can find primes $\ell_1, \dots, \ell_k$ satisfying the four conditions given above and such that

$$\bigcap_{1 \leqslant j \leqslant k} \mathcal{R}_{\ell_j}(F) \subset \{\pm 1 \bmod p\} \,,$$

and if we can do this for $F = F_1$, F_2, F_3, and F_4, we will have proved that $r = \pm 1$.

We wrote a short GP script to carry out the above calculations for $5 \leqslant p \leqslant 10^6$, and this took about three hours to run.

On the other hand, and this is outside the scope of this chapter, a careful application of linear forms in logarithms to this problem (see [Lau-Mig-Nes] and Chapter 12) shows that $p < 8200$ if $y \neq -1$. It follows that for any $p > 8200$ we have $y = -1$, so that $r = \pm 1$ as is easily seen, and for $p < 10^6$ we also have $r = \pm 1$ by the GP computation above, proving the proposition. $\square$

Note that the method of linear forms in logarithms has not only shown that $r = \pm 1$ for $p > 8200$, but has completely solved the equation $x^2 - 2 = y^p$ in that range. It remains to solve the finite number of remaining equations for $p < 8200$, knowing now that in that range we have $r = \pm 1$. Knowing this last fact we can improve on this range as follows.

Lemma 15.7.2. *If $y \neq -1$ then $y \geqslant (\sqrt{p} - 1)^2$.*

Proof. Note the trivial fact that if $y \neq -1$ then $y > 1$, and furthermore y is odd, since otherwise $x^2 \equiv 2 \pmod{4}$, which is absurd. Thus there exists an odd prime ℓ such that $\ell \mid y$. Since $N = 2^7 \operatorname{rad}(y)$ and $N_p = 128$ it follows from Proposition 15.2.3 that

$$\ell + 1 \equiv \pm a_\ell(F) \pmod{p} \,,$$

where F is one of the four curves F_i. However, by Hasse's theorem we know that $|a_\ell(F)| < 2\sqrt{\ell}$. It follows that

$$p \leqslant \ell + 1 + 2\sqrt{\ell} = (\sqrt{\ell} + 1)^2 \leqslant (\sqrt{y} + 1)^2 \,,$$

proving the lemma. $\square$

Using this lemma and once again a careful application of linear forms in logarithms one can prove that $p < 1237$; see [Lau-Mig-Nes] for a slightly weaker result.

On the other hand, for the small values of p we can try to solve the Thue equations that one obtains with $r = \pm 1$. In fact if we let $F_r(U, V) = ((1 + \sqrt{2})^r (U + V\sqrt{2})^p - (1 - \sqrt{2})^r (U - V\sqrt{2})^p)/(2\sqrt{2})$, so that the equations to be solved are $F_r(U, V) = 1$, we see that $F_{-1}(U, V) = F_1(U, -V)$, so it is sufficient to solve the Thue equation $F_1(U, V) = 1$. Using the built-in Thue equation solver in GP for $5 \leqslant p \leqslant 37$ we obtain that $(U, V) = (1, 0)$ is the only solution in that range, thus proving the following modest result.

Lemma 15.7.3. *If $5 \leqslant p \leqslant 37$ the only solutions to $x^2 - 2 = y^p$ are $(x, y) = (\pm 1, -1)$.*

15.7.2 Application to the SMK Equation

Thanks to the method of predicting exponents it is possible to give a useful strengthening of Proposition 15.6.3 that will allow us to solve further equations. We recall that E is the Frey curve associated with a nonzero coprime solution to $x^p + L^r y^p + z^p = 0$, and that by Ribet's theorem we know that $E \sim_p f$ for some newform f of level $2L$.

Proposition 15.7.4. *Let $p \geqslant 5$ be a fixed prime number, and let f be the newform of level $2L$ such that $E \sim_p f$. Assume that there exists a positive integer n satisfying the following conditions:*

(1) $\ell = np + 1$ *is prime.*
(2) $\ell \neq L$.
(3) $L^n \not\equiv 1 \pmod{\ell}$
(4) $p \nmid \mathcal{N}_{K/\mathbb{Q}}((\ell + 1)^2 - c_\ell^2)$.

Set

$$\mathcal{X}_\ell = \{\delta \in \mathbb{F}_\ell^*, \ \delta^n = 1, \ \delta \neq -1, \ p \mid \mathcal{N}_{K/\mathbb{Q}}(a_\ell(E_\delta)^2 - c_\ell^2)\} \,,$$

and

$$\mathcal{R}_\ell = \left\{ \frac{\Phi(-1 - \delta)}{\Phi(L)}, \ \delta \in \mathcal{X}_\ell \right\} \subset \mathbb{Z}/p\mathbb{Z} \,,$$

where, as in the preceding section, Φ denotes a group homomorphism from $\mathbb{F}_\ell^*$ to the additive group $\mathbb{Z}/p\mathbb{Z}$. Then

$$r \bmod p \in \mathcal{R}_\ell \,.$$

Proof. By Proposition 15.6.3, if (x, y, z) is a nontrivial solution then $\delta = \overline{z/x}^p \in \mathcal{X}_\ell$ as defined above. We now apply the homomorphism Φ to the identity $L^r y^p = x^p(-1 - (z/x)^p)$ and note that $\Phi(L) \not\equiv 0 \pmod{p}$ since otherwise $L^n \equiv 1 \pmod{\ell}$, which has been excluded, proving the proposition. $\square$

Remarks. (1) Proposition 15.6.3 says that if we omit the condition $L^n \not\equiv 1$ (mod ℓ) and if $\mathcal{X}_\ell = \emptyset$ then the SMK equation has no nontrivial solution.
(2) If for some finite set I of ℓ we have $\bigcap_{\ell \in I} \mathcal{R}_\ell = \emptyset$, we deduce that the SMK equation has no nontrivial solution.

We give two examples of the use of the above proposition. First, we can now solve the case $p = 7$ of the SMK equation for $L = 31$, which we had been unable to treat by a direct application of the method of Kraus in Theorem 15.6.4.

Corollary 15.7.5. *The SMK equation does not have any nontrivial solution for $L = 31$ and $p = 7$.*

Proof. We use $n = 6$ hence $\ell = 43$. An immediate computation shows that the assumptions of the lemma are satisfied, and we find that $\mathcal{X}_\ell = \{6, 36\}$, and then that $\mathcal{R}_\ell = \{0 \bmod 7\}$. It follows that $r \equiv 0 \pmod 7$, and since we can assume that $0 \leqslant r < p$ we have $r = 0$. We are thus reduced to the ordinary FLT equation, which has no nontrivial solution. Another way of finishing the proof is as follows: we choose also $\ell = 71$ and we find similarly that $\mathcal{R}_\ell = \{1, 4 \bmod 7\}$, which has empty intersection with $\mathcal{R}_{43}$.   □

As an additional example, we solve the case $L = 23$ of the SMK equation.

Corollary 15.7.6. *The SMK equation does not have any nontrivial solution for $L = 23$ and $p \geqslant 5$.*

Proof. By Ribet's theorem we know that if E is the Frey curve corresponding to a possible solution (x, y, z) then $E \sim_p f$ for a newform f of level 46. By Proposition 15.1.1 there is a single newform at that level, necessarily rational, and equal to

$$f = q - q^2 + q^4 + 4q^5 - 4q^7 + \cdots .$$

With the notation of Proposition 15.4.1 we compute that $B_3(f) = B_7(f) = 0$ but $B_5(f) = 240$, and since $p \geqslant 5$ this shows that our equation is impossible for $p \geqslant 7$. There remains the case $p = 5$, which we treat using the method of predicting exponents. Under the modularity theorem f corresponds to the elliptic curve 46A1 in [Cre2] with equation $Y^2 + XY = X^3 - X^2 - 10X - 12$. We first use $\ell = 31$ in Proposition 15.7.4, and we compute that $\mathcal{R}_\ell = \{2 \bmod 5\}$, so that $r \equiv 2 \pmod 5$. We then use $\ell = 181$, and we compute that $\mathcal{R}_\ell = \{0, 1, 4 \bmod 5\}$, which has empty intersection with $\mathcal{R}_{31}$, proving that the SMK equation has no nontrivial solutions.   □

15.8 Recipes for Some Ternary Diophantine Equations

We end this chapter by giving a number of recipes for ternary Diophantine equations, in other words for super-Fermat equations of the type $Ax^p + By^q + Cz^r = 0$, where (p, q, r) will be called the *signature* of the equation. How to associate to such an equation a Frey curve is detailed for the three important signatures (p, p, p), $(p, p, 2)$, and $(p, p, 3)$ respectively by Kraus

[Kra3], by Bennett–Skinner [Ben-Ski], and by Bennett, Vatsal, and Yazdani [Ben-Vat-Yaz]. For the convenience of the reader we reproduce the recipes for the Frey curves and levels appearing in these papers. However, we would like to emphasize the following points:

- There is much more in the above-mentioned papers than just the recipes, and the reader is particularly urged to pursue them.
- The choice of Frey curve given here is far from canonical. Sometimes it is possible to construct an alternative Frey curve that turns out to be more useful for the problem at hand. See for example [Ivo2] and [Ivo-Kra].
- For some problems it is possible to use several Frey curves and exploit the information obtained from these curves simultaneously. For an example of this "multi-Frey" approach see [BMS3].

This section is heavily influenced by Bennett's paper [Ben2].

15.8.1 Recipes for Signature (p, p, p)

Consider the equation

$$Ax^p + By^p + Cz^p = 0 \,,$$

where $p \geqslant 5$ is prime and where we assume that Ax, By, and Cz are nonzero and pairwise coprime. Setting $R = ABC$, we may clearly assume that $v_q(R) < p$ for all prime numbers q, since otherwise (A, B, and C being pairwise coprime) this means that one of A, B, and C is divisible by q^p, and this can be removed by dividing x, y, or z by p. Without loss of generality we may of course assume that $By^p \equiv 0 \pmod 2$ and that $Ax^p \equiv -1 \pmod 4$. The Frey curve is the curve E with equation

$$Y^2 = X(X - Ax^p)(X + By^p) \,.$$

Its minimal discriminant is given by

$$\Delta_{\min} = \begin{cases} 2^4 R^2 (xyz)^{2p} & \text{if } 16 \nmid By^p \,, \\ 2^{-8} R^2 (xyz)^{2p} & \text{if } 16 \mid By^p \,, \end{cases}$$

and the conductor is given by $N = 2^\alpha \operatorname{rad}_2(Rxyz)$, where

$$\alpha = \begin{cases} 1 & \text{if } v_2(R) \geqslant 5 \text{ or } v_2(R) = 0 \,, \\ 1 & \text{if } 1 \leqslant v_2(R) \leqslant 4 \text{ and } y \text{ is even} \,, \\ 0 & \text{if } v_2(R) = 4 \text{ and } y \text{ is odd} \,, \\ 3 & \text{if } 2 \leqslant v_2(R) \leqslant 3 \text{ and } y \text{ is odd} \,, \\ 5 & \text{if } v_2(R) = 1 \text{ and } y \text{ is odd} \,. \end{cases}$$

Theorem 15.8.1 (Kraus). *Under the above assumptions we have $E \sim_p f$ for some newform f of level $N_p = 2^\beta \operatorname{rad}_2(R)$, where*

$$\beta = \begin{cases} 1 & \text{if } v_2(R) \geqslant 5 \text{ or } v_2(R) = 0 \,, \\ 0 & \text{if } v_2(R) = 4 \,, \\ 1 & \text{if } 1 \leqslant v_2(R) \leqslant 3 \text{ and } y \text{ is even} \,, \\ 3 & \text{if } 2 \leqslant v_2(R) \leqslant 3 \text{ and } y \text{ is odd} \,, \\ 5 & \text{if } v_2(R) = 1 \text{ and } y \text{ is odd} \,. \end{cases}$$

Proof. The proof is left as an exercise to the reader, who is referred to [Kra3]. $\square$

For a much deeper study of the equation $Ax^p + By^p + Cz^p = 0$ including several variants of the modular approach, we heartily recommend the paper by Halberstadt and Kraus [Hal-Kra2]. For lack of space we do not explain these variants, but we mention just one of many interesting results proved using these methods.

Theorem 15.8.2 (Halberstadt–Kraus). *Let A, B, and C be odd pairwise coprime integers and set $N = 2\operatorname{rad}(ABC)$. Let r be the number of isogeny classes of elliptic curves of conductor N (over $\mathbb{Q}$) having full 2-torsion defined over $\mathbb{Q}$. There exists a set P of prime numbers having density $1/2^r$ such that for each prime $p \in P$ the equation $Ax^p + By^p + Cz^p = 0$ has no nontrivial solutions.*

15.8.2 Recipes for Signature $(p, p, 2)$

Consider the equation
$$Ax^p + By^p = Cz^2 \,,$$

where $p \geqslant 7$ is prime and where we assume that Ax, By, and Cz are nonzero and pairwise coprime. As usual we may assume that for all primes q we have $v_q(A) < p$ and $v_q(B) < p$, and that C is squarefree. Also it is easy to see that without loss of generality we may suppose that we are in one of the following situations:

(1) $ABCxy \equiv 1 \pmod 2$ and $y \equiv -BC \pmod 4$.
(2) $xy \equiv 1 \pmod 2$ and either $v_2(B) = 1$ or $v_2(C) = 1$.
(3) $xy \equiv 1 \pmod 2$, $v_2(B) = 2$, and $z \equiv -By/4 \pmod 4$.
(4) $xy \equiv 1 \pmod 2$, $3 \leqslant v_2(B) \leqslant 5$, and $z \equiv C \pmod 4$.
(5) $v_2(By^p) \geqslant 6$ and $z \equiv C \pmod 4$.

In cases (1) and (2) we consider the curve E_1 with equation

$$Y^2 = X^3 + 2CzX^2 + BCy^pX \,.$$

In cases (3) and (4) we consider the curve E_2 with equation

$$Y^2 = X^3 + CzX^2 + \frac{BCy^p}{4}X \ .$$

Finally, in case (5) we consider the curve E_3 with equation

$$Y^2 + XY = X^3 + \frac{Cz - 1}{4}X^2 + \frac{BCy^p}{64}X \ .$$

Theorem 15.8.3 (Bennett–Skinner [Ben-Ski]**).** *With the above assumptions and notation we have:*

(1) *The minimal discriminant $\Delta_{\min}$ of E_i is given by*

$$\Delta_{\min} = 2^{\delta_i} AB^2 C^3 (xy^2)^p \ ,$$

where

$$\delta_1 = 6, \quad \delta_2 = 0, \quad \delta_3 = -12 \ .$$

(2) *The conductor N of E_i is given by*

$$N = 2^\alpha C^2 \operatorname{rad}(ABxy) \ ,$$

where

$$\alpha = \begin{cases} 5 & \text{if } i = 1, \text{ case } (1) \ , \\ 6 & \text{if } i = 1, \text{ case } (2) \ , \\ 1 & \text{if } i = 2, \text{ case } (3), \ v_2(B) = 2, \text{ and } y \equiv -BC/4 \ (\mathrm{mod} \ 4) \ , \\ 2 & \text{if } i = 2, \text{ case } (3), \ v_2(B) = 2, \text{ and } y \equiv BC/4 \ (\mathrm{mod} \ 4) \ , \\ 4 & \text{if } i = 2, \text{ case } (4) \text{ and } v_2(B) = 3 \ , \\ 2 & \text{if } i = 2, \text{ case } (4) \text{ and } v_2(B) = 4 \text{ or } 5 \ , \\ -1 & \text{if } i = 3, \text{ case } (5) \text{ and } v_2(By^7) = 6 \ , \\ 0 & \text{if } i = 3, \text{ case } (5) \text{ and } v_2(By^7) \geqslant 7 \ . \end{cases}$$

(3) *Assume that E does not have complex multiplication (which is the case if we assume that $xy \neq \pm 1$) and that it does not correspond to the equation*

$$64 \cdot 1^7 + 1 \cdot (-1)^7 = 7 \cdot 3^2 \ .$$

Then $E_i \sim_p f$ for some newform f of level

$$N_p = 2^\beta C^2 \operatorname{rad}(AB) \ ,$$

where

$$\beta = \begin{cases} \alpha & \text{in cases } (1) - (4) \ , \\ 0 & \text{in case } (5) \text{ and } v_2(B) \neq 0, 6 \ , \\ 1 & \text{in case } (5) \text{ and } v_2(B) = 0 \ , \\ -1 & \text{in case } (5) \text{ and } v_2(B) = 6 \ . \end{cases}$$

(4) *The curves E_i have nontrivial 2-torsion.*

(5) *Assume that $E = E_i$ is a curve associated with some solution (x, y, z) satisfying the above conditions. Suppose that F is another curve defined over $\mathbb{Q}$ such that $E \sim_p F$. Then the denominator of the j-invariant $j(F)$ of F is not divisible by any odd prime $q \neq p$ dividing C.*

Remark. Part (4) is included to help with the application of Proposition 15.4.1, and part (5) is often very useful for eliminating rational newforms (which correspond to elliptic curves), see for example Exercise (2) below.

Exercises. (1) Determine all the solutions to the equation

$$x^p + 2^r y^p = 3z^2, \quad r \geqslant 2, \quad p \geqslant 7 \text{ prime}$$

in coprime nonzero integers x, y, and z.

(2) Let F_n and L_n be the Fibonacci and Lucas numbers (see Section 6.8.1). Noting that $L_n^2 = 5F_n^2 + 4(-1)^n$, using the above recipes and in particular part (5) of the above theorem, prove that the equation $L_n = y^p$ has no solution with n even.

15.8.3 Recipes for Signature $(p, p, 3)$

Consider the equation
$$Ax^p + By^p = Cz^3 \,,$$

where $p \geqslant 5$ is prime and where we assume that Ax, By, and Cz are nonzero and pairwise coprime. As usual we may assume that for all primes q we have $v_q(A) < p$ and $v_q(B) < p$, and that C is cubefree. Without loss of generality we may also assume that $Ax \not\equiv 0 \pmod 3$ and $By^p \not\equiv 2 \pmod 3$. Let E be the elliptic curve with equation

$$Y^2 + 3CzXY + C^2 By^p Y = X^3 \,.$$

Theorem 15.8.4 (Bennett–Vatsal–Yazdani [Ben-Vat-Yaz]). *With the above assumptions and notation we have:*

(1) *The minimal discriminant $\Delta_{\min}$ of E is given by*

$$\Delta_{\min} = 3^3 AB^3 C^8 (xy^3)^p$$

(2) *The conductor of E is given by*

$$N = 3^\alpha \operatorname{rad}_3(ABxy) \operatorname{rad}_3(C)^2 \,,$$

where

$$\alpha = \begin{cases} 2 & \text{if } 9 \mid (2 + C^2 By^p - 3Cz) \,, \\ 3 & \text{if } 3 \| (2 + C^2 By^p - 3Cz) \,, \\ 4 & \text{if } v_3(By^p) = 1 \,, \\ 3 & \text{if } v_3(By^p) = 2 \,, \\ 0 & \text{if } v_3(By^p) = 3 \,, \\ 1 & \text{if } v_3(By^p) \geqslant 4 \,, \\ 5 & \text{if } 3 \mid C \,. \end{cases}$$

(3) *Assume that the curve E does not correspond to one of the equations*

$$1 \cdot 2^5 + 27 \cdot (-1)^5 = 5 \cdot 1^3, \quad 1 \cdot 2^7 + 3 \cdot (-1)^7 = 1 \cdot 5^3,$$
$$2 \cdot 1^5 + 27 \cdot (-1)^5 = 25 \cdot (-1)^3, \quad or \quad 2 \cdot 1^7 + 3 \cdot (-1)^7 = (-1)^3 \,.$$

Then $E \sim_p f$ for a newform f of level

$$N_p = 3^\beta \operatorname{rad}_3(AB) \operatorname{rad}_3(C)^2 \,,$$

where

$$\beta = \begin{cases} 2 & \text{if } 9 \mid (2 + C^2 By^p - 3Cz) \,, \\ 3 & \text{if } 3 \| (2 + C^2 By^p - 3Cz) \,, \\ 4 & \text{if } v_3(By^p) = 1 \,, \\ 3 & \text{if } v_3(By^p) = 2 \,, \\ 0 & \text{if } v_3(B) = 3 \,, \\ 1 & \text{if } v_3(By^p) \geqslant 4 \text{ and } v_3(B) \neq 3 \,, \\ 5 & \text{if } 3 \mid C \,. \end{cases}$$

(4) *The curve E has a point of order 3, namely the point $(0,0)$.*

(5) *Suppose that F is an elliptic curve defined over $\mathbb{Q}$ such that $E \sim_p F$. Then the denominator of the j-invariant $j(F)$ of F is not divisible by any odd prime $q \neq p$ dividing C.*

16. Catalan's Equation

The present chapter gives a complete proof of Catalan's conjecture, now Mihăilescu's Theorem 6.11.1. As with the other chapters in the last part of this book it is not totally self-contained, but is sufficiently complete that the reader can read what is missing in the literature (essentially the proof of Thaine's theorem). I have followed notes of Yu. Bilu [Bilu] and R. Schoof, but for the most part this is a rewrite of notes of J. Boéchat and M. Mischler [Boe-Mis]. I claim entire responsibility for possible errors.

16.1 Mihăilescu's First Two Theorems

Recall that Catalan's equation is $x^p - y^q = 1$ with p and q greater than or equal to 2 and x, y nonzero integers. We have seen that we can reduce to p and q distinct primes, that there are no solutions for $q = 2$, and that the only solutions for $p = 2$ are $(x, y) = (\pm 3, 2)$ for $q = 3$, so that we can assume that p and q are distinct odd primes. In that case Cassels's Theorem 6.11.5, and more precisely Corollary 6.11.6, tells us that there exist nonzero integers a and b, and positive integers u and v, such that

$$x = qbu, \quad x - 1 = p^{q-1}a^q, \quad \frac{x^p - 1}{x - 1} = pv^q ,$$

$$y = pav, \quad y + 1 = q^{p-1}b^p, \quad \frac{y^q + 1}{y + 1} = qu^p .$$

The goal of this section is to prove the first two theorems of Mihăilescu on Catalan's equation, the first of which considerably strengthens Cassels's theorem (Theorem 16.1.3). We begin by introducing some notation that will be used in the rest of this chapter, and then prove two lemmas, the first of which which will be used at other places in the proof.

Notation. In this chapter, unless mentioned otherwise, p and q will always stand for distinct odd primes, and x and y for nonzero integers such that $x^p - y^q = 1$. We will often write ζ instead of ζ_p. However, it will also be convenient at times to use ζ for any conjugate of ζ_p, and in that case we will have to use the notation ζ_p explicitly to mean a fixed primitive pth root of unity. We set $K = \mathbb{Q}(\zeta)$, $\pi = 1 - \zeta$, we let $\mathfrak{p} = \pi\mathbb{Z}_K$ be the unique prime ideal of K above p, we let $G = \mathrm{Gal}(K/\mathbb{Q}) \simeq (\mathbb{Z}/p\mathbb{Z})^*$, and when $p \nmid t$ we denote by $\sigma_t \in G$ the automorphism of K sending ζ to ζ^t. Since we will extensively use complex conjugation, as in Chapter 3 we will denote it by ι (Greek iota), although it will sometimes be convenient to write as usual $\overline{u}$ instead of $\iota(u)$. Note that $\iota \in G$, so it commutes with all the σ_t, and that $\iota\sigma_t = \sigma_{p-t}$.

Lemma 16.1.1. *For all i such that $1 \leqslant i \leqslant p-1$ set $\beta_i = (x - \zeta^i)/(1 - \zeta^i)$. The β_i are algebraic integers not divisible by $\mathfrak{p}$ and the ideals that they generate are pairwise coprime and equal to qth powers of ideals.*

Proof. By Cassels's relations recalled above we have $p \mid (x - 1)$, hence $v_p(x-1) \geqslant p - 1 \geqslant 2$, hence $v_\mathfrak{p}(\beta_i - 1) = v_\mathfrak{p}(x-1) - v_\mathfrak{p}(1 - \zeta^i) \geqslant 1$. It follows that $v_\mathfrak{p}(\beta_i) = 0$, and since $(1 - \zeta^i)\mathbb{Z}_K = \mathfrak{p}$, that β_i is an algebraic integer coprime to $\mathfrak{p}$. Furthermore, $(1 - \zeta^i)\beta_i - (1 - \zeta^j)\beta_j = \zeta^j - \zeta^i$. Since we have $(\zeta^j - \zeta^i)\mathbb{Z}_K = \mathfrak{p}$ for all $i \not\equiv j \pmod{p}$, it follows that for $1 \leqslant i \neq j \leqslant p-1$ the ideals $\beta_i\mathbb{Z}_K$ and $\beta_j\mathbb{Z}_K$ are integral and coprime. Finally, using the polynomial equality $\prod_{1 \leqslant i \leqslant p-1}(X - \zeta^i) = (X^p - 1)/(X - 1)$ and Cassels's relations we have

$$\prod_{1 \leqslant i \leqslant p-1} \beta_i = \frac{\prod_{1 \leqslant i \leqslant p-1}(x - \zeta^i)}{\prod_{1 \leqslant i \leqslant p-1}(1 - \zeta^i)} = \frac{x^p - 1}{p(x-1)} = v^q \ .$$

Since the ideals $\beta_i\mathbb{Z}_K$ are pairwise coprime, it follows that each of them is the qth power of an ideal, proving the lemma. $\qquad\square$

For simplicity, in the sequel we set $\beta = \beta_1 = (x - \zeta)/(1 - \zeta)$, so that there exists an ideal $\mathfrak{b}$ such that $\beta\mathbb{Z}_K = \mathfrak{b}^q$.

16.1.1 The First Theorem: Double Wieferich Pairs

We can now continue the study of Catalan's equation in two different ways. The first is the classical one initiated by Kummer in the study of FLT I, assuming that q does not divide the class number h_p of K, and deduce a contradiction. Indeed, this is what we are going to do in the next subsection. A second method, however, is to use Stickelberger's Theorem 3.6.19 instead, and this leads to complementary and stronger results. In the present subsection we begin with this latter method.

Recall that we denote by $I_s(p)$ the Stickelberger ideal of $K = \mathbb{Q}(\zeta_p)$.[1] Recall also that we denote complex conjugation by $\iota \in G$.

Lemma 16.1.2. *For any $\theta \in (1 - \iota)I_s(p)$ the element $(x - \zeta)^\theta$ is a qth power in K.*

---

[1] There is a necessary notational confusion here: when we were studying the Stickelberger ideal of a general cyclotomic field $\mathbb{Q}(\zeta_m)$ the letter p stood for a prime number not dividing m. This notation will no longer occur in this context, so here p is the prime such that $K = \mathbb{Q}(\zeta_p)$.

Proof. Write $\theta = (1 - \iota)\theta_1$ with $\theta_1 \in I_s(p)$. By the preceding lemma we have $\beta \mathbb{Z}_K = \mathfrak{b}^q$ for some ideal $\mathfrak{b}$, while by Stickelberger's Theorem 3.6.19 we have $\mathfrak{b}^{\theta_1} = \alpha \mathbb{Z}_K$ for some $\alpha \in K$. It follows that $\alpha^q \mathbb{Z}_K = \mathfrak{b}^{q\theta_1} = \beta^{\theta_1} \mathbb{Z}_K$, so there exists a unit $u \in U(K)$ such that $\beta^{\theta_1} = u\alpha^q$. We can thus write

$$(x - \zeta)^\theta = \left(\frac{1 - \zeta}{1 - \iota(\zeta)} \right)^{\theta_1} \frac{u}{\iota(u)} \left(\frac{\alpha}{\iota(\alpha)} \right)^q .$$

Now $(1 - \zeta)/(1 - \iota(\zeta)) = -\zeta$, and by Lemma 3.5.19, $u/\iota(u)$ is also a root of unity. Thus the first two factors are roots of unity in K, in other words $(2p)$th roots of unity, and since q is coprime to $2p$, they are qth powers, proving the lemma. $\square$

Mihăilescu's first theorem is the following.

Theorem 16.1.3 (Mihăilescu). *If p and q are odd primes, and x and y are nonzero integers such that $x^p - y^q = 1$, then*

$$p^2 \mid y, \quad q^2 \mid x, \quad q^{p-1} \equiv 1 \ (\mathrm{mod} \ p^2), \quad p^{q-1} \equiv 1 \ (\mathrm{mod} \ q^2) .$$

Proof. If $\mathfrak{m}$ is an integral ideal we will use the standard notation $u \equiv v$ (mod $^*\mathfrak{m}$) to mean that $v_\mathfrak{q}(u - v) \geqslant v_\mathfrak{q}(\mathfrak{m})$ for all prime ideals $\mathfrak{q} \mid \mathfrak{m}$, which allows us to work with congruences between algebraic numbers that are not necessarily algebraic integers. Since $(1 - x\zeta^{-1}) = (-\zeta^{-1})(x - \zeta)$ and since $(-\zeta^{-1})^\theta$ is a $2p$th root of unity hence a qth power, it follows from the above lemma that $(1 - x\zeta^{-1})^\theta$ is a qth power for any $\theta \in (1 - \iota)I_s(p)$. Furthermore, by Cassels's relations we have $q \mid x$, hence $(1 - x\zeta^{-1})^\theta \equiv 1$ (mod $^*q\mathbb{Z}_K$), so by the easy but crucial Exercise 20 of Chapter 3, since q is unramified in K it follows that $(1 - x\zeta^{-1})^\theta \equiv 1$ (mod $^*q^2\mathbb{Z}_K$). On the other hand, if we write $\theta = \sum_{\sigma \in G} a_\sigma \sigma$ it is clear by expanding and using $q \mid x$ that

$$(1 - x\zeta^{-1})^\theta = 1 - xS \ (\mathrm{mod} \ ^*q^2\mathbb{Z}_K) \quad \text{with} \quad S = \sum_{\sigma \in G} a_\sigma \sigma(\zeta^{-1}) .$$

It follows from these two congruences that $xS \equiv 0$ (mod $^*q^2\mathbb{Z}_K$). Assume by contradiction that $q^2 \nmid x$. Then $S \equiv 0$ (mod $^*q\mathbb{Z}_K$), and since the $\sigma(\zeta^{-1})$ form a permutation of the ζ^j for $1 \leqslant j \leqslant p - 1$, they form a $\mathbb{Z}$-basis of $\mathbb{Z}_K$, so $q \mid a_\sigma$ for all $\sigma \in G$. However, this cannot be true for all $\theta \in (1 - \iota)I_s(p)$: for instance, if we choose $\theta = (1 - \iota)\Theta_2$ then it is clear that

$$\theta = - \sum_{1 \leqslant j \leqslant (p-1)/2} \sigma_j^{-1} + \sum_{(p+1)/2 \leqslant j \leqslant p-1} \sigma_j^{-1}$$

does not satisfy this condition. It follows that $q^2 \mid x$, which is the first important result of the theorem. We can now easily conclude: since $q^2 \mid x$, using Cassels's relations we have $p^{q-1}a^q = x - 1 \equiv -1$ (mod q^2), and since $p^{q-1} \equiv 1$

(mod q) we thus have $a^q \equiv (-1)^q$ (mod q), so using once again Exercise 20 of Chapter 3 we deduce that $a^q \equiv -1$ (mod q^2), and replacing back in Cassels's relation gives $p^{q-1} \equiv 1$ (mod q^2). Since p and q play symmetrical roles (since they are odd, recall that we can change (p, q, x, y) to $(q, p, -y, -x)$) the symmetrical results of the theorem follow. $\square$

Remarks. (1) Because of Wieferich's criterion for FLT I (Corollary 6.9.10), a pair (p, q) of primes such that $p^{q-1} \equiv 1$ (mod q^2) and $q^{p-1} \equiv 1$ (mod p^2) is called a *double Wieferich pair*. The only known such pairs are $(2, 1093)$, $(3, 1006003)$, $(5, 1645333507)$, $(5, 188748146801)$, $(83, 4871)$, $(911, 318917)$, and $(2903, 18787)$; see [Kel-Ric]. However, on reasonable probabilistic grounds it is expected that there exist infinitely many, in fact even for a given p.

(2) In view of the simplicity of the above proof, based only on Stickelberger's theorem from 1890 and on Cassels's relations from 1960, it is quite surprising that it was not found before 2001 (published in 2003). It is even more surprising in view of the fact that Wieferich's and Furtwängler's criteria for FLT I (Theorem 6.9.9 and Corollary 6.9.10) are exactly of the same type and are also proved using Stickelberger's theorem, via Eisenstein's reciprocity law, which is a consequence.

16.1.2 The Equation $(x^p - 1)/(x - 1) = py^q$

The initial proof of Mihăilescu's second theorem used a very technical study of the action of complex conjugation ι. Part of this study will be necessary in any case for the third and fourth theorems, but for the second theorem a much simplified version, proving in fact a stronger result, has very recently been found by B. Dupuy, and I thank him and Yu. Bilu for the possibility of including it here. We thus begin by proving Dupuy's theorem, part of which had already been proved by Mihăilescu.

Theorem 16.1.4 (Mihăilescu, Dupuy). *Let p and q be distinct odd primes with $p \geqslant 5$. If $q \nmid h_p^-$ the equation $(x^p - 1)/(x - 1) = py^q$ has no solutions with $x \neq 1$ and $y \neq 1$.*

We prove this theorem by starting from Lemma 16.1.1 in a Kummer-like manner instead of using the Stickelberger ideal. We will first prove a number of intermediate results. In this section, we will always assume without further mention that p and q are distinct odd primes such that $q \nmid h_p^-$. Recall that we have set $\beta = (x - \zeta)/(1 - \zeta)$, and that we denote complex conjugation by ι, considered as an element of $\mathrm{Gal}(K/\mathbb{Q})$.

Lemma 16.1.5. *There exists $\mu \in K$ such that $\iota(\beta)/\beta = \mu^q$. Furthermore, μ is unique and satisfies $\iota(\mu) = \mu^{-1}$.*

Proof. By Lemma 16.1.1 we know that β is an algebraic integer coprime to its conjugates, not divisible by $\mathfrak{p}$, and equal to a qth power of an ideal, so write $\beta\mathbb{Z}_K = \mathfrak{b}^q$ for some integral ideal $\mathfrak{b}$ of K. Recall from Proposition 3.5.21 that the natural map j from $Cl(K^+)$ to $Cl(K)$ is injective, and that by definition $h_p^- = h_p/h_p^+ = [Cl(K) : j(Cl(K^+))]$. It follows that $[\mathfrak{b}]^{h_p^-} \in Cl(K^+)$; in other words there exist an ideal $\mathfrak{c}_1$ of K^+ and an element $\gamma_1 \in K^*$ such that $\mathfrak{b}^{h_p^-} = \gamma_1 j(\mathfrak{c}_1)$. Since q and h_p^- are coprime we can write $uq + vh_p^- = 1$ for some integers u and v, so that $\mathfrak{b} = \beta^u\gamma_1^v j(\mathfrak{c}_1^v) = \gamma j(\mathfrak{c})$, where $\gamma \in K^*$ and $\mathfrak{c} \subset K^+$. It follows that $j(\mathfrak{c}^q) = \beta\gamma^{-q}\mathbb{Z}_K$ is a principal ideal of K, and since j is injective, $\mathfrak{c}^q$ itself is a principal ideal of K^+, say $\mathfrak{c}^q = \delta\mathbb{Z}_{K^+}$ with $\delta \in K^+$. Thus $\beta\mathbb{Z}_K = \gamma^q\delta\mathbb{Z}_K$, so there exists a unit $u \in K$ such that $\beta = u\gamma^q\delta$. Since $\delta \in K^+$ it follows that $\iota(\beta)/\beta = (\iota(u)/u)(\iota(\gamma)/\gamma)^q$. By Lemma 3.5.19 we know that $\iota(u)/u$ is a root of unity, hence a $2p$th root of unity in K, and since q is odd and distinct from p, it is coprime to $2p$, so it follows that $\iota(u)/u$ is a qth power, proving the first statement. Since K does not contain any qth roots of unity different from 1 the element μ is unique, and since $(\iota(\mu)\mu)^q = 1$, for the same reason we have $\iota(\mu) = \mu^{-1}$. $\qquad\square$

Lemma 16.1.6. *Denote by r an inverse of q modulo p. Then $\phi = \beta(\mu+\zeta^r)^q$ is a unit of K.*

Proof. By the binomial theorem we have $\beta(\mu + \zeta^r)^q = \sum_{0\leqslant n\leqslant q} c_n\beta\mu^n$ for some algebraic *integers* c_n. Since $(\beta\mu^n)^q = \beta^{q-n}\iota(\beta)^n$ and β is an algebraic integer, it follows that ϕ is an algebraic integer. Now set

$$\phi' = \beta^{q-1}\left(\sum_{0\leqslant k\leqslant q-1} \mu^k(-\zeta^r)^{q-1-k}\right)^q .$$

As above, we have $\phi' = \sum_{0\leqslant n\leqslant q(q-1)} d_n\beta^{q-1}\mu^n$ for some algebraic integers d_n, and since $(\beta^{q-1}\mu^n)^q = \beta^{q(q-1)-n}\iota(\beta)^n$ is an algebraic integer, ϕ' is also an algebraic integer. Finally, a small computation using $\zeta^{rq} = \zeta$ shows that $\phi\phi' = ((1-\zeta^2)/(1-\zeta))^q$, which is a cyclotomic unit, showing that ϕ is a unit. $\qquad\square$

We now work in $K_{\mathfrak{p}} = \mathbb{Q}_p(\zeta)$. By Lemma 6.11.3 we have $x \equiv 1 \pmod{\mathfrak{p}}$, so that $v_p(\lambda) \geqslant 1 - 1/(p-1)$, where we have set $\lambda = (x-1)/(1-\zeta)$. Note that $\iota(\lambda) = (x-1)/(1-\zeta^{-1}) = -\zeta\lambda$. Since $p \neq q$, it follows from Corollary 4.2.15 that the series $(1+x)^{1/q}$ converges in $K_{\mathfrak{p}}$ for $x = \lambda$ and $x = -\zeta\lambda$.

Lemma 16.1.7. *We have*

$$\mu = (1+\lambda)^{-1/q}(1-\zeta\lambda)^{1/q} ,$$

where $(1+\lambda)^{-1/q}$ (and similarly $(1-\zeta\lambda)^{1/q}$) means the power series $(1+x)^{-1/q}$ evaluated at $x = \lambda$.

Proof. Denote by ν the right-hand side. Since $1 + \lambda = (x - \zeta)/(1 - \zeta) = \beta$ and $\iota(\lambda) = -\zeta\lambda$, we have $\nu^q = \iota(\beta)/\beta = \mu^q$, so that $\mu = \eta\nu$ for some qth root of unity $\eta \in K_{\mathfrak{p}}$. Since $K_{\mathfrak{p}} \simeq \mathbb{Q}_p(\zeta)$, it follows from Exercise 20 of Chapter 4 that if $p \not\equiv 1 \pmod{q}$ we have $\eta = 1$, while if $p \equiv 1 \pmod{q}$ we have $\eta \in \mathbb{Q}_p$. In particular, we have $\iota(\eta) = \eta$, where we identify $\mathrm{Gal}(K_{\mathfrak{p}}/\mathbb{Q}_p)$ with $\mathrm{Gal}(K/\mathbb{Q})$. Applying ι to the equality $\mu = \eta\nu$ and using $\iota(\mu) = \mu^{-1}$, we obtain $\mu^{-1} = \eta\iota(\nu) = \eta\nu^{-1}$, so that $\eta^2 = 1$. Since $\eta^q = 1$ and q is odd we have $\eta = 1$ in all cases, as claimed. $\qquad\square$

As a last result, we need a congruence for ϕ. We will use the convenient notation $u = v + O(x)$ for some $x \in K_{\mathfrak{p}}$ to mean that $v_p(u - v) \geqslant v_p(x)$, and we recall that $v_p(\lambda) \geqslant 1 - 1/(p - 1)$.

Lemma 16.1.8. (1) *We have*

$$\frac{\phi}{(1 + \zeta)^q} = \left(1 - (x - 1)\frac{\zeta^r - \zeta}{(\zeta - 1)(\zeta^r + 1)} + O(\lambda^2)\right).$$

(2) *If $q \equiv 1 \pmod{p}$, we have*

$$\frac{\phi}{(1 + \zeta)^q} = \left(1 - \frac{q - 1}{2q}(x - 1)^2\frac{\zeta}{(\zeta - 1)^2} + O((q - 1)^2\lambda^2) + O((q - 1)\lambda^3)\right).$$

Proof. (1). By definition we have $\beta = (x - \zeta)/(1 - \zeta) = 1 + \lambda$. On the other hand, by the preceding lemma we have $\mu = (1 + \lambda)^{-1/q}(1 - \zeta\lambda)^{1/q}$. Since $1/q \in \mathbb{Z}_p$ we have $\binom{1/q}{k} \in \mathbb{Z}_p$ for all k, so that by expanding the power series we obtain

$$\phi = \beta(\mu + \zeta^r)^q = (1 + \zeta^r)^q(1 + a_1\lambda + O(\lambda^2))$$

with $a_1 = (\zeta^r - \zeta)/(\zeta^r + 1)$, and (1) follows after replacing λ by $(x - 1)/(1 - \zeta)$.

(2). Set $a = (q - 1)/q$, so that $v_p(a) \geqslant 1$. We write

$$\mu = (1 - \zeta\lambda)(1 + \lambda)^{-1}(1 + \lambda)^a(1 - \zeta\lambda)^{-a}.$$

By Lemma 4.2.12 applied to $a_n = a(a - 1) \cdots (a - n + 1)$, for $v_p(x) \geqslant 1/(p-3)$ we have

$$(1 + x)^a = 1 + ax + \frac{a(a - 1)}{2}x^2 + O(ax^3) = 1 + ax - \frac{a}{2}x^2 + O(a^2x^2) + O(ax^3).$$

Furthermore, we have $v_p(\lambda) \geqslant 1 - 1/(p - 1) > 1/(p - 3)$ since $p \geqslant 5$. Thus after a small calculation we obtain

$$\mu(1 + \lambda) = 1 + \lambda(a(1 + \zeta) - \zeta) - \frac{a\lambda^2}{2}(\zeta + 1)^2 + O(a^2\lambda^2) + O(a\lambda^3),$$

so that $(\mu + \zeta)(1 + \lambda) = (1 + \zeta)(1 + a\lambda - a\lambda^2(1 + \zeta)/2 + O(a^2\lambda^2) + O(a\lambda^3))$. It follows that

$$\phi = (1+\zeta)^q(1+\lambda)^{1-q}\left(1+aq\lambda - aq\frac{1+\zeta}{2}\lambda^2 + O(a^2\lambda^2) + O(a\lambda^3)\right)$$

$$= (1+\zeta)^q(1+\lambda)^{1-q}(1-\frac{a}{a-1}\lambda - a\frac{1+\zeta}{2}\lambda^2 + O(a^2\lambda^2) + O(a\lambda^3))$$

since $aq = q-1 = a/(1-a) = a+O(a^2)$. Applying once again Lemma 4.2.12 to $a_n = (1-q)(-q)\cdots(2-q-n)$ we obtain similarly

$$(1+\lambda)^{1-q} = 1 + \frac{a}{a-1}\lambda - \frac{a}{2(a-1)}\lambda^2 + O(a^2\lambda^2) + O(a\lambda^3)$$

$$= 1 + \frac{a}{a-1}\lambda + \frac{a}{2}\lambda^2 + O(a^2\lambda^2) + O(a\lambda^3),$$

and replacing gives

$$\phi = (1+\zeta)^q\left(1 - \frac{a\zeta}{2}\lambda^2 + O(a^2\lambda^2) + O(a\lambda^3)\right),$$

proving (2) after replacing a by $(q-1)/q$ and λ by $(x-1)/(1-\zeta)$.

Note that we have proved the lemma by working in $K_{\mathfrak{p}}$, but since all the quantities that are involved are in K, the result is also true in K. □

For simplicity, write $\mathcal{N}$ and Tr instead of $\mathcal{N}_{K/\mathbb{Q}}$ and $\mathrm{Tr}_{K/\mathbb{Q}}$.

Lemma 16.1.9. *For $\alpha \in \mathbb{Z}_K$ we have*

$$\mathcal{N}(\alpha) \equiv 1 + \mathrm{Tr}(\alpha-1) + O(\alpha-1)^2.$$

Proof. Write $\varepsilon = \alpha - 1$ and $k = (p-1)v_p(\alpha-1) = v_{\mathfrak{p}}(\alpha-1)$. For any $\sigma \in G = \mathrm{Gal}(K/\mathbb{Q})$ we have $\sigma(\alpha) = 1 + \sigma(\varepsilon)$, and $\sigma(\varepsilon) \in \mathfrak{p}^k$ since $\mathfrak{p}$ is stable by σ. Multiplying these equations together gives

$$\prod_{\sigma \in G}\alpha = 1 + \sum_{\sigma \in G}\sigma(\varepsilon) + O(\mathfrak{p}^{2k}),$$

proving the lemma. □

Lemma 16.1.10. *Denote by π_K the canonical reduction map from $\mathbb{Z}_K$ to $\mathbb{Z}_K/\mathfrak{p} = \mathbb{F}_p$, which is a ring homomorphism. For any $\alpha \in \mathbb{Z}_K$ we have $\mathrm{Tr}(\alpha) \equiv -\pi_K(\alpha) \pmod{p}$.*

Proof. We have $\mathrm{Tr}(\zeta^k) = \mathrm{Tr}(\zeta) = -1$ if $p \nmid k$, and $\mathrm{Tr}(\zeta^k) = \mathrm{Tr}(1) = p - 1 \equiv -1 \pmod{p}$ if $p \mid k$, so the result is true for $\alpha = \zeta^k$ since $\zeta \equiv 1 \pmod{\mathfrak{p}}$, therefore for any $\alpha \in \mathbb{Z}_K$ by linearity. □

Proof of Theorem 16.1.4. Set $b_1 = (\zeta^r-\zeta)/((\zeta-1)(\zeta^r+1))$. From Lemma 16.1.8 (1), the above lemma, and the fact that $(x-1)b_1/\lambda$ is integral we deduce that $\mathcal{N}(\phi/(1+\zeta)^q) = 1 - (x-1)\mathrm{Tr}(b_1) + O(\lambda^2)$, and in particular $\mathcal{N}(\phi/(1+$

$\zeta)^q) \equiv 1 \pmod{\mathfrak{p}}$. Since ϕ and $1+\zeta$ are units we have $\mathcal{N}(\phi/(1+\zeta)^q) = \pm 1$, and since $\mathfrak{p}$ is not above 2 the sign must be $+$, so that $v_p((x-1)\operatorname{Tr}(b_1)) \geqslant v_p(\lambda^2)$. Now by the above lemma we have

$$\operatorname{Tr}(b_1) \equiv -\pi_K(b_1) \equiv -\pi_K((\zeta^r - \zeta)/(\zeta - 1))\pi_K(\zeta^r + 1)^{-1} \pmod{p} \ .$$

From $(\zeta^r - \zeta)/(\zeta - 1) = \sum_{1 \leqslant j \leqslant r-1} \zeta^j$ we deduce that $\pi_K((\zeta^r - \zeta)/(\zeta-1)) = r - 1$, and $\pi_K(\zeta^r + 1) = 2$, so that $\operatorname{Tr}(b_1) \equiv (r-1)/2 \pmod{p}$. It follows that if $r \not\equiv 1 \pmod{p}$, in other words if $q \not\equiv 1 \pmod{p}$, we have $v_p(\operatorname{Tr}(b_1)) = 0$. Since $\lambda = (x-1)/(1-\zeta)$ we deduce from the inequality given above that $v_p(x-1) \geqslant 2(v_p(x-1) - 1/(p-1))$, so that $v_p(x-1) \leqslant 2/(p-1)$, giving a contradiction when $p \geqslant 5$ since $v_p(x-1) \geqslant 1$, and proving Theorem 16.1.4 when $q \not\equiv 1 \pmod{p}$.

Assume now that $q \equiv r \equiv 1 \pmod{p}$. In that case $b_1 = 0$, so by Lemma 16.1.8 (2) we have

$$\phi = (1+\zeta)^q(1 - (x-1)^2((q-1)/(2q))c + O((q-1)^2\lambda^2) + O((q-1)\lambda^3))$$

with $c = \zeta/(\zeta-1)^2$. As in the preceding case we must have $\mathcal{N}(\phi/(1+\zeta)^q) = 1$, and

$$\mathcal{N}(\phi/(1+\zeta)^q) = 1 - (x-1)^2((q-1)/(2q))\operatorname{Tr}(c) + O((q-1)^2\lambda^2) + O((q-1)\lambda^3) \ .$$

By Exercise 25 of Chapter 2 we have $\operatorname{Tr}(c) = (1 - p^2)/12$, which has p-adic valuation 0 for $p \geqslant 5$, so the above equalities together with $v_p(\lambda) = v_p(x-1) - 1/(p-1)$ imply that $2v_p(x-1) + v_p(q-1) \geqslant \min(A, B)$, with

$$A = 2v_p(q-1) + 2v_p(x-1) - 2/(p-1) \text{ and } B = v_p(q-1) + 3v_p(x-1) - 3/(p-1)) \ .$$

If the minimum is equal to A this gives the inequality $v_p(q-1) \leqslant 2/(p-1) < 1$ for $p \geqslant 5$, a contradiction since $p \mid (q-1)$. If the minimum is equal to B this gives the inequality $v_p(x-1) \leqslant 3/(p-1) < 1$, once again a contradiction since $p \mid (x-1)$, proving the theorem. $\qquad\square$

16.1.3 Mihăilescu's Second Theorem: $p \mid h_q^-$ and $q \mid h_p^-$

Mihăilescu's second theorem is now immediate.

Theorem 16.1.11 (Mihăilescu). *Let p and q be distinct odd primes. If $p \nmid h_q^-$ or $q \nmid h_p^-$ the equation $x^p - y^q = 1$ does not have any nonzero solutions.*

Proof. By Cassels's theorem, we know that a solution to $x^p - y^q = 1$ implies $(x^p - 1)/(x - 1) = pv^q$, which does not have any nontrivial solution for $p \geqslant 5$ by Dupuy's theorem, and for $p = 3$ by Nagell's Corollary 6.7.15. Symmetrically $((-y)^q - 1)/((-y) - 1) = qu^p$ also does not have any nontrivial solution if $p \nmid h_q^-$. $\qquad\square$

Corollary 16.1.12. *If p and q are distinct odd primes and p or q is less than or equal to 43 then the equation $x^p - y^q = 1$ does not have any nonzero solutions.*

Proof. Thanks to the above theorem it is sufficient to check that for any p and q with $\min(p, q) \leqslant 43$ we have $p \nmid h_q^-$ or $q \nmid h_p^-$. For this we need to compute h_p^- for small values of p, which is easily done thanks to Corollary 10.5.27. We find in completely factored form that $h_p^- = 1$ for $p \leqslant 19$, and $h_{23}^- = 3$, $h_{29}^- = 2^3$, $h_{31}^- = 3^2$, $h_{37}^- = 37$ (coming from the fact that 37 is an irregular prime), $h_{41}^- = 11^2$, and $h_{43}^- = 211$. By symmetry we may assume that $3 \leqslant p < q$. From this list we see that $q \nmid h_p^-$ since all the prime divisors of h_p^- are less than or equal to p, except when $p = 43$. In that case $q \mid h_{43}^-$ for $q = 211$. Thus we must check that $43 \nmid h_{211}^-$, which is the case since we compute that

$$h_{211}^- = 3^2 \cdot 7^2 \cdot 41 \cdot 71 \cdot 181 \cdot 281 \cdot 421 \cdot 1051 \cdot 12251 \cdot 113981701 \cdot 4343510221$$

(the fact that h_{211}^- has so many small factors comes from Corollary 10.5.28, and evidently we do not need this complete factorization just to check that $43 \nmid h_{211}^-$). □

Remarks. (1) The reason we stop at 43 is that for $p = 47$ and $q = 139$ we check that $q \mid h_p^-$ and $p \mid h_q^-$ so that we cannot apply Theorem 16.1.11 to this pair. In any case we have proved this result just for fun, because we will use it only up to 11.

(2) We now have two quite different criteria for proving that Catalan's equation does not have any nonzero solutions: Theorems 16.1.3 and 16.1.11. It is highly plausible that there are no pairs (p, q) satisfying both, but this is not known. However, thanks to the work of Baker and followers on linear forms in logarithms, it is not difficult to show that the above theorems of Mihăilescu prove Catalan's conjecture up to a finite and not too unreasonable amount of computer calculations, which have been started but not completed, first because they would be very long, but second, mainly because thanks to further theorems of Mihăilescu that we will see below they are not necessary.

16.2 The + and − Subspaces and the Group S

From now on we will have to consider carefully the action of complex conjugation on all the objects that we study. Thus as usual we denote by K^+ the subfield of K fixed by complex conjugation ι, in other words the maximal totally real subfield of K; see Section 3.5.4.

16.2.1 The $+$ and $-$ Subspaces

Let R be a commutative ring and M an $R[G]$-module. We define $M^{\pm} = \{x \in M, \iota(x) = \pm x\}$. If 2 is invertible in R we set $\varepsilon^{\pm} = (1 \pm \iota)/2 \in R[G]$. It is clear that $\varepsilon^{\pm}$ are *complementary projectors*, in other words that $(\varepsilon^{\pm})^2 = \varepsilon^{\pm}$, $\varepsilon^+ + \varepsilon^- = 1$, and $\varepsilon^+ \varepsilon^- = 0$. It is also clear that $M^{\pm} = \varepsilon^{\pm} M$ and that $M = M^+ \oplus M^-$. If 2 is not invertible in R (for instance in the important special case $R = \mathbb{Z}$) we set $\varepsilon^{\pm} = 1 \pm \iota$, and we have only the inclusions $\varepsilon^{\pm} M \subset M^{\pm}$ and $M^+ \oplus M^- \subset M$, the indexes being powers of 2. In the special case where $M = R[G]$ however, we have the following:

Lemma 16.2.1. $\varepsilon^{\pm} R[G] = R[G]^{\pm}$, and both are free R-modules of dimension $(p-1)/2$.

Proof. The left-hand side is always a submodule of the right-hand side. Thus let $x = \sum_{1 \leqslant t \leqslant p-1} a_t \sigma_t \in R[G]^{\pm}$. Since $\iota \sigma_t = \sigma_{p-t}$, this means that $a_{p-t} = \pm a_t$. Thus if we set $y = \sum_{1 \leqslant t \leqslant (p-1)/2} a_t \sigma_t$ it is clear that $x = \varepsilon^{\pm} y$. The last statement is clear since $a_{p-t} = \pm a_t$.   □

Exercise: show that the index of $\mathbb{Z}[G]^+ \oplus \mathbb{Z}[G]^-$ in $\mathbb{Z}[G]$ is equal to $2^{(p-1)/2}$.

Recall that we have defined the Stickelberger ideal $I_s = I_s(p)$ by $I_s = \mathbb{Z}[G] \cap \Theta \mathbb{Z}[G]$, where

$$\Theta = \frac{1}{p} \sum_{1 \leqslant t \leqslant p-1} t \sigma_t^{-1} .$$

We define

$$I = (1 - \iota) I_s = \varepsilon^- I_s \subset I_s^- = I_s \cap \mathbb{Z}[G]^- = I_s \cap \varepsilon^- \mathbb{Z}[G] ,$$

the last equality following from the lemma (we have already used I in Lemma 16.1.2). By Lemma 3.6.17, I_s is generated by Θ_{p+1} and by the Θ_b for $1 \leqslant b \leqslant p-1$. Set $g_b = -\Theta_b$ for $1 \leqslant b \leqslant p-1$, and $g_p = -\Theta_{p+1}$. By Lemma 3.6.16 we have $g_b = \sum_{1 \leqslant t \leqslant p-1} \lfloor bt/p \rfloor \sigma_t^{-1}$, including for $b = p$ since $\lfloor (p+1)t/p \rfloor = t$ for $1 \leqslant t \leqslant p-1$. Finally, for $1 \leqslant i \leqslant p-1$ set

$$f_i = g_{i+1} - g_i = \sum_{1 \leqslant t \leqslant p-1} \left(\left\lfloor \frac{t(i+1)}{p} \right\rfloor - \left\lfloor \frac{ti}{p} \right\rfloor \right) \sigma_t^{-1} ,$$

where we note that the coefficient of σ_t^{-1} is equal to 0 or 1. Since the g_i for $1 \leqslant i \leqslant p$ generate I_s and that we have $g_1 = 0$, it follows that the f_i for $1 \leqslant i \leqslant p-1$ also generate I_s. Furthermore, since $\lfloor tp/p \rfloor = t$ and $\lfloor t(p-1)/p \rfloor = t-1$ for $1 \leqslant t \leqslant p-1$ it follows that $f_{p-1} = \sum_{1 \leqslant t \leqslant p-1} \sigma_t$. This is the *norm* element $s(G) \in \mathbb{Z}[G]$ since $\alpha^{s(G)} = \mathcal{N}_{K/\mathbb{Q}}(\alpha)$.

Definition 16.2.2. If $f = \sum_{1 \leqslant t \leqslant p-1} a_t \sigma_t \in \mathbb{Z}[G]$ we define

$$\|f\| = \sum_{1 \leqslant t \leqslant p-1} |a_t| .$$

It is clear that $\|f\| \geqslant 0$, that $\|f\| = 0$ if and only if $f = 0$, and it is immediately checked that $\|fg\| \leqslant \|f\|\|g\|$, and it is clear as well that there is equality if all the coefficients of f and g are nonnegative.

Lemma 16.2.3. (1) *For* $1 \leqslant i \leqslant p - 2$ *we have* $\|f_i\| = (p-1)/2$.
(2) I_s *is a free $\mathbb{Z}$-module of rank* $(p+1)/2$ *generated by the* f_i *for* $1 \leqslant i \leqslant (p-1)/2$ *and by* $f_{p-1} = s(G)$.
(3) I *is a free $\mathbb{Z}$-module of rank* $(p-1)/2$ *generated by the* $e_i = \varepsilon^- f_i$ *for* $1 \leqslant i \leqslant (p-1)/2$.
(4) *For* $1 \leqslant i \leqslant (p-1)/2$ *the coefficients of* e_i *are equal to* ± 1, *and in particular* $\|e_i\| = p - 1$.

Proof. (1) and (4). For $1 \leqslant t \leqslant p-1$ and $1 \leqslant i \leqslant p-1$ we note that

$$\lfloor ti/p \rfloor + \lfloor (p-t)i/p \rfloor = \lfloor ti/p \rfloor + i - \lceil ti/p \rceil = i - 1$$

since $p \nmid ti$. It follows that

$$\sum_{1 \leqslant t \leqslant p-1} \lfloor ti/p \rfloor = \sum_{1 \leqslant t \leqslant (p-1)/2} (\lfloor ti/p \rfloor + \lfloor (p-t)i/p \rfloor) = (i-1)(p-1)/2 \,.$$

Since the coefficients of f_i are equal to 0 or 1, for $1 \leqslant i \leqslant p - 2$ we have $\|f_i\| = i(p-1)/2 - (i-1)(p-1)/2 = (p-1)/2$, proving (1) (note that this is false for $i = p-1$ since in that case the above computation is not valid for $i + 1 = p$, and in fact we know that $\|f_{p-1}\| = \|s(G)\| = p - 1$). The proof of (4) follows immediately from (1) and is left to the reader.

(2). Exchanging i and t in the first equality proved in (1) we see that

$$\lfloor it/p \rfloor + \lfloor (p-i)t/p \rfloor = t - 1 = \lfloor (i+1)t/p \rfloor + \lfloor (p-i-1)t/p \rfloor \,,$$

in other words that

$$\lfloor (p-i)t/p \rfloor - \lfloor (p-i-1)t/p \rfloor = \lfloor (i+1)t/p \rfloor - \lfloor it/p \rfloor \,.$$

It follows that $f_{p-1-i} = f_i$, so the f_i for $1 \leqslant i \leqslant (p-1)/2$ together with $s(G)$ generate I_s.

Let us set $e_i = \varepsilon^- f_i$. Since trivially $\varepsilon^-(s(G)) = 0$ and $I = \varepsilon^- I_s$, it follows that the e_i for $1 \leqslant i \leqslant (p-1)/2$ generate I. Assume that we have shown (3), in other words that the e_i form a $\mathbb{Z}$-basis of I. It is then clear that the f_i for $1 \leqslant i \leqslant (p-1)/2$ together with $s(G)$ form a $\mathbb{Z}$-basis of I_s: indeed if we had a relation $\sum_{1 \leqslant i \leqslant (p-1)/2} \lambda_i f_i + \lambda s(G) = 0$, then applying ε^- we would obtain $\sum_{1 \leqslant i \leqslant (p-1)/2} \lambda_i e_i = 0$; hence $\lambda_i = 0$, and hence $\lambda = 0$ also, and (2) follows.

(3). We will prove (3) indirectly. Since I_s is a finitely generated torsion-free $\mathbb{Z}$-module, it is free, as are its submodules. Proving (3) is thus equivalent to showing that the $\mathbb{Z}$-rank of I is equal to $(p-1)/2$. Now by Lemma 16.2.1, $\dim_{\mathbb{Z}} \mathbb{Z}[G]^- = (p-1)/2$. By Lemma 3.6.22, multiplication by $p\Theta$ is an injective map from $\mathbb{Z}[G]^-$ to $\mathbb{Z}[G]^-$, so $\dim_{\mathbb{Z}} p\Theta\mathbb{Z}[G]^- = (p-1)/2$. Now by definition $I_s = \Theta\mathbb{Z}[G] \cap \mathbb{Z}[G]$, so $I_s^- = \Theta\mathbb{Z}[G] \cap \mathbb{Z}[G]^-$. Since $p\Theta \in \mathbb{Z}[G]$ we thus have the chain of inclusions

$$p\Theta\mathbb{Z}[G]^- = p\Theta\mathbb{Z}[G]^- \cap \mathbb{Z}[G]^- \subset \Theta\mathbb{Z}[G]^- \cap \mathbb{Z}[G]^- \subset I_s^- \subset \mathbb{Z}[G]^-.$$

Since the extremities of this chain have $\mathbb{Z}$-rank equal to $(p-1)/2$, it follows that all the terms of the chain do, and in particular that $\dim_{\mathbb{Z}}(I_s^-) = (p-1)/2$. Finally, we note that if $x \in I_s^-$ then $\varepsilon^- x \in I$, but on the other hand, $\varepsilon^- x = x + x = 2x$. It follows that $2I_s^- \subset I \subset I_s^-$, so $\dim_{\mathbb{Z}}(I) = \dim_{\mathbb{Z}}(I_s^-) = (p-1)/2$, proving the lemma. $\qquad\square$

Remark. It follows from this lemma that the e_i for $1 \leqslant i \leqslant (p-1)/2$ are $\mathbb{Z}$-linearly independent. We leave as an exercise for the reader to show that this is equivalent to showing that the $((p-1)/2) \times ((p-1)/2)$ matrix $M = (m_{i,j})_{1 \leqslant i,j \leqslant (p-1)/2}$ defined by $m_{i,j} = \lfloor (i+1)(j+1)/p \rfloor$ has a nonzero determinant. This can be done without too much difficulty by showing that $\det(M)$ is equal to the determinant of the map multiplication by Θ from $\mathbb{C}[G]^-$ to itself multiplied by $p/(2^{(p-3)/2})$, which as remarked after Lemma 3.6.22 is up to sign equal to h_p^- (see Exercise 61 of Chapter 10).

16.2.2 The Group S

Recall that in this whole chapter q denotes an odd prime distinct from p.

Definition 16.2.4. (1) *We define*

$$E = \{u\pi^k, \ u \in U(K), \ k \in \mathbb{Z}\}.$$

(2) *We define V to be the group of elements $\alpha \in K^*$ such that $v_{\mathfrak{r}}(\alpha) \equiv 0$ (mod q) for all prime ideals $\mathfrak{r} \neq \mathfrak{p}$, and we set $S = V/K^{*q}$.*

Remarks. (1) It we set $T = \{\mathfrak{p}\}$, the group E is simply the group of T-units of K, while the group S is the so-called q-Selmer group of the ring $\mathbb{Z}_{K,T}$ of T-integers of K, but we will not use this terminology.
(2) The group V is also equal to the set of $\alpha \in K^*$ such that $\alpha\mathbb{Z}_K = \mathfrak{b}^q\mathfrak{p}^k$ for some ideal $\mathfrak{b}$ and some $k \in \mathbb{Z}$. In particular, Lemma 16.1.1 tells us that $x - \zeta \in V$, so that the class of $x - \zeta$ belongs to S.
(3) We could perform the entire proof that follows using $U(K)$ instead of E, and make all the corresponding changes to the groups S, C, etc., that we will define, but we have slightly more freedom by allowing arbitrary powers of π in the elements that we use. The price to pay is that we will work in $\mathbb{Z}[\zeta_p, 1/p]$ instead of $\mathbb{Z}[\zeta_p]$.

Proposition 16.2.5. (1) *E is a $\mathbb{Z}[G]$-module and $E = \mathbb{Z}[\zeta_p, 1/p]^*$.*
(2) *$\alpha \in V$ if and only if there exists an ideal $\mathfrak{a}$ and $k \in \mathbb{Z}$ such that $\alpha\mathbb{Z}_K = \pi^k\mathfrak{a}^q$.*

(3) S *is a* $\mathbb{Z}[G]$*-module annihilated by* $q\mathbb{Z}[G]$*, so* S *is an* $\mathbb{F}_q[G]$*-module.*

Proof. Immediate consequences of the definition and left to the reader. □

We set $G^+ = \mathrm{Gal}(K^+/\mathbb{Q}) = G/\langle \iota \rangle$, which has cardinality $(p-1)/2$. We denote as usual by $Cl(K)$ and $Cl(K^+)$ the class groups of K and K^+ respectively. The group $Cl(K)$ is a $\mathbb{Z}[G]$-module, so we can speak of $Cl(K)^{\pm}$. By definition $Cl(K)^+$ is the subgroup of ideal classes invariant by ι. This is in general *not* equal to $Cl(K^+)$, but by Proposition 3.5.21 the natural map from $Cl(K^+)$ to $Cl(K)^+$ is injective, so that $Cl(K^+)$ can be considered as a subgroup of $Cl(K)^+$ and in particular $h_p^+ \mid |Cl(K)^+|$. Furthermore, by the general considerations given at the beginning of Section 16.2.1 we have $Cl(K)^- \oplus Cl(K)^+ \subset Cl(K)$. It follows that there is a natural injection from $Cl(K)^-$ to $Cl(K)/Cl(K)^+$. In particular

$$|Cl(K)^-| \quad \text{divides} \quad \frac{h_p}{|Cl(K)^+|} \quad \text{divides} \quad \frac{h_p}{h_p^+} = h_p^- \ .$$

In the following lemma, recall that if A is an abelian group, then $A[q]$ denotes the set of elements $x \in A$ such that $x^q = 1$ (or $qx = 0$ in additive notation).

Lemma 16.2.6. *Keep all the above notation.*

(1) *We have an exact sequence of* $\mathbb{F}_q[G]$*-modules*

$$0 \longrightarrow E/E^q \longrightarrow S \longrightarrow Cl(K)[q] \longrightarrow 0 \ .$$

(2) E/E^q *is invariant by* ι*, so is an* $\mathbb{F}_q[G^+]$*-module.*
(3) *We have* $S^- \simeq Cl(K)[q]^- = Cl(K)^-[q]$ *and an exact sequence of* $\mathbb{F}_q[G^+]$*-modules*

$$0 \longrightarrow E/E^q \longrightarrow S^+ \longrightarrow Cl(K)[q]^+ \longrightarrow 0 \ .$$

(4) S *is annihilated by* I*.*

Proof. (1) is a general property of Selmer groups and is immediate to prove: if $\overline{\alpha} \in S$ then $\alpha\mathbb{Z}_K = \pi^k \mathfrak{a}^q$ for some ideal $\mathfrak{a}$, and we send $\overline{\alpha}$ to the ideal class of $\mathfrak{a}$. It is clear that this lands in $Cl[q]$, and that it is independent of the chosen representative of $\overline{\alpha}$ (changing α into $\alpha\gamma^q$ amounts to changing $\mathfrak{a}$ into $\gamma\mathfrak{a}$, which is in the same ideal class). Its kernel is the set of $\overline{\alpha}$ such that $\alpha\mathbb{Z}_K = \pi^k \gamma^q \mathbb{Z}_K$ for some element γ; hence $\alpha = \pi^k \gamma^q u$ for some $u \in U(K)$, so that $\alpha/\gamma^q \in E$ is such that $\overline{\alpha/\gamma^q} = \overline{\alpha}$. Finally, the map is surjective since if $\mathfrak{a}^q = \alpha\mathbb{Z}_K$ then the class of $\mathfrak{a}$ is the image of the class of α.

(2). Let $\alpha = \pi^k u \in E$. Since $\iota(\pi) = 1 - \zeta^{-1} = -z^{-1}\pi$ and $-\zeta^{-1}$ is a qth power since q and $2p$ are coprime, it follows that $\iota(\pi)/\pi \in E^q$. Furthermore, if $u \in U(K)$ then by Lemma 3.5.19, $\iota(u)/u$ is a $2p$th root of unity, so once again $\iota(u)/u \in E^q$, proving (2).

(3). Since q is odd, 2 is invertible in $\mathbb{F}_q$, so that for any $\mathbb{F}_q[G]$-module M we have $M = M^+ \oplus M^-$. In particular, taking $+$ and $-$ parts in an exact sequence of $\mathbb{F}_q[G]$-modules preserves exactness. Since by (2) we have $(E/E^q)^+ = E/E^q$ and $(E/E^q)^- = 0$, taking the $-$ part of the exact sequence of (1) gives $S^- \simeq Cl(K)[q]^-$, which is clearly equal to $Cl(K)^-[q]$, and taking the $+$ part gives the exact sequence of (3).

(4). By Stickelberger's theorem we know that I_s annihilates $Cl(K)$ hence $Cl(K)[q]$, and by (3) that ε^- annihilates E/E^q with $\varepsilon^- = 1 - \iota$. Since $I = \varepsilon^- I_s$ it follows that I annihilates both $Cl(K)[q]$ and E/E^q, so it annihilates S thanks to the exact sequence of (1).   □

16.3 Mihăilescu's Third Theorem: $p < 4q^2$ and $q < 4p^2$

This is not an important part of the proof of Catalan's conjecture, and was found only afterward, but it has the great advantage of completely avoiding the use of linear forms in logarithms and extensive computer calculations (as opposed to the straightforward proofs above).

In this section we let as before p and q be distinct odd primes and x and y be nonzero integers such that $x^p - y^q = 1$. To simplify notation we will write $\mathcal{N}$ instead of $\mathcal{N}_{K/\mathbb{Q}}$, where as usual $K = \mathbb{Q}(\zeta_p)$. Recall that by Lemma 16.1.1 we know that the class $[x - \zeta_p]$ of $x - \zeta_p$ modulo qth powers belongs to the group S.

Definition 16.3.1. *We denote by X the annihilator of $[x - \zeta_p]$ in $\mathbb{Z}[G]$, in other words the set of $\theta \in \mathbb{Z}[G]$ such that $(x - \zeta_p)^\theta = \alpha^q$ for some $\alpha \in K^*$.*

It is clear that X is an ideal of $\mathbb{Z}[G]$.

Lemma 16.3.2. *The map sending $\theta \in X$ to $\alpha \in K^*$ such that $(x - \zeta_p)^\theta = \alpha^q$ is a well-defined injective group homomorphism.*

Proof. The map is well defined since $K = \mathbb{Q}(\zeta_p)$ does not contain any other qth root of unity than 1. It is clear that it is a group homomorphism from the additive group X to the multiplicative group K^*. Let us show that it is injective: let $\theta \in X$ be such that $(x - \zeta_p)^\theta = 1$. For any $\sigma \in G$ we thus have $(x - \sigma(\zeta_p))^\theta = \sigma(1) = 1$, hence $\mathcal{N}(x - \zeta_p)^\theta = 1$. If $\theta = \sum_{\sigma \in G} a_\sigma \sigma$, since $\mathcal{N}(x - \zeta_p) \in \mathbb{Z}$ it follows that $\mathcal{N}(x - \zeta_p)^s = 1$, where $s = \sum_{\sigma \in G} a_\sigma$. Now recall that by Lemma 16.1.1 we have $(x - \zeta_p)/(1 - \zeta_p) \in \mathbb{Z}_K$, and so, since $\mathcal{N}(1 - \zeta_p) = p$, we have $p \mid \mathcal{N}(x - \zeta_p)$, and in particular $\mathcal{N}(x - \zeta_p) \geqslant p$. Thus we must have $s = \sum_{\sigma \in G} a_\sigma = 0$, so we can write

$$1 = \frac{(x - \zeta_p)^\theta}{(1 - \zeta_p)^s} = \prod_{\sigma \in G} \left(\frac{x - \sigma(\zeta_p)}{1 - \zeta_p} \right)^{a_\sigma} ,$$

and since $(1 - \sigma(\zeta_p))/(1 - \zeta_p)$ is a unit for all $\sigma \in G$ it follows that $\prod_{\sigma \in G} \sigma(\beta)^{a_\sigma}$ is a unit, where we have set $\beta = (x - \zeta_p)/(1 - \zeta_p)$. Now by Lemma 16.1.1

the ideals $\mathfrak{b}_\sigma = \sigma(\beta)\mathbb{Z}_K$ are (integral and) coprime. Since $\prod_{\sigma \in G} \mathfrak{b}_\sigma^{a_\sigma} = \mathbb{Z}_K$ it follows that $a_\sigma = 0$ for all $\sigma \in G$, in other words that $\theta = 0$, proving injectivity and the lemma. $\qquad \square$

Proposition 16.3.3. *Assume that $\min(p, q) \geqslant 11$. Let $\theta = \sum_{\sigma \in G} a_\sigma \sigma \in X \cap (1 - \iota)\mathbb{Z}[G]$, let $\alpha \in K^*$ be such that $(x - \zeta_p)^\theta = \alpha^q$, and assume that $\|\theta\| = \sum_{\sigma \in G} |a_\sigma| \leqslant 3q/(p-1)$. Then for all $\tau \in G$ we have*

$$|\operatorname{Arg}(\tau(\alpha)^q)| \leqslant \frac{\|\theta\|}{|x| - 1} \quad \text{and} \quad |\operatorname{Arg}(\tau(\alpha))| > \frac{\pi}{q},$$

where $\operatorname{Arg}(z)$ denotes the principal determination of the argument, i.e., such that $-\pi < \operatorname{Arg}(z) \leqslant \pi$.

Proof. Since $\theta \in (1 - \iota)\mathbb{Z}[G]$ we have $\iota\theta = -\theta$, so for all $\tau \in G$,

$$|\tau(\alpha)|^{2q} = |(x - \zeta_p)^{\tau\theta}|^2 = (x - \zeta_p)^{\tau\theta}(x - \zeta_p)^{\tau\iota\theta} = (x - \zeta_p)^{\tau\theta}(x - \zeta_p)^{-\tau\theta} = 1 \,,$$

so that $|\tau(\alpha)| = 1$. For the same reason we have $a_{\iota\sigma} = -a_\sigma$, hence $s = \sum_{\sigma \in G} a_\sigma = 0$. It follows that

$$\alpha^q = (x - \zeta_p)^\theta = \prod_{\sigma \in G}(x - \sigma(\zeta_p))^{a_\sigma}$$

$$= x^s \prod_{\sigma \in G}(1 - \sigma(\zeta_p)/x)^{a_\sigma} = \prod_{\sigma \in G}(1 - \sigma(\zeta_p)/x)^{a_\sigma} \,.$$

Fix some $\tau \in G$, and set $\zeta = \tau(\zeta_p)$. We thus have

$$\tau(\alpha)^q = \prod_{\sigma \in G}(1 - \sigma(\zeta)/x)^{a_\sigma} \,.$$

Denote by Log the principal branch of the complex logarithm, in other words such that $\operatorname{Log}(z) = \log(|z|) + i \operatorname{Arg}(z)$, and let f be some determination of the complex logarithm, so that $f(z) - \operatorname{Log}(z)$ is an integral multiple of $2i\pi$. We thus have

$$\sum_{\sigma \in G} a_\sigma \operatorname{Log}(1 - \sigma(\zeta)/x) = f(\tau(\alpha)^q) \,.$$

Now since $|x| > 1$ we have

$$|\operatorname{Log}(1 - \sigma(\zeta)/x)| = \left| \sum_{k \geqslant 1} \sigma(\zeta)^k/(kx^k) \right| \leqslant \sum_{k \geqslant 1} |x|^{-k} = 1/(|x| - 1) \,.$$

Note that for all z we have $f(z) = \log(|z|) + i(\operatorname{Arg}(z) + 2k\pi)$ for some $k \in \mathbb{Z}$, hence $|f(z)| \geqslant |\operatorname{Arg}(z) + 2k\pi|$. If $k = 0$ this gives $|f(z)| \geqslant |\operatorname{Arg}(z)|$, while if $k \neq 0$ this gives

$$|f(z)| \geqslant |2k\pi| - |\operatorname{Arg}(z)| \geqslant (2|k| - 1)\pi \geqslant \pi \geqslant |\operatorname{Arg}(z)|$$

since $|\operatorname{Arg}(z)| \leqslant \pi$, so that we always have $|f(z)| \geqslant |\operatorname{Arg}(z)|$. Thus

$$|\operatorname{Arg}(\tau(\alpha)^q)| \leqslant |f(\tau(\alpha)^q)| \leqslant \frac{1}{|x| - 1} \sum_{\sigma \in G} |a_\sigma| \leqslant \frac{\|\theta\|}{|x| - 1},$$

proving the first inequality. Now assume by contradiction that $|\operatorname{Arg}(\tau(\alpha))| \leqslant \pi/q$. It is immediately checked that in that case $|\operatorname{Arg}(\tau(\alpha)^q)| = q|\operatorname{Arg}(\tau(\alpha))|$, so that $|\operatorname{Arg}(\tau(\alpha))| \leqslant \|\theta\|/(q(|x|-1))$. Furthermore, if we set $\phi = \operatorname{Arg}(\tau(\alpha))$, since $|\tau(\alpha)| = 1$ we have $\tau(\alpha) = \cos(\phi) + i \sin(\phi)$, hence

$$\tau(\alpha) - 1 = 2 \sin(\phi/2)(-\sin(\phi/2) + i \cos(\phi/2)),$$

so that

$$|\tau(\alpha) - 1| = 2|\sin(\phi/2)| \leqslant |\phi| = |\operatorname{Arg}(\tau(\alpha))|.$$

We thus have $|\tau(\alpha) - 1| \leqslant \|\theta\|/(q(|x| - 1))$, so taking the product over all $\sigma \in G$ we obtain

$$|\mathcal{N}(\alpha - 1)| = |\tau(\alpha) - 1|^2 \prod_{\substack{\sigma \in G \\ \sigma \neq \tau, \, \sigma \neq \iota\tau}} |\sigma(\alpha) - 1| \leqslant \left(\frac{\|\theta\|}{q(|x| - 1)} \right)^2 2^{p-3},$$

since $|\sigma(\alpha) - 1| \leqslant |\sigma(\alpha)| + 1 = 2$.

Now set $\theta^+ = \sum_{\sigma \in G, \, a_\sigma \geqslant 0} a_\sigma \sigma$ and $\theta^- = \sum_{\sigma \in G, \, a_\sigma \leqslant 0} (-a_\sigma)\sigma$, so that $\theta = \theta^+ - \theta^-$. Since $a_{\iota\sigma} = -a_\sigma$ we have $\iota\theta^+ = \theta^-$ hence $\alpha^q = (x - \zeta_p)^\theta = \beta/\iota(\beta)$, where $\beta = (x - \zeta_p)^{\theta^+}$ is an algebraic *integer*. Now

$$\mathcal{N}(\beta^2) = \mathcal{N}(\beta)\,\mathcal{N}(\iota(\beta)) = \mathcal{N}(\beta\iota(\beta))$$

$$= \mathcal{N}_{K/\mathbb{Q}} \left(\prod_{\sigma \in G} (x - \sigma(\zeta_p))^{|a_\sigma|} \right) \leqslant (|x| + 1)^{\|\theta\|(p-1)},$$

so that $\mathcal{N}(\beta) \leqslant (|x| + 1)^{\|\theta\|(p-1)/2}$. Write $\alpha\mathbb{Z}_K = \mathfrak{a}/\mathfrak{b}$, where $\mathfrak{a}$ and $\mathfrak{b}$ are coprime integral ideals. We have $\mathfrak{a}^q/\mathfrak{b}^q = (\beta/\iota(\beta))\mathbb{Z}_K$, hence $\mathfrak{a}^q\iota(\beta) = \mathfrak{b}^q\beta$, and since $\mathfrak{a}$ and $\mathfrak{b}$ are coprime it follows that $\mathfrak{b}^q \mid \iota(\beta)\mathbb{Z}_K$. In particular, $\mathcal{N}(\mathfrak{b}^q) \leqslant \mathcal{N}(\iota(\beta)) = \mathcal{N}(\beta)$, so that $\mathcal{N}(\mathfrak{b}) \leqslant (|x| + 1)^{\|\theta\|(p-1)/(2q)}$. Now by Lemma 16.3.2, since we have chosen $\theta \neq 0$ we have $\alpha \neq 1$. Thus, since $\mathfrak{b}\alpha = \mathfrak{a}$ and $\mathfrak{b}$ are integral ideals, it follows that $\mathfrak{a}_1 = \mathfrak{b}(\alpha - 1) = \{x\alpha - x, \, x \in \mathfrak{b}\}$ is also an integral ideal, so that $1 \leqslant \mathcal{N}(\mathfrak{a}_1) = \mathcal{N}(\mathfrak{b})|\mathcal{N}(\alpha - 1)|$. Combining the inequalities that we have obtained above we thus have

$$1 \leqslant (|x| + 1)^{\|\theta\|(p-1)/(2q)} \left(\frac{\|\theta\|}{q(|x| - 1)} \right)^2 2^{p-3}.$$

This inequality is going to lead to a contradiction. Since $|x| \geqslant 6$ (see below) we have $(1 + |x|)^2 \leqslant 2(|x| - 1)^2$, hence

$$(1 + |x|)^{2 - \|\theta\|(p-1)/(2q)} \leqslant 2^{p-1}(\|\theta\|/q)^2 \,,$$

so by the assumption $\|\theta\| \leqslant 3q/(p-1)$ of the proposition and the fact that $p \geqslant 5$ we deduce that

$$(1 + |x|)^{1/2} \leqslant 2^{p-1}(3/(p-1))^2 \leqslant 2^{p-1} = 4^{(p-1)/2} \,.$$

Now by Proposition 6.11.15 (essentially Hyrrö's result), we have $|x| \geqslant q^{p-1} + q$ (which incidentally shows that $|x| \geqslant 6$). It follows that $q^{(p-1)/2} < (1 + |x|)^{1/2} \leqslant 4^{(p-1)/2}$, which is absurd since $q \geqslant 5$ by assumption.   □

To prove the next result (Proposition 16.3.6) we need several lemmas.

Lemma 16.3.4. *The number of k-tuples of nonnegative integers λ_i such that $\sum_{1 \leqslant i \leqslant k} \lambda_i \leqslant s$ is equal to $\binom{s+k}{s} = \binom{s+k}{k}$.*

Proof. The map that sends $(\lambda_i)_{1 \leqslant i \leqslant k}$ to the set of $\sum_{1 \leqslant i \leqslant j}(\lambda_i + 1)$ for $1 \leqslant j \leqslant k$ is easily seen to be a bijection from the set of k-tuples with sum s to the set of subsets of cardinality k of $[1, s + k]$, whose cardinality is equal to $\binom{s+k}{k}$.   □

Lemma 16.3.5. *Assume that $\min(p, q) \geqslant 11$ and that $q > 4p^2$. There exist at least $q + 1$ elements $\theta \in I$ such that $\|\theta\| \leqslant 3q/(2(p-1))$.*

Proof. Recall from Lemma 16.2.3 that I has a basis of elements e_i for $1 \leqslant i \leqslant (p-1)/2$ that are such that $\|e_i\| = p - 1$. Consider the set of $\theta = \sum_{1 \leqslant i \leqslant (p-1)/2} \lambda_i e_i$, where $\lambda_i \in \mathbb{Z}_{\geqslant 0}$ and $\sum_i \lambda_i \leqslant s = \lfloor 3q/(2(p-1)^2) \rfloor$. For such a θ we have

$$\|\theta\| \leqslant (p-1)\sum_i \lambda_i \leqslant (p-1)s \leqslant 3q/(2(p-1)) \,.$$

By the preceding lemma the number of such θ is equal to $\binom{s+(p-1)/2}{s}$. Since we can also consider $-\theta$ when $\theta \neq 0$, it follows that we construct in this way $2\binom{s+(p-1)/2}{s} - 1$ distinct elements θ. Let us show that this quantity is greater than or equal to $q + 1$. First note that

$$\frac{\binom{s+(p-1)/2}{s}}{p^2(s+1)} = \frac{\prod_{2 \leqslant j \leqslant (p-1)/2}(s+j)}{p^2((p-1)/2)!} \,,$$

which is evidently an increasing function of s. Since $q \geqslant 4p^2 \geqslant 4(p-1)^2$ we have $s \geqslant 6$, so that $\binom{s+(p-1)/2}{s}/(s+1) \geqslant \binom{6+(p-1)/2}{6}/(7p^2) = f(p)$, say. We compute that $f(p)/f(p-2) = (p+11)(p-2)^2/(p^2(p-1))$, and it is easily checked that this is greater than 1 as soon as $p \geqslant 5$. Thus $f(p)$ is an increasing function of p. In particular, we compute that $f(11) = 6/11 > 1/3$. Thus if $p \geqslant 11$ we have

$$\binom{s + (p-1)/2}{s} > \frac{p^2(s+1)}{2} > \frac{p^2 q}{2(p-1)^2} \geqslant \frac{q+2}{2} \,,$$

the last inequality being immediate since $q > 4p^2$. The number of distinct elements θ that we have constructed is thus greater than or equal to $q+1$, as claimed. □

Proposition 16.3.6. *Assume that* $\min(p,q) \geqslant 11$ *and that* $q > 4p^2$. *For all* $\tau \in G$ *there exists a nonzero* $\theta \in I$ *such that* $\|\theta\| \leqslant 3q/(p-1)$ *and such that* $|\operatorname{Arg}(\tau(\alpha))| \leqslant \pi/q$, *where* $\alpha \in K^*$ *is the element such that* $(x - \zeta_p)^\theta = \alpha^q$.

Proof. By the above lemma there exist at least $q + 1$ elements $\theta \in I$ such that $\|\theta\| \leqslant 3q/(2(p-1))$. For each such θ there exists a unique α such that $(x - \zeta_p)^\theta = \alpha^q$. Since $\theta \in I \subset (1 - \iota)\mathbb{Z}[G]$, by the first inequality of Proposition 16.3.3 we deduce that $|\operatorname{Arg}(\tau(\alpha)^q)| \leqslant \|\theta\|/(|x| - 1)$. Now note that $\operatorname{Arg}(\tau(\alpha)^q) = q\operatorname{Arg}(\tau(\alpha)) + 2k\pi$ for some k, so that $2k\pi = -q\operatorname{Arg}(\tau(\alpha)) + \operatorname{Arg}(\tau(\alpha)^q)$, and since Arg is always between $-\pi$ and π we have $2|k|\pi < (q+1)\pi$, hence $2|k| \leqslant q$, hence $|k| \leqslant (q-1)/2$ since q is an odd integer. Since there are exactly q integers k such that $-(q-1)/2 \leqslant k \leqslant (q-1)/2$ and we have at least $q + 1$ distinct θ, it follows from the pigeonhole principle that there exist θ_1 and θ_2 with $\theta_1 \neq \theta_2$, satisfying the given properties, with in addition the same value of k. For $i = 1, 2$ write $(x - \zeta_p)^{\theta_i} = \alpha_i^q$, $\theta = \theta_1 - \theta_2$, so that $(x - \zeta_p)^\theta = \alpha^q$ with $\alpha = \alpha_1/\alpha_2$, and evidently $\|\theta\| \leqslant \|\theta_1\| + \|\theta_2\| \leqslant 3q/(p-1)$. Since

$$\operatorname{Arg}(\tau(\alpha_i)) = \frac{\operatorname{Arg}(\tau(\alpha_i)^q)}{q} - \frac{2k\pi}{q}$$

we have

$$|\operatorname{Arg}(\tau(\alpha_2)) - \operatorname{Arg}(\tau(\alpha_1))| = \frac{1}{q}|\operatorname{Arg}(\tau(\alpha_2)^q) - \operatorname{Arg}(\tau(\alpha_1)^q)| \leqslant 2\pi/q < \pi \,,$$

hence $\operatorname{Arg}(\tau(\alpha)) = \operatorname{Arg}(\tau(\alpha_2)) - \operatorname{Arg}(\tau(\alpha_1))$. Using the inequalities $\|\theta\| \leqslant 3q/(2(p-1))$ and $|x| - 1 > q^{p-1}$ (Proposition 6.11.15) we thus have

$$
\begin{aligned}
|\operatorname{Arg}(\tau(\alpha))| &= |\operatorname{Arg}(\tau(\alpha_2)) - \operatorname{Arg}(\tau(\alpha_1))| \\
&\leqslant |\operatorname{Arg}(\tau(\alpha_2)) + 2k\pi/q| + |\operatorname{Arg}(\tau(\alpha_1)) + 2k\pi/q| \\
&\leqslant (|\operatorname{Arg}(\tau(\alpha_2)^q)| + |\operatorname{Arg}(\tau(\alpha_1)^q)|)/q \\
&\leqslant 2\|\theta\|/(q(|x| - 1)) \leqslant 3/((p-1)q^{p-1}) < \pi/q \,,
\end{aligned}
$$

proving the proposition. □

Mihăilescu's third theorem is now immediate.

Theorem 16.3.7. *Let* p *and* q *be odd primes such that* $\min(p,q) \geqslant 11$, *and let* x *and* y *be nonzero integers such that* $x^p - y^q = 1$. *Then* $p < 4q^2$ *and* $q < 4p^2$.

Proof. By symmetry, it is enough to prove that $q < 4p^2$. Assume by contradiction that $q > 4p^2$. By Proposition 16.3.6 for all $\tau \in G$ there exists a nonzero $\theta \in I$ such that $\|\theta\| \leqslant 3q/(p-1)$ and $|\operatorname{Arg}(\tau(\alpha))| \leqslant \pi/q$, where $(x - \zeta_p)^\theta = \alpha^q$. By Lemma 16.2.6 (4), S is annihilated by I; hence the class $[x - \zeta_p]$ is annihilated by I, so that $I \subset X$. Since by definition $I = (1 - \iota)I_s \subset (1 - \iota)\mathbb{Z}[G]$, it follows that $\theta \in X \cap (1 - \iota)\mathbb{Z}[G]$, and since $\|\theta\| \leqslant 3q/(p-1)$ we deduce from Proposition 16.3.3 that $|\operatorname{Arg}(\tau(\alpha))| > \pi/q$, which contradicts the inequality obtained from Proposition 16.3.6 and proves the theorem. $\square$

16.4 Mihǎilescu's Fourth Theorem: $p \equiv 1 \pmod{q}$ or $q \equiv 1 \pmod{p}$

This is the most subtle part of the proof. Up to now, we have used rather simple properties of cyclotomic fields, the essential tool being Stickelberger's theorem and the properties of the minus part of the class group. In contrast, Mihǎilescu's fourth theorem rests on properties of the plus part of the class group. This is much less well understood (think about real quadratic fields compared to imaginary quadratic fields), but a remarkable theorem has been proved by F. Thaine on the plus part, which in some sense is an analogue of Stickelberger's theorem. This theorem has had a number of very important applications, for instance in the proof of the finiteness of the Tate–Shafarevich group of elliptic curves of rank less than or equal to 1. It is also crucial in the present section. It would take us too long to give a proof of Thaine's theorem, so I refer to [Boe-Mis] or to the second edition of [Was].

In the first three subsections we prove some necessary results on the plus part, which are independent of Catalan's equation, assuming at a crucial point Thaine's theorem, which we will of course state. In the last subsection we give the proof of Mihǎilescu's fourth theorem.

16.4.1 Preliminaries on Commutative Algebra

Lemma 16.4.1. *Let R be a commutative ring, $\mathfrak{b}$ an ideal of R, M an R-module of finite type, and ϕ an R-endomorphism of M such that $\phi(M) \subset \mathfrak{b}M$. There exists a nonzero monic polynomial $P \in R[X]$ such that $P(\phi) = 0$, and such that all the coefficients of P other than the leading one belong to $\mathfrak{b}$.*

In the above, recall that $\mathfrak{b}M$ is the R-module of linear combinations of the product of an element of $\mathfrak{b}$ by an element of M, and that for any endomorphism ϕ, we let ϕ^0 be the identity.

Proof. Let $(m_i)_{1 \leqslant i \leqslant n}$ be an R-generating set for M, and let $b_{i,j} \in \mathfrak{b}$ be such that $\phi(m_j) = \sum_{1 \leqslant i \leqslant n} b_{i,j}m_i$ for $1 \leqslant j \leqslant n$. The module M can be considered as an $R[\phi]$-module through the map $A(\phi) \cdot m = A(\phi)(m)$ for $A \in R[X]$ and $m \in M$. If we set $B = (b_{i,j})_{1 \leqslant i,j \leqslant n}$ and if we denote by I_n the identity matrix of order n we can thus write in the ring of matrices with coefficients in $R[\phi]$ the equation $(\phi I_n - B)V = 0$, where V is the (column) vector of the m_i. Multiplying by the comatrix of $\phi I_n - B$, we deduce that $\det(\phi I_n - B)V = 0$, in other words that $\det(\phi I_n - B)m_i = 0$ for all i. Since the m_i generate M it follows that $\det(\phi I_n - B)M = 0$, hence that $\det(\phi I_n - B) = 0$ as an endomorphism of M, and this is clearly a monic polynomial in ϕ whose coefficients are in $\mathfrak{b}$ apart from the leading one.   □

Recall that one denotes by $\mathrm{Ann}_R(M)$ the annihilator of an R-module M, in other words the set of $x \in R$ such that $xM = 0$. It is evidently an ideal of R.

Lemma 16.4.2. *Let R be a commutative ring, $\mathfrak{b}$ an ideal of R, M an R-module of finite type, and denote by ψ the canonical surjection from R to $R/\mathfrak{b}$. If $R/(\mathrm{Ann}_R(M) + \mathfrak{b})$ has no nonzero nilpotent elements then*

$$\psi(\mathrm{Ann}_R(M)) = \mathrm{Ann}_{R/\mathfrak{b}}(M/\mathfrak{b}M) \,.$$

Proof. The inclusion $\subset$ is trivial, so let us show the reverse inclusion. Thus, let $\psi(\alpha) \in \mathrm{Ann}_{R/\mathfrak{b}}(M/\mathfrak{b}M)$; in other words, $\alpha \in R$ is such that $\alpha M \subset \mathfrak{b}M$. Applying the preceding lemma to the map multiplication by α, we deduce that there exist $b_i \in \mathfrak{b}$ such that the map multiplication by $\beta = \alpha^n + b_{n-1}\alpha^{n-1} + \cdots + b_0$ is the zero map in M, in other words such that $\beta \in \mathrm{Ann}_R(M)$. Since $b_i \in \mathfrak{b}$ it follows that $\alpha^n \in \mathrm{Ann}_R(M) + \mathfrak{b}$, and since $R/(\mathrm{Ann}_R(M) + \mathfrak{b})$ has no nonzero nilpotent elements we have $\alpha \in \mathrm{Ann}_R(M) + \mathfrak{b}$, hence $\psi(\alpha) \in \psi(\mathrm{Ann}_R(M))$.   □

Lemma 16.4.3. *Let H be a cyclic group of order n, and assume that $q \nmid n$. Set $s = \sum_{\sigma \in H} \sigma \in \mathbb{F}_q[H]$. The rings $\mathbb{F}_q[H]$ and $\mathbb{F}_q[H]/(s\mathbb{F}_q[H])$ have no nonzero nilpotent elements.*

Proof. Since H is cyclic we have $\mathbb{F}_q[H] \simeq \mathbb{F}_q[X]/((X^n - 1)\mathbb{F}_q[X])$, and $\mathbb{F}_q[H]/(s\mathbb{F}_q[H]) \simeq \mathbb{F}_q[X]/((X^{n-1} + \cdots + X + 1)\mathbb{F}_q[X])$, so that

$$\mathbb{F}_q[H] \simeq (\mathbb{F}_q[X]/((X - 1)\mathbb{F}_q[X])) \times \mathbb{F}_q[H]/((X^{n-1} + \cdots + X + 1)\mathbb{F}_q[H])$$
$$\simeq \mathbb{F}_q \times \mathbb{F}_q[H]/(s\mathbb{F}_q[H])$$

if $(X - 1)$ and $X^{n-1} + \cdots + X + 1$ are coprime, which is the case since $q \nmid n$. If η is a nilpotent element of $\mathbb{F}_q[H]/(s\mathbb{F}_q[H])$, then under this isomorphism $(0, \eta)$ will be a nilpotent element of $\mathbb{F}_q[H]$, so it is enough to prove that there are none in this ring. Since $\mathbb{F}_q[H] \simeq \mathbb{F}_q[X]/((X^n - 1)\mathbb{F}_q[X])$, it follows that if the class of $A(X) \in \mathbb{F}_q[X]$ is nilpotent then $(X^n - 1) \mid A(X)^k$ for some $k \geqslant 1$. However, since the derivative of $X^n - 1$ is equal to nX^{n-1} hence is nonzero since $q \nmid n$, it follows that the roots of $X^n - 1$ in an algebraic closure of $\mathbb{F}_q$ are all distinct. Thus $X^n - 1 \mid A(X)$, and so the class of A is equal to 0, as claimed.   □

We end this subsection by recalling without proof some basic facts on semisimple rings and modules that can be found in any good textbook.

Definition 16.4.4. (1) *A commutative ring R is semisimple if it is a finite product of fields.*

(2) *An R-module M is simple if its only submodules are 0 and M.*

(3) *An R-module is semisimple if it is a finite direct sum of simple modules.*

(4) *An R-module M is cyclic if it is generated over R by a single element, in other words if $M = aR$ for some $a \in M$.*

Lemma 16.4.5. *Let H be a cyclic group of order n, and assume that $q \nmid n$. Then $\mathbb{F}_q[H]$ is a semisimple ring.*

Proof. Let $X^n - 1 = \prod_{1 \leqslant i \leqslant g} P_i^{e_i}(X)$ be the decomposition of $X^n - 1$ as a power product of distinct monic irreducible polynomials in $\mathbb{F}_q[X]$. Since $q \nmid n$ the polynomial $X^n - 1$ has distinct roots in an algebraic closure of $\mathbb{F}_q$, hence $e_i = 1$ for all i. Thus by the lemma

$$\mathbb{F}_q[H] \simeq \mathbb{F}_q[X]/((X^n - 1)\mathbb{F}_q[X]) \simeq \prod_{1 \leqslant i \leqslant g} K_i \, ,$$

where $K_i = \mathbb{F}_q[X]/(P_i(X)\mathbb{F}_q[X])$ is a field, so $\mathbb{F}_q[H]$ is semisimple. $\square$

The following proposition summarizes the results that we need.

Proposition 16.4.6. *Let R be a semisimple ring. Then:*

(1) *Any R-module is semisimple.*

(2) *Every exact sequence of R-modules is split.*

(3) *For any R-module M there exists $\alpha \in M$ such that $\mathrm{Ann}_R(\alpha) = \mathrm{Ann}_R(M)$, so M contains the cyclic submodule aR isomorphic to $R/\mathrm{Ann}_R(M)$.*

(4) *If R and M are finite then $|M| \geqslant |R/\mathrm{Ann}_R(M)|$ with equality if and only if M is cyclic.*

(5) *Let M be a cyclic module. Every submodule M' of M is also cyclic, $\mathrm{Ann}_R(M) = \mathrm{Ann}_R(M') \cdot \mathrm{Ann}_R(M/M')$, and $\mathrm{Ann}_R(M)$ and $\mathrm{Ann}_R(M/M')$ are coprime ideals.*

16.4.2 Preliminaries on the Plus Part

Recall some notation. We let as always p and q be distinct odd primes, and we set $K = \mathbb{Q}(\zeta_p)$ and $G = \mathrm{Gal}(K/\mathbb{Q})$, which is canonically isomorphic to $(\mathbb{Z}/p\mathbb{Z})^*$. We let $K^+ = \mathbb{Q}(\zeta_p + \zeta_p^{-1})$ be the maximal totally real subfield of K, $G^+ = \mathrm{Gal}(K^+/\mathbb{Q}) = G/\langle \iota \rangle$. We recall from Propositions 3.5.20 and 3.5.21 that $U(K) = \langle \zeta_p \rangle U(K^+)$ and that the natural map from $Cl(K^+)$ to $Cl(K)$ is injective.

Lemma 16.4.7. *We have $Cl(K^+)[q] = Cl(K)[q]^+$.*

Proof. By Proposition 3.5.21 we can write by abuse of notation $Cl(K^+)[q] \subset Cl(K)[q]$, and since evidently $Cl(K^+)$ is invariant by ι we have $Cl(K^+)[q] \subset Cl(K)[q]^+$. Conversely, let $\mathfrak{a}$ be a representative of an element of $Cl(K)[q]^+$. Since $Cl(K)[q]$ is an $\mathbb{F}_q[G]$-module and 2 is invertible in $\mathbb{F}_q$, it follows that $Cl(K)[q]^+$ is equal to the kernel of multiplication by $(1-\iota)/2$ (or by $1-\iota$) from $Cl(K)[q]$ to itself. Thus there exist α and β in K^* such that $\mathfrak{a}\iota(\mathfrak{a})^{-1} = \alpha\mathbb{Z}_K$ and $\mathfrak{a}^q = \beta\mathbb{Z}_K$. Let $\mathfrak{b}$ be the ideal of K^+ defined by $\mathfrak{b} = \mathcal{N}_{K/K^+}(\mathfrak{a})$. We have $\mathfrak{b}\mathbb{Z}_K = \mathfrak{a}\iota(\mathfrak{a})$, hence $\mathfrak{b}^q\mathbb{Z}_K = \mathfrak{a}^q\iota(\mathfrak{a}^q) = \beta\iota(\beta)\mathbb{Z}_K = \mathcal{N}_{K/K^+}(\beta)\mathbb{Z}_K$; hence intersecting with K^+, we deduce that $\mathfrak{b}^q = \mathcal{N}_{K/K^+}(\beta)K^+$, so that the class of $\mathfrak{b}$ belongs to $Cl(K^+)[q]$. Furthermore, setting $m = (q+1)/2$ we compute that

$$\mathfrak{b}^m\mathbb{Z}_K = \mathfrak{a}^m\iota(\mathfrak{a})^m = \mathfrak{a}^m(\mathfrak{a}\alpha^{-1})^m = \mathfrak{a}^{q+1}\alpha^{-m} = \mathfrak{a}\beta\alpha^{-m} \ ,$$

so the class of $\mathfrak{a}$ is equal to the class of $\mathfrak{b}^m\mathbb{Z}_K$, proving the lemma.   □

Recall from Definition 16.2.4 that $E = \{u\pi^k, \ u \in U(K), \ k \in \mathbb{Z}\} = \mathbb{Z}[\zeta_p, 1/p]^*$. This is a $\mathbb{Z}[G]$-module, so that E/E^q is an $\mathbb{F}_q[G]$-module. By Lemma 3.5.19 and the fact that $\pi = 1 - \zeta_p$, for any $x \in E$ the expression $\iota(x)/x$ is a $2p$th root of unity, and since q is coprime to $2p$, it is a qth power. It follows that E/E^q is pointwise invariant by ι, so that it is in fact an $\mathbb{F}_q[G^+]$-module. The following lemma describes its structure very precisely when $p \not\equiv 1 \pmod{q}$.

Lemma 16.4.8. *Assume that $p \not\equiv 1 \pmod{q}$.*

(1) *We have $|E/E^q| = q^{(p-1)/2}$.*
(2) *If we set $W = U(K^+)/\{\pm 1\}$, then $\mathrm{Ann}_{\mathbb{Z}[G^+]}(W) = s\mathbb{Z}[G^+]$, where $s = \sum_{\sigma \in G^+} \sigma$.*
(3) *We have $\mathrm{Ann}_{\mathbb{F}_q[G^+]}(W/W^q) = s\mathbb{F}_q[G^+]$.*
(4) *We have $\mathrm{Ann}_{\mathbb{F}_q[G^+]}(E/E^q) = 0$.*
(5) *E/E^q is a free $\mathbb{F}_q[G^+]$-module of rank 1.*

Proof. (1). The map (u, k) from $U(K) \times \mathbb{Z}$ to E is an isomorphism since k is defined uniquely as the $\mathfrak{p}$-adic valuation of $u\pi^k$, hence by Dirichlet's theorem, as an abelian group $E \simeq \mu_{2p} \times \mathbb{Z}^{(p-1)/2}$, since the rank of the group of units of K is equal to $(p-3)/2$. Since $2p$ is coprime to q it follows that $E/E^q \simeq (\mathbb{Z}/q\mathbb{Z})^{(p-1)/2}$, proving (1).

(2). Let $\sum_{\sigma \in G^+} a_\sigma \sigma$ belong to $\mathrm{Ann}_{\mathbb{Z}[G^+]}(W)$, in other words be such that $\prod_{\sigma \in G^+} \sigma(\varepsilon)^{a_\sigma} = \pm 1$ for all $\varepsilon \in U(K^+)$. Let $(\varepsilon_i)_{1 \leqslant i \leqslant (p-3)/2}$ be a system of fundamental units of K^+. Taking logarithms we have $\sum_{\sigma \in G^+} a_\sigma \log(|\sigma(\varepsilon_i)|) = 0$ for all i. On the other hand, by Dirichlet's theorem the $((p-3)/2) \times ((p-1)/2)$ matrix of the $\sigma(\varepsilon_i)_{i \leqslant (p-3)/2, \sigma \in G^+}$ has rank $(p-3)/2$, so its kernel has dimension 1. Since $\sum_{\sigma \in G^+} \log(|\sigma(\varepsilon_i)|) = 0$, this kernel is generated over $\mathbb{R}$ by the column vector having all $(p-1)/2$ coordinates equal to 1. It follows that $a_\sigma = a$ for all σ, hence that $\sum_{\sigma \in G^+} a_\sigma \sigma = a \cdot s$, as claimed.

(3). By Lemma 16.4.3 applied to $H = G^+$, we see that if $p \not\equiv 1 \pmod{q}$ the ring $\mathbb{F}_q[G^+]/(s\mathbb{F}_q[G^+])$ has no nonzero nilpotent elements. Set temporarily $I = s\mathbb{Z}[G^+] + q\mathbb{Z}[G^+]$. It is clear that $\mathbb{Z}[G^+]/I \simeq \mathbb{F}_q[G^+]/(s\mathbb{F}_q[G^+])$, hence has no nonzero nilpotents. It is clear from (2) that $s\mathbb{F}_q[G^+] \subset \mathrm{Ann}_{\mathbb{F}_q[G^+]}(W/W^q)$, so let us show the reverse inclusion. Let $\theta \in \mathrm{Ann}_{\mathbb{F}_q[G^+]}(W/W^q)$; in other words, $\theta \in \mathbb{F}_q[G^+]$ is such that $W^\theta \subset W^q$. We apply Lemma 16.4.2 to $R = \mathbb{Z}[G^+]$, $\mathfrak{b} = q\mathbb{Z}[G^+]$, and $M = W$, where of course we recall that the action of R on M is multiplicative, while it is written additively in the lemma. Since by (2) we have $\mathrm{Ann}_R(M) = sR$, we see that since $R/(\mathrm{Ann}_R(M) + \mathfrak{b})$ has no nonzero nilpotent elements we have $\psi(\mathrm{Ann}_R(M)) = \mathrm{Ann}_{R/\mathfrak{b}}(M/\mathfrak{b}M)$. Translating into our context this means that $s\mathbb{F}_q[G^+] = \mathrm{Ann}_{\mathbb{F}_q[G^+]}(W/W^q)$, which is (3).

(4). Let us compute the image and kernel of the natural map from $U(K^+)$ to $U(K)/U(K)^q$. By Proposition 3.5.20 any $u \in U(K)$ has the form $u = \zeta\varepsilon$, where $\varepsilon \in U(K^+)$ and ζ is a $2p$th root of unity, hence a qth power. It follows that the class of u in $U(K)/U(K)^q$ is equal to the class of ε, so the map is surjective. Now let $\varepsilon \in U(K^+)$ be in the kernel, in other words be such that $\varepsilon = u^q$ for some $u \in U(K)$. Thus $\varepsilon = \iota(u)^q = u^q$, hence $\iota(u) = u$, so that $u \in U(K^+)$; hence the kernel is equal to $U(K^+)^q$. It follows from this that

$$U(K)/U(K)^q \simeq U(K^+)/U(K^+)^q \simeq W/W^q \;,$$

so that $E/E^q \simeq W/W^q \times \mathbb{Z}/q\mathbb{Z}$. Note that all of the above isomorphisms are canonical, and in particular are isomorphisms of $\mathbb{F}_q[G^+]$-modules. Thus it follows from (3) that

$$\mathrm{Ann}_{\mathbb{F}_q[G^+]}(E/E^q) \subset \mathrm{Ann}_{\mathbb{F}_q[G^+]}(W/W^q) \subset s\mathbb{F}_q[G^+] \;.$$

Now note that for any $\sigma \in G^+$ we have $s\sigma = s$. It follows that $s\mathbb{F}_q[G^+] = \mathbb{F}_q s$. Thus, let $\bar{a}s \in \mathrm{Ann}_{\mathbb{F}_q[G^+]}(E/E^q)$ with $a \in \mathbb{Z}$. Since $\pi = 1 - \zeta_p \in E$ we have $\pi^{as} \in E^q$, hence $v_\mathfrak{p}(\pi^{as}) \equiv 0 \pmod{q}$ by definition of E. On the other hand, for all $\sigma \in G$ we have $\pi^\sigma = u_\sigma \pi$ for some unit u_σ, hence $\pi^s = u\pi^{(p-1)/2}$ for some unit u. It follows that $v_\mathfrak{p}(\mathfrak{p}^{as}) = a(p-1)/2$. Since $q \nmid (p-1)/2$ we thus have $q \mid a$, hence $\bar{a} = 0$, proving (4).

(5). By Proposition 16.4.6 (3) applied to the semisimple ring $R = \mathbb{F}_q[G^+]$ and to $M = E/E^q$, there exists $\alpha \in M$ such that $\mathrm{Ann}_R(\alpha) = \mathrm{Ann}_R(M)$, hence $\mathrm{Ann}_R(\alpha) = 0$ by (4). This means that the map $x \mapsto x \cdot \alpha$ from R to M is an injective R-module homomorphism. However, by (1) we have $|M| = |E/E^q| = q^{(p-1)/2} = |\mathbb{F}_q[G^+]| = |R|$. It follows that the map is a bijection, so that R and M are isomorphic R-modules. $\square$

Definition 16.4.9. (1) *To simplify notation we set $R_p = \mathbb{Z}[\zeta_p, 1/p]$, so that $E = R_p^*$.*
(2) *Recall that we denote by $[\alpha]$ the class of α modulo qth powers in S. We define the group of q-primary elements of S by*

$$S_q = \{[\alpha] \in S, \ \alpha \equiv \beta^q \pmod{q^2 R_p}, \ \beta \ \text{invertible modulo} \ q^2 R_p\},$$

and $E_q = \{u \in E, \ [u] \in S_q\}.$

Lemma 16.4.10. *We have*

$$E_q = \{u \in E, \ u \equiv \beta^q \pmod{q^2 R_p}\}.$$

Proof. If u belongs to the right-hand side then $u \in E$, $u \equiv \beta^q \pmod{q^2 R_p}$, so β^q modulo q^2 is equal to u. Since elements of E are invertible in R_p, it follows that β^q modulo q^2 is invertible, hence so is β, so that $u \in E_q$. Conversely, let $u \in E_q$, so that $u \in E$ and $[u] \in S_q$. By definition of S_q there exist $\alpha \in K^*$ and $\beta, \gamma \in R_p$ such that $u\alpha^q = \beta^q + q^2\gamma$, and β is invertible modulo $q^2 R_p$. Let $\mathfrak{q}$ be a prime ideal of K different from $\mathfrak{p} = \pi \mathbb{Z}_K$. We thus have $v_\mathfrak{q}(u) = 0$, and since β, γ are in R_p and $\mathfrak{q} \neq \mathfrak{p}$ we have $v_\mathfrak{q}(\beta) \geqslant 0$ and $v_\mathfrak{q}(\gamma) \geqslant 0$. It follows that $v_\mathfrak{q}(\alpha) \geqslant 0$ for all prime ideals $\mathfrak{q} \neq \mathfrak{p}$; in other words, $\alpha \in R_p$. Now modulo $q^2 R_p$ we have $\overline{u}\overline{\alpha}^q = \overline{\beta}^q$. Since $\overline{\beta}$ is invertible it follows that $\overline{\alpha}$ is also invertible, and $\overline{u} = (\overline{\beta}\overline{\alpha}^{-1})^q$. Thus if $\beta_0 \in R_p$ is a representative of $\overline{\beta}\overline{\alpha}^{-1}$, we have $\overline{u} = \overline{\beta_0}^q$; in other words, $u = \beta_0^q + q^2\gamma_0$ for some $\gamma_0 \in R_p$, proving the reverse inclusion and the lemma. $\qquad\square$

16.4.3 Cyclotomic Units and Thaine's Theorem

Definition 16.4.11. *The group C of p-cyclotomic units of K is the multiplicative subgroup of K^* generated by the roots of unity and the $1 - \zeta_p^k$ for $k \in \mathbb{Z}$. We define $C_q = C \cap E_q$ and call the elements of C_q the q-primary p-cyclotomic units.*

Note that the group $C \cap U(K)$ is the group of cyclotomic units from Definition 3.5.16. Here we also allow powers of $\pi = 1 - \zeta_p$.

Lemma 16.4.12. *If p and q are distinct odd primes then $C = C_q$ implies that $p < q$.*

Proof. Let ζ be any primitive pth root of unity, not necessarily equal to ζ_p. Then $1 + \zeta^q = (1 - \zeta^{2q})/(1 - \zeta^q) \in C$, so that $1 + \zeta^q \in C_q$. Furthermore, I claim that $R_p/q^2 R_p \simeq \mathbb{Z}[\zeta_p]/q^2\mathbb{Z}[\zeta_p]$: indeed, let ϕ be the map sending $x \in \mathbb{Z}[\zeta_p]$ to its class in $R_p/q^2 R_p$. Its kernel is equal to $q^2 R_p \cap \mathbb{Z}[\zeta_p] = q^2\mathbb{Z}[\zeta_p]$, so it is enough to prove that ϕ is surjective. So let $y/p^n \in R_p = \mathbb{Z}[\zeta_p, 1/p]$, with $y \in \mathbb{Z}[\zeta_p]$. Since p^n and q^2 are coprime there exist u and v in $\mathbb{Z}$ such that $up^n + vq^2 = 1$. It follows that $y/p^n = uy + vyq^2/p^n$, and hence the class of y/p^n in $R_p/q^2 R_p$ is equal to the class of $uy \in \mathbb{Z}[\zeta_p]$, so it is in the image of ϕ, proving my claim.

Since $1 + \zeta^q \in C_q \subset E_q$ we can write $1 + \zeta^q = \beta^q + q^2\gamma$ with β and γ in R_p, and thanks to the above isomorphism, changing if necessary β and γ by an element of $q^2 R_p$ we may assume that β and γ belong to $\mathbb{Z}[\zeta_p]$. It follows that

$1+\zeta^q \equiv \beta^q \pmod{q^2 \mathbb{Z}[\zeta_p]}$. Thus, by the binomial expansion we have $(1+\zeta)^q \equiv 1 + \zeta^q \equiv \beta^q \pmod{q \mathbb{Z}[\zeta_p]}$. Since q is unramified in K it follows from Exercise 20 of Chapter 3 that $(1+\zeta)^q \equiv \beta^q \pmod{q^2 \mathbb{Z}[\zeta_p]}$. Thus $(1+\zeta)^q \equiv 1 + \zeta^q \pmod{q^2 \mathbb{Z}[\zeta_p]}$, so that $F(\zeta) \in q\mathbb{Z}[\zeta_p]$, where $F(X) = ((1+X)^q-1-X^q)/(qX)$, which is clearly a polynomial with integer coefficients of degree $q - 2$. Denote by $\overline{F} \in \mathbb{F}_q[X]$ the reduction of F modulo q. If $\mathfrak{q}$ is again a prime ideal above q then in the finite field $\mathbb{Z}[\zeta_p]/\mathfrak{q}$ we have $\overline{F}(\overline{\zeta}) = 0$, where $\overline{\zeta}$ is the image of ζ in $\mathbb{Z}[\zeta_p]/\mathfrak{q}$. Since this is true for all the $p - 1$ roots of unity ζ distinct from 1, and since these roots of unity are not congruent modulo $\mathfrak{q}$ since the norm of their difference is equal to p, it follows that $\overline{F}$ has at least $p-1$ distinct roots in $\mathbb{Z}[\zeta_p]/\mathfrak{q}$. Since $\deg(F) = q - 2$ we thus have $p - 1 \leqslant q - 2$, hence $p < q$. $\square$

We now state without proof the remarkable theorem of F. Thaine, referring for the proof to [Boe-Mis] or to the second edition of [Was]. We state only the special case of the theorem that will be needed.

Theorem 16.4.13 (Thaine). *Recall that C is the group of p-cyclotomic units of K. We have*

$$\mathrm{Ann}_{\mathbb{F}_q[G^+]}(E/CE^q) \subset \mathrm{Ann}_{\mathbb{F}_q[G^+]}(Cl(K^+)[q]) \,.$$

Note that this theorem is also valid for the cyclotomic units themselves, with the corresponding modification of E.

The main result of this section, which will be used to prove the fourth and last theorem of Mihăilescu, is the following.

Theorem 16.4.14. *Let p and q be odd primes such that $p > q$ and $p \not\equiv 1 \pmod{q}$. Then $\mathrm{Ann}_{\mathbb{F}_q[G^+]}(S^+ \cap S_q) \neq 0$.*

Proof. Set $R = \mathbb{F}_q[G^+]$, which is semisimple by Lemma 16.4.5. By Lemma 16.4.8 (5), E/E^q is a cyclic R module, hence by Proposition 16.4.6 (5) and (4), any submodule M of E/E^q is also cyclic, hence isomorphic to $R/\mathrm{Ann}_R(M)$. Since $R \simeq \mathbb{F}_q[X]/((X^{(p-1)/2} - 1)\mathbb{F}_q[X])$, any ideal of R is isomorphic to $f(X)\mathbb{F}_q[X]/((X^{(p-1)/2}-1)\mathbb{F}_q[X])$ for some $f(X) \in \mathbb{F}_q[X]$ dividing $X^{(p-1)/2} - 1$, which we may assume to be monic, so in particular $M \simeq R/\mathrm{Ann}_R(M) \simeq \mathbb{F}_q[X]/(f(X)\mathbb{F}_q[X])$. In particular, $\dim_{\mathbb{F}_q}(M) = \deg(f)$.

Now recall from Lemma 16.2.6 (3) that we have an exact sequence of R-modules $0 \longrightarrow E/E^q \longrightarrow S^+ \longrightarrow Cl(K)[q]^+ \longrightarrow 0$. By definition we have $E_q = \{u \in E, \ [u] \in S_q\}$, so under restriction, this exact sequence leads to an exact sequence $0 \longrightarrow E_q/E^q \longrightarrow S^+ \cap S_q \longrightarrow Cl(K)[q]^+$, where the last map is not necessarily surjective. Since R is semisimple, by Proposition 16.4.6 (2) every exact sequence is split, so in particular $S^+ \cap S_q$ is isomorphic to a submodule of $E_q/E^q \oplus Cl(K)[q]^+$, which we will write as $S^+ \cap S_q \hookrightarrow E_q/E^q \oplus Cl(K)[q]^+$.

Recall also that C is the group of p-cyclotomic units and that $C_q = C \cap E_q$. Consider the sequence of inclusions $0 \subset C_q E^q/E^q \subset CE^q/E^q \subset E/E^q$, and call E_1, E_2, and E_3 the successive quotients, so that $E_1 = C_q E^q / E^q$, $E_2 = C E^q / C_q E^q$, and $E_3 = E / C E^q$. Since R is semisimple, by Proposition 16.4.6 (1) and (2) every R-module is semisimple and every exact sequence is split. In particular, if $0 \subset A \subset B \subset C$ is a sequence of inclusions then $C \simeq B \oplus (C/B) \simeq A \oplus (B/A) \oplus (C/B)$. Since by Lemma 16.4.8 E/E^q is a free R-module of rank 1 we thus have an isomorphism

$$ E_1 \oplus E_2 \oplus E_3 \simeq R \simeq \mathbb{F}_q[X] / ((X^{(p-1)/2} - 1) \mathbb{F}_q[X]) \, . $$

It follows that the E_i are isomorphic to submodules of R which as before are isomorphic to $\mathbb{F}_q[X] / (e_i(X) \mathbb{F}_q[X])$ for some monic factors $e_i(X)$ of $X^{(p-1)/2} - 1$ such that $\dim_{\mathbb{F}_q}(E_i) = \deg(e_i)$. By the above isomorphism we have $e_1 e_2 e_3 = X^{(p-1)/2} - 1$.

By definition of S we have $E^q \subset E_q$, hence $C_q E^q \subset E_q$. We thus have an exact sequence

$$ 1 \longrightarrow C_q E^q / E^q \longrightarrow E_q / E^q \longrightarrow E_q / C_q E^q \longrightarrow 1 \, . $$

Since exact sequences are split it follows that $E_q / E^q \simeq E_1 \oplus E_q / C_q E^q$. On the other hand, it is clear that the kernel of the natural map from E_q to E/CE^q is equal to $E_q \cap C_q E^q$. Indeed, one inclusion is trivial. Conversely, if $x \in E_q$ has the form $x = ce^q$ with $c \in C$ and $e \in E$, then since $e^q \in E_q$ we have $c \in E_q \cap C = C_q$, hence $x \in C_q E^q$ as claimed. It follows that $E_q / C_q E^q$ is isomorphic to a subgroup of $E_3 = E/CE^q$. Putting everything together we obtain

$$ S^+ \cap S_q \hookrightarrow E_q / E^q \oplus Cl(K)[q]^+ $$
$$ \simeq E_1 \oplus E_q / C_q E^q \oplus Cl(K)[q]^+ \hookrightarrow E_1 \oplus E_3 \oplus Cl(K)[q]^+ \, . $$

Now by Thaine's theorem, any annihilator of $E_3 = E/CE^q$ also annihilates $Cl(K^+)[q]$, which is equal to $Cl(K)[q]^+$ by Lemma 16.4.7. Since e_i annihilates E_i by definition, it follows that $e_1 e_3$ annihilates $E_1 \oplus E_3$, and Thaine's theorem implies that e_3 annihilates $Cl(K)[q]^+$, so $e_1 e_3$ annihilates $S^+ \cap S_q$. Thus assume now by contradiction that $\mathrm{Ann}_R(S^+ \cap S_q) = 0$. We thus have $e_1 e_3 = 0$ in $\mathbb{F}_q[X] / ((X^{(p-1)/2} - 1) \mathbb{F}_q[X])$, in other words $X^{(p-1)/2} - 1 = e_1 e_2 e_3 \mid e_1 e_3$, so that $e_2 = 1$, hence $E_2 = 0$. By definition this means that $C_q E^q = CE^q$. We have already noted that $E^q \subset E_q$, hence $C_q \cap E^q = C \cap E^q$. I claim that we have $C = C_q$. Indeed, let $c \in C$. Since $c = c \cdot 1 \in CE^q = C_q E^q$ we can write $c = c_q e^q$ with $c_q \in C_q$ and $e \in E$. Thus $e^q = c/c_q \in C \cap E^q = C_q \cap E^q \subset C_q$, so that $c = c_q e^q \in C_q$ as claimed. Applying Lemma 16.4.12 we deduce that $p < q$, which contradicts the assumption of the proposition. $\square$

16.4.4 Preliminaries on Power Series

Recall that if R is a commutative ring we denote by $R[[T]]$ the ring of formal power series with coefficients in R.

Lemma 16.4.15. *Let R be a commutative ring of characteristic 0, let $f(T) = \sum_{k \geqslant 0} (a_k/k!) T^k$ and $g(T) = \sum_{k \geqslant 0} (b_k/k!) T^k$, and let $q \in R$. Assume that there exist a and b in R such that $a_k \equiv a^k \pmod{qR}$ and $b_k \equiv b^k \pmod{qR}$. Then we have $fg(T) = \sum_{k \geqslant 0} (c_k/k!) T^k$ with $c_k \equiv (a+b)^k \pmod{qR}$.*

Proof. Immediate and left to the reader. □

As always, in the sequel we assume that p and q are distinct odd primes.

Definition 16.4.16. (1) *If $F(T) = \sum_{k \geqslant 0} a_k T^k \in K[[T]]$ is a formal power series in T with coefficients in K, for any $\sigma \in G$ we let $F^\sigma(T) = \sum_{k \geqslant 0} \sigma(a_k) T^k$.*

(2) *If $F(T) = \sum_{k \geqslant 0} a_k T^k \in K[[T]]$ is a formal power series in T, for any integer $k \geqslant 0$ we denote by $F_k(T)$ the sum of the terms of degree less than or equal to k, in other words $F_k(T) = \sum_{0 \leqslant j \leqslant k} a_j T^j$.*

(3) *Let $\theta = \sum_{\sigma \in G} n_\sigma \sigma \in \mathbb{Z}[G]$. We define $F_\theta(T) \in K[[T]]$ to be the formal power series defined by the product*

$$F_\theta(T) = \prod_{\sigma \in G} (1 - \sigma(\zeta_p) T)^{n_\sigma/q},$$

where the power is obtained using the generalized binomial expansion.

Note that since $|\sigma(\zeta_p)| = 1$, if $z \in \mathbb{C}$ is such that $|z| < 1$ then the power series obtained by replacing T by z in $F_\theta(T)$ converges absolutely, and its sum will evidently be denoted by $F_\theta(z)$.

Definition 16.4.17. *Let*

$$F(T) = \sum_{k \geqslant 0} a_k T^k \in \mathbb{C}[[T]] \quad and \quad G(T) = \sum_{k \geqslant 0} b_k T^k \in \mathbb{R}[[T]].$$

We say that F is dominated by G if for all k we have $|a_k| \leqslant b_k$.

Proposition 16.4.18. *For simplicity, write F instead of F_θ.*

(1) *The coefficients of $F(T)$ are integral outside q, in other words have the form a/q^k for some $a \in \mathbb{Z}_K$ and $k \in \mathbb{Z}_{\geqslant 0}$.*

(2) *More precisely, if $\theta = \sum_{\sigma \in G} n_\sigma \sigma$ then $F(T) = \sum_{k \geqslant 0} (a_k/(q^k k!)) T^k$, where $a_k \in \mathbb{Z}_K$ satisfies*

$$a_k \equiv \left(-\sum_{\sigma \in G} n_\sigma \sigma(\zeta_p) \right)^k \pmod{q \mathbb{Z}_K}.$$

(3) *If $\tau \in G$ and $|t| < 1$ the series $F^\tau(t)$ converges. If, in addition, $0 \leqslant n_\sigma \leqslant q$ for all $\sigma \in G$, then if we set $m = (\sum_{\sigma \in G} n_\sigma)/q$ we have*

$$|F^\tau(t) - F_k^\tau(t)| \leqslant \binom{m+k}{k+1} \frac{|t|^{k+1}}{(1-|t|)^{m+k+1}}.$$

Proof. We have

$$(1 - \sigma(\zeta_p)T)^{n_\sigma/q} = \sum_{k \geqslant 0} \binom{n_\sigma/q}{k} (-\sigma(\zeta_p))^k T^k \,,$$

hence (1) follows from Lemma 4.2.8. More precisely, we have

$$\binom{n/q}{k} = \frac{n(n - q) \cdots (n - q(k - 1))}{q^k k!} \,,$$

so $(1 - q\sigma(\zeta_p)T)^{n/q} = \sum_{k \geqslant 0} b_k/k!$ with $b_k \equiv \sum_{k \geqslant 0}(-n\sigma(\zeta_p))^k \pmod{q\mathbb{Z}_K}$. It thus follows from Lemma 16.4.15 that

$$F(qT) = \prod_{\sigma \in G} (1 - \sigma(\zeta_p)T)^{n_\sigma/q} = \sum_{k \geqslant 0} (a_k/k!)T^k \,,$$

where

$$a_k \equiv \left(\sum_{\sigma \in G} (-n_\sigma \sigma(\zeta_p)) \right)^k \pmod{q\mathbb{Z}_K} \,,$$

proving (2). For (3) we note that when $0 \leqslant n \leqslant q$ we have

$$\left| \binom{n/q}{k} \right| = \left| \frac{n(n - q) \cdots (n - q(k - 1))}{k!} \right|$$

$$= \frac{n(q - n)(2q - n) \cdots (q(k - 1) - n)}{k!}$$

$$\leqslant \frac{n(n + q) \cdots (n + q(k - 1))}{k!} = \binom{-n/q}{k} \,.$$

It follows that the series $(1 - \sigma(\zeta_p)T)^{n/q}$ is dominated by the series $(1 - T)^{-n/q}$, so that $F(T)$ is dominated by $\prod_{\sigma \in G}(1 - T)^{-n_\sigma/q} = (1 - T)^{-m}$, and the same is evidently true for $F^\tau(T)$. It follows that for $|t| < 1$ we have

$$|F^\tau(t) - F_k^\tau(t)| \leqslant \left| (1 - |t|)^{-m} - \sum_{0 \leqslant j \leqslant k} \binom{-m}{j} (-|t|)^j \right| = |S(|t|) - S_k(|t|)| \,,$$

say, where we have set $S(T) = (1 - T)^{-m}$. Now by the Taylor–Lagrange theorem there exists $c \in [0, |t|]$ such that $S(|t|) - S_k(|t|) = (|t|^{k+1}/(k + 1)!)S^{(k+1)}(c)$. Since all the derivatives of S are evidently positive on $[0, 1[$, they are increasing, so that

$$S^{(k+1)}(c) \leqslant S^{(k+1)}(|t|) = m(m + 1) \cdots (m + k)(1 - |t|)^{-m-k-1}$$

$$= (m + k)!/((m - 1)!(1 - |t|)^{m+k+1}) \,,$$

and (3) follows.   □

Proposition 16.4.19. *Keep the same assumptions and notation, but assume in addition that $\theta \in (1 + \iota)\mathbb{Z}[G]$. Then*

(1) $F_\theta = F \in K^+[[T]]$.
(2) *Assume that $t \in \mathbb{Q}$ satisfies $|t| < 1$ and is such that there exists $\alpha \in K$ such that $(1 - t\zeta_p)^\theta = \alpha^q$. Then $\alpha \in K^+$, and for all $\sigma \in G$ we have $F^\sigma(t) = \sigma(\alpha)$.*

Proof. Since $\theta = \sum_{\sigma \in G} n_\sigma \sigma \in (1 + \iota)\mathbb{Z}[G]$ we have $\iota\theta = \theta$ hence $n_{\iota\sigma} = n_\sigma$ for all $\sigma \in G$. Thus if as usual P is a set of representatives of G modulo $\langle \iota \rangle$ we can write $F = F_1\overline{F_1}$, where F_1 is the same product as F but only over $\sigma \in P$, so the coefficients of F are real, hence in K^+, proving (1). For (2), the same reasoning shows that $(1 - t\zeta_p)^\theta \in \mathbb{R}$. It follows that $\overline{\alpha}^q = \overline{\alpha^q} = \overline{\beta} = \beta = \alpha^q$, hence $\overline{\alpha} = \alpha$ since qth roots are unique in K. It follows that $\alpha \in K^+$. Since G is abelian, it follows that $\sigma(\alpha) \in K^+$ for all $\sigma \in G$. In addition $\sigma(\alpha)^q = (1 - t\sigma(\zeta_p))^\theta = F^\sigma(t)^q$. However, since we have seen that $F^\sigma(t) \in \mathbb{R}$, it follows that $F^\sigma(t)/\sigma(\alpha)$ is a real qth root of unity in $\mathbb{C}$. Since q is odd, it must be equal to 1, proving the proposition. $\square$

Note that although not difficult, this last argument is one of the most subtle in the proof, and was in fact initially overlooked.

16.4.5 Proof of Mihăilescu's Fourth Theorem

In the above subsections we have studied properties of cyclotomic units, the plus part of cyclotomic fields, and power series, without any reference to Catalan's equation. We now begin the proof proper. We keep all of the above notation, in particular $R = \mathbb{F}_q[G^+]$.

Theorem 16.4.20. *Let p and q be distinct odd primes such that $\min(p, q) \geqslant 11$, and let x and y be nonzero integers such that $x^p - y^q = 1$. The submodule of S^+ generated by the class $[x - \zeta_p]^{1+\iota}$ is free; in other words, $\mathrm{Ann}_R([x - \zeta_p]^{1+\iota}) = 0$.*

Proof. Recall that $[x - \zeta_p] \in S$, so that we indeed have $[x - \zeta_p]^{1+\iota} \in S^+$. Thus let $\overline{\psi} = \sum_{\sigma \in G^+} \nu_\sigma \sigma \in \mathrm{Ann}_R([x - \zeta_p]^{1+\iota})$ with $\nu_\sigma \in \mathbb{F}_q$, so that $[x - \zeta_p]^{(1+\iota)\overline{\psi}} = 1$. Let P be a system of representatives in G of $G^+ = G/\langle \iota \rangle$, and by abuse of notation if $\sigma \in G^+$ denote again by σ the element of P whose class is σ. If we set $\psi = \sum_{\sigma \in P} \nu_\sigma \sigma$ we thus have $[x - \zeta_p]^{(1+\iota)\psi} = 1$. By definition of S it follows that for any $\theta \in \mathbb{Z}[G]$ whose reduction modulo q is equal to $\pm(1 + \iota)\psi$ we have $(x - \zeta_p)^\theta \in K^{*q}$. If for $\sigma \in P$ we set $\nu_{\iota\sigma} = \nu_\sigma$ we have $(1 + \iota)\psi = \sum_{\sigma \in G} \nu_\sigma \sigma \in \mathbb{F}_q[G]$. Let $\theta_1 = \sum_{\sigma \in G} n_\sigma \sigma \in \mathbb{Z}[G]$ be the lift of $(1 + \iota)\psi$ such that $0 \leqslant n_\sigma < q$, so that $\|\theta_1\| < (p - 1)q$. If for any integer n such that $0 \leqslant n < q$ we set $c(n) = q - n$ if $n \neq 0$ and $c(0) = 0$, then we again have $0 \leqslant c(n) < q$, and evidently $n + c(n) \leqslant q$. Thus $\theta_2 = \sum_{\sigma \in G} c(n_\sigma)\sigma$ is a lift of $-(1 + \iota)\psi$, $0 \leqslant c(n_\sigma) < q$, and $\|\theta_2\| + \|\theta_1\| \leqslant q(p - 1)$. It follows that for $i = 1$ or $i = 2$ we have $\|\theta_i\| \leqslant q(p-1)/2$, and we let θ be equal to the θ_i satisfying this inequality.

Let $\alpha \in K^*$ be such that $(x - \zeta_p)^\theta = \alpha^q$. By Lemma 16.1.1 we know that $\beta = (x - \zeta_p)/(1 - \zeta_p) \in \mathbb{Z}_K$, that $v_{\mathfrak{p}}(\beta) = 0$, and that the ideals generated by the conjugates of β are pairwise coprime. It follows that for all $\sigma \in G$ we have $v_{\mathfrak{p}}(x - \sigma(\zeta_p)) = 1$. Thus

$$\|\theta\| = \sum_{\sigma \in G} n_\sigma v_{\mathfrak{p}}(x - \sigma(\zeta_p)) = v_{\mathfrak{p}}\left(\prod_{\sigma \in G} (x - \sigma(\zeta_p))^{n_\sigma} \right)$$
$$= v_{\mathfrak{p}}\left((x - \zeta_p)^\theta \right) = q v_{\mathfrak{p}}(\alpha) \equiv 0 \pmod{q} .$$

Since $0 \leqslant \|\theta\| \leqslant q(p-1)/2$ it follows that there exists $m \in [0, (p-1)/2]$ such that $\|\theta\| = mq$. In addition, since n_σ and $n_{\iota\sigma}$ both reduce to ν_σ modulo q and are both in the interval $[0, q-1]$, they are in fact equal. It follows that $\theta = (1 + \iota)\phi$, where $\phi = \sum_{\sigma \in P} n_\sigma \sigma$ is a lift of ψ. In particular, for all $\sigma \in G$, $(x - \sigma(\zeta_p))^\theta = ((x - \sigma(\zeta_p))(z - \iota(\sigma(\zeta_p))))^\phi$ is a real number. Since qth roots are unique in K, when they exist, it follows that all the conjugates of α are real. Since for $x \in \mathbb{Q}$ we have $x^\theta = x^{\|\theta\|}$ it follows that for all $\sigma \in G$ we have $(1 - \sigma(\zeta_p)/x)^\theta = (\sigma(\alpha)/x^m)^q$. Since $1/x \in \mathbb{Q}$ and $|1/x| < 1$, we may apply Proposition 16.4.19 and deduce that for all $\sigma \in G$ we have $\sigma(\alpha) = x^m F^\sigma(1/x)$, where $F = F_\theta$. Set

$$I_\sigma = q^{m + v_q(m!)} |\sigma(\alpha) - x^m F_m^\sigma(1/x)| .$$

We are now going to use a Runge-type argument and show that $|I_\sigma| < 1$ and that $\prod_{\sigma \in G} I_\sigma \in \mathbb{Z}$. First, by Proposition 16.4.18 (3) we have

$$I_\sigma = q^{m + v_q(m!)} |x|^m |F^\sigma(1/x) - F_m^\sigma(1/x)|$$
$$\leqslant q^{m + v_q(m!)} \binom{2m}{m+1} |x|^{-1} (1 - 1/|x|)^{-(2m+1)}$$
$$\leqslant q^{m + m/(q-1) + m(\log(4)/\log(q))} |x|^{-1} (1 - 1/|x|)^{-(2m+1)} ,$$

where we have used $v_q(m!) \leqslant m/(q-1)$ and $\binom{2m}{k} \leqslant 2^{2m}$. Since $m \leqslant (p-1)/2$ and by Proposition 6.11.15 we have $|x| \geqslant q^{p-1}$, it follows that

$$I_\sigma \leqslant q^{((p-1)/2)(1 + 1/(q-1) + \log(4)/\log(q))} |x|^{-1} (1 - 1/|x|)^{-p}$$
$$\leqslant q^{((p-1)/2)(-1 + 1/(q-1) + \log(4)/\log(q))} (1 - 1/q^{p-1})^{-p} .$$

Now $I_\sigma < 1$ is equivalent to $\log(I_\sigma)/\log(q) < 0$, and we have

$$\frac{\log(I_\sigma)}{\log(q)} = \frac{p-1}{2}\left(-1 + \frac{1}{q-1} + \frac{\log(4)}{\log(q)} \right) - \frac{p}{\log(q)} \log(1 - 1/q^{p-1}) .$$

However, by the mean value theorem there exists $c \in [0, 1]$ such that

$$-\log(1-1/q^{p-1}) = \log(q^{p-1}) - \log(q^{p-1}-1) = \frac{1}{q^{p-1}-c} \leqslant \frac{1}{q^{p-1}-1} \leqslant \frac{1}{q^2-1}$$

since $p \geqslant 3$. Since we have assumed that $q \geqslant 7$ we immediately obtain

$$\frac{\log(I_\sigma)}{\log(q)} \leqslant \frac{p-1}{2}\left(-1 + \frac{1}{6} + \frac{\log(4)}{\log(7)}\right) + \frac{p}{48\log(7)} \leqslant -0.0497p + 0.061 \, ,$$

and this is strictly negative as soon as $p \geqslant 2$, proving that $I_\sigma < 1$.

Let us now look at the arithmetic properties of I_σ. By Proposition 16.4.18 we have $F_m^\sigma(T) = \sum_{0 \leqslant k \leqslant m} a_k/(q^k k!) T^k$ with $a_k \in \mathbb{Z}_K$. It follows that $q^{m+v_q(m!)} a_k/(q^k k!) \in \mathbb{Z}_K$, hence that $q^{m+v_q(m!)} x^m F_m^\sigma(1/x) \in \mathbb{Z}_K$ (note that there are no convergence problems here since we deal with polynomials). In addition, since $(x - \zeta_p)^\theta = \alpha^q$ and all the coefficients of θ are nonnegative, α^q is an algebraic *integer*, hence α also, so that $\alpha \in \mathbb{Z}_K = \mathbb{Z}[\zeta_p]$. It follows that $\gamma = q^{m+v_q(m!)}(\alpha - x^m F_m(1/x)) \in \mathbb{Z}_K$, hence that $\mathcal{N}_{K/\mathbb{Q}}(\gamma) \in \mathbb{Z}$. However, $|\mathcal{N}_{K/\mathbb{Q}}(\gamma)| = \prod_{\sigma \in G} I_\sigma < 1$ by what we have proved above. It follows that $\mathcal{N}_{K/\mathbb{Q}}(\gamma) = 0$, hence that $\gamma = 0$, in other words that

$$q^{m+v_q(m!)}\alpha = \sum_{0 \leqslant k \leqslant m} q^{m+v_q(m!)} \frac{a_k}{q^k k!} x^{m-k} \, .$$

Now all the terms occurring in the sum are divisible by q except the term with $k = m$. Thus $0 \equiv (q^{v_q(m!)}/m!)a_m \pmod{q\mathbb{Z}_K}$, so $a_m \equiv 0 \pmod{q\mathbb{Z}_K}$. On the other hand, by Proposition 16.4.18 we have $a_m \equiv s^m \pmod{q\mathbb{Z}_K}$, where $s = -\sum_{\sigma \in G} n_\sigma \sigma(\zeta_p)$. Thus $s^m \equiv 0 \pmod{q\mathbb{Z}_K}$, so for every prime ideal $\mathfrak{q}$ of K above q we have $s^m \in \mathfrak{q}$, hence $s \in \mathfrak{q}$, and since q is unramified, by the Chinese remainder theorem we deduce that $s \equiv 0 \pmod{q\mathbb{Z}_K}$, in other words that $\sum_{\sigma \in G}(n_\sigma/q)\sigma(\zeta_p) \in \mathbb{Z}_K$. Since the $\sigma(\zeta_p)$ are up to permutation the ζ_p^j for $1 \leqslant j \leqslant p-1$, which form a $\mathbb{Z}$-basis of $\mathbb{Z}_K$, it follows that $n_\sigma/q \in \mathbb{Z}$ for all σ, and since $0 \leqslant n_\sigma < q$ we deduce that $n_\sigma = 0$ for all σ. Thus $\theta = 0$, hence $\psi = 0$ and $\overline{\psi} = 0$, as was to be proved.   □

Mihăilescu's fourth theorem is now immediate.

Theorem 16.4.21. *Let p and q be odd primes such that $\min(p,q) \geqslant 11$, and let x and y be nonzero integers such that $x^p - y^q = 1$. Then $p \equiv 1 \pmod q$ or $q \equiv 1 \pmod p$.*

Proof. By Theorem 16.4.20, $\mathrm{Ann}_R([x - \zeta_p]^{1+\iota}) = 0$. By Mihăilescu's first Theorem 16.1.3 we know that $q^2 \mid x$, and as usual $(-\zeta_p)$ is a qth power since q and $2p$ are coprime. It follows that $x - \zeta_p \equiv \beta^q \pmod{q^2 R_p}$, hence $[x - \zeta_p] \in S_q$, so that $[x - \zeta_p]^{1+\iota} \in S_q \cap S^+$. Now by symmetry assume for instance that $p > q$, so that of course $q \not\equiv 1 \pmod p$. If we assume by contradiction that $p \not\equiv 1 \pmod q$ then Theorem 16.4.14 tells us that $\mathrm{Ann}_{\mathbb{F}_q[G^+]}(S^+ \cap S_q) \neq 0$, and in particular $\mathrm{Ann}_R([x - \zeta_p]^{1+\iota}) \neq 0$, a contradiction.   □

16.4.6 Conclusion: Proof of Catalan's Conjecture

We now summarize what we have done in Chapter 6 and in the present chapter, and finish the proof of Catalan's conjecture. Let x and y be nonzero integers and m, $n \geqslant 2$ such that $x^m - y^n = 1$. Lebesgue's Proposition 6.7.12 tells us that $n = 2$ and m prime is impossible, from which we deduce that the case n even is impossible. Similarly, Ko Chao's Theorem 6.11.8 tells us that $m = 2$ and n prime is impossible apart from $3^2 - 2^3 = 1$, so the case m even is also solved. Thus we may assume that m and n are odd, and it is sufficient to prove impossibility for $m = p$ and $n = q$ odd primes. In particular, the equation becomes symmetrical since we can change (p, q, x, y) into $(q, p, -y, -x)$. By Mihăilescu's second theorem (more precisely Corollary 16.1.12) we may assume that $\min(p, q) \geqslant 11$ (in fact 43, but 11 is sufficient). Thus, by Mihăilescu's fourth Theorem 16.4.21, exchanging p and q if necessary thanks to the above symmetry, we may assume that $p \equiv 1 \pmod{q}$. By the binomial theorem we have

$$p^q = (1+(p-1))^q = 1+q(p-1)+ \sum_{2 \leqslant i \leqslant q-1} \binom{q}{i}(p-1)^i+(p-1)^q \equiv 1 \pmod{q^2}.$$

On the other hand, by Mihăilescu's first Theorem 16.1.3 we have $p^{q-1} \equiv 1 \pmod{q^2}$, hence $p^q \equiv p \pmod{q^2}$, so that $p \equiv 1 \pmod{q^2}$. Finally, by Mihăilescu's third Theorem 16.3.7, we have $p < 4q^2$. It follows that $p = 1 + kq^2$ with $k = 1$, 2, or 3. Clearly $k = 1$ and $k = 3$ are impossible since otherwise p would be even, and $k = 2$ is impossible since $q^2 \equiv 1 \pmod{3}$ hence $1 + 2q^2 \equiv 0 \pmod{3}$, which is again impossible, finishing the proof of Catalan's conjecture. $\qquad\square$

Bibliography

[Abou]   M. Abouzaid, *Les nombres de Lucas et Lehmer sans diviseur primitif*, J. Théor. Nombres Bordeaux **18** (2006), 299–313.

[Abr-Ste]   M. Abramowitz and I. Stegun, *Handbook of Mathematical Functions*, Dover publications (1972).

[AGP]   R. Alford, A. Granville, and C. Pomerance, *There are infinitely many Carmichael numbers*, Ann. of Math. **139** (1994), 703–722.

[Ami]   Y. Amice, *Les nombres p-adiques*, SUP/Le Mathématicien **14**, Presses Universitaires de France (1975).

[Ang]   W. Anglin, *The square pyramid puzzle*, American Math. Monthly **97** (1990), 120–124.

[Ax]   J. Ax, *Zeroes of polynomials over finite fields*, Amer. J. Math. **86** (1964), 255–261.

[Bac]   G. Bachman, *Introduction to p-adic Numbers and Valuation theory*, Academic paperbacks, Acad. Press (1964).

[Bak1]   A. Baker, *Linear forms in the logarithms of algebraic numbers*, Mathematika **13** (1966), 204–216.

[Bak2]   A. Baker, *Transcendental Number Theory*, Cambridge University Press, 1975.

[Bak-Dav] A. Baker and H. Davenport, *The equations $3x^2-2 = y^2$ and $8x^2-7 = y^2$*, Quart. J. Math. Oxford Ser. (2) **20** (1969), 129–137.

[Bak-Wus] A. Baker and G. Wüstholz, *Logarithmic forms and group varieties*, J. reine angew. Math. **442** (1993), 19–62.

[BDD]   R. Balasubramanian, J.-M. Deshouillers, and F. Dress, *Problème de Waring pour les bicarrés 1 : schéma de la solution, 2 : résultats auxiliaires pour le théorème asymptotique*, C. R. Acad. Sc. Paris **303** (1986), 85–88 and 161–163.

[Bal-Dar-Ono] A. Balog, H. Darmon, and K. Ono, *Congruences for Fourier coefficients of half-integral weight modular forms and special values of L-functions*, Proceedings of a Conference in honor of H. Halberstam **1** (1996), 105–128.

[Bar]   D. Barsky, *Congruences de coefficients de séries de Taylor (Application aux nombres de Bernoulli–Hurwitz)*, Groupe d'Analyse Ultramétrique **3** (1975-1976), Exp. 17, 1–9, available on the NUMDAM archives.

[Bat-Oli]   C. Batut and M. Olivier, *Sur l'accélération de la convergence de certaines fractions continues*, Séminaire Th. Nombres Bordeaux (1979–1980), exposé **23**.

[Bel-Gan] K. Belabas and H. Gangl, *Generators and relations for $K_2\mathcal{O}_F$*, K-Theory **31** (2004), 195–231.

[BBGMS]   C. Bennett, J. Blass, A. Glass, D. Meronk, and R. Steiner, *Linear forms in the logarithms of three positive rational numbers*, J. Théor. Nombres Bordeaux **9** (1997), 97–136.

[Ben1]   M. Bennett, *Rational approximation to algebraic numbers of small height: The Diophantine equation $\mid ax^n - by^n \mid = 1$*, J. reine angew. Math. **535** (2001), 1–49.

[Ben2]   M. Bennett, *Recipes for ternary Diophantine equations of signature (p, p, k)*, Proc. RIMS Kokyuroku (Kyoto) **1319** (2003), 51–55.

[Ben3]   M. Bennett, *On some exponential Diophantine equations of S. S. Pillai*, Canad. J. Math. **53** (2001), 897–922.

[Ben-deW]   M. Bennett and B. de Weger, *The Diophantine equation $\mid ax^n - by^n \mid = 1$*, Math. Comp. **67** (1998), 413–438.

[Ben-Ski]   M. Bennett and C. Skinner, *Ternary Diophantine equations via Galois representations and modular forms*, Canad. J. Math. **56** (2004), 23–54.

[Ben-Vat-Yaz] M. Bennett, V. Vatsal, and S. Yazdani, *Ternary Diophantine equations of signature $(p, p, 3)$*, Compositio Math. **140** (2004), 1399–1416.

[Ber-Eva-Wil] B. Berndt, R. Evans, and K. Williams, *Gauss and Jacobi Sums*, Canadian Math. Soc. series **21**, Wiley (1998).

[Bha1]   M. Bhargava, *Higher composition laws I, II, and III*, Ann. Math. **159** (2004), 217–250, 865–886, 1329–1360.

[Bha2]   M. Bhargava, *The density of discriminants of quartic rings and fields*, Ann. Math. **162** (2005), 1031–1063.

[Bha-Han]   M. Bhargava and J. Hanke, *Universal quadratic forms and the 290-theorem*, Invent. Math., to appear.

[Bilu]   Yu. Bilu, *Catalan's conjecture (after Mihailescu)*, Séminaire Bourbaki **909** (2002–2003), 1–25.

[Bil-Han]   Yu. Bilu and G. Hanrot, *Solving Thue equations of high degree*, J. Number Th. **60** (1996), 373–392.

[Bil-Han-Vou] Yu. Bilu, G. Hanrot, and P. Voutier, *Existence of primitive divisors of Lucas and Lehmer numbers*, with an appendix by M. Mignotte, J. reine angew. Math. **539** (2001), 75–122.

[Boe-Mis]   J. Boéchat and M. Mischler, *La conjecture de Catalan racontée à un ami qui a le temps*, preprint available on the web at the URL http://arxiv.org/pdf/math.NT/0502350.

[Bom]   E. Bombieri, *Effective Diophantine approximation on $\mathbf{G}_m$*, Ann. Scuola Norm. Sup. Pisa Cl. Sci. (4) **20** (1993), 61–89.

[Bor-Bai]   J. Borwein and D. Bailey, *Mathematics by Experiment*, A. K. Peters (2004).

[Bor-Bai-Gir] J. Borwein, D. Bailey, and R. Girgensohn, *Experimentation in Mathematics*, A. K. Peters (2004).

[Bor-Sha]   Z. I. Borevitch and I. R. Shafarevitch, *Number Theory*, Academic Press, New York (1966).

[Bre-Cas]   A. Bremner and I. Cassels, *On the equation $Y^2 = X(X^2 + p)$*, Math. Comp. **42** (1984), 257–264.

[Bre-Mor]   A. Bremner and P. Morton, *A new characterization of the integer 5906*, Manuscripta Math. **44** (1983), 187–229.

[Bre-Tza1]   A. Bremner and N. Tzanakis, *Lucas sequences whose 12th or 9th term is a square*, J. Number Th. **107** (2004), 215–227.

[Bre-Tza2]   A. Bremner and N. Tzanakis, *On squares in Lucas sequences*, J. Number Th., to appear.

[Bre-Tza3]   A. Bremner and N. Tzanakis, *Lucas sequences whose nth term is a square or an almost square*, Acta Arith., to appear.

[BCDT]   C, Breuil, B. Conrad, F. Diamond, and R. Taylor, *On the modularity of elliptic curves over* $\mathbb{Q}$*: wild 3-adic exercises*, J. Amer. Math. Soc. **14** (2001), 843–939.

[Bri-Eve-Gyo] B. Brindza, J. Evertse, and K. Győry, *Bounds for the solutions of some Diophantine equations in terms of discriminants*, J. Austral. Math. Soc. (Series A) **51** (1991), 8–26.

[Bru1]   N. Bruin, *Chabauty Methods and Covering Techniques Applied to Generalized Fermat Equations*, CWI Tract **133**, CWI, Amsterdam (2002).

[Bru2]   N. Bruin, *The Diophantine equations* $x^2 \pm y^4 = \pm z^6$ *and* $x^2 + y^8 = z^3$, Compositio Math. **118** (1999), 305–321.

[Bru3]   N. Bruin, *Chabauty methods using elliptic curves*, J. reine angew. Math. **562** (2003), 27–49.

[Bru4]   N. Bruin, *Primitive solutions to* $x^3 + y^9 = z^2$, J. Number theory **111** (2005), 179–189.

[Bru-Kra]   A. Brumer and K. Kramer, *The rank of elliptic curves*, Duke Math. J. **44** (1977), 715–742.

[Bug]   Y. Bugeaud, *Bounds for the solutions of superelliptic equations*, Compositio Math. **107** (1997), 187–219.

[Bug-Gyo]   Y. Bugeaud and K. Győry, *Bounds for the solutions of Thue–Mahler equations and norm form equations*, Acta Arith. **74** (1996), 273–292.

[Bug-Han]   Y. Bugeaud and G. Hanrot, *Un nouveau critère pour l'équation de Catalan*, Mathematika **47** (2000), 63–73.

[Bug-Mig]   Y. Bugeaud and M. Mignotte, *On integers with identical digits*, Mathematika **46** (1999), 411–417.

[BMS1]   Y. Bugeaud, M. Mignotte, and S. Siksek, *Classical and modular approaches to exponential Diophantine equations I. Fibonacci and Lucas perfect powers*, Annals of Math. **163** (2006), 969–1018.

[BMS2]   Y. Bugeaud, M. Mignotte, and S. Siksek, *Classical and modular approaches to exponential Diophantine equations II. The Lebesgue–Nagell equation*, Compositio Math. **142** (2006), 31–62.

[BMS3]   Y. Bugeaud, M. Mignotte, and S. Siksek, *A multi-Frey approach to some multi-parameter families of Diophantine equations*, Canadian J. Math., to appear.

[Buh-Gro]   J. Buhler and B. Gross, *Arithmetic on elliptic curves with complex multiplication II*, Invent. Math. **79** (1985), 11–29.

[BGZ]   J. Buhler, B. Gross, and D. Zagier, *On the conjecture of Birch and Swinnerton-Dyer for an elliptic curve of rank 3*, Math. Comp. **44** (1985), 473–481.

[Cal]   E. Cali, *Points de torsion des courbes elliptiques et quartiques de Fermat*, Thesis, Univ. Paris VI (2005).

[Can]   D. Cantor, *Computing on the Jacobian of a hyperelliptic curve*, Math. Comp., **48** (1987), 95–101.

[Cas1]   J. Cassels, *Local Fields*, London Math. Soc. Student Texts **3**, Cambridge University Press (1986).

[Cas2]   J. Cassels, *Lectures on Elliptic Curves*, London Math. Soc. Student Texts **24**, Cambridge University Press (1991).

[Cas3]   J. Cassels, *On the equation* $a^x - b^y = 1$, *II*, Proc. Cambridge Phil. Soc. **56** (1960), 97–103.

[Cas-Fly]   J. Cassels and V. Flynn, *Prolegomena to a Middlebrow Arithmetic of Curves of Genus 2*, LMS Lecture Note Series **230**, Cambridge University Press (1996).

[Cas-Frö]   J. Cassels and A. Fröhlich, *Algebraic Number Theory*, Academic Press, London, New York (1967).

[Cat]   E. Catalan, *Note extraite d'une lettre adressée à l'éditeur*, J. reine angew. Math. **27**, (1844), 192.

[Cha]   C. Chabauty, *Sur les points rationnels des variétés algébriques dont l'irrégularité est supérieure à la dimension*, C. R. A. S. Paris, **212** (1941), 1022–1024.

[Coa-Wil]   J. Coates and A. Wiles, *On the conjecture of Birch and Swinnerton-Dyer*, Invent. Math. **39** (1977), 223–251.

[Coh0]   H. Cohen, *A Course in Computational Algebraic Number Theory (4th corrected printing)*, Graduate Texts in Math. **138**, Springer-Verlag (2000).

[Coh1]   H. Cohen, *Advanced Topics in Computational Number Theory*, Graduate Texts in Math. **193**, Springer-Verlag (2000).

[Coh2]   H. Cohen, *Variations sur un thème de Siegel et Hecke*, Acta Arith. **30** (1976), 63–93.

[Coh3]   H. Cohen, *Sums involving L-functions of quadratic characters*, Math. Ann. **217** (1975), 271–285.

[Coh4]   H. Cohen, *Continued fractions for gamma products and $\zeta(k)$*, unfinished postscript preprint available on the author's home page at http://www.math.u-bordeaux1.fr/~cohen/.

[Coh-Fre]   H. Cohen and G. Frey, eds., *Handbook of elliptic and hyperelliptic curve cryptography*, Chapman & Hall/CRC press, 2005.

[Coh-Fri]   H. Cohen and E. Friedman, *Raabe's formula for p-adic gamma and zeta functions*, submitted.

[Coh-Len]   H. Cohen and H. W. Lenstra, *Heuristics on class groups of number fields*, Springer Lecture Notes in Math. **1068** (1984), 33–62.

[Coh-Mar]   H. Cohen and J. Martinet, *Class groups of number fields: numerical heuristics*, Math. Comp. **48** (1987), 123–137.

[Coh-Rhi]   H. Cohen and G. Rhin, *Accélération de la convergence de certaines récurrences linéaires*, Séminaire Th. Nombres Bordeaux (1980–1981), exposé **16**.

[Coh-Vil-Zag]   H. Cohen, F. Rodriguez-Villegas, and D. Zagier, *Convergence acceleration of alternating series*, Exp. Math. **9** (2000), 3–12.

[Cohn1]   J. Cohn, *The Diophantine equation $x^2 + C = y^n$*, Acta Arith. **65** (1993), 367–381.

[Cohn2]   J. Cohn, *The Diophantine equation $x^2+C = y^n$, II*, Acta Arith. **109** (2003), 205–206.

[Col]   R. Coleman, *Effective Chabauty*, Duke Math. J., **52** (1985), 765–780.

[Colm]   P. Colmez, *Arithmétique de la fonction zêta*, Journées mathématiques X-UPS (2002), Publications de l'Ecole Polytechnique, 37–164.

[Con-Sou]   J. B. Conrey and K. Soundararajan, *Real zeros of quadratic Dirichlet L-functions*, Invent. Math. **150** (2002), 1–44.

[Con]   J.-H. Conway, *The Sensual (Quadratic) Form*, Carus Math. Monographs **26**, MAA (1997).

[Con-Slo]   J.-H. Conway and N. Sloane, *Sphere Packings, Lattices and Groups (3rd ed.)*, Grundlehren der math. Wiss. **290**, Springer-Verlag, New York (1999).

[Cre1]   J. Cremona, *Computing the degree of the modular parametrization of a modular elliptic curve*, Math. Comp. **64** (1995), 1235–1250.

[Cre2]   J. Cremona, *Algorithms for Modular Elliptic Curves (2nd ed.)*, Cambridge Univ. Press (1996).

[Cre-Pri-Sik]   J. Cremona, M. Prickett, and S. Siksek, *Height difference bounds for elliptic curves over number fields*, J. Number theory **116** (2006), 42–68.

[Dar]   H. Darmon, *Rational Points on Modular Elliptic Curves*, CBMS Regional Conference Series in Mathematics **101** (2004), American Math. Soc.

[Dar-Gra]   H. Darmon and A. Granville, *On the equations $z^m = F(x,y)$ and $Ax^p + By^q = Cz^r$*, Bull. London Math. Soc. **27** (1995), 513–543.

[Dar-Mer]   H. Darmon and L. Merel, *Winding quotients and some variants of Fermat's Last Theorem*, J. reine angew. Math. **490** (1997), 81–100.

[Dem1]   V. Dem'yanenko, *О Суммах четырех кубов (On sums of four cubes)*, Izv. Visch. Outch. Zaved. Mathematika **54** (1966), 64–69.

[Dem2]   V. Dem'yanenko, *Rational points on a class of algebraic curves*, Amer. Math. Soc. Transl. **66** (1968), 246–272.

[Dem3]   V. Dem'yanenko, *The indeterminate equations $x^6 + y^6 = az^2$, $x^6 + y^6 = az^3$, $x^4 + y^4 = az^4$*, Amer. Math. Soc. Transl. **119** (1983), 27–34.

[Den]   P. Dénes, *Über die Diophantische Gleichung $x^\ell + y^\ell = cz^\ell$*, Acta Math. **88** (1952), 241–251.

[DeW1]   B. de Weger, *Solving exponential Diophantine equations using lattice basis reduction algorithms*, J. Number Th. **26** (1987), 325–367.

[DeW2]   B. de Weger, *A hyperelliptic Diophantine equation related to imaginary quadratic number fields with class number 2*, J. reine angew. Math. **427** (1992), 137–156.

[Dia1]   J. Diamond, *The p-adic log gamma funnction and p-adic Euler constants*, Trans. Amer. Math. Soc. **233** (1977), 321–337.

[Dia2]   J. Diamond, *On the values of p-adic L-functions at positive integers*, Acta Arith. **35** (1979), 223–237.

[Dia-Kra]   F. Diamond and K. Kramer, *Modularity of a family of elliptic curves*, Math. Res. Lett. **2** (1995), No. 3, 299-304.

[Dok]   T. Dokchitser, *Computing special values of motivic L-functions*, Exp. Math. **13** (2004), 137–149.

[Duq1]   S. Duquesne, *Rational Points on Hyperelliptic Curves and an Explicit Weierstrass Preparation Theorem*, Manuscripta Math. **108:2** (2002), 191–204.

[Duq2]   S. Duquesne, *Calculs effectifs des points entiers et rationnels sur les courbes*, Thesis, Univ. Bordeaux I (2001).

[Edw]   J. Edwards, *Platonic solids and solutions to $x^2 + y^3 = dz^r$*, Thesis, Univ. Utrecht (2005).

[Elk1]   N. Elkies, *ABC implies Mordell*, Internat. Math. Res. Notices **7** (1991), 99–109.

[Elk2]   N. Elkies, $\mathbb{Z}^{28}$ in $E(\mathbb{Q})$, Internet announcement on the number theory listserver (May 3rd, 2006).

[Ell]   W. Ellison and M. Mendès France, *Les nombres premiers*, Hermann (1975).

[Erd-Wag]   P. Erdős and S. Wagstaff, *The fractional parts of the Bernoulli numbers*, Illinois J. Math. **24** (1980), 104–112.

[Eva]   R. Evans, *Congruences for Jacobi sums*, J. Number Theory **71** (1998), 109–120.

[Fal]   G. Faltings, *Endlichkeitssätze für abelsche Varietäten über Zahlkörpen*, Invent. Math. **73** (1983), 349–366.

[Fer-Gre]   B. Ferrero and R. Greenberg, *On the behaviour of p-adic L-functions at $s = 0$*, Invent. Math. **50** (1978), 91–102.

[Fly]   V. Flynn, *A flexible method for applying Chabauty's Theorem*, Compositio Math. **105** (1997), 79–94.

[Fly-Wet1]   V. Flynn and J. Wetherell, *Finding rational points on bielliptic genus 2 curves*, Manuscripta Math. **100** (1999), 519-533.

[Fly-Wet2]   V. Flynn and J. Wetherell, *Covering collections and a challenge problem of Serre*, Acta Arith. **98** (2001), 197–205.

[Fre]   E. Freitag, *Hilbert Modular Forms*, Springer-Verlag (1990).

[Frö-Tay]   A. Fröhlich and M. Taylor, *Algebraic Number Theory*, Cambridge Studies in Adv. Math. **27**, Cambridge Univ. Press (1991).

[Gel]   A. O. Gel'fond, *On the approximation of transcendental numbers by algebraic numbers*, Doklady Akad. Nauk SSSR **2** (1935), 177–182.

[Gou]   F. Gouvêa, *p-adic Numbers: An Introduction*, Universitext, Springer-Verlag (1993).

[Gra-Sou]   A. Granville and K. Soundararajan, *Large character sums: pretentious characters and the Polya–Vinogradov theorem*, Journal of the American Math. Soc., to appear.

[Gran]   D. Grant, *A curve for which Coleman's effective Chabauty bound is sharp*, Proc. Amer. Math. Soc. **122** (1994), 317–319.

[Gras]   G. Gras, *Class Field Theory: From Theory to Practice*, Springer monographs in mathematics (2003).

[Gre-Tao]   B. Green and T. Tao, *The primes contain arbitrarily long arithmetic progressions*, Ann. Math., to appear.

[Gri-Riz]   G. Grigorov and J. Rizov, *Heights on elliptic curves and the Diophantine equation $x^4 + y^4 = cz^4$*, Sophia Univ. preprint (1998).

[Gro]   B. Gross, *Heegner points on $X_0(N)$*, in Modular forms, edited by R. Rankin (1984), 87–105.

[Gro-Kob]   B. Gross and N. Koblitz, *Gauss sums and the p-adic Γ-function*, Ann. Math. **109** (1979), 569–581.

[Guy]   R. K. Guy, *Unsolved Problems in Number Theory (3rd edition)*, Problem books in math. **1**, Springer-Verlag (2004).

[Hal-Kra1]   E. Halberstadt and A. Kraus, *Sur les modules de torsion des courbes elliptiques*, Math. Ann. **310** (1998), 47–54.

[Hal-Kra2]   E. Halberstadt and A. Kraus, *Courbes de Fermat : résultats et problèmes*, J. reine angew. Math. **548** (2002), 167–234.

[Har-Wri]   G. H. Hardy and E. M. Wright, *An Introduction to the Theory of Numbers (5th ed.)*, Oxford University Press (1979).

[Hay]   Y. Hayashi, *The Rankin's L-function and Heegner points for general discriminants*, Proc. Japan. Acad. **71** (1995), 30–32.

[Her]   G. Herglotz, *Über die Kroneckersche Grenzformel für reelle quadratische Körper I, II*, Gesam. Schr. (ed. H. Schwerdtfeger), Vandenhoeck and Ruprecht (1979), 466–484.

[Hul]   W. Hulsbergen, *Conjectures in Arithmetic Algebraic Geometry*, Aspects of math., Vieweg (1992).

[Ire-Ros]   K. Ireland and M. Rosen, *A Classical Introduction to Modern Number Theory (2nd ed.)*, Graduate Texts in Math. **84**, Springer-Verlag (1982).

[Ivo1]   W. Ivorra, *Sur les équations $x^p + 2^\beta y^p = z^2$ et $x^p + 2^\beta y^p = 2z^2$*, Acta Arith. **108** (2003), 327–338.

[Ivo2]   W. Ivorra, *Equations diophantiennes ternaires de type $(p, p, 2)$ et courbes elliptiques*, Thesis, Univ. Paris VI (2004).

[Ivo-Kra]   W. Ivorra and A. Kraus, *Quelques résultats sur les équations $ax^p + by^p = cz^2$*, Can. J. Math., to appear.

[Iwa-Kow]   H. Iwaniec and E. Kowalski, *Analytic Number Theory*, Colloquium Publications **53**, American Math. Soc. (2004).

[Jan]   G. Janusz, *Algebraic Number Fields*, Pure and applied math. **55**, Academic Press (1973).

[Kap]   I. Kaplansky, *Ternary positive quadratic forms that represent all odd positive integers*, Acta Arith. **70** (1995), 209–214.

[Kat1]   N. Katz, *On a theorem of Ax*, Amer. J. Math. **93** (1971), 485–499.

[Kat2]   N. Katz, *The congruences of Clausen–von Staudt and Kummer for Bernoulli–Hurwitz numbers*, Math. Ann. **216** (1975), 1–4.

[Kea]   J. Keating, talk in Bordeaux, 2005.

[Kel-Ric]   W. Keller and J. Richstein, *Solutions of the congruence $a^{p-1} \equiv 1$ (mod p^r)*, Math. Comp. **74** (2005), 927–936.

[Kna]   A. Knapp, *Elliptic Curves*, Math. Notes **40**, Princeton University press (1992)

[Ko]   Ko Chao, *On the Diophantine equation $x^2 = y^n + 1$, $xy \neq 0$*, Sci. Sinica **14** (1965), 457–460.

[Kob1]   N. Koblitz, *p-adic Numbers, p-adic Analysis, and Zeta-Functions (2nd edition)*, Graduate Texts in Math. **58**, Springer-Verlag (1984).

[Kob2]   N. Koblitz, *An Introduction to Elliptic Curves and Modular Forms (2nd edition)*, Graduate Texts in Math. **97**, Springer-Verlag (1993).

[Kra1]   A. Kraus, *Sur l'équation $a^3 + b^3 = c^p$*, Experimental Math. **7** (1998), 1–13.

[Kra2]   A. Kraus, *On the equation $x^p + y^q = z^r$: a survey*, Ramanujan Journal **3** (1999), 315–333.

[Kra3]   A. Kraus, *Majorations effectives pour l'équation de Fermat généralisée*, Can. J. Math. **49** (1997), 1139–1161.

[Kra-Oes]   A. Kraus and J. Oesterlé, *Sur une question de B. Mazur*, Math. Ann. **293** (1992), 259–275.

[Kul]   L. Kulesz, *Application de la méthode de Dem'janenko–Manin à certaines familles de courbes de genre 2 et 3*, J. Number Theory **76** (1999), 130–146.

[Lan0]   S. Lang, *Algebra*, Addison-Wesley, Reading, MA (1965).

[Lan1]   S. Lang, *Algebraic Number Theory (2nd ed.)*, Graduate Texts in Math. **110**, Springer-Verlag (1994).

[Lau]   M. Laurent, *Linear form in two logarithms and interpolation determinants*, Acta Arith. **66** (1994), 181–199.

[Lau-Mig-Nes]   M. Laurent, M. Mignotte, and Yu. Nesterenko, *Formes linéaires en deux logarithmes et déterminants d'interpolation*, J. Number Theory **55** (1995), 255–265.

[Leb]   V. Lebesgue, *Sur l'impossibilité en nombres entiers de l'équation $x^m = y^2 + 1$*, Nouv. Ann. Math. **9** (1850), 178–181.

[Lem]   F. Lemmermeyer, *Kronecker–Weber via Stickelberger*, preprint.

[Ma]   D.-G. Ma, *An elementary proof of the solution to the Diophantine equation $6y^2 = x(x+1)(2x+1)$*, Sichuan Daxue Xuebao **4** (1985) 107–116.

[Man]   Yu. Manin, *The p-torsion of elliptic curves is uniformly bounded*, Izv. Akad. Nauk SSSR Ser. Mat. **33** (1969), 459–465; Amer. Math. Soc. Transl. 433–438.

[Mar]   J. Martinet, *Perfect Lattices in Euclidean Spaces*, Grundlehren der math. Wiss. **327**, Springer (2003).

[Marc]   D. A. Marcus, *Number Fields*, Springer-Verlag, New York (1977).

[Mart]   G. Martin, *Dimensions of the spaces of cusp forms and newforms on $\Gamma_0(N)$ and $\Gamma_1(N)$*, J. Number Theory **112** (2005), 298–331.

[Mat] E. M. Matveev, *An explicit lower bound for a homogeneous rational linear form in logarithms of algebraic numbers. II*, Izv. Ross. Akad. Nauk Ser. Mat. **64** (2000), 125–180. English transl. in Izv. Math. **64** (2000), 1217–1269.

[Maz] B. Mazur, *Rational isogenies of prime degree*, Invent. Math. **44** (1978), 129–162.

[McC] W. McCallum, *On the method of Coleman and Chabauty*, Math. Ann. **299** (1994), 565–596.

[Mes-Oes] J.-F. Mestre and J. Oesterlé, *Courbes de Weil semi-stables de discriminant une puissance m-ième*, J. reine angew. Math. **400** (1989), 173–184.

[Mig] M. Mignotte, *A note on the equation $ax^n - by^n = c$*, Acta Arith. **75** (1996), 287–295.

[Mig-Weg] M. Mignotte and B. de Weger, *On the Diophantine equations $x^2 + 74 = y^5$ and $x^2 + 86 = y^5$*, Glasgow Math. J. **38** (1996), 77–85.

[Mom] F. Momose, *Rational points on the modular curves $X_{split}(p)$*, Compositio Math. **52** (1984), 115–137.

[Mon-Vau] H. Montgomery and R. Vaughan, *Exponential sums with multiplicative coefficients*, Invent. Math. **43** (1977), 69–82.

[Mord] L. Mordell, *Diophantine Equations*, Pure and applied Math. **30**, Academic Press (1969).

[Mori] M. Mori, *Developments in the double exponential formula for numerical integration*, in Proceedings ICM 1990, Springer-Verlag (1991), 1585–1594.

[Morit1] Y. Morita, *A p-adic analogue of the Γ-function*, J. Fac. Sci. Univ. Tokyo Sect. IA Math. **22** (1975), 255–266.

[Morit2] Y. Morita, *On the Hurwitz–Lerch L-functions*, J. Fac. Sci. Univ. Tokyo Sect. IA Math. **24** (1977), 29–43.

[Nak-Tag] Y. Nakkajima and Y. Taguchi, *A generalization of the Chowla-Selberg formula*, J. reine angew. Math. **419** (1991), 119–124.

[New] D. Newman, *Analytic Number Theory (2nd corrected printing)*, Graduate Texts in Math. **177**, Springer-Verlag (2000).

[Pap] I. Papadopoulos, *Sur la classification de Néron des courbes elliptiques en caractéristique résiduelle 2 et 3*, J. Number Theory **44** (1993), 119–152.

[Poo-Sch-Sto] B. Poonen, E. Schaefer, and M. Stoll, *Twists of $X(7)$ and primitive solutions to $x^2 + y^3 = z^7$*, Duke Math. J., to appear.

[Poo-Wil] A. van der Poorten and K. Williams, *Values of the Dedekind eta function at quadratic irrationalities*, Canadian Jour. Math. **51** (1999), 176–224, corrigendum **53** (2001), 434–448.

[Rap-Sch-Sch] M. Rapoport, N. Schappacher, and P. Schneider, *Beilinson's Conjectures on Special Values of L-Functions*, Perspectives in Math. **4** (1988), Academic Press.

[Rib1] K. Ribet, *On modular representations of $\mathrm{Gal}(\overline{\mathbb{Q}}/\mathbb{Q})$ arising from modular forms*, Invent. Math. **100** (1990), 431–476.

[Rib2] K. Ribet, *On the equation $a^p + 2b^p + c^p = 0$*, Acta Arith. **LXXIX.1** (1997), 7–15.

[Rob1] A. Robert, *A Course in p-adic Analysis*, Graduate Texts in Math. **198**, Springer-Verlag (2000).

[Rob2] A. Robert, *The Gross–Koblitz formula revisited*, Rend. Sem. Math. Univ. Padova **105** (2001), 157–170.

[Rod-Zag]   F. Rodriguez-Villegas and D. Zagier, *Which primes are sums of two cubes?*, Canadian Math. Soc. Conference proceedings **15** (1995), 295–306.

[Ruc]   H.-G. Rück, *A note on elliptic curves over finite fields*, Math. Comp. **49** (1987), 301–304.

[Rud]   W. Rudin, *Real and Complex Analysis*, Mc Graw Hill (1970).

[Sam]   P. Samuel, *Théorie Algébrique des Nombres*, Hermann, Paris (1971).

[Sch]   E. Shaefer, *2-descent on the Jacobians of hyperelliptic curves*, J. Number Theory **51** (1995), 219–232.

[Sch-Sto]   E. Schaefer and M. Stoll, *How to do a p-descent on an elliptic curve*, Trans. Amer. Math. Soc. **356** (2004), 1209–1231.

[Scho]   R. Schoof, *Class groups of real cyclotomic fields of prime conductor*, Math. Comp. **72** (2003), 913–937 (see also the errata on Schoof's home page).

[Sel1]   E. S. Selmer, *The Diophantine equation $ax^3 + by^3 + cz^3 = 0$*, Acta Math. **85** (1951), 203–362.

[Sel2]   E. S. Selmer, *Completion of the tables*, Acta Math. **92** (1954), 191–197.

[Ser1]   J.-P. Serre, *Cours d'arithmétique*, P.U.F., Paris (1970). English translation: Graduate Texts in Math. **7**, Springer-Verlag (1973).

[Ser2]   J.-P. Serre, *Corps locaux (2nd ed.)*, Hermann, Paris (1968). English translation: Graduate Texts in Math. **67**, Springer-Verlag (1979).

[Ser3]   J.-P. Serre, *Sur les représentations modulaires de degré 2 de* $\mathrm{Gal}(\overline{\mathbb{Q}}/\mathbb{Q})$, Duke Math. J. **54** (1987) 179–230.

[Ser4]   J.-P. Serre, *Abelian ℓ-adic Representations and Elliptic Curves*, W. A. Benjamin, New York, 1968.

[Ses]   J. Sesanio, *Books IV to VII of Diophantus's Arithmetica in the Arabic Translation Attributed to Qusta ibn Luqa*, Sources in the History of Mathematics and Physical Sciences **3**, Springer-Verlag (1982).

[Shi]   G. Shimura, *Introduction to the Arithmetic Theory of Automorphic Functions*, Iwami Shoten (1971).

[Sho-Tij]   T. Shorey and R. Tijdeman, *Exponential Diophantine Equations*, Cambridge Tracts in Mathematics **87**, Cambridge University Press (1986).

[Sik]   S. Siksek, *On the Diophantine equation $x^2 = y^p + 2^k z^p$*, Journal de Théorie des Nombres de Bordeaux **15** (2003), 839–846.

[Sik-Cre]   S. Siksek and J. Cremona, *On the Diophantine equation $x^2 + 7 = y^m$*, Acta Arith. **109** (2003), 143–149.

[Sil1]   J. Silverman, *The Arithmetic of Elliptic Curves*, Graduate Texts in Math. **106**, Springer-Verlag (1986).

[Sil2]   J. Silverman, *Advanced Topics in the Arithmetic of Elliptic Curves*, Graduate Texts in Math. **151**, Springer-Verlag (1994).

[Sil3]   J. Silverman, *The difference between the Weil height and the canonical height on elliptic curves*, Math. Comp. **55** (1990), 723–743.

[Sil4]   J. Silverman, *Rational points on certain families of curves of genus at least 2*, Proc. London Math. Soc. **55** (1987), 465–481.

[Sil-Tat]   J. Silverman and J. Tate, *Rational Points on Elliptic Curves*, Undergraduate Texts in Math., Springer-Verlag (1992).

[Sim1]   D. Simon, *Solving quadratic equations using reduced unimodular quadratic forms*, Math. Comp. **74** (2005), 1531–1543.

[Sim2]   D. Simon, *Computing the rank of elliptic curves over a number field*, LMS J. Comput. Math. **5** (2002), 7–17.

[Sma]   N. Smart, *The Algorithmic Resolution of Diophantine Equations*, London Math. Soc. Student Texts **41** (1998).

[Sta1]   H. Stark, *Some effective cases of the Brauer–Siegel theorem*, Invent. Math. **23** (1974), 135–152.

[Sta2]   H. Stark, *Class numbers of complex quadratic fields*, in Modular Forms in One Variable I, Springer Lecture Notes in Math **320** (1973), 153–174.

[Sto]   M. Stoll, *Implementing 2-descent for Jacobians of hyperelliptic curves*, Acta Arith. **98** (2001), 245–277.

[Sug]   T. Sugatani, *Rings of convergent power series and Weierstrass preparation theorem*, Nagoya Math. J. **81** (1981), 73–78.

[Swd]   H.-P.-F. Swinnerton-Dyer, *A Brief Guide to Algebraic Number Theory*, London Math. Soc. Student Texts **50**, Cambridge University Press (2001).

[Tak-Mor]   H. Takashi and M. Mori, *Double exponential formulas for numerical integration*, Publications of RIMS, Kyoto University (1974), 9:721–741.

[Tay]   P. Taylor, *On the Riemann zeta function*, Quart. J. Math., Oxford Ser. **16** (1945), 1–21.

[Tay-Wil]   R. Taylor and A. Wiles, *Ring theoretic properties of certain Hecke algebras*, Annals of Math. **141** (1995), 553–572.

[Ten]   S. Tengely, *On the Diophantine equation $F(x) = G(y)$*, Acta Arith. **110** (2003), 185–200.

[Tij]   R. Tijdeman, *On the equation of Catalan*, Acta Arith. **29** (1976), 197–209.

[Tun]   J. Tunnell, *A classical Diophantine problem and modular forms of weight 3/2*, Invent. Math. **72** (1983), 323–334.

[Tza-Weg]   N. Tzanakis and B. de Weger, *On the practical solution of the Thue Equation*, J. Number Th. **31** (1989), 99–132.

[Vel]   J. Vélu, *Isogénies entre courbes elliptiques*, Comptes Rendus Acad. Sc. Paris Sér. A **273** (1971), 238–241.

[Wald1]   M. Waldschmidt, *Minorations de combinaisons linéaires de logarithmes de nombres algébriques*, Canadian J. Math. **45** (1993), 176-224.

[Wald2]   M. Waldschmidt, *Diophantine Approximation on Linear Algebraic Groups*, Grundlehren der math. Wiss. **326** (2000), Springer-Verlag.

[Wals]   P. G. Walsh, *A quantitative version of Runge's theorem on Diophantine equations*, Acta Arith. **62** (1992), 157–172.

[Was]   L. Washington, *Introduction to Cyclotomic Fields (2nd ed.)*, Graduate Texts in Math. **83**, Springer-Verlag (1997).

[Watk]   M. Watkins, *Real zeros of real odd Dirichlet L-functions*, Math. Comp. **73** (2004), 415–423.

[Wats]   G. Watson, *A Treatise on the Theory of Bessel Functions (2nd ed.)*, Cambridge Univ. Press (1966).

[Wet]   J. Wetherell, *Bounding the Number of Rational Points on Certain Curves of High Rank*, PhD thesis, Univ. California Berkeley (1997).

[Wil]   A. Wiles, *Modular elliptic curves and Fermat's last theorem*, Annals of Math. **141** (1995), 443–551.

[Yam]   Y. Yamamoto, *Real quadratic number fields with large fundamental units*, Osaka J. Math. **8** (1971), 261–270.

[Zag]   D. Zagier, *Modular parametrizations of elliptic curves*, Canad. Math. Bull. **28** (1985), 372–384.

Index of Notation

Page numbers in Roman type refer to the current volume, while italicized page numbers refer to the complementary volume.

Symbols

A

B

C

G

G	usually a group, also Catalan's constant, 127
g	sometimes the genus of a curve, *90*, 441
g	sometimes the number of prime ideals above $\mathfrak{p}$, *134*
G_0	group of points reducing to a nonsingular point, *507*
G_1	group of points reducing to $\overline{\mathcal{O}}$, *508*
$g_2(\Lambda)$	g_2-invariant of lattice Λ, *483*
$g_3(\Lambda)$	g_3-invariant of lattice Λ, *483*
γ	usually Euler's constant, 33
$\Gamma_p(s)$	p-adic gamma function at s, 368
$\Gamma_r(s,x)$	higher incomplete gamma function, *574*
$\gamma(s)$	$\pi^{-s/2}\Gamma(s/2)$, 172
$\Gamma(s,x)$	incomplete gamma function, *573*
$\Gamma(x)$	gamma function at x, 78
$\gamma_p(\chi),\ \gamma_p$	p-adic Euler constants, 308
$\gcd(a,b)$, GCD	greatest common dnivisor, viii, *viii*
$\gcd(a,b^\infty)$	limit of $\gcd(a,b^n)$, ix, *ix*
G_N	group of points of level $\geqslant N$, *508*
$G(\tau,s)$	nonholomorphic Eisenstein series, 211

H

$H^\perp$	orthogonal of H in V, *286*
$h(D)$	class number of quadratic order of discriminant D, *318*
$h(E)$	height of the elliptic curve E, *603*
H_k	$\sum_{1\leqslant j\leqslant k}1/j$, harmonic sum, 110
$h(K),\ h$	class number of K, *131*
H_n	$\sum_{1\leqslant k\leqslant n}1/k$, 85
H_n	harmonic sum $\sum_{1\leqslant j\leqslant n}1/j$, 128
HNF	Hermite normal form, *16*, *340*
$h(P)$	naïve height of a point $P\in E(\mathbb{Q})$, *530*
$\widehat{h}(P)$	canonical height of $P\in E(\mathbb{Q})$, *530*
h_p	class number of $\mathbb{Q}(\zeta_p)$, *432*
h_{p^k}	class number of $\mathbb{Q}(\zeta_{p^k})$, *148*
$h_{p^k}^-$	minus class number of p^kth cyclotomic field, *149*
$h_{p^k}^+$	class number of maximal totally real subfield, *148*
$H(p,t),\ H(t)$	conditions for $y^2=x^p+t$, *411*

I

$\Im(s)$	imaginary part of s, ix, *ix*
$I(\mathfrak{P}/\mathfrak{p})$	inertia group of $\mathfrak{P}/\mathfrak{p}$, *134*
$I_s(m),\ I_s$	Stickelberger ideal, *160*

N

$\lfloor x \rceil$	nearest integer to x, ix, *ix*
$\mathcal{N}\mathfrak{p}$	the absolute norm of a prime ideal $\mathfrak{p}$, *191*
$\left[{n \atop x}\right]$	$1 \cdot 2 \cdots n/(x(x+1) \cdots (x+n))$, *281*

O

$\mathcal{O}$	identity element of an elliptic curve, *473*
$\Omega(n)$	number of prime divisors of n with multiplicity, 156
ω	Teichmüller character, 391
$\omega(n)$	number of distinct prime divisors of n, 156
$\omega_{\mathfrak{P}}(x)$	$(q-1)$st root of 1 congruent to x mod $\mathfrak{P}$, *152*
$\omega_v(a)$	extension of Teichmüller character to $\mathbb{Q}_p^*$, 281
$\omega(x)$	Teichmüller character of x, *227, 228*
$\langle x \rangle$	$x/\omega(x) \in U_1$, *229*
ord_P	order of the point P, 443

P

$\Phi f(x,y)$	$(f(x) - f(y))/(x - y)$, 277
$\phi(n)$	Euler's ϕ function, *141*
Π	a uniformizer of a prime ideal in an extension, *432*
π	either a uniformizer of a prime ideal, or 3.14..., *432*
PID	principal ideal domain, *106, 129*
$\Pi^{(p)}$	product over integers prime to p, 302
ψ_b	$x \mapsto \zeta_p^{\mathrm{Tr}_{\mathbb{F}_q/\mathbb{F}_p}(bx)}$, *75*
$\psi_p(x)$	$\mathrm{Log}\Gamma'_p(x)$, Diamond's p-adic ψ function, 331
$\psi(x)$	logarithmic derivative of $\Gamma(x)$, 76
$\wp(z)$	Weierstrass $\wp$ function, *482*

Q

$\mathbb{Q}_p$	completion of $\mathbb{Q}$ at p, the field of p-adic numbers, *195*

R

r_1	number of real embeddings of a number field, *107*
r_2	half the number of nonreal embeddings of a number field, *107*
$R(A,B)$	resultant of polynomials A and B, *143*
$\mathrm{rad}(N)$	radical of the integer or polynomial N, 483
$R(E)$	regulator of the elliptic curve E, *522, 601*
$\Re(s)$	real part of s, ix, *ix*
$r_k(n)$	number of representations of n as a sum of k squares, *317*
$r_Q(n)$	number of representations of n by Q, 215

Index of Names

Page numbers in Roman type refer to the current volume, while italicized page numbers refer to the complementary volume.

A

Abel, N., 30, 200, 251, 256
Abouzaid, M., *417*
Adams, J., 67, 325
Alford, R., *94*
Almkvist, G., 69
Alpern, D., *384*
Amice, Y., 276
Apéry, R., 99, 141
Apostol, T., *94*
Arnold, V., 121
Artin, E., *70*, *115*, 167, *217*, 219
Atkin, O., *565*, *596*, *613*
Ax, J., *73*

B

Baker, A., viii, *viii*, 2, 411, 414, 424, *424*, *517*, *519*, 600, *603*
Balasubramanian, R., *4*
Balog, A., 133
Barnes, E., 135
Batut, C., 11, 99
Beck, M., *380*
Beilinson, A., 245
Belabas, K., x, *x*
Belyi, G., 478
Bender, C., 99
Bennett, M., x, *x*, *339*, 416, 423, 490, 523
Bernardi, D., x, *x*

Bernoulli, J., 3, 264
Bessel, F., 111
Beukers, F., 275, 400, 463
Beurling, A., 137
Bhargava, M., *107*, *313*
Bilu, Yu., viii, *viii*, 3, *413*, *417*, 436, *442*, 483, 529, 532
Binet, J., 125
Birch, B., vi, *vi*, 3, 245, *452*, *518*, *522*, *528*, *586*
Blichfeldt, H., *63*
Bloch, S., 245
Boéchat, J., viii, *viii*, *442*, 529
Borel, A., 244
Borevich, Z., x, *x*
Bourbaki, N., 21
Brauer, R., 242
Bremner, A., *410*, *462*, *608*, *614*
Breuil, C., *2*, 242, 498
Brindza, B., 437
Bruin, N., 456, 486, 489
Brumer, A., 501
Buchmann, J., *357*
Bugeaud, Y., viii, *viii*, 411, *424*, 518
Bump, D., 262

C

Cantor, D., 447
Cardano, G., *561*
Carlitz, L., 326

General Index

Page numbers in Roman type refer to the current volume, while italicized page numbers refer to the complementary volume.

Symbols

290-theorem, *313*

A

abc conjecture, 482
Abel–Plana formula, 30
Abelian extension, 167
abelian group
– finite, *14*
– finitely generated, *11*
abscissa
– of absolute convergence, 160
– of convergence, 162, 259
absolute norm, *109*
absolute trace, *109*
absolute value, *183*
– Archimedean, *184*
– equivalent, *184*
– extension, *237*
– non-Archimedean, *184*
– normalization, *191*
– trivial, *184*
absolutely irreducible, *468*
additive character, *74*
additive number theory, *4*
additive reduction, *472*
affine curve, *90*
AGM, *483, 486*
algebraic geometry, 7
algebraic integer, *126*

algebraic number theory, 6
algebraic rank, *522*
Almkvist–Meurman theorem, 70, 133, 327
analytic p-adic function, *205*
analytic element, *189*
analytic number theory, 151
analytic rank, *522*
approximate functional equation, 176
approximation of linear forms, *60*
Archimedean absolute value, *184*
arithmetic
– convolution, 152
– function, 151
arithmetic geometry, 7
arithmetic surface, 7
arithmetic–geometric mean, *483, 486*
Artin's conjecture, 167, 219
Artin–Hasse exponential, *217*
Artin–Schreier polynomial, *115*
Artin–Schreier subgroup, *70*
Artin–Schreier theory, *115*
asymptotic expansion, 19
Atkin–Lehner operator, *596*
automorphism
– Frobenius, *498*

B

baby-step giant-step algorithm, *357, 565*

Graduate Texts in Mathematics
(continued from page ii)

Graduate Texts in Mathematics

(continued from page ii)